V.E.A. Russo · S. Brody · D. Cove · S. Ottolenghi

DEVELOPMENT
The Molecular Genetic Approach

With 312 Figures

Springer-Verlag
Berlin Heidelberg New York London Paris Tokyo
Hong Kong Barcelona Budapest

Dr. VINCENZO E. A. RUSSO
MPI für Molekulare Genetik
Ihnestraße 73
W-1000 Berlin 33, FRG

Dr. STUART BRODY
Department of Biology
University of California
San Diego, La Jolla, CA 92093, USA

Dr. DAVID COVE
Department of Genetics
University of Leeds
Leeds LS2 9TJ, United Kingdom

Dr. SERGIO OTTOLENGHI
Dipartimento di Genetica e di Biologia
dei Microorganismi
Università degli Studi di Milano
Via Celoria 26
I-20133 Milano, Italy

ISBN-13: 978-3-642-77045-6   e-ISBN-13: 978-3-642-77043-2
DOI: 10.1007/978-3-642-77043-2

Library of Congress Cataloging-in-Publication Data. Development, the molecular genetic approach / V. E. A. Russo
... [et al.]. p. cm. Includes bibliographical references and index. ISBN-13: 978-3-642-77045-6 (Berlin: acid-free paper). –
1. Developmental biology. 2. Molecular genetics. I. Russo, V. E. A.
(Vincenzo E. A.) QH491.D424 1992   574.3 – dc20   91-45841

Typesetting: K+V Fotosatz GmbH, Beerfelden
31/3145-5 4 3 2 1 0 – Printed on acid-free paper

This Book Is Dedicated to
Ingela, Barbara, and Ann

Preface

Our understanding of the molecular basis of development is progressing at an explosive rate. This leads to a paradoxical situation. On the one hand, students wishing to enter this field need a text-book which contains up to-date accounts of the development of a wide range of different organisms, to allow them to choose an organism or system to study. On the other hand, it is almost impossible for one or few persons to write a textbook which does justice to a field which is progressing so fast. We offer here a solution to this paradox, a book written by 40 different laboratories on 35 different subjects, edited by us to give a uniform format and style. The goals of the book are to cover key concepts, key approaches, and many of the key systems. We have chosen a broad range of subjects relevant to the molecular genetic study of development, from virus assembly and microorganism development to the programming of flower structure, *Drosophila*, nematode, and mammal cell differentiation. We are proud to have been able to attract many of the best scientists in these different fields to contribute to this book.

We hope that we have produced a textbook which will be useful to both students and researchers. We are aware that in a field like the molecular genetics of development, any book will quickly become obsolete, which is why the authors were given the opportunity to update their chapters as late as possible (August 1991). We are also aware that there could be some errors or unclear pages in our book and we would be grateful to anyone who is so kind to point these out to us. We would like to thank the enthusiastic help of Uta Marchfelder as assistant to the Editor-in-Chief as well as the staff of Springer-Verlag for the accurate preparation of this book.

March 1992
V. E. A. Russo · S. Brody
D. J. Cove · S. Ottolenghi

Acknowledgements. We thank the following persons for reading the chapters in whole or in part: Chapter 2: J. Mandelstam (Oxford University); Chapter 4: D. A. Hopwood (John Innes Institute, Norwich); Chapter 5: A. Arthur (National Cancer Institute, Frederick); Chapter 6: D. Perkins, M. Springer, C. Yanofsky (Stanford University), J. F. Fincham (Cambridge University); Chapter 8: J. Martin, A. Metzenberg, G. Newport, T. White (University of California, San Francisco); Chapter 10: M. Mitchison (University of Edinburgh); Chapter 12: C. Knight (University of Leeds); Chapter 13: P. Quail (Plant Gene Expression Center, Albany), R. Stout (Montana State University); Chapter 17: P. Nevers (University of Kiel), P. Huiser (MPI für Züchtungsforschung, Cologne); Chapter 19: G. B. Ruvkun (Massachusetts General Hospital, Boston); Chapter 21: G. Jürgens (University of Munich); Chapter 23: P. Hardy (University of Cologne); Chapter 27: B. Herrmann (MPI für Entwicklungsbiologie, Tübingen), W. Alexander, R. Meehan, G. Tebb, A. Weith (Institute for Molecular Pathology, Vienna); Chapter 28: Q. Xu (National Institute for Medical Research, London); Chapter 30: F. Kiefer (Institute for Molecular Pathology, Vienna).

Contents in Brief

Contents

Chapter 6 **Development in *Neurospora crassa***
V. E. A. Russo and N. N. Pandit 88

Chapter 7 **Genetic Regulation of Sporulation in the Fungus**
Aspergillus nidulans
A. J. Clutterbuck and W. E. Timberlake 103

Chapter 8   Development of Trypanosomes

N. AGABIAN and S. METZENBERG 121

Chapter 13   Plant Photoperception: the Phytochrome System

Chapter 14   Exploration of *Agrobacterium tumefaciens*

Chapter 15   Exploitation of *Agrobacterium tumefaciens*

Section 3   Animals

Chapter 19   Embryogenesis in *Caenorhabditis elegans*

Chapter 20 **Generation of Temporal and Cell Lineage Asymmetry During *C. elegans* Development**

Chapter 21   Genetic and Molecular Analysis
of Early Pattern Formation in *Drosophila*

Chapter 22   Generation of Pattern in *Drosophila melanogaster*
Adult Flies

Chapter 23   Genetic Mechanisms in Early Neurogenesis
of *Drosophila melanogaster*

Chapter 24 *Xenopus* Embryogenesis

Chapter 25 The Developmental Regulation of the Genes Coding
for 5S Ribosomal RNA in *Xenopus laevis*

Chapter 28 The Use of in Situ Hybridisation
to Study the Molecular Genetics of Mouse Development
D.G. WILKINSON

Chapter 29 Insertional Mutagenesis and Mouse Development
M.R. KUEHN and J.P. STOYE

Chapter 30   **The Introduction of Genes into Mouse Embryos
and Stem Cells**
E.F. WAGNER and G. KELLER

List of Authors

NINA AGABIAN
University of California
Laurel Heights
Box 1204 Suite 150
3333 California Street
San Fransisco, CA 94143-1204
USA

BENOIT ARCANGIOLI
National Cancer Institut
Frederick Cancer Research
Facility
P.O. Box B, Bldg. 539
Frederick, MD 21702-1201, USA
Present Address:
Department de Biologie
Moleculaire
Institut Pasteur
25 rue du Docteur Roux
F-75724 Paris cedex 15

DENISE BARLOW
Research Institute of Molecular
Pathology
 Bohr-Gasse 7
A-1030 Wien

SALVO BOZZARO
Dip. di Scienze Cliniche e
Biologiche
Universitá di Torino
Ospedale San Luigi Gonzaga
Regione Gonzole 10
I-10043 Orbassano-Torino

DOUGLAS BRADLEY
Botany Department, KB-15
University of Washington
Seattle, WA 98195, USA

NICK J. BREWIN
John Innes Institute
Dept. of Genetics
Colney Lane
GB-Norwich NR4 7LH

STUART BRODY
Department of Biology
University of California
San Diego, La Jolla, CA 92093
USA

RUTH BRYAN
College of Physicians and
Surgeons of Columbia
University
Department of Microbiology
701 West 168th Street
New York, NY 10032, USA

JOSÉ CAMPOS-ORTEGA
Institut für
Entwicklungsphysiologie
Universität zu Köln
Gyrhofstraße 17
W-5000 Köln 41, FRG

KEITH CHATER
John Innes Institute
Dept. of Genetics
Colney Lane
GB-Norwich, NR4 7LH

A. JOHN CLUTTERBUCK
Dept. of Genetics
Glasgow University
Church Street
Glasgow G11 5JS
Scotland

DAVID J. COVE
Department of Genetics
University of Leeds
GB-Leeds, LS2 9JT

GERALD CRABTREE
Howard Hughes Medical
Institute
Research Laboratories,
Beckman Center
Unit in Molecular and Genetic
Medicine
Stanford University School of
Medicine
Stanford, CA 94305-5425, USA

J. ALAN DOWNIE
John Innes Institute
Dept. of Genetics
Colney Lane
GB-Norwich NR4 7LH

LOIS EDGAR
University of Colorado at
Boulder
Dept. of M.C.D.B
Porter Biosciences Building
Campus Box 347
Boulder, CO 80309-0347, USA

CHARLIE EMERSON
Institute for Cancer Research
Fox Chase Cancer Center
7701 Burholme Ave
Philadelphia, PA 19111, USA

JEFF ERRINGTON
Sir William Dunn School of
Pathology
Chemical Pathology Unit
University of Oxford
South Parks Road
GB-Oxford OX1 3RE

PETER FANTES
University of Edinburgh
ICMB
The Darwin Building
The King's Building
Mayfield Road
GB-Edinburgh EH9 3JR

ANTONIO GARCÍA-BELLIDO
Centro de Biología Molecular
Universidad Autonoma de
Madrid
Campus de Cantoblanco
E-28049 Madrid

J. GOLAY
Instituto di Ricerche
Farmacologiche "Mario Negri"
Via Eritrea, 62
I-20157 Milano

PETER GRUSS
M.P.I. für biophysikalische
Chemie, Abt. Molekul.
Zellbiologie
Postfach 2841
Am Faßberg
W-3400 Göttingen-Nikolausberg
FRG

BARBARA HOHN
Friedrich-Miescher-Institut
P.O. Box 2543
CH-4002 Basel

MARTINO INTRONA
Instituto di Ricerche
Farmacologiche "Mario Negri"
Via Eritrea, 62
I-20157 Milano

GORDON KELLER
National Jewish Center
1400 Jackson Street
Denver, CO 80206, USA

AMAR J.S. KLAR
National Cancer Institut
Frederick Cancer Research
Facility
P.O. Box B, Bldg. 539
Frederick, MD 21702-1201
USA

CSABA KONCZ
Max-Planck-Institut für
Züchtungsforschung
Carl von Linné-Weg 10
W-5000 Köln 30, FRG

ELISABETH KNUST
Institut für
Entwicklungsphysiologie
Universität zu Köln
Gyrhofstraße 17
W-5000 Köln 41, FRG

MICHAEL R. KUEHN
Dept. of Genetics
University of Illinois
College of Medicine
808 South Wood Street
Chicago, IL 60612, USA

CALVIN J. KUO
Howard Hughes Medical
Institute
Research Laboratories,
Beckman Center
Unit in Molecular and Genetic
Medicine
Stanford University School of
Medicine
Stanford, CA 94305-5425, USA

CORINNE LOBE
M.P.I. für biophysikalische
Chemie, Abt. Molekul.
Zellbiologie
Postfach 2841
Am Faßberg
W-3400 Göttingen-Nikolausberg
FRG

MANUEL MARÍ-BEFFA
Centro de Biología Molecular
Universidad Autonoma de
Madrid
Campus de Cantoblanco
E-28049 Madrid

STAN METZENBERG
University of California
Laurel Heights
Box 1204 Suite 150
3333 California Street
San Francisco, CA 93143-1204
USA

SERGIO OTTOLENGHI
Dipt. di Genetica e di Biologia
dei Microrganismi
Università degli Studi di
Milano
Via Celoria 26
I-20133 Milano

NIKETAN N. PANDIT
Max-Planck-Institut für
molekulare Genetik
Ihnestraße 73
W-1000 Berlin 33, FRG

LENNART PHILIPSON
EMBL
Postfach 102209
W-6900 Heidelberg, FRG

TOMAS PIELER
MPI für molekulare Genetik
Ihnestraße 73
W-1000 Berlin 33, FRG

DEBORAH PINNEY
Institute for Cancer Research
Fox Chase Cancer Center
7701 Burholme Ave
Philadelphia, PA 19111, USA

ROBERT E. PRUITT
University of Minnesota
Dept. of Genetics and Cell
Biology
250 Gortner Ave
St Paul, MN 55108-1095, USA

VINCENZO E. A. RUSSO
Max-Planck-Institut für
molekulare Genetik
Ihnestraße 73
W-1000 Berlin 33, FRG

GARY B. RUVKUN
Department of Molecular
Biology
Massachusetts General Hospital
Boston, MA 02114, USA

HANS SAEDLER
Max-Planck-Institut für
Züchtungsforschung
Carl von Linné-Weg 10
W-5000 Köln 30, FRG

ZUZSANNA SCHWARZ-
SOMMER
Max-Planck-Institut für
Züchtungsforschung
Carl von Linné-Weg 10
W-5000 Köln 30, FRG

JOZEF SCHELL
Max-Planck-Institut für
Züchtungsforschung
Carl von Linné-Weg 10
W-5000 Köln 30, FRG

ROBERT A. SHARROCK
Department of Biology
Montana State University
Bozeman, Montana 59717-0001
USA

HANS SOMMER
Max-Planck-Institut für
Züchtungsforschung
Carl von Linné-Weg 10
W-5000 Köln 30, FRG

VINCENZO SORRENTINO
EMBL
Postfach 102209
W-6900 Heidelberg, FRG

PETER STOCKLEY
Department of Genetics
University of Leeds
GB-Leeds LS2 9JT

J. P. STOYE
NIMR
Mill Hill
GB-London NW7 1AA

DIETHARD TAUTZ
Institut für Genetik und
Mikrobiologie der Universität
München
Maria-Wardstr. 1a
W-8000 München 19, FRG

WILLIAM E. TIMBERLAKE
Departments of Genetics and
Plant Pathology
University of Georgia
527 Biological Sciences
Building
Athens, GA 30602, USA

ERWIN F. WAGNER
Research Institute of Molecular
Pathology
Bohr-Gasse 7
A-1030 Wien

DAVE WILKINSON
Laboratory of Eukaryotic
Molecular Genetics
NIMR
The Ridgeway, Mill Hill
GB-London NW7 1AA

ALAN P. WOLFFE
Lab. of Molecular Biology
NIDDK, NIH
Bldg. 6, Rm. 131
Bethesda, MD 20892, USA

Introduction

David J. Cove, Stuart Brody, Sergio Ottolenghi, and Vincenzo E. A. Russo

1 Molecular Genetic Analysis Has Revolutionized the Study of Development

Genes and Their Regulatory Regions Can Be Isolated. Spectacular advances have been made in recent years in the techniques that permit genetic analysis to be carried out at the molecular level. Methods for isolation, cloning, and sequencing were first applied to genes with products that were well characterized and abundant. These methods have now been refined so that they can be extended to genes whose roles are developmental, and whose function in molecular terms is unknown. A battery of different techniques are now available to allow genes to be isolated. Once a gene or part of a gene has been isolated, it is not only possible to characterize the coding potential of the DNA, but the organization of the gene can also be investigated and neighboring sequences that are likely to be play significant roles in its regulation can be identified. These techniques have given the developmental biologist analytical power that was undreamed of previously, and as a result, our understanding of development is progressing at an accelerating pace.

The Pattern of Gene Expression in Time and Space Can Be Determined. Once a developmentally relevant gene has been isolated and cloned, the DNA can be used as a probe to investigate the pattern of expression of the gene. Messenger RNA isolated from different parts of an organism or a different times can be probed to establish the pattern of gene expression. Probes can also be used in in situ hybridization studies to establish which cell types are expressing a particular gene and at which developmental stage. In these ways, the patterns of gene regulation in time and in space can be established.

Gene Regulation Can Now Be Manipulated Experimentally. The techniques of DNA manipulation are now permitting developmental genetics to enter a new experimental phase. The role of possible regulatory sequences, associated with developmentally regulated genes, can be investigated directly by the techniques of genetic manipulation, which allow the sequences to be coupled to a reporter gene whose expression can be visualized. The *Escherichia coli* lac z gene is a favorite for this role since its product, the enzyme β-galactosidase, can utilize substrates which generate colored compounds on cleavage.

A developmentally relevant gene can also be uncoupled from its normal regulatory sequences, coupled to sequences which allow its expression to be controlled experimentally, and the engineered gene can then be re-introduced into an organism. The experimenter can now control whether the gene is active or inactive, and so investigate the effect of the expression of the gene at times and in tissues where it would not normally be active. Such investigations are only now beginning, and require a good basic knowledge of the molecular biology of the system under investigation, but there is no doubt that such techniques will have widespread applications.

Databases Allow Us to Guess Intelligently the Function of a Cloned Gene. Once a gene has been isolated and cloned, it is usual to determine its sequence. What information can be gained from this sequence information? Firstly it can be used to predict the amino acid sequence for which it might code. Unless information is also available on the direction of transcription of the DNA, the sequence's six possible reading frames (three in each direction) need to be examined. For eukaryotic DNA there is the added complication that coding sequences are likely to contain introns. Computer programs are now available that quickly identify the likely coding sequence. Such programs can identify stop codons, utilize previous knowledge of codon usage by the organism from which the DNA has been isolated, and search for consensus sequences such as those at intron-exon borders. Either the nucleic acid sequence or the deduced amino acid sequence can then be analyzed.

Two approaches can be exploited to analyze sequences. First, the physical properties of the deduced amino acid sequence can be assessed. If, for example, a region contains many hydrophobic amino acids, the protein may well be located, at least partly, in the membrane. Second, the nucleic acid or deduced amino acid sequence can be compared with known sequences stored in a computer database. The number of such sequences is growing rapidly and large databases already contain over 30000 amino acid sequences. Comparison of sequences is not simple. For example, when comparing amino acid sequences, two sequences differing from one another because there is a single extra amino acid inserted in one sequence half way through its length, would, on a crude comparison, appear to be similar only along half their length. The programs which compare sequences must allow for such features and also for amino acid differences that are "conservative", where it is unlikely that the properties of the protein products will be greatly affected by the difference since the amino acids concerned have similar properties. The program produces a list of sequences within the database, ranked in order of predicted homology to the sequence under investigation. In the chapters that follow there are many examples of the identification of homology using this approach, and often the results are surprising. Genes with similar sequences are found in species that are thought to be distantly related, both within and between the plant and animal kingdoms. The establishment of such homologies does not show that the two genes necessarily have a common evolutionary origin, but does show that their products are likely to play similar roles, e.g., as membrane receptors or in DNA binding.

The Function of Genes for Which No Mutants Have Been Isolated Can Be Determined. Classical developmental genetics investigates the programming of developmental processes by identifying the genes involved by mutation. For example, genes involved in the specification of flower or wing shape can be identified because they can mutate to affect these processes, and detailed analysis of mutant phenotypes can often indicate a more precise role for a gene. Although many examples of the power of this approach will be found in this book, molecular genetic analysis now allows the identification of genes by other means. These include genes that have developmentally regulated patterns of expression or that have sequences homologous to genes isolated in other organisms. Apart from analyzing the sequence of such genes, it is now possible to investigate their function in vivo. A number of approaches can be employed. First, the DNA sequence can be mutated in vitro, for example by engineering a small deletion which disrupts the reading frame, and the mutated gene can be introduced into the organism and exchanged for a resident functional allele. Sometimes the substitution of the mutant allele has no obvious effect. This may, indeed, be because the gene concerned plays a subtle role in development but it may instead indicate that the gene's function is duplicated by another gene, i.e., that there is redundancy of gene function, a not uncommon finding (see Sect. 2 below). An alternative approach introduces a so-called anti-sense version of the gene into the organism. This anti-sense version has a sequence complementary to the gene's coding sequence, and so leads to the production of an anti-sense RNA, which in some cases prevents the function of the normal gene. It is not yet clear how an anti-sense RNA interferes with the normal expression of the gene, nor does it in all cases have such an effect. It is possible that the sense and anti-sense transcripts hybridize and either the hybrid is destroyed or the translation of the sense transcript is blocked. The technique holds great promise but optimization of the ratio of anti-sense to sense message to effect maximum disruption of gene expression is not always straightforward. Using molecular techniques, it is possible not only to interfere with the expression of genes for which no mutant alleles are known, but also to study the effect of their overexpression. Such studies are often as revealing in establishing the role of the gene as studies which seek to block gene function.

Functional Homology Can Be Established by the Transfer of Genes Between Species. Genetic transformation allows the homologies of genes, established for example by sequence comparison, to be studied at the functional level. Instead of transforming a mutant strain with DNA coding for the wild-type allele of the gene concerned, a gene from another species, that is thought to have a similar function, can be introduced and its effect studied. This may confirm that the genes have homologous functions and, in turn, reveal the possible normal roles of the foreign genes. A few examples of the success of this strategy will be found in this book, for example, the human growth hormone gene can be transferred to genetically dwarf mice and stimulate their growth to that comparable to normal animals. This approach is likely to have an increasing impact on future developmental studies.

2 Common Themes

Developmental Programming Involves the Regulation of Genes in Time and Space. Living systems are immensely diverse, and it would be surprising if the mechanisms which program their development were not also diverse. Nevertheless, a number of common features are emerging as a result of the molecular genetic analysis of development. The first and most obvious finding is that development involves the regulation of gene expression. In a very wide range of organisms it can be shown that there is developmental regulation of gene expression. Some genes show tissue-specific patterns of activity, some are active only at specific times in development, others become active as a result of particular environmental stimuli such as light or heat shock. It is not always clear whether a gene that is developmentally regulated has a role in developmental programming or whether the developmental pattern of gene expression is a consequence of the developmental program (see Sect. 3).

Similar Mechanisms Are Used in the Developmental Regulation of Gene Expression. Similar mechanisms for the regulation of gene expression are found in a wide range of organisms. This may be because there are only a limited number of ways in which gene expression can be controlled or may indicate a common evolutionary origin. The mechanisms regulating developmental patterns of gene expression sometimes show detailed similarities between organisms that are thought bo be very distantly related and this situation suggests strongly that the regulatory mechanisms must have an ancient evolutionary origin.

The Mechanisms That Regulate Gene Expression in Development Are Seldom Simple, Often Involving Regulatory Cascades and Redundancy. Where details of the mechanisms that regulate the expression of genes in development have been characterized, these have usually been founded to be surprisingly complex. Cascades of gene regulation are often involved with genes regulating other genes which in turn regulate other genes and so on. These regulatory pathways may branch and so form networks. In a number of cases, it has been found that there are parallel regulatory pathways that apparently duplicate one another's function. It is likely that these complex patterns of gene regulation have evolved to achieve increasingly more effective mechanisms for the developmental control of gene expression.

Diverse Organisms Use Similar Cellular Signaling Pathways. Not only do distantly related organisms have similar mechanisms controlling gene expression but they also use similar mechanisms of signaling between and within cells. Mechanisms involving growth factors, membrane located signal receptors, reversable protein phosphorylation, and regulated channels for ion transport are found in a wide rage of organisms. The genes coding for these components often show considerable homology, for example, genes coding for GTP binding proteins, again suggesting that these cellular signaling pathways may have evolved early.

3 Operational Questions

Are There Developmental Genes? If a developmental gene is defined as any gene that can be shown to be important in development as a result of the study, for example, of the effect of its mutation, we have such a broad definition that it is of no practical use. Almost all genes, including those which are usually called "housekeeping genes", and which code for enzymes such as those involved in intermediary metabolism, would be regarded by this definition as developmental genes. Mutation in a gene coding for an enzyme of glycolysis, for example, is likely to be lethal and so could be said to block development completely. Clearly we require a more critical method of identifying genes more specifically involved in developmental processes.

We can try and narrow our search by considering whether there are genes that are involved uniquely in a specific developmental process. An example of this approach is provided by the study of spore development in the bacterium, *Bacillus subtilis*. A large number of genes have been identified that interfere with sporulation, and it can be argued that these have a specific developmental role since their mutation does not effect vegetative growth. Are there also genes in eukaryotes which are involved uniquely in a particular developmental pathway? Are there, for example, genes required only during leg or flower development, and is their control brought about by genes which deliver the message "make a leg" or "make a flower"?

Although it is improbable that structures as complex as legs or flowers have such simple programming, some genes have already been characterized, which can be regarded as master genes at the cellular level of development. The expression of such genes appears to be sufficient to induce cells to follow a pathway eventually leading to the development of a specific cell type. the activity of these master genes must itself be precisely controlled in time and space. Accounts of some such genes are to be found in this book, for example the expression of the *myo*D gene, which encodes a DNA-binding protein, in mammalian fibroblasts causes a cascade of gene expression which leads to fibroblasts differentiating into muscle cells. Other related genes are involved in directing the development of neural rather than epidermal tissue.

In other systems, the control of cell type and organ development seems to be more complex. In the chick embryo, a number of genes have been identified which appear to have central roles in regulating transcription during liver development. However, some of these regulatory genes are also expressed in other tissues, indicating that a specific combination of gene expression must be required for liver cell development. During the development of blood cells in mice and humans, the production of specific types of blood cells also appears to require combinations of regulatory genes to be expressed. Attempts to identify genes involved in floral induction in *Arabidopsis thaliana* by the direct method of isolating mutants unable to form floral primordia, have so far been unsuccessful, indicating that cellular developmental controls may be easier to unravel than those controlling organ formation. However, it is clear that homoeotic genes, genes which can mutate to switch development so that a structure is formed in an inappropriate position, may play master roles in development. Homoeotic genes have been best characterized in *Drosophila melanogaster*, where, for example, mutation can lead to the development of a leg instead of an antenna. Homologous genes have been identified in vertebrates but their roles are as yet less clear. Homoeotic genes have also been found in flowering plants, but these are structurally unrelated to those in animals. Loss of gene function in both *Arabidopsis thaliana* an *Antirrhinum majus* can, for example, lead to the production of petals instead of stamens during flower development.

Is a Developmentally Regulated Gene Necessarily Involved in the Programming of Development? Finding that a gene is developmentally regulated does not provide sufficient evidence to allow the conclusion that the gene is directly involved in the regulation of development. While the gene may indeed form part of a regulatory cascade, it may instead be at the end of the cascade, its pattern of expression being the result of developmental regulation rather than its cause. Even if this is the case, its further study may be worthwhile since we can work back from the regulated gene, finding what controls it and so come closer to the regulatory mechanisms. However, a gene which becomes active at a particular developmental stage may not be closely involved in that stage, since genes often become active well before the step at which their products are required. It is also possible that the gene's product could be involved in some process which occurs at the same time as the developmental step under study but is not directly connected with it. These considerations mean that caution must be exercized before concluding that a gene which shows a developmentally regulated pattern of expression is closely involved in the regulation of development.

How Do You Choose Which Genes and Which Organism to Study? This question cannot be answered without examining the different reasons why developmental research is carried out. If motivation is provided by natural curiosity, then the choice of genes and organism for study will be based on a number of factors, but experimental tractibility will be high on the list. If research is undertaken for utilitarian reasons, then an obvious approach is the direct one. If we wish to address an agricultural problem, the surely we should use the plant or animal concerned.

There are many reasons why this direct approach may not be best, and this is particularly true in the area of developmental biology. Because the mechanisms that we seek to analyze are bound to be complex, we are forced to attempt to tackle them a little at a time, using simpler systems than those which we might eventually wish to understand. Such an approach to developmental biological research is based on a belief that behind the multiplicity of forms of life there are mechanisms that are common to many living organisms. Particularly because of molecular genetic techniques, it is much faster and easier to check if a mechanism is present in an organism, than to characterize the mechanism in the first place. The logical consequence is that it is better to try to discover and characterize developmental mechanisms in organisms which are easier to study. Such organisms are often referred to as model systems. The bacterium, *Escherichia coli*, is an example of an organism of this type and its study has been spectacularly successful in establishing the general principles of molecular biology. The chapters that follow describe many examples of organisms and systems that have been chosen for their experimental tractability, and some illustrate how findings in one organism, for example *Drosophila melanogaster*, can be applied to another which is more difficult to study, for example, the mouse.

Experimental convenience is not the only factor that a scientist has to take into account in choosing a system to study. The economics of research must also be considered. Although we feel that our problems are greater today, research laboratories have always been confronted with limitations. Space is usually insufficient and there is seldom enough money for buying equipment or chemicals. Thought also has to be given to the availability of research staff and their experience.

Are We Studying Too Many Systems? In answering this question, there will be disagreement even among developmental biologists. There are clear advantages in having a large number of scientists studying the same organism or system. On the other hand, some of the most revealing discoveries have come unexpectedly as a result of research using systems that were sparsely studied. Despite the demands of politicians, it has to be accepted that it is difficult to plan the study of a complex process such as development, and serendipity is always likely to play an important role. A further factor is that knowledge of developmental biology is expanding rapidly. As more and more is learnt about development, new perspectives and new ideas emerge, and these in turn suggest that new or different systems should be studied to advance our knowledge further. Appreciation of this may lead to a co-ordinated effort to concentrate on a particular system, perhaps with a targeting of resources, as has been seen recently for research on *Arabidopsis thaliana*.

It also needs to be appreciated that developmental complexity has evolved independently many times. The major eukaryote groups, fungi, plants, and animals, probably evolved multicellularity after their evolutionary separation and, within these groups, multicellularity may have evolved a number of times independently. In additon, the process of development, being immensely complex, provides the opportunity for a great variety of control mechanisms. Only by studying a wide range of organisms will be establish the variety of solutions that have evolved to solve the problem of the programming of development.

Is Molecular Genetic Analysis Enough? The methods of molecular genetics have opened up developmental biology in a spectacular way. They can be used to identify and study developmentally regulated genes. They also allow the isolation and sequencing of genes which do not show developmental regulation but which mutate to affect development. Establishing the function of such genes may be difficult, especially if their sequence shows no homology to other known sequences, and yet these genes may include some of the functionally most important. It is therefore important that developmental studies should be multidisciplinary. Studies of the cell biology of developmental systems are as important as those on their molecular genetics and may lead us to establishing the function of genes that have not been established by molecular genetic methods. Our understanding of development will be much fuller when the discoveries of cell and molecular biology are connected and this is one of the major challenges confronting developmental biologists at the present time.

4 What Do We Need To Know in Order to Understand Development?

Molecular genetic analysis has revealed that the control of gene expression is central to the programming of development. However, to understand development more fully, it is essential to understand the function of the regulated genes. Many crucial questions remain to be answered, and some of these will be identified in this section.

How Is Asymmetry Achieved? One of the most fundamental developmental problems which the methods of molecular genetics have so far made little progress in solving is how asymmetrical structures are specified. The development of polar struc-

ture is widespread in living systems. Not only organisms but also cells and subcellular structures show regular patterns of asymmetry. How is this heterogeneity programmed? Asymmetrical virus particles provide a relatively simple system for the study of this problem at the molecular level, but do such structures represent the limit of complexity that can be achieved by self-assembly? Polarity at the cell level is widespread and the study of the mechanisms that bring this about is likely to be easiest in systems where individual cellular events can be studied directly. These include both prokaryotic and eukaryotic single-celled microorganisms such as *Caulobacter*, *Bacillus subtilis*, and yeast, but also more complex organisms where certain developmental stages such as egg development may be accessible to study.

At the level of the whole organism, our knowledge of how polarity is programmed is most advanced in *Drosophila melanogaster*, reflecting the supremacy ot this organism for developmental genetic analysis. We now know a great deal about how axes of asymmetry are interpreted during the early development of the *Drosophila* embryo, but the most likely explanation of how the axes are established is that they are dependent on the pre-existing asymmetry of the mother which produced the egg, her asymmetry having resulted from her mother's asymmetry and so on. Although it is likely that this type of maternal influence is central in establishing heterogeneity early in the development of many other organisms, e.g., *Xenopus*, it is not the only mechanism responsible. In mammals, although asymmetry in the embryo is likely to be generated by way of the mother, her role is less direct, generating the heterogeneous environment in which the embryo develops.

Many plants and fungi can be regenerated from almost any somatic cell. Here it is likely that environmental heterogeneities such as gravity and light may be exploited to achieve polarity but we remain almost wholly ignorant of the mechanisms that operate in such systems. Furthermore, there are structures in many multicellular organisms that show consistent asymmetry in three dimensions. Examples of these include the bilateral asymmetry shown by many organs in mammals, or the regular spiral origin of leaf primordia from the apex of a plant stem. Maternal and external environmental stimuli are unlikely to provide all the stimuli necessary to achieve such asymmetry.

How Is Cell Shape and Size Determined? This is another fundamental problem of developmental biology that has not yet been addressed to a sufficient extent, but is central to the achievement of regulated development. Again, most progress has been made in the study of micro-organisms. Research on yeasts is showing how cell size and the regulation of cell division are interlinked. Cell shape is studied extensively using the techniques of cell biology, and the role of the cytoskeleton in determining cell shape is well established. How cell shape is programmed is, however, much more obscure and this is an example of an area of research where molecular and cellular studies must be combined for progress to be made. Organisms such as *Physcomitrella patens* as well as better studied species such as *Drosophila melanogaster* and *Caenorhabditis elegans* should prove favorable for such studies.

How Is Cell Division Controlled? Cell number is intimately related to the final shape and size of an organism, and so the programming of development cannot be separated from the control of cell division. The control of the mitotic cell cycle has been studied extensively particularly in yeasts and genes that are central to the control of the cell cycle have been identified. These genes have homologs in many widely related organisms and so there may be common mechanisms for mitotic control. A number of interesting clues have also come from the study of viral oncogenes and mutated cellular oncogenes, some of which have arisen as a result of alterations to the pattern of expression of genes, the normal role of which is the regulation of cell proliferation. Much less is known about the control of cell division in a developmental context. How multicellular structures in adult organisms come to have a characteristic size and cell number is a fundamental problem of developmental biology. Studies on *Drosophila* and mammals are making progress in this area. Loss of function of the human Wilm's tumor suppressor gene, which has been cloned recently, results in kidney tumor formation in early childhood. This gene is expressed specifically in kidney cells, suggesting that it has a tissue-specific role in the regulation of cell proliferation. Little is known about the control of cell proliferation in other systems, particularly in plants, where cell immobility makes the control of cell division especially important in developmental regulation.

The initiation of cell differentiation may, in some systems, be linked to the cessation of cell division, and conversely the unregulated proliferation of tumor cells if often associated with a block in differentiation. A link between these processes may have been established in the development of mammalian muscle cells, where the product of the *myo*D gene may both inhibit the further proliferation of fibroblasts and induce their differentiation into muscle cell.

How Does Cell Lineage Influence Development? It has been known for many years that cell lineage is important in the development of some organisms. Experiments in which cells in the early embryo were removed or inactivated, resulted, in some species, in the absence of adult structures while in other species subsequent development could compensate for the deficiency and a normal adult would result. It was believed that the limited developmental potential shown by cells in some organisms might have been due to the elimination of blocks of genes from particular cell lineages, but there is no evidence that this is a general mechanism. Instead, the extent to which cell lineage is important in the development of different organisms is probably a reflection of differences in the diffusibility of signal molecules and in the role of cell to cell communication in early development. In organisms in which cell lineage plays a dominant role, it is likely that there are cytoplasmic heterogeneities and that cell division results in signal molecules being distributed unequally to daughter cells. These molecules must be long-lived or else be capable of triggering regulatory cascades that are localized within cells. In this way the ancestry of a cell will be important in determining its developmental fate. Examples will be found in the chapters that follow, of the varying extent to which cell ancestry is important in the development of different organisms. The nematode, *Caenorhabditis elegans*, provides an example of an organism in which cell lineage plays a dominant role in early development. In *Drosophila melanogaster*, groups of cells rather than specific cells have limited developmental capacity, but in vertebrates, and particularly in plants, cell lineage is much less important.

Evidence is now accumulating that cell lineage is important in another way. It is clear that in some organisms including humans, the parental origin of a gene, whether paternal or maternal, can lead to differences in the gene's expression, a phenomenon termed genomic imprinting. Similar mechanisms may be involved in the control of mating type switching in yeast. Extensive further study is required in order to understand the molecular basis of the mechanisms that are involved in developmental mechanisms which rely on cell lineage.

How Do Cells Know Their Position? There is a great deal of evidence, some from developmental genetic studies, some from experimental embryology, that gradients of signal molecules are established during development. The concentration of such molecules appears to be used as a method whereby cells can sense their position in a developmental field. In a number of systems the identity of the molecules that establish gradients is known, but details of the molecular mechanisms that interpret the gradient are unknown. Nor is it likely that this is the only mechanism that allows cells in a developing organism to sense their position. Work on *Drosophila* indicates that cell to cell communication is also important.

How Do Cells Communicate During Development? The ability of cells to signal to one another has been known for many years. Mechanisms involving extracellular molecules such as those involved in neuro-transmission or *Dictyostelium* aggregation are well characterized, and other mechanisms involving direct cell contact are also known. There is a great deal of evidence that inter-cell communication also plays an important role in development, and both diffusible signal molecules and mechanisms involving cell to cell contact are likely to be involved. Genes which have been identified by mutation as playing a role in development have been found to code for trans-membrane proteins that are likely to function in direct communication

between cells. Presumably such molecules interact with similar molecules on other cells and so establish communication networks, but this is another area of developmental biology where our knowledge is sketchy and in which more work is needed.

How Are Environmental Conditions Detected and Interpreted? The ability to modify development in order to respond to both internal and external environmental stimuli is important in most organisms, and response to the external environment is especially important in organisms with limited capacity to move. Many sessile animals such as sponges show remarkable variation in body form in response to the environment in which they develop. Plants and fungi also show such environmental responsiveness. Blue light alters the pattern of gene expression and development in bacteria, fungi, e.g., *Neurospora*, and plants. Most green plants also detect red light by way of the photoreceptor, phytochrome, and modify both gene expression and morphogenesis. Detection of changes in the environment are also crucial. Changes in day length trigger the induction of flowering in many plant species and impoverishment of nutrients leads to sporulation in many micro-organisms, including bacteria, e.g., *Bacillus subtilis*, *Streptomyces coelicolor*, slime molds, e.g., *Dictyostelium discoideum*, and fungi, e.g., *Saccharomyces cerevisiae* and *Neurospora crassa*. The mechanisms for detecting such environmental stimuli appear to be complex. Some examples of their study are contained in the chapters that follow, but this is yet another area where our knowledge remains sparse.

How Are Developmental Patterns of Gene Expression Achieved? The mechanisms responsible for developmental patterns of gene expression are especially amenable to study by the techniques of molecular genetic analysis. Progress is therefore likely to be rapid, but it is already obvious that mechanisms may be complex, often involving interlocking networks of gene regulation. Once more, work on the control of sporulation in *Bacillus subtilis* and on the early development of the *Drosophila melanogaster* embryo provide outstanding examples of the progress that is being made in this field, but systems as diverse as *Aspergillus nidulans*,

Xenopus laevis, Antirrhinum majus or the differentiation of mammalian cell types all promise to shed light on these processes.

Do Common Mechanisms Imply Common Origins? The finding that similar developmental mechanisms are found in species that are evolutionarily widely related is open to the interpretation that the mechanisms concerned may have arisen early in evolution. Alternatively, these observations may provide examples of convergent evolution. The evolution of developmental mechanisms is a subject of study that is only now beginning to gain momentum. Molecular genetic techniques are especially valuable in the area of study, enabling the homologies of gene function and structure to be established. Such studies have already revealed the unexpected finding that the divergence of the arthropod and vertebrate lines is likely to postdate the evolution of the mechanisms responsible for the differential development of body segments. There are likely to be many other similar discoveries as comparative developmental studies proceed.

ready been possible to engineer a male-sterile plant strain, an important tool in plant breeding, by coupling a regulatory sequence which results in the specific expression of a gene during pollen development, to a gene the product of which leads to cell death.

Understanding the mechanisms which control cell division is a central problem in developmental biology, but knowledge of these mechanisms will also have many practical applications, not least in combating "developmental diseases" such as cancer. Developmental studies are also essential if genetic diseases are to be treated by gene therapy, since supplying a functional copy of a defective gene is unlikely to be sufficient. It will be necessary to ensure that the gene is expressed at the right time and in the right tissue. Studies with the mouse, on the programming of the differentiation of liver, muscle, and blood cells all provide essential basic knowledge for the development of such strategies.

Developmental biology is undoubtedly one of the most exciting areas of science today. The questions posed in this section remain to be answered, but the tools already exist.

5 Does Developmental Biology Have Any Practical Benefit?

Although most of the studies described in the chapters of this book have been initiated because of the intrinsic interest of development, there is no doubt that a greater understanding of development will have widespread benefits, particularly in medicine and agriculture. Developmental mechanisms involved in the interaction between members of different species both in parasitism and symbiosis are of immediate relevance. Studies of parasites such as *Trypanasoma*, and of symbionts as *Rhizobium* are well advanced. A wider understanding of the molecular mechanisms involved in these relationships also opens the possibility of manipulating these mechanisms for practical purposes. Vectors developed from the T plasmid of the plant pathogen, *Agrobacterium tumefaciens*, are already used widely to transfer foreign genes into plants, and their exploitation in plant breeding has begun. It has al-

General References

Alberts B, Bray D, Lewis J, Raff M, Roberts K, Watson JD (1989) Molecular biology of the cell, 2nd edn. Garland, New York

Blackburn GM, Gait MJ (1990) Nucleic acids in chemistry and biology. IRL Press, Oxford University Press, London

Browder LN, Erickson CA, Jeffery WR (1991) Developmental biology, 3rd edn. Saunders, Philadelphia

Darnell J, Lodish H, Baltimore D (1990) Molecular cell biology. Scientific American Books, New York

Davidson EH (1986) Gene activity in early development, 3rd edn. Academic Press, New York

Gilberts S (1991) Developmental biology, 3rd edn. Sinauer Assoc, Sunderland, MA

Lewin B (1990) Genes IV. John Wiley and Sons, New York

Singer M, Berg P (1991) Genes and genomes, a challenging perspective. University Science Books, Blackwell, Oxford

Watson JD, Hopkins NH, Roberts JW, Steitz JA, Weiner AM (1987) Molecular biology of the gene, 4th edn. Benjamin/Cummings, Menlo Park, California

Section 1 Microbial Systems, Both Prokaryote and Eukaryote

Chapter 1 Virus Assembly and Morphogenesis

PETER G. STOCKLEY

1.1 Introduction[1]

1.1.1 Virusus as Model Systems for Development

This chapter will deal with the assembly of macromolecular complexes, in particular the assembly of viruses. At first glance this may seem an unusual topic for inclusion in a discussion of morphogenesis and development. However, as Monod described in his book *Chance and Necessity*, living things can be viewed as having two essential properties which separate them from the inorganic world. These are "reproduction invariance" and "structural teleonomy", or, more simply, the essentially faithful copying of genetic information in the form of nucleic acid sequences and the production of structures possessing specific functions. The interplay between these two properties is the essence of developmental studies.

Viruses are useful experimental systems, since their complete developmental repertoire must be completed before progeny viruses form. This greatly facilitates the application of genetic and biochemical techniques to the study of viral life cycles. The regulation of viral gene expression is as complex as many forms of cellular regulation and the end products of viral infection, i.e., new virus particles, are themselves teleonomic structures capable of specific encapsidation and protection of the viral nucleic acid, escape from the host cell and eventual re-entry into a target cell to allow the process to be repeated, an event usually coupled to specific release of the viral nuclei acid. The different steps in the viral life cycle can be viewed as developmental stages. Viruses have advantages over larger scale developmental systems because, in many cases, their genetic information content has been completely determined and the molecular structures of the resultant virus particles are known, in some cases to atomic resolution. The study of viruses, therefore, allows us to investigate some of the properties of developmental systems at the molecular level, with every confidence that, although chemical details will vary, similar principles will be seen to operate in more complex organisms.

1.1.2 Self-Assembling Systems: Molecular Ontogeny?

Development leads to the formation of novel biological entities, very often possessing properties and functions which appear to be more than the sum of their component parts. Viral systems display similar properties and for a few examples it can be shown that the properties of the whole are simply the result of a series of stereospecific recognition events between macromolecules. If this is true for viruses, it is potentially also true for the development of more complex systems such as subcellular organelles, and even the anatomies of higher organisms. The problem reduces to one of identifying the specific interactions involved in the formation of complex structures and how these lead to novel biological functions. The information required to build such complexes is clearly an inherent property of the macromolecules involved and, although it is entirely encoded within nulceic acid sequences is dependent on the interactions between macromolecules and the solvent, i.e., water. Biological structures have evolved and are expressed in defined solvent conditions. This is a point which is often ignored.

One of the most striking examples of "molecular development" in this sense is the spontaneous self-assembly of macromolecular complexes. The driving forces for such assembly reactions are identical to those causing proteins to fold into defined globular structures, namely the large number of noncovalent interactions between specific functional groups and small molecules, particularly water. Many viruses demonstrate spectacular abilities to self-assemble in vitro from the isolated and purified viral components. Furthermore, such reassembled structures appear indistinguishable in terms of physical, chemical and biological properties from those purified from the usual host. The simplest conclusion is that such structures normally self-assemble in vivo and that the presence of other cellular macromolecules does not interfere with this process. Assembled complexes very often have properties which the individual components do not and, therefore, it is not surprising that viruses have evolved mechanisms to control and regulate the timing of such assembly reactions, and hence the appearance of new functions. I shall use the term self-assembly in the strict sense that the components of the assembled complex are the only ones required for the process.

1.1.3 General Principles of Virus Organisation

As Crick and Watson pointed out, the essential feature of simple viruses is the protection of the viral nucleic acid between rounds of infections. The protective shells of many simple viruses, i.e., those containing only short nucleic acid molecules and, hence, few genes, are composed entirely of proteins; either a single polypeptide species or a relatively small number of different ones. Genetic economy would thus lead to viral shells (capsids) composed of multiple copies of this polypeptide(s), and this, in turn, implies that viral capsids should be highly regular in their structure. Implicit in this argument is the idea that the specific noncovalent contacts made by the viral coat proteins are identical throughout the viral shell. Caspar and Klug have investigated the structural principles of such viral capsids and have shown that only two geometrical forms allow such repeat uniform contacts to be made, i.e., structures in which the subunits are all equivalent. These are the *helix* and the *icosahedron*. The structures of simple viruses do indeed turn out to conform to one or other of these two

forms, and more complex viruses, such as those enclosed in membranes, have been shown to contain regular nucleocapsids based on helical or icosahedral symmetry. The limited number of morphological types of virus implies that there are a limited number of efficient designs for a biological container made from identical (or a limited number of different) subunits.

The fundamental principle of virus architecture is the construction of a large defined structure (virus capsid) from smaller subunits in a process similar to an industrial assembly line. This type of organisation allows a high degree of biological control at every level of construction so that if mistakes occur they can be rejected. The result is that very complex assemblies can be built with high efficiency. In a reguular structure all the subunits (or small subassemblies) are equivalent and each can be thought of as an asymmetric unit. Figure 1a illustrates the packing of a regular array of asymmetric units forming a close-packed plane net. The points of contact between the units can be considered equivalent to intersubunit bonds. Although there are six bonding sites (A–F) there are only three types of "bonds", namely, AD, BE, and CF. Figure 1b shows what happens when the paper containing the array is rolled into a cylinder. The result is a helical array of the asymmetric units which maintain their "bonding" interactions. In practice, when protein subunits form the bonding interactions the act of forming the helical array would deform the interactions slightly. Caspar and Klug have adopted the term "quasi-equivalent" to describe the subunits in such an arrangement. Although the symmetry may no longer conform to mathematical perfection, the structural principle of maximising identical inter-subunit contacts is preserved.

The second type of viral capsid is one which results in the formation of an essentially spherical shell. Caspar and Klug argued in favour of an icosahedral arrangement of protein subunits in such shells on the grounds of genetic economy, i.e., the more subunits allowed to enclose the viral nucleic acid the smaller they could be. As for a helical array, the construction of an icosahedral shell does not require any underlying symmetry in the viral subunit (see Fig. 2). In their attempts to understand

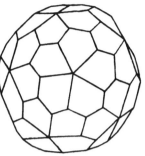

Fig. 2. Arrangement of nonsymmetrical polygons according to icosahedral symmetry. *Each unit makes identical contacts with its neighbours and there are 60 units in the shell.* (Caspar and Klug 1962)

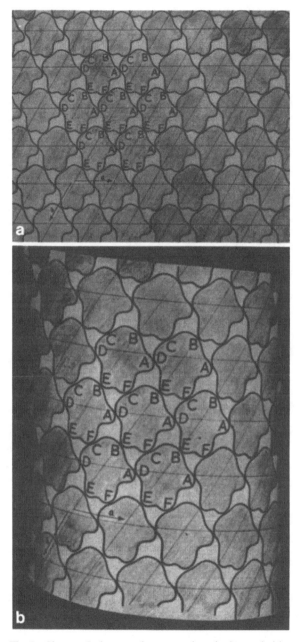

Fig. 1a Close-packed array of asymmetric units in a primitive plane net (i.e., without rotational symmetry but possessing translations in the a and b directions). Each subunit has six bonding sites A–F, but there are only three "bonds", *AD, BE, CF.* **b** The plane net of **a** can be rolled up to form a cylindrical surface which produces a helical array of subunits but does not alter the geometry of the "bonding" pattern. (Caspar and Klug 1962)

the implications of efficient packaging and genetic economy, Caspar and Klug realised that these were both aspects of the fundamental principles involved with the optimum design of a shell. The result of their studies is an elegant description of all possible viral capsids having icosahedral symmetry. For our purposes it is sufficient to know that the analysis results in a description of icosahedral shells in terms of their "triangulation number" (T). Shells are allowed which have multiples of 60 subunits according to the formula 60T, where T can be 1, 3, 4, 7... etc. Hence, Tomato Bushy Stunt Virus (TBSV), which has a shell composed essentially of 180 identical coat proteins subunits, is an example of a T = 3 capsid. This implies that the subunits are able to adopt three distinct quasi-equivalent conformations, a prediction spectacularly borne out in 1978 when Harrison and co-workers were the first to determine the structure of a virus to atomic resolution using X-ray crystallography (see Sect. 1.2.1.3).

1.2 Molecular Mechanisms of Viral Assembly

We will now discuss a number of examples of viral morphogenesis in detail. In each case the basic principles of shell construction discussed above can be seen to underly many of the processes involved.

1.2.1 T4 Bacteriophage, a Complex Virus[2]

The assembly of double-stranded (ds) DNA bacteriophages has been intensively studied. It is possible to isolate conditional lethal mutants of these phages, and this has allowed the details of their assembly pathways to be analysed despite their complexity. Figure 3 shows a schematic representation of the structure of bacteriophage T4. It is perhaps structurally the most elaborate of a group of dsDNA bacteriophages such as lambda, T2, P22, T3, and T7. We will discuss only T4 in detail since it illustrates very well the principles of building a complex structure via a series of ordered stereo-specific recognition events between macromolecules. The assembly pathways of the other bacteriophages have been studied and differ somewhat in detail

from T4 but the overall pattern of assembly is essentially the same.

T4 assembly occurs via a number of subassembly processes. The end result of one set processes is the production of the phage head, whilst the others lead to the production of the base plate and tail fibre. Each of these requires a specific set of virally encoded polypeptides, some of which become part of the assembled phage structure and some of which do not. The proteins which do not form part of the structure are often present in much smaller amounts than the structural components and so have historically been termed minor. This is a misnomer, since these proteins often repeatedly play crucial roles in the assembly process, whereas the final structural components, once assembled are essentially inert. Note that T4 also requires host fac-

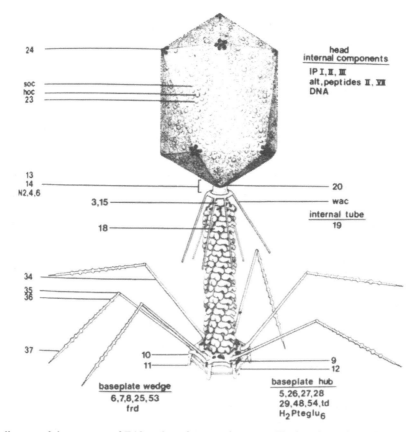

Fig. 3. Schematic diagram of the structure of T4 based on electron microscopy. The locations of protein components are indicated by gene number except for several unknown connector proteins, called *N2, 4,* and *6*

tors to complete the assembly process and therefore is not an example of self-assembly in the strict sense. At least 40% of the T4 genome (166 kbp) is required to supply components involved in the synthesis and assembly of the final bacteriophage structure. The timing of the assembly process is linked to the order and rate of viral gene expression.

Head Assembly and DNA Packaging. Twenty four genes are required for phage head morphogenesis, ten of which are structural components. Figure 4a shows an outline of the events occurring during T4 prohead assembly. The major head capsid protein is the product of the phage gene 23 (gp23). This is solubilised by association with gp31, and head assembly occurs after formation of an assembly initiation complex. The capsid assembles into an

icosahedral shell which is elongated along a five-fold axis. Once polymerisation is complete both the major capsid protein, gp23, and gp24, which is associated with the capsid vertices, undergo proteolytic maturation. These proheads undergo expansion during maturation and further proteins, such as gp*hoc* and gp*soc*, are added to the structure at this point. The neck and portal vertex of the head is the point of attachment to the tail fibre. This structure is complex and consists of the portal vertex formed by a disk of twelve gp20 proteins, which seem to be involved in DNA packaging, a series of six whiskers made up of 18 gp*wac* molecules, which serve as environmental sensors during binding of the target bacterium, and a number of other protein components.

A large number of mutants of T4 have been characterised which differ in the lengths of the phage head. These include the petite variant which

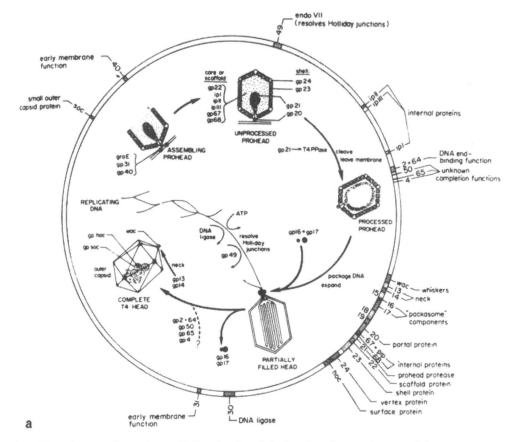

Fig. 4a. Assembly and maturation pathway of T4 proheads and the location of genes that control these processes

has a perfect, isometric T = 23 head, i.e., does not have the elongation along the fivefold axis. This variant appears only able to package 70% of the phage genome. A problem which arises from these observations is the molecular mechanism by which the head size is determined. A favoured hypothesis is that there is a "molecular ruler" which itself self-assembles to determine the overall length. The lengths of both lambda and P 22 heads are both determined by the self-assembly of a protein scaffold, which in the case of P 22 may assemble from an initiation complex formed from the scaffold protein and its messenger RNA, analogous to the assembly initiation complex formed by Turnip Crinkle Virus (see Sect. 1.2.1.3).

Replicating phage DNA exists as long multiples of a single genomic molecule, presumably generated by a rolling-circle type of replication mechanism. Phage proteins which associate with the par-

ticle, but are not structural components, function in part as nucleases ensuring the packing of single genomic copies. Once the head has been correctly packaged other proteins associate and make it competent for joining to tail assemblies.

Baseplate and Tail Assembly. The baseplate and tail fibre is apparently a unique structure restricted to the bacteriophages. However, its assembly represents an elegant example of how the regulation of protein self-assembly can control the formation of morphologically and functionally complex organelles.

Figure 4b shows the pathway of tail fibre morphogenesis. It is a sequential process and the proteins involved interact with each other only in a defined order. Mutations in proteins in the pathway at particular points result in the production of assembled precursors unable to interact any further with

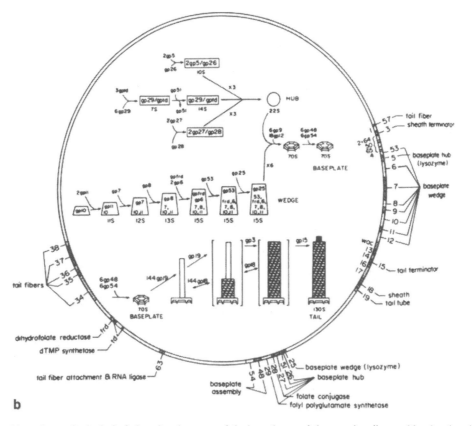

b

Fig. 4b. Assembly pathways for both the hub and wedge parts of the baseplate, and the steps in tail assembly, showing the positions of genes for tail and tail fibre synthesis and assembly components needed for later maturation, which thus remain unassembled and show no tendency to assemble in the absence of the preceding gene product. Addition of the missing component in vitro can often lead to rapid correct assembly.

Assembly of the baseplate and tail requires at least 31 phage gene products, 26 of which are found as structural components in the completed tail. Most of the proteins are required for baseplate assembly. The baseplate performs the vital biological function of triggering collapse of the tail fibre leading to injection of phage DNA into a target bacterium. During assembly of the baseplate five major structural proteins associate sequentially into a complex, characterised by its sedimentation coefficient of 15S, which is equivalent to a 1/6th wedge of the final hexagonal baseplate. Gp 53 converts this 15S complex into a form which polymerises into a 70S hexagon. The hexagon assembles around a central plug, which comprises a 22S complex of a number of viral proteins. Once the baseplate has reached this point gp 12 forms the short fibres which contact the target bacterium. At this point gp's 9, 48, and 54 add sequentially converting the baseplate into an assembly initiator for tail tube polymerisation.

The tail tube is formed by polymerisation of gp 19 and proceeds irreversibly until 24 annuli each comprising six subunits have been formed. Gp 3 then adds to the end of the assembled tube. The exact mechanism of tail tube length determination is unknown but it is thought that the structure of gp 19 is progressively distorted during polymerisation eventually leading to the loss of a site for further polymerisation. A tail sheath composed of gp 18 is then added on the outside of the tube. Assembly of the sheath can only begin after the tube polymerisation has started and terminates at the end of the tube. Sheath assembly is reversible until gp 15 functions by anchoring the ends of the tube and sheath together, producing the attachment site for head assembly.

1.2.1.1 Tobacco Mosaic Virus, a Helical Rod[3]

We now turn to an example of a much simpler virus. Tobacco Mosaic Virus (TMV) is a helical rod-shaped particle, 3000 Å long by 180 Å wide, with a central hole of 40 Å diameter. It comprises essentially a single coat protein subunit (MW 17.5 kDa) 2130 copies of which encapsidate a positive-sense, single-stranded RNA molecule (6390 bases). The virus particles are stable over periods of several decades during which time they remain infectious, implying that the RNA, which on its own is normally very sensitive to degradation, remains intact. The structure of the particle depends largely on the coat protein, and the protein alone can be induced to form helical rods of similar size and shape to the virus. However, these protein assemblies are not as stable as the intact virion, implying that in the virus, RNA-protein interactions act to stabilise the structure. The structure of the virus is shown schematically in Figure 5. As might be expected from the helical arrangement, there is essentially no secondary or tertiary structure in the RNA compo-

VIRAL RNA

PROTEIN SUBUNIT

Fig. 5. Schematic representation of Tobacco Mosaic Virus showing the helical arrangement of coat protein subunits which encapsidate the viral RNA (*dark grey*)

nent, rather the RNA exists as an extended helix buried in successive layers of the coat protein. Each protein subunit is associated with three nucleotides of the RNA and hence the RNA-protein interaction is not sequence-specific. The final structure of the virion, however, is clearly not the whole story since both in vivo and in vitro, only TMV RNA is packaged efficiently by the coat protein. Analysis of the RNA in vitro reveals extensive secondary and tertiary structure implying that during assembly of the virus, the viral RNA is effectively unfolded. The control and regulation of packaging specificity is a good example of molecular development.

In vitro the coat protein subunit self-assembles into a series of oligomeric structures. The exact nature of the aggregate present depends on the solvent conditions in a strictly defined way such that a phase diagram can be produced of the various forms (Fig. 6). A particularly important oligomeric form is the disk. The disk can be crystallised and

its structure has been determined to atomic resolution by X-ray diffraction techniques. It is a two-layered structure with seventeen coat protein subunits per layer. It is important because it approximates the organisation of the coat protein subunits in the virus itself.

Klug and colleagues have proposed that the disk can undergo a conformational change to form a lock-washer by a simple dislocation from the closed disk form. Such a structure then has growth points at both top and bottom which could be extended symmetrically to generate a helical array (Fig. 7). Dislocation of a disk to generate the lock washer form might then be imagined to act as an initiation event leading to the self-assembly of a helix. Specificity can be introduced into this scheme by supposing that the interaction between TMV disks and some element of secondary or tertiary structure in the viral RNA could trigger this conversion to the lock-washer and thus assembly initiation. Indeed,

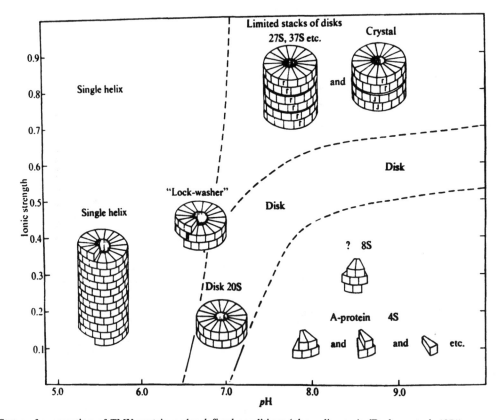

Fig. 6. States of aggregation of TMV protein under defined conditions (phase diagram). (Durham et al. 1971)

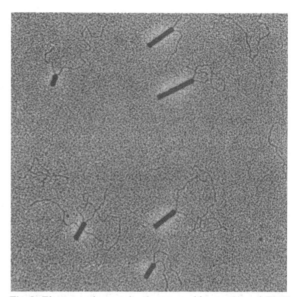

Fig. 8. Electron micrograph of an assembly reaction of TMV viewed in negative stain showing the two ends of the viral RNA emerging form the same end of the growing virus. (Lebeurier et al. 1977)

an assembly initiation sequence has been identified which is potenitally capable of forming a specific hairpin-loop structure. Incorporation of this sequence in recombinant RNA molecules leads to specific encapsidation into TMV-like structures supporting the idea of a sequence-specific assembly initiation followed by an encapsidation reaction which is not.

Once the assembly has been correctly initiated, helical symmetry allows for the production of any length of rod. Clearly the extent of polymerisation is determined by the length of the RNA being encapsidated. One problem the virus faces during assembly is the removal of secondary and tertiary interactions from the viral RNA. Detailed mechanistic studies show that both 5' and 3' terminii of the RNA project through the central hole of the growing virus to emerge on the same side of the helical

Fig. 7. Proposed mechanism of virus assembly involving dislocation of the symmetrical two-layer disk to form the lock-washer, which is then able to form nicked helices, which then anneal to form the final structure

rod (Fig. 8). Presumably one portion of the RNA is closely associated with the viral coat proteins, whilst the other passes through the central hole seen in the disks. The RNA in such a hole must essentially be single-stranded allowing the new RNA-protein contacts to be made at the growing assembly point without concomitant removal of other interactions, which must, rather, occur progressively as the RNA is taken up into the hole. Recently the three-dimensional structure of the entire virus particle has been determined to 2.9 Å resolution by X-ray fibre diffraction techniques. This has allowed the results described above to be understood in molecular detail. The protein subunits in the virus make essentially the same lateral protein-protein contacts as seen in the X-ray crystal structure of the disk, i.e., within one layer. However, the axial contacts are dramatically different confirming the idea that dislocation of the disk must occur during viral assembly. There are multiple protein-nucleic acid contacts as expected. Modelling of the different bases at the sites of interaction with protein has suggested that the basis of packing selectivity relies on favourable RNA-protein interactions between the coat protein and the loop residues of the assembly initiation site.

1.2.1.2 Tomato Bushy Stunt Virus, an Icosahedral Shell[4]

Tomato Bushy Stunt Virus has a T = 3 icosahedral shell comprising essentially 180 copies of a single polypeptide chain (MW 40 kDa) encapsidating a single-stranded RNA of ≈4000 bases, and as expected there are three quasi-equivalent conformations for the coat protein subunit. The structure of the virus was the first to be determined to atomic resolution by X-ray crystallography. Figure 9 shows a cartoon representation of the structure of the virus capsid as determined X-ray diffraction and a

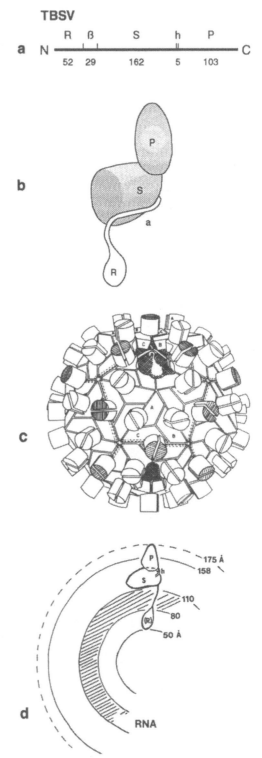

Fig. 9. Structure of TBSV. **a** Linear arrangement of domains in the polypeptide chain, *numbers* refer to amino acid sequence. **b** Overall fold of the subunit. **c** Packing of subunits in the viral shell. **d** Cartoon of a cross-section through the virus based on both X-ray and neutron diffraction results

cross-section of its organisation determined by neutron diffraction, which can identify regions of mostly protein or nucleic acid using the technique known as contrast matching.

In the viral capsid the coat protein subunits pack as noncovalent dimers, interacting with each other mostly via a hydrophobic interface between the projecting (P) domains. The P domain is connected to the rest of the polypeptide via a short hinge region, which leads to a second globular domain (S), which forms the shell of the virus. N-terminal to the S-domain is an extended region of polypeptide, the arm, which extends towards the centre of the virus particle and is lost from the electron density map. N-terminal to this region is a globular domain of 40 residues rich in positively charged amino acids, termed the R-domain because of its close association with the viral RNA. As predicted by quasi-equivalence, the coat protein subunits adopt three distinct conformations, A, B and C. In all these conformers the globular folding of the S or P domains is essentially identical; what does alter, however, is the orientation of the two domains relative to each other. This is achieved by conformational changes at the hinge region. The conformation adopted by the N-terminal arms is also dramatically different between the different quasi-equivalent subunits. In A/B type dimers the polypeptide comprising the arms leaves the underside of the S domain and immediately runs into the interior of the particle. However, in C/C type dimers the arms form an extra strand of β-sheet along the underside of the S-domain, running past each other to the particle threefold axis (where C/C dimers contact each other). Here they form an unusual element of secondary-structure with the arms from the symmetry related C/C dimers. This structure (Fig. 10) has been termed the β-annulus because the residues involved are essentially in a β-strand conformation. After the β-annulus the polypeptide chain again runs into the centre of the particle.

Assembly experiments with the close homologue Turnip Crinkle Virus (TCV) show that the purified viral components will self-assemble in vitro in an RNA sequence-specific fashion. The specificity is achieved via formation of an assembly initiation complex comprising three coat protein dimers which bind to specific RNA sequences distributed

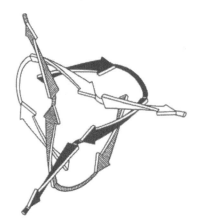

Fig. 10. Schematic representation of the β-annulus which forms at particle three-fold axes and may act as the switching signal for incoming coat protein dimers during assembly

throughout the viral genome. In essence the three-dimensional folding of the genomic RNA molecule generates a specific protein binding site. The proteins in this complex are of the C/C type interacting via a single β-annulus. In principle, this is all the information needed to ensure correct capsid formation, since the free N-terminal arms on each of the dimers of the complex are correctly positioned to form the next β-annulus in the structure as more coat protein dimers add to the growing shell. The region between the C/C dimers in the complex can accept dimers of the A/B type, thus correctly positioning curved or flat subunit dimers to create a shell of the correct size and symmetry (Fig. 11). After formation of the assembly initiation complex, all further RNA-protein interactions must essentially be nonsequence-specific.

Experimental support for this model comes from the use of coat protein subunits which have had their N-terminal arms removed by treatment with chymotrypsin, which cleaves close to the S-domain. Dimers can be prepared lacking either one or both arms. The assembly model suggests that dimers without arms would not really interfere with assembly since (a) they can not interact to form β-annuli and could thus dissociate readily from C/C positions (a form of editing) , and (b) could presumably be reasonably well tolerated at A/B positions. However, heterodimers ("one-arm bandits") can

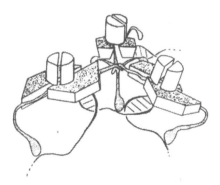

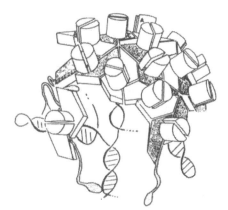

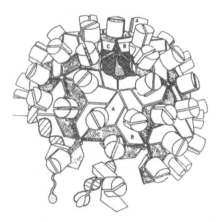

participate in forming a single β-annulus but would not be able to direct the formation of the next one, and hence, would poison the assembly process. Reassembly reactions spiked with proteolysed coat protein tolerate up to 40% by weight of homo-dimers but are completely poisoned by about 5% by weight of one-arm bandits.

1.2.1.3 Other Examples: Reovirus

A fascinating set of assembly problems are faced by viruses having segmented genomes. These include the rotaviruses which are a major cause of infant mortality in less developed areas of the world. The genomes of these viruses consist of between 10 to 12 dsRNA segments. Each virus particle contains a single copy of each segment, suggesting that there is a selective sorting mechanism for the genome fragments. The molecular details of this sorting remain unclear. However, there are some clues to the

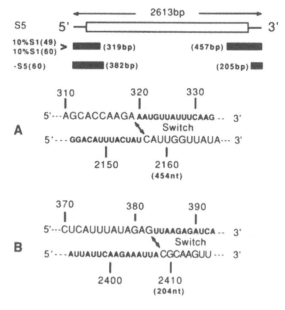

Fig. 11. Representation of the model for TCV self-assembly via formation of an assembly initiation complex, which consists of six coat protein subunits interacting via formation of a β-annulus in response to specific RNA sequences, followed by addition of further coat protein dimers from solution in a process akin to crystallisation. (Sorger et al. 1986)

Fig. 12 A, B. Sequence information retained by two different (**A** and **B**) DI RNA isolates of Wound Tumour Virus. In each case the bulk of the S 5 coding region is deleted and the remaining 5' fragments (319 nt for **A** and 382 nt for **B**) become joined to the corresponding 3' fragments (457 nt for **A** and 205 nt for **B**). The sequences around these switches are shown below. (Anzola et al. 1987)

mechanism which come from an analysis of some viral infections which slowly lose potency because of the build up of an abberant viral RNA known as defective interfering (DI) RNA. The sequences of DI RNAs show that they are deletions of the normal genomic segments. The deletions occur over the majority of the central section of the RNA's but always preserving the 5' and 3' terminal regions. The sequences within the terminal regions are able to form complementary base pairs between ends (Fig. 12).

The rules involving sorting appear to be: (1) that sequence information in the terminal domains is required for replication and packaging, (2) that 12 distinct segments are packaged irrespective of their size (i.e., DIs become packaged) and (3) that packaging of any particular segment excludes the packaging of a second copy of that segment. This suggest that the RNAs carry two sorting signals one which specifies viral RNA and an another which specifies genome segment number, the latter signal presumably requiring a unique recognition partner (either protein or RNA), which raises the problem of how these unique species are incorporated into the symmetrical viral capsid.

1.3 Structural Conservation in Icosahedral Viruses

1.3.1 A Conserved Protein Domain

A first determination of the three-dimensional structure of an icosahedral viral capsid occurred in 1978 when the work on Tomato Bushy Stunt Virus was completed. The techniques developed for this project were quickly applied to several other plant viruses which had been crystallised and, within a few years, two other viral capsid structures became known; namely Southern Bean Mosaic Virus (T = 3) and Satellite Tobacco Necrosis Virus (T = 1). The results revealed a striking conservation in the overall fold of the polypeptide forming the shell which encapsidates the viral RNA (S-domain for TBSV) (see Fig. 13). This fold consists of an eight-stranded β-barrel which presents a shallow,

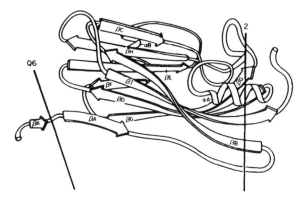

Fig. 13. Ribbon representation of the polypeptide fold of the S-domain of TBSV which has been conserved in the shell-forming portion of all RNA virus subunit structures determined to date with the exception of bacteriophage MS2

curving protein face to the inside of the viral particle, i.e., towards the RNA. This three-dimensional structural similarity had not been predicted on the basis of amino-acid sequence homology and suggested that this motif was involved in either RNA recognition or self-assembly or both.

Even more striking is the fact that since this period the structural homology has been shown to extend to simple RNA viruses from plants, insects and animals. These include polio and human rhino virus (HRV). With the exception of the MS2 RNA bacteriophage coat protein structure, which has recently been determined, the globular core of the polypeptide(s) enclosing the viral RNA in all simple spherical RNA viruses displays an essentially similar three-dimensional fold. Thus a protein motif involved in icosahedral shell formation and RNA encapsidation appears to have been evolutionarily conserved. Variations in structure are restricted to either end of this domain (e.g., the P domain of TBSV or the N-terminal arms and RNA-binding domains of some viruses) or to discreet insertions into the conserved motif such as the large antigenic loops of polio and HRV. Coat proteins from other RNA viruses of different symmetry (e.g., TMV) or from icosahedral DNA viruses such as polyoma or SV40 cores do not have this conserved motif.

1.3.2 Evolutionary Considerations[5]

Three-dimensional structures of biological macromolecules are generally conserved over far larger time spans than primary protein sequences. Many of the structural similarities described in Section 3.1 could not have been detected by protein homology searches. Three-dimensional protein structures can be used for alignment when amino acid sequences have diverged greatly. In homologous structures, differences tend to correspond to deletions and insertions, whilst the overall folding motif is preserved. The number of such changes is an indication of the degree of evolutionary divergence. Viral coat protein genes are usually the least conserved element of viral genomes in sequence comparisons, which makes the degree of similarity in the tertiary structure of icosahedral viral capsid proteins even more remarkable.

Even for nonicosahedral viruses which do not have the conserved protein fold, detailed structural analysis reveals an underlying homology suggesting that modern viruses could be descended from a common ancestor. Examination of sequence conservation among the most conserved viral elements, namely the RNA polymerase (replicase) genes can be used to construct an evolutionary tree for plant, bacterial and animal viruses. As more viral structures become available, these comparisons will be expanded, and should allow speculations to be made about the nature of the ancestral virus.

1.4 Outlook

It is widely assumed that the morphogenesis of plants and animals depends on the construction of a cytoskeleton, itself an assembly of macromolecules. Thus, even macroscopic events may ultimately be determined by such interactions. It is, therefore, important to extend the work on the systems described here in order to understand fully all their implications for the control of morphogenesis and to add to the list of available examples which can be studied at the molecular level, especially those that exhibit novel features. One area which will clearly advance rapidly is the determination of the structures of simple viruses by X-ray crystallography. Even during the preparation of this chapter the high resolution structures of TMV, MS2 and canine parvovirus were published for the first time and work on many other systems is in progress.

Important questions to be answered involve the evolution of macromolecular assemblies. How, for instance, do quasi-equivalent subunits evolve? Do such subunits have inherent structural flexibility so that they could perhaps form $T = 1$ or 3 or 4 etc. shells, depending on a few mutational changes? How is quasi-symmetry broken in a defined way in shells which contain deviations from ideal surface lattices? How can mutational change give rise to complicated "molecular machines" such as T4, which seem so much more than the sum of their parts? For some of these systems the combination of an X-ray structure and clones of the viral coat proteins should allow experiments aimed at addressing these questions to begin. Recent results on the chemical synthesis of complex organic molecules, including one containing the biological component poryphrin, suggest that self-assembly itself is not restricted to biological macromolecules, but is a more general property of interacting species.

To paraphrase Clausewitz, biology could be described as chemistry carried on by other means. Very often the chemistry of living things is hidden from view. The first secret of life was hidden in the three-dimensional structure of DNA. Monod described allosteric protein control as the second. The interactions between macromolecules to generate functional complexes, so beautifully illustrated by the simple viruses, is a third. It is an area of research where progress depends on synergic interaction between traditional scientific disciplines such as chemistry, genetics and cell biology, which makes it one of the most exciting areas for future endeavour.

1.5 Summary

We have seen that in a number of small viral systems, apparently purposeful behaviour is the direct result of well-defined interactions between macromolecules. Thus simple viruses, which exist on the borderline between the living world and the inorganic world, and yet perform many of the biological functions characteristic of living systems, such as replication and evolutionary change, operate solely in terms of the laws of physics and chemistry. Viruses are thus a reductionist's paradigm for the behaviour of more complex systems. The molecular mechanisms used by viruses apparently have close analogues in many areas of cellular functioning and it is perfectly possible for very complex functions to result directly from simple noncovalent interactions between macromolecules. Genetic information should therefore not be seen as merely a collection of linear nucleic acid and protein sequences rather it is a spatio-temporal interaction of these sequences with both solvent, i.e., water, and with each other. These ideas would not have surprised Monod. As one of the founders of the idea that allosterically regulated proteins could co-ordinate the chemical events of living cells, the regulations of complex and apparently purposeful morphogenetic events via a series of stereo-specific interactions between macromolecules would surely seem merely an extension of the same theme.

References

General

Knolle P, Hohn T (1975) Morphogenesis of RNA phages. In: Zinder ND (ed) 'RNA phages'. The Cold Spring Harbor Laboratory, New York, pp 147–201

Monod J (1972) Chance and necessity. Collins

Rossmann MG, Johnson JE (1989) Icosahedral RNA virus structure. Annu Rev Biochem 58:533–573

[1] Caspar D, Klug A (1962) Physical principles in the construction of regular viruses. In: Cold Spring Harbor Symposium, vol XXVII. The Cold Spring Harbor Laboratory, New York

[2] Mosig G, Eiserling F (1988) In: Calender R (ed) The bacteriophages, vol 2, Plenum Press, New York, pp 521–606

[3] Butler PJG (1984) The current picture of the structure and assembly of tobacco mosaic virus. J Gen Virol 65:253–279

Durham ACH, Finch JT, Klug A (1971) States of aggregation of Tobacco Mosaic Virus protein. Nature New Biol 229:37–50

Hirth L, Richards KE (1981) Tobacco mosaic virus: model for structure and function of a simple virus. Adv Virus Res 26:145–199

Lebeurier G, Nicolaieff A, Richards KE (1977) Inside-out model for self-assembly of tobacco mosaic virus. Proc Natl Acad Sci USA 74:149–153

Turner DR, Butler PJG (1986) Essential features of the assembly origin of tobacco mosaic virus RNA as studied by directed mutagenesis. Nuc Acids Res 14:9229–9242

[4] Anzola JV, Xu Z, Asamiza T, Nuss DL (1987) Segment-specific inverted repeats found adjacent to conserved terminal sequences in wound tumor virus genome and defective interfering RNAs. Proc Natl Acad Sci USA 84:8301–8305

Harrison SC, Olson AS, Schutt CE, Winkler FK, Bricogne G (1978) Tomato bushy stunt virus at 2.9 Å resolution. Nature (London)276:368–373

Hogle JM, Maeda A, Harrison SC (1986) Structure and assembly of turnip crinkle virus. 1. X-ray crystallographic structure analysis at 3.2 Å resolution. J Mol Biol 191:625–638

Sorger PK, Stockley PG, Harrison SC (1986) Structure and assembly of turnip crinkle virus. II. Mechanism of reassembly in vitro. J Mol Biol 191:639–658

Stockley PG, Kirsch AL, Chow EP, Smart JE, Harrison SC (1986) Structure and assembly of turnip crinkle virus. III. Identification of a unique coat protein dimer. J Mol Biol 191:721–725

Wei N, Heaton LA, Morris TJ, Harrison SC (1990) Structure and assembly of turnip crinkle virus. VI. Identification of coat protein binding sites on the RNA. J Mol Biol 214:85–95

[5] Rao TVS, Lawrence DS (1990) Self-assembly of a threaded molecular loop. J Am Chem Soc 112:3614–3615

Chapter 2   *Bacillus subtilis* Sporulation: a Paradigm for the Spatial and Temporal Control of Gene Expression

JEFF ERRINGTON

2.1 Introduction

2.1.1 *B. subtilis:* One of the Best-Known of All Organisms[1]

B. subtilis was one of the first species of bacteria to be characterized, as long ago as 1876. Very soon it was recognised as being unusual among bacteria because it could exist in two forms with very different morphological and physiological properties. Under nutritionally rich conditions it grew as a typical rod-shaped bacterium, elongating and then dividing by binary fission. However, on starvation it formed an unusual life form now known as an endospore. Endospores are very remarkable structures. They are extremely resistant to treatments such as boiling, that would rapidly kill more conventional forms of life, including most bacteria. They are also dormant forms, and can remain so for inestimable periods, certainly many thousands of years, if necessary.

At about the beginning of the 1960s, it was recognised that spore formation was a simple form of cellular differentiation. Being much less complex organisms than eukaryotes, and easy to manipulate, bacilli soon attracted the attention of developmental biologists, particularly those interested in studying the genetic and molecular aspects of development. From about the same time, the spectacular successes of studies on another bacterium, *Escherichia coli*, were probably a driving force behind the popularity of sporulation. The second crucially important reason for the emerging interest in *B. subtilis* was the discovery that this organism was naturally transformable with naked double-stranded DNA. This discovery has had far-reaching consequences, both in the early days of classical genetic analysis of sporulation and also in the modern era of molecular genetics. The transformation system has been exploited with immense success to study and manipulate the genes controlling development and it can be argued that the tools available to the *B. subtilis* geneticist are now more versatile and powerful than those of any other organism, including *E. coli*!

2.1.2 The *B. subtilis* Life Cycle[2]

The life cycle of *B. subtilis* comprises two alternative phases, vegetative growth and sporulation (Fig. 1). The former can be considered as a proliferative phase in which each cell divides to give two identical daughter cells. The latter comprises a developmental process in which two sister cells differentiate. Vegetative proliferation continues for as long as the state of the external medium can sustain growth. Depletion of an essential nutrient, usually limitation for the source of carbon, nitrogen or phosphorous, initiates the developmental phase of the life cycle.

The first morphological feature characteristic of sporulating cells is typical of many differentiating systems – an asymmetric cell division. Each of the two unequal cells that arise then undergoes its own distinct series of developmental changes, culminating with the release of the mature endospore by lysis of its enveloping mother cell. Of the two sister cells, then, one is chosen to be the "germ line"; the other, a "somatic cell", undergoes terminal differentiation and is discarded. The development of the two cells is highly coordinated, not least because the developing endospore, or prespore as it is known in its immature state, is engulfed inside the mother cell cytoplasm at an early stage of the process. The prespore thus depends on the mother cell for the generation of energy and is afforded a degree of protection. The larger resources of the

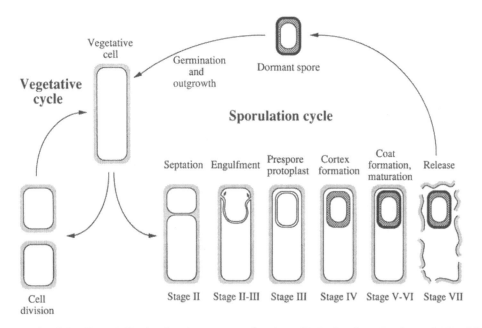

Vegetative cycle

Sporulation cycle

Vegetative cell

Germination and outgrowth

Dormant spore

Septation   Engulfment   Prespore protoplast   Cortex formation   Coat formation, maturation   Release

Cell division

Stage II   Stage II-III   Stage III   Stage IV   Stage V-VI   Stage VII

Fig. 1. The life cycle of *Bacillus subtilis*, showing the separate phases of vegetative growth and sporulation. The vegetative cycle is proliferative and comprises alternating rounds of growth by elongation and central division. On starvation, the sporulation cycle is chosen. It begins with a modified cell division, in which an asymmetric septum is produced. A series of intermediate morphological states then ensues, leading to the production of a mature spore. Completion of the septum is defined as *Stage II*. The edges of the septal disc, which comprises a pair of lipid bilayers, migrate towards the pole of the cell, engulfing the smaller cell, known now as the prespore, within the cytoplasm of the larger

"mother cell". During *Stage IV*, a layer of cell-wall-like material, known as the cortex is synthesised in the space between the membranes surrounding the prespore. In *Stage V*, a multilayered proteinaceous coat is deposited on the outside surface of the spore. During the final period of development, known as "maturation", the extreme resistance and dormancy properties that typify the mature spore develop, with little overt change in morphology. Finally, the completed spore is released by lysis of the mother cell (*Stage VII*). The cycle is completed by spore germination and outgrowth of a new vegetative cell, which occurs when conditions suitable for the resumption of growth recur

mother cell are mainly spent in contributing to the protective layers known as the cortex and coat, which are deposited around the prespore as development proceeds.

Once completed, the prespore has remarkable resistance properties against heat, desiccation, organic solvents, lytic enzymes, ionizing radiation, and the like. Its dormancy is tightly controlled so that the germination response, which is followed by the resumption of vegetative growth, occurs only under appropriate nutritional conditions.

2.1.3 Advantages and Disadvantages of *B. subtilis* as an Experimental System [3]

A wealth of previous work and background information is not always a good reason for continued study of a system that has outlived its usefulness.

But in the case of *B. subtilis* the enormous investment in prior work is a priceless asset. The fact that such an amount of painstaking work has been done on the basic morphological, biochemical and physiological events accompanying sporulation has greatly facilitated the yield of useful information from the more fashionable molecular genetic approaches applied recently. In organisms in which there is little or no such background information the payoff from molecular genetics is relatively poor. Moreover, the incentive to initiate such background work is low, as such work is less fashionable and not as overtly productive as molecular biology.

What are the specific experimental advantages of sporulation? In general, bacterial cells have a much simple organization than eukaryotic cells and their basic biochemistry and physiology are much better understood. Bacterial cells are easy to grow and they can grow rapidly, doubling in as little as 20 min. Sporulation is easy to induce and it involves only two differentiating cells. Nevertheless, it exhibits many of the hallmarks of more complex developmental systems. Some of the most basic questions of development are thus readily accessible in this organism and they are considered in more detail below.

Another major reason for studying *B. subtilis*, however, lies in its sophisticated genetics. *B. subtilis* cells are naturally transformable with DNA. They require no manipulations other than culturing in a simple growth medium. The cells then take up DNA, in either linear or cirucular double-stranded form, and highly active recombination systems direct crossing over between the incoming DNA and homologous DNA resident in the cell. Integration

of the transforming DNA is efficient so long as there is a stretch of at least partially homologous DNA in the cell. Some specific applications of the phenomenon, such as gene replacement and plasmid integration, will be illustrated below. *B. subtilis* also has a number of useful temperate bacteriophages, of which ϕ105 and the larger SPβ, are particularly important. Like phage lambda of *E. coli*, they can exist in a stable chromosomally integrated state known as a prophage, or undergo lytic growth producing large numbers of progeny virus. ϕ105 and SPβ have been instrumental in the cloning of developmental genes and the analysis of their regulation, as we shall see.

B. subtilis does have some limitations as a model system for development. Clearly, it does not undergo multicellular development, and cell-cell commu-

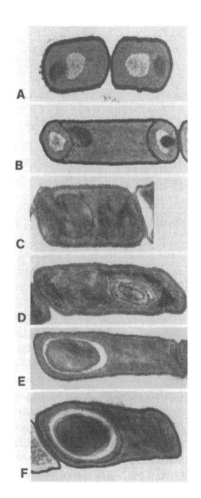

Fig. 2 A–F. Examples of the morphological properties of mutants defective in sporulation. Electron micrographs of the ultrastructural morphologies of several mutants defective in sporulation are shown. Some of the mutants are arrested in forms that are observed as normal intermediates in development. Others produce aberrant morphologies unlike anything observed with wild-type cells. **A** *spoOA*. The mutation blocks sporulation at the earliest stage. The mutant fails to make an asymmetric septum and goes on to complete an additional central septum instead, leading to the production of unusually short cells. **B** *spoIIA*. The mutant completes an apparently normal sporulation septum but proceeds no further in normal development. Instead, it has produced a second "symmetry restoring septum", giving an aberrant phenotype known as "abortively disporic". **C** *spoIID*. This mutant proceeds to an intermediate stage of prespore engulfment and is then completely blocked. The bulging of the prespore cytoplasm into the mother cell is normal and observed at an intermediate stage of development in wild-type cells. **D** *spoIVA*. The mutant has a normal phenotype until stage III (engulfment) is completed. However, cortex synthesis is reduced and although some spore coat proteins are synthesised, assembly is aberrant and the proteins aggregate in the mother cell rather than on the surface of the spore. **E** *spoIVB*. Cortex synthesis is almost complete but the spore coat layers are virtually absent. **F** *spoVF*. This mutant produces apparently normal cortex and coat layers but the final maturation of the prespore is defective. **C** and **D** reprinted from Piggot, PJ and Coote, JG (1976) *Bacteriol Revs.* **40**:908–982 (with permission). **E** and **F** from Coote, JG (1972) *J Gen Microbiol* **71**:1–16 (with permission)

nication, other than between prespore and mother cell, does not seem to play a prominent role. Also, the small size means that the details of morphological events can often only be resolved by electron microscopy.

2.2 The Powerful Molecular Genetic Approach

2.2.1 Defining the Developmental Genes[4]

Many of the genes involved in the regulation of sporulation were identified during the 1960s and 1970s by pleiotropic mutations called *spo*, which had no detectable effect on vegetative growth but caused a more or less complete block in sporulation. Often these mutations produced a blocked phenotype that was similar to an intermediate step in normal development. Others produced aberrant phenotypes in which development proceeded normally to a certain stage but abnormal changes then occurred. Some examples of the ultrastructural phenotypes of *spo* mutants are shown in Fig. 2. During normal sporulation, a number of easily measured biochemical events, such as the synthesis of specific enzymes, were known to occur at specific and reproducible times, so that they could be used as temporal "marker events" for progress through the developmental programme. Generally, both the morphological and biochemical events were blocked at a similar stage in each *spo* mutant. This suggested that the *spo* genes encoded regulatory proteins and that they acted in a linear-dependent sequence. Each gene in the sequence would be responsible for controlling both morphological and biochemical events at a particular stage of development, and for the activation of the next regulatory gene in the cascade. Unfortunately, at that time it was not possible to deduce the precise functions determined by the various *spo* gene products. This had to await the development of molecular genetic techniques and DNA sequencing.

In addition to the *spo* genes, developmental genes have been identified by various other kinds of mutations. One important class are the *ger* genes, which are needed for the germination response of the mature spore. Since spore germination does not require *de novo* gene expression, the proteins needed for this response must be built into the spore during its development. Thus, the *ger* genes are actually expressed during sporulation and the distinction between *ger* and *spo* genes is often an arbitrary one. For example, some of the genes originally identified by *ger* mutations have been found, on subsequent analysis, to exhibit defective sporulation.

The largest classes of gene that have been identified more recently encode the major structural components of the spore; the spore coat proteins (*cot* genes), and the small acid-soluble proteins of the spore core (*ssp* genes). These were isolated by using N-terminal amino acid sequence data from the purified proteins to design oligonucleotide probes for the genes. Surprisingly, few, if any, of the latter genes correspond to previously identified *spo* genes, despite the fact that they encode some of the most abundant proteins of the spore. It is now clear that individual mutations in these genes do not lead to a Spo phenotype because there is functional redundancy within these families of proteins; loss of a single member of a family has relatively little effect on the overall structure. The *ssp* and *cot* genes have been useful in studying the regulation of gene expression during sporulation. Not surprisingly, however, in most cases these experiments have led no further than to the *spo* genes already identified as important regulators.

2.2.2 Cloning and Physical Characterization of *spo* Genes[5]

The key to unravelling the complex interactions between the products of the *spo* genes, and the details of their functions, lay in molecular cloning. The most successful approach, used to clone virtually all of the *spo* genes, was based on the use of the temperate phage $\phi 105$ and genetic complementation. Because *spo* mutations abolish spore formation, and spores are extremely resistant to treatments with heat, organic solvents etc., rare sporulating cells containing a recombinant phage with a fragment of DNA that complements a particular *spo* mutation can be selected directly from

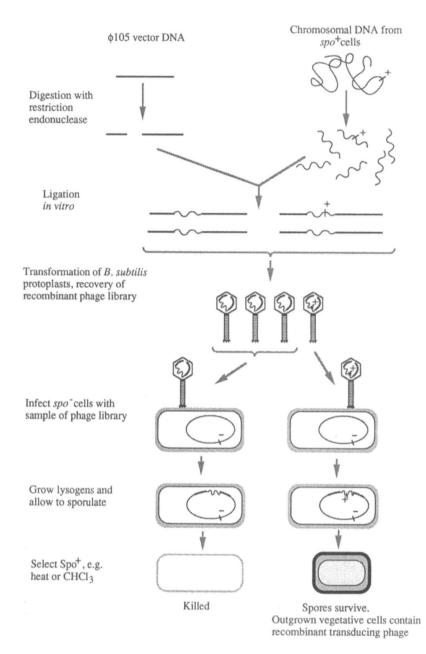

Fig. 3. Cloning *spo* genes with bacteriophage ɸ105. Vector and chromosomal DNA samples are digested with a restriction endonuclease and ligated in vitro. Transformation of a sensitive strain of bacteria results in lytic phage growth and the release of a large number of recombinant phage progeny, which constitute a genomic library. Recombinants containing any *spo* gene can be selected directly from a sample of the library because they will restore the ability to sporulate to an otherwise asporogenic mutant. The symbols + and − refer to wild-type and mutant alleles of a *spo* gene

large numbers of nonsporulating mutant cells (Fig. 3). Thus, from only two genomic libraries, a series of ϕ105 derivatives collectively capable of complementing virtually any of the known *spo* mutations were isolated. To complement a *spo* mutation the DNA insert in a recombinant phage must contain both the complete functional *spo* gene, together with the *cis*-acting regulatory signals necessary for correct expression during development. Thus, the recombinant phages provided DNA that could be used to study both the functions of the *spo* gene products, and the nature of the regulatory mechanisms determining the timing and spatial localization of gene expression during sporulation.

Various techniques taking advantage of the *B. subtilis* transformation system greatly facilitate the analysis of the boundaries of each developmental gene or operon. Techniques such as the one illus-

trated in Fig. 4 are important for accurate DNA sequence interpretation. By application of these methods of cloning and physical analysis, many of the genes involved in the regulation of sporulation have been identified and partially characterised.

2.2.3 Pathways of Gene Expression[6]

The powerful methods of gene manipulation also facilitate the analysis of gene expression. A particularly important procedure is based on the use of gene fusions (Fig. 5), in which the *spo* gene of interest is fused by recombinant DNA methods to a "reporter gene", usually the *lacZ* gene of *E. coli*. The reporter gene encodes an easily assayed product, in this case the enzyme β-galactosidase, the activity of which then provides a convenient indicator of gene

Fig. 4A–C. Defining the boundaries of the transcription unit (TU) of a sporulation operon by integration plasmid analysis. **A** and **B** show the results of transformation of two different integrative plasmids into wild-type *B. subtilis*. *Thin lines* denote sequences of plasmid origin, *thick lines* are used for the host chromosome and the homologous segment of DNA cloned in the plasmid. The *box* indicates a *spo* TU (comprising both structural genes and promoter). Integrative plasmids are maintained and manipulated in *E. coli* by standard methods but they cannot replicate autonomously in *B. subtilis*. Selection for chloramphenicol resistance results in transformants in which the plasmid has integrated into the host chromosome by a single crossover homologous recombination event as shown. Integration by this mechanism generates structures in which the plasmid sequences are flanked by directly repeated sequences of host origin. In **A** segment of DNA cloned in the plasmid is internal to the TU, as denoted by the *zig-zag ends* to the box. Integration results in two inactive copies of the TU: one is truncated at the promoter proximal end, the other at the promoter distal end. The transformants are therefore phenotypically Spo. In **B** the insert in the plasmid extends beyond one end of the TU. Insertion of this plasmid again results in a partial duplication of the TU but this time one of the copies is functional and the transformants remain Spo$^+$. **C** summarizes the results obtained by transforming a series of integration plasmids into wild-type *B. subtilis* and observing the resultant phenotypes. Insertions internal to the TU generated a mutant phenotype, as indicated by the *solid bars*. Insertions *extending beyond one or other end* of the TU result in a wild-type (Spo$^+$) phenotype. Comparison of the physical maps of the cloned DNA segments with the phenotypic effects of integration allowed the ends of the TU to be defined

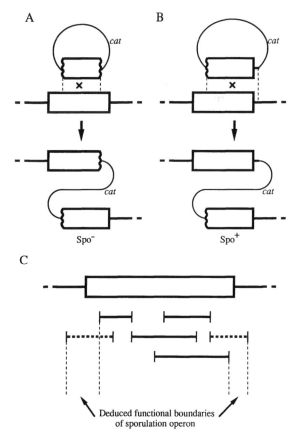

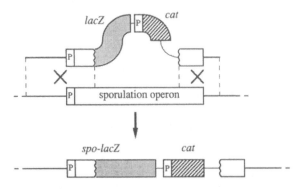

Fig. 5. Disruption of a sporulation gene and construction of a *spo-lacZ* fusion. Fragments of DNA comprising the promoter proximal and promoter distal segments of a sporulation operon are ligated in vitro to a *lac cat* cartridge. The latter consists of a selectable chloramphenicol resistance gene and a promoterless *lacZ* gene encoding the "reporter" enzyme β-galactosidase. Transformation of wild-type *B. subtilis* with selection for chloramphenicol resistance results in transformants in which the cartridge has integrated by a double-crossover homologous recombination event as indicated by the crosses. The cartridge has replaced part of the sporulation operon, which is therefore disrupted and non-functional. However, the promoter (*P*) of the operon is still functional and can drive transcription of the previously inactive *lacZ* gene. β-galactosidase synthesis can now be used as a convenient measure of the activity of the promoter during growth and development

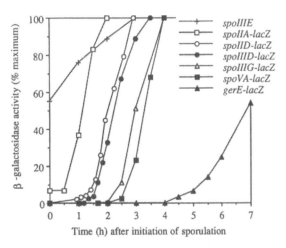

Fig. 6. Time courses for the expression of various *spo* genes during sporulation as measured with a *lacZ* reporter gene. Various *spo* genes were fused to *lacZ* and the production of β-galactosidase was used to follow their expression. Each of the genes shows its own characteristic time of induction during development. *spoIIIE* is unusual in that it is expressed during both growth and sporulation. The total period during which changes in gene expression occur occupies at least 4 h

expression. Note that in this case integration of the heterologous DNA occurs by double crossover homologous recombination, which deletes and replaces part of the *spo* operon. Gene fusions have been used to derive three important kinds of information about the regulation of *spo* gene expression.

1. Timing of Gene Expression. Virtually all of the *spo* genes required for the intermediate steps of development (i.e., other than the *spo0* genes described below) are expressed only during development. Figure 6 shows that each gene displays a characteristic temporal pattern of expression. Many different temporal classes have been identified, showing that the regulation of expression is both complex and diverse. The ordered activation of successive sets of genes is one of the hallmarks of cellular development in all organisms.

2. Spatial Localization of Expression. Division of the prespore and mother cell occurs about an hour after the induction of sporulation. Genes that begin to be expressed after this time have the possibility of being expressed in different cellular compartments. Remarkably, all such genes that have been studied so far seem to show cell-specific expression. Thus, from the time of septation onwards, different programmes of gene expression occur in the prespore and mother cell.

3. Dependence Relationships. By taking a fusion of *lacZ* to gene *spoX* and introducing it into a strain with a mutation in gene *spoY* it is possible to determine whether the product of gene *spoY* is needed for expression of gene *spoX*. By repeating this experiment with a series of different *spo* mutations, the "dependence pattern" for gene *spoX* can be determined. This kind of analysis has now been applied to many different *spo-lacZ* fusions. What has emerged is a picture of the overall pattern of gene expression, as shown in Fig. 7. It is clear that the genes are arranged in a complex interacting hierarchy, just as in many eukaryotic developmental systems. The pattern of interactions branches at about the time of septations. Thereafter, separate programmes of gene expression occur in the two cells. The apparent linear nature of the interactions predicted from earlier analysis of the morphological properties of *spo* mutations (see above) can be explained by intercellular interactions that operate at several key points in the process. Some of these interactions, such as the requirement of the *spoIVB* gene in mother cell development, will be considered in more detail below.

Elucidation of this pattern of interactions has been an important achievement because it has led to the identification of *spo* genes that control key steps in the developmental programme. Some of these key steps will be discussed in detail later in this chapter.

2.2.4 Transcriptional Control of *spo* Gene Expression [7]

Dependence patterns provide information about the pattern of interactions between *spo* genes but they do not lead directly to the functions of their products. However, some of the mechanisms underlying these interactions have been deduced by various means. Thus, the solid arrows in Fig. 7 indicate direct transcriptional effects. A particularly important class of molecules known to be involved in the regulation of sporulation are the sigma factors. These are subunits of RNA polymerase that

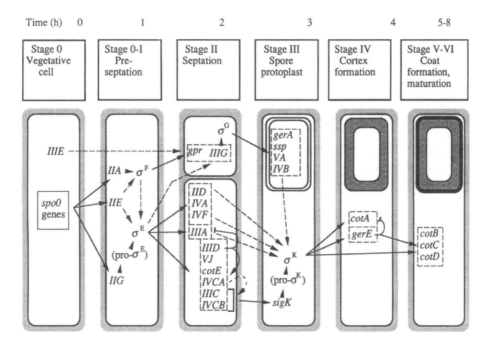

Fig. 7. Pattern of interactions between the genes controlling spore formation in *B. subtilis*. The basic morphological stages are shown with an approximate time scale. Sporulation genes are placed on the figure according to the time at which they begin to be expressed, and after septation, in the appropriate cellular compartment. For clarity, the *spo* gene designations have been abbreviated. Four sigma factors are known to be involved, of which two are synthesised as precursors (pro-σ^E and pro-σ^K). Genes with promoters recognised by the same sigma factor are *grouped together in dotted boxes*. The *solid arrows* indicate direct effects on transcription, either positive regulation (*lines ending in an arrowhead*), or negative regulation (*lines ending in*

a bar). *Dotted arrows* indicate other kinds of effect on gene expression, mainly of an as yet unknown nature. The *bracket* indicates a remarkable site-specific recombination event that brings together the two separate segments of the hybrid *sigK* gene encoding the precursor of σ^K. This event occurs only in the mother cell relatively late in development. Its general significance is unclear because molecular genetic experiments indicate that the recombination per se is not essential for sporulation. Indeed, closely related spore forming bacilli have an undivided *sigK* gene throughout growth and sporulation. No attempt has been made to represent the complex series of interactions between the *spo0* genes (see Fig. 9)

determine promoter specificity. That is, each sigma factor is responsible for the recognition of a particular class of promoter sequences by RNA polymerase. Thus, the appearance of a new form of sigma factor during development can switch transcriptional activity so that a whole new set of genes become expressed. "Consensus" sequences for the promoters recognised by the sigma factors involved in sporulation are shown in Fig. 8.

The functions of the sporulation-specific sigma factors are apparent from an appraisal of Fig. 7. For example, *spoIIIG* encodes σ^G, which directs expression of many of the prespore-specific genes. Thus, the arrows leading from *spoIIIG* to the collection of prespore-specific genes can be explained by a direct transcriptional activation mediated by σ^G. The four sigma factors synthesized during sporulation clearly effect major switches in the spectrum of genes transcribed during development and differentiation.

The sigma factors were first identified by in vitro experiments using "core" RNA polymerase, devoid of any sigma factor, which has catalytic activity but does not initiate transcription at defined sites. Addition of purified sigma factor can direct accurate transcription initiation on a defined promoter template. Different sigma factors can be purified by using a series of promoter containing DNAs. Although the first few sigma factors were defined in this way, they are now known to comprise a large family of proteins with closely related amino acid sequences. Thus, σ^F and σ^G were first identified by DNA sequence analysis of *spo* genes.

Although sigma factors are undoubtedly important regulators of gene expression during sporulation, and they control key steps in the process, as we will see below, the story does not end there. It is clear that many of the arrows in Fig. 7 are mediated by other kinds of regulatory proteins. In a few cases the nature of the mechanism is known. Thus, the Spo0A, SpoIIID and GerE proteins are transcription factors that work in conjunction with the RNA polymerase holoenzyme. The mechanisms underlying several of the other steps in the regulatory cascade are still unresolved. In the next few sections some of the more interesting and better understood features of the developmental process in *B. subtilis* are discussed in more detail.

2.3 Stage 0: Proliferation or Development[3]

2.3.1 What Factors Influence the Decision to Initiate Sporulation?[8]

The decision to sporulate or to continue proliferating can be crucial to the fate of a cell or a population of cells. Vegetatively growing cells face such a decision as they begin to run out of nutrients. Continued growth, allowing perhaps one more division would be futile if both daughter cells ultimately starved. *B. subtilis* cells, perhaps because their natural habitat in the soil is so variable and generally poor, have a wide repertoire of adaptive responses available to them as they approach the stationary phase of growth. They can begin to produce a variety of extracellular hydrolytic enzymes that may release a new source of energy. They can become motile and swim off to a more profitable micro-environment. They can take up DNA and so acquire genes that would increase their chances of surviving. They can produce antibiotics in the hope of elimi-

	−35			−10
	·			·
σ^A	TTGACA	17–18 bp		TATAAT
σ^E	GT-ATA	16–17 bp		ATACAAT
σ^F	TGCAT-	17–18 bp		-A-A-T
σ^G	TGAATA	17–18 bp		CATACTA
σ^K	GTCACA	15–16 bp		CATAA-ATA

Fig. 8. Comparison of the promoter sequences recognised by RNA polymerase containing the major vegetative sigma factor σ^A and the sporulation specific sigma factors. As in most other bacterial promoters two blocks of conserved sequences centred at about −10 and −35 with respect to the start site of transcription (+1) are probably involved in promoter recognition by RNA polymerase. The most frequent base at each position is shown; a hyphen if no single base is prefered. The σ^F recognition sequence contains several hyphens because it is based on a relatively small number of promoter sequences. Although some of the pairs of recognition sequences are quite similar, particularly σ^E and σ^K, and σ^F and σ^G, there are sufficient differences for the proteins to discriminate between different types of promoter

nating competitors and releasing nutrients. Finally, when vegetative growth becomes too difficult, they sporulate. But this decision is not taken lightly, as it commits at least one cell (the mother cell) to death. Also, if conditions favouring growth recur, the cell committed to sporulation is at a considerable disadvantage because of the time lost before it can resume growth. Not surprisingly, then, the decision to initiate sporulation is controlled by a very elaborate network of signal transduction pathways, which integrate the information from various sensory devices monitoring both internal and external conditions. In addition to the nutritional stimuli, it is also clear that the cell can monitor progress through its cell division cycle, and that initiation of sporulation can only occur at a specific point in the cycle. Extracellular communication may also be involved. The cells appear to secrete an extracellular factor that is used to determine the density of the population. Sporulation is induced poorly on starvation of cells below a threshold culture density.

Little is yet known about the sensory mechanisms involved in monitoring the various parameters to which the efficiency of sporulation is linked. However, some progress has been made on the genetics and biochemistry of the computational machinery that integrates the various sensory inputs and controls the decision to sporulate.

2.3.2 The Roles of the *spo0* Gene Products[9]

The products of at least seven genes (called *spo0* genes) are involved in controlling the decision to sporulate, of which the key player is encoded by *spo0A*. The Spo0A protein is a transcriptional activator or repressor of many genes. The key to its function lies in its state of phosphorylation (Fig. 9). It is a member of a family of bacterial proteins that are involved in environmental signalling and responsiveness. Like other related proteins, its transcriptional activity is modulated by phosphorylation, and it can receive phosphate groups from at least two different histidine kinase proteins called KinA and KinB. Both kinases probably respond to information from the internal or external environment, on the nutritional state of the cell, although the precise parameters they measure are not yet clear. The process is further complicated by the existence of phosphorelay proteins encoded by the *spo0F* and *spo0B* genes. These act as intermediate carriers transferring phosphate groups between the

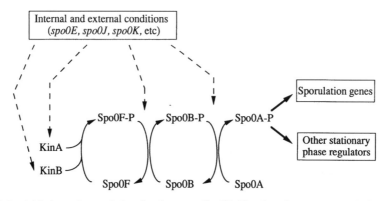

Fig. 9. Regulation of the initiation of sporulation involves a phosphorylation/desphosphorylation cascade. The key effector of early sporulation gene expression is a transcriptional activator/repressor protein, Spo0A. The activity of this protein is influenced by its state of phosphorylation, and the probability of the cell entering sporulation is proportional to the concentration of the phosphorylated form (Spo0A−P). Phosphorylation of Spo0A occurs via a series of phosphotransferase reactions involving intermediate phosphate carrier proteins, Spo0F and

Spo0B. The phosphate groups are derived originally from either of at least two different protein kinases, KinA and KinB. At each step in the cascade, information from intracellular or extracellular monitoring devices may be introduced, so as to stimulate or inhibit the production of Spo0A−P and thus modulate the probability of the cell initiating stationary phase adaptive responses, such as sporulation. The precise nature of the response shown by the cell is influenced by a further complex set of interactions involving several "transition state" regulatory proteins protein kinases and Spo0A. Both steps in the relay are also probably modulated by internal or external environmental cues; for example, the state of the cell division cycle or the density of the population of cells. Clearly this decision is an important feature of many developmental systems, where the first step in differentiation involves cell division.

The decision as to which of the various stationary phase responses should occur is partly determined by the interactions between Spo0A protein and a collection of additional regulatory proteins called "transition state" regulators. The products of these genes form another complex interacting hierarchy which is only partially understood. Clearly, the regulation of gene expression as cells enter stationary phase is very complex and it will not be understood in detail for some time to come.

In addition to the Spo0A protein, at least one "minor" sigma factor, σ^H, encoded by the *spo0H* gene, plays an important role in stationary phase gene expression. Some but not all of the genes involved in the early stages of sporulation are transcribed by the σ^H form of RNA polymerase. The overall role of this protein is not yet clear.

2.4 Stages II to III: Generation of Cellular Asymmetry [4]

2.4.1 Cellular Asymmetry [10]

About 40 min after a vegetatively growing *B. subtilis* culture is starved, some of the cells begin to complete their final rounds of DNA replication. About 30 min later they divide, and the two recently completed daughter chromosomes are segregated. Unlike vegetative cell division, during sporulation, division is asymmetric; the division septum being placed near one end of the cell. Such cellular asymmetry is characteristic of many developmental systems but in none is its molecular basis understood. *B. subtilis* sporulation would seem to be a good model system in which to approach the problem of asymmetric cell division. Unfortunately, the mechanism of symmetric septation in vegetatively growing *B. subtilis*, or in fact in any rod-shaped

bacterium, is not yet understood. Unlike eukaryotes, bacteria have a single circular chromosome. There is no known equivalent of the eukaryotic centromere / spindle mechanism in bacteria. Thus, the mechanism for segregating daughter chromosomes at cell division is unclear, as is the way in which the cell "measures" its length so that it can place the division septum precisely at the mid point of the cell. The most attractive proposition at the moment is that the daughter chromosomes segregate from each other by simple electrostatic repulsion and that the septum forms in the space between the separated chromosomes. Because of the sheer bulk of the chromosomes, they do not move far from the mid point of the cell.

A tentative model for asymmetric septation in *B. subtilis* based on an extension of this proposal for vegetative septation is shown in Fig. 10. The proposal is based on two assumptions: firstly, that the sporulating cell is derived from four potential "cousins"; secondly, that the state of asymmetric septation (Stage II in Fig. 10) is reached by inhibition of one of two potential symmetric division events. The final structure thus has three chromosomes in the mother cell compartment and one in the prespore.

To arrive at such a structure reliably the cell must be able to choose between the two partially replicated chromosomes at point 3 in Fig. 10. It is at this point that the asymmetry becomes apparent. How might this be achieved? One obvious answer lies in the asymmetry of the DNA molecule itself. The "Watson" and "Crick" strands are complementary but chemically different, and hence asymmetric. Whenever a DNA molecule is replicated, each daughter chromosome receives one old strand and one newly synthesised strand. One daughter receives a new "Watson"; the other a new "Crick". In principle, the cell could choose the daughter chromosome containing a specific newly synthesisied strand and use the "information" to generate a grossly asymmetric structure via a mechanism such as that shown in Fig. 10. Similar mechanisms converting the asymmetry of the DNA molecule into cellular asymmetry have been proposed in other developmental systems (see Arcangioli and Klar, Chap. 5, this Vol.).

1. Starvation stimulus occurs

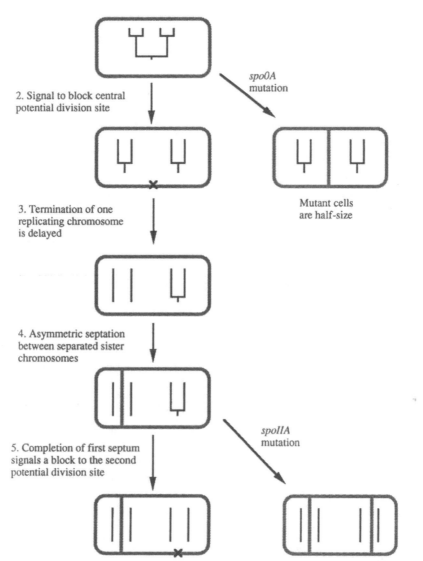

2. Signal to block central potential division site

spo0A mutation

Mutant cells are half-size

3. Termination of one replicating chromosome is delayed

4. Asymmetric septation between separated sister chromosomes

spoIIA mutation

5. Completion of first septum signals a block to the second potential division site

Abortively disporic mutant

Fig. 10. A model for the generation of the asymmetric spore septum. The first cell shown will soon complete a round of DNA replication and hence is about to become committed to a central (vegetative) cell division. For clarity, only one half of the circular chromosome is shown. The cells are growing rapidly, so a second round of DNA replication is already in progress. Starvation leads to a signal that blocks the forthcoming central division, as indicated by the *cross*. Replication of one of the chromosome pairs is then delayed relative to the other, generating a transient block in septation at one end of the cell. This allows a single asymmetric septum to form between the sister chromosomes that have separated. Completion of the asymmetric septum then leads to a signal that permanently blocks the second septum, again indicated by a *cross*. The consequences of mutations preventing the signal to block the first central septum (e.g., *spo0A*) or the final signal to block the second "symmetry restoring" division (e.g., *spoIIA*) are also indicated. Electron micrographs of such mutants are shown in Fig. 2. For simplicity, growth of the cell is not shown

2.4.2 Differential Gene Expression[11]

The above discussion indicates how cellular asymmetry could be achieved. The obvious question that follows concerns the establishment of asymmetric gene expression. Immediately after septation, it is clear from Fig. 7 that separate programmes of gene expression are established in the prespore and mother cell. The genomes in the two cells are identical but different sets of genes are activated in each. Again, a complete answer to this basic problems is not yet forthcoming, but some of the components of the mechanism have been identified. Two sigma factors appear to play crucial roles. The genes encoding σ^E and σ^F are expressed before septation, in the pre-divisional cell. Thus, their products might be expected to be distributed between the prespore and the mother cell at septation. However, the genes transcribed by the σ^E and σ^F forms of RNA polymerase appear to be expressed in a cell specific manner, rather than in both cell types (see Fig. 7). This implies the existence of a mechanism that restricts the activity of each sigma factor to a specific cell. The simplest explanation would be that the two proteins are physically segregated at septation. However, the way in which this is achieved remains obscure. There are also several other possibilities, in addition to the direct physical segregation of the proteins. For example, the proteins could be differentially degraded in the two cells, or they could be present in both cells but the genes controlled by a secondary level of regulation. However, in either case, there is still a requirement for some component to be differentially partitioned or activated in the two cells. Whatever the mechanism, it is likely to be complex as a number of other genes involved have been identified by mutations that abolish one or both of the cell-specific pathways. Where the mutation lies in a gene that is expressed *before* septation, it is likely that the gene's product contributes in some way to the spatial separation of the sigma factor activities. Although the roles of these various gene products are not yet known, it seems likely that the problem of the establishment of asymmetric gene expression should now be tractable to analysis at the molecular level.

2.5 Stages IV to V: Differential Morphogenesis[5]

2.5.1 Synthesis of the Spore Cortex and Coat Layers[12]

Electron microscopy of cells in stages IV and V of sporulation has revealed the construction of two major macromolecular structures, the spore cortex and the spore coat (Fig. 1). The cortex is made up of peptidoglycan, only slightly different in composition from that of the wall of vegetative cells. Presumably cortex synthesis requires the action of some dual-purpose proteins that are involved in both processes. However, the products of some *spo* genes must participate to produce the modified structure found in the spore. Little is known about the mechanism of cortex synthesis or its regulation.

Rather more is known about the synthesis of the spore coat. It is a multilayered structure built up by the successive deposition of 12 or more different types of protein. The order of synthesis of the proteins loosely follows the order of their assembly into the coat with certain exceptions. Self-assembly processes akin to the assembly of viral capsids undoubtedly play a significant part. The synthesis and assembly of this complex structure is a classic example of macromolecular morphogenesis, and study of its regulation is likely to provide important insights into the general principles of developmental regulation. Although little is yet known about the nature of the assembly process, significant progress has been made in understanding the mechanisms controlling the timing of gene expression during mother cell development.

2.5.2 Temporal Control of Gene Expression[13]

Three types of mechanism that participate in controlling the timing of gene expression and coordinating this with progress in the morphogenic process have been identified.

1. The Switch from σ^E to σ^K. The early phase of mother-cell-specific transcription is directed by the σ^E form of RNA polymerase. Most probably, this sigma factor is synthesised specifically in the mother cell and its presence is the dominant factor determining the programme of gene expression in the early stages of mother cell development (see above). One of the proteins under the ultimate control of σ^E is another sigma factor σ^K. Because the synthesis of σ^K is complex, being regulated at three different levels (see Fig. 7), its activity appears in the mother cell at least an hour after that of σ^E (see below). Thus, genes encoding proteins needed in the early stages of spore coat synthesis are expressed from σ^E-dependent promoters. Synthesis of proteins required for the outer layers of the coat is delayed because their genes are transcribed from σ^K-dependent promoters.

Switching from σ^E to σ^K would obviously turn off genes expressed during the early stages of mother cell development. What about genes that need to be expressed for a relatively protracted period, during both phases of development? The answer seems to be that such genes can be transcribed by both the σ^E and σ^K forms of RNA polymerase. There are at least two ways in which this has been

achieved. Firstly, the recognition properties of σ^E and σ^K are sufficiently similar (Fig. 8) for some promoter sequences to be recognised by both forms of polymerase. Alternatively, some genes, such as *spoVJ* and *cotE*, actually have two promoters upstream of their coding sequences. One of them directs expression by the σ^E form of RNA polymerase, the other by σ^K. As the rate of transcription from this promoter falls off, σ^K appears and activates the second promoter sequence. A similar strategy is used in eukaryotic systems where a single complex promoter region can be controlled by multiple regulatory effectors.

2. The SpoIIID Time Switch. Within the regulon of genes controlled by each of these σ factors, individual genes do not always show identical patterns of expression. Other factors can fine-tune the time at which gene expression begins and the period for which its expression is maintained. The SpoIIID protein, for example, appears to act as a time switch that modulates the expression of some of the genes in the σ^E regulon (Fig. 11). The SpoIIID protein is

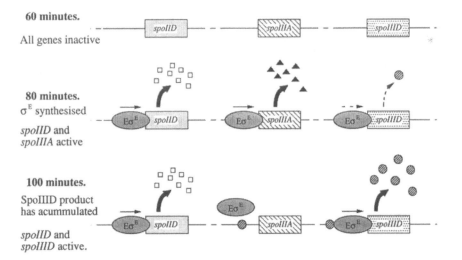

Fig. 11. The SpoIIID protein acts as a time switch during mother cell development. The expression of three unlinked genes at different stages of sporulation is shown. All of the genes are transcribed by the σ^E form of RNA polymerase (Eσ^E) and so are inactive until this enzyme appears, after about 80 min. The *spoIID* and *spoIIIA* genes have promoters that are recognised well by Eσ^E and so are expressed strongly (*thick arrows*). The *spoIIID* promoter is recognised only poorly by Eσ^E, and so is expressed only weakly at this stage (*thin dotted arrow*). It re-

quires its own product, SpoIIID, for full activity. When sufficient SpoIIID protein has accumulated, after about 100 min, it binds to the promoter of its own gene and strongly stimulates transcription. However, it also binds to the promoter of the *spoIIIA* gene in such a way as to prevent transcription. So, after a transient period of expression, the *spoIIIA* promoter is shut down. SpoIIID does not bind to the *spoIID* promoter and so this gene continues to be expressed while Eσ^E is active

a small DNA-binding protein that can either stimulate or repress gene expression, perhaps depending on the precise location of its binding sites in the regulatory region of a gene. The key to its action as a time switch lies in the interactions of the protein with the regulatory region of its own gene. The *spoIIID* gene is transcribed by the σ^E form of RNA polymerase but full expression also requires the SpoIIID product itself. Thus, the main phase of *spoIIID* expression does not begin at the same time as that of some other σ^E-dependent genes, such as *spoIID*, but rather after a short delay, during which the SpoIIID protein builds up slowly. Thus, *spoIIID* expression is temporally delayed compared with σ^E-dependent genes that do not require the SpoIIID protein. Other genes that are completely dependent on SpoIIID also show delayed expression (Fig. 7). A third class of genes do not require SpoIIID for expression and so are activated relatively early, but their expression is repressed by SpoIIID. Consequently, after a transient period of

rapid expression, these genes are shut down, even befor σ^E is replaced by σ^K. It seems likely that another small DNA-binding protein, GerE, could have a similar function to that of SpoIIID later in mother cell development (see Fig. 7).

Thus, during mother cell development the transcriptional apparatus undergoes a series of programmed changes. Each individual gene, by subtle adjustment of its upstream regulatory sequences, has had its timing and level of expression finely tuned according to the requirements of its product.

3. Feedback regulation from morphogenic structure to gene expression. So far, the temporal control of gene expression has been viewed as an automatic process. A series of regulatory effectors appear in sequence, the synthesis of each depending on the action of the previous factor. This form of temporal control operates simply on the basis of the time required for each successive product to build up, which might not suffice to ensure that each protein is available in the required amount at the required time. However, this could be achieved by feedback regulation, and there is mounting evidence for the existence of such mechanisms.

The σ^K protein, which has already been considered in terms of temporal control, is synthesised as an inactive precursor with an N-terminal amino acid extension that is cleaved off by a specific protease. It seems that the protease becomes active only when a certain stage of morphological development is reached (Fig. 12). Thus, mutations in several *spo* genes that are needed for the earlier stages of morphogenesis cause the precursor of σ^K to accumulate and prevent expression of the genes needed for completion of the spore coat. Because of the pleiotropic effects of the mutations, these *spo* genes had been thought to act as regulators in the conventional sense, but the products predicted from DNA sequence analysis are more reminiscent of structural effectors, needed for the fabrication of the spore structures or its metabolic or physiological maturation. The precise signal detected by the pro-σ^K protease is not known, but it is thought to reflect some physical property of the developing spore.

Two important principles thus emerge. Firstly, the developmental programme incorporates feed-

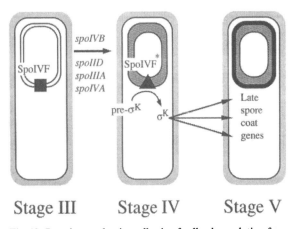

Stage III   Stage IV   Stage V

Fig. 12. Putative mechanism allowing feedback regulation from morphological development to gene expression. The genes encoding the proteins that form the outermost layers of the spore coat are transcribed from promoters recognised by the σ^K form of RNA polymerase. This sigma factor is synthesised as a precursor that remains inactive until cleaved by a protease, possibly encoded by the *spoIVF* operon. The SpoIVF protease, which is synthesised during an earlier stage of sporulation, remains inactive until the prespore reaches the appropriate stage of development (Stage IV). The mechanism would serve to coordinate expression of the late spore coat genes with progress in prespore morphogenesis. Mutations in any of the genes contributing to prespore morphogenesis, such as *spoIVB*, can thus prevent expression of the late spore coat protein genes

back regulation to keep genetic events in register with morphological events. Secondly, mutations with a pleiotropic effect, having a severe effect on gene expression, are not necessarily in genes encoding regulatory proteins. They could lie in genes needed for morphological transitions to which subsequent gene expression is coupled.

2.6 Outlook and Summary

In the space available it has been possible to do little more than give a superficial (and slightly personalized) overview of some of the more interesting areas of research on sporulation. It should, however, be clear that the system is surprisingly complex. Even a simple two-cell differentiation system in a bacterium is a complex process involving many genes. Fortunately, the tools for analysing the system are powerful and it is likely that answers to some of the most basic questions relevant to developmental biology will soon be available.

In fact, a great deal has already been achieved. Many, perhaps most, of the genes involved in regulation of sporulation have now been identified, cloned and partially characterized in terms of their DNA sequences and regulation. The basic pattern of interactions between these genes and their proteins products has been worked out. The molecular basis for some of the interactions, such as the action of the various sigma factors, have been resolved, although many of the details remain to be worked out.

Several of the more interesting and challenging problems have not yet been solved, but much of the preliminary work needed to make them accessable to molecular analysis has now been done. Thus, the details of the decision-making process controlling the onset of sporulation should soon be resolved satisfactorily. The mechanism of the establishment of differential gene expression, may also be understood in detail before long. However, the latter problem may depend on first understanding the basis for the asymmetry of the prespore septum. This may in turn hinge on whether progress can be made in the poorly understood and relatively intractable problems of bacterial cell division and cell cycle regulation.

Other areas of interest, such as the mechanism of prespore engulfment and the evolution of sporulation have only just begun to be considered, but they clearly represent challenging problems for the future. It is clear from this short survey that bacterial systems have much to offer to a general understanding of developmental processes and that they will remain a fruitful area of research for many years to come.

References

[1] Keynan A, Sandler N (1983) Spore research in historical perspective. In: Hurst A, Gould GW (eds) The bacterial spore, vol 2. Academic Press, London, pp 1–48

[2] Fitz-James P, Young E (1969) Morphology of sporulation. In: Gould GW, Hurst A (eds) The bacterial spore. Academic Press, London, pp 39–72

[3] Hurst A, Gould GW (eds) (1983) The bacterial spore, vol 2. Academic Press, London
Smith I, Slepecky RA, Setlow P (eds) (1989) Regulation of prokaryotic development. American Society for Microbiology, Washington DC

[4] Moir A, Smith D (1990) The genetics of bacterial spore germination. Annu Rev Microbiol 44:531–553
Piggot PJ (1989) Revised genetic map of *Bacillus subtilis* 168. In: Smith I, Slepecky RA, Setlow P (eds) Regulation of prokaryotic development. American Society for Microbiology, Washington DC, pp 1–42
Piggot PJ, Coote JG (1976) Genetic aspects of bacterial endospore formation. Bacteriol Rev 40:908–962

[5] Harwood CR, Cutting SM (eds) (1990) Molecular biological methods for *Bacillus*. Wiley, Chichester

[6] Losick R, Kroos L (1989) Dependence pathways for the expression of genes involved in endospore formation in *Bacillus subtilis*. In: Smith I, Slepecky RA, Setlow P (eds) Regulation of prokaryotic development. American Society for Microbiology, Washington DC, pp 223–241
Losick R, Kroos L, Errington J, Youngman P (1989) Pathways of developmentally regulated gene expression in *Bacillus subtilis*. In: Hopwood DA, Chater KF (eds) Genetics of bacterial diversity. Academic Press, London, pp 221–242

[7] Moran CP Jr (1989) Sigma factors and the regulation of transcription. In: Smith I, Slepecky RA, Setlow P (eds) Regulation of prokaryotic development. American Society for Microbiology, Washington DC, pp 167–184

[8] Freese E (1981) Initiation of bacterial sporulation. In: Levinson HS, Sonenshein AL, Tipper DJ (eds) Sporulation and germination. American Society for Microbiology, Washington DC, pp 1–12

[9] Burbulys D, Trach KA, Hoch JA (1991) Initiation of sporulation in *B. subtilis* is controlled by a multicomponent phosphorelay. Cell 64:545–552

Grossman A (1991) Integration of developmental signals and the initiation of sporulation in *B. subtilis*. Cell 65:5–8

Smith I (1989) Initiation of sporulation. In: Smith I, Slepecky RA, Setlow P (eds) Regulation of prokaryotic development. American Society for Microbiology, Washington DC, pp 185–210

[10] Errington J (1991) A model for asymmetric septum formation during sporulation in *Bacillus subtilis*. Mol Microbiol 5:785–789

[11] Errington J, Illing N (1992) Establishment of cell-specific transcription during sporulation during sporulation in *Bacillus subtilis*. Mol Microbiol (in press)

[12] Jenkinson HF, Kay D, Mandelstam J (1980) Temporal dissociation of late events in *Bacillus subtilis* sporulation from expression of genes that determine them. J Bacteriol 141:793–805

Jenkinson HF, Sawyer WD, Mandelstam J (1981) Synthesis and order of assembly of spore coat in *Bacillus subtilis*. J Gen Microbiol 123:1–16

[13] Cutting S, Oke V, Driks A, Losick R, Lu S, Kroos L (1990) A forespore checkpoint for mother cell gene expression during development in *B. subtilis*. Cell 62:239–250

Foulger D, Errington J (1991) Sequential activation of dual promoters by different sigma factors maintains *spoVJ* expression during successive developmental stages of *Bacillus subtilis*. Mol Microbiol 5:1363–1373

Kunkel B, Losick R, Stragier P (1990) The *Bacillus subtilis* gene for the developmental transcription factor 6^K is generated by excision of a dispensable DNA element containing a sporulation recombinase gene. Genes Devel 4:525–535

[14] Zheng L, Losick R (1990) Cascade regulation of spore coat gene expression in *Bacillus subtilis*. J Mol Biol 212:645–660

Chapter 3 Development in *Caulobacter crescentus*

Ruth Bryan

3.1 Introduction[1]

Caulobacters are dimorphic bacteria, with a motile phase in which the cell body is flagellated, and a sessile stage, when the cell grows a stalk. Development involves the asymmetric division of the parent stalked cell to yield one flagellated swarmer cell and one stalked cell, followed by differentiation of the swarmer cell into a stalked cell. Some of these events are the subject of this chapter. The study of *Caulobacter* development is facilitated by the simplicity of its cell cycle, the polar location of the organelles, and the ability to obtain synchronous swarmer cell populations.

Caulobacters are common micro-organisms in nutrient poor environments. They have been found in freshwater, salt water, in soils, and even in the gut of the millipede. *Caulobacter crescentus*, shown in Fig. 1, is a freshwater organism and the most extensively studied member of the group. *C. crescentus* is distinguished from other caulobacters by its crescent shape.

As Gram-negative bacteria, caulobacters possess an outer membrane, a peptidoglycan layer, and a cytoplasmic membrane. The membranes of the stalk are apparently continuous with the cellular membranes, and the inside of the stalk probably contains cytoplasm. Survival in nutrient poor environments may be easier because the stalk greatly increases the cell's surface area and may allow more efficient uptake of nutrients. The stalk grows at its site of attachment to the cell body. Its growth depends upon the synthesis of phospholipids. The stalk contains crossbands which are made of protein and which are apparently located between the outer and inner membrane.

The genetics of *C. crescentus* has been well developed. Cells can be mutagenized by UV irradiation, chemical mutagenesis, and by insertion of trans-

posable elements such as Tn5 or derivatives. Cell division mutants, mutants unable to make flagella, stalk-less mutants, other developmental mutants, and a variety of auxotrophic mutants have been isolated and the mutations mapped. Mapping of insertion mutations is facilitated by the fact that a circular map of the chromosome has been constructed using pulsed field-gel analysis of DNA fragments. In addition, genetic exchanges can be carried out by generalized phage transduction or by RP4 mediated conjugation. Stalk and flagellum formation, which are dependent upon cell cycle events, have been the most intensely studied events. Many mutations have been found which affect these steps. Broad host range plasmids are capable

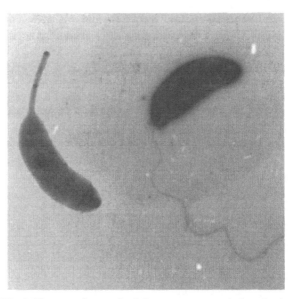

Fig. 1. Electron micrograph of *C. crescentus*, negatively stained with uranyl acetate. The cell on the *left* contains a stalk, the *right one* contains a flagellum. The cell body is about 2 µm in length, the stalk diameteris about 909 nm, and the flagellar diameter is about 15 nm. (Bryan et al. 1990, see [4])

of replication in *C. crescentus*. This has made it possible to reintroduce cloned genes to complement mutant genes or to study the control of gene expression.

3.2 Each Cell Division Cycle Produces a Stalked and a Swarmer Cell

3.2.1 Stalked Cell and Swarmer Cell Carry Out Different Programs[2]

The *C. crescentus* cell cycle consists of stalked cell, predivisional cell, and swarmer cell phases. The stalked cell divides repeatedly, giving rise to one motile, flagellated, daughter and one stalked daughter (Fig. 2). After cell division, the different daughter cells follow different developmental pathways and have different types of cell cycle. The daughter stalked cell immediately begins DNA replication, while the daughter swarmer cell delays replication until it differentiates into a stalked cell.

Following cell division the daughter swarmer cell is fully motile and undergoes chemotaxis by reversal of the direction of rotation of its flagellum in response to environmental stimuli. The swarmer cell also assembles pili, at the same pole as the flagellum. When grown under conditions in which nutrients are not limited, the swarmer cell spends about one third of its life cycle swimming. It then undergoes a major morphogenetic change, in which several enzyme activities disappear, the flagellum is ejected, and the cell gets ready to grow a stalk and replicate its DNA. The ejection of the filament must involve some internal rearrangement to prevent the cell contents, which are under high pressure, from leaking out.

Two other genera of stalked, dimorphic bacteria, *Hyphomicrobium* species and *Rhodomicrobium* species, also have swarmer cell stages in which no DNA replication occurs. This stage has been referred to as a shutdown, or growth precursor stage. The energy required for swimming is fairly low, estimated to be about 0.1% of a growing cell's total energy expenditure. Under certain starvation conditions these organisms persist in the swimming phase, avoiding DNA replication and not growing.

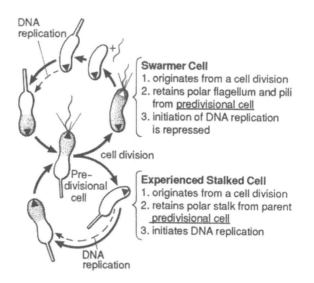

Inexperienced Stalked Cell
1. originates from a swarmer-to-stalked cell transition
2. initiates DNA replication
3. nascent stalk formed at cell pole

Swarmer Cell
1. originates from a cell division
2. retains polar flagellum and pili from predivisional cell
3. initiation of DNA replication is repressed

Experienced Stalked Cell
1. originates from a cell division
2. retains polar stalk from parent predivisional cell
3. initiates DNA replication

Fig. 2. *C. crescentus* cell cycle, showing sites of polar organelle development and times of DNA synthesis. *Shaded areas* exhibit swarmer cell characteristics. *Filled-in triangles* mark the site where flagellum synthesis will occur. (Shapiro 1985)

3.2.2 Methods for Obtaining Synchronous Cell Populations

Several methods are available for the isolation of synchronous swarmer cells. In one method, growing stalked cells are attached to the surface of a petri plate. Newly divided swarmers, which do not stick to the plate, are collected by washing the plate. Swarmers can also be isolated by sedimentation of a special strain of *Caulobacter* with altered buoyancy through a gradient of Ludox, a silica suspension. Purified swarmers can also be obtained by filtration of mixed populations through sterile cheesecloth. The stalked cells stick to the cheesecloth, and the swarmers can be eluted. The first method allows the isolation of a small number of highly synchronized cells. The other methods produce populations which contain swarmer cells of all ages, some of them newly divided and others almost ready to become stalked cells.

The synchronizability of the organism allows the biochemical characterization of the internal changes which occur as a function of the cell cycle. Pulse labeling experiments and nuclease S1 experiments allow the identification of changes in protein and RNA synthesis that occur throughout the cell cycle.

3.2.3 DNA Replication Is Under Cell Cycle Control[3]

Chromosomal replication occurs only in the stalked cell. Replication begins at the same time as the growth of the stalk, in the newly differentiated stalked cell. In the "experienced" stalked cells, replication begins following the cell division event. Replication occurs bidirectionally, starting from a unique origin. In *C. crescentus*, there is only one round of DNA replication per cell division, whether the cells are growing quickly or slowly. This is unlike DNA replication in *E. coli*, where rapidly growing cells contain multiple replicating forks. As shown in Fig. 2, there is no chromosomal replication in the swarmer cell stage (G1), and there is a post-synthetic gap (G2), following the completion of chromosome replication before cell division. The two daughter cells do not differ in their inheritance of chromosomes: there is no segregation of chromosomes on the basis of the age of the template DNA strand. Labeling experiments demonstrate that the daughter chromosomes whose templates were one generation old segregated equally to the two types of daughter cells.

What distinguishes the swarmer-cell chromosome from the stalked cell chromosome, and what is responsible for the fact that only the stalked cell chromosome is replicated immediately following cell division? Differences in structure between the two daughter chromosomes or differences in availability of replication enzymes in the two daughter cells might both play a role in controlling the time of replication. The chromosome in the daughter swarmer is different physically from the stalked cell chromosome. Membrane-associated chromosomes, called nucleoids, isolated from swarmer cells have a high sedimentation value, implying that they are condensed. Nucleoids isolated from stalked cells, which are replicating their DNA, have a sedimentation value that is less than half in the swarmer cell. Cells which have finished replication but have not yet divided contain two types of nucleoids, one with a sedimentation value that is the same as the nucleoid in the swarmer cell, the other with an s-value the same as the nucleoid present during the replication stage.

To test for the presence of the machinery for DNA replication in stalked and swarmer cells, broad host range plasmids, whose replication depends on the presence of host-encoded replication enzymes, were introduced into a synchronizable *Caulobacter* strain. The cells were synchronized and the time of chromosome and plasmid replication was examined. Most of the plasmid replication occurred during the stalked cell stage, at the time of chromosome replication. However, some replication also occurred during the swarmer cell stage, when there was no chromosomal replication. This indicates that the machinery for DNA replication is present in the daughter swarmer cell, although probably at a very low level. Other evidence that the concentration of at least one of the replication enzymes in the swarmer cells is low is that DnaK, the product of the *dnaK* gene, which is required for initiation of DNA replication and made in the predivisional cell, is preferentially segregated to the daughter stalked cell (see Sect. 5.4).

The absolute lack of chromosomal replication in swarmer cells thus cannot be explained by a complete lack of replication machinery. The unequal fates of the two daughter cells is probably a consequence both of the difference in folding between the two newly replicated chromosomes and the differences in enzymes present in the two daughters. It is not clear whether these factors are independent; in fact, they probably influence each other.

3.2.4 Membrane Growth and Cell Division

The *C. crescentus* membrane phospholipid contains primarily cardiolipin and phosphatidylglycerol. The membrane composition does not change throughout the cell cycle, as measured by steady-state labeling with ^{32}P and separation of the membrane components by two dimensional thin layer chromatography. Pulse labeling of synchronous cultures at different stages of the cell cycle shows that synthesis of phospholipids is constant throughout swarmer, stalked cell, and early predivisional stages. Close to the time of cell division, a change in synthesis of fatty acids and phospholipids occurs. Synthesis of cardiolipin and an abundant precursor of phosphatidylglycerol ceases, and the synthesis of phosphotidylglycerol phosphate increases dramatically. This may represent a temporary slowdown in phospholipid synthesis at the time of cell separation.

Synthesis of at least 15 different membrane proteins is periodic, occuring at specific portions of the cell cycle. Six of these are made immediately before cell division. Synthesis of some membrane proteins is dependent upon ongoing membrane synthesis.

Caulobacter crescentus divides by a gradual pinching in of the cell membranes. Electron microscopic examination shows that there no septum is formed. The close relative of the caulobacters, *Asticcaulus*, which is not tapered at the ends, does form a septum. In this organism, the appendages are located subpolarly, indicating that the method of cell division may determine both the location of polar organelles and the cell shape.

3.3 The *C. crescentus* Flagellum Is Similar to Other Bacterial Flagelli[4]

3.3.1 The Structure of the Flagellum is Complex

Bacterial flagella are made up of several components. Many of the individual parts are made up of multimeric assemblies of individual proteins. The *C. crescentus* flagellum, like other bacterial flagella, is made up of a basal body, a hook, and a flagellar filament (Fig. 3). The basal body, which attaches the flagellum to the cell, consists of ring shaped assemblies of proteins embedded in the cell envelope. *C. crescentus* basal bodies contain five rings, compared to four found in *E. coli* and other Gram-negative organisms. (The basal bodies in Gram-positive bacteria, which lack an outer membrane, contain only two rings.) In Gram-negative

organisms, the innermost ring, the M-ring, lines up with the cytoplasmic membrane. The S, or supra membrane, ring lies adjacent to the M-ring, on the periplasmic surface. The P (periplasmic) ring seems to be in contact with the peptidoglycan layer. The L ring is in the outer membrane. The E (extra) ring in *C. crescentus* is located between the S and P rings. The location of the E ring, at the site on the rod where filament ejection takes place, suggests that it may play a role in the ejection process.

The rings are threaded on a rod, which in *E. coli* is made up of four different proteins. The rod in *C. crescentus* apparently separates into two parts, because when the flagellar filament is shed into the medium during *C. crescentus* development, it is attached to the hook and to a portion of the rod which extended into the basal body. The basal body is not ejected with the rest of the flagellum.

The hook is made up of several intersecting families of helices composed of 70 000 MW hook monomers. It is external to the cell and is attached to the basal body by the rod. The cell proximal end of the hook is tapered and appears small enough to fit partially inside the outermost ring of the basal body. The distal end of the hook forms a cup-shaped structure onto which the flagellar filament attaches.

The flagellar filament is approximately 6 microns in length, about four times the length of the cell. Unlike most simple flagellar filaments, the *C. crescentus* filament is made up of more than one flagellin; in fact, there are at least three different flagellins in each flagellar filament. Although they are closely related and immunologically cross-reactive, when they are assembled their available epitopes are not identical. Antibody decorations, performed by lysing cells on electron microscope grids and treating with various antiflagellin antibodies show that filaments contain four distinct regions. Immediately adjacent to the hook is a tiny region whose composition is undefined. The next region depends on the presence of the gene encoding a flagellin of 29 000 MW. The third region contains flagellins of 27 500 MW and comprises about 1/4 of the filament. The most distal 3/4 of the filament is composed of 25 000 MW flagellin. The minor flagellins apparently facilitate the assembly process, but it is possible to assemble active fila-

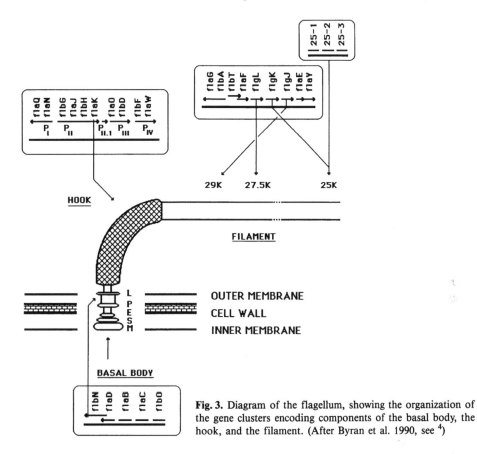

Fig. 3. Diagram of the flagellum, showing the organization of the gene clusters encoding components of the basal body, the hook, and the filament. (After Byran et al. 1990, see [4])

ments in the absence of either of the minor flagellins. Interruption of the genes encoding both the 29K and the 27.5K flagellins impairs motility more than deletion of either gene alone.

Rotation of the flagellum depends on the presence of Mot (motor) proteins. Rotation of the flagellum in a clockwise direction propels the cell forward; counterlock-wise rotation pulls it backwards. The direction of rotation depends on interaction of the motor with a "switch complex" thought to be on the cytoplasmic surface of the basal body. The state of the switch complex is in turn controlled by the activities of several chemotaxis proteins, molecules involved in sensing changes in the environment and directing rotation accordingly. Some of the chemotaxis proteins are membrane bound and may be positioned close to the flagellum, others, including methylesterase and methyltransferase, are soluble, cytoplasmic proteins.

3.3.2 Flagellar Gene Organization[5]

Flagellar genes are genes required for the formation of the flagellum. Strains containing mutations in these genes were isolated by loss of motility, and then screened for loss of flagellum formation. Some mutations were spontaneous, others were the result of transposon mutagenesis. Flagellar genes may encode structural components of the flagellum, proteins necessary for its assembly, or regulatory proteins. Some flagellar genes have pleiotropic effects, altering other polar structures and interfering with cell division (see section on pleiotropic mutations).

Flagellar genes (*fla*, *flb*, and *flg* genes) are arranged in four clusters, as shown in Fig. 3. An additional 19 individual genes, not shown, are scattered on the genetic map. The gene locations were found by transductions, matings, pulse field gel analysis of Tn5 insertions, and cloning and DNA sequencing. There is evidence that some of the individual genes are actually small clusters, and there are a total of at least 44 genes required. The *flg* genes, encoding the flagellins, are found in two clusters. One cluster contains three genes for the 25K flagellin, and apparently no other flagellar genes. The other cluster contains a 25K flagellin gene, and the genes for the 27.5 and 29K flagellin. The second cluster of flagellin genes also includes a gene essential to turn off flagellin synthesis, and several other genes which play a role in flagellin assembly and perhaps in ejection of the filament. The basal body cluster contains five flagellar genes and one *mot* (motor) gene. One of these genes, *flb*N, encodes the protein which makes up the L-ring, the outermost ring of the basal body. The sequence of this protein is 31% identical to that of the L-ring protein from *S. typhimurium*. Mutations in any of the flagellar genes in this cluster prevent synthesis of a complete basal body.

The largest gene cluster is refered to as the hook gene cluster because it contains the structural gene for the hook protein. This cluster contains one operon encoding four genes, and six or seven other individual genes in five transcription units. The hook operon also contains a gene which in some way controls the length of the hook itself. Several genes in this cluster have regulatory effects on the expression of flagellar genes in other locations. (see Sect. 4.2)

Within the gene clusters, there is an interesting overlapped gene structure connecting some genes. *fla*E and *fla*Y, for example, are arranged so that the promoter for *fla*Y is within the coding sequence for *fla*E. The stop codon for *fla*E and the start codon are both located on what should be a stable stem and loop structure. Although both the *fla*E and the *fla*Y promoters are apparently used, mutations in *fla*E have a polar effect on expression of *fla*Y. This arrangement, which is somewhere between an operon and separate genes, certainly has an impact on the regulation of these genes, but exactly what it is is not clear.

The arrangement of flagellar genes in *C. crescentus* bears only a superficial resemblance to that from *E. coli* and *S. typhimurium*. The flagellar genes in *E. coli* and *S. typhimurium* contain three gene clusters, while *C. crescentus* contains four, and the clusters in *E. coli* and *S. typhimurium* are larger. The latter organisms do not have isolated flagellar genes. The flagellar operons in *C. crescentus* are smaller than those in the other organisms. The gene encoding the basal body L-ring protein in *S. typhimurium* is part of an operon, while in *C. crescentus* it has its own promoter. There does not appear to be conservation of the arrangement of genes within the clusters. For example, in *E. coli* and *S. typhimurium*, the basal body and hook genes are in the same cluster, while they are separated in *C. crescentus*. In *E. coli* and *S. typhimurium*, the chemotaxis genes are located within the flagellar gene clusters; in *C. crescentus*, they are located in their own clusters. Overall, the flagellar genes of *C. crescentus* are more dispersed than those of the enteric bacteria.

3.3.3 Flagellar Assembly Proceeds from the Inside to the Outside

Mutants blocked at different stages of flagellar assembly define the genes required for flagellar assembly and the steps in the assembly process. Mutants in 18 of the 44 known flagellar genes have a phenotype which has been defined by electron microscopic examination of the whole cell, the ejected filament, or the basal body purified from the membrane. The assembly phenotypes of these mutants are shown in Fig. 4. Mutations in nine additional genes, not shown in the figure, prevent the formation of the hook and flagellar filament, but have not been tested for the presence of basal bodies. The assembly phenotypes define the direction of assembly of the flagellum. Extensive studies in *E. coli* and *S. typhimurium* indicate that assembly proceeds from the innermost structures toward the outside of the cell. Similar analysis of *C. crescentus* mutations indicates that the direction of assembly is the same in *C. crescentus*. Mutation of *flb*N, the structural gene for the L-ring of the basal body, and mutations in three genes whose structural function is not known completely abolish flagellar formation. A strain containing a mutation in the

I. No basal body assembled
    fla B, fla C, flb N, flb O

II. Partial basal body

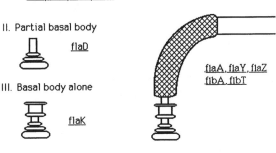

    fla D

III. Basal body alone

    fla K

IV. Basal body plus hook

V. Basal body, hook and partial
    filament

    fla A, fla Y, fla Z
    flb A, flb T

VI. Normal or paralyzed flagellum

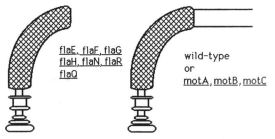

    fla E, fla F, fla G
    fla H, fla N, fla R
    fla Q

    wild-type
    or
    mot A, mot B, mot C

Fig. 4. Partial flagellar products made by various mutants. The basal bodies are oriented with the inner ring, which sits in the inner membrane, on the *bottom* of each figure

fla D gene of *C. crescentus* makes a partial structure containing only the two inner rings. No mutants were found containing an outer but not inner rings, so it appears assembly of the inner rings is necessary for assembly of the outer rings. *fla* D may encode the third, E-ring. This protein coding sequence predicts the presence of a signal sequence, so the protein is apparently transported through the cytoplasmic membrane and then cleaved and assembled. Similarly, sequence analysis indicates that, in both *C. crescentus* and *S. typhimurium*, the protein which makes up the L-ring contains a cleavable signal sequence. On the other hand, the protein making the M-ring in *E. coli* does not have a signal sequence or membrane spanning sections, indicating that it is assembled by a different mechanism from the outer rings. This is probably true for *C. crescentus* as well, as the genes for the basal body proteins appear to be highly conserved

among the organisms examined. Following the assembly of the basal body, the hook is assembled outside the cell, by threading of the hook monomers through the hollow core of the rod. The hook protein does not contain a signal sequence.

Twelve different genes seem to be required for the assembly of flagellins on to the hook, as compared with only four genes in *E. coli* and *Salmonella*. *C. crescentus* strains with mutations in seven of these genes assemble a basal body and a hook, but no flagellar filament. Mutations in the other five genes cause a phenotype in which a partial flagellar filament is assembled. The products of some of these genes may function as adaptors to join the dissimilar helices of the hook and the flagellins, as is thought to be the case for the hook-associated proteins in *S. typhimurium* and *E. coli*. Perhaps some are enzymes for post-translational modification of the flagellins, which do contain methylated residues.

Similar classes of mutations have been described for *E. coli* and *S. typhimurium*. In those organisms there are 23 genes in which mutation leads to a complete block in appearance of any structure, and only six genes which are blocked at the stage of addition of flagellin. The overall structure of the flagellum in these organisms is very similar to that found in *C. crescentus*. It is not clear whether the differences in the numbers of classes of mutations found reflects differences in methods, differences in stability of the partial structures, or real differences in the assembly pathway.

Flagellar biogenesis involves the interaction of many gene products whose order of assembly is apparently carefully controlled. One possible mechanism for controlling assembly order could be to control temporally the synthesis of the component parts. It seems as if the time of expression of a gene and the order of assembly of the gene product are correlated. For example, the basal body L-ring protein is assembled before the hook protein, which is assembled before the flagellins, and *fla* N (encoding the L-ring protein) is expressed before *fla* K (hook gene), which is expressed before the flagellins. There are many genes whose function in assembly is not known. If their time of expression can be found, it may help to make their role clear.

3.4 Transcription Controls Timing and Level of Flagellar Gene Expression

3.4.1 Alternate Promoters and a Variety of Regulatory Factors [6]

The ability of bacterial RNA polymerase to recognize promoters is conferred by the sigma factor, a subunit of the holoenzyme. Most bacterial promoters share common recognition elements; promoters containing these common recognition elements are recognized by the most abundant sigma factor within a cell. Other sigma factors allow recognition of different families of promoter sequences. This is one mechanism by which bacteria are able to regulate synthesis of groups of related genes. The flagellar genes in *C. crescentus* have at least two different types of promoters and are apparently transcribed under the direction of at least two different sigma factors. Neither type of promoter sequence resembles the consensus sequence of the most common bacterial promoters. One type of flagellar promoter looks like those used by flagellar genes in *B. subtilis* and *E. coli*. In fact, an *E. coli* chemoreceptor gene is transcribed in a cell cycle dependent manner in *C. crescentus*. Another type of promoter is quite similar to the promoters found upstream of genes for nitrogen metabolism in *E. coli*, and upstream of flagellar genes in *Pseudomonas*. The *E. coli* RNA polymerase containing σ^{54}, used for transcribing *ntr* genes in *E. coli*, also transcribes *C. crescentus* flagellar genes which contain this type of promoter, designated a σ^{54} promoter.

Upstream from the flagellar genes are different types of conserved sequences which play a role in regulation of transcription. One protein which binds to a conserved sequence upstream or within flagellar genes is structurally and functionally similar to integration host factor (IHF) from *E. coli*. IHF in other organisms has been found to bind to specific DNA sequences, inducing bends which influence recombination of DNA molecules as well as transcription of specific genes, including some genes with *ntr* promoters. In *C. crescentus*, there are IHF binding sites between the σ^{54} promoters and the *ftr*-elements (see below) upstream of the hook operon, the 25 K flagellin gene, and the 27.5 K

flagellin gene. Specific deletions or alterations of the IHF binding site upstream from the hook operon prevent transcription from that promoter.

At least six *C. crescentus* flagellar genes, *fla*E, *fla*Y, *flg*J, *flb*N, *fla*D, and *fla*K contain a partially conserved "13-mer" sequence in the untranscribed region upstream from their promoter. The role of this sequence is not known. A *ftr* element, a conserved 19 base pair sequence, is located about 100 nucleotides upstream from the *ntr*-like promoters in front of the hook operon and the genes encoding the 25 K and 27.5 K flagellins, and within the *fla*N coding sequence. The sequence of the *ftr* element is not the same as that of elements found upstream from *ntr* genes in *E. coli*, but it apparently serves the same purpose, and is essential for achieving normal levels of transcription of *fla* genes.

Some of the sequence elements are binding sites for temporally controlled proteins, which play roles in controlling the time and/or level of transcription of the *fla* genes. FlbD, the product of a gene at or near the top of the regulatory hierarchy controlling flagellar gene expression, is expressed briefly, before the start of transcription of the major flagellins and the hook gene. This protein is quite similar to NtrC protein from *Klebsiella pneumonia*. In the domain of the protein thought to be involved in DNA binding and interaction with RNA polymerase to promote open complex formation, 55% of the amino acids are identical.

3.4.2 Regulatory Hierarchy [7]

The level of expression of flagellar genes is controlled by a complex regulatory network, including both positive and negative interactions which are diagrammed in Fig. 5. Positive interactions are shown by arrows, negative interactions by bars. Positive interactions are inferred from experiments in which a mutation in a gene at the tail of the arrow leads to a decrease in expression of the gene at the tip of the arrow. Negative interactions mean that a mutation in the gene at the tail leads to an increase in the level of expression of the genes at the tip. It is not known which of these interactions are direct. This regulatory hierarchy is similar in underlying principal to hierarchies of flagellar gene expression eluci-

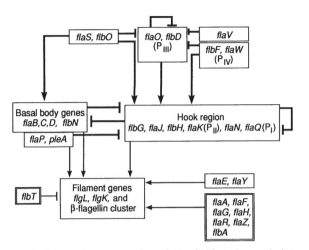

Fig. 5. Schematic representation of genetic hierarchy regulating expression of flagellar genes. *Arrows* indicate so-called positive interactions, in which the target gene appears to require the gene(s) at the other end of the line for full expression. *Lines ending in a bar* indicate what appear to be negative interactions. The *heavy lines* represent experiments in which transcription was shown to be altered. The expression of the genes at the end of the light lines was measured at the protein level. (Byran et al. 1990, see [4])

dated in *S. typhimurium* and *E. coli*, but there are several differences in design. The underlying principle in both organisms seems to be that expression of genes encoding proteins which are assembled at early stages in assembly is required for expression of genes whose products are assembled later.

The light arrows pointing to the flagellin (*flg*) genes encoding the 25 K and 27.5 K flagellins, at the bottom of the hierarchy, indicate that a mutation in almost any *fla* or *flb* gene leads to a large decrease in the level of flagellin synthesis. Flagellin synthesis was measured by pulse labelling cultures and immunoprecipitating cell extracts with antiflagellin antibodies. Mutations in flagellin genes do not alter the level of expression of other flagellar genes. In all the mutants that have been examined, the flagellins are made at the normal time in the cell cycle, but at low levels, implying that there are different controls on timing and level of flagellin synthesis.

At the middle level of the hierarchy are the hook gene operon and the genes in the basal body cluster. The heavy lines represent interactions in which changes in transcription level occur. The regulation

of these genes was found by putting transcription fusion reporter genes into *C. crescentus fla* genes, then putting the chimeric gene into different mutant backgrounds and measuring the effect on transcription of the reporter gene. These genes require other genes, the genes at the top of the hierarchy, for their expression. In addition, these genes apparently exert negative regulatory effects on each other. Mutations in the hook gene lead to increased levels of transcription of both the hook operon and the basal body genes. Similarly, mutations in the basal body genes lead to increases in transcription of the basal body and the hook transcriptional units. Apparently the hook and basal body are regulated as a unit. Complete assembly of the hook-basal body complex is necessary to turn off expression of all of the hook and basal body structural genes. Mutations that prevent assembly prolong and enhance transcription of these genes.

Another class of genes is located at the lower right side of Fig. 5. These genes are required for flagellin synthesis and filament assembly, but not for hook and basal body assembly, and they may not be controlled by the genes at the top of the hierarchy.

The genes at the top of the hierarchy, *fla*S, *fla*O, and *flb*O, *flb*D, and *fla*W are required for transcription of the basal body, hook, and flagellin genes. Although *fla*K mutations elevate transcription of the *fla*K gene, strains with both *fla*S and *fla*K mutations do not make *fla*K mRNA, indicating that *fla*S is epistatic to *fla*K, and may be absolutely required for transcription of *fla*K.

A similar, but more completely nested, hierarchy controls flagellar gene expression in *E. coli* and *S. typhimurium*. In *E. coli*, the genes which are at the top of the regulatory cascade are thought to encode an alternate sigma factor which is required for transcription of all the other flagellar genes. Expression of these top genes in *E. coli* is under catabolite repression; *E. coli* does not assemble flagelli and swim when grown in glucose. Swimming in *C. crescentus* is not repressed when the cells are grown in glucose. There may be other genes which are higher in the regulatory cascade, but which have not been found yet. In addition, not all the possible experiments to order these genes at the top of the hierarchy have been done, so it is possible that one of the current top genes will emerge as the absolute top gene. On the other hand, it is possible that there is more than one "top" of the *C. crescentus* flagellar gene hierarchy, reflecting the diversity of flagellar gene promoters described above. It is also possible that the "top" gene has not been found because it has a function in the cell cycle that makes it indispensible.

3.4.3 Timing[8]

All flagellar genes whose timing has been measured are expressed for a small portion of the cell cycle. The cell must therefore have mechanisms both for turning on and for turning off the expression of these genes; proper timing may be important to the assembly process. For flagellar genes that have been cloned and whose products have been identified, nuclease protection assays show that RNAs are present only at the times when the protein is being made. This indicates that the timing is controlled at the level of transcription.

3.4.3.1 Methods for Studying Timing of Synthesis of Flagellar Genes

The time of expression of flagellar genes, diagramed in Fig. 6, has been measured in a variety of ways. The synthesis of flagellins can be measured by pulse-labeling synchronized populations of cells at different stages of development. The labeled proteins are then precipitated with anti-flagellin antibody, and the precipitated proteins are separated by SDS polyacrylamide gel electrophoresis and visualized by autoradiography. 29K flagellin synthesis begins before DNA replication is finished, followed by synthesis of the 27.5K flagellin, and then the 25K flagellin. 25K flagellin synthesis continues in the swarmer cell for a short period of time before being shut off. The timing of expression of the hook gene has been measured in the same ways. The hook protein is made in late predivisional cells, starting at about the same time as the synthesis of the 27.5K flagellin, but continuing for a longer period. The times of synthesis of these proteins overlap. The flagellins and the hook protein are all reg-

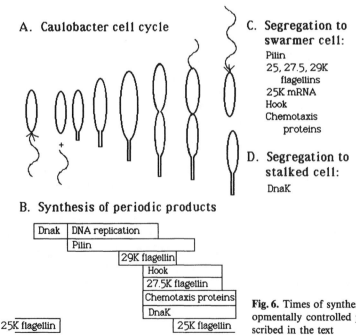

Fig. 6. Times of synthesis and patterns of segregation of developmentally controlled proteins. Experiments were done as described in the text

ulated transcriptionally. The synthesis of chemotaxis proteins is also periodic, occuring in predivisional and newly divided swarmer cells. Transcription has been measured using reporter genes and nuclease protection assays.

3.4.3.2 DNA Replication Plays a Role in Controlling Timing of Flagellar Gene Expression

Expression of the major flagellins and the hook protein is dependent upon DNA replication, and requires that DNA synthesis pass a certain "execution point". Experiments in which DNA replication was delayed for a specific time showed that flagellin synthesis was delayed for an equal period. It therefore seems as if DNA replication may be acting as a clock. The method by which the ticks of the clock are transmitted to the genes is not clear. Flagellar genes located on plasmids which replicate throughout the cell cycle are still subject to cell cycle control of initiation of transcription, which means that it is not replication of a specific gene which controls its time of expression.

It is possible that the condensation of the DNA may play a role in controlling time of synthesis of certain genes. Differences in the packing of the swarmer cell chromosome may be responsible for controlling both the time of synthesis of flagellar genes and the time of chromosome replication. The turn on of expression of flagellar genes on plasmids is like that on chromosomes, but the time of turn off seems to be delayed. Similarly, plasmid replication peaks at the same time as chromosomal replication, but some replication also occurs at the "wrong" time.

3.4.3.3 The Role of Promoters and Upstream Regulatory Regions in Controlling Timing

The promoters for flagellar genes seem to be necessary but not sufficient to control the time of synthesis of these genes. For example, the hook gene, one of the genes encoding the major flagellin, and the gene encoding the 27.5K flagellin are all expressed under the direction of promoters recognized by RNA polymerase containing the sigma

factor which reads *ntr* genes in *E. coli*. The time of expression of these genes, however, is not identical. *ntr*-type promoters have been found to direct transcription of a wide variety of specialized genes in various bacteria. By themselves, the promoter and the RNA polymerase holoenzyme, containing the *ntr*-sigma factor, are not able to direct transcription from these promoters. They require the presence of additional factors, which bind to conserved DNA sequences. In other systems, a variety of different factors have been found which bind to different upstream sequences. These factors are themselves regulated. For example, the regulatory factor NtrC must be phosphorylated in order to bind to upstream sequences and allow transcription of *ntr* genes by RNA polymerase containing the *ntr* sigma. The phosphorylation of NtrC by NtrB is itself controlled by the response of other enzymes to the available nitrogen.

Similarly, upstream sequences essential for proper expression of flagellar genes have been found in *C. crescentus*. The *ftr* element upstream from the hook gene and two of the flagellin genes is essential for the proper level of expression of these genes, and may play a role in regulating timing as well, but while the sequences are conserved among the three genes, the timing of expression of these genes is not identical.

3.4.3.4 Does the Hierarchy Control Timing?

It is tempting to think that the hierarchy described in Section 4.2 could be used to control the time of synthesis of flagellar genes. The product of gene 1 could turn on expression of genes 2, 3 and 4, one of whose products could turn on expression of genes 5, 6, 7 and 8; and so on. It is possible that the hierarchy could play a role in timing, but it cannot be as simple as outlined above. Mutations which lead to drastically lowered expression of flagellin genes do not interfere with the timing of expression of these genes. This implies that control of timing is independent of the hierarchy controlling level of gene expression. It is possible that a given gene is regulated by several factors, which are part of the hierarchy and which specify both timing and level of expression. A mutation which removes one of these factors would leave timing of the target gene unchanged, but lower its level of expression. The presence of many conserved sequences upstream of flagellar genes makes this interpretation seem feasible. Although many promoter and upstream regulatory mutations have been introduced to try to change the regulation of flagellar genes, alterations which preserve full expression tend to preserve correct timing as well. DNA replication may interact with these factors, either by changing the structure of DNA, or by allowing transcription of a factor at the top of the hierarchy.

The mutations, described in the section on hierarchy, which lead to increased synthesis of other flagellar genes, may indicate the presence of genes whose function is to turn off synthesis of other flagellar genes, or even themselves. However, even in the case of elevated synthesis, timing is preserved, again indicating that there may be multiple controls on timing. For example, cells with mutations in the hook gene overproduce hook mRNA, to such an extent that the message for the hook protein persists until the swarmer cell stage. However, no hook mRNA is made in the swarmer cells. Apparently, the swarmer cells are missing a required factor essential for transcription of the hook gene.

In summary, DNA synthesis may be the overriding force that controls timing, perhaps through interaction with several DNA binding factors.

3.5 Gene Products Are Segregated to the Appropriate Daughter Cell

What directs proteins to go where they belong? Is the signal for targeting within the protein, or is it in an upstream regulatory region or in a leader or signal sequences? Theoretically, proteins could be made at the site where they are needed, because of the localization of either transcription or translation factors. Alternatively, their localization could be driven by assembly. A third possibility is that turnover of proteins that are out of place could account for apparent segregation. These questions have been addressed for flagellins, chemotaxis pro-

teins, and heat shock proteins. The results are summarized in Fig. 6.

3.5.1 Flagellins[9]

Flagellin messenger RNA is made in the predivisional cell. It is segregated to the daughter swarmer cell, as shown by the following experiment. Predivisional cells, swarmer and stalked cells were isolated, and their RNA was hybridized to DNA encoding the 25 and 29K flagellin. The swarmer cell ended up with four times more flagellin mRNA than did the daughter stalked cell, and with more flagellin RNA than the predivisional cell had, indicating that the RNA had been segregated to the swarmer cell, and not preferentially degraded in the stalked cell. mRNA segregation may occur either because of localization of ribosomes containing nascent protein, or because of localized transcription.

A different type of result was obtained using *flg* J, the gene for the 29K flagellin. A transcriptional fusion was made by placing a promoterless neomycin phosphotransferase gene within the protein coding region of *flg* J. Predivisional cells containing both the wild-type *flg* J gene and the interrupted gene on a plasmid were pulse labeled with ^{35}S methionine. The cells were allowed to divide, and swarmers were separated from stalked cells. Under these conditions, the 29K flagellin was segregated to the daughter swarmer cell, even in *fla*-strains which cannot assemble a filament. NPTII produced from the transcriptional fusion to the 29K flagellin gene, however, was found in the daughter stalked cell as well as in the daughter swarmer. This indicates that the 29k protein, or sequences within the coding region for that protein, are responsible for the segregation of the 29K protein. Regulatory sequences upstream from that gene are not sufficient to target a chimeric mRNA to the swarmer cell. This implies that, for the 29K gene at least, localized transcription is not the method by which the protein reaches its target.

The same kind of experiment was done labeling predivisional cells in a hook mutant strain to see if the 29K segregation was dependent upon assembly of the flagellin into the flagellar filament. The 29K flagellin segregated exclusively to the daughter swarmer, even though it could not be assembled because there was no hook. So it appears that there is a method for flagellin localization that does not depend on either the site of transcription, or on the assembly of the protein outside the cell. It is, of course, possible that even in a hook mutant strain the components of the flagellum responsible for flagellin export are still present, and that these proteins allow proper segregation of the flagellins.

3.5.2 Chemotaxis Proteins[10]

Like flagellins, the chemotaxis proteins are made in the predivisional cell and preferentially found in the daughter swarmer after cell division. The methylated chemotaxis proteins (MCPs) contain two membrane-spanning domains and are located in the inner membrane. In order to measure the location of the MCPs, membrane vesicles containing filaments were isolated on an affinity column containing anti-flagellin antibodies. When vesicles were isolated from swarmer cells, both flagellated and non-flagellated vesicles contained MCPs, indicating that the MCPs were not preferentially localized to the area immediately adjacent to the flagellum. Larger vesicles, isolated from predivisional cells showed a predominance of MCPs in the flagellated vesicles, indicating that the MCPs were positioned at the daughter swarmer pole before cell division. Membrane fractions from swarmer cells were able to act as methyl acceptors, membranes from stalked cells lacked acceptor function. Similarly, soluble fractions from swarmer but not stalked cell populations were found to contain methylesterase and methyltransferase activities. The fact that soluble proteins are segregated means that there is a mechanism for segregation that does not depend on assembly in a structure.

3.5.3 Pilin[11]

At the same pole as the filament, the swarmer cell assembles pili, structures composed of pilin monomers. Unlike the flagellar genes, which are expressed only briefly, the pilin gene is expressed throughout the stalked cell stage, and its expression does not depend upon DNA synthesis. Because pilin is made throughout the stalked cell stage, unlike the structural proteins for the flagellum, there is apparently not a common mechanism controlling both time of synthesis and localization of proteins for these two organelles. The pilus structure does not form until the time of cell division. The pili remain until the ejection of the filament, at which time they disappear from the cell surface. They may be retracted into the cell, because phage adsorbtion to the pili prevents their loss from the cell surface.

3.5.4 Heat-Shock Proteins[12]

Other proteins are segregated to the daughter stalked cell. Several heat shock proteins are made at defined stages in the cell cycle, under non-heat shock conditions. These include proteins which cross-react with antibodies to Lon protease, Gro EL, and DnaK from *E. coli*, and to RNA polymerase from *C. crescentus*. The DnaK made in the predivisional cell is segregated predominantly to the daughter stalked cell, where DNA replication will begin following cell division. Similarly, the putative Lon protease and the heat shock sigma factor, when made in predivisional cells, are segregated preferentially to the daughter stalked cell. GroEL is evenly divided among the two cell types. The putative Lon protease and the RNA polymerase subunit are made predominantly in the stalked cell, while the protein which cross-reacts with Gro EL is preferentially made in the swarmer cell. DnaK is synthesized at two times in the cell cycle: just prior to the start of DNA replication in the swarmer cell, as it is becoming a stalked cell, and just prior to cell division. The diversity of segregation and of time of synthesis of this set of heat shock proteins makes it seem unlikely that timing of synthesis and segregation are controlled by the same signals.

3.6 Mutations and Antibiotics with Pleiotropic Effects Define Dependent Pathways and Common Developmental Steps [13]

Mutations and antibiotics which block either metabolic or developmental pathways often have pleiotropic effects. Analysis of these effects has led to the following conclusions:

1. Phospholipid synthesis is required for stalk formation, DNA synthesis is not required.
2. Membrane synthesis is required for DNA replication.
3. Completion of a round of DNA replication is required for cell separation; substantial DNA synthesis is required for initiation of cell division.
4. Completion of a round of DNA replication without reinitiation allows cell viability in the absence of membrane synthesis.
5. DNA synthesis is required for flagellum formation, but not for pilin production.
6. Some steps in polar organelle development are shared by several organelles.
7. Polar organelles are not obligatory steps in the cell cycle.

3.7 Environmental Influences on *C. crescentus* [14]

While the *C. crescentus* cell cycle is invariant under conditions where the nutrients are not limited, when cells are starved they respond in a variety of ways. G1 and S phase increase in length with increasing doubling times, but G2 does not. Nitrogen limitation triggers the accumulation of ppGpp (magic spot) and causes the cells to remain in the swarmer cell stage. Phosphate limitation in the presence of nonlimiting carbon leads to accumulation of carbon as poly-β-hydroxybutyrate, as well as to formation of very long (up to 20 µm) stalks. Exogenous cGMP also stimulates stalk elongation, and mutants which have abnormally long stalks, in the presence of phosphate, also have abnormally low ATP and GTP pools.

Shifting from glucose to lactose as a carbon source, or depletion of the glucose from media containing glucose plus lactose, causes a growth lag, in which the cells arrest at the nonmotile, predivisional stage for several hours, with the accumulation of cGMP. Addition of dibutryl cyclic AMP to cells in this condition allows the cells to escape from cell cycle arrest.

Several *C. crescentus* enzymes that degrade fatty acids have been identified, and their metabolic regulation studied. Like similar enzymes from *E. coli*, these enzymes are under catabolite control in *C. crescentus*, and are repressed when cells are grown in the presence of glucose. In *C. crescentus*, the level of glucose repression is only about two- to threefold, compared with 50–300-fold repression in *E. coli*.

Some starvation conditions lead to cell death, apparently because the cell is unable to reach a stage where it can complete DNA synthesis and shut itself down. This may be why *Caulobacter* replicates its DNA only once per cell cycle: replicating cells may be more vulnerable to environmental insults and to starvation. Keeping the replication stage to a minimum would therefore be protective.

3.8 Outlook

The combination of tractable genetics and a simple cell cycle with polar organelles will hopefully allow elucidation of the mechanisms which control placement of the polar organelles, including stalk, pili, and flagellum. Placement of the filament is probably dependent upon the assembly of a first structure, whose nature is not known. Presumably this organelle leaves a trace of itself following ejection, which could then be a marker for stalk location. Deciphering the role of genes which alter both flagellum and stalk formation may answer the question of how these locations are marked.

What controls the timing of expression of the genes required for polar organelle formation? Many details of the answer to this question are emerging; timing seems to depend in part on DNA replication and in part on a cascade of DNA-binding proteins and alternative transcription factors. The source of the cascade may be hard to find, however. Altering time of expression of these timed genes and looking at the effects both on expression of other genes and on the ability of the cell to assemble a flagellum will test the importance of timing on the assembly process.

Understanding the difference between the daughter swarmer and stalk cell in DNA replication ability is crucial to understanding differentiation in *C. crescentus*, for two reasons. First, it is an essential part of understanding how a stem cell can give rise to two different progeny. In addition, the state of replication seems to be essential to controlling flagellar development. Characterization of the enzymes of DNA replication, examining their synthesis, activity, and distribution in developing cells, would be an important step in answering these questions.

Answers to these questions will certainly overlap; for example, the polar location of organelles may reflect the presence of structures required for the replication of the DNA.

3.9 Summary

The study of *C. crescentus* has demonstrated the following:

1. A complex structure, namely the flagellum, is made from component parts that are made at specific, brief times during the cell cycle. The time of synthesis of these components is controlled in part by the actions of DNA-binding proteins and alternative RNA polymerase sigma factors. Ongoing DNA synthesis is both a prerequisite and a timing device for flagellum biosynthesis.
2. The level of synthesis of the flagellar proteins depends on a complex genetic regulatory hierarchy, with expression of genes at the botton of the hierarchy dependent on expression of genes at the top of the hierarchy.
3. Several cell cycle controlled products are made in the predivisional cell and preferentially segregated to either the daughter swarmer or the daughter stalked cell. In some cases, there is apparently segregation of the mRNA. In other cases, the protein product may be segregated.
4. The polar organelles are not obligatory components of the cell cycle, making the isolation of defective mutants relatively easy. On the other

hand, complete polar organelle development depends on membrane synthesis, DNA synthesis, and cell division steps.

References

[1] Ely B, Shapiro L (1984) Regulation of cell differentiation in *Caulobacter crescentus*. In: Losick R, Shapiro L (eds) Microbial development. Cold Spring Harbor, New York, pp 1–26

Poindexter JS (1964) Biological properties and classification of the *Caulobacter* group. Bacteriol Rev 28:231–295

Pointdexter JS (1972) The caulobacters: ubiquitous unusual bacteria. Microbiol Rev 45:123–179

[2] Dow CS, Whittenbury R, Carr NG (1983) The "shut down" or "growth precursor" cell – an adaptation for survival in a potentially hostile environment. In: Slater JH, Whittenbury R, Wimpenny JWT (eds) Symposium of the society for general microbiology, vol 34. Microbes in their natural environments. Cambridge University Press, Cambridge, England; New York, NY, USA, pp 187–247

Ely B, Gerardot CJ, Fleming DL, Gomes SL, Frederikse P, Shapiro L (1986) General nonchemotactic mutants of *Caulobacter crescentus*. Genetics 114:717–730

Macnab RM (1987) Motility and chemotaxis. In: Neidhardt FC, Ingraham J, Low KB, Magasanik B, Schaecter M, Umbarger HE (eds) *Escherichia coli* and *Salmonella typhimurium*: Cellular and molecular biology. American Society for Microbiology, Washington DC, pp 732–759

Shapiro L (1985) Generation of polarity during *Caulobacter* differentiation. Annu Rev Cell Biol 1:173–207

[3] Degnen ST, Newton A (1972) Chromosome replication during development in *Caulobacter crescentus*. J Mol Biol 64:671–680

Dingwall A, Shapiro L (1990) Chromosome replication rate, origin and bidirectionality as determined by pulsed field gel electrophoresis. Proc Natl Acad Sci USA 86:119–123

Evinger M, Agabian N (1979) *Caulobacter crescentus* nucleoid: analysis of sedimentation behavior and protein composition during the cell cycle. Proc Natl Acad Sci USA 76:175–178

Marczynski G, Dingwall A, Shapiro L (1990) Plasmid and chromosomal replication and partitioning during the *Caulobacter crescentus* cell cycle. J Mol Biol 212:709–722

Poindexter JS, Hagenzieker JG (1981) Constriction and septation during cell division in caulobacters. Can J Microbiol 27:704–719

[4] Bryan R, Glaser D, Shapiro L (1990) A genetic regulatory hierarchy in *Caulobacter* development. In: Wright TRF (ed) Advances in genetics, vol 27. Academic Press, London, pp 1–31

Driks A, Bryan R, Shapiro L, DeRosier DJ (1989) The organization of the *Caulobacter crescentus* flagellar filament. J Mol Biol 206:627–636

Macnab RM (1987) Flagella. In : Neidhardt FC, Ingraham J, Low KB, Magasanik B, Schaecter M, Umbarger HE (eds) *Escherichia coli* and *Salmonella typhimurium*: Cellular and molecular biology. American Society for Microbiology, Washington, D.C., pp 70–83

Minnich SA, Ohta N, Taylor N, Newton A (1988) Role of the 25-, 27- and 29-kDa flagellins of *Caulobacter crescentus* in cell motility; a method for the construction of the Tn*5* insertion and deletion mutants by gene replacement. J Bacteriol 170:3953–3960

Newton A (1989) Differentiation in *Caulobacter* flagellum development, motility and chemotaxis. In: Chater K, Hopwood DA (eds) Genetics of bacterial diversity. Academic Press, London, pp 199–222

Stallmeyer MJB, Hahnenberger KM, Sosinsky GE, Shapiro L, DeRosier DJ (1989) Image reconstruction of the flagellar basal body of *Caulobacter crescentus*. J Mol Biol 205:511–518

Wagenknecht T, DeRosier D (1981) Three-dimensional reconstruction of the flagellar hook from *Caulobacter crescentus*. J Mol Biol 151:439–465

[5] Dingwall A, Gober JW, Shapiro L (1990) Identification of a *Caulobacter* basal body gene and a *cis*-acting site required for activation of transcription. J Bacteriol 172:6066–6076

Ely B, Croft RH, Gerardot CJ (1984) Genetic mapping of genes required for motility in *Caulobacter crescentus*. Genetics 108:523–532

Ely B, Ely TW (1989) Use of pulsed field gel electrophoresis and transposition mutagenesis to estimate the minimal number of genes required for motility in *Caulobacter crescentus*. Genetics 123:649–654

Hahnenberger K, Shapiro L (1987) Identification of a gene cluster involved in flagellar basal body biogenesis in *Caulobacter crescentus*. J Mol Biol 194:91–103

Johnson RC, Ely B (1979) Analysis of non-motile mutants of the dimorphic bacterium *Caulobacter crescentus*. J Bacteriol 137:627–635

Johnson RC, Ferber DM, Ely B (1983) Synthesis and assembly of flagellar components by *Caulobacter crescentus* motility mutants. J Bacteriol 154:1137–1144

Kaplan JB, Dingwall A, Bryan R, Champer R, Shapiro L (1989) Temporal regulation and overlap organization of two *Caulobacter* flagellar genes. J Mol Biol 205:71–83

[6] Frederikse PH, Shapiro L (1989) An *Escherichia coli* chemoreceptor gene is temporally controlled in *Caulobacter*. Proc Natl Acad Sci USA 86:4061–4065

Gober JW, Shapiro L (1990) Integration host factor is required for the activation of developmentally regulated genes in *Caulobacter*. Genes Dev 4:1494–1499

Mullin DA, Newton A (1989) Ntr-like promoters and upstream regulatory sequence *ftr* are required for transcription of a developmentally regulated *Caulobacter crescentus* flagellar gene. J Bacteriol 171:3218–3227

Ninfa AJ, Mullin DA, Ramakrishnan G, Newton A (1989) *Escherichia coli* σ^{54} RNA polymerase recognizes *Caulobacter crescentus* *flb*G and *flb*N flagellar gene promoters in vitro. J Bacteriol 171:383–391

Ramakrishnan G, Newton A (1990) FlbD of *Caulobacter crescentus* is a homologue of the NtrC(NR$_1$) protein and activates σ^{54}-dependent flagellar gene promoters. Proc Natl Acad Sci USA 87:2369–2373

[7] Champer R, Dingwall A, Shapiro L (1987) Cascade regulation of *Caulobacter* flagellar and chemotaxis genes. J Mol Biol 194:71–80

Newton A, Ohta N, Ramakrishnan G, Mullin D, Raymond G (1989) Genetic switching in the flagellar gene hierarchy requires negative as well as positive regulation of transcription. Proc Natl Acad Sci USA 86:6651–6655

Schoenlein PV, Ely B (1989) Characterization of strains containing mutations in the contiguous *flaF*, *flbT*, or *flbA-flaG* transcription unit and identification of a novel Fla phenotype in *Caulobacter crescentus*.. J Bacteriol 171:1554–1561

Xu H, Dingwall A, Shapiro L (1989) Negative transcriptional regulation in the *Caulobacter* flagellar hierarchy. Proc Natl Acad Sci USA 86:6656–6660

[8] Bryan R, Champer R, Gomes S, Ely B, Shapiro L (1987) Separation of temporal control and trans-acting modulation of flagellin and chemotaxis genes in *Caulobacter*. Mol Gen Genet 206:300–306

Loewy ZG, Bryan RA, Reuter SH, Shapiro L (1987) Control of synthesis and positioning of a *Caulobacter crescentus* flagellar protein. Gene Dev 1:626–635

Minnich SA, Newton A (1987) Promoter mapping and cell cycle regulation of flagellin gene transcription in *Caulobacter crescentus*. Proc Natl Acad Sci USA 84:1142–1146

Sheffery M, Newton A (1981) Regulation of periodic protein synthesis in the cell cycle: control of initiation and termination of flagellar gene expression. Cell 24:49–57

[9] Millhausen M, Agabian N (1983) *Caulobacter* flagellin mRNA segregated asymmetrically at cell division. Nature 302:630–632

[10] Gomes SL, Shapiro L (1984) Differential expression and positioning of chemotaxis methylation proteins in *Caulobacter*. J Mol Biol 177:551–568

Nathan P, Gomes SL, Hahnenberger K, Newton A, Shapiro L (1986) Differential localization of membrane receptor chemotaxis proteins in the *Caulobacter* predivisional cell. J Mol Biol 191:433–440

[11] Smit J, Agabian N (1982) *Caulobacter crescentus* pili: analysis of production during development. Dev Biol 89:237–247

[12] Reuter S, Shapiro L (1987) Asymmetric segregation of heat-shock proteins upon cell division in *Caulobacter crescentus*. J Mol Biol 194:653–662

[13] Driks A, Schoenlein PV, DeRosier DJ, Shapiro L, Ely B (1990) A *Caulobacter* gene involved in polar morphogenesis. J Bacteriol 172:2113–2123

Newton A, Ohta N, Huguenel E, Chen LS (1985) Approaches to the study of cell differentiation in Caulobacter crescentus. In: Setlow P, Hock J (eds) The molecular biology of microbial differentiation. American Society for Microbiology, Washington DC, pp 267–276

[14] O'Connell M, Henry S, Shapiro L (1986) Fatty acid degradation in *Caulobacter crescentus*. J Bacteriol 168:49–54

Poindexter JS (1987) Bacterial responses to nutrient limitation. In: Fletcher M, Gray TRG, Jones JG (eds) Symposium of the society for general microbiology, vol 41. Ecology of microbial communities. Cambridge University Press, Cambridge, pp 283–317

Chapter 4 *Streptomyces coelicolor*: a Mycelial, Spore-Bearing Prokaryote

KEITH F. CHATER

4.1 Introduction[1]

Many bacteria grow rapidly by binary fission, with cell populations that are usually presumed to be virtually identical to each other except in cell cycle-related processes or very rare events such as transposition or other DNA rearrangements. Generally, they do not have intracellular compartments. These qualities, along with their simple nutrition and exceptional amenability to a wide range of types of genetic analysis, make them wonderful objects for the investigation of many basic biological questions, for which they may act as models for higher organisms. On the other hand, many important aspects of higher organisms are a consequence of multicellularity and structural complexity. Some bacteria do indeed also show cellular differentiation, the formation of endospores by *Bacillus subtilis* discussed in the chapter by Errington being the best-known example; and a few are truly multicellular organisms – perhaps most notably the Gram-negative myxobacteria, in which unicellular motile cells cooperate to produce sporulating fruiting bodies, and the Gram-positive actinomycetes, many of which form mycelial colonies on which specialized spore-bearing organs develop. How is such multicellular organisation controlled and coordinated? How great are the genetic changes needed to account for the evolution of multicellular differentiation? Can students of eukaryotic developmental biology learn anything from prokaryotic developmental systems and vice versa? This chapter discusses current knowledge of the developmental biology of the most intensively studied actinomycetes, members of the genus *Streptomyces*. Much of this work has focused on *S. coelicolor* A3 (2), which is genetically the most studied species, though some important discoveries have also been made in other species, especially *S. griseus* and *S. antibioticus*, which sporulate particularly abundantly.

Figure 1 shows scanning electron micrographs of the development of a streptomycete colony on agar medium. In a suitable environment, colonies are initiated by the germination of a spore giving rise, after a number of hours, to a mycelium of branching hyphae. Occasional crosswalls divide the cytoplasm of the hyphae into compartments that contain many copies of the genome. During the early stages of growth the mycelium seems to colonise the medium relatively efficiently, hyphae rarely crossing each other. Soon, however, the colony centre becomes crowded with hyphae, and overgrowth occurs, giving a three-dimensional colony in which growth of the upper layers must depend on nutrient diffusion either within hyphae, or through the aqueous matrix held by capillarity between hyphae. Typically, after 2–4 days the surface of the colony changes, with the emergence of aerial hyphae. These differ from the vegetative or substrate hyphae in possessing a hydrophobic outer layer. The aerial hyphae grow quickly, at the expense of nutrients derived from the substrate mycelium. Eventually, hyphal tips several tens of microns long begin to metamorphose into spore chains, with specialized crosswalls forming at every 1–2 µm to divide the cytoplasm into compartments. Typically, each compartment contains a single genome. The cylindrical cells so formed then round up to produce chains of ovoid or spherical spores. The sporulating aerial hyphae and spore chains of *S. coelicolor* are helical. In some other species they are straight. As the spores mature, their walls thicken and they often acquire a species-specific pigment (grey in the case of *S. coelicolor*). While these events take place on the colony surface, cells in the lower part of the colony swich to secondary metab-

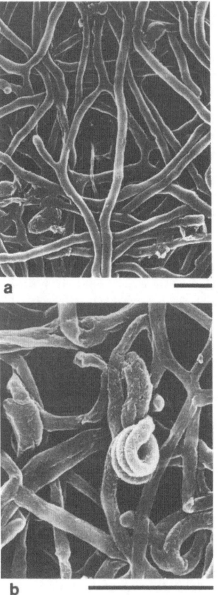

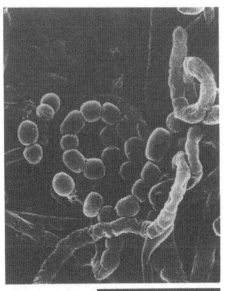

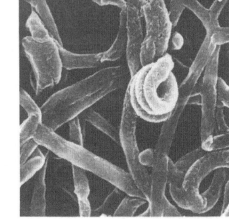

Fig. 1a–c. Stages in the formation of *Streptomyces* spore chains. **a** Branching substrate mycelium (48-h culture). **b** Curling tip of an aerial hypha emerging from substrate mycelium (72-h culture). **c** Spore chains at different stages of development (96 h). *Bars* 5 μm. (Originals from J. Burgess, Norwich)

olism, during which they produce a variety of biologically active chemicals, including antibiotics. This delayed production of antibiotics may at first sight seem paradoxical, if a strain is to gain advantage over its competitors during growth. However, the main adaptive value of antibiotics may be to protect colonies at this late stage of their development, when many compartments of the substrate hyphae are undergoing senescence and lysis, releasing the nutrients that support aerial growth: at such a stage, a colony might potentially be vulnerable to invasion, especially by motile coinhabitants of the soil (Fig. 2). Overall, a maturing *Streptomyces* colony may be regarded as a multicellular organism containing morphologically and physiologically differentiated tissues (Fig. 2).

4.2 Physiological Aspects of Early Events in *Streptomyces* Differentiation

4.2.1 Is there an Universal First Signal for Differentiation in *Streptomyces*?[2]

The formation of aerial mycelium and spores may be seen as an adaptation by which mycelial organisms encountering nutritional stress convert a relatively large amount of the biomass that has been accumulated in nutritionally favourable conditions, into a maximal number of readily dispersed viable units. Conceptually, then, the simplest signal for differentiation is an intracellular "alarm bell" set off by nutrient limitation. The nutrients that are most likely to limit growth are assimilable carbon, nitrogen and phosphate. The work of Freese and his collaborators on sporulation in *Bacillus subtilis* and budding yeast has suggested that the common intracellular change triggered by starvation for each of these components is a fall in the level of GTP. This is probably also the case for several *Streptomyces* species. Moreover, the compound decoyinine – which rather specifically inhibits the formation of GMP from IMP – elicits sporulation in some species.

4.2.2 Pheromones and Differentiation in *Streptomyces* spp.[3]

Evolutionary adaptations sometimes depend on correlations that give advance warning of an approaching new situation or hazard (for example, many plants respond to day length in anticipating the changing seasons). Some streptomycetes appear to have evolved to respond to a situation – high cell density – that correlates with approaching nutrient limitation, either in addition to or perhaps even supplanting its direct perception. In the most-studied example of this, some strains of *Streptomyces griseus* secrete a diffusible extracellular compound, A-factor, the extracellular concentration of which must bear some proportionality to the hyphal density for any particular set of conditions. Provided that *S. griseus* can recognise a certain threshold level of extracellular A-factor, then it can

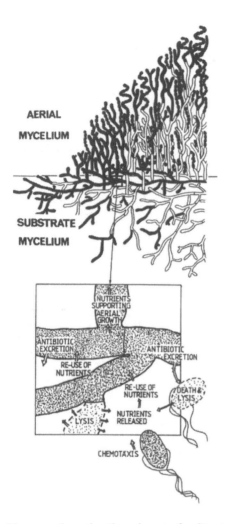

Fig. 2. Diagrammatic section through part of a *Streptomyces* colony. The *upper part* of the figure shows viable cells in black, and lysing or dead cells in white. (After Wildermuth 1970, see [1]). The *lower part* shows a more enlarged section of the region at the base of an aerial hypha, and illustrates a model in which cell lysis supports aerial growth, and antibiotic production plays an important role in colony defence against invading motile microorganisms

In this chapter, I shall discuss *Streptomyces* colony development from several standpoints: the underlying program of sequential gene activation; the intracellular and extracellular signals that trigger the switch from vegetative growth to the formation of aerial hyphae, the roles of metabolism, and the determinants of form.

gauge cell density, allowing its colonies to "antici-pate" the need to sporulate and to produce antibi-otic (in this case, streptomycin), and to co-ordinate the initiation of these processes over large areas of its mycelium. Some progress has been made in un-derstanding how the level of this hormone is per-ceived by hyphae. Mutants defective in A-factor production cannot sporulate or produce streptomy-cin in pure culture, but they can do so when cul-tured close to an A-factor-producer, or when A-fac-tor is added at a concentration as low as 10^{-9} M. A cytoplasmic protein binds A-factor efficiently with a k_D of about this concentration, and mu-tants lacking the protein sporulate and produce streptomycin prematurely and abundantly even when A-factor is absent. The conclusion is that morphological and physiological differentiation in this organism is repressed by the A-factor-bind-ing-protein, and that repressor activity is eliminat-ed when the protein binds A-factor. Presumably, other regulatory elements must also be involved, or the binding-protein mutants would sporulate con-stitutively and therefore be unable to grow at all. It is interesting to note a certain similarity between A-factor and signalling molecules involved in the de-velopment of disparate organisms (Fig. 3). A com-mon theme is of a lipophilic side chain attached to a simple ring structure. Presumably the side chains may help in the free diffusion of the molecules through cell membranes.

Although production of A-factor-like substances is widespread among streptomycetes, some of the producers do not appear to use them as autoregula-tors, and their function remains a mystery in those strains. Perhaps in the natural environment they may cause premature sporulation of competing streptomycetes that have A-factor-dependent spor-ulation. Extracellular autoregulators other than A-factor, both of related and of quite distinct types, occur in different streptomycetes. For example, in *S. virginiae* an A-factor-like compound (Fig. 3) reg-ulates production of virginiamycin antibiotic, and is also perceived by a cytoplasmic high-affinity binding protein; but there is no cross-recognition between the *S. griseus* and *S. virginiae* binding pro-teins and their cognate factors. Extracellular pro-teins or even antibiotics may induce differentiation in other species.

4.3 Genes Involved in *Streptomyces* Differentiation – an Overview[4]

4.3.1 The Use of *Streptomyces coelicolor* for Genetic Studies

In order to exploit the power of genetics in analys-ing complex processes, it is necessary to have access to systems for natural genetic exchange and gene cloning, and to mutants affected in the process of interest. In *Streptomyces coelicolor* A3(2), proce-dures for genetic analysis are well established. Moreover, many cloning vectors based on indige-nous low and high copy number plasmids and tem-perate bacteriophages are available, all readily in-troduced as DNA into *Streptomyces* protoplasts with the help of polyethylene glycol. Mutants de-fective in aerial mycelium formation and sporula-tion can readily be isolated, since these processes

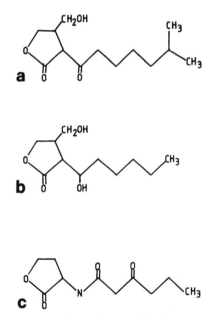

Fig. 3a–c. Autoregulators that coordinately control popula-tions of cells in *Streptomyces* spp. and other microorganisms. **a** A-factor from *Streptomyces griseus*. **b** A γ-butyrolactone compound from *Streptomyces virginiae*. **c** Compound that in-duces expression of luminescence genes in the marine bacterium *Vibrio fischeri*

are not obligatory for viability: vegetative hyphae can be maintained by sequential subculture. Genetic analysis has been carried out on about 50 mutants unable to make aerial mycelium on normal minimal medium (*bld* mutants: Fig. 4) and on a similar number with aerial hyphae that fail to produce normal grey spores (*whi* mutants: Fig. 5). This analysis revealed the presence of (generally several) representative mutants for each of six *bld* and eight *whi* genes (Fig. 6). Several of these genes have been cloned by introducing genetic libraries into representative mutants and then recognising relevant clones by their ability to restore the normal morphology.

Although it seems from this genetic analysis that only a small number of genes are specific to, and indispensable for, normal differentiation, many of the processes necessary for differentiation must also be necessary for vegetative growth (for example, macromolecular synthesis and intermediary metabolism); and some – perhaps many – genes specific to differentiation may show sufficient functional overlap with others to mask the pheno-

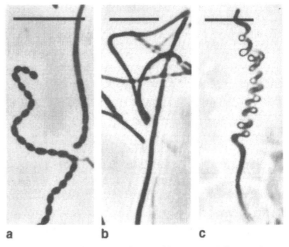

Fig. 5a–c. Aerial hyphae from *whi* mutants (phase-contrast microscopy). **a** Wild-Type, with spore chains. **b** *whiG* mutant. **c** *whiB* mutant. *Bar* 10 μm

typic effects of mutations. In the latter case, the uncovering of these genes and their roles may require biochemical procedures involving enrichment for developmentally regulated proteins or RNA, and the eventual design of molecular probes for the relevant gene(s), as in the case of spore-associated proteins (Saps). An exciting alternative approach is to use vectors or transposons containing reporter genes, such as the *lux* genes for light emission, to detect promoters active at particular times or locations in colonies.

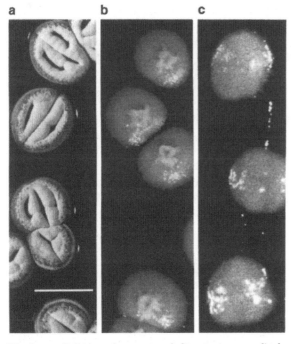

Fig. 4a–c. Colonies of mutants of *Streptomyces coelicolor* lacking aerial mycelium. **a** Wild-type. **b** *bldA* mutant. **c** *bldC* mutant. *Bar* 0.25 cm

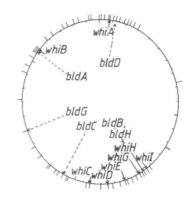

Fig. 6. Genetic map of *Streptomyces coelicolor* genes for aerial mycelium formation (*bld*) and for the metamorphosis of aerial hyphae into spores (*whi*). The *marks outside the circle* indicate the positions of various other mapped genes, principally involved in primary metabolism

4.3.2 What Genes Are Switched On in Response to the Signals for Initiation of Differentiation?[5]

Most of the *bld* mutations that block the start of morphological differentiation also block antibiotic production. Such *bld* genes are likely to play a regulatory role and to act early in differentiation, and might therefore be among the first differentiation-specific genes to be activated (though direct evidence of this is currently lacking). At the time of writing, three *bld* genes (other than those involved in A-factor production or response in *S. griseus*) have been cloned: two from *S. coelicolor* and one from *S. griseus*.

The most intelligible, but perhaps particularly surprising, finding is that the *bldA* gene of *S. coelicolor* specifies a tRNA for the leucine codon UUA, which occurs (as TTA) extremely rarely in *Streptomyces* genes. Genetic evidence, based on the *bldA*-dependence of genes containing this codon, suggests that the *bldA*-specified tRNA is the only efficient means of translating UUA codons (although some mRNAs containing UUA codons can be expressed as functional proteins in *bldA* mutants, albeit at much reduced efficiency). It follows, since *bldA* mutants grow vegetatively with normal vigour, that genes essential for vegetative growth are unlikely to contain TTA codons – a deduction

Table 1. Of the first 100 *Streptomyces* genes sequenced, only eight contain TTA codons

	No. of TTA codons
Resistance genes for:	
Streptomycin (*S. glaucescens, strA*)	1
Hygromycin (*S. hygroscopicus, hyg*)	1
Carbomycin (*S. thermotolerans, carB*)	2
Actinorhodin (*S. coelicolor, act*II ORF2)	1
Regulatory genes for production of:	
Streptomycin (*S. griseus, strR*)	1
Actinorhodin (*S. coelicolor, act*II ORF4)	1
Genes for differentiation	
S. griseus bld gene	1
Putative transposase gene	
IS*466* (*S. coelicolor*)	1

so far entirely borne out by sequence analysis. The presence of the *bldA* tRNA apparent only in older cultures led to a model in which translation of one or more crucial genes for differentiation and antibiotic production is limited by the presence of at least one UUA codon in the mRNA. This model has yet to be verified.

A distinct gene has been cloned from *S. griseus* by its ability to restore wild-type morphology to an *S. griseus bld* mutant. Its sequence suggests that it encodes a protein able to interact with DNA, since it contains a region resembling the helix-turn-helix motif of "classical" bacterial sequence-specific DNA-binding proteins. Mutations in this *bld* gene are again pleiotropic, resulting in failure to produce antibiotics and pigments as well as loss of aerial mycelium. This gene contains a TTA codon, tempting speculation that it is one of the targets of *bldA* action (but note that *bldA* mutants have yet to be studied in species other than *S. coelicolor* A3(2) and its close relative *Streptomyces lividans* 66).

The idea that *bldA* mediates its effects on genes for differentiation and antibiotic production by translational control of genes that encode transcriptional regulators receives some further support from studies of genes involved in antibiotic production: the transcription of genes encoding biosynthetic enzymes for different antibiotics in *S. coelicolor* is greatly reduced in *bldA* mutants, and several pathway-specific regulatory and resistance genes associated with antibiotic biosynthesis contain TTA codons (Table 1). Clearly, however, this view is over-simplified, because it fails to take into account any of the six or more other *bld* genes of *S. coelicolor*, or the ways in which they might interact with several other pleiotropic genes that affect antibiotic production without being required for morphogenesis.

4.4 Development and Metamorphosis of the Aerial Mycelium[6]

4.4.1 The Emergence of Aerial Hyphae[6]

One might have expected to find *bld* genes that are specifically and unconditionally needed for aspects of aerial mycelium growth, yet none of the at least seven *bld* genes currently known in *S. coelicolor* fulfils this expectation: not only are most of the *bld* mutants also deficient in antibiotic production, but also most of them can be phenotypically restored to the normal morphology by physiological means, such as changing the carbon source or growing them next to a wild-type strain. As already mentioned, this can be interpreted in three alternative ways: there may be many relevant genes, but with functional overlap such that mutants have no ob-

servable phenotype; the genes involved may also be essential for vegetative growth; or there may be very few genes involved. Figure 7 proposes a speculative hypothesis that would necessitate relatively few genes and which deals with the requirements of aerial growth for nutrients, turgor, directionality and protection against desiccation. Nutrients are supplied by the re-use of macromolecules stored in the ageing substrate hyphae: proteins, nucleic acids, and compounds such as glycogen and poly-β-hydroxybutyrate that accumulate in *Streptomyces*, as in many other bacteria, in certain conditions of nutrient limitation. Limitation for nutrients would be short-lived in a *Streptomyces* colony, because a fresh supply would become available through the lysis of many hyphae. These nutrients would allow growth to be reinitiated, and in activated hyphae there would be breakdown of storage polymers, causing an increase in osmotic pressure and hence of turgor inside the hyphae. This could provide a physical driving force for the emergence of new hyphal branches. In circumstantial support for this model, glycogen accumulation is very low in the surface hyphae of *bldA* mutants. Directionality – the recognition of the hyphal surface in contact with air – could be achieved by the inability of certain secreted proteins or other compounds to diffuse away from this face, so that aerial branches would automatically be covered in a distinctive surface (perhaps with potential for protection against dehydration). Some of the "spore-associated proteins" – Saps – that are found on the surface of aerial hyphae may conceivably fulfil this role. To account for the paucity of mutants that are specifically and unconditionally deficient in aerial mycelium, it is proposed that there is some degree of redundancy for the genetic determinants of nutrition and turgor control, and of the specialized surface components.

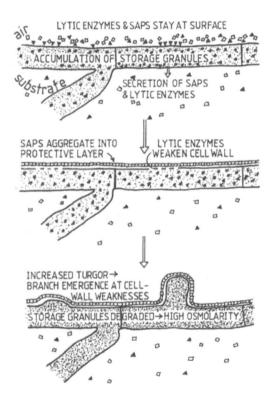

Fig. 7. Model to explain the physiological control of aerial mycelium formation in *Streptomyces*. The model is based on observations suggesting that nutrient-limited hyphae at the substrate-air interface accumulate osmotically inactive cytoplasmic polymeric storage granules (*asterisks*), and secrete lytic enzymes (*black triangles*) and so-called "spore-associated proteins" (*Saps, white rectangles*). It is proposed that rapid degradation of storage granules leads to increased osmolarity and turgor pressure, forcing branches to form at points of cell-wall weakness (*dashed lines*); and that aerial branches become coated in an organised surface layer of Saps that protects them against desiccation

4.4.2 Overview of Sporulation of Aerial Hyphae and its Genetic Control in *S. coelicolor*[6,7]

Once formed, aerial hyphae retain a connection with an aqueous external environment only through their basal contact with the substrate mycelium. Probably, therefore, the metamorphosis

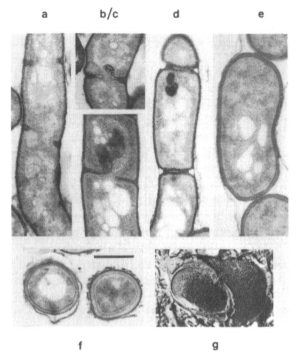

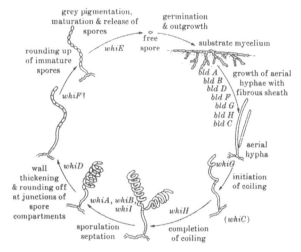

Fig. 9. Genes required to achieve specific stages in the differentiation of *Streptomyces coelicolor*. (After Chater and Merrick 1979, see[4])

Fig. 8a–g. Formation of spore chains from aerial hyphae in *Streptomyces coelicolor*. The electron micrographs of thin sections show **a** initiation of septation, **b, c** ingrowth of septa, **d** spore compartments completed, **e** rounding off of spore compartments, **f** mature spore chains. In addition, a freeze-fractured view of the spore surface is shown in **g**. *Bar* 1 µm. (McVittie 1974, see[7])

of aerial hyphal tips into chains of spores is largely independent of changes in the medium. The sequence of intracellular morphologically observable events during sporulation is shown in Fig. 8. The hope is that such morphological studies will become interpretable in terms of gene function because of the availability of *whi* mutants blocked at various stages in the process.

Six *whi* loci (*A, B, C, G, H, I*) have been recognised that are needed for the formation of regularly spaced, specialised sporulation septa. Further *whi* loci are involved in spore wall thickening (*whiD*) and spore pigment formation (*whiE*) (Fig. 9). No doubt many other genes are involved in sporulation; but probably in most cases mutations in these genes do not give the white colony phenotype either because of functional redundancy, or

because they are needed for vegetative growth, or because some genes essential for, and specific to, sporulation may not be needed for grey pigment to form. Most probably, the *whi* genes include a (the?) major set of regulatory genes. If so, some of the genes determining the structural features of sporulation will probably eventually be recognised because of their regulation by *whi* genes, rather than by the isolation of mutants.

4.4.3 The Crucial Role of a Transcription Factor in the Onset of Sporulation[8]

Early light microscope studies of recombinants carrying mutations in two different *whi* genes showed that the morphology of the double mutants was in each case the same as that of one of the single mutant parent strains, indicating a simple hierarchy of epistatic interactions. Thus, morphologically, *whiG* was epistatic to *whiH*, *A*, *B* and *I*; *whiH* to *whiA*, *B* and *I*; and *whiA* and *whiB* to *whiI*. This suggested that *whiG* might determine a very early event in sporulation of aerial hyphae, perhaps even the primary "decision" to sporulate. Accordingly, a fragment of *S. coelicolor* DNA able to complement *whiG* mutants was cloned (using a phage vector to guard against problems that might be caused by multicopy cloning of a key morphogenetic gene). It was soon seen that even one extra copy of *whiG* mutants gave a substantial increase in the greyness of colonies: a reflection of hypersporulation. When *whiG* was introduced on a multicopy plasmid, the sporulation was increased so much that the colonies grew very slowly, probably because hyphae in the substrate mycelium were also sporulating instead of continuing to grow and branch. The cause of this phenotype was pin-pointed to the *whiG* gene product by finding that a small (42 bp) in-frame deletion from the centre of the *whiG* coding sequence eliminated both hypersporulation and the complementation of *whiG* mutations.

The suggestion from these results that *whiG* might specify a regulatory protein, was substantiated when analysis of the *whiG* nucleotide sequence revealed that its deduced gene product was related to a well-known family of bacterial regulatory proteins, the sigma factors. Upon interaction with the catalytic core RNA polymerase enzyme, sigma factors direct it to particular kinds of promoters. A central role for sigma factors in major regulatory switching points has often been observed in other bacteria, perhaps most pertinently in controlling successive waves of transcription during endospore formation in *Bacillus subtilis*, as described by Errington (Chap. 2, this vol.), and in the life cycles of some bacteriophages. There is now very good evidence that most sigma factors, when associated with RNA polymerase, recognise specific features of their cognate promoters at DNA regions localised about 10 and 35 base pairs upstream of transcription start sites. Moreover, these contacts at the so-called -10 and -35 regions involve aminoacids in two of the characteristic regions — the 2.4 and 4.2 regions respectively — that are conserved in their general features among different sigma factors. Sigma-WhiG is expected to recognise -10 and -35 regions that are different from those recognised by the principal bacterial sigma factors that direct most transcription during growth, because the regions 2.4 and 4.2 of Sigma-WhiG are markedly different from those of the principal sigma factors, resembling instead those of a particular minor sigma factor — Sigma D — of *Bacillus subtilis*. A consensus sequence has been derived for promoters recognised by Sigma D RNA polymerase holoenzyme, and it seems probable that a sequence similar to this is recognised by the Sigma-WhiG holoenzyme. In support of this, a ca. 500 bp fragment of *B. subtilis* DNA containing a Sigma D-dependent promoter, when present at high copy number in *S. coelicolor*, causes a *whiG* phenocopy. Presumably, the Sigma-WhiG holoenzyme is being sequestered by the excessive number of cognate promoters, so that transcription of its correct targets — some unknown sporulation genes — is greatly reduced. Homology between Sigma-WhiG and Sigma D is surprising, because Sigma D does not play a major part in sporulation in *B. subtilis*: most of the genes that it controls are concerned with motility and chemotaxis.

These molecular studies of *whiG* give rise to a simple hypothesis in which *whiG* expression is activated in aerial hyphae by some unknown mechanism, and Sigma-WhiG then binds to RNA polymerase core enzyme (perhaps effectively competing with other sigma factors). RNA polymerase containing Sigma-WhiG then transcribes genes that carry out the first steps in sporulation and also some regulatory gene (another *whi* gene?) that takes the sporulation process on to the next stage. As we shall see, this hypothesis needs to be modified.

4.4.4 Not All Sporulation Genes Depend on *whiG* for Expression[9]

The results with *whiG* described in the previous section lead us to expect that the expression of genes such as *whiH, A, B* and *I*, to which *whiG* is epistatic at the morphological level, should be switched off in a *whiG* mutant. One of these four genes, *whiB*, has been cloned and used as a hybridisation probe to measure the level of *whiB* mRNA in RNA samples prepared from cultures of different ages and genotypes. The *whiB* mRNA was not present in cultures on solid medium unless they were producing aerial hyphae, in line with the simplest expectation for a gene specifically involved in sporulation. Surprisingly, however, *whiB* transcripts with the sam 5' end were present in mutants representative of all the *whi* genes, including *whiG*. Thus the *whiG*-dependent cascade proposed in the preceding section is certainly an oversimplification: more than one sequence of regulatory events must be occurring in parallel early in sporulation. These parallel pathways must converge in the course of sporulation for mutations in either *whiB* or *whiG* to prevent later sporulation events such as grey pigment deposition from taking place.

4.4.5 RNA Polymerase Diversity in *S. coelicolor* and its Relevance to Differentiation[10]

We have seen that *whiB* is transcribed independently of Sigma-WhiG. The *whiB* promoter shows a marked resemblance to the majority class of currently known *Streptomyces* promoters, which are grouped together because of their similarity to the consensus sequence derived from "typical" promoters of enteric bacteria and bacilli. Thus, *whiB* is probably transcribed by a form of RNA polymerase containing a "principal sigma factor". However, although in *E. coli* and *Bacillus subtilis* there seems to be a single principle sigma factor, *S. coelicolor* contains no fewer than four genes (*hrd* genes, for "homologue of *rpoD*", the gene for the major *E. coli* sigma factor) whose deduced products closely resemble principle sigma factors. This similarity is most marked in the regions believed to contact the -10 and -35 regions of promoters. Does this apparent profligacy reflect the complexity of *Streptomyces*, for example with one specific *hrd* gene perhaps being used during differentiation to transcribe *whiB*? This question was addressed by analysing the effects of disrupting *hrd* genes. However, when the *hrdA*, *C*, or *D* genes were disrupted, the capacity of *S. coelicolor* to grow and sporulate normally was not destroyed. Thus, none of these three gene products is uniquely capable of directing transcription of *whB*. The fourth *hrd* gene, *hrdB*, seems to be essential for normal growth, because attempts to disrupt it have been unsuccessful. This leaves an open question about whether *whiB* can be transcribed by alternative RNA polymerase forms, or only by the form containing Sigma-HrdB.

4.4.6 The Spore Pigment Locus *whiE* As a Potential Tool in Analysing Spore Development[11]

Since the regulatory pathways involving *whi* genes must successfully converge to allow spores to become mature and pigmented, the regulation of the gene(s) for spore pigment is of considerable interest. *S. coelicolor* DNA complementing a *whiE* mutant (potentially specifically defective in pigment formation) has been cloned, and sequencing has revealed that *whiE* consists of at least seven genes closely related to genes involved in the biosynthesis of certain polyketide antibiotics. These antibiotics share a common biosynthetic origin in that their carbon chains are formed by the condensation of carboxylic acid residues in a process closely similar to the synthesis of fatty acids. The cloned DNA also contains promoters, and perhaps other cis-acting information, that ensure correct timing and localisation of *whiE* expression. These promoters are dependent on *whiB*, but not on *whiG*, for activity. Examination of the promoter-containing regions reveals no sequences resembling known promoters. Probably, therefore, another — as yet unidentified — form of RNA polymerase is needed for *whiE* transcription.

4.4.7 A Role for Sigma-WhiG in Physiological Regulation?[9]

The limited information so far available on sporulation of aerial hyphae has presented us with an intriguing paradox about *whiG*. On the one hand, Sigma-WhiG levels appear to be a crucial determinant of sporulation; and on the other, *whiB* and *whiE* promoters are transcribed in a *whiG* mutant. What genes are controlled by *whiG*? An interesting avenue has been opened up by the study of glycogen accumulation during sporulation. Glycogen granules accumulate abundantly in the compartments formed by sporulation septation (Fig. 10), even during sporulation induced in the substrate mycelium by the presence of a high copy number of *whiG*. On the other hand, the granules are absent from the aerial hyphal tips of *whiG* mutants. Possibly, this phase of glycogen accumulation is controlled by Sigma-WhiG. Thus, the effect of *whiG*

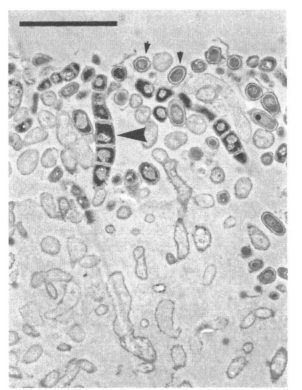

Fig. 10. Glycogen accumulates specifically in the sporulating parts of aerial hyphae. Glycogen is stained to give black areas. These are seen only in immature spore chains (*large arrowheads*). Mature spores (*small arrowheads*) do not contain glycogen. (Original from K.A. Plaskitt.) *Bar* 5 μm

on sporulation might have more to do with changing the physiology of hyphae than with a developmental regulatory cascade of gene expression. Here, as with studies of earlier stages in colony differentiation, the cloning of genes for glycogen metabolism would greatly aid future studies.

4.5 Implications [12]

Aerial mycelium formation on streptomycete colonies is typically initiated more or less synchronously over a large area containing many hyphal compartments. At least in species such as *S. griseus*, this probably involves perception of an extracellular hormone-like substance, together with response to an intracellular signal such as a fall in GTP levels. Responses to these two very different kinds of signals must somehow be appropriately integrated to allow differentiation to proceed. Possibly, the integration involves different levels of information transfer in the cell: for example, A-factor receptor protein appears to regulate transcription in *S. griseus*, whereas the *bldA* tRNA of *S. coelicolor* exerts translational effects (this begs the question of the regulation of *bldA* itself, but it is worth pointing out that formation of a functional tRNA offers a wealth of potentially regulatable steps including RNA processing, modification of bases, and the charging of the tRNA with its cognate amino acid, as well as regulation of the primary transcription of the gene). On the hypothesis advanced in Section 4.1, the emergence of aerial hyphae may involve mainly physiological changes, i.e., in turgor, controlled by storage compound metabolism, and in the "sidedness" imparted by secretion of proteins from hyphae at the interface between air and an aqueous environment. A model such as this has the attraction of seeming to require rather few evolutionary steps: most simple bacteria can accumulate glycogen during stationary phase and degrade it when growth begins again, and many produce surface proteins for specialised functions. The subsequent use of aerial hyphae as sporophores may be little more complex than the ability of many bacteria — especially actinomycetes — to form filaments that subsequently undergo septation and separation. Once formed, aerial hyphae are removed from direct contact with solutes in the extracellular environment. In this situation, the major "external" signals capable of influencing the fate of aerial hyphae are transmitted from the substrate hyphae (though it is by no means excluded that the aerial hyphae might also perceive molecules in the gaseous phase). These intra-hyphal signals could be of two main kinds, neither of which has been experimentally addressed: nutrient limitation and, perhaps, the abundance of specific regulatory molecules. Presumably in response to these signals, at least two sets of genes are transcriptionally activated: those dependent on *whiG*, and those, including *whiB*, that are *whiG*-independent. On the unsubstantiated assumption that these two sets of transcripts are each part of multistep cascades of gene activation, then perception by each cascade of the achievement of steps in the other could be used for sensitive coordination of development. In *Bacillus subtilis*, a beautiful example of such coupling is the dependence of activation of mother-cell coat-protein genes on the achievement of forespore-specified events (see Errington Chap. 2, this Vol.).

The production of spores in *Streptomyces* spp. reveals many differences from endospore formation in *Bacillus subtilis*, at the molecular level as well as at the morphological level. For example, although specific sigma factors play a role in both processes, it is interesting to note that Sigma-WhiG is most closely related to a *B. subtilis* sigma factor (Sigma D) that plays no role in endospore formation, but instead is needed for the expression of genes involved in motility. An independent evolutionary origin for the two kinds of sporulation is therefore likely.

gene that is itself dependent on *brlA* have an analogue in *S. coelicolor*? Mutual awareness of developmental processes in these evolutionarily distant organisms by their investigators is at least likely to provide extra imaginative insights.

In addition to these wider questions, many more specific questions are raised by the progress made so far: for example, how is *bldA* transcription regulated? What are its targets? What are the mechanisms that regulate glycogen metabolism in relation to development? If glycogen metabolism controls turgor, does turgor play a part in controlling developmental genes? How are *whiG* and *whiB* switched on? What are the products of the other *whi* genes? How do sporulation septa form, and how is their formation coordinated with genome segregation? Clearly, *Streptomyces* development can provide exciting research possibilities for a number of years to come.

4.6 Outlook

Molecular analysis has yet to reveal clear-cut homologies between prokaryotic and eukaryotic developmental systems. For example, the homeobox proteins crucial to many eukaryotes are absent from prokaryotes, while sigma factors play a central role in prokaryotic developmental gene regulation but do not appear to occur in eukaryotes. Probably, then, informative comparisons between prokaryotic and eukaryotic systems are likely to involve analogy rather than homology. In one such analogy, described in Chapter 7 (this Vol.) on *Aspergillus*, sporulation (conidiation) on aerial hyphae of *Aspergillus nidulans* is controlled by a transcriptional regulatory protein, the product of the *brlA* locus, untimely expression of which causes ectopic conidiation. This is closely similar to the observations with the *whiG* gene of *Streptomyces coelicolor* described above. Does *A. nidulans* also have an analogue of the *whiG*-independent regulatory hierarchy that operates in parallel with *whiG* early in *S. coelicolor* sporulation? And does the feedback circuitry by which *brlA* is regulated by a

4.7 Summary

The application of molecular genetics is beginning to reveal how multicellular differentiation of mycelial *Streptomyces* colonies is controlled. In certain species, the development of aerial mycelium and spores depends on the accumulation of a threshold level of an extracellular phenomone, and its perception by a regulatory protein. Some of the genes involved in early stages of aerial mycelium formation contain a codon, TTA, that is apparently absent from vegetative genes, and which is possibly used to regulate development. Aerial mycelium formation, and the paucity of mutants specifically defective in it, are accounted for by a speculative model involving macromolecular storage compounds, turgor, and spore-surface-associated proteins. Once aerial hyphae have formed, sporulation is activated by a specific RNA polymerase sigma factor related to the motility sigma factors of other bacteria. However, the expression of other genes involved in sporulation does not require this sigma factor, implying that parallel regulatory cascades are taking place.

References

General

Hopwood DA (1988) Towards an understanding of gene switching in *Streptomyces*, the basis of sporulation and antibiotic production. Proc R Soc Lond B 235:121–138

[1] Ensign J (1978) Formation, properties and germination of Actinomycete spores. Annu Rev Microbiol 32:185–219

Hardisson C, Manzanal B, Méndez C, Braña A (1986) Colony development in *Streptomyces*. In: Szabó G, Biró S, Goodfellow M (eds) Biological, biochemical and biomedical aspects of actinomycetes. Akadémiai Kiadó, Budapest, pp 433–442

Kaiser D (1989) Multicellular development in Myxobacteria. In: Hopwood DA, Chater KF (eds) Genetics of bacterial diversity. Academic Press, London, pp 243–263

Kalakoutskii LV, Agre NS (1976) Comparative aspects of development and differentiation in Actinomycetes. Bacteriol Rev 40:469–524

Kendrick KE, Ensign JC (1983) Sporulation of *Streptomyces griseus* in submerged culture. J Bacteriol 155:357–366

Losick R, Kroos L, Errington J, Youngman P (1989) Pathways of developmentally regulated gene expression in the spore-forming bacterium *Bacillus subtilis*. In: Hopwood DA, Chater KF (eds) Genetics of bacterial diversity. Academic Press, London, pp 221–242

Shapiro JA (1988) Bacteria as multicellular organisms. Sci Am 258:62–69

[2] Freese E (1982) Initiation of bacterial sporulation. In: Levinson HS, Sonenshein AL, Tipper DJ (eds) Sporulation and germination. American Society for Microbiology, Washington DC, pp 1–12

Ochi K (1987) Metabolic initiation of differentiation and secondary metabolism by *Streptomyces griseus*: significance of the stringent response (ppGpp) and GTP content in relation to A-factor. J Bacteriol 169:3608–3616

Ochi K (1988) Nucleotide pools and stringent response in regulation of *Streptomyces* differentiation. In: Okami Y, Beppu T, Ogawara H (eds) Biology of Actinomycetes '88. Japan Scientific Societies Press, Tokyo, pp 330–337

[3] Biró S, Békési I, Vitális S, Szabó G (1980) A substance effecting differentiation in *Streptomyces griseus*. Purification and properties. Eur J Biochem 103:359–363

Horinouchi S, Beppu T (1988) Regulation of secondary metabolism in *Streptomyces*. In: Okami Y, Beppu T, Ogawara H (eds) Biology of Actinomycetes '88. Japan Scientific Societies Press, Tokyo, pp 71–75

Khohklov AS, Anisova LN, Tovarova II, Kleiner FM, Kovalenko IV, Krasilnikova OI, Kornitskaya EY, Pliner SA (1973) Effect of A-factor on the growth of asporogenous mutants of *Streptomyces griseus*, not producing this factor. Z Allg Mikrobiol 13:647–655

Kondo S, Yasni K, Natsume M, Katayama M, Marumo S (1988) Isolation, physico-chemical properties and biological activity of panamycin-607, an aerial mycelium-inducing substance from *Streptomyces alboniger*. J Antibiot 41:1196–1204

Miyake K, Yoshida M, Chiba N, Mori K, Nogawa N, Beppu T, Horinouchi S (1989) Detection and properties of the A-factor-binding protein from *Streptomyces griseus*. J Bacteriol 171:4298–4302

Silverman M, Martin M, Engebrecht J (1989) Regulation of luminescence in marine bacteria. In: Hopwood DA, Chater KF (eds) Genetics of bacterial diversity. Academic Press, London, pp 71–86

Szabó G, Szeszák F, Vitális S, Tóth F (1988) New data on the formation and mode of action of factor C. In: Okami Y, Beppu T, Ogawara H (eds) Biology of Actinomycetes '88. Japan Scientific Societies Press, Tokyo, pp 324–329

[4] Chater KF, Merrick MJ (1979) *Streptomyces*. In: Parish JH (ed) Developmental biology of prokaryotes. Blackwell, Oxford, pp 93–114

Guijarro J, Santamaria R, Schauer A (1988) Promoter determining the timing and spatial localization of transcription of a cloned *Streptomyces coelicolor* gene encoding a spore-associated polyketide. J Bacteriol 170:1895–1901

Hopwood DA, Bibb MJ, Chater KF, Kieser T, Bruton CJ, Kieser HM, Lydiate DJ, Smith CP, Ward JM, Schrempf H (1985) Genetic manipulation of *Streptomyces* – a laboratory manual. The John Innes Foundation, Norwich

Hopwood DA, Bibb MJ, Chater KF, Kieser T (1987) Plasmid and phage vectors for gene cloning and analysis in *Streptomyces*. In: Wu R (ed) Methods in enzymology, vol 153. Recombinant DNA, part D. Academic Press, San Diego, pp 116–166

Schauer A, Rancs M, Santamaria R, Guijarro J, Lawlor E, Mendez C, Chater K, Losick R (1988) Visualizing gene expression in time and space in the morphologically complex, filamentous bacterium *Streptomyces coelicolor*. Science 240:768–772

[5] Babcock MJ, Kendrick KE (1990) Unusual transcriptional and translational features of a developmental gene of *Streptomyces griseus*. Gene 95:57–63

Champness WC (1988) New loci required for *Streptomyces coelicolor* morphological and physiological differentiation. J Bacteriol 170:1168–1174

Chater KF (1990) The improving prospects for yield increase by genetic engineering in antibiotic-producing streptomycetes. Bio/Technology 8:115–121

Fernández-Moreno MA, Caballero JA, Hopwood DA, Malpartida F (1991) The *act* cluster contains regulatory and antibiotic export genes, direct targets for translational control by the *bldA* tRNA gene of *Streptomyces*. Cell 66:1–20

Harasym M, Zhang L-H, Chater K, Piret J (1990) The *Streptomyces coelicolor* A3(2) *bldB* region contains at least two genes involved in morphological development. J Gen Microbiol 136:1543–1550

Lawlor EJ, Baylis HA, Chater KF (1987) Pleiotropic morphological and antibiotic deficiencies result from mutations in a gene encoding a tRNA-like product in *Streptomyces coelicolor* A3(2). Genes Dev 1:1305–1310

Leskiw BK, Lawlor EJ, Fernández-Abalos JM, Chater KF (1991) TTA codons in some genes prevent their expression in a class of developmental, antibiotic-negative *Streptomyces* mutants. Proc Natl Acad Sci USA 88:2461–2465

Merrick MJ (1976) A morphological and genetic mapping study of bald colony mutants of *Streptomyces coelicolor.* J Gen Microbiol 96:299–315

[6] Chater KF (1989a) Sporulation in *Streptomyces.* In: Smith I, Slepecky RA, Setlow P (eds) Regulation of procaryotic development. American Soc Microbiol, Washington DC, pp 277–299

Chater KF (1989b) Multilevel regulation of *Streptomyces* differentiation. Trends Genet 5:372–376

[7] Chater KF (1972) A morphological and genetic mapping study of white colony mutants of *Streptomyces coelicolor.* J Gen Microbiol 72:9–28

Hardisson C, Manzanal MB (1976) Ultrastructural studies of sporulation in *Streptomyces.* J Bacteriol 127:1443–1454

McVittie A (1974) Ultrastructural studies on sporulation in wild-type and white colony mutants of *Streptomyces coelicolor.* J Gen Microbiol 81:291–302

Wildermuth H, Hopwood DA (1970) Septation during sporulation in *Streptomyces coelicolor.* J Gen Microbiol 60:57–59

Williams ST, Sharples GP (1970) A comparative study of spore formation in two *Streptomyces* species. Microbios 5:17–26

[8] Chater KF (1975) Construction and phenotypes of double sporulation deficient mutants in *Streptomyces coelicolor* A3(2). J Gen Microbiol 87:312–325

Chater KF, Bruton CJ, Plaskitt KA, Buttner MJ, Méndez C, Helmann J (1989) The developmental fate of *Streptomyces coelicolor* hyphae depends crucially on a gene product homologous with the motility sigma factor of *Bacillus subtilis.* Cell 59:133–143

Helmann JD, Chamberlin MJ (1988) Structure and function of bacterial sigma factors. Annu Rev Biochem 42:839–872

Losick R, Pero J (1981) Cascades of sigma factors. Cell 25:582–584

[9] Chater KF, Bruton CJ, Davis NK, Plaskitt KA, Soliveri J, Tan H (1990) Gene expression during sporulation in *Streptomyces coelicolor* A3(2). In: Noack D, Krügel H, Baumberg S (eds) Genetics and product formation in *Streptomyces.* Plenum Press, New York, pp 3–9

Chater KF, Brian P, Davis NK, Leskiw BK, Plaskitt KA, Soliveri J, Tan H (1991) Developmental pathways in *Streptomyces*: a comparison with endospore formation in *Bacillus.* In: Heslot H, Davies J, Florent J, Bobichon L, Durand G, Pénasse L (eds) Genetics of Industrial Microorganisms. Proc 6th Int Symp on Genetics of industrial microorganisms. Vol I. Société Française de Microbiologie, Paris, pp 373–378

[10] Buttner MJ (1989) RNA polymerase heterogeneity in *Streptomyces coelicolor* A3(2). Mol Microbiol 3:1653–1659

Buttner MJ, Chater KF, Bibb MJ (1990) Cloning, disruption and transcriptional analysis of three RNA polymerase sigma factor genes of *Streptomyces coelicolor* A3(2). J Bacteriol 172:3367–3378

Hopwood DA, Bibb MJ, Chater KF, Janssen GR, Malpartida F, Smith CP (1986) Regulation of gene expression in antibiotic-producing *Streptomyces.* In: Booth I, Higgins C (eds) Regulation of gene expression – 25 years on. 39th Symp of Soc Gen Microbiol. Cambridge University Press, Cambridge, pp 251–276

Tanaka K, Shiina T, Takahashi H (1988) Multiple principal sigma factor homologs in eubacteria: identification of the "*rpoD* box". Science 242:1040–1042

[11] Davis NK, Chater KF (1990) Spore colour in *Streptomyces coelicolor* A3(2) involves the developmentally regulated synthesis of a compound biosynthetically related to polyketide antibiotics. Mol Microbiol 4:1679–1691

[12] Cutting S, Oke V, Driks A, Losick R, Lu S, Kroos L (1990) A forespore checkpoint for mother cell gene expression during development in *Bacillus subtilis.* Cell 62:239–250

Gehring WJ, Müller M, Affolter M, Percival-Smith A, Billeter M, Qian YA, Otting G, Wüthrich W (1990) The structure of the homeodomain and its functional implications. Trends Genet 6:323–329

Mirabito PM, Adams TH, Timberlake WE (1989) Interactions of three sequentially expressed genes control temporal and spatial specificity in *Aspergillus* development. Cell 57:859–868

Chapter 5 The Programme of Cell Type Determination in Fission Yeast

Benito Arcangioli and Amar J. S. Klar

5.1 Introduction

Any mechanism of cellular development and differentiation must explain how two daughters of a cell are produced in such a way that their developmental fates differ. The studies of developmental biology in the past decade have led to the model of cellular differentiation based on the judicious expression of cell-type-specific genes. It is thought that differential gene expression, in turn, is controlled by modulated expression of key regulatory genes. The major question that remains unanswered is: what regulates these cell-type-specific regulators? Perhaps another regulator does, and still another, and so on. Although the regulated gene expression does explain execution of the programme of cell-type determination rather well, the model fails to explain how the programme was initially started in one cell and not in its sister cell.

It is easier to imagine that environmental cues can dictate initial events required for differentiation in procaryotes (e.g., sporulation of *Bacillus* and *Myxococcus*) or single-celled eucaryotes (e.g., meiosis and sporulation of yeast). The problem becomes much more complex when many different cell types are required for multicellular development. Varied systems presumably employ different strategies. For example, major differences in plant and animal development are known to exist. In animals, reproductive and somatic cell lineages diverge early in embryogenesis. In contrast, reproductive and vegetative structures arise from the same meristem in plants, showing that these structures share a common lineage until late in development. Although in lower plants, such as ferns and mosses, the shoot arises from a single large cell at the apex of the meristem, in higher plants the multicellular structures generally originate from a large progenitor cell population of 30–100 cells. In other words,

cell lineage plays little or no role in plant development. Also, as most plant cells are not terminally differentiated, since they can be induced to give rise to an entire plant, the concept of stem cell lineage is not of critical importance in plant development.

Stem cells are progenitor cells whose division produces one daughter like the parent cell, while the other daughter is advanced in its developmental programme. Stem cells occupy central importance in regenerating tissues such as those of skin, alimentary canal, hemopoietic and immune systems, etc., in animals. The mechanism of producing daughters differing in their developmental programme is not known in any system.

To address these issues, one can use a simpler, one-celled organism which undergoes differentiation by environmental signals as well as follows a cell-autonomous programme analogous to the mammalian stem cell lineage. The fission yeast, *Schizosaccharomyces pombe*, is an ascomycetes fungus that can initiate a developmental programme resulting in meiosis and sporulation under nutritional limitation conditions. Secondly, the two mating cell types, called *P* (for *P*lus) and *M* (for *M*inus), spontaneously interchange in an efficient manner by rigidly following the pattern of stem cell lineage. This chapter summarizes our understanding of the mechanism of the mating type switching pertaining to the programme of cell type interchange. The reason for two daughters being developmentally different in their fates is best understood in this system, and it does not depend upon assuming prior differential gene regulation!

5.2 Life Cycle[1]

S. pombe has two distinct life cycles, asexual and sexual, which involve three different cell types

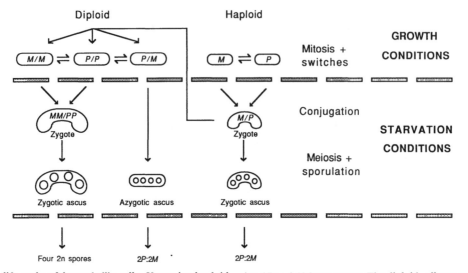

Fig. 1. The life cycle of homothallic cells. Vegetative haploid (*middle part*) and diploid (*left part*) cells grow mitotically and switch mating types in favourable growth conditions. Starvation induces conjugation between cells of opposite mating type, ultimately leading to meiosis producing zygotic asci, each containing *2P* and *2M* ascospores. The diploid cells can undergo meiosis without conjugation, producing azygotic asci when they are heterozygous at *mat 1* (*P/M*). The diploid 2n spores result from a diploid by diploid cross. The *horizontal arrows* indicate interchange of cell type between *P* and *M*

(Fig. 1). Under growth conditions, *S. pombe* cells reproduce by mitotic division. When nutritional conditions become limited, sexual activities are induced where cells of the opposite mating type undergo sexual fusion. Normally, the zygotic cells immediately enter into meiotic cycle and give rise to four haploid spores contained within a structure called the ascus.

5.2.1 Asexual Life Cycle

Cells of *S. pombe* grow as a cylinder approximately 7 µm long and 3 µm wide and divide by fission thus the prefix *Schizo* (for fission). Mitotic growth follows a typical eukaryotic cell cycle comprised of discrete G_1, S, G_2, and M phases. The generation time is about 2.5 h. The G_2 phase is about 70% of the cell cycle, and the remaining G_1, S, and M phases each occupy about 10% of the cell cycle length. When sexual development is not allowed under nutritional limitation conditions, stationary-phase cells can enter from either the G_1 or G_2 phase into a dormant state, called "GO". The two major cell cycle events are S-phase when chromosomes are replicated and M-phase when replicated chromosomes are segregated into daughter cells. Many recent studies have established the existence of a universal control mechanism for cell growth and division, called the cell cycle, common to all eukaryotic cells. The *cdc2* gene has a central role in advancing the cell cycle. This gene codes for a 34-kilodalton (kDa) protein kinase which is required for traverse of the cell cycle from G_1 to S and from S into M phase.

5.2.2 Sexual Life Cycle: Evidence for Three Cell Types

In response to starvation, which includes oxygen, glucose, or nitrogen deprivation, sexual activities are induced in *S. pombe*. The ability to engage in such activities is determined by the mating type of the cells. Based on their mating behaviour and their ability to undergo meiosis and sporulation, three different types of yeast cells have been defined. The *P* and *M* cell types are specialized for mating with cells of opposite mating type, whereas diploid heterozygous cells (*P/M*) do not mate but can undergo

Table 1. Three cell types of *S. pombe*

Cell type	Genotypes	Properties	
		Conjugation	Meiosis
P	*mat 1-P* or *mat 1-P/mat 1-P*	Yes	No
M	*mat 1-M* or *mat 1-M/mat 1-M*	Yes	No
P/M	*mat 1-P/mat 1-M*	No	Yes

meiosis and sporulation. Table 1 summarizes the properties of the three cell types.

The mating type of the cell is determined by the allele contained in the mating type (*mat 1*) locus located in chromosome II. Cells with the *mat 1-P* allele exhibit the *P* mating type, and those with the *mat 1-M* allele exhibit the *M* mating type. These alleles encode for key regulatory genes whose expression is induced by starvation. Under these conditions, cells of the opposite mating type undergo sexual fusion forming diploid zygotes. This so-called conjugation is restricted to a short period in late G_1 phase of the cell cycle. Conjugation result in producing diploid zygotic cells which immediately undergo conventional meiosis, followed by sporulation giving rise to asci containing four haploid spores (Fig. 1). The spores are called ascospores. Two of the four spores are of the *P* type and the other two are of the *M* type contained in each ascus. When the ascospores encounter favourable growth conditions, they germinate to restart the vegetative asexual life cycle.

Since conjugation and meiosis are induced by starvation, the diploid state is only transitory. Thus, the haploid state is the predominant life form in *S. pombe*. However, if soon after conjugation the diploid zygote is transferred to growth conditions which inhibit sporulation, then it can restart the vegetative cell cycle instead of undergoing meiosis. Depending on the combination of the mating type alleles, diploid cells behave differently in starvation conditions. Diploids which become homozygous at the *mat 1* locus (*mat 1-P/mat 1-P* or *mat 1-M/mat 1-M*) by mating type switching (see below for switching) will react like their haploid counterparts in their ability to mate with cells of

the opposite mating type and in their inability to undergo meiosis and sporulation (Fig. 1). When the diploid is heterozygous at *mat 1* (i.e., *mat 1-P/mat 1-M*), it can sporulate without conjugation, giving rise to an azygotic ascus containing four spores (note the difference in shape between the zygotic and azygotic ascus in Fig. 1).

5.3 *mat 1* Locus and the Cell Types [2]

DNA cloning and sequencing of the *mat 1-P* and *mat 1-M* loci have indicated that both alleles contain two open reading frames [Pc, 118 amino acids (aa); Pi, 159 aa; Mc, 181 aa; Mi, 42 aa; Fig. 2]. In growth medium, only *Pc* and *Mc* are weakly transcribed. However, in starvation conditions, all four genes are transcriptionally induced. Mutational

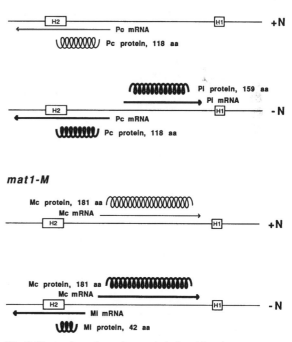

Fig. 2. Expression of *mat 1* genes is induced by nitrogen starvation. Each *mat 1* allele encodes two genes. Under the growth condition (+N), only *Pc* and *Mc* genes are partially expressed. Under the nitrogen starvation regime (−N), all four genes are induced. *H1* and *H2* are sequence homologies with the donor loci (see Fig. 5)

Table 2. Requirement of mating type genes

Gene	Conjugation	Meiosis
M_c	Yes	Yes
M_i	No	Yes
P_c	Yes	Yes
P_i	No	Yes

analysis has shown that Pc and Mc are required for conjugation and meiosis, whereas Pi and Mi are required only for meiosis (Table 2). The mechanism by which these key regulatory factors control sexual functions is not known. However, analysis of the predicted amino acid sequence of the Pi product indicated the presence of a canonical homeodomain characteristic of DNA-binding proteins. In addition, the predicted amino acid sequence of Mc shows homologies with the Tdy and TDF genes located in the testis-determining region in the Y chromosome of mice and humans, respectively. Furthermore, this conserved domain showed homologies with the DNA-binding motif present in the chromosomal high-mobility group proteins HMG 1 and HMG 2 thought to be acting as transcription factors. These results strongly suggest that Pi and Mc gene products are involved in controlling specific gene expression possibly as specific transcription factors.

5.4 Mating Type Switching: Homothallism Versus Heterothallism[4]

A remarkable feature found in many fungi is the instability of their mating types. Such strains are called homothallic, whereas those variants or mutants which cannot interchange their mating type efficiently are called heterothallic. As a consequence of homothallism in *S. pombe*, a colony grown from a single cell of either P or M type contains nearly equal proportions of P and M cells. On the other hand, colonies originating from a single heterothallic cell of either P or M type contain only P or M cells, respectively. This cell type switching has no practical physiological consequences until

IODINE STAINING OF *S. POMBE* COLONIES ON SPORULATION MEDIA

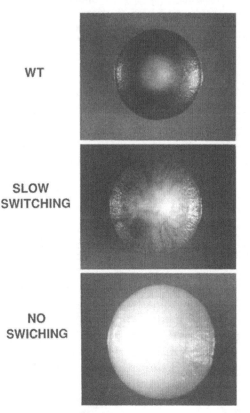

Fig. 3. Iodine vapour staining of *S. pombe* colonies of a homothallic (wild-type and slow switching) and heterothallic (no switching) strains. Dark iodine staining streaks in the slow switching colony result from mating and sporulation of region in the adjoinging sectors

cells in the colony experience starvation. When this occurs, cells of opposite mating type contained in a homothallic colony undergo conjugation and meiosis followed by sporulation. In contrast, cells of a colony from a heterothallic strain lacking cells of opposite sexual type accumulate in G_0 phase. Homothallic and heterothallic strains can be easily identified while growing on a plate by the iodine vapour staining procedure. By this procedure, colonies of yeast growing on solid medium are indirectly exposed to iodine vapours for about 1 min. The procedure stains switching-proficient homothallic colonies black because they contain spores, but switching-deficient heterothallic colonies do not stain (Fig. 3). The black stain is due to a starch-like compound present in spores. From a Darwinian point of view, it is not surprising that wild-type *S. pombe* strains are mostly homothallic, since the ascospores are a fairly resistant dormant life form that can withstand adverse conditions.

5.5 The Programme of Switching

The *S. pombe* cell biology requires a combination of cell-autonomous capability of switching mating type and an appropriate response to adverse environmental conditions for undergoing sexual development. We will see how the cell with a single genome can produce, in a highly regulated fashion, different cell types; that is one of the central questions which remains unanswered in classical developmental biology.

How can one determine that a yeast cell has a switchable mating type and when this mating type actually switches in a cell lineage? One can make a pedigree of cell lineage taking into account two properties:

1. One constructs diploid cells where one of the mating type alleles never switches because of cis-acting genetic mutations while the other allele may switch. In the example of Fig. 4, the diploid is *mat 1-P/mat 1-P^{Ns}* (*Ns* = never switchable).
2. On a sporulation medium, the cell which has switched the wild-type *mat 1-P* into *mat 1-M* will not divide anymore and will make an ascus owing to *mat 1-M/mat 1-P* constitution of the diploid. The ascus is easily distinguishable by watching them through the microscope from a

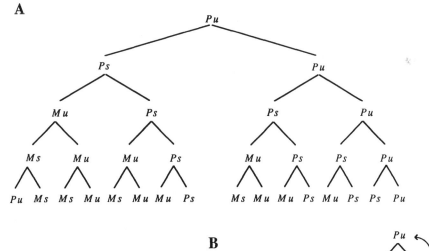

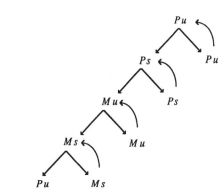

Fig. 4A, B. Pattern of switching mating type in a cell lineage. **A** Diagrammatic representation of a pedigree of mating type interconversion illustrating the rules of switching. The postscript *s* indicates *s*witching while *u* indicates *u*nswitchable cells. **B** Switching pattern demonstrating the stem cell lineage. Note that each cell division produces one daughter identical to the parental cell, while the other daughter is advanced in the programme. Actual pedigrees are conducted with diploid cells which contain nonswitchable allele in one homologue. Only pattern of switching of the wild-type allele residing in the other homologue of the diploid is diagrammed

nonsporulating cell, the latter of which is capable of growing and switching in subsequent generations.

From a pedigree analysis one obtains a cell lineage of a type of Fig. 4A. Then one sees a sporulated cell, one can say that this cell has actually switched its mating type of the wild-type locus to *M*. Its mother cell must have a switchable mating type *P* per definition and can be called *Ps* (*s* = switchable). The grandmother cell did not have any of the two daughters switched, therefore, her mating type *P* was unswitchable. One calls this cell *Pu* (*u* for *u*nswitchable). From the analysis of these pedigrees, the following rules can be formulated.

Rule 1. Among a pair of sister cells, only one cell switches her mating type in about half of pairs of cells, and simultaneous switches of both sisters are never observed — the single daughter switching rule.

Rule 2. When mating type switching is accomplished in one pair of cells by Rule 1, it will not occur in the other half of the lineage, i.e., in their first cousin cells — the one-in-four granddaughters switching rule.

Rule 3. The sister of the recently switched cell is switchable, as she produces one switched and one unswitched daughter in the following cell division in about 90% of the consecutive cell divisions — the consecutive switching rule.

Rule 4. The switchable cell switches to the opposite allele in about 90% of cell divisions. Thus, cells are directed to switch primarily (or only) to the opposite allele — the directionality of switching rule.

From these observations, the following conclusions can be reached:

- The switching is formally a G_2 event, as only one of the two daughters of a switching-competent cell actually switches.
- The one-in-four switching rule implies that two consecutive developmentally asymmetric cell divisions are required to restrict switching of one cell among four related cousin cells. The first asymmetric division produces two cells which differ in their potential of switching. The second asymmetric division executes the switching

event such that only one in four granddaughters of a cell ever switches.
- The consecutive switching rule indicates that the switching potential is inherited by the unswitched daughter, which can repeat the cycle of switching again and again, producing a chain of recurrent switching. In summary, all four granddaughters of a cell possess distinct developmental identities; two are switching-competent, the other two are switching incompetent and one among the latter was newly switched.
- The directionality rule indicates existence of a mechanism ensuring switching to the opposite *mat1* allele.

The observed pattern of switching is analogous to the mammalian stem cell lineage, where a cell produces one daughter like itself while the other daughter is advanced in the developmental programme.

Before we address the question of how the programme of switching is controlled by the cell, we will describe the molecular structure of the mating type locus of *S. pombe* and the *cis/trans* elements required for switching.

5.6 The Mating Type Region[2,5]

The existence of switching suggests that all cells contain the potential information coding for *mat1-P* and *mat1-M*, but express only one type at the *mat1* locus at a time. This was first shown by genetic analysis of the mutant in either type of *mat1* allele. Switching allowed the mutant allele to be replaced ("healed") by either a wild-type *mat1-P* or *mat1-M* gene. This result suggested the existence of unexpressed mating type information elsewhere in the genome. Molecular cloning of the *mat1* locus and the silent storage loci, i.e., *mat2-P* and *mat3-M*, confirmed the transposition model. The cells switch by transposing a copy of either *mat2* or *mat3* "controlling elements" or "cassettes" into *mat1*, replacing the previously existing information by the process of unidirectional gene conversion (Fig. 5).

Genetic analysis has mapped the mating type locus on the right arm of chromosome II between the

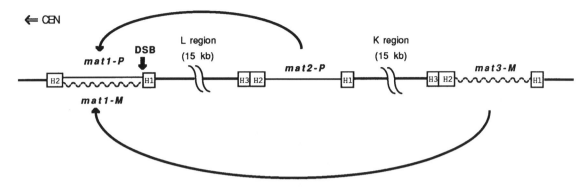

Fig. 5. Schematic representation of the mating type region showing the *mat 1*, *mat 2*, and *mat 3* loci (cassettes) separated by about 15 kb each on chromosome II. *mat 1* can contain either *P* or *M* information which is interchanged by unidirectional gene conversion as indicated. *DSB* represents the position of the double-stranded break, which initiates the process of recombination. The position of the centromere with respect to *mat 1* is indicated by an *open arrow*

his 7 and *his 2* markers. Molecular analysis has defined the structure of the mating type locus. The *mat 1*, *mat 2*, and *mat 3* cassettes are organized as direct repeats. The *mat 1* locus exists in two forms, one containing the *p* allele and the other containing the *M* allele, while the *mat 2* donor cassette contains the *P* information and the *mat 3* donor cassette contains the *M* information. *mat 1*, *mat 2*, and *mat 3* are separated by two spacer regions of roughly 15 kilobases (kb) each, termed L and K (Fig. 5). It is interesting to note that meiotic crossing-over in the K regions is totally absent. The DNA sequence determination of all three cassettes showed that the *P*-specific region is 1113 base pairs (bp) long (present both at *mat 1-P* and *mat 2-P*), while the *M*-specific region is 1127 bp long (present both at *mat 1-M* and *mat 3-M*). These specific *P* and *M* sequences are flanked by homology boxes: H1, 59 bp long, is located to the right and H2, 135 bp long, is located to the left of all the three cassettes. A third homology box, the 57 bp H3 box, is present to the left of H2 only at the two silent cassettes, *mat 2* and *mat 3*. The different functions attributed to these homology boxes for mating type expression and for switching will be discussed later in the chapter.

Southern blot analysis of genomic DNA prepared from homothallic wild-type strains revealed that a constant proportion (approximately 25%) of cells contain a double-stranded break (DSB) at the *mat 1* locus. Surprisingly, the DSB is persistent and

is maintained at a constant level during the entire length of the cell cycle. It is not known how the chromosome is held together. The other problem is to determine how the cleaved chromosome duplicates. Genetic evidence has suggested that the chromosome is sealed, replicated and cleaved again in a switchable cell. The site of the cleavage is at the junction of the *mat 1* allele-specific and H1 sequences and is found in both *mat 1* alleles (see Fig. 5). By analogy with the budding yeast, *Saccharomyces cerevisiae*, *MAT*-switching system and by employing several switching-defective mutants containing a reduced level of DSB and reduced efficiency of switching, it was suggested that the DSB constitutes the recombination-initiation event required for *mat 1* switching. This scenario implies that the programme of mating-type switching is controlled by the generation of the DSB at *mat 1*. The molecular aspect of this cell-autonomous decision-making event will be developed in a later section. We will first describe several *mat 1 cis*- and *trans*-acting mutations which adversely affect the rate of switching.

5.7 "Position-Effect" Control for Expression and Switching of Cassettes

Although the *mat 1* cassettes and the respective donor loci contain the same H2, allele-specific, and

H1 sequences, only *mat1* is transcriptionally active. The problem is particularly interesting and enigmatic since both mRNA transcripts of both *mat1* alleles initiate from within the cassette and extend outwards (Fig. 2). Therefore, identical promoter sequences present in expressed and unexpressed loci promote transcription only when located at *mat1*. In higher eukaryotes, cytosine methylation is thought to inhibit gene expression (the issue is controversial, though). Such a mechanism is not operative to keep the donor loci unexpressed as base modification does not occur in yeast. It is suggested that this so-called transcriptional position-effect is mediated through "silencer" elements found at *mat2* and *mat3* keeping them unexpressed. The most likely candidate for a silencer element consists of the 57 bp H3 sequence present on the left of the *mat2* and *mat3* H2 regions (Fig. 5). It is likely that some *trans*-acting functions are also required to keep the donor loci unexpressed.

Another important position-effect control regulates switching of a particular cassette. Although all cassettes contain identical sequences at the cut site, only cassettes containing *mat1*-distal sequences are cleaved in vivo. Further deletion analysis of the *mat1*-distal sequence has identified two *cis*-acting sites, called *SAS1* (switching activating sequence) and *SAS2*, located within the 200 bp region distal to the DSB.

Thus, the two position-effect controls for keeping *mat2* and *mat3* cassettes silent and for producing a DSB only at the *mat1* cassette are regulated independently through respective *cis*-acting sequences. Clearly, these *cis*-acting sequences must not act alone, *trans*-acting protein functions must also be present to promote these processes.

5.8 *Trans*-acting *swi* (switch) Genes[6]

In addition to the *cis*-acting mutations described above, at least 11 complementation groups have been defined genetically as *trans*-acting functions required for switching. Based on their requirement for particular steps of recombination, these genes have been classified into three groups (Table 3). The *swi1*, *swi3*, and *swi7* genes are required for the for-

Table 3. Role of *swi* genes in switching

Swi loci	DSB	Switching	Rearrangement
Swi 1, 3, 7	Reduced	Reduced	No
Swi 2, 5, 6	Yes	Reduced	No
Swi 4, 8, 9, 10	Yes	Yes	Yes

mation of the DSB at *mat1*, as is evidenced by a reduced level of cut DNA found in strains harbouring mutations in these genes. Mutations in *swi2*, *swi5*, and *swi6* do not affect the level of the DSB but cause defects in switching, presumably because they are defective in the processing and/or utilization of the DSB for recombination. Mutations in *swi4*, *swi8*, *swi9*, and *swi10* and in *rad22* result in a high frequency of rearrangement of the mating type region thereby a copy of *mat2*, K region, and *mat3* is transferred to *mat1* (Fig. 5). Such strains are called heterothallic as they rarely switch ($\sim 10^{-4}$). Thus, this class of mutants appears to be deficient in the resolution of recombination intermediates. Clearly, some of these genes are involved in other functions, as *swi5* mutants are defective in general recombination and several of the *swi* mutants are UV or γ-ray sensitive. Furthermore, some *swi* mutant combinations such as those of *swi1*, *swi7* and *swi3*, *swi7* mutants are lethal, suggesting a role for these genes in other cellular functions.

5.9 The Programme of Switching Is Dictated by Parental DNA Chain Inheritance[4,7]

Assuming that the DSB initiates gene conversion/recombination which is required for switching, it follows that the pattern of switching in a cell lineage should be dictated by the regulation of generating the DSB. In other words, the question of the programme of switching could be restated: what causes the generation of the DSB in some cells and not in their immediate sister cells such that only one in four granddaughters of a cell ever switches? The usual expectation is that some factors, such as those encoded by *swi* genes, are unequivalently distributed or that those genes are unequally expressed in some cells and not in their sister cells. The question then is, what are those critical factors and how are they so precisely and unequivalently distributed to progeny cells. Another possibility is that the switching potential is segregated to progeny cells in a genetical way, via the DNA template. In this model, some semi-heritable chromosomal modification, called imprinting, occurs in some chromosomes and not in their sister chromatids causing developmental asymmetry between sister cells. Imprinting is defined as a non-mutational alteration of a chromosome affecting its behaviour later in development.

In pedigrees of diploid cells homozygous for the h^{90} configuration at *mat 1* but defective at *mat 2*-P in one of the chromosomes, it was observed that the asymmetric segregation of switching potential on one chromosome occurs essentially independently of the asymmetric segregation on the other chromosome. This observation indicated that switching competence segregates in *cis* with *mat 1* and most likely it is the long-lived DSB.

What, however, regulates the DSB such that it is present in some cells and not in their sister cells? Whatever the mechanism, the decision for switching only a single granddaughter cell must have been made two generations earlier (that is, by the grandparent cell) and that two consecutive developmentally asymmetric cell divisions must be required to produce the observed pattern of switching. To explain this programme of differentiation, a specific strand segregation model (Fig. 6A) was suggested in which the Watson and Crick strands of DNA molecule are considered nonequivalent in their ability to acquire the developmental potential for switching in each cell division. Specifically, another sequence-specific imprinting event, acting as a precondition for producing the DSB, is hypothesized to occur in the parent cell but only on one specific strand, say, the Watson (W) strand. Replication of the semi-modified chromosome generates unequivalent sister chromatids. The chromatid possessing the old imprinted W strand is cleaved, perhaps during replication (which is inherited by the *Ps* daughter in Fig. 6A), while the unimprinted (*Pu*)

chromatid cannot be cleaved, as it lacks the imprinted event on the newly synthesized W strand. The *Pu* unimprinted chromatid will be imprinted later in the cell cycle and will be inherited by the *Pu* daughter. The *Ps* cell generates two types of chromatids: the chromatid containing the original imprinted strand remains unchanged, has the DSB, and thus is switchable (*Ps*), while the sister chromatid is healed by switching (*Mu*). The key idea is that segregation of the parental and grandparental DNA strands confers developmental asymmetry to progeny cells.

Genetic and molecular tests of the strand-segregation model are based on determining the efficiency of cleavage and the switching pattern of cells containing a tandem inverted duplication of *mat 1* (Fig. 6B). The additional cassette was inserted proximal to the *mat 1* cassette approximately 4.5 kb away by DNA-mediated transformation. A specific molecular prediction is that, as compared to wild-type strains which contain only one *mat 1* gene, twice as many chromosomes should be cleaved in duplication-containing strains. This is predicted because both strands of a parental cell should be imprinted, one strand in one cassette and the other strand in the second cassette.

Consequently, both daughters of that cell should have the DSB, one daughter in one cassette and its sister in the other cassette. Another testable prediction is that both cassettes should never be cleaved simultaneously in the same chromosome as, at a maximum, only one cassette should be cleaved in any chromosome. DNA analysis by Southern transfer verified the predictions, since twice as many chromosomes were cleaved, each site was as well cleaved as in wild-type strains, and both sites were not cleaved simultaneously in the same chromosome.

Simpler and even more important genetic predictions of the model are that in *mat 1*-inverted duplication-containing strains, two cousins in four related granddaughter cells should switch, each using a different *mat 1* locus as the target for switching, and the additional cassette in the duplication itself should switch by the one-in-four rule (Fig. 6B). Results exactly satisfying these predictions were obtained.

In summary, the major tenant of the model arguing that the two DNA strands confer developmen-

A: Strand Segregation Model

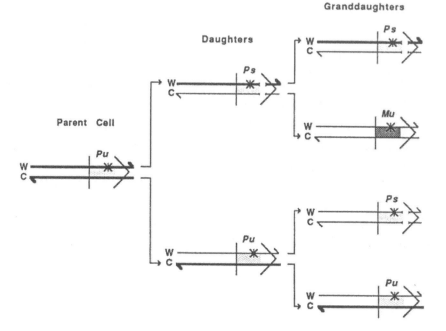

B: Strand Segregation of *mat1* Duplication

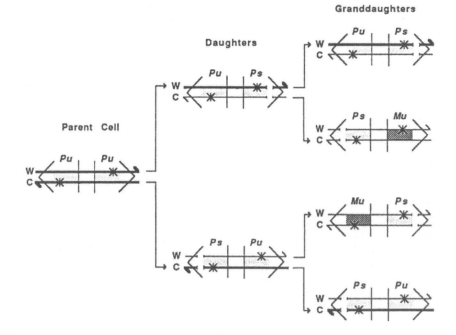

tal asymmetry to daughter cells in each of two consecutive generations is fulfilled. These results show that the cell-autonomous potential of differentiation in this system is not through a diffusible factor; rather, there is a structural basis for its segregation that is based on the two strands of DNA possessing a complementary but not identical DNA sequence, and on their semiconservative replication and segregation to progeny cells. Two types of imprinting events segregating in *cis* with *mat1* have been suggested. One of these is a hypothetical precursory state of DNA making the locus cleavable, while the other one is surely a DSB making the locus switchable. What physically might be involved in the first type of genome imprinting can only be speculated at present. It may be a protein complex that segregates with a specific strand, DNA methylation or some other base modification, a site-specific single-stranded nick that becomes a double-stranded break in the next round of replication, or an unrepaired RNA primer of Okazaki fragment. The major lesson of general interest is that the double-helical nature of the chromosome dictates the programme of cellular differentiation and it does not depend upon cell-cell interaction or on differential gene expression of sister cells.

5.10 Comparison of *S. pombe* with *S. cerevisiae* Mating Type Interconversion[8]

An exciting story has also developed for mating type (*MAT*) switching in the evolutionarily distantly related budding yeast, *S. cerevisiae*. Although

---

Fig. 6A, B. The pattern of switching is dictated by the segregation of parental DNA strands. **A** Strand segregation model. The *large arrows* show the orientation of the *mat1* locus, the *small arrows* show segregation of specific DNA strands to progeny cells, the *asterisk* (*) indicates the imprinted strand, and the discontinuity in *mat1* represents the DSB. **B** Strand segregation in the *mat1* inverted duplication containing cells provides the molecular and cellular tests of the model in **A**. Note that the switching programme is predicted to be modified in duplication-containing cells such that two (cousins) in four related granddaughters should switch, instead of the usual one in four

switching analogously involves transposition of a copy of silent information into the expressed *MAT* locus, almost all the details of the switching process are very different in both organisms. The cell lineage determination of *MAT* switching of *S. cerevisiae* is based on transcriptional regulation of the *HO* gene which encodes the endonuclease required for cleaving *MAT* locus. The *HO* is expressed in parental "mother" and not in the first born "daughter" cell, the basis for differential regulation is not known. Perhaps another regulator is involved in the process. It is important to realize that the mating type cassettes of both yeast do not have significant DNA sequence homology with each other. Thus, it is not surprising that many of the details of the transposition process are very different. The phenomenon of sexual differentiation by homothallism is very widespread in fungi. It will be highly rewarding to learn details of many more systems of homothallism before we can fully address how these systems might have evolved.

5.11 Outlook

It is important to realize that homothallism promotes genetic crossing between sibs and thus goes against the useful concept of outbreeding in evolution. Perhaps the selective advantage for homothallism is provided by the ability of switched progeny to mate and undergo meiosis, resulting in producing ascospores which can better weather adverse conditions. We can also imagine that evolution never stands still and this organism has not yet benefited from all the imaginable parameters of evolution.

The general outline of the programme of mating type switching is apparently clear now, but none of the molecular details is known. Important mechanistic questions must be addressed in the future. It is certain that transposition does not occur through diffusible intermediates. How does the interaction between the donor and the acceptor loci occur? Is this a gene conversion of an unusual type, as it seems to be catalyzed by the DNA replication process? Does transposition occur by copying a specific DNA strand from the donor or by copying both strands simultaneously? What is the nature of the DNA imprinting event hypothesized to be required for catalyzing the break at *mat 1*? What keeps the cut ends protected from degradation given that the cut persists for the entire length of the cell cycle? Once the DSB is produced at *mat 1*, what dictates the donor choice such that 90% of the switches occur to the opposite *mat1* allele? What is the molecular basis for the lack of recombination between *mat2* and *mat3* and what keeps these loci transcriptionally inactive? Furthermore, how prevalent is this type of mechanism of generating developmental asymmetry between sister cells? In this context, it should be stressed that generation of developmental asymmetry need not necessarily require recombination. The process of DNA replication is inherently asymmetric as leading and lagging strands are replicated by different enzymatic activities. Any mechanism which assembles chromatin structure of a developmental gene in a strand-specific fashion during replication, in principle, can modulate gene expression making the sisters developmentally different. In other words, the important general lesson to learn from the yeast system is that the act of DNA replication and the double helical nature of DNA causes the developmental asymmetry in sister cells. These are some of the obvious and important questions which remain unanswered. As this system is amenable to sophisticated tools of molecular genetics and molecular biology, answers to some of these questions should be forthcoming.

5.12 Summary

As is clear from the forgoing discussion, the initial critical event conferring developmental asymmetry on sister cells is now understood in *S. pombe*. The pattern of switching is not dictated by a diffusible element, and it follows the classical model of semiconservative DNA replication and segregation/inheritance of parental DNA chains to progeny cells. In no other system is the initial event so clearly understood. More importantly, this mechanism does not depend on differential regulation of the determinant of differentiation. This mechanism underscores the importance of DNA replication for differentiation, a feature not fully recognized as important for development and differentiation.

References

[1] Egel R (1990) Mating-type genes, meiosis and sporulation. In: Nasim A, Johnson B, Young P (eds) Molecular biology of the fission yeast. Academic Press, San Diego, pp 32–73

Fantes P (1990) Cell cycle control. In: Nasim A, Johnson B, Young P (eds) Molecular biology of the fission yeast. Academic Press, San Diego, pp 127–204

Nurse P (1990) Universal control mechanism regulating onset of *M*-phase. Nature 344:503–507

[2] Kelly M, Burke J, Smith M, Klar A, Beach D (1988) Four transcriptionally regulated mating type genes control sexual differentiation in the fission yeast. EMBO J 7:1537–1547

Sinclair AH, Berta P, Palmer MS, Hawkins JR, Griffiths BL, Smith MJ, Foster JW, Frischauf A, Lovell-Badge R, Goodfellow PN (1990) A gene from human sex-determining region encodes a protein with homology to a conserved DNA-binding motif. Nature 346:240–244

[3] Egel R (1977) Frequency of mating-type switching in homothallic fission yeast. Nature 266:172–174

[4] Egel R, Eie B (1987) Cell lineage asymmetry in *Schizosaccharomyces pompe*: unilateral transmission of a highfrequency state of mating-type switching in diploid pedigrees. Curr Genet 12:429–433

Klar AJS (1990) The developmental fate of fission yeast cells is determined by the pattern of inheritance of parental and grandparental DNA strands. EMBO J 9:1407–1415

Miyata H, Miyata M (1981) Mode of conjugation in homothallic cells of *Schizosaccharomyces pompe*. J Gen Appl Microbiol 27:365–371

[5] Beach DH (1983) Cell type switching by DNA transposition in fission yeast. Nature 305:682–687

Beach DH, Klar AJS (1984) Rearrangements of the transposable mating-type cassettes of fission yeast. EMBO J 3:603–610

Egel R (1984) Two tightly linked silent cassettes in the mating-type region of *Schizosaccharomyces pombe*. Curr Genet 8:199–203

Egel R, Gutz H (1981) Gene activation by copy transposition in mating-type switching of a homothallic fission yeast. Curr Genet 3:5–12

Klar AJS, Bonaduce MJ, Cafferkey R (1991) The mechanism of fission yeast mating type interconversion: seal/replicate/cleave model of replication across the double-stranded break site at *mat1*. Genetics 127:489–496

Nielsen O, Egel R (1989) Mapping the double-strand-breaks at the mating type locus in fission yeast by genomic sequencing. EMBO J 8:269–276

[6] Egel R, Beach DH, Klar AJS (1984) Genes required for initiation and resolution steps of mating-type switching in fission yeast. Proc Natl Acad Sci USA 81:3481–3485

Gutz H, Schmidt H (1985) Switching genes in *Schizosaccharomyces pombe*. Curr Genet 9:325–331

Schmidt H, Kapitza P, Gutz H (1987) Switching genes in *Schizosaccharomyces pombe*: their influence on cell viability and recombination. Curr Genet 11:303–308

[7] Egel R (1984) The pedigree pattern of mating-type switching in *Schizosaccharomyces pombe*. Curr Genet 8:205–210

Klar AJS (1987) Differentiated parental DNA strands confer developmental asymmetry on daughter strands in fission yeast. Nature 362:466–470

Klar AJS, Miglio LM (1986) Initiation of meiotic recombination by double-stranded DNA breaks in *Schizosaccharomyces pombe*. Cell 46:725–731

[8] Klar AJS (1987) Determination of the yeast cell lineage (mini-review). Cell 49:433–435

Klar AJS (1989) The interconversion of yeast mating type: *Saccharomyces cerevisiae* and *Schizosaccharomyces pombe*. In: Berg DE, Howe MW (eds) Mobile DNA. American Society for Microbiology, Washington DC, pp 671–691

Chapter 6 Development in *Neurospora crassa*

Vincenzo E. A. Russo and Niketan N. Pandit

6.1 Introduction[1]

Neurospora crassa is a eukaryotic organism belonging to the group of fungi called Ascomycetes. In nature, it grows on decaying or burnt vegetation in tropical or subtropical areas. It was a serious contaminant of bakeries in the past and was known under the trivial name of red bread mold. The genus *Neurospora* comprises ten species, of which *N. crassa* is by far the best studied. *N. crassa* became very famous when Beadle and Tatum isolated the first biochemical mutants in any organism in 1941 and suggested the "one gene-one enzyme" hypothesis which led to the explosive development of biochemical genetics and molecular biology. Many important contributions, e.g., the first demonstration of gene conversion that was generally believed, the first demonstration of interallelic complementation, and the discovery of the genes which regulate recombination in specific regions of the chromosomes, have come from the studies on *N. crassa*.

Today, a large amount of information is available on the genetic, biochemical, ultrastructural, and cytogenetic aspects of this organism. There are more than 6000 auxotrophic, morphological and developmental mutant strains which are maintained at the Fungal Genetic Stock Center (FGSC) in Kansas City (USA). The center also publishes the Fungal Genetic Newsletter, an annual publication which contains short articles on techniques, new strains, genetic maps, and recent bibliography. Mutants available so far define more than 600 different genes. About half of these genes have effects on morphology. The strain collection also has a group of *Neurospora* isolates collected from nature.

Besides the well advanced classical genetics and molecular biology, there are many features that make *Neurospora* an attractive model system to study morphogenesis. The first is that this fungus is structurally much simpler than multicellular eukaryotes, producing only about a dozen different morphological structures in its life cycle. Sufficient quantities of several of these structures can be produced in the laboratory for biochemical or molecular biological work. Thus, it is one of the simplest among the complex eukaryotic organisms. Secondly, it grows on minimal medium, the only requirement for vigorous growth in addition to inorganic salts and a carbon source being biotin. The ability to grow on simple, well-defined medium is very useful for studies relating to biochemistry and nutritional effects on morphogenesis. Thirdly, it grows very fast and under laboratory conditions, the entire life cycle can be completed in 2 weeks if one wants to make sexual crosses, while the vegetative cycle requires only a few days.

6.2 Life Cycle[2]

N. crassa has two distinct life cycles, the asexual and the sexual cycle (Fig. 1). In the asexual cycle, a conidium or an ascospore germinates on a suitable substrate and forms mycelium. The mycelium is a mat of branched intertwined hyphae. The hyphae have perforated cross walls which allow organelles such as nuclei or mitochondria to pass through. The mycelium is therefore coenocytic. Under optimal conditions, the mycelial front advances on solidified medium at the rate of more than 10 cm per day. After a few days, aerial hyphae and macroconidia are formed. Macroconidia have a variable number of nuclei, from one to five. A different class of spores, microconidia, are formed directly from mycelia in old cultures. Microconidia are smaller than macroconidia and have only one nucleus.

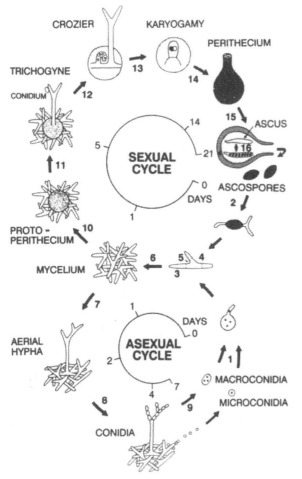

Fig. 1. Life cycles of *N. crassa*. The time scale is shwon for the periods usually observed in laboratory. *Numbers near arrows* refer to the morphogenetic steps described in Table 1. Sexual cycle begins with the formation of protoperithecia followed by the fertilization with the conidia of opposite mating tpye. The asexual cycle begins with germination of conidia or ascospores. The drawings are not to scale

Some species of *Neurospora*, including *N. crassa*, are heterothallic and have two mating types. The two mating type *A* and *a*, which are determined by two alternative DNA sequences at the same chromosomal locus, are identical in morphology and distinguishable only in their mating reaction. The length of unique *A* DNA is 5 kb while that of unique *a* DNA is 3 kb. The experimenter can choose to inoculate either mating type in suitable conditions (see later) and wait for 5 days till the

protoperithecia are formed. These female structures are fertilized by contact with the cells of the other mating type. Usually the conidia provide the fertilizing nuclei of the opposite mating type although the mycelium is also quite effective in doing so. Protoperithecia are regarded as female structures because they provide the cytoplasm of the ascospores including mitochondria. After 24 h the fertilized protoperithecia already appear bigger and browner, and they develop into perithecia with developing asci in a few days. The mature fruiting body, the perithecium, harbors 200–400 asci containing eight ascospores each. The ascospores, once they are mature, are ejected vigorously through the ostiole of perithecial neck. About 2 weeks from fertilization the perithecia start shooting ascospores in groups of eight. This process continues for several days.

6.3 Microbiological and Genetic Techniques[3]

Neurospora tissues are haploid except for a transient phase which precedes meiosis during the sexual cycle. The haploid genome contains about 4×10^7 base pairs roughly corresponding to ten times the size of the *E. coli* genome. The young ascospores have two genetically identical nuclei. Most of the genetical or biochemical work is done using conidia or mycelia grown from them and is technically similar to that with a prokaryotic organism like *E. coli*. Four days after inoculation, it is possible to obtain abundant conidia. A typical culture obtained from 20 ml of agar medium in a 250-ml Erlenmeyer flask, contains about 10^9 conidia. A conidial suspension at a titer higher than 10^7 conidia per ml in distilled water can be stored frozen at $-30\,°C$ for more than a year. For longer preservations up to 10 years, silica gel stocks are recommended. It is also possible to freeze the mycelium for many months without losing viability.

For studies requiring large masses of the fungus, as needed routinely for biochemical analyses, suitable liquid medium can be inoculated with 10^5 to 10^6 conidia per ml and incubated at optimal conditions to yield about 2 g of cell mass per 100 ml culture medium after reaching the stationary phase. Tough cell walls of fungi presented a major technical problem in the past in obtaining high yields of intact organelles. However, gentle methods are now available to disrupt the cells of *N. crassa* and it is relatively easy to isolate functional nuclei, plasma membrane, or mitochondria from the mycelium.

Conidia generally serve as the starting material for genetical studies. It is possible to induce mutations, to map and complement mutant alleles and to construct multiple mutants by recombination. Several mutagens can be used to mutagenize the nuclei of macroconidia (or microconidia). The mutagenized conidia are analysed by growth on plates containing the desired medium and a certain amount of sorbose. The presence of sorbose induces colonial growth, keeping the individuality of mycelia derived from different conidia. It is possible to have hundreds of colonies on each plate. For a complementation test, heterokaryons can easily be constructed because of the coenocytic nature of mycelia and the ability of hyphae to fuse with each other. A forced heterokaryon can be made by co-inoculating different strains on a medium on which neither parent alone can grow. After 1–2 days the germinated conidia have fused and the forced heterokaryotic mycelium grown. The only requirement is that the two strains be isogenic for ten different genes called *het* (heterokaryon incompatibility). When a pure genotype is desired, it is possible to eliminate known or hypothetical heterokaryosis simply by crossing the strain in question with a suitable parent. The ascospores are normally homokaryotic.

Molecular genetic techniques which have been applied to *Neurospora* include the separation of chromosomes using pulse field gel electrophoresis, which should soon allow the construction of specific chromosomal banks, thereby further simplifying the cloning of mapped genes. *N. crassa* protoplasts can be prepared and transformed efficiently to yield about $10^4 - 10^5$ transformants per μg DNA. The transformants arise by integration of incoming DNA into the chromosome. Recent studies show that irrespective of the number of integrated DNA, the majority of transformants arise by integration of incoming DNA into a single nucleus of

the protoplasts. In several transformants containing more than one copy of the transforming foreign gene hygromycin phosphotransferase (*hph*), cytosine methylation is induced, which may result in reversible inactivation of the gene function during the asexual cycle. Except for the mitochondrial plasmids, no other autonomous plasmid replicon is known in *Neurospora*. Although several other fungi can be transformed with DNA preparations, high frequency of integrative transformation is obtained only with *N. crassa*. This permits the screening of gene libraries by looking for complementation of a genetic marker. Furthermore, mutations can be introduced in DNA sequences which have been cloned. This is feasible because of a process termed RIP (*R*epeat *I*nduced *P*oint mutations) which operates during the sexual cycle of *N. crassa* and introduces GC to AT transitions in DNA segments which are present in the cell at the copy number of 2 or more. This finding should help define the functions of genes which have been cloned. In addition, a transposon has been identified in some strains of *Neurospora*. It should be possible to use this transposon for insertional mutagenesis.

The 600 genes which have already been mapped define seven linkage groups corresponding to seven chromosomes which can be observed using cytological staining techniques. More than 80 genes have already been cloned and a cosmid library and a genomic bank in a lambda vector is available from the FGSC. Any chromosomal segment, even when cloned only as cDNA, can be mapped with a restriction polymorphism kit also available with FGSC. The idea is based on the use of restriction enzyme recognition site(s) as genetic markers and determination of their linkage to already known markers. The mapping kit essentially consists of the progeny of a cross between parents that are sufficiently different from each other in having several RFLPs (*R*estriction *F*ragment *L*ength *P*olymorphism), thus obviating the need to perform such a cross in different laboratories. However, because of the small number of progeny, the resolution achieved by this technique in mapping is limited to those segments which show more than 5% recombination, or in other words are separated by about 80 or more genes on the genome of *N. crassa*.

6.4 Morphological Studies

The importance of light and electron microscopic studies in relation to genetic and molecular biological studies cannot be over-emphasized. It is clear that the description of various morphological structures helps define the changes which take place during morphogenesis. This in turn, helps tentative identification of stages on which particular genes may act. This section is confined to a description of the major stages in the life cycles of *N. crassa*.

6.4.1 From Spore to Mycelium [4]

N. crassa can produce three different types of spores, asexual microconidia and macroconidia, and sexual spores ascospores. The ascospores are dormant and resistant to a number of environmental stresses. Asexual spores will grow at once when put into growth medium; they are not long-lived at normal temperatures and conditions. Ascospores, on the other hand, will remain viable in the dormant state for many years, and do not grow simply in response to availability of nutrients. They are protected by their thick black coats. The ascospores can be germinated by a heat shock at 60 °C for 30 to 60 min or by furfural treatment. Sequence of visible morphological events is the same as that for conidia once dormancy is broken. A germ tube is formed after about 3 h and elongates by apical extention (Fig. 2). After a few hours, branching and continued elongation results in fully grown multinuclear mycelial mass (Fig. 3a, b).

Fig. 2. Germinating ascospore. The ascospore is 25 μm long. (Courtesy of Rudi Lurz and Wolfgang Heimler)

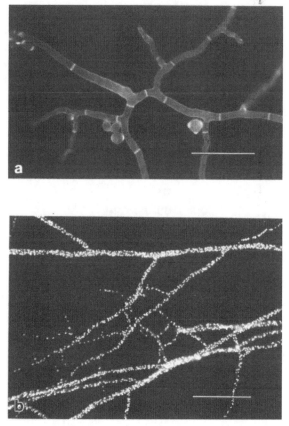

Fig. 3a, b. Fluorescence micrographs of mycelial hyphae. **a** Hyphal cross walls visualized with Calcofluor. **b** Multinucleate hyphae stained with Hoechst 33258. *Bars:* **a** 50 μm, **b** 50 μm. (After Springer and Yanofsky 1989, see [5])

CONIDIOPHORE TIMELINE

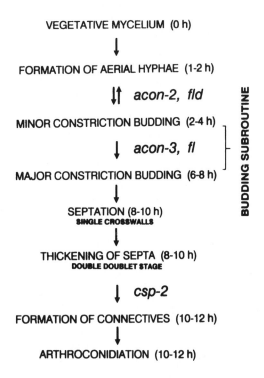

VEGETATIVE MYCELIUM (0 h)

↓

FORMATION OF AERIAL HYPHAE (1-2 h)

↕ *acon-2, fld*

MINOR CONSTRICTION BUDDING (2-4 h)

↓ *acon-3, fl*

MAJOR CONSTRICTION BUDDING (6-8 h)

↓

SEPTATION (8-10 h)
SINGLE CROSSWALLS

↓

THICKENING OF SEPTA (8-10 h)
DOUBLE DOUBLET STAGE

↓ *csp-2*

FORMATION OF CONNECTIVES (10-12 h)

↓

ARTHROCONIDIATION (10-12 h)

BUDDING SUBROUTINE

BUDDING SUBROUTINE

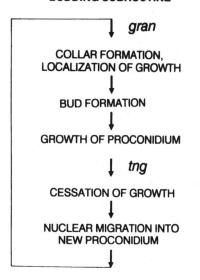

↓ *gran*

COLLAR FORMATION,
LOCALIZATION OF GROWTH

↓

BUD FORMATION

↓

GROWTH OF PROCONIDIUM

↓ *tng*

CESSATION OF GROWTH

↓

NUCLEAR MIGRATION INTO
NEW PROCONIDIUM

6.4.2 From Mycelium to Conidia[5]

Conidiation, the formation of asexual spores, typically takes place on solid medium or above a stationary liquid culture, but can occur in liquid under certain conditions (see below). The most complete studies on conidiation using a wild type and several conidiation defective mutants support a timeline of events shown in Fig. 4. During conidiation, the aerial hyphae cease to elongate by apical extension, and instead grow by successive apical budding. This results in the formation of proconidial chains (Fig. 5a). Cross walls are laid down between each proconidial element (Fig. 5b). These cross walls then thicken and separate, leaving the newly formed conidia held together by a fragile connective (Fig. 5c), which can break and release windborne conidia at the slightest disturbance.

6.4.3 The Sexual Cycle[6]

A prerequisite for the sexual cycle is the production of protoperithecia. These spherical structures, roughly 50 micrometer diameter have several hyphae protruding from their surface (Fig. 6). One of these hyphae is called the trichogyne and is connected with the ascogonial cell within the protoperithecium. Protoperithecia can be formed only on mycelia that are not submerged. Fertilization begins when a conidium or a piece of mycelium of the opposite mating type comes in contact with the trichogyne. The trichogyne grows towards the cell of the opposite mating type with the help of attractants which are not yet defined. Nuclei of opposite mating type pass into the trichogyne by wall fusion and are then transported to the ascogonium. Here, nuclei of both mating types undergo several divi-

Fig. 4. Developmental timeline of conidiation. The timeline is divided into two parts: the conidiophore timeline, which applies to the entire conidiophore as it is developing, and the budding subroutine, which occurs every time a proconidial chain buds. The approximate time at which each step in the conidiophore timeline begins after induction is indicated in parentheses. The step at which conidiation is blocked in the mutant strains *acon-2, acon-3, csp-2, fl, fld, gran,* and *tng* is indicated. (After Springer and Yanofsky 1989)

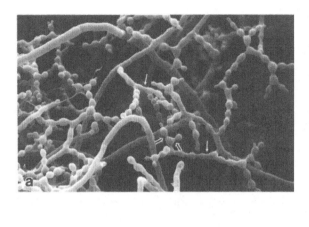

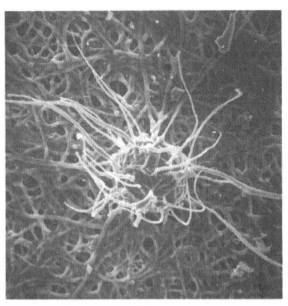

Fig. 6. Scanning electron micrograph of a protoperithecium. (Courtesy of Rudi Lurz and Wolfgang Heimler)

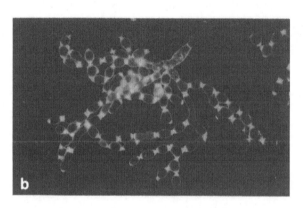

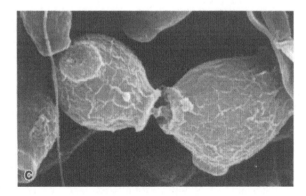

Fig. 5a–c. Scanning electron micrographs of developing wild-type conidiophores. **a** Major constriction chains. *Double line arrows* point to major constrictions; *small arrows* point to minor constrictions, **b** Combined dark field and Calcofluor fluorescence images of a conidiophore with thickened cross walls (double-doublet stage). **c** Connective between two mature conidia. (After Springer and Yanofsky 1989)

sions before karyogamy (Fig. 7). Nuclei of opposite mating type finally fuse in the crozier, a hook shaped structure composed of three cells, and immediately undergo meiosis. In the meantime, the protoperithecium becomes browner, bigger, and develops a beak-shaped structure with a aperture (ostiole) through which the ripe ascospores are eventually discharged (Fig. 1). The resulting structure, now called perithecium, contains hundreds of asci each one coming from a different crozier. Each ascus has eight ascospores representing an outcome of meiotic divisions. Of the eight ascospores, each pair of adjacent ascospores has identical genotype which may be different from the neighboring pair of ascospores. Each young ascospore has two nuclei formed by mitotic division inside the ascospore; the genotype of the two nuclei is therefore identical. The four potentially different genotypes of the ascospore pairs are the result of two meiotic divisions. The ascospores of one ascus are ejected together through the ostiole of the perithecium. Because whole meiotic tetrads could be analyzed, *N. crassa* was the first organism in which gene conversion was demonstrated.

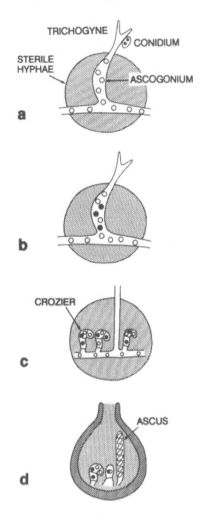

Fig. 7 a – d. Schematic representation of the major stages in the development of a perithecium. **a** Conidium attaches to the trichogyne. *Circles* represent nuclei. **b** Nuclei of the conidium go inside ascogonium. **c** Croziers are formed. **d** Asci are formed from croziers. The drawings are not to scale

6.5 Influence of the Environment on Development

6.5.1 Environmental Factors Identified So Far[7]

Normally, mycelium does not differentiate in submerged culture. A mycelium can be induced to conidiate by filtering it onto filter paper and letting it

start to dry. With this method, aerial hyphae and conidia are obtained in about 6 h. The actual stimuli are not known. Using this technique it is possible to show that humidity and to some extent light have an influence on conidiation. It is possible to induce conidiation up to the point of Fig. 5 a in liquid culture by nitrogen or carbon limitation. In such conditions the conidia are formed directly on the hyphae with a lag time of two hours. Conidiogenesis on solid medium is inhibited by as little as 0.2% CO_2. Aging of mycelium induces formation of microconidia on vegetative hyphae. Variation in temperature has no effect on conidiation between 4 and 37 °C the temperature range in which the fungus can grow. In contrast, production of protoperithecia is sensitive to temperature, the optimum being between 18 and 26 °C. Nitrogen starvation is essential for this differentiation process while a short pulse of blue light in nitrogen limiting conditions can accelerate the process by a factor of 30. Blue light is a signal which induces a variety of physiological responses not only in *N. crassa* but also in several microorganisms, plants, and animals. The major physiological responses include carotenoid production, phototropism, and developmental changes.

The fact that it is possible to induce differentiation is very important for biochemical and molecular biological studies of morphogenesis. This is because comparison of induced and uninduced cultures of the same age then helps separate the events that could be specifically involved in morphogenesis from those which occur simply as temporally regulated coincidences.

6.5.2 The Biological Meaning of Environmental Influence on Development[8]

The micro-environment in which microorganisms grow can change rapidly in time. It is therefore understandable that microorganisms have evolved mechanisms which sense a changing environment. Nutrient depletion, for example, is a stimulus for differentiation in a variety of organisms including *Bacillus subtilis, Streptomyces coelicolor, Dictyostelium discoideum,* yeasts, *Aspergillus,* plants, and animals. Two main strategies have evolved to survive adverse conditions. First, the ability to produce dormant cells which resist environmental stresses and initiate growth once favorable conditions are available. Secondly, the ability to sense the conditions which would favor wide dispersion of such spores thereby escaping from a potentially unfavorable environment. Compared to mycelium, both conidia and ascospores are more resistant to a variety of environmental stresses such as adverse temperature, chemicals and starvation. It is therefore understandable that starvation is a signal which induces both sexual as well as asexual differentiation. What is the advantage of the sensitivity of mycelium to blue light? The natural habitat of *Neurospora* is burnt vegetation or wood that has been heated. Thus, if the mycelium is in confined areas with low air convection for the dispersion of the spores, the chances that protoperithecia are fertilized and the ascospores dispersed by wind or insects are low. Thus, the ability of the organism to sense open fields increases the possibility of wider dispersal of the dormant structures. How does *Neurospora* sense the open fields? Light would certainly seem to be a good signal. In line with this interpretations is the fact that the perithecial necks display a phototropic response, suggesting that when the ascospores are discharged from perithecia, they are likely to get into the open, with higher chances of dispersal.

Sensing of blue light is consistent with the widely accepted belief that life started in water and the fact that at depths of 30 m or more in water, mainly the blue component of sunlight is present. This may also explain why blue light is sensed by a large number of organisms. The ability of organisms to sense blue light might have been the first photoperception in evolution.

6.6 Developmental Genes of *Neurospora*[9]

More than 250 genes which affect one or more of the morphological processes in *N. crassa* are already known. A list of phase-specific and pleiotropic genes for each morphogenetic process is given in Table 1. Analysis of these data needs care for several reasons. First, not all the mutants are isogenic. Although, the standard wild-type strain used in several laboratories these days is St. Lawrence (SL), at least a dozen other wild-type strains have been used in the long history of research with *N. crassa*. As a result, about 60% of the mutant strains of today have a mixed genetic background. Second, although several of the mutants were isolated specifically on the basis of their morphogenetic changes, the others were not, and therefore the classification of such alleles is often based on marginal comments of the authors. In spite of these limitations, it is possible to draw several conclusions from these data.

For most of the morphogenetic steps, at least one gene is known. In cases such as sexual development, many genes are known. On the basis of the distribution of mutations over loci, the number of genes needed for protoperithecia formation is estimated as around 100, while for perithecia formation probably 200 genes are necessary. Considering that the number of genes already known to be involved in sexual development is about 80, it can be seen that about 25% of the expected number of genes are already known (from protoperithecia to the female functions in Table 1).

As far as we know, there is no clustering of genes. In other words, the genome does not have regions specifically dedicated to developmental genes. These data also show that more than 50% of the genes are pleiotropic; the most pleiotropic is the gene R (round spore) which influences five different morphogenic steps (Table 1).

6.6.1 Biosynthetic and Developmental Pathways Have Little in Common

The absence of any correlation between auxotrophy and morphogenesis is striking. Out of the 130 genes known to be involved in various biosynthetic pathways, there are only 17 which affect morphogenesis. Of these, 12 block ascospore germination when the mutant alleles are homozygous in crosses and probably represent a special case. Thus, only five genes are known which affect both metab-

Table 1. Genes affecting development of *Neurospora crassa*

1. Conidial germination
i. *pdx-1*
ii. *age-1, age-2, moe-3, mus-9, mus-11, nd, psi-1, scon, sf, slo-1, slo-2, so, uvs-3, uvs-6*

2. Ascospore germination
i. *ad-1, ad-7, arg-1, arg-10, cys-3, leu-5, lys-3, lys-5, mtr, pan-2, pyr-1, ser-1, ser-3, spe-1, ws-2*
ii. *ace-1, col-6, col-12, fl, gsp, gul-3, gul-4, gul-5, gul-6, le-1, le-2, mei-1, mei-2, mo-2, R, scon, sf, sg, slo-1, slo-2, smco-5, ts, ty-2, tyr-2, un-17, uvs-1, uvs-4, ws-1*

3. Hyphal elongation
i. Not known
ii. *st*

4. Septation
i. Not known
ii. *cot-1, cwl*

5. Branching
i. *bal, chol-2, col-9, col-16, col-17, com, cot-3, csh, do, dot, lp, mat, md, mel-1, mel-3, moe-2, pf, smco-1, smco-3, smco-4, smco-6, smco-8, sn, spco-6, spco-8, spco-9, spco-10, spco-12, spco-13, su(bal), su(col-2), ti*
ii. *amyc, bn, chol-1, cl, col-1, col-2, col-4, col-5, col-6, col-8, col-10, col-12, col-13, col-15, cot-1, cot-2, cot-4, cot-5, cr-1, cum, del, erg-4, fr, fs-2, gpi-1, gran, inl, ipa, le-1, le-2, moe-3, pk, rg-1, scot, scr, sh, smco-5, smco-7, smco-9, sp, spco-5, spco-7, spco-11, spco-14, ti, vel*

6. Mycelial morphology
i. *car, cel, coil, cy, (cyh-2, cyh-3), fls, fz, glp-2, gul-1, gul-2, mo-1, mod(sc), mus(SC17), os-2, os-4, os-5, rol-1, rol-2, rol-3, spg, un-8*
ii. *amyc, atr-1, cfs(OY305), cfs(OY306), cfs(OY307), col-2, cpk, dn, ff-3, fs-1, gul-3, gul-5, gul-6, med, mo-2, mo-4, mo-5, moe-1, nd, oak, os-1, pat, pi, psi-1, R, rg-1, rg-2, ro-1, ro-2, ro-3, ro-4, ro-6, ro-7, ro-10, sc, sg, sh, spco-4, st, ta, tng, tyr-2, un-17, uvs-4, uvs-5, uvs-6*

7. Aerial hyphae
i. *cut, pl*
ii. *age-1, age-2, bd, cl, cot-5, cpt, cr-1, cr-4, cum, del, dow, dr, eas, erg-3, ff-1, fr, fs-5, pe, pk, sc, scr, sk, smco-5, so, sp, ty-1, ty-2, un-20, var-1, wa*

8. Conidia
i. *ace-3, acon-2, adh, an, at, cr-2, cr-3, da, dir, fld, gap, (gray, cr-1), su(pe), vis(3717), ylo-2, ylo-3*
ii. *acon-3, amyc, bd, bn, chol-1, col-1, col-4, col-5, col-10, col-13, col-15, cot-1, cot-4, cpt, cr-1, cr-4, cwl, cya-8, dow, dr, eas, erg-1, erg-3, fl, fr, gran, inl, le-1, mo-2, mo-4, mo-5, moe-1, oak, os-1, pat, pe, pi, pk, rg-1, rg-2, ro-1, ro-2, ro-3, ro-4, ro-6, ro-7, ro-10, sc, scot, shg, sk, slo-1, slo-2, smco-7, so, sp, spco-4, spco-11, spco-14, ta, tng, un-20, vel, wa*

9. Conidial separation
i. *csp-1, csp-2*
ii. Not known

10. Protoperithecia
i. *ff-2, ff-6, fs-3, fs-4, fs-6, fs-n, stopper(mito), wc-1, wc-2*
ii. *ban, erg-3, ff-1, ff-3, ff-5, fs-2, fs-5, var-1*

11. Trichogyne
i. Not known
ii. *ff-5*

Perithecia (female) [a]
i. *per-1*
ii. *fmf-1, fs-1, pen-1*

Perithecia (male) [b]
i. Not known
ii. *fmf-1, mb-1, mb-2, mb-3*

Perithecia (interallelic) [c]
i. *A/a, arg-13*
ii. *col-8, cot-4, mus-9, spco-7*

Female function(s) [d]
i. *acr-7, erg-2, leu-1, mms-5, mus-10, mus(SC10), pyr-3, pyr-4, (rg-1, cr), ssu-1, ssu-2, ssu-3, ssu-4, ssu-5, ssu-6, ssu-7, ssu-8, ssu-9, ssu-10*
ii. *acon-3, atr-1, cot-5, cya-1, cya-2, cyb-2, cyt-1, cyt-2, cyt-12, erg-1, erg-4, fr, gul-3, gul-4, R, ro-1, ro-2, ro-3, ro-4, ro-6, ro-7, ro-10, sk, so, tng, ty-1, ty-2, tyr-2*

12. Crozier
i. *mei-4*
ii. *Prf, uvs-6*

13. Karyogamy
i. Not known
ii. *uvs-3, uvs-5*

14. Perithecial neck
i. Not known
ii. *mb-1, mb-2, mb-3, pen-1*

15. Ascus
i. *asc(K1897), Iasc*
ii. *cl, col-2, col-10, cr-1, mei-1, mei-2, pk, Prf, R, sc, scr, smco-9, spco-5*

16. Ascospores
i. *fsp-1, fsp-2, mei-3, mms-6, mms-7, mus-7, mus-8, Prf*
ii. *ace-1, ban, cot-2, cr-1, gsp, mus-11, R, scon, scr, sk, spco-7, ts, un-17, uvs-5, uvs-(SA3B), ws-1*

The developmental genes involved in various stages of morphogenesis. The numbers from 1 to 16 correlate with those which define morphogenetic steps as shown in Fig. 1. The genes grouped in (i) are known to affect only one morphogenetic step while those grouped in (ii) are known to affect two or more morphogenetic steps.
[a] Perithecia are not formed when the mutants are used as the female parent.
[b] Perithecia are not formed when the mutants are used as the male parent.
[c] Perithecia are not formed in a specific, interallelic cross.
[d] Ascospores are not produced. The differentiation step blocked is not known.

olism and development, indicating that the biosynthetic and morphogenetic pathways are quite separate from each other.

In the Section 6.7 we will discuss some colonial mutants with altered Km values for some enzymes in the carbohydrate pathways (probably because of missense mutations). This observation opens a possibility that several other morphological mutants may also have missense mutations in the genes encoding enzymes involved in metabolic pathways. If this is the case, then overlap between the morphogenetic and metabolic pathways may be more extensive. The interpretation would then be that an enzyme per se is not necessary for differentiation as long as the product(s) of enzymatic reactions(s) can be or have to be supplied from the outside. However, if the mutation does not lead to a total loss of the function of an enzyme but only reduces its activity, then the internal balance between the different intermediate metabolites can be so disturbed that the results is abnormal development. Cloning of the morphogenetic genes and determination of the function of their products can throw some light on the link between metabolic intermediates and morphogenetic pathways.

6.6.2 About the Pleiotropism of Developmentally Relevant Genes [10]

We have seen that about 50% of all the morphogenetic genes influence two or more of the 16 morphogenetic steps of *Neurospora* (Table 1). The highest pleiotropism is between production of protoperithecia/perithecia and the production of hyphae or conidia. About one third of the genes affecting the former morphogenetic processes also affect the other two (19 genes). This is an indication that there is much in common between the pathways of sexual and asexual differentiations. In asexual differentiation, aerial hyphae have to redirect cell wall synthesis in order to make septa and produce conidia. In sexual differentiation also cell wall synthesis must be modulated in order to make the ascogonium. The modulation of cell wall synthesis necessary in both cases requires that information and/or building blocks from within the cell

must come through the plasma membrane to the cell wall. One possibility is that some of the 19 genes necessary for both types of morphogenesis may be involved in this kind of information flow or in cell wall synthesis. For instance, the *erg* mutants which have cell membranes deficient in ergosterol, display impaired morphogenesis. The *erg-1, erg-2, erg-3,* and *erg-4* mutants function poorly or are sterile as a female parent, and *erg-1* and *erg-3* mutants make fewer conidia and uneven aerial hyphae. Another evidence is that the protein components of the plasma membrane show different patterns in *R* and *Iasc* mutants.

6.6.3 On the Role of Mitochondria in Development [11]

Virtually nothing is known about how the mitochondrial genetic system influences nuclear genes. In *Neurospora* there are two lines of evidence that mitochondria are important for morphogenesis. The more direct one is the discovery that one class of mitochondrial mutants called *stopper* are unable to make protoperithecia. As male parents they are normal but the genetic trait cannot be transmitted by crosses because in *Neurospora,* as in all animals and plants, the mitochondria are transmitted through the female gamete. All the *stopper* mutants have a deletion of 350 to 5000 bp in the mitochondrial genome, depending on the allele. One of these mutants has been shown to lack a subunit of the NADH dehydrogenase.

Indirect evidence showing that mitochondria are important for morphogenesis is the fact that six genes are necessary both to make normal cytochrome complements in the mitochondria and to form normal perithecia. The evidence is indirect because other genes affecting the same cytochromes are normal for sexual differentiation. Therefore normal complements of cytochromes per se are not necessary for differentiation but something which interferes with their function (the mitochondrial membrane?) can also influence morphogenesis. In several plants like *Vicia faba,* maize, and petunia, the genome of the mitochondrion is important for male fertility.

6.7 The Carbohydrate Pathways and Development[12]

Because the cell wall is made of polysaccharide, glucans, and chitin, one might expect that carbohydrate biosynthetic pathways are important in development. The detailed study of the biochemical genetics of several morphological mutants made by Tatum and his colleagues has revealed new and unexpected results. The *rg-1*[+] and *rg-2*[+] genes probably encode two isoenzymes of phosphoglucomutase which synthesize glucose-1-phosphate from glucose-6-phosphate. The evidence is based on temperature sensitivity of the mutants, different Kms and other characteristics. A mutant in the gene *smco-9* has higher amounts of an inhibitor for the enzyme alpha-1,4 glucan-6-glucosyltransferase which synthesizes the 1−6 link of the glucan chains. This should have a direct effect on the synthesis of glucan and may therefore affect the shape of mycelia. In the same region of the carbohydrate pathway there are two other enzymes, glucose-6-phosphate dehydrogenase and 6-phosphogluconate dehydrogenase, which are tetrameric and dimeric respectively; the genes encoding them are known and designated *col-2*[+], *bal*[+], and *fr*[+] for the former and *col-3*[+] and *col-10*[+] for the latter. Mutants in any of these genes have abnormal morphology (Fig. 8). The evidence that these five genes encode subunits of the two enzymes is again based on thermal stability and other parameters, which, before the cloning era, was the best evidence that a given mutation is in the coding gene of an enzyme and not in a regulatory gene.

An interesting observation is that the mutants which alter glucose-6-phosphate dehydrogenase or 6-phosphogluconate dehydrogenase have lower NADPH level than WT. On the other hand, WT grows semi-colonialy and has low levels of NADPH if grown on minimal medium with acetate as the sole source of carbon. There is a good correlation between NADPH and linolenic acid levels in that low NADPH levels are found whenever the level of linolenic acid is low. At least in the case of the *frost (fr)* mutant, the defect in growth can be alleviated by addition of linolenic acid to the medium. The sequence of events thought to take place is the

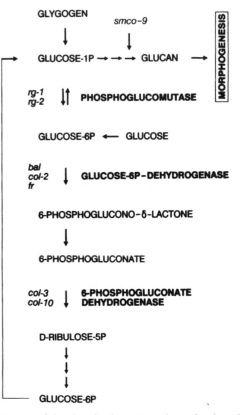

Fig. 8. A part of the phosphogluconate pathway showing which morphogenetic genes encode which enzymes. Phosphoglucomutase is a heterodimer encoded by *rg-1* and *rg-2*, glucose-6P dehydrogenase is a heterotetramer encoded by *bal, col-2,* and *fr,* and 6-phosphogluconate dehydrogenase is a heterodimer encoded by *col-3* and *col-10*

following: low NADPH → low linolenic acid → abnormal morphology. No data are available on the rescue of *col-2, bal, col-3, col-10* mutants by linolenic acid.

It is striking that three enzymes of the phosphogluconate pathway are so important for normal morphology. A primary purpose of this pathway in most cells is to generate reducing power in the extramitochondrial cytoplasm in the form of NADPH. The role of NADPH and linolenic acid as key molecules in regulation of morphogenesis could be excluded by experiments which demonstrate that low NADPH or linolenic acid levels do not change the morphology of mycelium.

6.8 Molecular Biology

6.8.1 The Mating Type Locus[13]

The mating type locus of *N. crassa,* is important not only for the sexual cycle but also for heterokaryon incompatibility. If cells of opposite mating type fuse on medium inappropriate for mating, the resulting heterokaryotic cells die. Mutational analysis of the mating type locus shows that all the mutants that confer sterility will also nullify the heterokaryon incompatibility. No mutations converting one mating type to the other or to a state of self-fertility have been observed. The situation in *Neurospora* is therefore very different from that of *Saccharomyces cerevisiae* (see Arcangioli and Klar, Chap. 5, this Vol.). The genomic cloning of both the *A* and *a* mating type loci has shown that the mating type activity is included in a stretch of DNA about 3 to 5 kb long. The *A* and *a* specific DNAs do not hybridize with each other. Each mating type has one and only one of these two pieces of DNA; there is no "silent" cassette as found in yeast. When a strain carrying a mutation in the mating type locus which leads to sterility is transformed with an appropriate stretch of DNA, both the fertility and the heterokaryon incompatibility are restored simultaneously. It is reasonable to assume that in *N. crassa* the mating type locus is a master gene, which has a regulatory effect on many genes necessary for both sexual fusion and heterokaryon compatibility. Surprisingly, most homothallic species of *Neurospora* have only the *A* counterpart of the mating type locus; one has both *A* and *a* genetic material.

6.8.2 Gene Regulation During Conidiation[14]

In experimental conditions which induce conidiation, there are dozens of proteins which are regulated. Through 2-D-gel analysis it is possible to see that the concentration of 14 polypeptides increases and of 38 polypeptides decreases. The number of resolved polypeptides was 300, which means that 20% of the major proteins are differentially regulated. Several genes which are transcriptionally regulated at different times after the conidiation

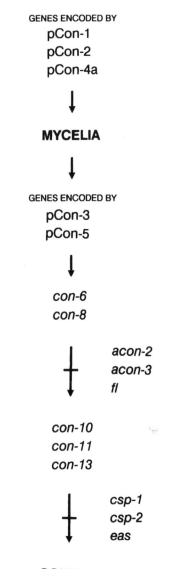

Fig. 9. Diagram representing the proposed order of *con* gene expression and the stages at which various mutants are blocked in the conidiation pathway. (After Roberts and Yanofsky 1989). The term *con* implies a gene where as pCon refers to the plasmid harboring the *con* gene

stimulus, have been cloned. Some of these genes are regulated in mutants known to affect conidiation. All the information so far available can be summarized in Fig. 9. However, the drawing does not imply that all the regulated genes such as *con-8*[+] are necessary for conidiation. It is possible that its expression merely represents a side effect. There are no data yet to show that the disruption of any of these genes would make *Neurospora* aconidial. In *Aspergillus nidulans,* deletion of genes which are actively expressed during conidiation does not affect conidiation, at least under laboratory conditions.

6.8.3 Blue Light Regulates Many Genes[15]

In *Neurospora* and many other microorganisms, blue light regulates a series of physiological responses. In *N. crassa,* blue light responses include carotenoid production, the circadian rhythm of conidiation (in the *bd* mutant it can be synchronized and shifted), phototropism of perithecial necks, and protoperithecia production. Two genes, *wc-1*[+] and *wc-2*[+] are necessary for all these responses to blue light. Blue light is a signal which can be given in a short pulse (60 s) to induce protoperithecia formation. It is therefore an ideal case for studying gene regulation during differentiation because the illuminated mycelium can be compared with a dark-grown control.

In *N. crassa* where regulation of transcription by blue light has been extensively studied, 3 to 4% of genes that are expressed during vegetative growth are blue light-regulated within 30 min following irradiation. This represents 70–200 of the 2000–6000 genes estimated to be expressed in the vegetative mycelium. The regulation is at the level of more translatable mRNA in vitro, which means that either more mRNA is made or more is unmasked. The regulation is time-staggered, the first gene being regulated after a 2-min lag, the last one after 20 min. Eight genes (*al-1*[+], *al-2*[+], *al-3*[+], *bli-3*[+], *bli-4*[+], *bli-7*[+], *bli-13*[+], and *bli-75*[+]), which are blue light-regulated, have already been cloned. For four of these, *al-1*[+], *al-2*[+], *bli-3*[+], and *bli-4*[+], the regulation is at the level of mRNA transcriptional initiation. The lag time of mRNA increase is 2 min, and after 30 min exposure to light there is about 100 times more mRNA than in the dark. White collar mutants *wc-1* and *wc-2* are insensitive to light. It is known that *al-1*[+], *al-2*[+], *al-3*[+] genes are not necessary for development. No experiments have yet been done to abolish gene

function for the other five genes. It is not known therefore if any of these genes is necessary for development. It is not yet known whether the time-staggered regulation is an indication of a cascade regulation or not.

6.9 Outlook

We have learned so far from studies of *N. crassa* that morphogenetic pathways can be activated through external stimuli and the same stimuli regulate several genes. For example, filtering the mycelium induces conidiation and regulates expression of many genes. Similarly, blue light induces protoperithecia formation and turns on dozens of genes. We know further that hundreds of genes are necessary and specific for these unknown pathways. In the example of the sexual cycle, it is estimated that 300 genes are necessary for morphogenesis and already 80 genes are mapped. With a few exceptions, the functions of most of the genes are not yet known.

Are all or some of those regulated genes necessary for development? There is so far no proof that any of the regulated genes are indeed necessary for morphogenesis, but this should come soon. Very little is known about which morphogenetic genes are necessary for which gene regulation.

The complete picture of which genes are regulated and which genes are specific for this regulation will be difficult to obtain in the near future because of the huge number of genes involved and because it is difficult to look for regulation of minor genes. However, with more clones of the morphogenetically relevant genes, clear information about morphogenetic pathways should accumulate. Even an incomplete picture should throw some light on the regulatory mechanisms involved. A major contribution could come from the studies of the regulation of morphogenetic pathways by nitrogen, carbon source, and by blue light.

Another field in which *Neurospora* is an ideal model system is the study of the role of organelles, such as mitochondria, in morphogenesis. *Neurospora* is one of the few eukaryotes in which mutants in the mitochondrial genomes result in abnormal sexual morphogenesis.

6.10 Summary

N. crassa is a filamentous fungus with both a sexual and an asexual cycle. In total there are more than a dozen different morphological structures. The nucleus is haploid and contains seven chromosomes with a DNA content of 4×10^7 bp, about ten times more than *E. coli*. Microbiological and genetic methods used to study *N. crassa* are similar to those used for prokaryotic organisms. Differentiation of both cycles can be regulated by environmental stimuli, i.e., nitrogen and glucose limitation, blue light. More than 600 genes are mapped and about 250 nuclear genes influence one or more steps in morphogenesis. A few mitochondrial mutants are available which are defective in protoperithecia production. There is apparently very little overlap between the biosynthetic and morphogenetic pathways. Some morphogenetic genes code for enzymes in the carbohydrate pathway (phosphogluconate pathway). Some environmental signals which induce conidiation or protoperithecia formation regulate hundreds of genes. The cloning of more than a dozen of those genes shows temporal regulation at the transcriptional level. This regulation can be very fast (2 min) and extensive (a factor 100).

N. crassa is a very good model for studying the organization and regulation of morphogenetic pathways and their relationship to metabolic pathways. It could be a good model also for studying the interaction between the mitochondrial genome and the nuclear genome in differentiation.

References

[1] Beadle GW, Tatum EL (1941) Genetic control of biochemical reactions in *Neurospora*. Proc Natl Acad Sci USA 27:499–506

Catalogue of Strains, vol 35 (1988) Fungal Genetics Stock Center, University of Kansas City, Kansas

Fungal Genetics Stock Center. Department of Microbiology, University of Kansas City, Kansas

Perkins DD, Turner BC, Barry G (1976) Strains of *Neurospora* collected from nature. Evolution 30:281–313

[2] Fincham JRS, Day PR, Radford A (1979) Fungal genetics. Blackwell Scientific Publications, Oxford

Perkins D, Barry EG (1977) The cytogenetics of *Neurospora*. Adv Genet 19:133–285

[3] Cambareri EB, Jensen BC, Schabtach E, Selker EU (1989) Repeat-induced G-C to A-T mutations in *Neurospora*. Science 244:1571–1575

Davis RH, de Serres FJ (1970) Genetic and microbiological research techniques for *Neurospora crassa*. Methods Enzymol 27:79–143

Fincham JRS (1989) Transformation in fungi. Microbiol Rev 53:148–170

Kinsey JA, Helber J (1989) Isolation of a transposable element from *Neurospora crassa*. Proc Natl Acad Sci USA 86:1929–1933

Orbach MJ, Vollrath D, Davis RW, Yanofsky C (1988) An electrophoretic karyotype of *Neurospora crassa*. Mol Cell Biol 8:1469–1473

Pandit NN, Russo VEA Reversible inactivation of a foreign gene (*hph*) during the asexual cycle of *N. crassa* transformants. Mol Gen Genet (submitted)

[4] Brody S (1981) Genetic and biochemical studies on *Neurospora* conidia germination and formation. In. Turian G, Hohl H (eds) The fungal spore: morphogenetic controls. Academic Press, New York, pp 605–626

Farkas V (1979) Biosynthesis of cell walls of fungi. Microbiol Rev 43:117–144

Garnjobst L, Tatum EL (1967) A survey of new morphological mutants in *Neurospora crassa*. Genetics 57:579–604

Seale T (1973) Life cycle of *Neurospora crassa* viewed by scanning electron microscopy. J Bacteriol 113:1015–1025

[5] Rossier C, Oulevey N, Turian G (1973) Electron microscopy of selectively stimulated microconidiogenesis in wild type *Neurospora crassa*. Arch Microbiol 91:345–353

Springer ML, Yanofsky C (1989) A morphological and genetic analysis of conidiophore development in *Neurospora crassa*. Genes Dev 3:559–571

[6] Bistis GN (1983) Evidence for diffusible, maiting-type-specific trichogyne attractants in *Neurospora crassa*. Exp Mycol 7:292–295

Harris JL, Howe HB jr, Roth IL (1975) Scanning electron microscopy of surface and internal features of developing perithecia of *Neurospora crassa*. J Bacteriol 122:1239–1246

Raju NB (1980) Meiosis and ascospore genesis in *Neurospora*. Eur J Cell Biol 23:208–223

[7] Guignard R, Grange F, Turian G (1984) Microcycle conidiation induced by partial nitrogen deprivation in *Neurospora crassa*. Can J Microbiol 30:1210–1215

Müller BT, Russo VEA (1989) Nitrogen starvation or glucose limitation induces conidiation in constantly shaken liquid cultures of *Neurospora crassa*. Fungal Genet Newsl 36:58–60

Rossier C, Ton-That T-C, Turian G (1977) Microcyclic microconidiation in *Neurospora crassa*. Exp Mycol 1:52–62

Sommer Th, Degli Innocenti F, Russo VEA (1987) Role of nitrogen in the photoinduction of protoperithecia and carotenoids in *Neurospora crassa*. Planta 170:205–208

[8] Russo VEA, Chambers JAA, Degli Innocenti F, Sommer Th

(1985) Photomorphogenesis in microorganisms. In: Colombetti G, Lenci F, Song P-S (eds) Sensory perception and transduction in aneural organisms. Plenum Press, New York, pp 231–249

[9] Johnson TE (1978) Isolation and characterization of perithecial development mutants in *Neurospora*. Genetics 88:27–47

Perkins DD, Radford A, Newmeyer D, Björkman M (1982) Chromosomal loci of *Neurospora crassa*. Microbiol Rev 46:426–570

Perkins DD, Pollard VC (1987) Newly mapped chromosomal loci of *Neurospora crassa*. Fungal Genet Newsl 34:52–53

[10] Bowman J, Srb AM (1983) Two-dimensional electrophoresis of plasma membranes, showing differences among wild-type and abnormal ascospore mutant strains of *Neurospora crassa*. J Bacteriol 155:1393–1398

[11] Attardi G (1988) Biogenesis of mitochondria. Annu Rev Cell Biol 4:289–333

Bertrand H, Pittenger TH (1972) Complementation among cytoplasmic mutants of *Neurospora crassa*. Mol Gen Genet 117:82–90

Bertrand H, Nargang FE, Collins RA, Zagozeski CA (1977) Nuclear cytochrome-deficient mutants of *Neurospora crassa*: isolation, characterization, and genetic mapping. Mol Gen Genet 153:247–257

de Vries H, Alzner-De Weerd B, Breitenberger CA, Chang DD, de Jonge JC, RajBhandary UL (1986) The E35 stopper mutant of *Neurospora crassa*: precise localization of deletion endpoints in mitochondrial DNA evidence that the deleted DNA codes for a subunit of NADH dehydrogenase. EMBO J 5:779–785

[12] Brody S, Tatum EL (1966) The primary biochemical effect of a morphological mutation in *Neurospora crassa*. Proc Natl Acad Sci USA 56:1290–1297

Brody S, Tatum EL (1967) Phosphoglucomutase mutants and morphological changes in *Neurospora crassa*. Proc Natl Acad Sci USA 58:923–930

Brody S (1970) Correlations between reduced nicotinamide adenine dinucleotide phosphate levels and morphological changes in *Neurospora crassa*. J Bacteriol 101:802–807

Brody S, Nyc JF (1970) Altered fatty acid distribution in mutants of *Neurospora crassa*. J Bacteriol 104:780–786

Brody S (1972) Regulation of pyridine nucleotide levels and ratios in *Neurospora crassa*. J Biol Chem 247:6013–6017

Mishra NC, Tatum EL (1970) Phosphoglucomutase mutants of *Neurospora sitophila* and their relation to morphology. Proc Natl Acad Sci USA 66:638–645

Nyc JF, Brody S (1971) Effects of mutations and growth conditions on lipid synthesis in *Neurospora crassa*. J Bacteriol 108:1310–1317

Scott WA, Tatum EL (1970) Glucose-6-phosphate dehydrogenase and *Neurospora* morphology. Proc Natl Acad Sci USA 66:515–522

Scott WA, Abramsky T (1973) *Neurospora* 6-phosphogluconate dehydrogenase. II. Properties of two purified mutant enzymes. J Biol Chem 248:3542–3545

[13] Glass NL, Grotelueschen J, Metzenberg RL (1990) *Neurospora crassa* A mating-type region. Proc Natl Acad Sci USA 87:4912–4916

Griffiths AJF (1982) Null mutations of the *A* and *a* mating type alleles of *Neurospora crassa*. Can J Genet Cytol 24:167–176

Staben C, Yanofsky C (1990) *Neurospora crassa a* mating-type region. Proc Natl Acad Sci USA 87:4917–4921

[14] Aramayo R, Adams TH, Timberlake WE (1989) A large cluster of highly expressed genes is dispensable for growth and development in *Aspergillus nidulans*. Genetics 122:65–71

Berlin V, Yanofsky C (1985) Isolation and characterization of genes differentially expressed during conidiation of *Neurospora crassa*. Mol Cell Biol 5:849–855

Roberts AN, Yanofsky C (1989) Genes expressed during conidiation in *Neurospora crassa*: characterization of *con-8*. Nucl Acids Res 17:197–214

[15] Degli Innocenti F, Russo VEA (1984) Genetic analysis of blue light induced response in *Neurospora crassa*. In: Senger H (ed) Blue light effects in biological systems. Springer, Berlin Heidelberg New York, pp 214–219

Lauter FR, Russo VEA (1990) Light-induced dephosphorylation of a 33 kDa protein in the wild-type strain of *Neurospora crassa*: the regulatory mutants *wc-1* and *wc-2* are abnormal. J Photochem Photobiol 5:95–103

Nawrath C, Russo VEA (1989) Fast regulation of translatable mRNAs by blue light in *Neurospora crassa wt*: the *wc-1* and *wc-2* mutants are blind. J Photochem Photobiol 4:261–271

Nelson MA, Morelli G, Carattoli A, Romano N, Macino G (1989) Molecular cloning of a *Neurospora crassa* carotenoid biosynthetic gene (*albino*-3) regulated by blue light and the products of the white collar genes. Mol Cell Biol 9:1271–1276

Schmidhauser TJ, Lauter FR, Russo VEA, Yanofsky C (1990) Cloning, sequence, and photoregulation of *al-1*, a carotenoid biosynthetic gene of *Neurospora crassa*. Mol Cell Biol 10:5064–5070

Sommer Th, Chambers JAA, Eberle J, Lauter FR, Russo VEA (1989) Fast light-regulated genes of *Neurospora crassa*. Nucl Acid Res 17:5713–5723

Chapter 7 Genetic Regulation of Sporulation in the Fungus *Aspergillus nidulans*

A. JOHN CLUTTERBUCK and WILLIAM E. TIMBERLAKE

7.1 Introduction

The name *Aspergillus nidulans* means "holy-water sprinkler making small nests," the first part referring to the brush-like form of the asexual reproductive apparatus, and the second to the sexual fruiting bodies and their surrounding cells. *A. nidulans* is a filamentous fungus related to many industrially important species such as *Penicillium chrysogenum* and *Aspergillus niger*. The genetics of *A. nidulans* have been extensively explored following the lead of Pontecorvo and his colleagues. It is this organism in which the parasexual (mitotic) genetic cycle was first described, and it has frequently been used as a model for the investigation of parasexual genetics. Even so, *A. nidulans* possesses an active sexual system which is also used extensively in genetic analysis. The life cycle of *A. nidulans*, illustrated in Fig. 1, has three phases: vegetative mycelium, sexual cycle, and asexual cycle. The asexual cycle is the topic of this chapter.

7.2 Conidiation in the Wild Type [2]

7.2.1 General Considerations

Vegetative growth occurs by apical extension and branching of hyphae (Fig. 2). Polarized growth is a characterstic of fungi and is accomplished by selective deposition of cell wall precursors and polymerization enzymes at the apex of the growing cell. Hyphae are divided into compartments by perforated cross walls (septa). *A. nidulans* colonies in petri dishes grow at a constant rate by radial extension, thus forming circular colonies (Fig. 3). Cleistothecia and conidiophores are supported by

the physiologically interconnected substrate mycelium embedded in the growth medium, which acts much like a higher plant root system. Growth of *A. nidulans* in submerged culture is usually aconidial and the normal stimulus to conidiate is exposure of hyphae to an aerial surface.

7.2.2 Induction of Conidiation

In addition to exogenous stimuli, there are endogenous requirements for sporulation. Cultures grown as submerged shaking cultures are not competent to respond to induction of conidiation by exposure to air until a defined amount of growth has occured. Mutants having reduced growth intervals required for competence acquisition may identify genes controlling this process. The *A. nidulans* wild type conidiates freely only after exposure of hyphae to red light and that induction is partially reversible by exposure to far red light, similar to phytochrome-mediated responses in higher plants. This requirement went undetected for many years because most laboratory strains carry the mutation *veA1*, which abolishes light dependence. The velvet mutation is a recessive loss of function mutation, indicating that *veA* is a negative regulator of conidiation. The identity of the primary photoreceptor(s) and the mechanisms through which *veA* controls sporulation are unknown.

7.2.3 Morphological Alterations

Conidiophores begin to develop 5 h after induction of a competent mycelium, at a density of approximately 450 per mm^2, facilitating biochemical analysis. Conidiophores grow at right angles to the sur-

LIFE CYCLE OF <u>ASPERGILLUS NIDULANS</u>

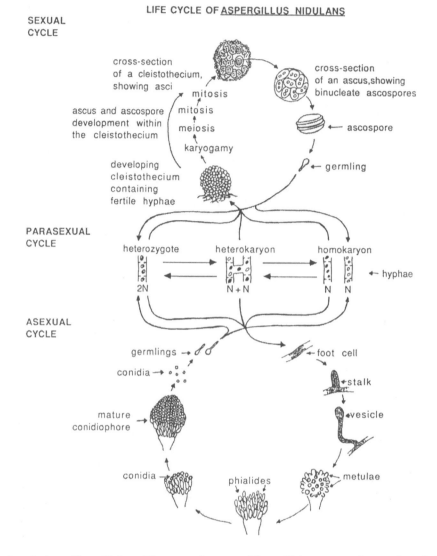

Fig. 1. Life cycle of *Aspergillus nidulans*. The vegetative mycelium consists of an aggregate of multinucleate, tubular cells, called hyphae, that grow by apical extension and sub-apical branching (*central portion*). A mycelium may contain nuclei that are genetically identical, in which case it is referred to a homokaryotic. Mycelia containing genetically differentiated nuclei may form by fusion of two genetically distinct strains; such mycelia are said to be heterokaryotic. Genetically distinct nuclei may fuse to form heterozygous diploids. Spontaneous or induced chromosome loss rapidly leads to reformation of haploids containing a mixture of chromosomes from each parental genome, a process described as the parasexual cycle that is frequently used in assigning genes to linkage groups. Sexual reproduction (*upper portion*) involves formation of asci containing eight unordered ascospores in fuiting bodies, called cleistothecia. Single cleistothecia contain thousands of spores, providing sufficient progeny for genetic analysis. *A. nidulans* is homothallic and essentially all laboratory strains have been derived from a single common ancestor. This means that virtually any laboratory strain can be crossed with any other strain. Asexual sporulation (*lower portion*) involves formation of mitotically derived, uninucleate spores, called conidia, on a specialized reproductive apparatus, called the conidiophore. Conidiation is particularly suitable for genetic and molecular analysis: it is dispensable in the laboratory, so that nonconditional, viable mutants can be obtained, and it involves a very simple, but precise, series of developments. Successive stages of differentiation are accompanied by apical growth and budding; as far as is known cell-cell interactions are not important. This simplicity makes conidiation particularly well suited for unraveling the regulatory network involved in a defined developmental sequence. (Timberlake and Marshall 1988)

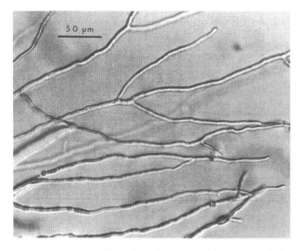

Fig. 2. Vegetative hyphae of *Aspergillus nidulans*. The tubular cells are approximately 6 µm in diameter and grow by apical extension. The cell walls are continuously synthesized near the growing tip and consist of chitin, glucans, and proteins. Septa divide the hyphae up into compartments, but are perforate, permitting controlled movement of nutrients and organelles between compartments. (Timberlake and Marshall 1989)

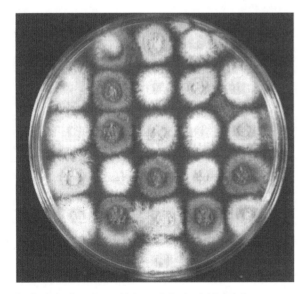

Fig. 3. Colonies of *A. nidulans*. Colonies initiated from spores increase their size at a constant rate radial extension and are usually covered with conidiophores and conidia that give them a velvet texture. Colonies can be made more compact by inclusion of detergents in the growth medium so that hundreds of colonies can be examined in a single Petri dish. (Timberlake and Marshall 1988, see[1])

face of the medium, independent of light and gravity. The conidiophore wall is thicker and more hydrophobic than that of the vegetative cells. The compartment from which a conidiophore arises is therefore distinguishable as a "foot cell". Conidiophores probably lose their ability to take up nutrients and lack the necessary metabolic machinery to utilize nutrients even if they could be acquired from the environment. Thus, the connection of the foot cell to the vegetative mycelium may provide a crucial source of cellular materials. Commitment to development is reversible at early stages, but irreversible by the phialide stage. The major stages in the development of the conidiophore are shown in Fig. 4.

Formation of the specialized conidial wall appears to involve deposition of materials by both the phialide and the spore protoplast itself. The phialide continuously extrudes primary wall layers and is also involved in formation of conidial pigment. Conidia contain a dark green pigment in their walls that protects them from ultraviolet light. Although development of the spore pigment only occurs in older spores, and the enzymes needed for pigment synthesis are present in the spore walls themselves, the genes concerned with pigmentation are transcribed in the phialide and the transcripts are not found in mature spores. Thus, pigmentation enzymes must by synthesized in the phialide and secreted into the developing spore wall along with substrates. This observation shows that the phialide acts to some extent like the mother cell in *Bacillus subtilis* sporulation, providing for the synthesis of outer wall layers.

Very late during spore maturation, just before the conidia become impermeable, an inner wall layer forms under the direction of the conidial protoplast. This wall layer is probably responsible for sealing the spores off from the environment. Its formation requires production of some spore specific mRNAs. Thus, the *Aspergillus* conidium, like the *Bacillus* endospore, depends partially on itself and partially on a terminally differentiated mother or nurse cell for wall formation. Conidia are long-lived, respirationally inactive, and impermeable. Nevertheless, they germinate readily upon addition of only water and glucose and are not markedly resistant to toxic agents or heat treatments.

Fig. 4 A–E. Scanning electron micrographs showing stages of conidiophore development. **A** The conidiophore stalk(s) emerges as a branch from the foot cell and grows by apical extension, very much as do hyphae. **B** At a height of approximately 100 µm polarized growth ceases and the tip of the cell begins to swell to from a terminal "vesicle" (*V*). **C** Up to about 60 buds, called metulae (*M*), form simultaneously in close spacing on the upper surface of the conidiophore vesicle. The conidiophores, like the vegetative mycelium, are coenocytic, and a series of nuclear divisions culminates in synchronous division of nuclei in the neck of each metular bud so that one daughter nucleus enters each metula. The metula is then cut off by a septum at its base. **D** The metulae grow to about 5 µm in length and then bud and undergo nuclear division to give rise to a second tier of cells, called phialides (*P*). Metulae and phialides are referred to collectively as sterigmata. The septa delimiting metulae and phialides are perforate, so there is cytoplasmic continuity between all the cells of the conidiophore and the supporting mycelium. **E** conidia are formed by repeated mitotic divisions of the phialide nucleus. These divisions are oriented along the long axis of this vase-shaped cell so that one daughter nucleus enters the tip of the cell while the other is retained in its base. A spore wall is formed around the distal, G1-arrested nucleus. The proximal nucleus undergoes another division and the previously formed spore is displaced by the newly formed spore. As this process continues, a long chain of clonally derived spores forms. Thus, conidia are budded from the phialide in basipetal succession so that the oldest spore is at the top and the youngest at the bottom. The phialide is a stem cell that retains its own differentiated characteristics while producing many other differentiated cells with a completely different set of characteristics. Conidia undergo maturation as they are displaced from the phialide. While spores are being formed, the metulae bud again so that in the mature conidiophore each metula bears three or four phialides. (Mims et al. 1988)

7.3 Molecular Genetic Facilities[3]

One of the principle reasons for studying development in *A. nidulans* is that it has an exceedingly well-characterized genetic system. Similarly, it has a DNA-mediated transformation system that permits many of the genomic manipulations best known in prokaryotes and yeast. Some of the basic techniques that are utilized in the analysis of development are as follows.

7.3.1 Mutation

Very large numbers ($>10^9$) of haploid, uninucleate conidia can be readily obtained and subjected to mutagenasis. Mutants defective for conidiation are expected to result from loss of function mutations in structural or positive regulatory genes for conidiation and are very readily isolated. Mutants with any serious modifications or deficiency in conidiation stand out as abnormally pigmented or colorless colonies in a background of wild-type green colonies. Other selection strategies for loss of function mutations in negative regulators of development and for dominant mutations are available.

7.3.2 Complementation

Complementation tests are used to determine if new mutations occur in known or previously identified genes. Heterokaryons or heterozygous vegetative diploids are readily constructed in this species for complementation tests.

7.3.3 Strain Construction and Genetic Mapping

Construction of strains having desired genotypes is accomplished through use of both the sexual and parasexual cycles. Whole chromosomes are reassorted by forming diploids and haploidizing them. Markers on the same chromosome are usually recombined by standard sexual crossing. Meiotic recombination rates are fairly high (1 map unit = 5 – 10 kbp) so that even physically closely associated genes can be readily recombined.

There is a well-marked genetic map of the *A. nidulans* chromosomes, and mapping of new mutations is routine. The parasexual cycle is used to locate a mutation to a chromosome and to establish its position on the chromosome arm relative to widely spaced markers. Meiotic analysis is then used to map its precise position. More recently, electrophoretic karyotyping has permitted resolution of all eight chromosomes. Thus, for cloned genes, chromosome assignments are most easily made by Southern blot analysis of resolved chromosomes. In addition, two reference *A. nidulans* cosmid libraries have been divided into chromosomespecific sub-collections.

7.3.4 DNA-Mediated Transformation

A variety of systems have been developed for transformation of *A. nidulans*. Transformation often results in homologous integration of transforming DNA into the chromosome. Integration at a particular locus can also be forced by use of an incomplete selective marker which can only regenerate the complete gene by recombination with the chromosomal copy. Developmental genes can most readily be cloned by complementation of mutants with gene banks in plasmid or cosmid vectors. Integrated vectors can then be recovered from uncut or restricted DNA of transformants. Rapid methods are available for physically mapping complementing activities within cloned DNA fragments, and ho-

mologous integration allows specific gene replacement and disruption strategies. Thus, the *A. nidulans* transformation and genetic systems provide convenient and rapid methods for cloning a gene of interest, proving formally that the correct gene has been cloned, physically mapping the gene, and testing the outcome of gene manipulations. Few multicellular eukaryotes permit such a variety of approaches.

7.4 How Many Genes Are Needed for Growth and Development?[4]

The principle thrust of *Aspergillus* developmental genetics is to understand how the products of genes are utilized to give rise to the complex conidiophore. What are the particular molecular characteristics of conidiophore cells and conidia that impart upon them their specialized functions? How are these characteristic acquired in a genetically pre-determined, orderly series of cell differentiations leading to formation of the conidiophore? Genes important conidiophore development have been divided into four conceptual categories for the sake of discussion and hypothesis building: (1) genes whose products are needed for both vegetative growth and development, the "housekeeping" or "support" genes; (2) "strategic" genes that regulate the conversion from vegetative growth to conidiation; (3) "tactical" genes that regulate the orderly assembly of the conidiophore; and (4) "auxiliary" genes that directly determine the structure and physiological characteristics of the conidiophore or spores. Expression of this class of genes is presumed to be regulated in space and time by the tactical loci. The scope of the problem of understanding the genetic control of conidiophore development can be assessed by estimating the number of genes in each class involved in the process. Molecular and genetic estimates indicate that about 1000 genes are selectively activated during conidiation. However, a significant proportion of these genes are probably not essential for spore production, and some may be needed for wild-type vegetative growth as well as for development.

7.5 Analysis of Development with Stage-Specific Mutants[5]

7.5.1 Mutants of Interest

Several developmental mutants have proved to be deficient in important steps in conidiation and to define pivotal regulatory loci. Mutants with lesions at the *brlA* and *abaA* loci are fully aconidial. Mutants defective at the *stuA* and *medA* loci have a reduced conidial yield on a distorted conidial apparatus. Mutants with lesions at the *wetA* locus have a reduced conidial yield on a fairly normal conidiophore due to the production of autolytic spores.

Other mutants produced conidia with abnormal colors. In addition, a class of mutants designated as "ivory" were identified that lacked a grey-brown, melanin-like pigment in the conidiophore, as opposed to the conidia themselves.

7.5.2 Mutants Found but Ignored

A large proportion of conidiation defective mutants are also subnormal for vegetative growth, implying that many standard metabolic processes are more critical for conidiation than for hyphal growth. A second large class of mutants were put aside because, even though they were specifically affected in conidiation, the stage at which the defect manifested itself was indistinct. A third large, and potentially very interesting, class of mutants are defective at the early stages of development before overt conidiophore development begins. These comprise as much as 83% of conidiation-less mutants. The probable genetic complexity of this stage may reflect the variety of options open to a fungal colony at the time the decision to conidiate is made. The colony can continue with vegetative growth or it can go for sexual reproduction. Interconvertible chemical morphogens produced in *A. nidulans* colonies that stimulate the asexual reproductive phase simultaneously inhibit the sexual reproductive phase and vice versa. This suggests a model in which the colony initially produces one morphogen (in this case called psiA) that stimulates conidiation and inhibits cleistotheciation. During colony development psiA is converted to the alternative morphogen psiC that inhibits conidiation and stimulates cleistotheciation. Clearly mutations that interfered with morphogen metabolism would have major effects on the choice to conidiate versus cleistotheciate.

7.5.3 Mutants Expected but Not Found

The morphological description of the steps of conidiation given above suggests that the biological process of development would proceed by similar steps and deficiency mutants would be found for each. In actuality, this idea proved wrong, implying that morphological steps may not coincide with developmental steps.

Unless a specially designed search was made, one would not expect to pick up mutants for a number of conidial functions. For example, there may be many biochemical aspects of development that are not essential for apparently normal conidial development under laboratory conditions, but that might be important in the wild. In this vein, it has been shown that deletion of a large cluster of genes that are transcribed at high levels specifically in differentiating spores has no detectable effect on conidium morphology or viability in the laboratory. On the other hand, mutations leading to expression of conidiation genes during vegetative growth are likely to be lethal (see Sect. 7.4) and isolation of such mutants would require special selection procedures.

7.6 Characteristics of Important Developmental Mutants[6]

7.6.1 General Considerations

The many developmental mutants isolated fall into a limited number of classes that provide some important clues to the functions of the identified wild type genes. In addition, a number of developmentally important genes have now been cloned and partially characterized, providing further insights into their functions. Descriptions of some of the most important mutant classes follow.

7.6.2 Velvet Mutants

Velvet (veA) mutants conidiate profusely, a frequently desirable property that makes the colonies easier to replicate. Most A. nidulans genetic stocks carry the veA1 mutation, to the extent that the mutation is often not even listed in a strain's genotype. veA+ strains can be induced to conidiate at levels equivalent to veA1 strains by exposure to red light whereas red light has no effect on conidiation in veA1 mutants. veA is probably a repressor of conidiation that is inactivated directly or indirectly by red light. Red light induction is partially reversible by far-red light, suggesting a phytochrome-mediated response.

7.6.3 Fluffy Mutants

Fluffy (flu) mutants form an excessive aerial mycelium, giving them a powder puff appearance, and produce few or no conidiophores. Fluffy mutants are often described as invasive because they can overgrow adjacent, wild-type colonies. This has been interpreted to imply that they are unable to respond to normal signals that inhibit hyphal extension as the growing fronts of independent colonies approach one another. However, overgrowth of colonies by fluffy strains does not occur when the colonies are grown under cellophane, which prevents development of the aerial mycelium. Thus, the ability to invade other colonies may depend only on the proliferation of aerial hyphae which can literally overgrow and fall down upon neighboring colonies.

7.6.4 Aconidial Mutants

For many developmental mutants the stage of development at which the normal gene acts can be inferred from the morphology of the abnormal conidiophore produced. However, conidial induction and the early, committed steps leading to overt conidiation are unmarked by morphological events. Thus, the analysis of these processes requires a somewhat different approach from that used to detect genes involved in the later stages of development. Thermosensitive aconidial mutants and time-shift experiments have been used to determine the critical period of action for 11 of the conditional aco genes.

Some of these mutants accumulate phenolic compounds, some of which are antibacterial. Others appear to be colony morphogens, called psi factors (for precocious sexual induction). Three psi-active compounds (psiA, B, and C) have been isolated and identified as C-18 hydroxy fatty acids. PsiA stimulates conidiation at fairly high concentrations. By contrast, psiC strongly inhibits conidiation and stimulates cleistotheciation at low concentrations. PsiC is 5,8-dihydroxy linoleic acid, whereas psiA is the delta lactone derivative of psiC. PsiB is probably a transient intermediate in the synthesis of psiC from psiA. PsiA and psiC are interconvertible in vitro, and it has been proposed that psiA is produced in the colony at a relatively high concentration, which stimulates conidiation, and is slowly converted to psiC, which inhibits conidiation and stimulates cleistotheciation. This could explain why cleistothecia usually form behind the area of a colony where conidiophores are being actively produced.

7.6.5 Bristle Mutants

All characterized mutants with bristle phenotypes map at the brlA locus. Null mutants develop undifferentiated bristles in place of conidiophores (Fig. 5, 6). These grow up to ten times the normal conidiophore height and may eventually branch, much like vegetative hyphae. However, bristles are distinguishable as conidiophores, rather than ordinary aerial hyphae, by their thickened walls.

Leaky brlA mutants can be arranged in a morphological series (Fig. 5) that gives a more complex and interesting picture, suggesting that this locus controls a variety of functions. Slightly leaky mutants develop forked bristles with two or more branches arising at one point. Leakier mutants form a rudimentary vesicle and multiple branches, each of which may rebranch after a short distance. Additional brlA+ activity leads to shortening of the branch intervals to form multiple tiers of metula-like cells. The bristles of brlA null mutants are unpigmented, but leakiness is accompanied by

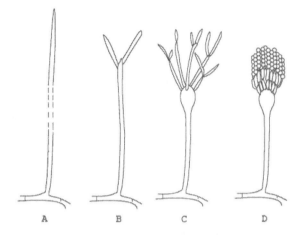

Fig. 5 A – D. Structures of bristle conidiophores. Alleles of *brlA* are highly variable, ranging from null (**A**), in which case mutants produce no structures beyond conidiophore stalks, to partially leaky (**B**), to very leaky (**C**), in which case mutants produce conidiophores with layers of sterigmata, similar to the wild type (**D**). (After Clutterbuck 1990)

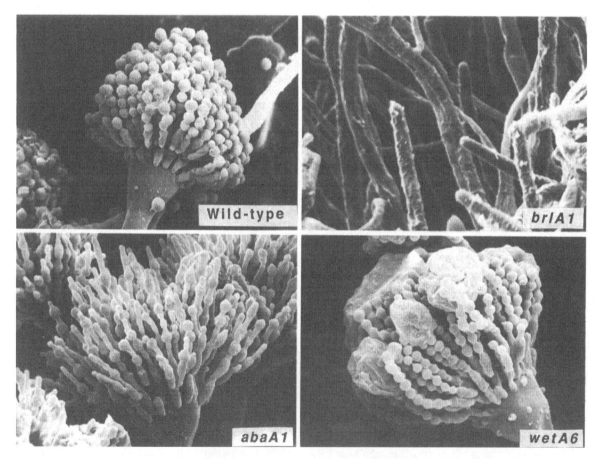

Fig. 6. Scanning electron micrographs of wild type and mutant conidiophores. *brlA* null mutants produce only conidiophore stalks. *abaA* mutants produce initially normal conidiophores, but metulae fail to produce functional phialides and instead form chains of abnormal cells having metula-lika ultrastructure.

wetA mutants form normal conidiophores and initially normal conidia that autolyse instead of undergoing the final stages of maturation. Lysed spores coalesce into the structures seen at the ends of conidial chains. (Boylan et al. 1987)

conidiophore pigmentation and associated activation of the *ivoA* and *ivoB* loci (see below). Given that *brlA* appears to encode a positively acting transcription factor (see below), the activation of some loci, but not others, in leaky bristle mutants may imply differential affinity of the BrlA protein for developmentally regulated promoters. Alternatively, there might be different, but related, *brlA*-encoded proteins responsible for activation of different subsets of developmentally regulated genes.

7.6.6 Abacus Mutants

Abacus mutants have nearly normal conidiophores in which the phialides and conidia are replaced by beaded, abacus-like structures (Fig. 6). *abaA* mutants bear some resemblance to very leaky *brlA* mutants and to medusa mutants (see Sect. 7.6.8) in that they produce supernumerary, albeit morphologically aberrant, layers of sterigmata. As with bristle mutants, abacus mutants fail to accumulate transcripts for a number of late developmental genes.

7.6.7 Stunted Mutants

Stunted (*stuA*) mutants produce diminutive conidiophores with unthickened walls, but these bear conidia that are apparently normal in pigmentation and storage properties. Metulae and phialides are not distinctive in *stuA* mutants and may even be absent, in which case the conidia must be budded straight from the conidiophore vesicle. *stuA* mutations modify the abnormal structures formed by other mutants.

7.6.8 Medusa Mutants

Medusa (*medA*) mutants have phenotypes reminiscent of the leakiest *brlA* mutants, characterized by a normal conidiophore vesicle bearing multilayered metulae and all characterized medusa mutants map at one locus, *medA*. Unlike the leakiest *brlA* mutants, medusa conidiophores are only lightly pigmented which correlates with a deficiency in the

pigment precursor that is the product of *ivoA* locus activity (Sect. 7.6.9). However, activity of the related *ivoB* locus is normal or possibly elevated in medusa mutants.

7.6.9 Ivory (Conidiophore Pigment) Mutants

Complementation tests divide *ivo* mutants into two groups identifying the *ivoA* and *ivoB* loci. *ivoA* mutants lack the conidiophore pigment precursor 6-hydroxy-N-acetyltryptophan, whereas *ivoB* mutants lack a phenol oxidase that is specific for this substrate. *ivoA* and *B* are under the direct or indirect control of *brlA* and *medA*.

7.6.10 Rodletless Mutants

Conidia and probably conidiophores contain an outer hydrophopic protein layer, the rodlet layer, that may help prevent water loss from these cells which are exposed directly to air. Mutants lacking this layer have been induced by site-directed inactivation of a gene that is turned on during conidiation. This gene, designated *rodA*, has been mapped and shown not to correspond to any previously identified gene.

7.6.11 Benomyl-Resistant Mutants

Hyphal growth and conidiation are sensitive to the potent fungicide benomyl, which acts by disrupting microtubules. Hyphal resistance is conferred by *benA*^{R} mutations that modify β-tubulins. conidiation in these mutants, however, remains sensitive to the antibiotic, suggesting the involvement of conidiation-specific β-tubulins. A further mutation in a different locus, *tubC*, confers resistance during conidiation and does so by eliminating a developmentally regulated species of β-tubulin (β3). Loss of β3 does not disrupt the activities of the fungus under any conditions yet discovered. Even though the product of *tubC* appears only to make conidiation sensitive to benomyl, β3 presumably has a function in conidiation that can be replaced by *benA*^{R}-encoded β-tubulins. An involvement of the cytoskeleton in morphogenesis is implied by these results.

7.6.12 Wet-White Mutants

Wet-white (*wetA*) mutants produce unpigmented conidia that autolyse soon after formation due to a failure to form an inner wall layer, designated C4. Examination of spore-specific mRNAs from *wetA* mutants indicates that this locus is regulatory in that mutants are deficient for a number of late-stage conidiation functions.

7.6.13 Spore Color Mutants

White-spored (*wA⁻*) and yellow-spored (*yA⁻*) mutants were some of the earliest used in *A. nidulans* genetics. Many other genes have been identified since. These include *yB* and *ygA* (yellow-green); *chaA* (chartreuse); *bwA* (brown); *fwA* (fawn); *pA* (pale color); *dilA, B* (dilute color); *drkA, B* (dark color); and *wB*.

So far *yA* is the only color gene whose product has been characterized. *yA* encodes a *p*-diphenol oxidase (laccase) that is deposited by the phialide into the conidium wall where it converts a yellow pigment intermediate into the mature green form. The *yA* gene has been cloned, sequenced, and shown to be developmentally regulated. The *wA* gene has also been cloned and shown to be developmentally regulated.

7.7 Molecular Analysis of Developmental Regulation[7]

7.7.1 General Considerations

The *A. nidulans* DNA-mediated transformation system makes it possible to clone any gene that has been identified by mutation, and many of the genes discussed in the previous section have been cloned and partially characterized. Studies with cloned genes can be divided into two general categories. In one, genes presumed to be regulators of develop-

ment are studied with the aim of understanding how they control the entry into the conidiation pathway or the steps of conidiophore development. In the other, genes whose products contribute directly to conidiophore form and function are studied with the aim of understanding how their expression is regulated and how their products work at the biochemical and physiological levels.

7.7.2 Isolation and Characterization of Regulatory Genes

Three genes that are good candidates for pivotal regulators of development once it is manifest, *brlA, abaA,* and *wetA*, have been investigated in some depth. All three genes were cloned by complementation of appropriate mutants. The putative genes were subsequently molecularly mapped, inactivated by site-directed mutagenesis, and shown to correspond to those originally identified by mutation. These experiments in addition confirmed the nature of null mutations. Transcript and reporter gene fusion analyses showed that all three genes are inactive in vegetative cells. *brlA* is activated a few hours after inducing conidiation and *abaA* is activated

Table 1. Time and sites of expression of *Aspergillus* developmental control genes

Gene or class	*brlA*	*abaA*	*wetA*	Early genes	Phialide genes	Spore genes
Time of expression:	10 h	15 h	25 h	10 h	15 h	25 h
Sites						
Hyphae	–	–	–	–	–	–
Vesicles	+	–	–			
Metulae	+	+	–	} +	} +	} +
Phialides	+	+	+			
Conidia	–	–	+	–	–	+

The site and time of expression of the *brlA, abaA,* and *wetA* were determined by using fusions of each gene to the *E. coli lacZ* gene and detecting the resulting activity using X-gal. The time of expression is recorded in hours after induction by exposure to air. Transcripts of the early, phialide-specific and spore-specific genes were determined in Northern blots of mRNAs from vegetative hyphae, surface cultures (containing conidiophores at all stages of development, vesicles, metulae, phialides) and from isolated conidia shortly thereafter (Table 1). Both transcripts accumulate in conidiating cultures but not in mature conidia. By contrast, the *wetA* transcript, which begins to accumulate last, is present in mature spores. Consistent with results from morphological epistasis tests, *abaA* and *wetA* transcripts fail to accumulate in *brlA1* (null) strains and *wetA* transcript fails to accumulate in *abaA1* (null) strains. *wetA* transcript also fails to accumulate in *wetA6*[ts] strains, raising the possibility that formation of viable *wetA* product is required for *wetA* transcription, that is, *wetA* is positively autoregulatory. These results are summarized under "normal development" in Table 2.

The DNA sequences of these three genes have been determined. The predicted BrlA polypeptide contains near its carboxyl terminus repeated sequences closely resembling the cys_2-his_2 Zn(II) coordination sites first recognized in *Xenopus laevis* transcription factor IIIA (TFIIIA), suggesting that it is a nucleic acid binding protein. Disruption of either Zn finger by site-direct mutations that convert finger cysteines to serines results in complete loss of *brlA* activity. The sequences of *abaA* and *wetA* have provided few clues to their functions.

All three of the mRNAs from these putative regulatory genes have long untranslated regions in comparison to many other *A. nidulans* nonregulatory gene transcripts. It is possible that the leaders and trailers have important functions, for example in stabilizing the transcripts during development. This notion is made more attractive by the observation that massive RNA degradation occurs during sporulation. An interesting possibility is that RNAs not needed for development are rapidly degraded, while those that are needed carry specific molecular tags that protect them from this generalized turnover.

7.7.3 Isolation and Characterization of Responder Genes

The most interesting developmental genes identified by mutation have turned out to be regulatory. There is a deficiency of mutations in genes believed to respond to these regulators, and in fact the only genes of this type cloned by complementation of mutants are those defective for spore or conidiophore pigmentation. Failure to find mutants in morphogenetic responder genes leaves us with two problems. One is a lack of information on the biochemistry of morphogenesis, and the other is the need for an explanation for the absence of the expected mutants. There are two types of explanations for failure to find mutants: one is that the predicted mutants have no detectable phenotype, and the other is that the phenotypes are lethal. Undetectable phenotypes would be likely if the relevant genes were redundant, i.e., present in multiple copies such that loss of function of any one copy would not be noticed. A related explanation is suggested by *medA* and *sthA* and *B* mutants, each of which has only a partially deficient phenotype on its own, but together mimic *brlA* mutants. These suggest that rather than multiple redundant genes determining morphogenesis, *brlA* may control a polygenic series of genes with distinct but partially overlapping functions.

Developmental mutants would only be lethal if the activities affected were essential for vegetative

Table 2. Genes expressed in normal and artificially induced *Aspergillus* developmental cultures

Genotype	mRNAs expressed					
	brlA	*abaA*	*wetA*	Early genes	Phialide genes	Spore genes
Normal development						
Wild type	+	+	+	+	+	+
brlA⁻	+	−	−	−	−	−
abaA⁻	+	+	−	+	−	−
wetA⁻	+	+	−	+	−	−
Artificially induced development						
brlA'	+	+	+	+	+	+
brlA'abaA⁻	+	+	−	+	−	−
brlA'wetA⁻	+	+	−	+	−	−
abaA'	+	+	+	+	+	+
abaA'brlA⁻	+	+	+	+	−	+
abaA'wetA⁻	+	+	−	+	−	−
wetA'	−	−	+	−	−	+

brlA⁻, *abaA*⁻, and *wetA*⁻ mutants all give null phenotypes. *brlA'*, *abaA'*, and *wetA'* are constructs in which genes are fused to the inducible *alcA* promoter. Note that while null mutants of *brlA* and *abaA* score positive for their own (inactive) mRNAs, no *wetA* mRNA was detected from the *wetA*⁻ mutant (see Sect. 7.7.2)

growth as well as conidiation. Such activities might be either increased or decreased to produce the distinctive conidiophore morphology. In the extreme case, development would depend on switching off hyphal growth functions rather than switching on specific conidiophore growth determinants.

The pigmentation gene clones, although perhaps unhelpful on the subject of morphogenesis, provide markers for successive developmental stages and have been used to investigate interactions of the regulatory genes which control them. Other clones valuable for this purpose have been obtained by isolating cDNA clones corresponding to mRNAs which are present preferentially during conidiation. These genes, although quite various in their time of expression and pattern of regulation, are summarized in Tables 1 and 2 and Fig. 8 as "early", "phialide-specific", or "spore-specific" genes. For only one of these cDNA clones (*rodA*) has a function so far been determined. The mutant phenotype determined by integration of a disrupted copy of this gene is subtle, and failure to pick up *rodA* mutants in standard mutant screening programs is not surprising.

In an earlier screening, this time of phage clones containing more than one gene per clone, it was found that of approximately 200 mRNAs expressed selectively in conidia, 81% occur in clusters with an average of 3.3 genes per cluster (excluding solitary genes); this implies that there are about 50 such clusters. One large cluster (SpoC1) has been analyzed in some detail by transferring a component gene (SpoC1C) elsewhere and by insertions of the *argB* gene into the cluster. It was concluded that transcription depends both on elements closely associated with the individual genes and on components of the cluster acting at a distance. Most genes in the SpoC1 cluster are in the spore-specific category and are dependent on *wetA* for their transcription. The function of the cluster is not known, but it seems rational that in transcriptionally inactive spores, genes required only during spore maturation or germination should be segregated into chromosomal domains which are accessible to polymerase at this stage of development and perhaps inaccessible during vegetative growth.

The *yA* and *wA* genes both fall into the phialide-specific category of responder genes, whereas the

ivoA and *B* genes are expressed early and are good candidates for direct targets of regulation by *brlA*. In the wild type, both *ivo* genes are activated at about the time of vesicle formation. In *brlA* null mutants, both activities are absent. On the other hand, leaky *brlA* mutants can be arranged in a series with increasing morphological development, and increased *brlA* activity is correlated with increased *ivoA* and *B* activities. We propose that *brlA* directly regulates both of the *ivo* genes as well as unknown loci responsible for morphogenesis.

7.7.4 Controlled Induction of Regulatory Genes

An experimental approach that has been particularly useful for understanding the modes of action of the *brlA*, *abaA*, and *wetA* regulatory genes involves fusion of the structural components of the genes to a promoter [*alcA(p)*] that allows their controlled activation in hyphae under conditions that normally suppress conidiation ("Artificially Induced Expression" in Table 2). The fusion genes were introduced into the genome by transformation. Induction of *brlA* caused the hyphal tips to differentiate into phialide-like structures that gave rise to conidia (Fig. 7). Thus, *brlA* expression is sufficient to direct the essential developmental pathway leading to spore differentiation. Further, *brlA* activation turned on the *abaA* and *wetA* regulatory genes, the *yA* and *ivoB* structural genes, and many other developmentally regulated genes of the "early" class (Table 2) whose functions are unknown. These results indicate that the product of *brlA* is a positively acting molecule that controls the central pathway leading to conidium formation. Thus, *brlA* is both necessary and sufficient to direct conidiation even in vegetative cultures.

Forced activation of *abaA* in hyphae growing in submerged culture inhibited hyphal elongation and caused formation of abnormal vacuoles and thickened septa, but did not lead to spore differentiation. *abaA* activation also led to activation of *wetA* and, surprisingly, *brlA*, showing that *brlA* and *abaA* are mutually inductive, that is, *abaA* is a positive feedback regulator of *brlA* (Fig. 8). The positive feedback loop formed by *brlA* and *abaA* could serve to amplify the primary signals that normally

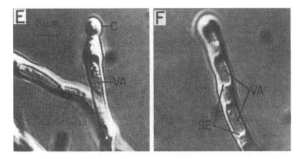

Fig. 7. *brlA*- and *abaA*-induced cellular differentiation. The *brlA* or *abaA* coding region was fused to the vegetatively inducible *alcA* promoter and introduced into an appropriate *A. nidulans* strain by transformation. Light micrographs of induced cells (**E** *brlA*; **F** *abaA*) cells are shown. *V* vacuoles; *C* conidia; *SE* septa. (Mirabito et al. 1989)

turn on *brlA* and could even make expression of the conidiation pathway independent of the signals that initially activated it. *wetA* induction required both *abaA*⁺ and *wetA*⁺, but not *brlA*⁺ activities. The requirement for *wetA*⁺ for *wetA* mRNA accumulation raises the possibility that *wetA* is positively autoregulatory.

Induction of *brlA* in vegetative cells leads to formation of functional spore-producing cells, and spore production is *abaA*-dependent. On the other hand, forced expression of *abaA* fails to induce spore formation, even though it does lead to accu-

mulation of *brlA, abaA,* and *wetA* transcripts to the same levels as those caused by induction of *brlA*. These results indicate that the correct order of *brlA* and *abaA* activation is essential for the formation of sporogenous cells, that is, *brlA* expression conditions the cells so that they can respond appropriately to *abaA*.

wetA induction in hyphae inhibits vegetative growth and leads to some morphological changes, including increased branching and hyphal contortions. *wetA* activation does not induce either *brlA* or *abaA*, but does cause accumulation of spore-specific transcripts. Phialide-specific transcripts require the activities of all three regulatory loci.

These results have led to the formulation of the regulatory model shown in Fig. 8. In this model, *aco* genes are responsible for conversion of a metabolite "X", which inhibits conidiation, to "Y", which is not inhibitory. Thus, loss of function mutations in these genes would lead to accumulation of X and inhibit conidiation. The observation that *veA1* is a suppressor of *aco* pre-induction mutations indicates that X exerts its effect through the product of *veA*. *veA*⁺ is required for light control of conidiation; *veA1* mutants conidiate in complete darkness, whereas *veA*⁺ strains do not. Thus *veA* acts as a red-light-inactivatable repressor of conidiation. It is possible that fluffy loci have pre-induc-

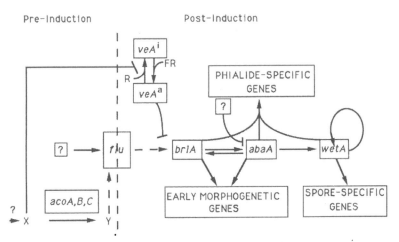

Fig. 8. A model for genetic interactions controlling conidiophore development. *Solid arrows* and *bars* indicate positive and negative interactions, respectively, inferred from genetic analyses. *Dashed arrows* are hypothetical. The *veA* product is proposed to exist in active (*a*) and inactive (*i*) forms that are inter-

convertible by exposure to red (*R*) or far red (*FR*) light. There is no evidence that VeA is itself a chromophore, so other gene products may be involved in light reception. *flu* genes are placed over the line separating pre- and post-induction controls because the developmental times at which they act are unknown

tion and post-induction functions and are involved in *brlA* activation. However, given the lack of data concerning the regulation and biochemical activities of fluffy loci, their position in the model is highly tentative.

abaA and genes whose products are needed early during development are initially activated by *brlA* following environmental induction of conidiation; *brlA* can only be activated when pre-induction functions have been completed, corresponding to competence acquisition. *abaA* induction may in part involve displacement of an unidentified repressor. *abaA* further stimulates expression of these early genes, reinforces expression of *brlA*, and activates *wetA*. *wetA* activates spore-specific genes and, in conjunction with *brlA* and *abaA*, phialide-specific genes. In addition, *wetA* stimulates its own expression, perhaps so that it can continue to be transcribed in differentiating conidia that have separated from the phialide, the closest site of *abaA* expression. This model is certainly incomplete and over-simplified, but does provide a framework for further studies.

7.8 Comparisons with Other Systems

Many of the processes revealed by careful examination of conidiation in *A. nidulans* are reminiscent of related processes in prokaryotes, some of which are discussed in other chapters in this Volume. For example, the phialide can be compared to the *B. subtilis* spore mother cell (see Errington, Chap. 2, this Vol.) and, like that cell, may become terminally differentiated through essentially irreversible genetic processes. In many sporulating prokaryotes, a good example being *Myxococcus xanthus*, a substantial proportion of the population of cells contribute to the formation of spores by a minority, much as the *A. nidulans* mycelium appears to support sporulation by a fraction of the cells. Programmed cellular asymmetries are a property of some prokaryotes, notably *Caulobacter* (see Bryan, Chap. 3, this Vol.), and similar asymmetries are exhibited during *A. nidulans* development in formation of the phialides from metulae and in production of spores from phialides. Some similarities are quite striking. For example, misscheduled expres-

sion of the *whiG* gene of *Streptomyces coelicolor* (see Chater, Chap. 4, this Vol.) leads to abnormal sporulation much as *brlA* expression does in *A. nidulans*. In each case, the relevant gene codes for a factor directly involved in controlling transcription of genes in the sporulation pathway. How mechanistically related the other similarities are remains to be seen.

The *brlA*, *abaA*, and *wetA* mutants were originally chosen as more interesting than a variety of other mutants because their morphologies appeared to reflect a diversion of development into an abnormal form, without any effect on growth. These mutants can be compared with homeotic mutants in which an organ expresses a morphology appropriate to another part of the body. In the case of *A. nidulans*, however, we have so far identified only heterochronic mutants that continue to express what can be interpreted as a previous stage of development: *brlA* mutants show hyphal characteristics, *abaA* mutants repeat metula-like structures, and *wetA* mutants may activate hyphal metabolic systems instead of instituting dormancy. In this respect the fungal mutants parallel those in the flowering plant *Antirrhinum* (see Schwarz-Sommer et al., Chap. 17, this Vol.). The classes of mutants picked up may, of course, reflect a limitation of the system that can be expected to be remedied as more is learned. For example, we are at present ignorant of sexual developmental stages to the extent that mutants exhibiting homeotic transformations between sexual and asexual sporulation might well be missed.

In *Drosophila melanogaster*, the prototype for homeotic studies, the segments that can exhibit homeotic effects are developed simultaneously, both at the embryonic stage and at the time of imaginal disk development in the pupa. However, it is postulated that dipteran flies are evolved from ancestors in which segments developed sequentially (Akam 1987). Thus, even in *D. melanogaster*, the bithorax and antennapedia complexes are seen as specifying modifications from a ground state plan which is retained only in the thoracic segments. Homeotic mutants are divided into loss-of-function recessive mutants that confer more thoracic characteristics, and dominant, hypermorphic mutants conferring inappropriate distal morphologies.

In *A. nidulans*, the morphogenetic mutants initially isolated were all recessive, and most likely due to loss of gene function, whereas hypermorphic mutants have been constructed by fusion of wild-type coding regions of the morphogenetic loci to a vegetatively inducible promoter. These constructed mutant genes are lethal or severely deleterious when expressed in hyphae and so would not have been picked up in normal mutant searches. They can be described as dominant as the constructs are expressed in the presence of a normal allele of the gene which has been manipulated. It is interesting to note that in the hypermorphic *brlA* mutant the specialized vegetative tips are clearly transformed into abnormal phialide-like cells that produce spores, reminiscent of a homeotic transformation in an animal.

7.9 Outlook

One of the striking features of the conidiophore development mutants of *A. nidulans* is that only one locus has been detected for each phenotype, even though more than 30 mutants at each of the *brlA* and *abaA* loci have been isolated. Given the complexity of the developmental process, it might have been expected that many more critical loci would have been found. In addition, it might be predicted that mutants with similar phenotypes due to lesions at both regulatory and structural genes would occur. There may be several explanations for the absence of such mutants. A large number of conidiation-defective mutants have been ignored because they are also affected in vegetative functions. Systematic searches for mutations in genes that repress development have not yet been made. The availability of cloned regulatory and structural genes and the sophisticated molecular genetic system of *A. nidulans* make it possible to make mutations in preselected genes to see the effects on development and to set up novel genetic selections and screens for new classes of mutants. Both types of approaches can be expected to expand our understanding of the control of development in this organism.

It is likely that although conidiation appears to be a simple developmental sequence, fairly complex feedback regulatory systems act to control it. In addition to the *brlA*-*abaA* feedback loop and *wetA* autoregulatory loop described above, there is some evidence that *brlA* is autoregulatory in that there is a clear correlation between *brlA* mutant leakiness and *brlA* mRNA levels. An understanding of these and other feedback regulatory systems will be key to understanding development. In addition, the biochemical functions of the central regulators of conidiation have yet to be determined. Knowledge of these functions will certainly open up whole new areas of investigation.

The problem of how biochemistry relates to morphogenesis remains difficult and requires some imaginative leaps even to decide on the questions to ask. It is a much repeated assumption that the hyphal wall is the main determinant of fungal morphology, but analysis of wall composition and biosynthesis still begs the question of spatial distribution of shape determinants. We can guess that the answers will involve interactions of the cytoskeleton, enzyme compartmentation, secretion, and site-specified activation, which have already been invoked to explain specific cases, but which in no case can yet provide a convincing picture of a morphogenetic process. Conidiophore development in *A. nidulans* perhaps provides a uniquely suitable system for addressing the most basic questions about the mechanisms controlling fungal morphogenesis.

7.10 Summary

Conidiation in *A. nidulans* is amenable to analysis by traditional and molecular genetics. Numerous genes that are specifically activated during development of the multicellular conidiophore have been identified by mutation or molecular hybridization. Most of these genes encode products that are involved in morphogenesis or the specialized physiological functions of conidiophores or spores. A few are regulatory genes whose combined activities control expression of target genes in space and time. Future studies are expected to reveal the molecular mechanisms through which the regulatory genes exact their activities and the precise developmental functions of their target genes.

References

[1] Clutterbuck AJ (1974) *Aspergillus nidulans*. In: King RC (ed) Handbook of genetics. Plenum, New York, pp 447–510

Pontecorvo G, Roper JA, Hemmons LM, MacDonald KD, Bufton AWJ (1953) The genetics of *Aspergillus nidulans*. Adv Genet 5:141–238

Timberlake WE (1991) Molecular genetics of *Aspergillus* development. Annu Rev Genet 24:5–36

Timberlake WE, Marshall MA (1988) Genetic regulation of development in *Aspergillus nidulans*. Trends Genet 4:162–169

[2] Adams TH, Timberlake WE (1990) Developmental repression of growth and gene expression in *Aspergillus*. Proc Natl Acad Sci USA 87:5405–5409

Axelrod DE (1972) Kinetics of differentiation of conidiophores and conidia by colonies of *Aspergillus nidulans*. J Gen Microbiol 73:181–184

Axelrod DE, Gealt M, Pastushok M (1973) Gene control of developmental competence in *Aspergillus nidulans*. Dev Biol 34:9–15

Clutterbuck AJ (1972) Absence of laccase from yellow-spored mutants of *Aspergillus nidulans*. J Gen Microbiol 70: 423–435

Clutterbuck AJ (1990) *Aspergillus nidulans*. In: O'Brien SJ (ed) Genetic maps; maps of complex genomes, vol 3. Lower eukaryotes. Cold Spring Harbor Laboratory, New York, pp 3.97–3.108

Gooday GW (1983) The hyphal tip. In: Smith JE (ed) Fungal differentiation. A contemporary synthesis. Marcel Dekker, New York, pp 315–356

Mayorga ME, Timberlake WE (1990) Isolation and molecular characterization of the *Aspergillus nidulans* wA gene. Genetics 126:73–79

Mims CW, Richardson EA, Timberlake WE (1988) Ultrastructural analysis of conidiophore development in the fungus *Aspergillus nidulans* using freeze-substitution. Protoplasma 44:132–141

Mooney JL, Yager LN (1990) Light is required for conidiation in *Aspergillus nidulans*. Genes Dev 4:1473–1482

O'Hara EB, Timberlake WE (1989) Molecular characterization of the *Aspergillus nidulans* yA locus. Genetics 121:249–254

Pontecorvo G, Roper JA, Hemmons LM, MacDonald KD, Bufton AWJ (1953) The genetics of *Aspergillus nidulans*. Adv Genet 5:141–238

Sewall T, Mims CW, Timberlake WE (1990) Conidial differentiation in wild-type and wetA-strains of *Aspergillus nidulans*. Dev Biol 138:499–508

Timberlake WE, Marshall MA (1989) Genetic engineering of filamentous fungi. Science 244:1313–1317

[3] Brody H, Carbon J (1989) Electrophoretic karyotype of *Aspergillus nidulans*. Proc Natl Acad Sci USA 86: 6260–6263

Brody H, Griffith J, Cuticchia AJ, Arnold J, Timberlake WE (1991) Chromosome-specific recombinant DNA libraries from the fungus *Aspergillus nidulans*. Nucl Acids Res 19:3105–3109

Clutterbuck AJ (1974) *Aspergillus nidulans*. In: King RC (ed) Handbook of genetics. Plenum, New York, pp 447–510

Rambosek J, Leach J (1987) Recombinant DNA in filamentous fungi: progress and prospects. CRC Crit Rev Biotechnol 6:357–393

Timberlake WE (1991) Cloning and analysis of fungal genes. In: Bennet JW, Lasure L (eds) More gene manipulations in fungi. Academic Press, Orlando, pp 51–85

Timberlake WE, Marshall MA (1988) Genetic regulation of development in *Aspergillus nidulans*. Trends Genet 4:162–169

[4] Clutterbuck AJ (1977) The genetic of conidiation in *Aspergillus nidulans*. In: Pateman JA, Smith JE (eds) Genetics and physiology of *Aspergillus*. Academic Press, New York, pp 305–317

Martinelli SD, Clutterbuck AJ (1971) A quantitative survey of conidiation mutants in *Aspergillus nidulans*. J Gen Microbiol 69:261–268

Timberlake WE (1980) Developmental gene regulation in *Aspergillus nidulans*. Dev Biol 78:497–510

[5] Champe SP, Simon LD (1990) Cellular differentiation and tissue formation in the fungus *Aspergillus nidulans*. In: Rossomando E, Alexander S (eds) Morphogenesis: an analysis of the development of biological structures. Marcel Dekker, New York (in press)

Clutterbuck AJ (1969) A mutational analysis of conidial development in *Aspergillus nidulans*. Genetics 63:317–327

Martinelli SD, Clutterbuck AJ (1971) A quantitative survey of conidiation mutants in *Aspergillus nidulans*. J Gen Microbiol 69:261–268

[6] Aramayo R, Timberlake WE (1990) Sequence and molecular structure of the *Aspergillus nidulans* yA (laccase I) gene. Nucl Acids Res 18:3415

Birse CE, Clutterbuck AJ (1990) N-acetyl-6-hydroxytryptophan oxidase, a developmentally controlled phenol oxidase from *Aspergillus nidulans*. J Gen Microbiol 136: 1725–1730

Boylan MT, Mirabito PM, Willett CE, Zimmermann CR, Timberlake WE (1987) Isolation and physical characterization of three essential conidiation genes from *Aspergillus nidulans*. Mol Cell Biol 7:3113–3118

Butnick NZ, Yager LN, Hermann TE, Kurtz MB, Champe SP (1984) Mutants of *Aspergillus nidulans* blocked at an early stage of sporulation secrete an unusual metabolite. J Bacteriol 160:533–540

Butnick NZ, Yager LN, Kurtz MB, Champe SP (1984) Genetic analysis of mutants of *Aspergillus nidulans* blocked at an early stage of sporulation. J Bacteriol 160:541–545

Champe SP, El-Zayat AAE (1989) Isolation of a sexual sporulation hormone from *Aspergillus nidulans*. J Bacteriol 171:3982–3988

Champe SP, Simon LD (1990) Cellular differentiation and tissue formation in the fungus *Aspergillus nidulans*. In: Rossomando E, Alexander S (eds) Morphogenesis: an analysis of the development of biological structures. Marcel Dekker, New York (in press)

Clutterbuck AJ (1969) A mutational analysis of conidial development in *Aspergillus nidulans.* Genetics 63:317–327

Clutterbuck AJ (1972) Absence of laccase from yellow-spored mutants of *Aspergillus nidulans.* J Gen Microbiol 70:423–435

Clutterbuck AJ (1990) The genetics of conidiophore pigmentation in *Aspergillus nidulans.* J Gen Microbiol 136:1731–1738

Dorn GL (1970) Genetic and morphological properties of undifferentiated and invasive variants of *Aspergillus nidulans.* Genetics 66:267–279

Law DJ, Timberlake WE (1980) Developmental regulation of laccase levels in *Aspergillus nidulans.* J Bacteriol 144:509–517

Martinelli SD (1979) Phenotypes of double conidiation mutants of *Aspergillus nidulans.* J Gen Microbiol 114:277–287

May GS, Morris NR (1988) Developmental regulation of a conidiation specific β-tubulin in *Aspergillus nidulans.* Dev Biol 128:406–414

Mayorga ME, Timberlake WE (1990) Isolation and molecular characterization of the *Aspergillus nidulans* wA gene. Genetics 126:73–79

Mazur P, Meyers HV, Nakanishi K, El-Zayat AAE, Champe SP (1990) Structural elucidation of sporogenic fatty acid metabolites from *Aspergillus nidulans.* Tetrahedron Lett 31:3837–3840

McCorkindale NJ, Hayes D, Johnston GA, Clutterbuck AJ (1983) N-acetyl-6-hydroxytryptophan a natural substrate of a monophenol oxidase from *Aspergillus nidulans.* Phytochemistry 22:1026–1028

Miller KY, Toennis TM, Adams TH, Miller BL (1991) Isolation and transcriptional characterization of a morphological modifier: the *Aspergillus nidulans* stunted (*stuA*) locus. Mol Gen Genet 227:285–292

Mooney JL, Yager LN (1990) Light is required for conidiation in *Aspergillus nidulans.* Genes Dev 4:1473–1482

O'Hara EB, Timberlake WE (1989) Molecular characterization of the *Aspergillus nidulans* yA locus. Genetics 121:249–254

Roper JA (1958) Nucleo-cytoplasmic interactions in *Aspergillus nidulans.* Cold Spring Harbor Symp Quant Biol 23:141–154

Sewall T, Mims CW, Timberlake WE (1990) Conidial differentiation in wild type and *wetA⁻* strains of *Aspergillus nidulans.* Dev Biol 138:499–508

Sewall TC, Mims CW, Timberlake WE (1990) abaA controls phialide differentiation in *Aspergillus nidulans.* Plant Cell 2:731–739

Stringer MA, Dean RA, Sewall T, Timberlake WE (1991) *Rodletless,* a new *Aspergillus* developmental mutant induced by directed gene inactivation. Genes Dev 5:1161–1171

Tamame M, Antequerra F, Villanueva JR, Santos T (1983) High-frequency conversion to a "fluffy" developmental phenotype in *Aspergillus* spp. by 5-azacytidine treatment: evidence for involvement of a single gene. Mol Cell Biol 3:2287–2297

Weatherbee JA, May GS, Gambino J, Morris NR (1985) Involvement of a particular species of β tubulin β-3 in conidial development in *Aspergillus nidulans.* J Cell Biol 101:706–711

Yager LN, Kurtz MB, Champe SP (1982) Temperature-shift analysis of conidial development in *Aspergillus nidulans.* Dev Biol 93:92–103

[7] Adams TH, Timberlake WE (1990) Upstream elements repress premature expression of an *Aspergillus* developmental regulatory gene. Mol Cell Biol 10:4912–4919

Adams TH, Timberlake WE (1990) Developmental repression of growth and gene expression in *Aspergillus.* Proc Natl Acad Sci USA 87:5405–5409

Adams TH, Boylan MT, Timberlake WE (1988) *brlA* is necessary and sufficient to direct conidiophore development in *Aspergillus nidulans.* Cell 54:353–362

Aguirre J, Adams TH, Timberlake WE (1990) Spatial control of developmental regulatory genes in *Aspergillus nidulans.* Exp Mycol 14:290–293

Birse CE, Clutterbuck AJ (1990) N-acetyl-6-hydroxytryptophan oxidase, a developmentally controlled phenol oxidase from *Aspergillus nidulans.* J Gen Microbiol 136:1725–1730

Birse CE, Clutterbuck AJ (1991) Isolation and developmentally regulated expression of an *Aspergillus nidulans* phenol oxidase encoding gene, *ivoB.* Gene 98:69–76

Boylan MT, Mirabito PM, Willett CE, Zimmermann CR, Timberlake WE (1987) Isolation and physical characterization of three essential conidiation genes from *Aspergillus nidulans.* Mol Cell Biol 7:3113–3118

Clutterbuck AJ (1990) The genetics of conidiophore pigmentation in *Aspergillus nidulans.* J Gen Microbiol 136:1731–1738

Gems DH, Clutterbuck AJ (1991) *sthA* and *sthB*: polygenes involved in conidiophore morphogenesis in *Aspergillus nidulans.* (in preparation)

Johnstone IL, Hughes SG, Clutterbuck AJ (1985) Cloning an *Aspergillus nidulans* developmental gene by transformation. Embo J 4:1307–1311

Marshall MA, Timberlake WE (1991) *Aspergillus nidulans wetA* regulates spore-specific gene expression. Mol Cell Biol 11:55–62

Miller BL, Miller KY, Roberti KA, Timberlake WE (1987) Position-dependent and -independent mechanisms regulate cell-specific expression of the SpoC1 gene cluster of *Aspergillus nidulans.* Mol Cell Biol 7:427–434

Mirabito PM, Adams TH, Timberlake WE (1989) Interactions of three sequentially expressed genes control temporal and spatial specificity in *Aspergillus* development. Cell 57:859–868

Orr WC, Timberlake WE (1982) Clustering of spore-specific genes in *Aspergillus nidulans.* Proc Natl Acad Sci USA 79:5976–5980

Timberlake WE (1991) Cloning and analysis of fungal genes. In: Bennett JW, Lasure L (eds) More gene manipulations in fungi. Academic Press, Orlando (in press)

Zimmermann CR, Orr WC, Leclerc RF, Barnard EC, Timberlake WE (1980) Molecular cloning and selection of genes regulated in *Aspergillus* development. Cell 21: 709–715

[8] Akam M (1987) The molecular basis for metameric pattern in the *Drosophila* embryo. Development 101:1–22

Chater KF, Bruton CJ, Plaskitt KA, Buttner MJ, Mendez C, Helmann JD (1989) The development fate of *S. coelicolor* hyphae depends upon a gene product homologous with the motility sigma factor of *B. subtilis*. Cell 59:133–143

Coen ES, Romero JM, Doyle S, Elliot R, Murphy G, Carpenter R (1991) *floricaula:* a homeotic gene required for flower development in *Antirrhinum majus*. Cell 63:1311–1322

Kaiser D (1989) Multicellular development in Myxobacteria. In: Hopwood DA, Chater KF (eds) Genetics of bacterial diversity. Academic Press Inc, San Diego, CA, pp 243–263

Chapter 8   Development of Trypanosomes

Nina Agabian and Stan Metzenberg

8.1 Introduction

8.1.1 Why Study Parasites?[1]

Parasites are organisms which, during the course of their life cycle, dwell within or upon another living species. The relationship between a parasite and its host is often complicated; however, the parasite usually derives some benefit from its host, and the health of the host is usually compromised in some obvious way by the parasite. As an example, infection by the malaria protozoan *Plasmodium falciparum* causes clinical symptoms of fever, anemia, and sometimes death, and so it is considered to be a parasite.

Parasites are found among all kingdoms, and possibly there are more parasitic organisms in the world than nonparasitic ones. This chapter describes features of the protozoan parasites belonging to the genera *Trypanosoma* and *Leishmania* which are relevant to the study of differentiation and development. Some of these features are inextricably linked to the ability of these organisms to cause disease; however, there are many fundamental discoveries in cell biology which have been made through the study of these organisms. Because of the interaction between host and parasite, one of the challenges in parasitology is to understand the complex ecology between the parasite and its host at the physical, biochemical, immunological, and behavioral level.

8.1.2 *Trypanosoma* and *Leishmania* Species[2]

Trypanosoma and *Leishmania* are genera of protozoa (in the family Trypanosomatidae) which have complex life cycles and are obligate parasites, occupying, at different stages of their life cycles, either vertebrate or invertebrate hosts. Other related trypanosomatids are either free-living or found in association with one or more plant or animal hosts. The morphologies of trypanosomatids are complex, and change with life cycle stages and host. A few of the characteristic morphologies adopted by most species are shown in Fig. 1. For example, *Leishmania* species are found as amastigotes (Fig. 1d) in their mammalian host, and as promastigotes (Fig. 1c) in their insect host. This system of morphological classification is derived from the position of the flagellum (if any) and basal body with respect to the nucleus and anterior end of the cell. In many cases, these terms (trypomastigote, epimastigote, promastigote, and amastigote) do not adequately distinguish between the differentiated stages of the various developmental forms, and adjectives are used to increase the precision of the nomenclature. Some strains of African trypanosomes, especially those isolated from the wild or recently passaged through the tsetse fly vector, are so variable in their morphology that they are called pleomorphic, while those expressing a more uniform morphology in bloodstream stages are called monomorphic. This complicated nomenclature is related to specific properties of the organisms at each life cycle stage, and will become useful as our discussion of differentiation and development of trypanosomes continues.

While many of the medically and economically important species in the family Trypanosomatidae have a similar overall morphology, they may frequently be distinguished by their restricted host cell or species ranges. In addition, trypanosomes may differ with respect to the complexity of their life cycles and developmental stages, sites of proliferation, and development within their respective hosts (for example, hindgut vs. foregut in an insect host), and the spectrum of diseases they cause. Species

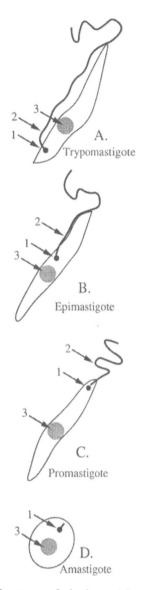

Fig. 1A–D. Structures of developmental stages of trypanosomatid organisms. The relative positions of the flagellar basal body (*arrow 1*), flagellum (*arrow 2*), and nucleus (*arrow 3*) are shown for four developmental stages of trypanosomes. In all four drawings (**A–D**), the anterior end of the cell is oriented at the top and the posterior end at the bottom

and subspecies are also now commonly distinguished by combinations of their isoenzyme patterns, restriction fragment length polymorphisms (RFLPs), or karyotypes determined by pulse-field gel electrophoresis.

8.1.3 The Life Cycle of *Trypanosoma brucei*[3]

Although there is recent evidence for genetic recombination of *T. brucei* in the insect vector, the asexual replication cycles of Trypanosomatidae species are understood best. For *Trypanosoma* and *Leishmania* species, the life cycle represents a developmental program which permits survival and replication within both mammalian and insect hosts, and ensures efficient transmission between them. Trypanosomatid parasites with life cycles split between two types of hosts are said to be digenetic, while those developing in one type of host are said to be monogenetic.

There are three subspecies of *T. brucei*, called *T. brucei rhodesiense, T. brucei gambiense,* and *T. brucei brucei,* but the latter subspecies is not of medical importance as it does not survive in the human bloodstream. In nature, *T. brucei brucei* infects certain species of wild game in Africa, as well as domestic livestock introduced into endemic areas. The *rhodesiense* and *gambiense* subspecies are distinct geographically and in their clinical presentation in humans. For most of the remainder of chapter, these three subspecies will simply be discussed as a single species, *T. brucei.*

Infecting *T. brucei* organisms enter a mammalian bloodstream through the saliva of a biting insect, *Glossina* (commonly called tsetse flies). The parasites have undergone a complicated developmental program within the insect host (also called, the "vector"), one salient feature of which is the development of a variant surface glycoprotein (VSG) coat on the surface of the parasite. The VSG coat, established in advance of their entry into the bloodstream, protects the metacyclic trypanosome from the immune system of the mammalian host (at least temporarily).

Upon entry into a mammal, the parasites rapidly develop into proliferating bloodstream forms (Fig. 2). The differences between these stages are poorly understood but they may be distinguished at least at the level of VSG gene transcription. In the bloodstream the parasites are free-swimming organisms (unlike *T. cruzi* and *Leishmania,* which become intracellular), and several significant events occur during the first few weeks following infection. First, the immune system of the infected

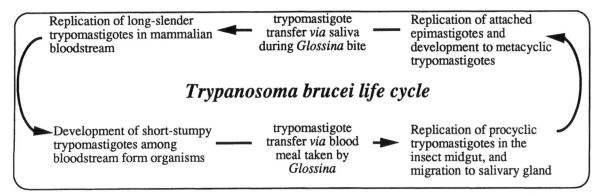

Replication of long-slender trypomastigotes in mammalian bloodstream ← trypomastigote transfer *via* saliva during *Glossina* bite — Replication of attached epimastigotes and development to metacyclic trypomastigotes

Trypanosoma brucei life cycle

Development of short-stumpy trypomastigotes among bloodstream form organisms — trypomastigote transfer *via* blood meal taken by *Glossina* → Replication of procyclic trypomastigotes in the insect midgut, and migration to salivary gland

Fig. 2. Important features of the life cycle of *T. brucei*. The *arrows* show the direction of the developmental program for *T. brucei*, with the mammalian host stage on the *left* and insect host stage on the *right*. The *horizontal arrows* indicate the passage of infective organisms from the insect host to the mammalian host, or vice versa

mammal eventually develops an antibody against the VSG coat initially expressed by the infecting trypanosomes, and these parasites are killed by the body's antibody-mediated complement fixation mechanism. By the time this initial immune response has been mounted, a small fraction of the replicating population will have switched their expression of VSG to an antigenic type that is distinct, and therefore resistant to the current immune response of the host. These parasites will replicate during the time it takes for the host immune system to recognize this new VSG antigenic type, and will

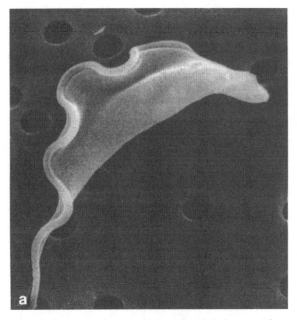
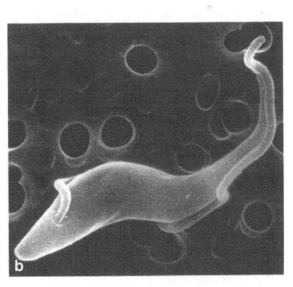

Fig. 3a, b. Scanning electron micrographs of *T. brucei*. A long slender (bloodstream form) trypomastigote (**a**). A metacyclic trypomastigote from *Glossina morsitans* (**b**). Each of these trypomastigotes has a flagellum that exits the cell body near the posterior end of the cell and traverses the outside of the cell body towards the anterior end. Only the bloodstream form cell (**a**) shows the characteristic of an undulating membrane connecting the flagellum to the body of the cell. (Electron micrographs kindly provided by Dr. Keith Vickerman. ×7000)

be almost entirely killed by the subsequent immune response. This process of immune selection and emergence of VSG variants leads to a series of "waves" of parasitemia in the bloodstream, during which time (a period of weeks or months) the titer of parasites increases and decreases dramatically.

A second significant event which occurs during infection of the mammalian host is the development of some trypomastigotes from a long slender (see Fig. 3a) to a short stumpy morphology. The short stumpy trypomastigotes have a more developed mitochondrion than the long slender trypomastigotes. These short stumpy organisms appear to be infectious to *Glossina*, and non-dividing in the bloodstream of the mammal; however, there is considerable disagreement about this in the literature.

Ultimately *T. brucei* returns to *Glossina* in a blood meal taken by the insect, and begins the second part of its life cycle. The VSG coat, which had protected the parasite in the mammalian bloodstream, is replaced by a surface coat composed of the glycoprotein procyclin. These trypomastigotes replicate in the midgut of the *Glossina*, and then migrate as nonreplicating organisms by a circuitous route to the salivary gland of the fly, where they differentiate to an epimastigote form (see Fig. 1b), become anchored to the salivary gland epithelium, and begin to replicate once again. These epimastigotes divide asymmetrically, producing metacyclic trypomastigotes (see Fig. 3b), which are nondividing forms in the insect, but infectious to the mammalian host.

8.2 Trypanosomes as an Experimental System

8.2.1. *Trypanosoma* and *Leishmania* – in Vitro and in Vivo Cultivation are Practical in the Laboratory [4]

Amongst the protozoan parasites, those of the genera *Trypanosoma* and *Leishmania* have received the most attention, not only because of the importance of the diseases they cause, but also because they are comparatively easy to maintain in the laboratory and to cultivate in relatively large quantity. For these reasons, molecular and biochemical studies have been possible to a greater extent with these organisms than with other parasitic protozoa, and these studies have revealed many new strategies for processing molecular information. Culture systems have been developed to grow *T. brucei*, *T. cruzi*, and *L. major* in the laboratory, and under conditions where most life cycle stages may be isolated, cultured in vitro, and induced to differentiate between the forms found in insect and mammalian hosts.

The in vitro culture systems can be complemented and validated through various animal models of trypanosomiasis which accurately reflect the course of disease in humans. Small laboratory mammals (mice, rats, and rabbits), themselves susceptible to infection by *Trypanosoma* and *Leishmania* species, often follow a course of disease similar to that found in humans and thus provide especially useful models for human pathogenesis. Since this is the case, the potential exists to address experimentally such questions as the role of immunity in the course of disease and the role of parasite components and other factors which contribute to pathogenesis. In the case of *Leishmania*, the full power of mouse genetics has provided some elegant data concerning resistance to parasite growth.

Although *T. brucei brucei* typically infests wild game species in Africa, the bloodstage parasites can be readily grown by serial passage in mice and rats. An average adult rat may yield $10^9 - 10^{10}$ bloodstream form *T. brucei* organisms, or 10^8 parasites per cubic centimeter of blood. With some difficulty, the bloodstream form of *T. brucei* may also be propagated in vitro. The differentiation of the bloodstream form to a form similar to that found in the insect (a procyclic culture) may be induced by placing the bloodstream parasites in suitable media (usually one supplemented with mammalian serum and high levels of several amino acids as carbon source) and decreasing the culture temperature to approximately 27°C (close to the body temperature of the invertebrate host *Glossina*). Once established under these conditions, the culture can be passaged indefinitely, and may reach a level of 10^{10} to 10^{11} cells per liter, making it relatively easy to obtain large quantities of viable *T. brucei* organisms. These procyclic cultures are generally not infective in mammals. Recently however, it has become possible to induce such cultures of *T. b. brucei* or *T. b. rhodesiense* to develop into epimastigotes, and subsequently metacyclic trypomastigote forms which are able to infect mice. The addition of invertebrate tissue explants to the culture medium, in the form of either abdomen or head tissues dissected from *Glossina morsitans*, appears to be sufficient to induce this differentiation. The in vitro differentiation of procyclic cultures into mammalian infective forms holds great promise for the study of differentiation and development since, in principle, the entire life cycle may be reproduced in vitro. The use of insect tissue overcomes the need for an intact (i.e., buzzing) *Glossina* host, which for reasons of both difficulty in maintenance and safety are unavailable in many parts of the world.

8.2.2 Energy Metabolism and Respiration in *Trypanosoma* and *Leishmania* Species[5]

Clearly, the life cycles of trypanosome species are complicated not only by their requirement for passage between vertebrate and invertebrate hosts, but also by the restrictions imposed at different times in their proliferation and development as they occupy specific body compartments within each host. Thus their stereotyped development and differentiation must also include a sufficient level of versatility to allow adaptation to a variety of biochemical and physical conditions. *T. brucei*, for example, is exposed to both high oxygen tension in the mammalian bloodstream, and low oxygen tension within the insect gut. In addition, the ambient temperature is different in the two hosts (27 °C in the insect vs. 37 °C in the mammal), as are the available carbon sources (predominantly amino acids in the insect and glucose in the mammal). *T. cruzi* and *Leishmania* species are similarly subjected to vast changes in their biochemical environment during the course of their life cycle.

Trypanosoma and *Leishmania* species contain an unusual analog of glutathione, which consists of a pair of glutathione molecules linked to the terminal amino groups of a spermidine molecule [as N^1,N^8-bis(glutathionyl)spermidine]. This molecule,

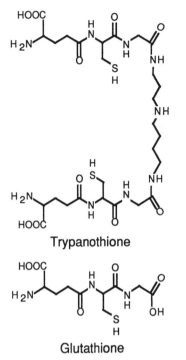

Fig. 4. Two potential carriers of redox potential in trypanosomatid species. The chemical structures of trypanothione [N1,N8-bis(glutathionyl)spermidine] and glutathione are shown in their reduced (thiol) forms

only recently discovered, has been given the trivial name trypanothione (shown in Fig. 4), and is probably found in all species of Trypanosomatidae. Trypanosomatids specifically reduce trypanothione to a dithiol through the activity of a novel enzyme, trypanothione reductase (a homolog of glutathione reductase). Trypanothione may also participate in peroxide removal by a trypanothione-dependent peroxidase activity. It is probable that trypanothione and its associated enzymes provide an important line of defense for these organisms against oxidative stress, and the unique features of this system are likewise promising as targets for the development of specific drugs for disease control.

8.2.3 The Nuclei of *Trypanosoma* and *Leishmania* Species Have Unusual Properties[6]

Variations in nuclear organization and chromosome structure are one of the hallmarks of protozoan species, and while it may seem problematic to the beginning protozoology student, these permutations of nuclear architecture can provide a real advantage to the research scientist. The structure of telomeres in eukaryotes, for example, was first discovered in the ciliated protozoan *Tetrahymena*, primarily because its macronucleus contains an enormous number of chromosomes (and hence telomere copy number). *Trypanosoma* and *Leishmania* species contain only a single type of nucleus; however, for the African trypanosomes, which undergo a process of antigenic variation, there is a subset of "mini-chromosomes" which is quite large in number (about 50–200), their complexity being roughly correlated with the antigenic repertoire of the species/subspecies.

Trypanosoma and *Leishmania*, like many other unicellular organisms, contain nuclear envelopes that remain intact throughout the cell division process. The nucleus undergoes binary fission prior to cytokinesis, and each of the nuclei are segregated to the daughter cells. Research interest in trypanosomatid molecular genetics was coincident with the development of pulse field gel electrophoresis and its use in molecular karyotyping. These studies provide a great deal of information about chromosome numbers and genetic translocations, especially when combined with DNA hybridization studies. In general, with organisms whose genomes are as plastic as those of the trypanosomatids, a basis has not been developed with which to correlate the relevance of these alterations in chromosome number to speciation. Nevertheless, these studies indicate that the chromosome number is highly variable in all of the designate species.

Leishmania species have approximately 20–30 stable chromosomes, as do African trypanosomes, but the latter may have an additional 100–200 mini-chromosomes as well. *T. cruzi* appears to have a relatively stable chromosome number, but its populations are naturally heterogeneous, as judged by a variety of other biochemical, immunological, and cell biological parameters.

While the genetics of trypanosomatids have not been studied in great detail, it is generally believed on the basis of nuclear DNA content that the proliferating blood and insect stages are diploid. Genetically this is harder to establish, since many of the genes in these organisms are found in multiple copies. In several cases where two copies of a gene are present per nucleus, the DNA sequences are located on chromosomes of different sizes. This evidence suggests that trypanosomes are at least partially diploid, or perhaps more accurately "allodiploid" (the situation in which alleles are not in equivalent chromosomal loci).

8.2.4 Trypanosomes Have a Special Mechanism of RNA Processing[7]

Another distinguishing feature of the genomes of trypanosomatids is the organization of their protein-encoding sequences. Instead of being interrupted by introns, as is common in most eukaryotes, the genes of *Trypanosoma* and *Leishmania* species are composed of single, uninterrupted exons (at least this has been the case in all examples studied to date). Quite frequently these genes appear as tandem arrays within the chromosome, and are expressed as long polycistronic pre-mRNAs which are processed into monocistronic units by several RNA maturation events (as outlined in Fig. 5).

The processing of pre-mRNAs involves an endonucleolytic cleavage at the 3'-end of each coding sequence, which may also serve in trypanosomes to subdivide polycistronic pre-mRNAs into monocistronic transcripts. The nucleotide sequences and enzymes that direct the 3'-end cleavage reaction have not been determined for trypanosomatids, as they have been in many higher eukaryotes, and it is not known what factors are required for polyadenylation of the RNA. At the 5'-end of each pre-mRNA, an unusual type of splicing reaction occurs where a 5'-capped 39 nucleotide RNA is added to each pre-mRNA transcript. This small spliced leader (SL) RNA is transcribed separately as a 139 base transcript, but only the 5'-most 39 bases are joined at a splice-acceptor site on the pre-mRNA. The bimolecular reaction which covalently links these two pre-mRNAs is given the name *trans*-splicing, to distinguish it from *cis*-splicing reactions which result in the removal of intervening sequences from the pre-mRNAs in most other eukaryotes (see Fig. 6). The prototypic trypanosome mRNA consists of a coding exon, with a capped SL RNA at its 5'-end

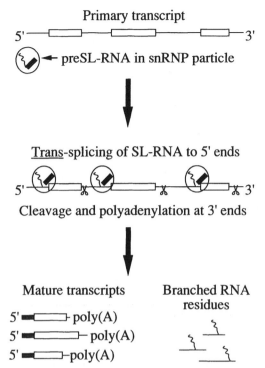

Primary transcript

5' ──────□□□──────□□□────────□□□──── 3'

← preSL-RNA in snRNP particle

Trans-splicing of SL-RNA to 5' ends

5' ────────────────────────────── 3'

Cleavage and polyadenylation at 3' ends

Mature transcripts   Branched RNA
                              residues

5' ▄▄▄□□□□ poly(A)

5' ▄▄▄□□□□ poly(A)

5' ▄▄▄□□□□ poly(A)

Fig. 5. Processing of pre-mRNA molecules in the nuclei of trypanosomes. Three stages in the processing of pre-mRNA are shown, separated by *arrows* that show the direction of the processing reactions. At the *top* is shown a primary transcript with three cistrons (exons), and a preSL-RNA molecule (140 bases, including the capped SL portion shown as a *stippled rectangle*) contained within an elliptical SL-small nuclear ribonucleoprotein (snRNP) particle. The next stage of processing (*middle*) is the assembly of a splicing complex at the 5' end of each exon (as represented by the SL-snRNP) and specific cleavage and polyadenylation (as represented by a *pair of scissors*) at the 3' end of each exon. The result of these reactions (*bottom*) are three *trans*-spliced (SL RNA-containing) and polyadenylated mature transcripts, and several branched RNA residues of splicing

and a poly(A) tail at its 3'-end (detailed in Fig. 7). It is not known whether *trans*-splicing precedes 3'-end cleavage and polyadenylation, or viceversa, nor has it been determined whether the 3'- and 5'-end processing reactions are coordinated in some manner.

A comparison between the factors which participate in *trans*-splicing in trypanosomes with those of *cis*-splicing in yeast and animal cells (see Fig. 6), suggest that the two processes are related evolutionarily. Trypanosomes contain homologs of the mammalian U2, U4, and U6 small nuclear RNAs

(in the form of ribonucleoprotein particles RNPs), which like their *cis*-splicing counterparts are necessary components of the *trans*-splicing reaction. Additionally, the intron-lariat intermediate of *cis*-splicing has a homolog in the *trans*-splicing reaction in the form of a discontinuous lariat, or Y-shaped intermediate. This intermediate consists of the 3'-terminal 100 nucleotides of the SL-RNA transcript (the portion not included in the mRNA), covalently bound via a 2'−5' phosphodiester linkage to a branchpoint nucleotide upstream of the splice acceptor site.

Counterparts of the U1 and U5 RNAs of the *cis*-splicing system have not been found in trypanosomes, although there is some suggestion that the SL-RNA may have a U1-like function in splicing. The precise nucleotide and protein constituents of the *trans*-splicing system in trypanosomes are being actively studied and should contribute much to our understanding of how RNA processing mechanisms have developed through evolution.

8.2.5 The Expression of VSG Proteins Is Vital to the Bloodstream Form of African Trypanosomes [8]

The organization and expression of the variant surface glycoprotein (VSG) genes of *T. brucei*, and by analogy of all other species and subspecies of African trypanosomes which undergo antigenic variation, were among the first gene families shown, at the molecular level, to be regulated by transposition and chromosomal position effects. As judged by hybridization analysis, the *T. brucei* genome is estimated to contain between 200 and 1000 VSG genes and pseudogenes. Some of these VSG gene DNA sequences reside in chromosomal loci at or near a telomere, while others are located at internal chromosomal sites. Within a bloodstream-form trypanosome, only one VSG gene, a so-called expression-linked copy (ELC), is expressed (at any time), and that expressed gene is always located proximal to a telomere. The structure of VSG expression sites can be complicated, with the transcriptional promoter regions located 50 kbp upstream of the VSG gene. Additional non-VSG genes

Stages in RNA-splicing

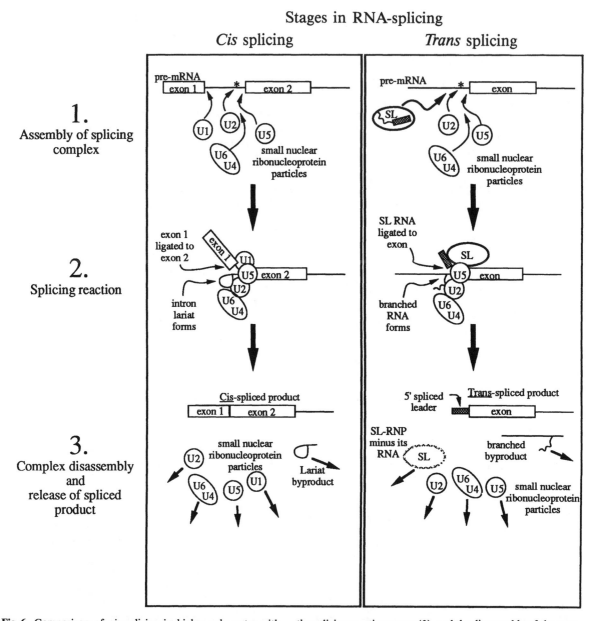

Fig. 6. Comparison of *cis* splicing in higher eukaryotes with *trans* splicing in trypanosomes. In each reaction, the ultimate goal is the covalent connection of two RNA sequences, whether it occurs by intron removal (*cis* splicing, *left*) or spliced leader (*SL*) RNA addition (*trans* splicing, *right*). The two types of splicing reactions are divided into three stages each, with the different stages connected by vertical arrows to show the direction of the reaction. These stages, as indicated by the *labels on the left side* of the figure, include assembly of a splicing complex (*1*), the splicing reaction per se (*2*), and the disassembly of the complex (*3*). Within each panel, exons are indicated by rectangles, nonexon sequences (introns or intergenic regions) by *horizontal lines,* branch points by *asterisks,* and small nuclear ribonucleoproteins by *circles* or *ellipses.* In the *trans* splicing panel (*right*), the SL-RNA is indicated by a *stippled rectangle,* and the SL-snRNP by an *ellipse labeled SL.* The representation of a U5 snRNP in the panel at right is speculative, since no analogous particle has yet been identified in trypanosomes

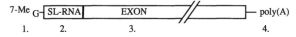

Fig. 7. The prototypical mRNA from a trypanosomatid organisms contains the following components (listed in the order of their appearance, from 5' to 3'). *1* A 7-methyl guanosine cap residue, connected to the 5' terminal adenosine of the SL-RNA by a ^{7Me}G 5'-ppp-5' A triphosphate ester linkage. *2* A 39 base SL-RNA, which begins with the nucleotide sequence AACUAA.... The first four of these nucleotides are modified, respectively, to N6,N6,2'-O-trimethyl A, 2'-O-methyl A, 2'-O-methyl C, and N3,2'-O-dimethyl U. The functions of this unusual 5' end are not know. *3* A coding exon, containing an AUG initiator codon, an open reading frame encoding a protein product, and a codon specifying termination of translation. This portion of the RNA is contiguous with its DNA template (i.e., the gene was not interrupted by introns). *4* A 3' poly(A) tail, added post-transcriptionally by a poly(A) polymerase enzyme

located within an expression site are called ESAGs (for Expression Site Associated Genes). While the identity of some of these ESAGs has been deduced from sequence homology with other known protein sequences, no consistent set of ESAG functions have been realized from one expression site to another. ESAGs and their downstream VSG genes appear to be transcribed as a single, multicistronic pre-mRNA unit.

One unusual feature of transcription within expression sites is that it is remarkably resistant to the drug alpha-amanitin, an inhibitor of eukaryotic RNA polymerase II. Since all known protein-encoding genes in eukaryotes are transcribed by an alpha-amanitin-sensitive RNA polymerase II, the apparent resistance of a protein-encoding gene such as the VSG gene to this drug is unprecedented and it has been suggested that the VSG genes and ESAGs are transcribed by either a modified RNA polymerase II or an entirely different RNA polymerase enzyme. Transcription by this "special" enzyme may then be one level at which VSG gene expression is regulated.

The VSG protein product is crucial to *T. brucei*, because it provides the organism with protection from the immune system of its mammalian host. Each trypanosome is covered with approximately 10^7 copies of the ~60 kDa VSG protein, forming a dense monolayer which coats the entire cell surface. It is thought that this VSG coat may prevent anti-

bodies and complement factors from reaching the cell membrane. The VSG proteins expressed at the cell surface have a number of common structural features, including at least one N-linked oligosaccharide, several conserved cysteine residues near their amino terminus, and a number of similarities within their carboxy terminal 50 to 100 amino acid residues. X-ray crystallography has shown us that the VSG must be able to form homodimers as part of its assembly process at the cell surface.

As one would expect, the expression of a particular VSG only provides the trypanosome with a temporary reprieve from the mammalian immune system. Eventually, as the levels of circulating antibody against that VSG increase, the only trypanosomes which survive the onslaught of VSG specific antibodies will be those that have, by chance, switched their expression to an antigenically different VSG; one not already recognized by circulating host antibodies. In some chronic infections, the expression of over 50–100 different antigens have been documented. The probability of a trypanosome switching from one VSG to another is only about 10^{-6} to 10^{-7} per cell division, but that rate is sufficient to ensure the survival of the organism through recurrent challenges from the immune system.

Some VSGs appear to have a higher probability of being expressed during the early stages of infection than others, a phenomenon which may be related to the expression of that antigen at the time of ingestion by a tsetse vector as well as to the molecular details of the switching event per se. All expressed VSG genes (ELCs) are found on telomeres and it appears, from molecular genetic analysis, that most, if not all of the telomeres in the trypanosome nucleus are occupied by copies of VSG genes. It is not known how switching between ELCs occurs, nor whether expression of a particular VSG is maintained by genetic or epigenetic modifications of the telomeric sites; nevertheless, expression of one VSG gene is always seen to the exclusion of all others. Thus it is a necessary, but not sufficient condition for a VSG gene to be in a telomeric location for transcriptional activation. There are several schemes by which the VSG gene which is being expressed (as outlined in Fig. 8) may be switched. First, the trypanosome may shut off

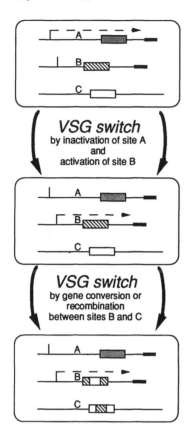

Fig. 8. Methods by which the expression of variant surface glycoprotein (VSG) genes are switched in bloodstream form *T. brucei*. Each of the *three rounded rectangles* shows a different state of VSG expression in bloodstream form *T. brucei*. In an initial state (*top panel*), three VSG genes are shown (labeled *A, B,* and *C,* and represented by *stippled, hatched,* and *open rectangles* respectively) at three chromosomal loci represented by *horizontal lines.* Sites *A* and *B* are associated with telomeres (*vertical hatch marks*), and transcriptional promoters (*vertical lines*); however, in the top panel only site A is an expression site (as indicated by a *horizontal dashed line and arrow*). One method of switching of VSG types involves an epigenetic or genetic change, such that a new telomeric expression site is activated and the former expression site is deactivated. This is shown in the *middle panel,* where site *B* is active and *A* is inactive. An alternate means of VSG switching involves a structural change in the VSG coding sequence by gene conversion or recombination with a different telomeric (or nontelomeric) VSG sequence. Recombination may result in the transplantation of an entire structural gene into an expression site, or the creation of a hybrid gene containing sequences derived from both VSG genes. As shown in the example in the *bottom panel,* the VSG gene in site B is transcriptionally active (as it was in the middle panel), but it has recombined with site C to create a hybrid structural gene (as indicated by the *partially hatched rectangle*)

transcription of its expression linked copy of VSG, and activate a VSG from an antigenically different VSG gene family which likewise is proximal to a different telomere. Alternatively, VSG genes may be switched through gene conversion or recombination mechanisms that alter the coding sequence of the VSG genes. Since approximately 10% of the trypanosome's genome consists of VSG sequences (telomeric and otherwise) it is not hard to imagine that this process could lead to nearly endless variations in VSG type.

The switching of the VSG genes does not occur as a consequence of the parasite's interaction with the host immune system since parasites cultured in vitro have been shown to express alternative VSGs. The effect of the immune system is rather to destroy the prevailing VSG in the population, thus favoring the appearance of a new VSG by immune selection. This process leads to the persistence of infection as one trypanosome population is immunologically destroyed while another, which bears an antigenically distinct VSG coat, proliferates.

8.2.6 Differentiation of the Mitochondrion of *T. brucei* Occurs During Development[9]

Many scientists believe that mitochondria may have originated as symbiotic intracellular prokaryotes in early eukaryotic cells. It is likely that acquisition of mitochondria a billion or so years ago was beneficial, since nearly every eukaryotic cell today still maintains these organelles. Despite their central position in eukaryotic cell biology however, the regulatory interplay between mitochondrial and nuclear encoded functions is not well characterized, no doubt because of the degree of interdependence between these cellular compartments in most eukaryotes. In this respect, the study of *T. brucei* may yield some clues, since mitochondrial development and function can be induced in response to environmental cues and is therefore more tractable experimentally.

The long slender bloodstream form of *T. brucei* feeds primarily on glucose in the blood, catabolizing it to pyruvate and glycerol. The single mitochondrion in each cell at this stage of development is rudimentary in both form and enzymatic repertoire. There is essentially no oxidative phosphorylation system nor are there significant levels of the enzymes of the tricarboxylic acid (TCA) cycle. In contrast, the mitochondrion of procyclic forms of *T. brucei*, are well developed, containing large plate-like cristae. The coordination of mitochondrial differentiation with the transition of the trypanosome between its vertebrate and invertebrate hosts is one of the unusual features of the *T. brucei* developmental program.

When the bloodstream forms of *T. brucei* are placed in tissue culture media at 26 °C, they develop into organisms which resemble the form of *T. brucei* found in the midgut of the *Glossina* host. Presumably, this in vitro development mimics the normal developmental program initiated when trypanosomes are ingested by flies during a blood meal. The effect of in vitro culture conditions on the trypanosome mitochondrion is outlined in Fig. 9. In terms of gross morphology (see Fig. 9a,b), it is evident that the mitochondrion of the bloodstream form organism is rudimentary, with a diameter of 0.1 to 0.3 microns, and containing few cristae. In cultured procyclic organisms, the

mitochondrion is branched and involuted with numerous plate-like cristae. This process of mitochondrial development occurs rapidly in culture and is accompanied by an increase in levels of enzymes and cytochrome proteins specific to the mitochondrion.

Interestingly, this induction of high levels of mitochondrial proteins is not always accompanied by an equivalent increase in the levels of the transcripts which encode them. In the case of cytochrome c protein, the protein levels increase several-hundred-fold while there is only a two- to five fold increase in cytochrome c mRNA. The levels of mitochondrial transcripts such as those encoding cytochrome b, and cytochrome oxidase subunits I and II, are induced several-hundred fold during mitochondrial development, while others such as NADH dehydrogenase subunit 5 remain essentially unchanged. It appears that the expression of some of these enzymes is regulated at the level of transcription, since more than one site of initiation of transcription has been mapped on the mitochondrial genome. Post-transcriptional editing of mitochondrial transcripts to yield translatable RNA species (see below) is also developmentally

Bloodstream form

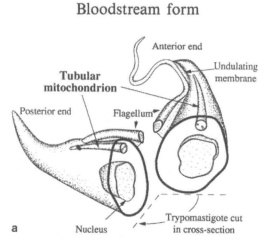

a

Procyclic form

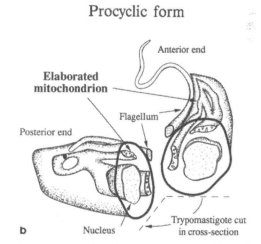

b

Fig. 9a, b. The mitochondria of bloodstream form *T. brucei* are elaborated, both at the morphologic and molecular levels, as these organisms develop towards the procyclic form. A long-slender trypomastigote (**a**), and a procyclic trypomastigote (**b**) are shown in perspective and cross-section (bisecting the nucleus) to demonstrate their mitochondrial volumes and structures. In **a** the mitochondrion is tubular and contains only a few tubular cristae. In **b** the mitochondrion is more sheet-like and contains many plate-like cristae. The *black ovoid body* towards the posterior end of each mitochondrion represents the kinetoplast. As bloodstream form *T. brucei* develop into procyclic form organisms in culture (at 26 °C) over a period of several days, their relative levels of the mitochondrial cytochrome c increase by roughly 50-fold

regulated and thus can affect the rate of expression of specific protein products.

The regulation of mitochondrial gene expression during development must be closely coordinated with the continually changing demands of a cell undergoing complex differentiation events. For example, ultimately, the frequency and efficiency of transcription, the rate of editing, and the half-lives of the specific transcripts must together control the levels of each mature transcript and of protein synthesis during the developmental process. Additionally, the levels of mitochondrial ribosomal RNA increase 30-fold during development suggesting that the efficiency of translation of a given mitochondrial mRNA might increase dramatically during this period, in proportion to the increase in ribosome levels. Alternatively, the increase in mitochondrial mRNA transcription may be triggered by the developmental requirement for more gene product, which may upregulate ribosome synthesis. Finally, the architecture of the mitochondrion changes during development, and proteins destined to be localized there must be sufficiently versatile to assure the correct assembly of enzyme systems such as those for oxidative phosphorylation.

Fig. 10. Electron micrograph of the concatenated mini- and maxicircles contained within the kDNA of *Leishmania tarentolae*. (Electron micrograph kindly provided by Dr. Larry Simpson)

8.2.7 The Mitochondrial DNA of *Trypanosoma* and *Leishmania* Are Organized into Kinetoplasts[10]

Organisms within the order Kinetoplastidae (which includes the family of Trypanosomatidae) derive their name from the unique organization of their mitochondrial genome in a structure called a kinetoplast. The kinetoplast is a highly catenated mass of DNA (kDNA) which contains multiple copies of two classes of circular elements, maxicircles and minicircles (see the electron micrograph in Fig. 10). The maxicircular genome (22 kbp) is analogous to the mitochondrial genome of other eukaryotes. About 50 copies of the maxicircular genome, all of which appear to represent a single sequence class, are found in the kDNA of an individual mitochondrion. The minicircles range from 1−2.5 kbp in length, and number approximately 5000 copies in the kDNA (by mass, about 95% of the total kDNA). They are heterogeneous in se-

quence and the complexity of this part of the mitochondrial genome varies between trypanosomatids. Together, the maxi and minicircles account for 10−30% of the total DNA in *T. brucei, T. cruzi,* or *Leishmania* species. The different populations of sequences within the kDNA can only be released from one another through selective endonuclease cleavage since both maxi and minicircles are inextricably catenated within the structure (Fig. 10). The function of this linking of minicircles and maxicircles is unknown, but it may represent a strategy for kDNA organization or packaging. Clearly this kind of organization imposes a substantial molecular dilemma during DNA replication and cell division, and many researchers have wondered how the kDNA is duplicated and segregated to daughter cells. Some experimental data suggest that circular DNA elements may be released occasionally from their catenated state by the enzyme topoisomerase II, replicated as free structures, and then re-linked to the kDNA mass. A 100–200 bp conserved sequence in each minicircle may serve as an origin of DNA replication. Replication of kDNA is roughly concurrent with that of the nuclear DNA (based on experiments in which cells are synchronized by a temporary block of cell cycle progression), and each DNA sequence is replicated only once. Segregation of kDNA sequences to progeny cells during mitosis may either be precise, ensuring that each daughter cell receives an identical complement of kDNA, or somewhat random, allowing for drift in the sequence composition and overall copy numbers of kDNA maxi and minicircles. The selective pressure for maintaining kDNA during the bloodstream stages of the life cycle are minimal, since the enzyme and proteins which it encodes are not highly expressed until the organism enters the invertebrate host. Indeed, strains of *T. brucei* lacking kDNA exist and grow well as bloodstream form organisms. These are not, however, infective to *Glossina,* underscoring the view that functional mitochondria are required in fly-borne trypanosomes and thus for parasite transmission. Parenthetically, not all species of Kinetoplastidae catenate their kDNA (a notable exception is *Bodo caudatus*), so it may be possible to understand the role of specific replicases and helicases in this process by comparing the activities and roles of these enzymes in kDNA structures in different trypanosomatids.

8.2.8 kDNA Sequences Participate in RNA Editing[11]

Until recently, kDNA was viewed as an oddity of trypanosome organellar structure and although many possible roles for the kDNA had been postulated, the function of the minicircles in particular was not understood. With the discovery that some of the maxicircle transcripts are subject to RNA editing, and later that the minicircle sequences participate in this process, a function could be assigned to some of the components of the kDNA genome. RNA editing can occur by more than a single mechanism and has been described for a number of organisms, but no species performs this feat with the panache of the Trypanosomatidae. RNA editing of

Trypanosoma and *Leishmania* mitochondrial transcripts is manifest as uridine addition, and more rarely, subtraction, at various sites along the pre-mRNA. As a result, the coding sequence of the mature edited transcript may be fundamentally different from that transcribed from the maxicircle template. In the case of the *T. brucei* mitochondrial transcript for cytochrome oxidase subunit III protein (coIII) for example, almost 60% of the RNA coding sequence consists of uridines added during the editing process. The name "pan editing" (*pan,* taken from Greek, meaning all or universal) has been suggested for these extreme examples of RNA sequence modification, and the gene on the maxicircle that encodes the unedited transcript has been labeled a cryptogene, to indicate that its coding of the mature transcript is cryptic, or hidden. Several other mitochondrial cryptogenes have pan-edited transcripts, and these include the NADH dehydrogenase subunit 7 (ND7) gene, and the ATPase subunit 6 gene (formerly known by the name MURF4). Not all *T. brucei* mitochondrial transcripts are pan-edited; a notable exception is the transcript encoding cytochrome b, which is edited only near its 5′ terminus, generating an AUG start codon in the process. Whether a mitochondrial transcript is edited to a greater or lesser extent appears to be a species- or genus-specific trait, since many of the cryptogene transcripts that are pan-edited in *T. brucei* receive many fewer modifications in *Leishmania tarentolae*. A comparison of the extent of RNA editing of the same gene in different genera is shown in Fig. 11, where 12 amino acid residues, internal to the coIII polypeptide of *L. tarentolae* (top) and *T. brucei* (bottom), are shown along with the RNA and DNA sequences that encode them. In the case of *L. tarentolae*, the 12 amino acids are encoded by 36 nucleotides in the RNA and DNA, and no editing is apparent in this portion of the transcript (although the *L. tarentolae* coIII transcript is edited in other regions). Conversely, in *T. brucei* this same region of the coIII polypeptide (with 9 out of 12 amino acids matching) is encoded by only 18 nucleotides in the primary transcript, which is edited by the addition of 20 uridines and the removal of two uridines.

Although the system of RNA editing used by *Trypanosoma* and *Leishmania* species appears ar-

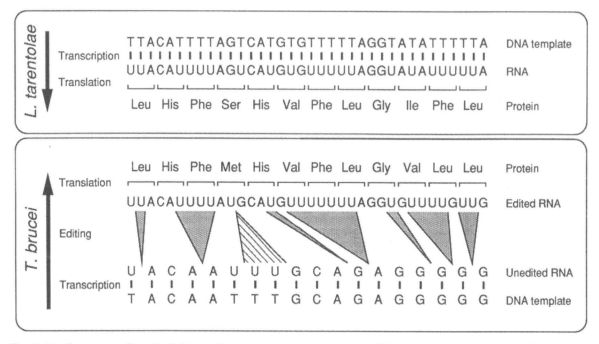

Fig. 11. To edit or not to edit – that is the question. An internal segment of the cytochrome oxidase III (coIII) gene of *L. tarentolae* (*top panel*) and *T. brucei* (*bottom panel*) is shown, along with the protein sequences that are ultimately generated from each (the segment of coIII shown in each panel is 66 amino acids from the C-terminus of the protein). In *L. tarentolae*, the 12 amino acids are encoded by 36 nucleotides in the mRNA and corresponding DNA template (*arrows* show the direction of expression, from DNA template to protein sequence in each panel). No editing events occur in this segment, although other regions of the *L. tarentolae* coIII transcript are edited. Conversely in *T. brucei*, the homologous sequence of 12 amino acids is encoded by only 18 nucleotides in the DNA template, and RNA editing modifies the transcript by the addition of 20 nucleotides (*stippled wedges*) and the removal of two nucleotides (*hatched wedge*)

cane, it is very likely that it provides some advantage to the cells which employ it; however, for the moment the purpose of RNA editing within this system remains to be discovered.

8.3 Outlook

8.3.1 Mitochondrial Development – How Is It Triggered?

An important and as yet unsolved question, is what triggers mitochondrial development in bloodstream form cells. In the tissue culture model, the effects of temperature or ingredients in the culture media are obvious candidates, but this in vitro system may not duplicate the process in vivo. The stumpy bloodstream form of *T. brucei* cells, for example, have more developed mitochondria than do the long slender bloodstream forms and contain some of the enzymes of oxidative metabolism found in the procyclic insect forms. Populations of cells which contain stumpy forms have a higher efficiency of *Glossina* infection, suggesting perhaps that this limited development of a procyclic-like mitochondrion represents a preadaptation to the insect midgut. It is generally believed that stumpy forms do not proliferate in the mammalian host. Thus the early stages of mitochondrial differentiation may either be neutral or perhaps harmful to a stumpy organism while still in the mammal, while at the same time providing a selective advantage to an organism which finds itself in its insect host.

8.3.2 How Is RNA Editing Developmentally Regulated? [12]

A perplexing aspect of RNA editing has been developing a model which would account for the source of information provided in the nucleotide additions and subtractions that occur during RNA maturation. Early experiments indicated that pan-edited transcripts such as coIII (of *T. brucei*) do not contain complete DNA or RNA templates which encode them in the cell, at least based on the rules of Watson-Crick base pairing. The nucleotide sequences of minicircles revealed, however, that small regions of kDNA complementary to edited RNAs exist, provided that one allows guanidine-uridine base pairing. These regions of the minicircles are expressed as small (approximately 50–100 base) complementary RNA transcripts, called guide RNAs (gRNAs). The typical gRNA contains a 3' anchor sequence, consisting of approximately ten nucleotides at its 5' end which may base-pair with the unedited RNA; approximately 20–30 internal nucleotides that may base-pair with the edited RNA and act as a template for editing; and finally a 3' terminal sequence of variable length, and containing no similarity to edited or unedited sequences. Based on information obtained from cDNA copies of putative editing intermediates, it has been postulated that editing may initiate at the 3' end of the cryptogene transcript, and proceed towards the 5' end. The enzymatic steps would consist of specific cleavage of the transcript, addition (or removal) of uridine residues to bring the transcript into complementarity with the coding region of the guide RNA, and lastly RNA ligation. At present, the precise mechanism of editing has not been determined, but in vitro RNA editing systems may reveal some of the details in the future.

8.4 Summary

The study of parasitic protozoa has provided scientists with a fascinating look at some of earth's most primitive eukaryotes, and their relationship and dependence on more complex host organisms. In particular, the unusual life cycles of *Trypanosoma* and *Leishmania* species have permitted the analysis of protozoan cell development at the level of morphology, biochemistry, and molecular biology.

The trypanosomatid organisms discussed in this chapter are digenetic, in that one complete life cycle takes them from an insect host to a mammalian host, and then back again to an insect. Extensive differentiation of the trypanosome takes place at each stage. This may occur for the purpose of adaptation to the current host (for example, the mitochondrion develops in procyclic *T. brucei*, as discussed in Sect. 8.2.6), or it may indicate some preparation for entry into the next host (for example, expression of a VSG surface coat in metacyclic trypomastigotes prepares the parasite for future exposure to the mammalian immune system, as discussed in Sect. 8.1.3). The molecular details of these events are largely unknown, but there is hope that current studies will provide some clues as to how these developmental processes are regulated.

One avenue of research that has been illuminating is the study of how *Trypanosoma* and *Leishmania* species generate their mature mRNA molecules. In the nucleus, a capped SL-RNA is added to the 5' ends of transcribed coding sequences, by a process called trans-splicing (Sect. 8.2.4) that is mechanistically related to the *cis*-splicing process of higher eukaryotes. In the mitochondria of these parasites, many mature mRNAs are only partly encoded by the maxicircle genome (these incompletely coding genes have been termed "cryptogenes"). The remainder of the sequence information is contained in small "guide RNAs", which are used during the process of RNA editing (as described in Sects. 8.2.8 and 8.3.2). These uncommon features of RNA expression have indicated in recent years that trypanosomatid organisms differ substantively from their hosts in the processing of genetic information. In future research, these differences between trypanosomatid species and higher eukaryotes may lead to the rational design of drugs that combat trypanosomiasis and leishmaniasis.

References

[1] Markell EK, Voge M, John DT (1986) Medical parasitology. W. B. Saunders, Philadelphia

Mulligan HW (1970) The African trypanosomiases. George Allen & Unwin, London

[2] Lumsden WHR, Evans DA (1976) Biology of the Kinetoplastida, vol 1. Academic Press, London

[3] Vickerman K, Tetley L, Hendry KAK, Turner CMR (1988) Biology of African trypanosomes in the tsetse fly. Biol Cell 64:109–119

Seed JR, Sechelski JB (1989) Mechanisms of long slender (LS) to short stumpy (SS) transformation in the African trypanosomes. J Protozool 36:572–577

[4] Kaminsky R, Beaudoin E, Cunningham I (1988) Cultivation of the life cycle stages of Trypanosoma brucei sspp. Acta Trop (Basel) 45:33–43

Blackwell JM, Toole S, King M, Dawda P, Roach TIA, Cooper A (1988) Analysis of Lsh gene expression in congenic B10.L-Lsh^r mice. Curr Top Microbiol Immunol 137:301–309

[5] Gutteridge WE, Coombs GH (1977) Biochemistry of parasitic protozoa. University Park Press, Baltimore

Henderson GB, Fairlamb AH (1987) Trypanothione metabolism: a chemotherapeutic target in trypanosomatids. Parasitol. Today 3:312–315

[6] Woodward R, Gull K (1990) Timing of nuclear and kinetoplast DNA replication and early morphological events in the cell cycle of Trypanosoma brucei. J Cell Sci 95:49–57

[7] Agabian N (1990) Trans-splicing of nuclear pre-mRNAs. Cell 61:1157–1160

Michaeli S, Roberts TG, Watkins KP, Agabian N (1990) Isolation of distinct small ribonucleoprotein particles containing the spliced leader and U2 RNAs of Trypanosoma brucei. J Biol Chem 265:10582–10588

Tschudi C, Ullu E (1990) Destruction of U2, U4, or U6 small nuclear RNA blocks trans-splicing in trypanosome cells. Cell 61:459–466

[8] Borst P (1986) Discontinuous transcription and antigenic variation in trypanosomes. Annu Rev Biochem 55:701–732

Pays E, Coquelet H, Tebabi P, Pays A, Jefferies D, Steinert M, Koenig E, Williams RO, Roditi I (1990) Trypanosoma brucei: constitutive activity of the VSG and procyclin gene promoters. EMBO J 9:3145–3151

[9] Brown RC, Evans DA, Vickerman K (1973) Changes in oxidative metabolism and ultrastructure accompanying differentiation of the mitochondrion in Trypanosoma brucei. Int J Parasitol 3:691–704

[10] Ray DS (1987) Kinetoplast DNA minicircles: high-copy-number mitochondrial plasmids. Plasmid 17:177–190

Silver LE, Torri AF, Hajduk SL (1986) Organized packaging of kinetoplast DNA networks. Cell 47:537–543

[11] Feagin JE, Abraham JM, Stuart K (1988) Extensive editing of the cytochrome oxidase III – transcript in Trypanosoma brucei. Cell 53:413–422

Sturm NR, Simpson L (1990) Partially edited mRNAs for cytochrome b and subunit III of cytochrome oxidase from Leishmania tarentolae mitochondria: RNA editing intermediates. Cell 61:871–878

[12] Feagin JE, Jasmer DP, Stuart K (1986) Differential mitochondrial gene expression between slender and stumpy bloodforms of Trypanosoma brucei. Mol Biochem Parasitol 20:207–214

Sturm NR, Simpson L (1990) Kinetoplast DNA minicircles encode guide RNAs for editing of cytochrome oxidase subunit III mRNA. Cell 61:879–884

Chapter 9   *Dictyostelium:* From Unicellularity to Multicellularity

SALVATORE BOZZARO

9.1 Introduction[1]

Differentiation and morphogenesis of multicellular organisms depend on the coordinated interaction of intra- and intercellular signals that act together to mold their overall shape. The regulatory role of intercellular signals can be conveniently studied in the slime mold *Dictyostelium discoideum*, where both diffusible signals and cell-cell adhesion molecules have been identified.

D. discoideum is a soil-living amoeba, which propagates asexually, feeding on bacteria. When starved, the cells initiate a developmental process which involves aggregation, formation of a motile multicellular organism and, finally, a fruiting body containing spores (Fig. 1). The cells are haploid, with a small genome (50 000 kb), are easily transformed, and gene disruption by homologous recombination occurs with a very high frequency.

A key feature of all slime mold species is that cell division and development are temporally separated: starvation stops growth and initiates development. Developmental mutants which divide normally can, therefore, be easily isolated, and the entire developmental process can be selectively studied independently of cell growth. Synchronous development is initiated by the removal of food (bacteria or defined nutrient media). In less than 24 h, the genetically uniform amoebae differentiate into two cell types, spores and stalk cells. The terminally differentiated stalk cells are nonviable, and their only function is to form a rigid support for a small ball of spores. Cells which will later form spores and stalk cells are called "pre-spore" and "pre-stalk" cells, respectively, and are present at the so-called "slug" stage in the constant ratio of 70% pre-spore and 30% pre-stalk cells. In addition to this proportioning, the two cell types are localized in different regions of the slug, and resemble, in their regulation of differentiation and sorting, animal embryos.

9.2 Development of *Dictyostelium discoideum*

The development of *D. discoideum* can be roughly divided into two phases, the aggregation phase, during which individual, free-living amoebae collect into multicellular mounds, and the differentiation phase, which takes place within the aggregate, and results in the formation of two cell types, which originate from a common population of unspecialized cells. Cell differentiation is accompanied by a sequence of morphogenetic changes of the whole aggregate, as outlined in Fig. 1.

Aggregation is the result of two forms of cellular interactions: cell-cell communication between distant cells, which leads to amoebae gathering in clusters of several thousand cells, and cell-cell adhesion which renders the amoebae cohesive to each other, transforming heaps of cells into true, albeit primitive, organisms.

9.2.1 Cell-Cell Communication[1,2]

Within 3 to 4 h from the end of growth, starving cells develop the ability to aggregate by chemotaxis. Foci of aggregation arise in the initially uniform cell layer and the amoebae collect into aggregation territories (Fig. 2). *D. discoideum* is perhaps the best understood example of eukaryotic chemotaxis. Individual amoebae in the homogeneous cell population start to synthesize and secrete cyclic AMP in discrete pulses, giving rise to circular waves of this chemoattractant which propagate outwardly. Cells

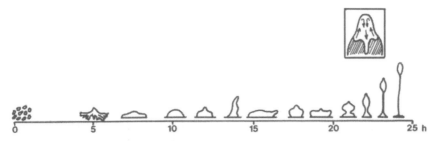

Fig. 1. The development of *Dictyostelium discoideum*. *Dictyostelium* cells start to aggregate by chemotaxis at about 5 h. In their chemotactic movement they form streams that flow into aggregation centers. By about 10 h, mounds form, in which the cells maintain their individual identity, though adhering in a tissue-like structure. Shortly thereafter, a "tip" is formed on top of the mound, the mound elongates and form a motile organism, called "slug", that migrates toward light or heat. The tip and the anterior part of the slug contains pre-stalk cells, whereas the posterior contains pre-spore cells. Culmination, the final stage of differentiation, initiates with the slug settling back and elongating slowly. During culmination, the pre-stalk cells become vacuolated and invaginate through the mass of pre-spore cells (*inset*). The stalk is formed by secretion of a cellulose sheat and vacuolization of the differentiating stalk cells, whereas the spore cells differentiate by loosing water and secreting a rigid spore coat made of glycoproteins and polysaccharides

in the immediate vicinity sense the gradient formed by the leading edge of the passing wave by means of cAMP membrane receptors, and respond both by moving towards cells which released cAMP first and secreting their own pulse of cAMP. In this way the chemotactic signal is amplified and relayed from the center to the periphery of an aggregation territory.

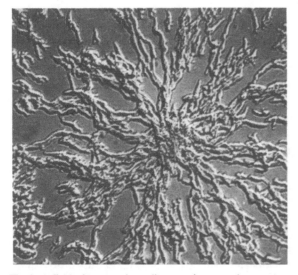

Fig. 2. A field of aggregating cells streaming towards an aggregation center

The chemotactic response involves several elements: adenylate cyclase for the synthesis of cAMP, cell surface receptors for binding, and extracellular and membrane-bound phosphodiesterase for degradation of the signal molecule. Signal transduction pathways, which are only partially understood, control oriented cell motility and the cAMP relay (Fig. 3).

9.2.2 Cell-Cell Adhesion[3]

In their chemotactic movement towards the aggregation centers, the cells adhere to each other at their ends, giving rise to aggregating streams (Fig. 2). Later, they form tight aggregates. Cohesiveness is a developmentally regulated property, which starts to appear either by the end of exponential growth or shortly after starvation, depending on the strain and the experimental conditions. During the first 3–4 h of development cells adhere only through EDTA-sensitive contacts, but shortly before the onset of aggregation new contact sites accumulate on the cell surface which are resistant to 10 mM EDTA.

A 24 kDa and an 80 kDa glycoprotein mediate, respectively, the EDTA-sensitive and EDTA-resistant form of adhesion. The 80 kDa glycoprotein, which has been called contact site A (csA), is an integral membrane protein modified with glycosidic

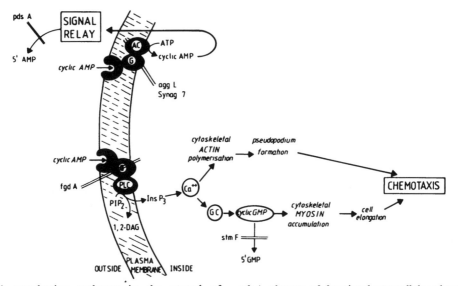

Fig. 3. Signal transduction pathways in the control of chemotaxis. Cyclic AMP binding to a membrane receptor activates via G protein the adenylate cyclase, stimulating further release of cAMP (*top*). Cyclic AMP interaction with the same or a similar class of membrane receptors activates the phosphoinositol cycle, with increase of intracellular Ca^{2+} that leads to actin polymerization and myosin accumulation (*bottom*). Mutations in several elements of the cAMP signaling system are shown: *pdsA*, absence of functional extracellular phosphodiesterase; *aggL*, *Synag 7* inability to relay due to deficiency in the G protein/adenylate cyclase coupling; *fgdA* nonfunctional $G\alpha_2$ subunit of G protein; *stmF* deficiency in cGMP phosphodiesterase. Other abbreviations: *G* G protein; *AC* adenylate cyclase; *PIP$_2$* phosphatidylinositol 4,5-biphosphate; *1,2-DAG* 1,2-diacylglycerol; *InsP3* inositol (1,4,5)trisphosphate; *GC* guanylate cyclase. (Newell P et al. 1990)

chains, which are necessary for its transport to the cell surface, stability to proteolytic cleavage and biological function. CsA-mediated adhesion is homotypic, i.e., two identical molecules on opposing surfaces bind to each other. Whether gp24 also binds homotypically or rather heterotypically, i.e., to a complementary binding site, is not known.

Accumulation of csA is subject to a bimodal control, being induced by starvation and strongly enhanced by cAMP pulses (see Sect. 9.3.1), whereas gp24 appears to be only under the control of signals linked to starvation. After the cells have aggregated, transcription of the csA gene is repressed, the glycoprotein slowly disappears and its function is taken over by other membrane components. Antibodies against slug membranes block EDTA-resistant adhesion at the slug, but not at the aggregation stage. The target antigens have not yet been unequivocally identified.

9.2.3 Multicellular Development[1]

At the end of aggregation, the initially loose aggregates undergo compaction, forming hemispherical mounds, which are enveloped in a sheath composed of cellulose and proteins. Each mound develops a distinct tip, it elongates slightly and then, under conditions favoring migration, falls on its side to form a migrating slug (Fig. 4), with the tip at the anterior end. The slug will migrate a considerable distance continuously synthesizing new slime sheat.

The slug has a polarity, with the tip and the anterior one-fifth containing pre-stalk cells and the posterior four-fifths containing pre-spore cells. The tip can be considered a sort of "sensor" and "organizer" of the entire organism. It is the tip that is sensitive to light and chemical signals, controlling slug movement. If a slug is cut in two, the portion with the tip will differentiate rapidly, whereas the

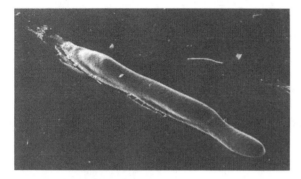

Fig. 4. A migrating slug

tipless portion first regenerates a new tip. Similarly, if a tip of one slug is grafted onto the flank of another it captures part of the host and splits off to form a separate slug. Despite its importance for size regulation and possibly for proportioning of the pre-stalk versus pre-spore region, the tip does not seem to be necessary for cell differentiation (see Sect. 9.5).

When conditions are changed to those favoring sporulation (i.e., low humidity and removal of ammonia), the slug settles back with the tip upright and begins to form a fruiting body, a process called culmination (Fig. 1). During culmination, the pre-stalk cells invaginate and form the stalk of the fruiting body, while the mass of pre-spore cells is lifted off the substrate up to the top of the mature fruiting body. Morphologically differentiated stalk and spore cells first become visible during fruiting body formation. Up to this stage, the masses of pre-stalk and pre-spore cells have a tissue-like structure in which individual cells are undetectable.

9.3 Molecular Analysis of Development[4]

The fact that *Dictyostelium* differentiates in basically two cell populations, and that pattern formation is simple compared to animal embryos, makes it an attactive model for molecular studies of development. Nevertheless, the complexity of the organism should not be underestimated. It has been calculated that between 3000 and 5000 distinct mRNA sequences are expressed during the growth phase, which is similar to the number found in cells of

higher organisms. Estimates of distinct mRNA sequences expressed during development oscillates between 800 and 2000. Genetic analysis, based on the rate of lethal mutations versus mutations in developmental genes (keep in mind that developing *Dictyostelium* cells have ceased to grow) has led to the conclusion that genes essential for development range between 200 and 400. Of these, between 40 and 100 would be involved in aggregation.

These are still large numbers, if an understanding of development requires the mutual interactions of all these genes to be determined. However, studies at both mRNA and protein level have shown that major changes in gene expression occur at a few distinct stages of development. These changes occur in response to distinct signals, which are not environmental but generated by the cells as they proceed through development. This has allowed genes to be sorted into classes which are coordinately expressed or repressed by the action of a few regulatory signals.

The first class is represented by genes expressed at the beginning of development under the control of pre-starvation and starvation signals. A pre-starvation factor (PSF) is secreted by the cells during growth, accumulates as a function of cell density and is believed to act as a signal for initiating development. Some of the genes controlled by this signal are necessary for establishing the cAMP relay system (cAMP receptor, phosphodiesterase, adenylate cyclase, and G protein α_2 subunit).

Once the cAMP relay system becomes functional, pulsed production of cAMP begins, which further stimulates expression of genes involved in chemotaxis and cell adhesion. Two hours later aggregation occurs. By the end of aggregation, the Differentiation Inducing Factor (DIF) and its receptor accumulate and, shortly thereafter, the third major class of genes, most of which are cell-type specific, is activated, apparently in response to cAMP and DIF (see Sect. 9.5.1).

Finally, during culmination, a small number of new genes are expressed which are necessary for terminal differentiation and fruiting body formation. It is unclear whether ammonia, which accumulates during development as the final product of protein degradation (the primary energy source in starving amoebae), is an inducer of late gene expression. A decrease in ammonia concentration, due to its dispersal in the air, triggers culmination of the slug as well as terminal stalk cell differentiation, apparently by a mechanism involving a drop in intracellular pH.

This brief description suggests that *Dictyostelium* development is controlled by a limited number of sequentially ordered signals, so that one signal triggers a class of genes, some of which are required to build up the machinery for the next signal, and so forth. Despite the limited number of signals and the temporal linearity of their production, the regulatory circuits in the control of gene expression are complex because of three factors:

1. Each signaling system stimulates or inhibits genes, depending on how the signal is provided. The csA gene, for example, is stimulated by pulsed production of cAMP, but repressed by constant amounts of cAMP. High concentration of cAMP in the aggregate could be the signal that down-regulates csA at the end of aggregation.
2. Different types of signals act together, either additively, synergistically, or antagonistically. Thus, (pre)-starvation signals induce transcription of several genes, which is further enhanced by cAMP. cAMP and DIF act antagonistically in the induction of prespore gene expression, but synergistically, at least in part, in prestalk cell differentiation (see Sect. 9.5.1).
3. Once aggregates are formed, formation of short-range signal gradients as well as responsiveness of cell groups may be affected by cell adhesion, though we know little about how this occurs (see Sect. 9.5.2).

9.3.1 Cyclic AMP: a Morphoregulatory Signal[2,5]

The best characterized signals in *D. discoideum* are cAMP and DIF. cAMP is necessary throughout development, both as a chemoattractant and a regulator of gene expression. The expression of all genes involved in chemotaxis and cell adhesion is enhanced by cAMP, which represses genes specific for growth. Cyclic AMP also controls postaggregative gene expression, regulating cell differentiation in combination with signals, such as DIF. Consistent with a key morphoregulatory role of cAMP, mutants which fail to produce or to respond to cAMP are blocked at the beginning of development and do not express any postaggregative gene. Altering the cAMP signaling system late in development, for example by adding extracellular phosphodiesterase to slugs, disrupts any subsequent development (see also Sect. 9.4.1).

An interesting feature of the cAMP signaling system is its autocatalytic control. The ability to produce, detect and degrade cAMP is present in *D. discoideum* at the beginning of development. However, at this stage, cells possess few cAMP membrane receptors, little adenylate cyclase, little membrane phosphodiesterase, and a high level of a phosphodiesterase inhibitor, which down-regulates an extracellular phosphodiesterase expressed highly during growth. Within a few hours, all these gene products accumulate dramatically, with the exception of the extracellular phosphodiesterase inhibitor, which declines to background levels.

This coordinated accumulation/decline is regulated by cAMP, which therefore controls the structural components of its own signaling system, generating an autocatalytic loop. Binding of cAMP to membrane receptors activates, via G protein, the adenylate cyclase, which is deactivated by desensitation of the receptor (Fig. 3). Degradation of bound cAMP by the membrane phosphodiesterase restores the excitability of the receptor, whereas the extracellular phosphodiesterase destroys the cAMP gradient around the cell. Activation/deactivation of adenylate cyclase results in pulsed production and secretion of cAMP. Cyclic AMP pulses are the stimuli that enhance at the transcriptional level further expression of adenylate cyclase, cAMP membrane receptors and membrane phosphodiesterase. The extracellular phosphodiesterase, instead, accumulates, and its inhibitor declines, in response to a net increase in cAMP concentration. Each component of the signaling system is clearly necessary for the entire system to function well, and thus for full expression of the other components.

The constituents of the cAMP signaling system belong to a larger class of gene transcripts which are subject to cAMP regulation. The function of most transcripts is unknown, but one which is well defined is the gene encoding the csA glycoprotein.

Cyclic AMP exerts its control on gene expression by binding to membrane receptors. This has been shown by analysis of mutants defective in cAMP receptor-mediated activation of adenylate cyclase, which can be induced to develop by stimulation with exogenous cAMP. Pharmacological studies with cAMP analogs which have selective affinity for either membrane receptors or intracellular binding sites support the same conclusion. The transduction pathway leading from the membrane receptor to gene expression is still unknown. However, a start has been made to identify *cis*-acting regulatory elements in cAMP-inducible genes and transacting factors.

9.3.2 DIF: a Histospecific Regulatory Signal[6]

DIF is a chlorinated hexaphenone (Fig. 5), which is released externally, it permeates the cell membrane and is able to induce stalk cell differentiation of isolated cells in vitro. Consistent with DIF's ability to stimulate stalk-cell differentiation, pre-stalk-specific genes have been identified which are rapidly induced by DIF. DIF also acts as a transcriptional repressor of at least some pre-spore-specific genes. Mutants have been selected which do not produce DIF and are blocked at the loose aggregate stage, unless DIF is added.

DIF is assumed to act by binding to intracellular receptors, in a steroid hormone-like manner, and indeed a preliminary characterization of a cytoplasmic DIF-binding activity has been reported. This activity appears to be developmentally regulated, being detectable only after the cells have aggregated.

DIF

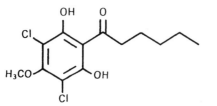

Fig. 5. Structure of DIF, the pre-stalk-specific differentiation factor

The biosynthetic pathway of DIF and the enzymes involved are unknown. DIF itself is degraded through a complex route involving 11 steps. A DIFase which degrades DIF in a monochlorinated derivative has been recently identified. The enzyme is enriched in pre-stalk cells and is induced by DIF itself.

9.4 Molecular Genetic Analysis of Aggregation[7]

Mutants blocked at different stages of development are easily isolated in *Dictyostelium*, and they have helped to better define the temporal sequence of differentiation. Classical genetic analysis has been used to determine the genetic complexity of phenotypes and to establish the linkage groups in the genome. The construction of successful transformation vectors has however led to current genetic research largely using molecular techniques. The most powerful tool for a molecular genetic analysis of development is the high recombinogenic capacity of *Dictyostelium*. After transformation with sequences identical to a resident gene, homologous recombination occurs with high frequency and leads to disruption of the native gene. As an alternative to gene disruption, indirect gene inactivation with cDNA coding for antisense RNA has been used to study the function of given gene products. These approaches, in combination with biochemical and cell biology studies, have allowed the molecular genetic analysis of the three primary processes involved in aggregation, cell-cell communication, cell adhesion and motility.

9.4.1 Essential Genes in Cell-Cell Communication[2,5,8]

The cAMP membrane receptor is present in two kinetically distinct forms, which are believed to interact with either adenylate cyclase or phospholipase C (Fig. 3). They may be different proteins or just different states of the same receptor protein. A cAMP receptor gene (CAR 1) has been isolated and shown to encode a typical G protein-linked receptor with seven transmembrane segments and several phosphorylatable serines in the cytoplasmic C-terminal region. This receptor oscillates between two interconvertible forms, depending on phosphorylation of the serine residues. Phosphorylation appears to be the mechanism by which the receptor is desensitized. Cloning analysis has revealed the existence of three other receptor subtypes, CAR2 to 4. The four subtypes exhibit temporal and spatial specificity of expression, with CAR1 being the major form expressed before and during aggregation.

The key role of CAR1 for intercellular communication and development has been confirmed by cell transformation with cDNA coding for antisense RNA. Transformed strains do not express any form of the receptor, fail to produce cAMP and cGMP, lose chemotactic sensitivity to cAMP, and do not develop at all.

Another essential enzyme for aggregation is the cAMP phosphodiesterase. It exists in two forms: one membrane-bound to the cell surface, and the other extracellular, which eliminates the cAMP gradient around the cells. Mutants which fail to produce the phosphodiesterase are unable to aggregate and develop. They can be rescued by transformation with a cloned phosphodiesterase gene. Late in development, the phosphodiesterase is required for proper formation of fruiting bodies, since absence or overproduction of the enzyme blocks morphogenesis.

The enzyme is subject to multiple control, both at transcriptional and posttranslational level. At transcriptional level, three different promoters, which are located at different distances from the coding region, allow a temporal and tissue-specific regulation of the enzyme (Fig. 6). The three promoters are specific for growth, aggregation and development, and the three mRNA transcripts produced differ only in their 5' noncoding sequence. The late promoter is responsible for tissue-specific production of the phosphodiesterase, which is found only in pre-stalk cells, and is induced by DIF. The aggregation-specific promoter is under the control of cAMP. At this stage, the enzyme exists in a membrane-bound and an extracellular form. Expression of the membrane-bound phosphodiesterase is stimulated by cAMP pulses, whereas the extracellular form accumulates in response to an increase of cAMP concentration. How both forms are regulated, being encoded by the same gene, is not clear. The cDNA-derived sequence reveals a long hydrophobic leader, that could anchor the phosphodiesterase or, if cleaved, give rise to the extracellular form. Modulation of the extracellular

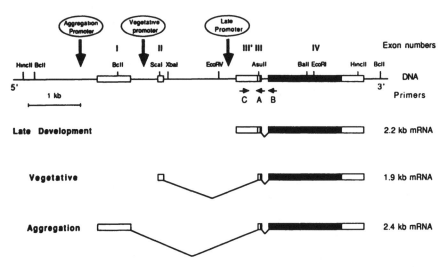

Fig. 6. Three different promoters control the developmental expression of the phosphodiesterase gene. Transcribed but untranslated regions are shown by *white bars*, and coding regions are shown by *black bars*. The splicing patterns of the three mRNAs are shown. Other explanations in the text (Faure et al. 1990)

phosphodiesterase activity is also obtained by down-regulation of a phosphodiesterase inhibitor, which is induced by starvation signals and repressed by cAMP pulses.

The central event in cAMP amplification and propagation from cell to cell is the periodic activation of adenylate cyclase. Mutants defective in cAMP receptor-mediated activation of adenylate cyclase do not aggregate by chemotaxis. Coupling of the cAMP receptor to adenylate cyclase occurs through G proteins. Three genes coding for Gα subunits have been cloned and show a high degree of homology in certain conserved regions and also with mammalian genes coding for Gα subunits. The cloned genes are differently regulated during development, with Gα1 and Gα3 being expressed in growth-phase cells. In contrast Gα2 is absent during growth, is induced by starvation and its expression during development is strongly regulated by cAMP pulses. Mutants defective in Gα2 are blocked in chemotaxis, cannot be rescued by cAMP, and fail to develop.

It is remarkable that defects in any one of the genes which code for the cAMP receptor, phosphodiesterase or G protein, not only block chemotaxis but also late gene expression, confirming the key regulatory role of cAMP-mediated signaling not only for aggregation but also for multicellular development.

9.4.2 Cell Adhesion and Motility[3,9]

In contrast to early expectations, csA, the glycoprotein-mediating EDTA-resistant cell contacts in the aggregation phase, has proven not to be essential for aggregation and development. A spontaneous structural mutant as well as csA-null transformants obtained by gene disruption are able to aggregate and form fruiting bodies in the complete absence of csA. Interestingly, expression of the cs-A gene in growth phase cells, achieved by placing the gene under the control of an actin promoter, leads to formation of EDTA-resistant aggregates, thus confirming the involvement of csA in this type of adhesion.

Sorting out experiments between transformants expressing csA constitutively, csA-null mutants and cells of the related slime mold species *Polysphondylium pallidum* have recently indicated that csA is probably necessary to prevent interspecific adhesion during the aggregation phase. Different slime mold species coexist in the same environment feeding on the same bacterial strains. During aggregation, different species sort out to form species-specific aggregates. Chemotaxis toward species-specific chemoattractants is not sufficient to guarantee species specificity, since some species share cAMP as a chemoattractant. Others such as *P. pallidum* produce cAMP, though they do not respond chemotactically to it. Thus, selective cell-cell adhesion can be considered a double-insurance mechanism for establishing species-specific multicellular aggregates during development.

Cell motility is essential for both aggregation and the morphogenetic events occurring in the aggregate and in the slug. The molecular basis of motility has been clarified by biochemical studies, which have led to the identification of several proteins which in vitro regulate actin filament formation and acto-myosin complexes. The use of antisense RNA or gene disruption has, however, revealed a higher degree of redundancy in the molecular basis of motility than previously expected. Antisense or null mutants for myosin and other cytoskeletal components, such as α-actinin, severin and gelation factor, are able to aggregate by chemotaxis, though the rate of movement is reduced and chemotactic orientation is somewhat impaired. In favor of the notion that motility is characterized by a high degree of redundancy is the finding that *D. discoideum* cells possess genes coding for myosin head-like proteins, in addition to single-copy genes encoding myosin I and II. Actin was already known to be coded by several genes.

9.5 Multicellular Development: Determining the Time of Cell Divergence[10]

To explain how cell interactions control differentiation and pattern formation, it is necessary to determine when cells start to diverge into the pre-spore or pre-stalk pathway. In the past, vital dyes or anti-bodies against pre-spore cells have been used as a tool to identify different cell populations. Identification of pre-spore and pre-stalk-specific gene transcripts and characterization of the genes which code for them has made a more sensitive approach possible, involving the use of "reporter" genes. The promoter of a pre-stalk or pre-spore gene is fused to sequences for β-galactosidase or SV40 T antigen, and the expression of the gene is followed in situ by an enzymatic or immunological reaction. In this way, it has been possible to determine with a greater resolution the temporal and spatial differentiation of both pre-spore and pre-stalk cells during development.

The most relevant result of these studies is that differentiation of both pre-spore and pre-stalk cells can be traced back to the loose aggregate stage, where both cell types are found homogeneously scattered within the aggregate. Following compaction, the number of both cell types increases and they sort out in different areas of the aggregate.

The random distribution observed in the early stage of differentiation suggests that short-range interactions are the relevant factor in determining the differentiation pathways of cell groups, and this must be taken into account by models involving positional signaling as the mechanism of cell differentiation.

Whether cells are determined prior to aggregation is a matter of debate. A dependence of cell-type proportioning on the phase of the cell cycle has been suggested by the finding that cells which are in early G2 at the start of development become aggregation-competent hours before cells in late G2, which in turn preferentially form spores. However, this result has not been unequivocally confirmed. It has also been shown that cells with considerable glycogen stores preferentially form spores, when mixed with less well-fed cells. On the other hand, however, when the two populations develop separately, both the timing of development and cell proportioning are identical, and therefore independent of glycogen content.

Even though heterogeneities in the cell population at the beginning of development cannot be excluded, the fact remains that cells are plastic until late in development, and their committment is modulated and modified by cellular interactions.

9.5.1 Diffusible Signals, Cell Differentiation and Pattern Formation

Assuming totipotency up to the end of aggregation, it is necessary to explain how cells diverge into the pre-spore and pre-stalk pathway and how pattern formation in the compact aggregate is achieved.

There is wide agreement, as was already mentioned, that cell differentiation is regulated by extracellular signals of cAMP and DIF, acting, at least in part, antagonistically. The evidence in favor of this hypothesis can be summarized as follows:

1. Cyclic AMP favors expression of pre-spore-specific genes. When compact aggregates are mechanically disaggregated, exogenous cAMP stabilizes pre-spore mRNA transcripts which display shorter half-lives following disaggregation. Pre-spore genes, which are repressed by disaggregation, can also be re-induced by cAMP. Degradation of cAMP by treatment of slugs with cAMP phosphodiesterase results in dedifferentiation of pre-spore cells. Isolated cells can be induced to differentiate into spore cells by using a membrane-permeant cAMP analog to elevate the concentration of intracellular cAMP.

2. Isolated cells are induced to differentiate into stalk cells by a sequential treatment with cAMP and DIF. Thus, cAMP is necessary for both differentiation pathways, however, if added at the time when only DIF is required, cAMP inhibits stalk cell formation. DIF, in contrast, inhibits pre-spore cell differentiation, while inducing stalk cell formation of isolated cells. Pre-stalk-specific genes are induced by DIF, but remarkably the requirement for DIF changes with time for different subclasses of pre-stalk-specific genes, suggesting that they are controlled by a combination of more signals, one of which is certainly cAMP. Deletions in the promoter sequences of these genes will help in dissecting the regulatory components and explaining their synergistic or inhibitory interactions.

Although these results clearly establish a role for cAMP and DIF in regulating cell differentiation, we do not know how they interact to bring about cell divergence in vivo. To explain the localized, scattered differentiation of pre-spore and pre-stalk cell (sub)populations in the aggregates, it would be necessary to determine the short-range distribution of both signals as well as the degree of responsiveness of cell groups, which very likely depends on their position in the aggregate, intracellular parameters notwithstanding.

9.5.2 Cell Adhesion and the Control of Development

Cell-cell adhesion is obviously a pre-requisite for the multicellular state to be established and, in this sense, it is crucial for generation of spatial inhomogeneities, setting the conditions for cell divergence and pattern formation. It is currently believed that cell contacts play only the passive role of keeping cells in close proximity, allowing local high concentrations of diffusible regulatory signals (cAMP and DIF) to be formed.

It is possible, however, that close cell-cell contacts are required, in the first place, for generating a diffusible signal system. We do not know, for example, how the synthesis of DIF and its receptor are regulated, and there is some evidence that full expression of the cAMP signaling system is regulated by cell adhesion.

Starving *Dictyostelium* cells are able to develop also in shaken cultures. Under these conditions, genes involved in chemotaxis and cell adhesion are fully expressed with the same timing and sequence as in cells developing on a solid substratum. Shaking does not prevent formation of small cell clumps, which are initially formed through EDTA-sensitive contacts. Some strains, however, do not aggregate at all in these conditions, probably because of reduced EDTA-sensitive contacts. These strains are unable, in shaking cultures, to relay cAMP, and fail to develop, unless treated with exogenous cAMP pulses. This suggests that cell adhesion, in addition to starvation factors, is likely to be required for the cAMP signaling system to be fully established.

In agreement with these results, selective blockage with antibodies of gp24, the glycoprotein which mediates EDTA-sensitive cell adhesion, inhibits aggregation and all subsequent development.

If this result can be confirmed by using the gene disruption or antisense approach, it would strengthen the hypothesis that cell adhesion molecules act as morphoregulatory signals.

Furthermore, expression of the csA glycoprotein in growth phase cells results in formation of aggregates and expression of some developmentally regulated genes, while cells are still in nutrient medium. Even if the stimulatory effect on gene expression could be a secondary effect of reduced nutrient uptake by the inner cell mass of aggregates, this result shows that cell adhesion affects the cellular environment in such a way that developmental gene expression is induced.

It is likely that cell adhesion plays a major role in pattern formation. As already mentioned, pre-spore and pre-stalk cells differentiate in regions of the aggregate that are different from their final location. At the compact aggregate stage, pre-spore cells congregate in the central area of the aggregate, while pre-stalk cells are enriched at the periphery of the aggregate, and later in the tip. Pre-spore and pre-stalk cells have been shown to be differentially cohesive, the pre-spore cells being much less sensitive to dissociation with EDTA. This differential cohesiveness could explain the pattern observed in compact aggregates and fits well with sorting out models, in which the more cohesive cells are localized in the interior of co-aggregates.

9.6 Outlook

Considerable progress has been made in the definition of the molecular basis of cell-cell communication and cell adhesion in *Dictyostelium*. The regulation of aggregation by cAMP appears to be largely understood. However, the transmembrane and intracellular signal elements linking the cAMP receptor to gene expression as well as the regulatory elements at the gene level are still obscure. The structure of DIF and its metabolism are known, and attempts to characterize its receptor are under way. Both DIF and the receptor are regulated, but we do not have knowledge of the enzymes involved in DIF synthesis and of how they are regulated.

Cell divergence into pre-spore- and pre-stalk pathways can be accounted for by short-range interactions of DIF and cAMP within the aggregate. The isolation of pre-spore and pre-stalk-specific genes which are controlled by cAMP and DIF should facilitate identification of the different regulatory sequences at gene level and clarify their mutual interactions. A major task in the future will be to dissect the mechanisms that give rise to cell sorting out during pattern formation. The involvement of cell adhesion in this process could be investigated at molecular level by transforming cells with the csA gene under the control of pre-spore- or pre-stalk-specific promoters, thus altering the adhesive capacities of a specific cell population.

Pattern formation is a dynamic process involving extensive movements of individual cells as well as whole tissues during slug formation and, in particular, during culmination, when the pre-stalk region invades the pre-spore region of the forming fruiting body. Little is known about the mechanisms controlling these morphogenetic movements, but it is reasonable to assume as a working hypothesis that they are regulated at least in principle by the same cell interactions which are active during aggregation, i.e., oriented cell motility, cell-cell adhesion, and cell adhesion to the substratum (the extracellular matrix of the stalk tube). For an understanding of these mechanisms it is therefore important to define the molecular basis of adhesion at this stage and the potential role of chemotactic motility in the culmination process.

9.7 Summary

Because of the features of its development, *Dictyostelium* is a convenient model to illustrate how cell interactions lead to the construction of a multicellular organism, to cell differentiation, and morphogenesis. These processes involve about 2000 genes, which are coordinately expressed in clusters at different times of development, under the control of cell-cell signaling systems which are themselves developmentally regulated. Analysis of the best characterized signal systems in *Dictyostelium*, cAMP and DIF, reveals complex interactions between the signals, which work in parallel as well as synergistically and antagonistically to bring about cell differentiation and pattern formation.

Cyclic AMP also acts as chemoattractant and is required, together with cell-cell adhesion and cell motility, for transition from unicellularity to multicellularity. DIF is a unique effector of cell-type specific gene expression, which regulates stalk-cell differentiation. Cell adhesion and motility are important primary processes for transition from unicellularity to multicellularity and for both pattern formation as well as morphogenetic shape changes that occur in the multicellular aggregate. A combination of biochemical, molecular genetic, and cell biology approaches has made it possible to unravel the molecular basis of cell communication, cell adhesion, and motility, to identify structural genes involved in these processes, to study their regulation, and to better determine the role of each gene during development.

References

[1] Bonner JT (1967) The cellular slime molds. Princeton University Press, Princeton

Loomis WF (1975) *Dictyostelium discoideum*. A developmental system. Academic Press, New York

Raper KB (1984) The dictyostelids. Princeton University Press, Princeton

[2] Devreotes PN, Zigmond SH (1988) Chemotaxis in eukaryotic cells: a focus on leukocytes and *Dictyostelium*. Annu Rev Cell Biol 4:649–678

Janssens PMW, Van Haastert PJM (1987) Molecular basis of transmembrane signal transduction in *Dictyostelium discoideum*. Microbiol Rev 51:396–418

McRobbie SJ (1986) Chemotaxis and cell motility in the cellular slime molds. CRC Crit Rev Microbiol 13:335–375

Newell PC, Europe-Finner GN, Liu G, Gammon B, Wood CA (1990) Signal transduction for chemotaxis in *Dictyostelium* amoebae. Semin Cell Biol 1:105–113

[3] Beug H, Katz FE, Gerisch G (1973) Dynamics of antigenic membrane sites relating to cell aggregation in *Dictyostelium discoideum*. J Cell Biol 56:647–658

Bozzaro S, Merkl R, Gerisch G (1987) Cell adhesion: its quantitation, assay of the molecules involved and selection of defective mutants in *Dictyostelium* and *Polysphondylium*. In: Spudich JA (ed) *Dictyostelium discoideum*: Molecular approaches to cell biology. Methods in Cell Biology, vol 28. Academic Press, New York, pp 359–385

Gerisch G (1986) Interrelation of cell adhesion and differentiation in *Dictyostelium discoideum*. J Cell Sci (Suppl) 4:201–219

Loomis WF (1988) Cell-cell adhesion in *Dictyostelium discoideum.* Dev Genet 9:549–559

Müller K, Gerisch G (1878) A specific glycoprotein as the target site of adhesion-blocking Fab in aggregating *Dictyostelium* cells. Nature 274:445–449

Siu CH, Cho A, Choi AHC (1987) The contact site A glycoprotein mediates cell-cell adhesion by homotypic binding in *Dictyostelium discoideum.* J Cell Biol 105:2523–2533

Wilcox D, Sussman M (1981) Serologically distinguishable alterations in the molecular specificity of cell adhesion during morphogenesis in *Dictyostelium discoideum.* Proc Natl Acad Sci USA 78:358–362

4 Cardelli JA, Knecht DA, Wunderlich R, Dimond RL (1985) Major changes occur during at least four stages of development of *Dictyostelium discoideum.* Dev Biol 110:147–156

Clarke M, Yang J, Kayman SC (1988) Analysis of the pre-starvation response in growing cells of *Dictyostelium discoideum.* Dev Genet 9.315–326

Kessin RH (1988) Genetics of early *Dictyostelium discoideum* development. Microbiol Rev 52:29–49

Loomis WF (1990) Essential genes for development of *Dictyostelium.* In: Jeanteur P, Kuchino Y, Müller WEG, Paine PC (eds) Progress in molecular and subcellular biology, vol 35. Springer, Berlin Heidelberg New York, pp 159–183

Mangiarotti G, Bozzaro S, Landfear S, Lodish HF (1983) Cell-cell contact, cyclic AMP and gene expression during development of *Dictyostelium discoideum.* In: Moscona AA, Monroy A (eds) Curr Top Dev Biol, vol 18. Academic Press, New York, pp 117–154

5 Gerisch G (1987) Cyclic AMP and other signals controlling cell development and differentiation in *Dictyostelium.* Annu Rev Biochem 56:853–879

Haribabu B, Dottin RP (1986) Pharmacological characterization of cAMP receptors mediating gene regulation in *Dictyostelium discoideum.* Mol Cell Biol 6:2402–2408

Hjorth A, Datta S, Khanna NC, Firtel RA (1988) Analysis of cis- and trans-acting elements involved in cAMP-inducible gene expression in *Dictyostelium discoideum.* Dev Genet 9:435–454

Mann SKO, Firtel RA (1989) Two-phase regulatory pathways control cAMP receptor-mediated expression of early genes in *Dictyostelium.* Proc Natl Acad Sci USA 86:1924–1928

May T, Kern H, Müller-Taubenberger A, Nellen W (1989) Identification of a cis-acting element controlling induction of early gene expression in *Dictyostelium discoideum.* Mol Cell Biol 9:4653–4659

Williams JG (1989) Extracellular signals and intracellular transduction pathways regulating *Dictyostelium* development. Curr Opinions Cell Biol 1:1132–1138

6 Insall R, Kay RR (1990) A specific DIF binding protein in *Dictyostelium.* EMBO J 9:3323–3328

Morris H, Taylor G, Masento M, Jermyn K, Kay RR (1987) Chemical structure of the morphogen differentiation inducing factor from *Dictyostelium discoideum.* Nature 328:811–814

Town CD, Gross JG, Kay RR (1976) Cell differentiation with-

out morphogenesis in *Dictyostelium discoideum.* Nature 262:717–719

Williams JG, Ceccarelli A, McRobbie S, Mahbubani H, Kay RR, Early A, Berks M, Jermyn KA (1987) Direct induction of *Dictyostelium* prestalk gene expression by DIF provides evidence that DIF is a morphogen. Cell 49:185–192

7 De Lozanne A (1987) Homologous recombination in *Dictyostelium* as a tool for the study of developmental genes. In: Spudich JA (ed) *Dictyostelium discoideum:* molecular approaches to cell biology. Methods Cell Biol, vol 28. Academic Press, New York, pp 489–495

Nellen W, Datta S, Reymond C, Sivertsen A, Mann S, Crowley T, Firtel RR (1987) Molecular biology in *Dictyostelium:* tools and applications. In: Spudich JA (ed) *Dictyostelium discoideum:* molecular approaches to cell biology. Methods Cell Biol, vol 28. Academic Press, New York, pp 67–100

Newell PC (1982) Genetics. In: Loomis WF (ed) The development of *Dictyostelium discoideum.* Academic Press, New York, pp 35–70

8 Faure M, Franke J, Hall AL, Podgorski GI, Kessin RH (1990) The cyclic nucleotide phosphodiesterase gene of *Dictyostelium discoideum* contains three promoters specific for growth, aggregation and late development. Mol Cell Biol 10:1921–1930

Klein PS, Sun TJ, Saxe CL, Kimmel AR, Johnson RL, Devreotes PN (1988) A chemoattractant receptor controls development in *Dictyostelium discoideum.* Science 241:1467–1472

Kumagai A, Pupillo M, Gundersen R, Miake-Lye R, Devreotes PN, Firtel RA (1989) Regulation and function of Gα protein subunits in *Dictyostelium.* Cell 57:265–275

9 De Lozanne A, Spudich JA (1987) Disruption of the *Dictyostelium* myosin heavy chain gene by homologous recombination. Science 236:1086–1091

Faix J, Gerisch G, Noegel AA (1990) Constitutive overexpression of the contact site A glycoprotein enables growth-phase cells of *Dictyostelium discoideum* to aggregate. EMBO J 9:2709–2716

Knecht DA, Loomis WF (1987) Antisense RNA inactivation of myosin heavy chain expression in *Dictyostelium discoideum.* Science 236:1081–1086

Noegel AA, Schleicher M (1990) The contractile system in nonmuscle cells: involvement of actin and actin-binding proteins. In: Dedmann JR, Smith VL (eds) Stimulus response coupling: the role of intracellular calcium. CRC Press, Boca Raton, pp 57–82

Titus MA, Warrick HM, Spudich JA (1989) Multiple actin-based motor genes in *Dictyostelium.* Cell Regulation 1:55–63

10 Ceccarelli A, Mahbubani HJ, Williams JG (1991) Positively and negatively acting signals regulating stalk cell and anterior-like cell differentiation in *Dictyostelium.* Cell 65:14–20

Gomer RH, Firtel RA (1987) Cell-autonomous determination of cell-type choice in *Dictyostelium* development by cell-cycle phase. Science 237:758–762

MacDonald SA, Durston AJ (1984) The cell cycle and sorting behaviour in *Dictyostelium discoideum*. J Cell Sci 66:195–204

Maeda Y, Ohmori T, Abe T, Abe F, Amagai A (1989) Transition of starving cells to differentiation phase at a particular position of the cell cycle. Differentiation 41:169–175

Schaap P (1986) Regulation of size and pattern in the cellular slime molds. Differentiation 33:1–6

Tasaka M, Takeuchi I (1981) Role of sorting out in pattern formation in *Dictyostelium discoideum*. Differentiation 18:191–196

Williams JG, Duffy KT, Lane DP, McRobbie SJ, Harwood AJ, Traynor D, Kay RR, Jermyn KA (1989) Origin of the prestalk-prespore pattern in *Dictyostelium* development. Cell 59:1157–1163

Chapter 10 Control of the Cell Cycle in Yeasts

Peter Fantes

10.1 Introduction

The basic eukaryotic cell cycle comprises the series of processes by which the cell's genetic material is duplicated, and the two new copies segregated to daughter nuclei. In eukaryotes, the two processes of DNA replication and chromosome segregation are temporally separated: the former taking place during the S (synthetic) phase, the latter during mitosis. Except in a few special cell types, there are intervals between S phase and mitosis: G1 is the period (gap 1) between mitosis and S phase, while S phase is followed by G2, which leads to mitosis (Fig. 1). The S phase and mitosis alternate under most circumstances, although there are naturally occurring situations where this rule is broken, such as during polytenisation of dipteran chromosomes.

The "core" cycle of G1-S-G2-M is usually accompanied by other processes such as cell division, which although closely associated with mitosis in most cells, is not an essential part of the core cycle. Early *Drosophila* embryos undergo a number of rounds of DNA replication and mitosis before cellularisation. Plasmodia of the slime mould *Physarum polycephalum* consist of a single multinucleate cell or syncytium, in which there is no cell division.

The ways in which the cells of plants, animals and fungi divide are diverse, but the basic processes of DNA replication and mitosis appear to be very similar in all eukaryotic cells. This chapter will concentrate on the central DNA synthesis-mitosis cycle, only referring to other processes where they are particularly significant.

A number of cell types have been particularly informative about certain aspects of the cycle (Fig. 2). *Physarum polycephalum,* for instance, has been used because it is a convenient system in which all the nuclei in a single plasmodium pass through mitosis and replicate their DNA with remarkable nat-

ural synchrony. The rapid cycles of early embryos (fertilised eggs) consist of only S and M, and have been considered to represent the "pure" cell cycle since G1 and G2 are not present. In addition to their convenience, it has recently become clear that observations on these model systems often have much general relevance.

This chapter concentrates on studies of the cell cycle in the two yeasts *Saccharomyces cerevisiae* and *Schizosaccharomyces pombe.* These species are ideally suited to genetical investigations: they have small haploid genomes and so it is relatively easy to obtain mutants defective in particular aspects of the cell cycle; mutants can be analysed and combined through meiotic crosses; and in the past decade sophisticated methods have been developed for manipulation of their genomes at the molecular level. Another chapter (see Philipson and Sorrentino, Chap. 36, this Vol.) deals with the related topic of how mammalian cell proliferation is controlled.

10.2 Yeast Cell Cycles[1]

As *Saccharomyces cerevisiae* cells progress through the cell cycle, they undergo characteristic morphological and biochemical changes. The most prominent events observable by the light microscope are the emergence and subsequent expansion of a bud, the extension into the bud neck of the nucleus, the division of the nucleus, and finally the cleavage of the bud from the mother cell. Normally, the nuclear DNA starts to replicate at the time of bud emergence (Fig. 3).

In *S. pombe,* morphological changes during the cell cycle are less striking than in *S. cerevisiae.* The cells increase in length during the first 3/4 of the

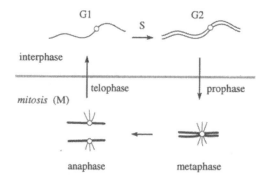

Fig. 1. The chromosome cycle. Uncondensed chromosomes in G1 are replicated during the S phase and remain uncondensed through G2. At prophase of mitosis the chromosomes condense, and the nuclear envelope breaks down. At metaphase the chromosomes align on the metaphase plate, attached by their centromeres to the spindle microtubules. The centromere splits at anaphase and the daughter chromosomes are segregated to daughter nuclei by the force of the spindle microtubules. During telophase the chromosomes decondense as they are enclosed in the daughter nuclear envelopes. Only a single chromosome is shown for clarity. The nuclear envelope is not shown.

cycle, at which point elongation ceases, and the cells enter mitosis. A septum is laid down centrally, and this is the only prominent cell cycle landmark visible under low power microscopy. The septum is cleaved with the formation of two new daughter cells. The S phase is short and takes place very early in the cycle (Fig. 4).

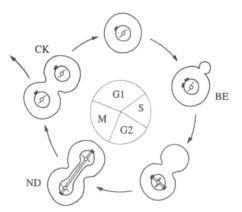

Fig. 3. *Saccharomyces cerevisiae* cell cycle. Progress of a cell through the cycle, indicating the progress of the bud, and the nucleus, spindle pole body and spindle. Approximate cycle phases are also shown. The progress of a single chromosome during the cycle is shown.

10.2.1 Cell Cycle Mutants

The procedure originally used to isolate *cdc* (cell division cycle) mutants of *S. cerevisiae* started with the isolation of a large number of temperature-sensitive lethal mutants, able to grow at 23 °C but not at 36 °C. Temperature-sensitive (*ts*) mutants were used because the inability to progress through the cell cycle is lethal, and a mutant with an absolute

Cell type	Relative cycle phases				Total cycle time
cultured fibroblast	G1	S	G2	M	24 hr
Saccharomyces cerevisiae	G1	S	G2	M	1.5 hr
Schizosaccharomyces pombe	G1 S		G2	M	2.5 hr
Physarum polycephalum	S		G2	M	12 hr
Fertilised egg *Xenopus*		S	M		0.5 hr

Fig. 2. Durations of the cycle phases. The approximate relative durations of the cycle phases are shown, along with typical total cycle times. Cycle times can be substantially greater than those shown, and this can affect the relative lengths of the cycle phases

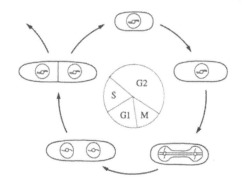

Fig. 4. *Schizosaccharomyces pombe* cell cycle. Progress of a cell through the cycle, showing the behaviour of the nucleus, the septum and cell division. The spindle pole body duplicates shortly before mitosis (not shown). Note: the durations of the cycle phases are deliberately not drawn to scale: G2 occupies 75% of the cycle, with the three other phases occupying roughly equal periods of the remainder. G1 and S phase are very short and are essentially completed during the interval between M and cell division. Hence, newborn cells are already in G2

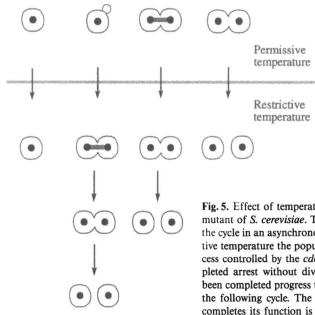

Fig. 5. Effect of temperature shift on temperature-sensitive *cdc* mutant of *S. cerevisiae*. The *top line* shows cells at all stages of the cycle in an asynchronous population. On shift to the restrictive temperature the population is split. Cells in which the process controlled by the *cdc* gene in question has not been completed arrest without division; cells in which this process has been completed progress through the cycle, divide, and arrest in the following cycle. The point at which the *cdc* gene product completes its function is termed its execution point

(temperature-independent) defect in cycle progress would never progress beyond a single defective cell. To identify those mutants specifically defective in cell cycle progress, use was made of cycle-related changes in cellular morphology.

Any growing *S. cerevisiae* culture contains a mixture of unbudded cells and cells with small and large buds. When growing cells of a *ts cdc* mutant are shifted to the restrictive temperature, the ratios of the various morphological types will change. For instance, the effect of a temperature shift on a mutant conditionally defective in bud emergence is as follows (Fig. 5). Cells early in the cycle will be unable to initiate bud emergence and will arrest as unbudded, while budded cells will complete their current cycles but will arrest before bud initiation in the next cycle. After a few generations all the cells will be unbudded. Other types of *cdc* mutant will arrest with only large buds, and so on. Therefore mutants arresting with uniform morphology after shift to the restrictive temperature were picked as prospective *cdc* mutants. Other temperature-sensitive lethals would be unlikely to lead to uniform arrest morphology.

The original definition of a cell cycle mutant in *S. pombe* was rather different. The criterion for de-

ciding whether a particular mutant had a cell cycle defect was the formation of abnormally elongated cells at the restrictive temperature. The rationale for this was based on the previous observation that inhibition of the *S. pombe* cell cycle using chemical inhibitors permitted cell elongation, and oversize cells were produced as biomass continued to increase in the absence of division. Mutants showing elongation at the restrictive temperature were analysed, and found to be defective in DNA replication, mitosis, or septum formation (Table 1).

The difference in the criteria used to define cell cycle mutants in the two yeasts means that the spectrum of mutants isolated do not completely overlap. While the majority of *S. cerevisiae* mutants increase in biomass during the cycle block, a minority do not. This class was, of course, not obtained, in *S. pombe* because of the imposed requirement for cell elongation.

Other approaches to isolating cell cycle mutants have been taken; for instance, many cold-sensitive *cdc* mutants have been isolated in both yeasts. Often cold-sensitive mutations have identified new genes in which no heat-sensitive mentations have been obtained. Other cell cycle genes have been identified by more focussed searches; for instance, mutations with specific biochemical defects, or new

Table 1. cdc Mutants discussed in the text[a]

Gene	Role in cycle	Biochemical activity	Comments
Saccharomyces cerevisiae			
CDC28	Start	Protein kinase	May have a role later in cycle
CLN3	Start	Cyclin homologue	Also called *WHI1*, *DAF1* affects cell size
CLN1,2	Start	Cyclin homologue	
CKS1	?Start	Homologue of *S. pombe* *suc1*	Interacts with *CDC28*
MIH1	G2	Homologue of *S. pombe* *cdc25*	Not essential
RAD9	G2	Unknown	Part of G2 checkpoint system
Schizosaccharomyces pombe			
cdc10+	Start	Unknown	
cdc2+	Start G2-M	p34 protein kinase	Can mutate to *cdc⁻* by loss of activity, or to *wee* phenotype
wee1+	G2-M	Predicted protein kinase	Negative regulator of *cdc2+*
cdc25+	G2-M	Unknown	Activator of *cdc2+*
nim1+/cdr1+	G2-M	Predicted protein kinase	Negative regulator of *wee1+*
cdc13+	G2-M	Cyclin homologue	Protein interacts with p34 protein kinase
suc1+	Start G2-M	Unknown	Protein interacts with p34 protein kinase
nda2+	M	α-Tubulin	Cold-sensitive mutants defective in spindle
nda3+	M	β-Tubulin	Cold-sensitive mutants defective in spindle
top2+	M	Topoisomerase II	Required for chromosome condensation and separation at mitosis
cut1+	M	Unknown	Mutants block nuclear division but not cytokinesis

[a] The conventions for naming normal and mutant genes (strictly, genetic alleles) are different for the two yeasts. In *S. cerevisiae*, the wild-type allele is normally given in capital letters, and mutants in lower case. In *S. pombe*, the wild-type allele is indicated by a superscripted " + " sign

mutations that interact genetically with known mutants. About 70 cdc and related genes are known in *S. cerevisiae*; for *S. pombe* the number is about 50. Estimates suggest that 1–10% of genes are concerned with cell cycle progress.

The availability of conditional cdc mutants has extended the range of tools for cell cycle analysis considerably. Far more cell cycle processes can be directly manipulated than by using chemical or physical agents. A further advantage in using mutants is that the primary defect in a mutant is confined to a single gene, and therefore to a single protein. In contrast, inhibitors often have an unknown primary target within the cell, and many have side-effects.

There are, however, a few drawbacks to the use of cdc mutants.

1. Only genes that can be mutated to temperature-sensitive function can be readily accessed genetically by the procedures described above. Furthermore, some genes are present in more than one copy in the genome, and mutations in one of them may not lead to an observable cellular phenotype.
2. Complete absence of function can never be guaranteed with a temperature-sensitive mutant: such mutations are always of the "missense" type, and residual function at the restrictive temperature can lead to misleading conclusions. In a similar way, a mutant grown at its permissive temperature may not behave as would a wild-type cell.
3. Some genes which are required for cell cycle progress may not give a cdc phenotype when mutated, because the gene product is also required for some other cellular process.
4. Occasional cell cycle mutants reflect not loss but alteration of function: this is not easy to detect without a full molecular analysis.

cdc mutants can be used in a number of ways as physiological tools to dissect the workings of the cell cycle, and can also be used to obtain clones of the genes for molecular analysis. The following short section is included to provide an outline of the molecular techniques available in yeasts.

10.2.2 Genes and Clones

Exogenous DNA can be introduced into yeasts by transformation, which can be used in a variety of ways. To isolate a plasmid from a genomic library that carries a functional copy of a gene, the library is transformed into a strain carrying a lethal mutation in the gene. The presence of a selectable marker in the library vector allows transformants to be selected. Among these, transformants containing the gene of interest on a plasmid will be able to grow at the restrictive temperature, while others will generally not. Plasmid is recovered from such strains for analysis.

Plasmids carrying sequences other than the authentic cdc^+ gene are sometimes isolated by this procedure; they carry instead extragenic suppressor genes. Extragenic suppressors are usually effective only when overexpressed, such as when present on multi-copy plasmids. They can be a nuisance, since the aim is generally to clone the cdc^+ gene itself, but many new genes have been identified as suppressors which have later been shown to be functionally related to the cdc gene initially sought.

Integrative transformation is a powerful tool in the molecular genetic analysis of yeasts, because integration of a plasmid into the genome usually occurs by homologous recombination. This can be used to "tag" a chromosomal locus, or often to generate a deletion allele of a cloned gene after in vitro modification.

In general, any genes derived from the host species' own genome will be expressed and regulated by the cells' own machinery. Heterologous genes can also be expressed under appropriate conditions.

10.3 Start[2]

10.3.1 Pathways in the *S. cerevisiae* Cell Cycle

The cell cycle of *S. cerevisiae* involves several prominent events: bud emergence (BE) and growth, replication of the nuclear DNA (DS), extension of the nucleus into the bud neck (nuclear migration, NM), nuclear division (ND), and separation of the bud from the mother cell (cytokinesis, CK). By arresting the cycle with a mutant defective in one event and observing the effect on others, it has been possible to elucidate the relationships between the various processes. This approach led to the model of the *S. cerevisiae* cell cycle shown in Fig. 6, with DS and ND forming one dependent sequence, and BE, bud expansion, and CK the other.

In order to complete CK, both ND and the budding sequence must have taken place, providing a point of convergence between the pathways. Evidence for another connection is provided by mutants which arrest as unbudded G1 cells, and initiate neither bud formation nor the DS-ND pathway. Mutants that arrest in this way define a unique event early in the cycle, whose completion is necessary for initiation of both dependent pathways. This event is termed "start".

10.3.2 Start and Commitment

"Start" in *S. cerevisiae* is the point in the cycle at which the cell takes decisions about its future development: whether to embark on a new mitotic cycle, or to enter the meiotic pathway, or to conjugate with a cell of opposite mating-type, or to remain in G1 if nutrients are limiting. Once the decision has been taken to initiate a new mitotic cycle, that decision is irrevocable, and no switch into any other developmental route can be made. As shown in Fig. 6, three independent intracellular pathways are initiated by passing start: one leads to DS and thence to mitosis; another allows the spindle pole body to duplicate, which subsequently initiates formation of the mitotic spindle; and a third leads to bud emergence and subsequent expansion.

When *S. cerevisiae* start mutants arrest, they are uncommitted to any developmental pathway. In particular, start mutants are able to conjugate directly from the arrested state, which distinguishes them from other *cdc* mutants. The corresponding property was used to investigate start in *S. pombe,* where it was found that only *cdc2* and *cdc10* mutants were able to conjugate from their block points, suggesting that these genes control start in this yeast.

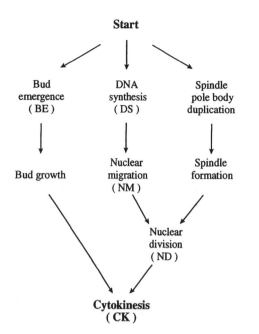

Start

Bud
emergence
(BE)

DNA
synthesis
(DS)

Spindle
pole body
duplication

Bud growth

Nuclear
migration
(NM)

Spindle
formation

Nuclear
division
(ND)

Cytokinesis
(CK)

Fig. 6. "Start" and developmental pathways in *S. cerevisiae*. The three main pathways in the *S. cerevisiae* cell cycle are shown. The initiation of each requires completion of start; nuclear division requires the DNA replication and spindle pathways; cytokinesis requires nuclear division and bud formation

The *CDC28* gene is known to play a central role in the execution of start in *S. cerevisiae*. The gene was cloned by complementation; its product was predicted from the DNA sequence to be a protein kinase, a prediction subsequently confirmed directly. The kinase is active against a number of substrates in vitro, and putative in vivo substrates have been identified.

Proteins believed to interact with the *CDC28* kinase have been identified by an indirect genetic route. Plasmids able to suppress *ts* mutants of *CDC28* were isolated, and the sequences of the genes determined. Two of these, *CLN1* and *CLN2*, were found to show significant homology at the protein level with cyclins. Cyclins had previously been identified as concerned with the G2-M transition (see Sect. 10.5). In contrast, the products of the *CLN* genes may be primarily concerned with start. A third cyclin homologue that also acts in G1 is discussed below. Recently, G1-acting cyclins have been identified in *S. pombe* and other organisms.

10.3.3 Growth Control

Mutants such as *cdc28* continue to accumulate biomass after cycle arrest. There are other mutants that arrest at start but do not show net growth at the restrictive temperature. Whether they should properly be classified as cell cycle mutants is debatable, but some are clearly involved in controlling cell development in G1; some are defective in initiating meiosis, while others initiate meiosis under inappropriate environmental conditions. A number of nongrowing start mutants are defective in a complex growth control system in which the *S. cerevisiae* homologues of the *ras* genes regulate the activity of adenylate cyclase, and thereby the level of cAMP and the activity of cAMP-dependent protein kinases (see Philipson and Sorrentino, Chap. 36, this Vol.). The relationship between the growth control system and cell cycle control at start is unclear.

10.3.4 "Start" in Other Cells?

The existence of the start control in G1 of *S. cerevisiae* is beyond doubt, but its exact nature is obscure, and cells arrested at start by different means are not exactly equivalent. In neither *S. cerevisiae* nor *S. pombe* is the biochemistry of start well understood, and this will remain a topic of great interest for future investigation.

The existence of controls in the G1 of mammalian fibroblasts has long been known. The "R" or restriction point may define a unique cell cycle stage equivalent to start. The restriction point is the major control point in actively cycling mammalian cells, although cells emerging from quiescence and entering the cell cycle must traverse several controls en route (see Philipson and Sorrentino, Chap. 36, this vol.).

10.4 Coordination of Growth and Division[3]

Cells grown in culture tend to maintain a characteristic size under given conditions. For yeasts and some other microorganisms, this has been shown to reflect the operation of a control that regulates cell size. Cell size control has been investigated both genetically and by manipulation of the growth conditions.

10.4.1 *S. cerevisiae*

In *S. cerevisiae,* the major cell size control mechanism acts at start. The size at which a mother cell initiates a bud (BE) is characteristic of the genotype and the growth conditions. When the medium is poor and growth slow, the critical size for BE is reduced. By shifting cells between poor and rich media, the point in the cycle at which the critical size is "set" has been shown to occur shortly before BE itself, at the time of start.

Mutants with altered cell size at BE have been isolated. The most interesting of these map at the locus variously referred to as *WHII, DAFI* or (now) *CLN3.* As the third name suggests, the predicted product of the gene shows homology to cyclins, as do *CLN1* and *CLN2* genes. The *CLN* mutants show a reduced critical size, and some exhibit a higher level of resistance to α-factor than normal cells (α-factor is a mating pheromone secreted by cells of mating-type *MATa* which arrests cells of mating-type *MATa* at start). *CLN3* mutants, at least, have a significantly reduced G1, suggesting that start is advanced. A surprising observation is that deletion of the functional *CLN3* allele leads to an *increase* in size, rather than the size reduction in the original mutants, suggesting that the protein products of small-cell alleles are hyperactive. A likely explanation for this is that the hyperactive alleles have greater activity because their products are poorly degraded by intracellular proteases. As discussed in Section 10.5, the degradation of G2 cyclins plays a central role in mitosis. By analogy, an alteration in the degradability of the *CLN* start cyclins might influence their net activity within the cell, and advance start.

10.4.2 *S. pombe*

The control of cell size in *S. pombe* is similar to that in *S. cerevisiae,* but an important difference is

that regulation occurs at the G2-M transition rather than in G1 at start. Size control is better understood in *S. pombe,* at least in part because its linear growth pattern makes changes in size easier to detect. This has facilitated both direct observation and the isolation of cell size mutants. Time-lapse observations on growing *S. pombe* cells show that random deviations from the average size at one division are almost completely corrected by the next, indicating tight control over size.

The growth of *S. pombe* on poor media reduces cell size, and shifts between poor and rich media are accompanied by a resetting of the critical size, but in *S. pombe* this is a critical size for mitosis (Fig. 7). Cell size mutants are particularly well characterised in *S. pombe,* and have provided the genetic route into the control regulating entry into mitosis, as discussed in detail in Section 10.5.

Point mutants in two genes, *wee1* and *cdc2,* can lead to small cell phenotype. Some *wee1* mutants show temperature-sensitive expression of the small cell phenotype, so that at 25 °C their size is similar to wild-type, while at 35 °C it is reduced to about half the normal. Transfers between the two temperatures show resetting of the critical cell size required to initiate mitosis, in a similar way to that observed for nutrient shifts. The *wee1*[+] gene encodes a negative regulator or inhibitor of mitosis, since null *wee1* mutations are advanced into mitosis.

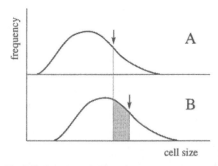

Fig. 7 A, B. Cell size control. The diagrams show the size distributions of populations of cells: in **A**, small cells; in **B**, large cells. Within each population larger cells are older, i.e., closer to division. Attainment of a critical size (*arrows*) commits cells to division. On shift from conditions shown in **B** to **A**, a proportion of the cells (*shaded*) are oversize for the new conditions, and are stimulated to divide

10.4.3 Size Controls in G1 or G2

The cell size controls in *S. cerevisiae* and *S. pombe* (Fig. 8) are similar in some respects, for instance in being nutritionally sensitive, despite acting at different stages in the cycle. This similarity extends to the genetic level, since the cloned *S. cerevisiae*

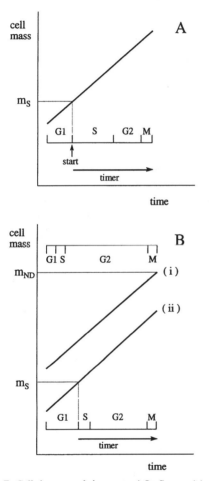

Fig. 8 A, B. Cell size controls in yeasts. **A** In *S. cerevisiae* the major cell size control acts in G1, and newborn cells must attain a critical mass (m_s) to complete start. The time from the initiation of DNA replication to division is relatively constant, irrespective of the generation time. This period may be controlled by a specific mechanism referred to as a "timer", or may simply reflect the time it takes for a cell to complete the events of S phase, G2 and mitosis. **B** In *S. pombe* the major size control acts at entry to mitosis (i) for which a critical mass (m_{nd}) is required. In undersize cells (ii) a back-up control operates similar to initiation of DNA replication limiting overall cycle progress. A timer may control the period from S phase to division.

CDC28 and *S. pombe cdc2$^+$* genes are highly homologous and functionally interchangeable (Sect. 10.5). Both genes are capable of functioning at G1 and at G2/M.

The requirement for *cdc2$^+$* function at start in *S. pombe* has already been mentioned, and may be related to a second size control mechanism. Normally G1 is very short in *S. pombe*, but when cells are unusually small, for example in *wee* mutants or after nitrogen deprivation, it is prolonged. Such small cells need to grow and attain a critical size before S phase can be initiated. This takes time and hence G1 is extended. Under normal conditions, cells are born larger than the critical size, and S phase starts very soon after mitosis. The G1/S phase size control is therefore cryptic in *S. pombe* under many conditions.

There is little evidence for the occurrence in *S. cerevisiae* of a cell size-related control after start. However the cross-complementation of the *S. pombe cdc2$^+$* and *S. cerevisiae CDC28* genes, and other evidence, shows that the latter has at least the potential to act in a G2 role (Sect. 10.5).

10.4.4 Size Controls in Other Cells

In question of whether cell size is actively controlled in other cell types, especially mammalian cells, is of long standing. In prokaryotes such as *Escherichia coli*, both the length and the mass of the cell are important in regulating the key cell cycle events of the initiation of DNA synthesis and cell division. Among eukaryotes, some form of size control has been observed in *Amoeba proteus*, *Physarum polycephalum*, and other "lower" eukaryotes. The cycles of the green algae *Chlamydomonas reinhardtii* and *Chlorella fusca* are subject to size control in a different way. Their cycles are split into growth and division phases, corresponding to the light and dark parts of the day. During the dark phase, between one and three divisions may occur in the absence of growth, and cell size is thought to determine the number of divisions. Eggs and early embryos are nongrowing systems whose cycles are very short and lack G1 and G2. These special circumstances may abolish the need for any size-monitoring mechanism.

In cultured mammalian cells the size control question is unresolved, despite extensive investigation. However genes concerned with size control in yeasts such as *cdc2/CDC28* have homologues in many and perhaps all eukaryotes, showing that the potential for a size control mechanism is very widespread.

10.5 The G2-Mitosis Transition[4]

The transition from G2 to mitosis has been intensively studied in the past few years, from a genetical standpoint in yeasts, and biochemically in other cells such as early embryos. The recent convergence of the two approaches has led to some of the most exciting recent insights in cell biology. This section will describe the work on *S. pombe* and indicate parallels with other systems.

10.5.1 Regulation of the *cdc2* Protein Kinase p34

Entry into mitosis in *S. pombe* is regulated by a cell size control mechanism, in which the *cdc2*, *wee1* and *cdc25* genes are of central importance. Other genes are involved, and the combination of classical and molecular genetics has led to the model in Fig. 9. The protein kinase encoded by the *cdc2* gene, p34, acts on several substrates, whose phosphorylation initiates the various pathways in mitosis (see Sect. 10.7). The key question is how p34 is regulated so that it triggers mitosis at the correct time and/or the correct cell size. So far this question remains unanswered, but recent rapid progress in understanding the behaviour of p34 at the molecular level will enable the more subtle timing controls to be unravelled.

The p34 protein kinase is phosphorylated at several sites, and loss of phosphorylation on one particular tyrosine residue is thought to activate it. Replacing this tyrosine by a phenylalanine residue, which cannot be phosphorylated, results in the loss of negative regulation and *wee* phenotype. During the cell cycle, the tyrosine residue is transiently dephosphorylated at mitosis coincident with high p34 protein kinase activity. The *wee1*[+] and

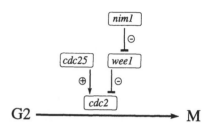

Fig. 9. Mitotic control in *S. pombe*. The pathways involved in controlling the G2-M transition in *S. pombe* are shown. *Arrows* and "+" symbols indicate positive regulation (activation); *bars* and "–" symbols, negative regulation (inhibition). It is not known whether there is direct contact between the gene products indicated. Other genes such as *suc1* are known to be involved, but their precise roles are unknown, and are not shown

cdc25[+] gene products are both rate-limiting for initiating mitosis and act through p34, but have opposing effects. *wee1*[+] inhibits mitosis, and loss of *wee1*[+] function advances mitosis. Recently it has been shown that the *wee1*[+] product has protein kinase activity: unusually, it can phosphorylate both serine and tyrosine residues. It is able to phosphorylate p34 on the critical tyrosine under appropriate conditions. The product of a newly identified gene, *mik1*[+], is a protein kinase also able to phosphorylate the critical tyrosine of p34.

In contrast, *cdc25*[+] function is normally essential for activation of p34, and increased *cdc25*[+] expression leads to advanced mitosis. A homologue of the *cdc25*[+] product has been shown to act as a tyrosine phosphatase with activity against p34.

The *suc1* gene plays a major role in the control of mitosis. Genetic evidence suggested that its product, p13, might physically interact with the p34 kinase, and this has been confirmed directly at the protein level. How the interaction affects p34 kinase activity in vivo, and what the role of p13 is in mitosis, are still unknown. Homologues of *suc1* and its product have been identified in several cell types. In *S. cerevisiae* the *CKS1* gene, identified as a suppressor of *cdc28* mutations, encodes a p18 protein highly homologous to p13.

10.5.2 MPF and Cyclins[5]

In *Xenopus laevis* eggs, a complex called maturation promoting factor (MPF) was identified some years ago as the control element that brought about maturation of the eggs. Maturation involves a transition from interphase to metaphase in this case, the metaphase of the second meiotic division. MPF may be an immediate activator of metaphase, since its action is independent of protein synthesis. In contrast, other activators such as the hormone progesterone require protein synthesis in order to be effective. Subsequently, factors with activity similar to MPF have been identified in many cell types. It is now referred to by some investigators as mitosis- or metaphase-promoting factor, to reflect that its activity is not restricted to the maturation of *Xenopus* eggs. A striking feature of MPF is its ability to auto-activate, so that injection of a small portion of an egg with high MPF activity into a second egg with low activity stimulates an increase in MPF activity in the second.

MPF has now been purified and shown to consist of at least two polypeptides: a cyclin, and the p34 protein kinase. A likely model for the changes that occur during mitosis is shown in Fig. 10. Before mitosis, the p34 component is phosphorylated and complexed to unphosphorylated cyclin in a complex called pre-MPF. Shortly before mitosis the p34 component is dephosphorylated and the cyclin becomes phosphorylated, generating active MPF. This triggers a set of processes in mitosis: nuclear envelope breakdown, chromosome condensation and spindle assembly. At the same time, the cyclin is specifically degraded and free p34 released. Cyclin degradation is necessary for exit from metaphase. In the following cell cycle, cyclin must be resynthesised de novo, and accumulates until mitosis is once again activated. There is some evidence that cyclin accumulation is rate-limiting for the interphase to metaphase transition at least in eggs and early embryos.

In *S. pombe*, the *cdc13* gene product is homologous to cyclins, and has been shown by genetical and direct biochemical means to interact with p34. The general scheme in Fig. 10 seems to hold for *S. pombe*, but there are two important differences from eggs. First, *cdc13* is not rate-limiting for entry into mitosis, since overexpression of *cdc13*[+] does not accelerate mitosis. However, the product of *cdc13*, p56, does show cyclical changes in level during the cell cycle in the way expected for a cyclin. Second, the nuclear envelope remains intact during mitosis of *S. pombe*, as for fungi in general. This presents a paradox, since chromosome condensation and spindle assembly are intranuclear events, and the factors responsible for these processes need to be imported into the nucleus. Recent work has shown that p56 accumulates within the nucleus during the cell cycle, and abruptly disappears at the time of mitosis. Furthermore the nuclear level of p34 increases in the same way, although the total cellular level of p34 does not vary significantly during the cycle. A likely explanation is that p56 is targetted to the nucleus, and p34 complexed to it enters at the same time. In this way a sufficient amount of p56−p34 complex (pre-MPF?) is accumulated within the nucleus in time for mitosis. The precise timing of these events may be regulated by changes in the phosphorylation state of p34. This in turn is likely to be controlled by a balance of tyrosine phosphorylating and phosphatasing activities, by the gene products of *wee1*[+] (and perhaps *mik1*[+]) and *cdc25*[+] respectively (Fig. 9).

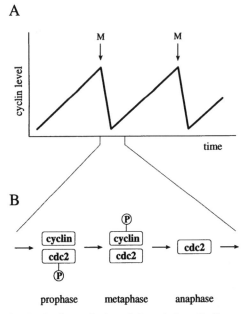

A

B

prophase   metaphase   anaphase

Fig. 10A, B. Cyclin synthesis and degradation. Cyclin accumulates during interphase **A**; when it reaches a critical level mitosis is initiated and the events shown in **B** take place. Destruction of cyclin is needed for the metaphase-anaphase transition

10.6 Dependency of Mitosis on Prior DNA Replication[6]

With very few exceptions, S phase and mitosis alternate under natural conditions. Several recent studies have shed some light on the question of how this alternation is maintained. This section deals with the dependency of mitosis on prior DNA replication, while the converse requirement, that mitosis must precede S phase, is discussed in Sect. 10.7.

Several situations have been described where mitosis can take place without prior DNA replication. For instance, the protraction of S phase in *Xenopus* eggs with aphidicolin, a DNA polymerase inhibitor, does not prevent chromosome condensation, and continued DNA replication has been observed under these conditions. In *Aspergillus nidulans* and mammalian cells there are mutants that allow mitosis in cells blocked in DNA synthesis. In the yeasts, recent genetical studies have opened new routes to the molecular analysis of the dependency.

10.6.1 *S. pombe*

When the cell size control over mitosis is defective, as in *wee1* mutants, cell size at mitosis is nevertheless subject to backup regulation. Unless this were so, cells would divide at ever-decreasing size at succeeding generations. There is some evidence that the size control over DNA replication might indirectly control the time of mitosis (Fig. 8). This pathway might also be involved in the normal dependency of mitosis on prior DNA replication.

As mentioned earlier, there are two regulatory controls influencing the p34 protein kinase that controls entry into mitosis (Fig. 9). Inhibition of DNA replication in wild-type cells or *wee1* mutants also blocks entry into mitosis. But if the *cdc25* pathway is artificially activated, for example by *cdc25*[+] overexpression, blocking DNA replication does not prevent cells from initiating mitosis (Fig. 11). This suggests that completion of start normally activates the *cdc25* pathway, which provides the link between start, DNA replication and mitosis.

10.6.2 *S. cerevisiae*

Another mechanism that couples DNA replication to mitosis has been identified by a quite different method. DNA damage, caused for instance by X- or UV-irradiation, delays mitosis in many cell types, presumably to allow time in G2 for the cell to repair the damage. Treating mammalian cells with caffeine drastically reduces the G2 delay, and the cells enter mitosis on schedule and die due to uncorrected chromosome damage.

A parallel phenomenon is observed in *S. cerevisiae*. Among a collection of radiation-sensitive mutants, those in the *RAD9* gene are unusual in failing to arrest in G2 after irradiation. Instead, they attempt to undergo mitosis, and die. It seems that the *RAD9* gene encodes a product that is somehow involved in monitoring the state of the chromosomes. In normal cells DNA damage may be detected at a G2 checkpoint by the *RAD9* system, which generates a signal to delay mitosis (Fig. 12). In *rad9* mutant cells this does not occur,

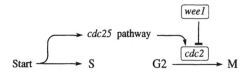

Fig. 11. Two mitotic pathways in *S. pombe* and possible relationship to S phase. Completion of start commits cells to S phase. It also activates the *cdc25* pathway. Normally mitosis does not occur unless S phase has been completed, but hyperactivation of the *cdc25* pathway alleviates this requirement

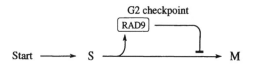

Fig. 12. Role of *RAD9* in coupling DS and mitosis. The completion of S phase is normally required for mitosis. DNA damage is sensed by the *RAD9* system which prolongs G2. In the absence of *RAD9*, mitosis is initiated irrespective of DNA damage

and mitosis starts on schedule. It is not known yet whether the *RAD9* system acts through the *CDC28* product or another mitotic control element.

10.7 Pathways in Mitosis[7]

The successful completion of mitosis requires that the behaviour of the chromosomes, mitotic spindle and nuclear envelope be coordinated. The chromosomes condense during prophase and assemble at the metaphase plate, after which the centromeres split and the daughter chromosomes segregate to the mitotic poles. Concurrently, the cytoplasmic microtubules disassemble and the spindle is formed. The spindle then undergoes two sorts of movements; the pole-to-pole distance increases while the pole-to-chromosome distances decrease. Finally, the spindle dissociates and the cytoplasmic microtubule array is reestablished. In most cell types (but not yeasts) the nuclear envelope breaks down at prophase and is absent throughout mitosis until telophase, when new envelopes are formed around the daughter nuclei. These three processes involve different cellular components, and can to a great degree take place independently of one another. The early assembly of the *S. cerevisiae* spindle relative to nuclear division (see Sect. 10.5) shows that these two processes do not need to take place simultaneously, although in most cell types they do so.

10.7.1 Pathways Within Mitosis

In *S. pombe*, many mutants, including mutations in the control gene *cdc2*, block the cycle before mitosis is underway, and neither spindle formation nor chromosome condensation are observed. However, mutants with specific defects in chromosome or spindle behaviour have been identified, and are particularly informative about the mechanisms of mitosis.

There are three tubulin genes in *S. pombe*: one encoding β-tubulin, and two α-tubulin. Cold-sensitive lethal mutations in the β-tubulin gene (*nda3*) and one of the α-tubulin genes (*nda2*) have been

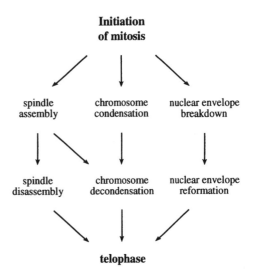

Fig. 13. Pathways in mitosis. The initiation of mitosis starts three trains of events, concerned with the spindle, chromosomes and nuclear envelope. These pathways are largely independent but blocking spindle formation prevents chromosome decondensation. The three pathways meet again at telophase

isolated. The mutations lead to arrest in mitosis, and the chromosomes condense in a similar way to those of wild-type cells treated with the microtubule agent thiabendazole. This parallels the condensation of mammalian chromosomes after cells are treated with colcemid. In all these cases, chromosome condensation continues in the absence of a spindle (Fig. 13). In these and other *S. pombe* mutants defective in spindle function, there is no further progress in morphological aspects of the cell cycle after arrest; specifically, septum formation is defective or absent.

The chromosomes do not decondense again in cells without spindles, showing that the inability to complete the spindle pathway prevents progress of the chromosome pathway. How the cell senses that the spindle is absent or defective is unknown, though feedback via MPF has been suggested.

A number of mutants (*top2, cut*) have been isolated which despite producing apparently normal spindles, are specifically defective in chromosome behaviour. Septum formation occurs in such mutants despite the failure to complete mitosis, and is therefore not dependent on the chromosome pathway. In contrast, septum formation is dependent on completion of the spindle pathway.

10.7.2 Dependency of S Phase on Prior Mitosis

Perhaps one of the most surprising recent findings is that DNA synthesis continues in arrested *top2* and *cut* mutants despite failure of the chromosome pathway in mitosis. More than one extra round of DNA replication takes place, suggesting that the core S-M cycle is trying to continue, although overt mitosis is prevented. In contrast, tubulin mutants and others defective in the spindle pathway do not undergo subsequent DNA replication. This is an unexpected finding; intuitively, the behaviour of the *chromosomes*, rather than the spindle, might have been expected to control subsequent DNA replication.

10.8 Outlook

The rate of recent progress in many areas has been so rapid that it is difficult to make predictions about the direction of future work. Perhaps the major lesson from recent years is that bringing together observations on different cell types can expand our understanding greatly, and lead to new insights about the individual systems. So any predictions about how the *yeast* cell cycle will develop in the future must also consider progress in other cell types. Areas in which yeast research should be at the forefront are:

- the nature of "start",
- how chromosomes and spindles interact in mitosis,
- further details of the mitotic control.

Yeasts are ideal genetic organisms, and it by using this approach that most progress must be expected. New mutants and genes will be identified with cell cycle roles, and unravelling their functions will lead to new insights. Doubtless homologues of these genes will be found in other cell types, but only in yeasts can the effect of manipulating them be readily investigated. On the less positive side, yeasts have been found to be poorly suited to biochemical work, and cytological work is hampered by the small size of the cells. Technical advances that would improve these approaches, or that would al-

low cell-free yeast systems to be developed, would revolutionise study of the yeast cell cycle.

Molecular biology and allied techniques have demonstrated how widely spread across the eukaryotes some of the key elements that control the cycle are. It is likely that the fundamental processes that regulate entry into mitosis or S phase, or that coordinate the events of mitosis, will turn out to be very similar in a wide range of eukaryotic cell types.

A word of warning is due here: the striking conservation of many key components suggests that the way these components function within the cell may also be conserved. There are, however, basic differences among the cycles of different cell types. One example is that the nuclear envelope of animal and plant cells breaks down in the prophase of mitosis, while in fungi and *Physarum*, the nuclear envelope remains intact throughout. It is not yet clear how important to the overall control of the cycle such differences might be, but in principle a control mechanism that evolved to include nuclear envelope breakdown might have diverged considerably through evolution from one that did not.

The operation of major regulatory systems may also differ considerably from one cell type to the next. The absence of G1 and G2 from the cycles of eggs of many species may reflect a fundamental difference between these and other cycles. There is little evidence in such cells for a G1 control akin to start in yeasts, and while MPF controls the interphase-mitosis transition, its regulation may be different from that in other cells. The evidence that cyclin accumulation is critical for initiating mitosis in eggs is strong, but the *S. pombe* cyclin homologue *cdc13* appears not to be rate-limiting. On the other hand, the *cdc25* product of *S. pombe* is rate-limiting, as appears to be so for a *Drosophila melanogaster* homologue of *cdc25*, the *string* gene.

It has also been suggested that certain checkpoints within the cycle may be absent from the early cycles of fertilised eggs. Evolution has selected for very short cycles in many early embryos, and this may have resulted in the loss of controls present in other cells. For instance any DNA introduced into *Xenopus* eggs can be replicated, irrespective of sequence, while yeasts have stringent requirements for an ARS (autonomous replication sequence) to initiate replication. Other controls may also be absent in eggs, such as the equivalent of the *RAD9* system in *S. cerevisiae*, and hence *Xenopus* eggs may continue replication even in mitosis.

10.9 Summary

The cell cycle can be viewed as the overt expression of a cellular oscillator. Equally, and more appropriately in the context of this book, it can be thought of as a complex developmental process that takes place within a single cell. The cycle might appear at first sight to consist of a single linear pathway starting with a new-born cell in G1, passing through S, G2 and mitosis, and finally ending at cell division. More thorough analysis of the cycle has shown this view to be an oversimplification, since several parallel pathways must be completed by the cell in order to proceed from birth to division. The cell cycle is nevertheless subject to overall coordination so that individual events take place at the appropriate time and in the correct order. To ensure this, parallel pathways come together at particular key control points in the cycle.

Some of the key controls acting in the cell cycle are now well-established. In most cell types there is a major control point in G1 at which the decision whether to embark on a new cycle is made: in yeasts this is called "start". Both start and a similar control in mammalian cells, the restriction point, have been known for some time, but it remains to be seen whether they are related at more than a superficial level. The process of "start" is poorly understood, although the recent discovery of cyclin homologues that act in G1 in *S. cerevisiae* and *S. pombe* is a major step forward.

The existence of a G2-M control in some cells, particularly *S. pombe* and *Physarum*, has also been known for a considerable time. The observation that the cycle of fertilised eggs consisted of an alternation of interphase and mitosis placed emphasis on entry to mitosis as a key event. The realisation that other cells possess a G2 control arose from combining observations from three sources: biochemical studies on MPF in *Xenopus* eggs, the discovery of cyclins whose levels rose and fell in a periodic manner in invertebrate eggs, and genetical

studies on the *cdc2* gene of *S. pombe*. The identification of one component of MPF as the *cdc2* product was soon followed by the identification of cyclin as another. Since then, control components first discovered in one system have been identified in others, to the point where the fundamental similarities between cell types probably outweigh the differences.

References

General

Baserga R (1985) The biology of cell reproduction. Harvard University Press, Cambridge, Massachusets

Brooks R, Fantes P, Hunt T, Wheatley D (eds) (1989) The cell cycle. J Cell Sci 12 (Suppl)

John PCL (ed) (1981) The cell cycle. Cambridge University Press, Cambridge, UK

Prescott DM (1976) Reproduction of eucaryotic cells. Academic Press, New York

Pringle JR (1978) The use of conditional lethal cell cycle mutants for temporal and functional sequence mapping of cell cycle events. J Cell Physiol 95:393–405

Pringle JR (ed) (1989) Current opinion in cell biology, vol 1, No 2. Current Science, London

[1] Fantes PA (1989) Cell cycle controls. In: Nasim A, Young P, Johnson BF (eds) Molecular biology of the fission yeast. Academic Press, San Diego, California, pp 128–204

Hartwell LH (1974) *Saccharomyces cerevisiae* cell cycle. Bacteriol Rev 38:164–198

Hirano T, Yanagida M (1989) Controlling elements in the cell division cycle of *Schizosaccharomyces pombe*. In: Walton EF, Yarranton GT (eds) Molecular and cellular biology of yeasts. Blackie, Glasgow, UK, pp 223–245

Pringle JR, Hartwell LH (1981) The *Saccharomyces cerevisiae* cell cycle. In: Strathern JN, Jones EW, Broach JR (eds) The molecular biology of the yeast *Saccharomyces*. Cold Spring Harbor Press, Cold Spring Harbor, NY, pp 97–142

Reed SI (1984) Genetic and molecular analysis of division control in *Saccharomyces*. In: Nurse P, Streiblova E (eds) The microbial cell cycle. CRC Press, Boca Raton, pp 89–107

[2] Mendenhall MD, Jones CA, Reed SI (1987) Dual regulation of the yeast *CDC28* protein kinase: cell cycle, pheromone, and nutrient limitation effects. Cell 50:927–935

Nurse P (1981) Genetical control of the yeast cell cycle: a reappraisal of "start". In: Gull K, Oliver S (eds) The fungal nucleus. Cambridge University Press, London, pp 331–345

Nurse P, Bisset Y (1981) Gene required in G1 for commitment to cell cycle and in G2 for control of mitosis in fission yeast. Nature 292:558–560

Pardee AB (1989) G1 events and regulation of cell proliferation. Science 246:603–608

Reed SI (1980) The selection of *S. cerevisiae* mutants defective in the START event of cell division. Genetics 95:561–577

Reed SI, Hadwiger JA, Lorincz AT (1985) Protein kinase activity associated with the product of the yeast cell division cycle gene *CDC28*. Proc Natl Acad Sci USA 82:4055–4059

[3] Berger JD (1989) The cell cycle in lower eukaryotes. Curr Opinion Cell Biol 1:256–262

Carter BLA (1981) The control of cell division in *Saccharomyces cerevisiae*. In: John PCL (ed) The cell cycle. Cambridge University Press, Cambridge, UK, pp 99–117

Cross FR (1989) Further characterization of a size control gene in *Saccharomyces cerevisiae*. J Cell Sci 12:117–127 (Suppl)

Fantes PA (1977) Control of cell size and cycle time in *Schizosaccharomyces pombe*. J Cell Sci 24:51–67

MacNeill SA, Nurse P (1989) Genetic interactions in the control of mitosis in fission yeast. Curr Genet 16:1–6

McAteer M, Donnan L, John PCL (1985) The timing of division in *Chlamydomonas*. New Phytol 99:41–56

Nash R, Tokiwa G, Anand S, Erickson K, Futcher AB (1988) The *WHI1*+ gene of *Saccharomyces cerevisiae* tethers cell division to cell size and is a cyclin homolog. EMBO J 7:4335–4346

Nurse P, Thuriaux P (1977) Controls over the timing of DNA replication during the cell cycle of fission yeast. Expl Cell Res 107:365–375

Nurse P, Thuriaux P (1980) Regulatory genes controlling mitosis in the fission yeast *Schizosaccharomyces pombe*. Genetics 96:627–637

Piggott J, Rai R, Carter B (1982) A bifunctional gene product involved in two phases of he yeast cell cycle. Nature 298:391–393

Reed SI, Hadwiger JA, Richardson HE, Wittenberg C (1989) Analysis of the CDC28 protein kinase complex by dosage suppression. J Cell Sci 12:29–37 (Suppl)

[4] Beach D, Durkacz B, Nurse P (1982) Functionally homologous cell cycle control genes in budding and fission yeast. Nature 300:706–709

Featherstone C, Russell P (1991) Fission yeast p107^wee1 mitotic inhibitor is a tyrosine/serine kinase. Nature 349:808–811

Gould KL, Nurse P (1989) Tyrosine phosphorylation of the fission yeast *cdc2*+ protein kinase regulates entry into mitosis. Nature 342:39–45

Hayles J, Nurse P (1989) A review of mitosis in the fission yeast *Schizosaccharomyces pombe*. Exp Cell Res 184:273–286

Lee M, Nurse P (1988) Cell cycle control genes in fission yeast and mammalian cells, Trends Genet 4:287–290

Lundgren K, Walworth N, Booher R, Dembski M, Kirschner M, Beach D (1991) mik1 and wee1 cooperate in the inhibitory tyrosine phosphorylation of cdc2. Cell 64:1111–1122

MacNeill SA, Nurse P (1989) Genetic interactions in the control of mitosis in fission yeast. Curr Genet 16:1–6

Moreno S, Hayles J, Nurse P (1989) Regulation of the cell cycle timing of mitosis. J Cell Sci 12:1–8 (Suppl)

Parker LL, Atherton-Fessler S, Lee MS, Ogg S, Falk JL, Swenson KI, Piwnica-Worms (1991) Cyclin promotes the tyrosine phosphorylation of p34^cdc2 in a wee1+ dependent manner. EMBO J 10:1255–1263

[5] Alfa CE, Booher R, Beach D, Hyams JS (1989) Fission yeast cyclin: subcellular localisation and cell cycle regulation. J Cell Sci 12:9–19 (Suppl)

Minshull J, Pines J, Golsteyn R, Standart N, Mackie S, Colman A, Blow J, Ruderman J, Wu M, Hunt T (1989) The role of cyclin synthesis, modification and destruction in the control of cell division. J Cell Sci 12:77–97 (Suppl)

Murray AW (1989) Cyclin synthesis and degradation and the embryonic cell cycle. J Cell Sci 12:65–76 (Suppl)

Murray AW, Kirschner MW (1989) Dominoes and clocks: the union of two views of the cell cycle. Science 246:614–621

Russell P, Moreno S, Reed SI (1989) Conservation of mitotic controls in fission and budding yeasts. Cell 57:295–303

[6] Bueno A, Richardson H, Reed SI, Russell P (1991) A fission yeast B-type cyclin functioning early in the cell cycle. Cell 66:149–159

Enoch T, Nurse P (1990) Mutation of fission yeast cell cycle control genes abolishes dependence of mitosis on DNA replication. Cell 60:665–673

Forsberg SL, Nurse P (1991) Identification of a G1-type cyclin *puc1*+ in the fission yeast *Schizosaccharomyces pombe*. Nature 351:245–248

Hartwell LH, Weinert TA (1989) Checkpoints: controls that ensure the order of cell cycle events. Science 246:629–634

Hutchison CJ, Brill D, Cox R, Gilbert J, Kill I, Ford CC (1989) DNA replication and cell cycle control in *Xenopus* egg extracts. J Cell Sci (Suppl) 12:197–212

Osmani SA, Engle DB, Doonan JH, Morris NR (1988) Spindle formation and chromatin condensation in cells blocked at interphase by mutation of a negative cell cycle control gene. Cell 52:241–251

[7] Blow JJ (1989) DNA replication and its control. Curr Opinion Cell Biol 1:263–267

Hayles J, Nurse P (1989) A review of mitosis in the fission yeast *Schizosaccharomyces pombe*. Exp Cell Res 184:273–286

Hirano T, Yanagida M (1989) Controlling elements in the cell division cycle of *Schizosaccharomyces pombe*. In: Walton EF, Yarranton GT (eds) Molecular and cellular biology of yeasts. Blackie, Glasgow, UK, pp 223–245

Chapter 11   Circadian Rhythms of *Neurospora*

STUART BRODY

11.1 Introduction[1]

The phenomena of circadian rhythms, sometimes called biological clocks, may seem out of place for a book devoted to development. The reasons are clear: biological clocks are often characterized in adult organisms and not in developing systems; and biological clocks show periodic functions whereas developmental systems are basically aperiodic. Both types of phenomena are temporal, however. This chapter, although basically devoted to the facts and strategies in the clock field, will also attempt to illustrate the connection between clocks and development. It will point out how an understanding of the mechanism of the clock and/or its input and output pathways, can be useful to the understanding of developmental events, cell cycle control, and responses to various environmental stimuli. It will also attempt to show how genetic and biochemical techniques, and combinations thereof, have been employed to probe the mechanism that is responsible for a given phenotype.

Circadian rhythms are defined as biological rhythms with innate periodicities of roughly one day. The term comes from *circa diem*, meaning "about a day". Such rhythms are widespread throughout the plant and animal kingdoms and among eukaryotic microorganisms. The manifestations of this rhythmicity can be found in single cells where cell division is confined to particular times in a 24 h cycle, and it can be found in multi-cellular organisms as complex behavioral changes. Sexual and asexual reproduction, bioluminescence, hormonal levels, body temperature and activity/rest cycles are a few of the many additional biological phenomena under the control of the biological clock. With respect to developmental events, numerous systems have been characterized for their circadian rhythms. The pupation of *Drosophila* larvae is an excellent example, as is the spore-forming (conidiation) rhythm of *Neurospora*. Both of these events can occur in the absence of an obvious circadian rhythm, i.e., the clock mechanism synchronizes these events, rather than initiating them. This idea that the circadian rhythm mechanism is some type of a "master clock" has also been suggested with respect to cell division control, where the circadian rhythm mechanism restricts cytokinesis to certain times of day. Numerous other events are affected by this master clock and many are listed in the books on circadian rhythms cited at the end of this chapter. An impressive literature also exists on the manifestation of the biological clock in humans, i.e., the phenomenon of jet-lag, shift-work disorders, seasonal affective disorders, etc. Recent work has shown the similarities in some of the properties of the primate biological clock, based in the central nervous system, to that of other organisms such as *Neurospora* and *Drosophila*.

Despite the large amount of phenomenological data about observed rhythms, there is little biochemical knowledge about the underlying mechanism. This chapter focuses primarily on the circadian clock in the ascomycete fungus, *Neurospora crassa*, and also discusses various approaches being employed in this field. Certain basic properties are common to the circadian rhythms of most organisms: (1) the rhythm is endogenous and self-sustaining; (2) the period of the rhythm is close to but not equal to 24 h; (3) the period of the rhythm varies little with temperature, a property called temperature-compensation; (4) the phase of the rhythm can be changed by pulses of light or temperature; and (5) the rhythm can be altered to follow light/dark or temperature cycles with periods other than 24 h.

The circadian rhythm mechanism is, in a sense, a biological sensory apparatus able to receive input from the environment, process it through a central oscillator, and generate an output. It is different from other biological sensors in that it distinguishes changes, in the environment. Similarly, the output is not just a biochemical, developmental, or behavioral response, but the rhythmic restriction of such a response, a "window", to a particular, appropriate time of day.

11.1.1 The *Neurospora* Conidiation Rhythm – an Example of a Synchronized Developmental Event

The circadian rhythm in *Neurospora* is routinely assayed by its expression as a conidiation (asexual spore formation) rhythm on agar media (Fig. 1). This rhythm has a period of about 22 h. This complex developmental process is allowed to occur by the clock mechanism at a particular phase of the

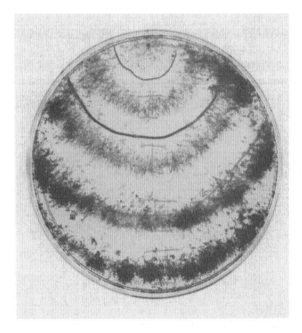

Fig. 1. The *Neurospora* conidiation rhythm. The rhythm is expressed as arcs of conidiation by a culture which had been inoculated on agar medium at the perimeter of a large (150 mm diameter) Petri plates, and then grown in constant darkness. The *more dense areas* are the "band" regions, and the *lines drawn on the bottom of the plate* are the edges of the growing front on successive days

circadian cycle. Although the conidiation process is closely tied to the oscillator mechanism, it can occur in the absense of the rhythm, and vice versa. The observations that suggest that this linkage is one-way are: (1) biochemical rhythms persist in the absence of the conidiation rhythm; (2) cultures growing in liquid medium, which conidiate poorly, if at all, are apparently rhythmic, based on the phase of the conidiation rhythm which starts upon transfer from liquid culture to agar medium; and (3) morphological mutations affecting conidiation do not change the underlying periodicity of the circadian rhythm. Therefore, the conidiation process itself is not part of the mechanism involved in generating the oscillation. The conidiation process involves the formation of new structures, such as aerial hyphae and conidia, and as expected, a considerable number of genes are activated in this process. The region of conidiation is also characterized by more dense mycelial mass (measured as mass per unit area) than the non-conidiating region, increased branching of the mycelia, and aerial outgrowths from the mycelia from which conidia are formed. These changes result in a clearly discernable area called a "band." The regions between the bands, called "interbands," are characterized by a less dense mycelial growth and few conidia. Although the interband areas continue to increase in mass, conidiation does not occur in these regions. These alternating band and interband regions leave a "fossil" record of the rhythm (see Fig. 2).

In practice, cultures for monitoring rhythmicity are grown on agar media in Petri plates or on cylindrical glass growth tubes. Cultures growing in constant darkness are monitored for growth at frequent intervals, and the position of the growing edge of each culture is marked on the bottom of the petri dish or race tube under red light, which has no effect on rhythmicity in *Neurospora*. When the cultures have finished growing, the positions of the clearly visible conidiation bands, relative to the marked growth fronts, allow the calculation of the period and phase of the rhythm. For monitoring rhythmicity, a mutant strain of *Neurospora,* band (*bd*), which allows the expression of the conidiation rhythm in closed Petri dishes or growth tubes is used. Although the primary biochemical defect is not known, the physiological basis of the mutant

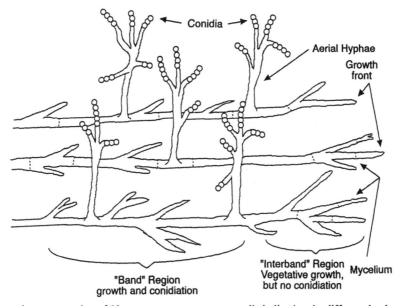

Fig. 2. A highly schematic representation of *Neurospora* grown on agar media indicating the different developmental regions of the culture

phenotype appears to be a resistance of growth and conidiation to CO_2: Developmental mutants that have altered conidia such as *csp*-1 or *csp*-2, are also often employed in monitoring rhythmicity. These strains are deficient in the formation of cross walls separating the conidia from aerial hyphae and from each other; consequently, conidia remain attached to aerial hyphae, and the plates or tubes are unable to self-inoculate during handling.

11.2 Input[2]

Study of environmental input to the circadian oscillator is of interest not only in itself, but also for the clues which may be provided as to the nature of the oscillator. Starting with a given environmental input, it is potentially possible to trace the input pathway to the oscillator itself by identifying the receptor for the input, second messengers triggered by the activated receptor, other messengers triggered by these and so on, although it is possible that second messengers may so multiply that the oscillator mechanism becomes hidden in the altered physiology of the organism. In nature, the most obvious circadian-rhythmic environmental variables are temperature and illumination, and both of these variables have been extensively studied with respect to their effects on rhythmicity in *Neurospora* and *Drosophila*. The environment also presents numerous other daily rhythms, such as of

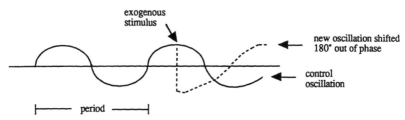

Fig. 3. A schematic diagram of an oscillation and an effect that leads to a stable phase change

humidity, ultraviolet irradiation, color of ambient illumination, magnetic field strength, etc. Although changes in magnetic field strength apparently do not affect the *Neurospora* rhythm, it is not known which of the other environmental signals above may affect circadian rhythmicity.

In discussing the effects of environmental variables on rhythmicity in the following sections, the concept of a phase shift will occur repeatedly. Many treatments and insults, when applied as discrete pulses, induce a stable change in the phase of observed rhythms (Fig. 3), presumably either through specific signal transduction mechanisms or through direct effects on the oscillator. In either case, the central oscillator itself is altered, and the new phase to which the oscillator is shifted by a pulse depends on the old phase at which the pulse was applied.

11.2.1 Responses to Light

There are many different in vivo responses to light which have been described in *Neurospora* (see Russo and Pandit, Chap. 5, this Vol.). All are classified as blue-light responses since they are induced by blue (or white) light but not by red light, and in cases where an action spectrum has been determined, the major peak is around 465 nm. Although these responses are similar to each other with respect to their sensitivity to blue light, there may be more than one blue light receptor in *Neurospora*. There may also be more than one intracellular signal produced by a single photoreceptor, and more than one target for each signal. Studies on the *wc* mutant (described below) indicate that the photoreceptor and signal transduction pathway leading to the oscillator may be shared with other blue light responses.

Light phase-shifting in *Neurospora* is similar to that in many other organisms. This property allows the clock to be synchronized (entrained) by 24-h light-dark cycles, and by single light pulses every 24 h. It can also be entrained to cycles of other than 24 h, but the limits of entrainment have not been determined. Constant light obscures or suppresses the rhythm in that constant conidiation is observed.

11.2.1.1 Light-Insensitive Mutants

Several mutants with reduced sensitivity to light have been identified: [*poky*], *rib* (riboflavin-requiring), *lis* (light-insensitive) and *wc* (white-collar). The first such mutant to be reported was [*poky*], which is resistant to light-induced suppression of banding. It was demonstrated that [*poky*] continues to express circadian conidiation under light intensities two orders of magnitude greater than required to suppress the rhythm in wild type. This mutant is defective in mitochondrial protein synthesis, and is therefore deficient in some cytochromes and in mitochondrial respiration. The insensitivity to light is not a consequence of [*poky*]'s respiratory deficiency, as a respiratory mutant *rsp*-2 shows normal light suppression of the rhythm. Phase shifts by light are smaller in [*poky*] than in wild type, although phase-shifting by pulses of cycloheximide, a protein synthesis inhibitor, was normal. This indicates that the oscillator itself is not affected, but only the light signal transduction pathway. Other mutants showing altered light responses are the riboflavin-requiring mutants (*rib*) which showed reduced responses to light pulses, but normal circadian rhythm when flavin deficiency is induced; and three nonallelic light-insensitive (*lis*) mutants.

Results reported with the white-collar (*wc*) mutants indicate that at least some component(s) of the phototransduction mechanism must be common to most blue-light responses in *Neurospora*. These mutants were originally described as deficient in light-induced carotenoid synthesis in the hyphae, but not deficient in constitutive synthesis in conidia, and were proposed to be regulatory mutants blocked in photoinduction. A total of 11 *wc* mutants have been isolated and all fall into only two complementation groups. These mutants have also been shown to be insensitive to the effects of light on other processes (see Russo and Pandit, Chap. 5, this Vol.).

11.2.2 Temperature Effects

With few exceptions, organisms are subjected to large variations in environmental temperature, often within a single day. There is a twofold relationship between this variation in environmental temperature and circardian rhythmicity. On the one hand, the period of the rhythm must be conserved despite temperature variation since a clock which runs faster or slower depending on the temperature would not keep accurate time. This conservation of period length despite variation in temperature is termed temperature compensation. On the other hand, changes in temperature can provide time cues to synchronize biological rhythms to environmental rhythms. That organisms take advantage of these cues is indicated by phase-shifting responses to temperature changes and by entrainment to temperature cycles. *Neurospora* shows both temperature compensation of period and responsiveness to temperature changes.

11.2.2.1 Temperature Compensation

The term temperature compensation was chosen to describe the conservation of period despite changes in temperature in order to emphasize that rhythmicity is not temperature-independent, since temperature changes can phase shift the rhythm. Mutations can also partially or completely abolish temperature compensation. Mutations in the frequency (*frq*) gene that lead to a lengthened period exhibit weaker temperature compensation at lower temperatures than does the wild type. Another allele at the *frq* locus, *frq*-9, has lost temperature compensation altogether. The *cel* (Chain Elongation) mutant, which is defective in fatty acid synthesis, also has altered temperature compensation. Although temperature compensation of the conidiation rhythm is normal at temperatures from 22° to 30 °C, below about 21 °C there is no temperature compensation (Table 1). The *frq*-9 and *cel* mutants of *Neurospora* are the only known mutants which have essentially lost the property of temperature compensation. Both, however, have a normal circadian period at a temperature near the middle of the temperature range permitting growth, around 20° to 26 °C. Both mutants have rhythms which are light-sensitive, and both can be entrained by light/dark cycles. Hence, in both mutants, temperature compensation has been lost independently of other properties of rhythmicity.

Table 1. The effect of temperature and exogenous fatty acids on the period of the *cel*+ and *cel*- strains

Temperature (°C)	Supplement	Period of the strain	
		cel+ (h)	*cel*- (h)
18	None	21 h	32 h
22	None	21 h	21 h
26	None	20 h	20 h
30	None	19 h	18 h
22	16:0	21 h	21 h
22	18:2	20 h	40 h
22	16:0, 18:2	21 h	21 h

11.2.2.2 Phase Shifting by Temperature Steps and Pulses

Although the periods of circadian rhythms are generally temperature-compensated, their phases can be shifted by changes in temperature, such as a temperature step, in which temperature is abruptly raised or lowered to a new value. It was found that both temperature increases and temperature decreases phase-shift the rhythm of *Neurospora*. The size of the observed phase shift is a function of the time in the cycle at which the temperature step occurred, of the magnitude and direction of the temperature change, and of the starting temperature. It was also found that temperature pulses, in which the temperature is raised or lowered for only a few hours and then returned to its original value, phase-shift the rhythm of *Neurospora*. Again, the size of the phase shift is a function of the time of the pulse, the magnitude and direction of the temperature change, and the duration of the pulse.

11.3 The Strategies and Experiments Pursued Towards Understanding the Oscillator[3]

To date, there is no biochemical data which would allow identification of the components of the oscillator mechanism in *Neurospora*, or in any other organism. Therefore, this chapter can only focus on strategies and techniques, while discussing experiments which may or may not have any direct bearing on the identity of these unknown components. These strategies and tactics may also be applicable to studying complex developmental events as well. The basic strategies have been: (1) perturbing an input pathway and tracing this path to the oscillator; (2) perturbing an output pathway and tracing this path back to the oscillator; or (3) perturbing the oscillator directly. The strategies can be further classified according to whether the perturbation is by mutation (the genetic approach) with drugs or inhibitors (the biochemical approach), or with light and temperature (the physical approach).

11.3.1 The Genetic Approach: Strategies

The basic genetic strategies for studying rhythmicity have been the isolation of mutants with altered periods or the screening of existing mutants for alterations in period, since any mutants which alters the period must have some effect, direct or indirect, on the oscillator mechanism. Mutations which directly affect clock components might alter either the actual molecules whose levels oscillate, or components necessary for the oscillations, but whose levels do not actually oscillate. Mutations which indirectly affect the oscillator would include any that alter gene products which normally are not part of the clock mechanism but, because their levels or activities are changed by the mutation, now have some effect on the clock. Once isolated, mutants can be studied biochemically, or with classic genetic techniques, such as determination of dominance and studies of interactions between them.

11.3.1.1 Clock-Affecting Mutations

Mutations Detected Based on Changes in Period (Function Mutants). The isolation of clock-affecting mutants provides just one half of the structure-function relationship, in that these mutants have alterations in a known function but a defect in an unknown structure. Techniques for tracking down the unknown primary defects in such mutants can involve several different kinds of screening procedures. In addition to cloning and sequencing the

genes in question, one could screen particular kinds of low molecular weight compounds for changes in their levels, or screen the protein population by two-dimensional electrophoresis. The use of several approaches is desirable, and it should be kept in mind that proteins communicate with each other through small molecules, so that even after the successful identification of a primary biochemical defect, one must analyze the resulting changes in metabolism.

In *Neurospora,* seven genes have been identified based on their ability to alter the period of the free-running rhythm: *frq* (frequency), *chr* (chrono), *prd*-1 (period), *prd*-2, *prd*-3, *prd*-4, and *cla*-1 (clock-affecting). These map at seven different locations in the genome of *Neurospora*. With the exception of the *frq* gene, the genes are known by one mutant allele each. Except for *cla*-1, which is a chromosomal rearrangement, it is not known whether these mutations are point mutations, deletions, insertions, or other rearrangements. Some properties of these mutants are listed in Table 2.

Table 2. Summary of period changes in *Neurospora* due to mutation

Shorter mutants	Period (h)
arg-13	19
cys-4	19
cys-12	19
frq-1	16
frq-2	19
frq-9 (>26 °C)	17−21
glp-3	19
oli^r	18−19
phen-1	19
prd-4	18
[MI-3] and other mitochondrial mutants	19

Longer mutants	
cel (<22 °C)	25−40
cel	42[a]
chr	23.5
frq-3	24
frq-7	29
frq-9 (<26 °C)	21−35
prd-1	26
prd-2	25
prd-3	25

[a] In the presence of linoleic acid.

The most intensively studied clock-affecting gene in *Neurospora* is the *frq* gene. Eight independent *frq* mutants have been isolated (Table 2), and the evidence is convincing that the different isolates are allelic to each other. The *frq*-9 allele is unusual in that its period is dependent on the composition of the growth medium, and under certain conditions, this mutant appears arrhythmic.

The *frq* gene has been cloned employing a chromosomal walk from a closely linked gene. Cosmids containing the *frq* region transform strains containing the recessive *frq*-9 allele to normal period and normal temperature compensation. The sequence of the genomic region encoding the *frq* gene has one large open reading frame (ORF) of approximately 800 amino acids, and two transcripts, of approximately 1.5 and 5 kb, are found which hybridize to DNA from this region. Although the amino acid sequence predicted from the ORF in the *frq* region does not match any currently known protein sequence, a region of glycine-threonine repeats has been noted which shows some sequence similarity to a region in the *Drosophila* clock gene, *per*. The possibility that the *Neurospora frq* gene and the *Drosophila per* gene might code for similar gene products has been suggested previously. There are, however, differences in the properties of the organisms; the *per* gene is thought to play a role in a rhythm of its own transcription, and possibly affect gap junctions between cells. If the *per* gene product played a role in coupling rhythmic cells into a rhythmic tissue, then it would be a component of what has been termed the intercellular clock, as distinct from the intracellular clock, which generates rhythmicity within single cells.

Clock-Affecting Mutations Discovered by Screening Known Mutants (Structure Mutants). The rationale of this strategy is straightforward: to screen mutants with known biochemical defects for any altered clock function. Two types of information can be obtained from this approach: either the pathway affected by the mutation has an influence on the clock, directly or indirectly, or the flux through the pathway does not affect the clock mechanism. Although the screening of biochemical and auxotrophic mutants appears to be a type of "fishing expedition," it has eliminated a number of metabolic pathways from consideration as components of the oscillator, and it has identified mutations which lead to significant clock effects (see studies on the *cel* strain, below).

A total of 85 biochemical and morphological mutants, representing a significant portion of the known mutants in *Neurospora*, have been screened for changes in period; some of these are listed in Table 2. Seventy-three of these show no changes in period at the screening temperature and these are affected in a wide range of metabolic reactions, including amino acid synthesis, lipid synthesis, assimilation of alternate carbon and nitrogen sources, and mitochondrial metabolism, as well as mycelial morphology and resistance to inhibitors (see Table 3). In the case of many auxotrophs, the mutants could only be screened in the presence of the required supplement because growth or conidiation was poor without supplement. One cannot conclude, therefore, that rhythmicity does not require the end product of the mutated pathway, but only that the flux through that pathway is unnecessary for normal period. The studies on the morphological mutants indicated that slowing the growth rate by mutation did not change any clock property, and that defects in the conidiation process also did not lead to changes in any clock property.

Of the 12 mutations which lead to an altered period, three affect amino acid synthesis, two affect lipid synthesis, one affects carbon assimilation, and six affect mitochondrial energy metabolism. The *cel* mutant, which requires saturated fatty acids for growth, has lost temperature-compensation,

Table 3. Summary of period studies on biochemical mutants

Pathways of potential interest
1. *Biosynthesis of*: saturated, but not unsaturated, fatty acids (*cel* mutant)
2. Enzymes or a pathway increased by increased mitochondrial dosage (*oli*ʳ and *MI* mutants)

Steps not required for a normal period
1. *Biosynthesis of*: 13 of the 13 amino acids tested; 4 of the 4 vitamins tested; some lipid components; carotenoids
2. *Pathways*: glyoxylate shunt; catabolism of basic amino acids
3. *Enzymes or enzyme complexes*: pyruvate dehydrogenase; nitrate reductase; NADase; invertase

as discussed above. The period of *cel* is also altered by the addition of naturally occurring fatty acids to the growth medium, and effects of these compounds are relatively large. This mutant has a deficiency in fatty acid synthetase, and is therefore blocked in the synthesis of saturated fatty acids. The *cel* mutant was first isolated based on its requirement for the fatty acid 18:0. Fatty acids are designated in this section by the x:y notation where x = number of carbons and y = number of double bonds. It was shown to require saturated fatty acids for optimal growth; however, the requirement is not absolute, as it grows, albeit slowly, without supplement. The mutant exhibits reduced fatty acid synthesis resulting from very low levels (2% of wild-type levels) of 4'-phosphopantetheine cofactor bound to the fatty acid synthetase. The period of the conidiation rhythm of the *cel* mutant is sensitive to the addition of certain fatty acids to the growth medium, while the period of the wild type is not affected by the addition of any fatty acid (Table 1). Although saturated fatty acids such as 16:0 and 18:0 permit optimal growth of the mutant without affecting the period at 22 °C, the unsaturated fatty acids, 18:1, 18:2, and 18:3, lengthen the period to as long as 26 h, 40 h, and 33 h, respectively, at 22 °C. This observation is particularly interesting, since the active compounds are not exotic antibiotics or inhibitors, but normal constituents of *Neurospora* hyphae. This initial observation has been extended by the finding that saturated fatty acids, 8 to 13 carbons long, also lengthen the period of *cel* at 22 °C. Fatty acids 14 or 16 carbons long, however, not only do not lengthen the period, but also reduce or cancel the effects of period-lengthening fatty acids like 18:2 when the two species are added to the growth medium together.

The *cel* mutant is the only mutant with a known primary defect in which temperature compensation is lost. If the nature of this loss can be elucidated, it would provide invaluable insight into the mechanism of circadian rhythmicity. Research with the *cel* mutant has implicated the participation of membrane properties in maintaining temperature-compensation. Evidence for this includes the altered phospholipid fatty acid composition of the mutant, and the lability of the phospholipid fatty acid composition to supplemental fatty acids.

Despite the evidence implicating membranes, the mechanism of temperature-compensation in *Neurospora* remains unknown. Although a wealth of data has been collected on the *cel* mutation, as yet no testable hypothesis has been developed which is consistent with all of this data. Much of the data is phenomenological, i.e., measurement of period, rather than biochemical. Nonetheless, it is research on the lipid and mitochondrial biochemistry of *cel* that promises the greatest chance of understanding altered temperature compensation in this mutant.

11.3.2 The Biochemical Approach: Strategies

11.3.2.1 The Use of Chemicals to Probe the Oscillator

The rationale for this approach is straightforward, even though the interpretation of the results may not be. The objective is to try to identify biochemical reactions or pathways that are components of the biological clock by treating *Neurospora* with specific chemical effectors of specific reactions or pathways and studying the effect on the clock. If these compounds have effects on rhythmicity, the affected area of metabolism would be implicated in the mechanism of circadian rhythmicity. The compounds may be applied continuously and the rhythm monitored for changes in period, since such a change would signal an effect on the oscillator itself. Alternatively, the compounds may be applied

Table 4. Summary of period changes in *Neurospora* due to chemicals

	Period (h)
SHORTER	
Antimycin A	19
Aurovertin	19
Chloramphenicol	19
2-phenylethanol	18
LONGER	
D_2O	24
Lithium	24
Theophylline and related compounds	21 – 25[a]

[a] Concentration-dependent.

in discrete pulses, and the rhythm monitored for changes in phase, since a stable change in phase would also signal an effect on the oscillator. Some of the compounds tested for effects on period in *Neurospora* are listed in Table 4.

11.3.2.2 Biochemical Characterization of Clock-Affecting Mutants

An alternative biochemical approach is to compare the metabolism of clock-affecting mutants to that of the wild type in hopes of correlating changes in metabolism with changes in rhythmicity. An example of this approach is the research on the *cel* mutant. Another area of interest is that of protein synthesis. Protein synthesis inhibitors have been widely studied for effects on rhythmicity in many organisms. In *Neurospora,* the inhibitor, cycloheximide, has strong phase-shifting effects, and inhibition of protein synthesis by this drug has been quantitated. Cycloheximide-resistant mutants have been used to demonstrate that the phase-shifting effects are mediated by the primary effect of the drug on protein synthesis on 80S ribosomes. A more recent discovery that protein synthesis in the *frq*-7 strain is sensitive to cycloheximide, but little or no phase shifts are seen, is not yet fully understood.

11.3.2.3 Membranes

Membranes have been frequently proposed as components of the circadian clock. In *Neurospora,* the evidence that membranes play a role in rhythmicity consists of the phase-shifting effects of ionophores; the effects of substances, such as D_2O, which affect membrane properties; the loss of temperature-compensation in the *cel* mutant, which is defective in fatty acid synthesis; the sensitivity of the period of this mutant to exogenous fatty acids; the altered phospholipid fatty acid composition of *cel* and of the clock-affecting mutant *prd*-1; and the rhythm of fatty acid composition of the phospholipids of *Neurospora.* These various lines of evidence involving the *cel* mutant and the altered fatty acid composition remain currently the strongest correlation between rhythmicity and any biochemical function.

In particular, evidence from studies on the *cel* mutant, discussed in detail above, strongly implicates membranes in the mechanism of temperature-compensation.

11.4 Output[4]

The function of a circadian oscillator, if it has a function at all, must be to supply timing information to other processes in the cell, and therefore either to "drive" these processes in a rhythmic manner or to allow the expression of the processes. In the absence of any information about the identity of the oscillator itself, we can only observe its workings indirectly by observing its effects on these rhythms; therefore all information about the oscillator and its input comes from observing its output. The most obvious and most easily assayed output in *Neurospora* is the conidiation rhythm, and by using this assay the fundamental properties of the *Neurospora* oscillator have been described. Any other rhythmic process in the cell which is driven by the oscillator will share the same properties as the conidiation rhythm and before any fluctuating variable can be classified as genuine circadian output, it must have at least some of these same properties.

11.4.1 Rhythmic Variables

The conidiation rhythm is expressed only at the growing front of cultures on solid medium, although it has been shown that the circadian oscillator continues to run in old areas of the culture behind the growing front. The phase of the oscillator determines which developmental program will be followed by the new growth at the front, and once the binary decision has been made (to conidiate or not to conidiate), that area of the culture carries out the program and becomes permanently differentiated. The oscillation between conidiating and non-conidiating growth at the growing front produces the visible morphological rhythms in production of aerial hyphae, production of conidiospores, and synthesis of carotenoids. The developmental program leading to conidiation clearly requires changes in the activities of a number of enzymes. The activities of enzymes related to conidiation at the growing front of cultures growing on solid medium were assayed. Rhythmic activity was found during the first 2 days of growth for certain enzymes, such as citrate synthase, isocitrate lyase, and others. A rhythm in total soluble protein was also found. Two of the enzymes were assayed again over days 4 to 6, and the rhythms were shown to persist, thus satisfying the first criterion for establishing the circadian nature of the rhythms. None of the other criteria was demonstrated, but the close association of these rhythms with the conidiation rhythm makes it likely that they represent genuine circadian output. These rhythms were shown to be related to conidiation by two criteria: (1) the activities do not oscillate in old areas of the culture, but instead reflect the permanent developmental state (band or interband), and (2) none of these activities oscillates in the wild-type or fluffy (*fl*) strains, which undergo little or no conidiation.

Recent work has demonstrated the application of molecular biology techniques to the analysis of circadian output. Clock-controlled genes have been isolated by screening both cDNA and genomic libraries with cDNA probes enriched in poly $(A)^+$ RNA sequences expressed at two different circadian times ("morning" and "evening"). Two "morning-specific" genes were identified, and their expression was shown to be rhythmic when assayed across two circadian cycles by Northern analysis of total cellular RNA extracted from cultures at various phases. Expression was assayed in two strains, frq^+ and *frq*-7, with conidiation rhythm periods of 21.5 and 29 h, respectively. These two morning-specific genes are rhythmically expressed with periods appropriate to the strains from which the RNA was derived, thus providing strong evidence that these genes are controlled by the circadian oscillator. Expression of these genes is apparently controlled at the level of transcription. There was no evidence of circadian changes in total RNA, or in rRNA, although total RNA decreases with age. Other rhythms observed in *Neurospora* are: a rhythm in CO_2 production; oscillations in both K^+ and Na^+ ion content; and an oscillation in adenine nucleotides. No evidence was provided to indi-

cate whether these rhythms were associated with conidiation or if they were independent of morphological changes.

A rhythm in fatty acid composition was described by using cultures on solid medium. This rhythm was seen in the mole percentages of two unsaturated fatty acids, linoleic (18:2) and linolenic (18:3), in both the total lipids and in the phospholipids of the mycelia. The peak in the level of one fatty acid coincides with the trough in the level of the other. The oscillation has a period of about 20 h, and the phase of the oscillation is set by the light-to-dark transition rather than the total growth time, indicating a true circadian rhythm. The rhythm persisted in the bd^+ *csp*-1 strain, which shows no gross morphological rhythmic changes.

11.5 Outlook – Possible Insights into Developmental Events from Studies on the Clock Mechanism

Although this section is clearly quite speculative, it seems reasonable to outline some of the possible "take-home" lessons for developmental biology from studies on the oscillator mechanism. At the level of gene activation, an understanding of the details of the clock mechanism could lead one to some of the early initiating events for those developmental processes controlled by the clock. More specifically, the knowledge of which genes and their control elements are involved in this oscillator, and in regulating the transcriptional rhythms driven by the clock, should indicate directly the "downstream" genes and their controlling elements. Perhaps some of these genes will have promoter elements which respond to the circadian clock "signal", in addition to their other derepression signals. It is also likely that the gene products that show large diurnal variations will have to be ones with short half-lives, otherwise there could not be the "decay" part of the rhythm. This feature of a clock makes additional predictions then about mRNA and protein half-lives for products of these driven genes. It should be kept in mind, however, that the clock mechanism may be more of a synchronizer than initiator of developmental events, and would not necessarily be the trigger, but more of a modulator for a cascade of gene activation.

The biochemistry of the clock mechanism could lead us to a fuller understanding of the effects of environmental signals, such as light and temperature, on developmental events. The effects of light on microbial and plant differentiation are too numerous to list, but it is apparent that many of these processes are triggered by the same wavelengths of light that provoke large changes in the oscillator mechanism. If some part of this mechanism involves the level of an ion, then it is expected that there will be some "fallout" for the study of ion fluxes, etc. and their role in development from understanding how light pulses shift the oscillator. The "second message" ideas should be quite pertinent here. Likewise, since cell division in many organisms is restricted by the clock mechanism to certain times of day, there should be some direct biochemical link from the clock mechanism to the key controlling elements in cell division. These may be both at the level of gene activation and secondary messages in the cell.

The approaches used for studying clocks are also illustrative in a general way. The genetic approach again showed that each of the individual properties of a complex mechanism such as light response or temperature-compensation could be dissected out by examining appropriate mutants. It has also indicated the advantages and disadvantages in employing mutants isolated on the basis of function (clock mutants) or isolated on the basis of structure (auxotrophs or drug-resistant mutants). In combination with the studies on such physical effectors as light and temperature, these mutants have been very helpful in elucidating some of the properties of the clock, and in eliminating many areas of cellular biochemical function from consideration. These approaches should be applicable, in a general way, to studies on developmental processes and phenotypes.

11.6 Summary

The *Neurospora* oscillator sends periodic signals to the mycelium to induce it to conidiate. This oscillator mechanism is sensitive to and can be phase-shifted by both light pulses and temperature changes. The *Neurospora* oscillator is therefore similar to many other circadian clocks. The action spectrum for phase-shifting by light and genetic evidence implicates a flavin/cytochrome complex as the photoreceptor. Mutants lacking the light response ("blind") indicate that this input pathway can be disrupted without affecting the oscillator itself, and must therefore be separate from the oscillator mechanism. Mutants have been found that have apparently lost temperature-compensation, indicating that temperature-compensation can also be disrupted without destroying rhythmicity. On the other hand, the oscillator can be altered without altering the growth rate as seen from the studies of the *frq* mutants that have either shorter or longer periods than wild type, but normal growth rates. The *frq* gene has been cloned and partially sequenced, but its primary cellular biochemical activity is still not known. The cloning of clock-affecting genes paves the way for the use of molecular biology techniques to discover the biochemical functions of these genes. A number of other biochemical pathways have been eliminated as components of the oscillator. The absence of clock defects in a number of mutants defective in amino acid synthesis, some aspects of lipid synthesis, vitamin synthesis, and several other pathways implies that flux through these pathways is not essential to the clock mechanism. However, 12 mutants with known biochemical defects have been shown to affect the period. These mutants have alterations in amino acid synthesis, lipid synthesis, glycerol utilization, and mitochondrial functions. A number of clock-affecting compounds have been shown to alter period or phase. These compounds affect a variety of cellular processes: protein synthesis, mitochondrial energy metabolism, cyclic AMP, membranes, ion fluxes, and calcium metabolism. There is evidence that the maintenance of normal membrane composition is important for normal clock functioning. The *cel* mutant, deficient in fatty acid synthesis, has an altered membrane fatty acid composition and is defective in temperature-compensation.

There are a number of rhythmic variables which show true circadian rhythmicity and which may be driven by the oscillator. Although the majority of these variables appear to be linked to conidiation, a few have been shown to persist in its absence. Two genes showing rhythmic transcription have been cloned. These genes could provide tools for using molecular biology techniques to dissect output pathways.

References

[1] Edmunds JLN (1988) Cellular and molecular bases of biological clocks. Springer, Berlin Heidelberg New York

Feldman JF, Dunlap JC (1983) *Neurospora crassa:* a unique system for studying circadian rhythms. Photochem Photobiol 7:319–368

Hall JC, Rosbash M (1987) Genes and biological rhythms. Trends Genet 3:185–191

Lakin-Thomas P, Coté G, Brody S (1990) Circadian rhythms in *Neurospora crassa:* biochemistry and genetics. Crit Rev Microbiol 17:365–416

Morse DS, Fritz L, Hastings JW (1990) What is the clock? Translational regulation of circadian bioluminescence. TIBS 15:262–265

Sweeney BM (1976) Circadian rhythms, definition, and general characterization. In: Hastings JW, Schweiger H (eds) The molecular basis of circadian rhythms. Dahlem Konferenzen, Berlin, pp 77–83

[2] Borgeson CE, Bowman BJ (1985) Blue light-reducible cytochromes in membrane fractions from *Neurospora crassa.* Plant Physiol 78:433–437

Francis CD, Sargent ML (1979) Effects of temperature perturbations on circadian conidiation in *Neurospora.* Plant Physiol 64:1000–1004

Gardner GF, Feldman JF (1981) Temperature compensation of circadian period length in clock mutants of *Neurospora crassa.* Plant Physiol 68:1244–1248

Lakin-Thomas P, Brody S (1985) Circadian rhythms in *Neurospora crassa:* interactions between clock mutations. Genetics 109:49–66

Muñoz V, Brody S, Butler WL (1974) Photoreceptor pigment for blue light responses in *Neurospora crassa.* Biochem Biophys Res Commun 58:322–327

Paietta J, Sargent ML (1983) Isolation and characterization of light-insensitive mutants of *Neurospora crassa.* Genetics 104:11–21

Russo VEA (1988) Blue light induces circadian rhythms in the *bd* mutant of *Neurospora:* double mutants *bd, wc-1* and *bd, wc-2* are blind. J Photochem Photobiol 2:59–65

Sargent ML, Briggs WR (1967) The effects of light on a circadian rhythm of conidiation in *Neurospora.* Plant Physiol 42:1504–1510

[3] Coté GG, Brody S (1987) Circadian rhythms in *Neurospora crassa:* membrane composition of a mutant defective in temperature compensation. Biochim Biophys Acta 898:23–36

Dunlap JC, Feldman JF (1988) On the role of protein synthesis in the circadian clock of *Neurospora crassa.* Proc Natl Acad Sci USA 85:1096–1100

Loros JJ, Feldman JF (1986) Loss of temperature compensation of circadian period length in the *frq*-9 mutant of *Neurospora crassa.* J Biol Rhythms 1:187–198

Loros JJ, Richman A, Feldman JF (1986) A recessive circadian clock mutation at the *frq* locus of *Neurospora crassa.* Genetics 114:1095–1110

Mattern DL, Forman LR, Brody S (1982) Circadian rhythms in *Neurospora crassa:* a mutation affecting temperature compensation. Proc Natl Acad Sci USA 79:825–829

McClung CR, Fox BA, Dunlap JC (1989) The *Neurospora* clock gene *frequency* shares a sequence with the *Drosophila* clock gene *period.* Nature 339:558–562

[4] Hochberg ML, Sargent ML (1974) Rhythms of enzyme activity associated with circadian conidiation in *Neurospora crassa.* J Bacteriol 120:1164–1175

Loros JJ, Donome SA, Dunlap JC (1989) Molecular cloning of genes under control of the circadian clock in *Neurospora.* Science 243:385–388

Roeder PE, Sargent ML, Brody S (1982) Circadian rhythms in *Neurospora crassa:* oscillations in fatty acids. Biochemistry 21:4909–4916

Section 2 **Plants**

Chapter 12 Regulation of Development in the Moss, *Physcomitrella patens*

DAVID J. COVE

12.1 Introduction

12.1.1 The Suitability of *Physcomitrella patens* for the Study of Plant Development

The study of the molecular basis of development should ideally be carried out in an organism which can be grown under controlled environmental conditions and which is amenable to both traditional and molecular genetic analysis. In addition, development events should be directly observable, preferably at the level of the individual cell. Few organisms and no plants measure up to this ideal specification, but mosses have a number of advantages which gives them considerable potential as systems for understanding plant development at the molecular level. Many moss species, including *P. patens*, can be cultured and can complete their life cycle on a defined growth medium which need contain no organic compounds, using essentially microbiological methods. Light intensity and quality, temperature and humidity can all be controlled, and their effects on development are therefore straightforward to study.

The principal phase of the moss life cycle, the gametophyte, is haploid, and as a result, mutants with altered gametophyte morphology can be isolated with relative ease. Haploidy also makes genetic analysis simple (see Sect. 12.2.2), but also has disadvantages since it is possible that mutation in many genes involved in developmental programming will lead to lethality. An approach commonly used in viruses and micro-organisms to overcome this problem, is to isolate conditional mutants. This solution is certainly available in mosses, although it has not yet been exploited extensively. The mosses, and particularly *P. patens* and its near relatives, offer an alternative way of coping with this aspect of haploidy. Mosses show vigorous powers of vegetative regeneration. Almost any tissue of *P. patens* is capable of regeneration, behaving in much the same way as a germinating spore. The normal sexual life cycle is therefore not obligatory. Even if a mutation results in a block in development comparatively soon after spore germination, lethality does not occur and the mutant strain can easily be propagated.

Mosses have other advantages for the study of development. Many mosses, including *P.* patens, have a filamentous phase consisting of cells which can be observed directly in living plants. The cellular events which occur during this phase are described in Section 12.1.2 and are particularly suitable for the detailed study of plant development. The advantages which make *P. patens* suitable for developmental studies are shown by many moss species. One reason for choosing *P. patens* is that this species grows rapidly, growing in its natural environment on newly exposed soil. Another reason is chance; *P. patens* was one of the first species in which genetic analysis of induced mutations was shown to be possible, and since then work on it has gathered momentum. Other species may prove worthwhile subjects for the molecular genetic analysis of development and molecular genetic techniques will make it easier to use the knowledge acquired in the study of one species to the study of another.

12.1.2 Development of *Physcomitrella patens*

Descriptions of normal wild-type development often make two assumptions. The choice of a strain as a wild type for a particular species may be made uncritically, and different experimenters may use different strains which may show natural genetic variation. This can lead to difficulties in reconciling the findings of one group of research workers with those of another. For *P. patens*, almost all recent work has been carried out on a strain which has been designated a standard wild type and which was isolated by Harold Whitehouse in 1962 as a single spore from a plant near Cambridge. The use of this strain by most research groups who work on *P. patens* makes comparison between their work easier, but this single genotype has no particular claim to represent the wild type of the species. The second assumption is that there is a set of environmental conditions that can be regarded as normal. There have been no studies of development of *P. patens* in its natural environment and since it will grow in a range of environments, such studies would have no greater claim to describing normal development. It has become customary to describe as normal, development under a set of convenient and defined conditions, far removed from those experienced by plants growing in their natural environment. These conditions include the use of a simple, defined culture medium containing inorganic salts, with nitrate as nitrogen source and solidified with agar. Cultures are lit continuously with white light, produced by fluorescent electric lights, at intensities from 5 to $20\,W\,m^{-2}$ and are grown at $25\,°C$. These conditions appear to be near-optimal for gametophyte development.

Under these standard conditions, wild-type spores germinate about 48 h after inoculation, giving rise to a system of cell filaments. The first filaments produced are composed of primary chloronemal cells. Chloronemal filaments are ca. $18\,\mu m$ in diameter and the sub-apical cells are ca. $115\,\mu m$ long. Chloronemal filaments are very green, the cells containing many large chloroplasts, and they have cross walls which are perpendicular to the filament axis. Filaments grow by apical cell division, which occurs every about 20 h. The sub-apical cells of chloronemal filaments can also divide, giving rise to side branches from the original filament axis. About 3 days after germination, some chloronemal apical cells undergo a developmental transition to give rise to the second cell type, caulonema.

The newly formed caulonemal apical cell divides with a cell cycle time of about 8 h, giving rise to a filament of caulonemal cells. As a result of the shorter cell cycle time, caulonemal filaments spread

out from the original mass of tissue. Caulonemal filaments are a little larger in diameter than chloronemal filaments. The sup-apical cells are also longer (from 130 to $180\,\mu m$), and the cross walls between them are oblique to the filament axis.

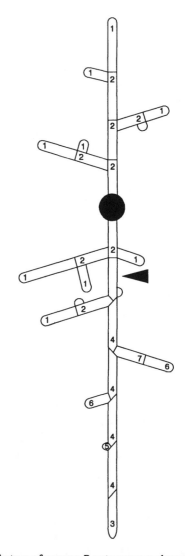

Fig. 1. Cartoon of a young *P. patens* gametophyte showing the relationship of the various cell types. The cartoon represents a gametophyte approximately 5 days after spore germination. *1* primary chloronemal apical cell; *2* primary chloronemal sub-apical cell; *3* caulonemal apical cell; *4* caulonemal sub-apical cell; *5* caulonemal side branch initial; *6* secondary chloronemal apical cell; *7* secondary chloronemal sub-apical cell. The *arrow* indicates the position at which the apical cell of the filament changed from chloronemal to caulonemal. Figure not drawn to scale

Caulonemal cells are less green than chloronemal cells, since they are more vacuolated and contain fewer chloroplasts. These chloroplasts are spindle-shaped and smaller than those in chloronemal cells.

Under standard conditions, almost all caulonemal sub-apical cells divide, giving rise initially to single-celled side branches. A few side-branch initials divide no further but most adopt one of three developmental fates, and it is the balance between these which results in the characteristic pattern of gametophyte development (Figs. 1 and 2). About 90% of initials develop into filaments of secondary chloronemal cells. These are morphologically similar to primary chloronemal filaments but show some differences in developmental behaviour (see Sect. 12.5.1). The next most common developmental fate of side-branch initials is to generate further caulonemal filaments. This allows the growing plant to fill in the spaces between the caulonemal filaments as they spread out to colonise more surface area. The remaining developmental fate of side branch initials, bud formation, is only adopted by about 3% of initials but is nevertheless the key to further development. Each of these fates involves different cellular events which may be readily observed in living tissue as early as the first division of the side-branch initial (Fig. 3).

Buds develop to give rise to structures called gametophores, the leafy shoots which are the most familiar phase of the moss gametophyte generation (Fig. 4). Gametophores show little cell specialisation and contain no vascular tissue. Growth of the gametophore axis occurs by division of a tetrahedral apical cell (Fig. 3, 4) which cuts off new cells from each of its three basal surfaces in turn. Sup-apical meristematic tissue gives rise to leaves which are simple structures, one cell thick. Rhizoids, filaments which anchor the gametophore and keep it upright, are produced from the gametophore base. Rhizoids are morphologically similar to caulonemal filaments but produce few side branches. As their name implies, gametophores are the organs which produce gametes. Gametogenesis requires temperatures below 19 °C, and so sexual reproduction does not occur in the standard conditions described. The environmental conditions required for gametogenesis have not been investigated in detail and the control of the onset of gamete formation

Fig. 2. Caulonemal filaments from a 3-week-old culture of *P. patens*, which has grown under the standard conditions described in Section 2.1.2. The main filament axes are composed of caulonemal cells. Almost every sub-apical caulonemal cell has divided to produce one or more side branches. Most of these have developed into filaments of secondary chloronema. A few have produced buds, which may be seen at various stages of development. *Bar* 1 mm

may be an aspect of development which would be worth studying.

P. patens is monoecious, i.e., male and female gametes are both produced on the same gametophore. Following fertilisation, zygotes give rise to sporophytes, the diploid phase of the moss life cycle.

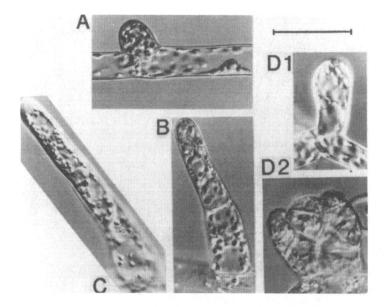

Fig. 3A–D. Early stages in the alternative developmental fates which can be adopted by caulonemal side branches of *P. patens.* **A** Single celled side branch initial. **B** Secondary chloronemal side branch. **C** Caulonemal side branch. **D1** Early stage in bud formation. **D2** Later stage in bud formation similar to that shown in the top left scanning electron micrograph in Fig. 4. *Bar* 50 μm

Fig. 4. Scanning electron micrographs of gametophore development. *Above* four stages in early bud development. *Bar* 60 μm; *below* mature gametophores. *Bar* 1 mm

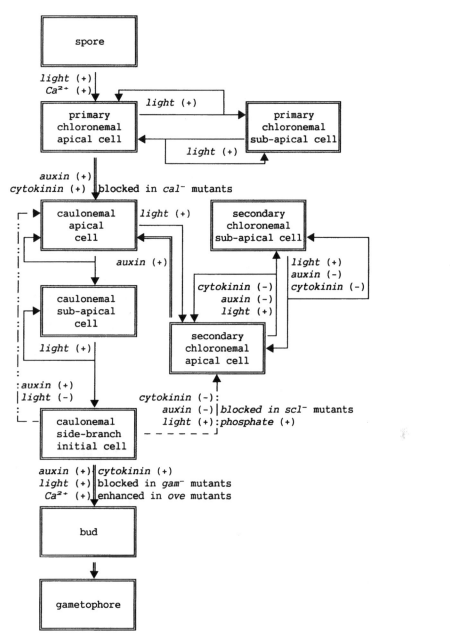

Fig. 5. Cell lineages in the development of the gametophyte of *P. patens*. Transitions between stages which are connected by a *broken arrow* do not require cell division. The *branched arrows with single lines* represent developmental steps involving a cell division, the *arrows* indicate the two products of the division. Stages connected by a *double-lined arrow* require more than one cell division. The (+) sign beside a signal indicates that it is required for or enhances the frequency of the transition, the (−) sign, that it decreases the frequency of the transition

Sporophytes of *P. patens* are structurally simple and consist of a very short stalk and a spore capsule which has no special structure to facilitate spore release. Each capsule contains about 4000 spores, produced in tetrads by meiosis of spore mother cells. *P. patens* has no mating types, and sporophytes can arise as a result of either cross- or self-fertilisation.

The description of the pathway of development given here started arbitrarily with spore germination. It may, however, be that this start point can be regarded as having some developmental significance since tissue regeneration follows a similar developmental path, although not with precisely the same timing. Whatever tissue is used as a start point, whether gametophytic or sporophytic, regeneration begins by the production of chloronemal filaments. Chloronemal filaments are also formed following protoplast regeneration (see Sect. 12.2.2.3). No extensive study has been made of the factors which trigger regeneration. It is likely that tissue damage is an important factor but that the signalling is more complex than this alone.

Figure 5 represents diagrammatically the lineages in the development of the gametophyte of *P. patens*. By representing development in this way, it is possible to see clearly the developmental switches which must occur and whose mechanisms need to be elucidated. Figure 5 also includes information about some of the transitions for which environmental effects have been identified, and the stages at which different classes of mutants are likely to be blocked. Further explanation of both these topics is given in Sections 12.3 and 12.4.

A cell lineage diagram is a helpful simplification, but it fails to emphasise that some of the processes represented in the lineage occur repeatedly while others occur only a limited number of times. For example, the apical cells of both chloronemal and caulonemal filaments divide repeatedly (but at different rates), and can therefore be regarded as stem cells which give rise to sub-apical cells. On the other hand, sub-apical cells divide only a few times. Some other transitions shown do not even involve cell division. Another shortcoming of the cell lineage diagram is that it does not represent spatial or temporal relationships between developmental events.

12.2 Techniques[1]

12.2.1 Culture Methods

P. patens is usually grown on simple solid medium, containing only inorganic salts and agar. Although spores may be used as an inoculum, it is more common to use fragments of tissue. Because growth occurs within, as well as on the surface of the medium, it is difficult to harvest tissue from cultures growing on solid medium. Small quantities of tissue can be obtained by inoculating a suspension of tissue fragments, obtained by gentle blending, onto cellophane placed on the surface of solid medium. Fragments regenerate and give rise to a mat of tissue. After culture for 7 d about 500 mg of tissue, composed mostly of chloronemal cells, can be harvested from each 90-mm diameter Petri dish. For larger quantities of tissue, *P. patens* may be grown in a fermenter. Again, tissue fragments are used as an inoculum. Remarkably, development within a fermenter occurs in a similar way to that on solid medium.

A further method of culture, which is valuable for the study of hormone action, uses continuous medium feeding. Tissue is inoculated onto a nylon support net, placed on cellophane, overlaying solid medium and incubated for 7 d while regeneration occurs. The culture is then transferred, on its net, to a special apparatus where fresh liquid medium is pumped onto it. It is also possible to culture *P. patens* to allow continuous observation under a microscope. Here the culture is grown in a very thin layer of solid medium, which is prevented from drying out by passing a slow stream of liquid medium through it. In this way, cultures have been observed and developmental processes recorded using time lapse video recording equipment, for periods of over 48 h.

Growth with repeated sub-culture is not the only method of maintaining cultures. Spores from fertile strains can be preserved for long periods but since many developmentally abnormal strains are sexually sterile, they can only be propagated vegetatively. Fortunately, a method of strain maintenance, using cryopreservation, has been devised, since the maintenance of strains as vegetative cultures is not only expensive, but also runs the risk of strains reverting or accumulating new mutations. Cryopreservation involves freezing tissue at a controlled rate in the presence of a protective chemical, until it reaches a temperature of $-20\,°C$. The tissue can then be transferred into liquid nitrogen, where it may be preserved in a viable state for at least 5 years. When a stored culture is required for use, it is warmed to room temperature, washed and used to inoculate fresh medium.

12.2.2 Genetic Methods

12.2.2.1 Mutant Isolation

Both spores and tissue can be mutagenised. Tissue can be treated with enzymes, after mutagenesis, to obtain uninucleate isolated protoplasts which regenerate to give clones for screening. Large numbers of mutants, including many with auxotrophies or developmental abnormalities, have been isolated by carrying out non-selective screens of strains produced following mutagenesis.

12.2.2.2 Genetic Analysis Using the Sexual Cycle

In *P. patens*, the sexual cycle is completed in about 10 weeks. Because the gametophyte is haploid, characters which can be scored in the gametophyte, segregate at each generation and with gametic ratios. Genetic analysis is thus both simpler and quicker than in organisms which are predominantly diploid. However, this form of conventional genetic analysis is only possible for mutants which do not lead pleiotropically to sexual sterility.

12.2.2.3 Somatic Hybridisation and Parasexual Genetic Analysis

An alternative method of genetic analysis involves artificial hybridisation of mutant strains by way of protoplast fusion. Protoplasts can be isolated in large numbers from *P. patens* chloronemal tissue by enzyme treatment. The hybridisation of two developmentally abnormal strains is possible if each carries a different auxotrophy. Protoplasts are isolated

from the two strains, mixed, subjected to a treatment which causes protoplast fusion, and then plated onto medium which is not supplemented for either auxotrophy. Only hybrids will grow on this medium. These hybrids produce gametophytes which are similar to haploid strains although they develop more slowly. They can be used for genetic analysis, but somatic hybrids are usually used simply to carry out complementation tests.

12.2.2.4 Genetic Transformation

As with other plants, the use of transformation for genetic analysis of *P. patens* is in its infancy, but is now developing rapidly. Transformants for antibiotic resistance can be selected following treatment of isolated protoplasts with plasmid DNA. Initial selection leads to the regeneration of large numbers of resistant plants. However, many of these express antibiotic resistance only transiently and lose plasmid DNA rapidly. Some are able to replicate the plasmid DNA and retain it through many cell divisions providing selection is maintained. At present, only in a few plants does the gene coding for antibiotic resistance become stably integrated into the *P. patens* genome, where it can be transmitted through sexual crosses. It should not be difficult to increase the frequency of recovery of stable transformants as our knowledge of transformation improves.

12.3 The Environmental Modification of Development[2]

12.3.1 The Effect of Light

Spore germination only occurs in the light and phytochrome is likely to be involved (see Chap. 13, this Vol.). The relationship between light intensity and time of irradiation is not, however, straightforward, and the developmental mechanisms involved are probably complex. Mutants affected in spore germination have not been isolated, and the genes involved may have their effects in the diploid spore mother cells and so be similar to maternal effect genes in, for example, *Drosophila melanogaster* (see Chap. 21, this Vol.).

Chloronemal cell division is light-dependent. A culture consisting of chloronemal cells cannot grow upon transfer to darkness, even if supplied with a carbon source. Chloronemal apical cells also show tropic responses to light, their direction of growth being influenced by light direction (see Sect. 12.5.1).

Caulonemal apical cells can divide in darkness. When a culture containing both chloronemal and caulonemal filaments is transferred to darkness, only the caulonemal filaments continue to grow, showing a negative gravitropic response (see Sect. 12.5.2). The production of side-branch initials by caulonemal sub-apical cells is light-dependent, but only very low levels of light are required to saturate this response. The developmental fate of initials also shows light dependence. Figure 6A shows the morphology of a caulonemal filament cultured at a low light intensity and is markedly different from the morphology of filaments grown at high light levels (Fig. 2). Detailed studies of the influence of light on side-branch fate have yet to be undertaken, but phytochrome is likely to be involved in many of the phenomena described here. Light also influences the direction of growth of caulonemal apical cells (see Sect. 12.5).

Gametophore development is also influenced by light. In darkness and at low light levels, gametophores etiolate, producing very elongated stems, and leaf development is much reduced. Gametophores also show phototropism. This is the only light-mediated effect for which mutants showing an abnormal response have so far been obtained (*ptr* mutants, see Sect. 12.5.1).

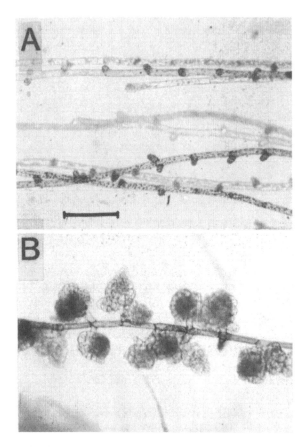

Fig. 6. A Morphology of filaments growing under standard conditions (see Sect. 12.1.2) except that level of white light (100 mW m^{-2}) is low. **B** As **A** except that the cytokinin, 6-benzyl adenine, was added to the medium at a concentration of 100 nM. *Bar* 200 μm

12.3.2 The Effect of Nutritional Status

The development effects of nutritional status are probably complex. Only two effects are well documented. The morphology of cultures grown with nitrate as nitrogen source differs from that of cultures supplied with ammonium. On ammonium, secondary chloronemal side branches contain more cells and branch more often that those produced with nitrate. This may not be a specific effect of nitrate and ammonium, but a more general reflection of nutrient status. Starvation for other nutrients,

with the exception of phosphate, results in morphogenetic effects which resemble growth on nitrate.

The effect of phosphate starvation on development is distinct. Both caulonemal production and apical cell growth is more or less unimpaired. Gametophores are produced but these are smaller and less numerous than on cultures produced under phosphate-sufficient conditions. The most dramatic effect of phosphate starvation is that secondary chloronemal filament production is blocked; caulonemal filaments produce large numbers of single-celled side branch initials and resemble filaments grown at low light levels (Fig. 6A). Phosphate limitation therefore appears to have a specific effect on the developmental transition of side-branch initials to secondary chloronemal filaments. The basis of this effect is unknown. Mutants which have a morphology resembling that of the wild type grown under conditions of phosphate limitation (*scl⁻* mutants) have been isolated, but these are rare and difficult to culture and therefore to investigate. In liquid medium, more drastic phosphate starvation can be achieved and side-branch initial production is also impaired, but caulonemal apical cell growth still occurs. Again, there is a parallel with the effect of light, since caulonemal filaments grown in darkness are unbranched but show normal apical cell growth.

There are therefore two different patterns of developmental response to starvation. The effect of phosphate starvation parallels that of light limitation and blocks the formation of caulonemal side-branch initials and their development into secondary chloronemal filaments. The effect of starvation for nitrogen and most other nutrients is to limit the luxurience of secondary chloronemal development. More drastic starvation for nitrogen does not lead to a morphology similar to that obtained with phosphate or light limitation, but reduces culture vigour more generally.

12.3.3 The Effect of Phytohormones (Plant Growth Regulators)

The discovery that treatment with cytokinin led to a greatly increased production of buds in a number of moss species was one of the earliest reports of a specific developmental effect for a phytohormone. This response has allowed mosses to be used as a sensitive bioassay for cytokinins. Detailed studies of the effects of cytokinins on *P. patens* development have shown that not only do such compounds lead to a massive switch of the developmental fate of caulonemal side branch initials from secondary chloronemal filaments to buds (Fig. 6B) but they also inhibit secondary chloronemal cell division, and at higher concentrations, interfere with normal bud development. The sensitivity of the response varies for different cytokinins and for different moss species. In *P. patens*, an enhancement of bud production can be detected in wild-type cul-

tures treated with 6-benzyl adenine (BA) at concentrations greater than 5 nM. BA has not been found as a natural product, but is used experimentally because it is more stable than many natural cytokinins. *P. patens* has been shown to produce two cytokinins, N^6-(^{2}-isopentenyl)adenine and zeatin, and its sensitivity to these natural cytokinins is similar to that for BA. Evidence from the analysis of mutant phenotypes, which will be reviewed in Section 12.4, has indicated that cytokinins also play a role in the transition of primary chloronemal apical cells to caulonemal apical cells, but treatment of wild-type cultures with cytokinins does not lead to a detectable effect at this stage of development.

Treatment of *P. patens* with auxins leads to an enhanced production of caulonemal filaments. An increased number of both primary chloronemal apical cells and caulonemal side branch initials give rise to caulonemal filaments. Auxins have an effect similar to cytokinins on the development of secondary chloronemal filaments, reducing both their length and complexity. As with cytokinins, higher concentrations of auxin inhibit bud development, but auxins also lead to increased rhizoid production. No specific effects of other substances known to regulate growth in flowering plants have been identified in *P. patens*, but studies of such effects have not been extensive. Physiological analysis of wild-type strains has therefore shown that auxins enhance caulonemal production and cytokinins enhance bud production, but such studies are not able to show whether these substances are required for these stages of development. The establishment that they were indeed required at these stages required the study of developmentally abnormal mutants, which will be dealt with in the next section.

12.4 The Roles of Auxin and Cytokinin in Moss Development[3]

12.4.1 Mutants Altered in Hormone Synthesis and Response

Three classes of developmentally abnormal mutants have been shown to involve alterations in synthesis of or response to auxins or cytokinins. The first mutant class (*cal⁻*) is blocked early in development and produces few if any caulonemal filaments. Under standard conditions, mutant plants of this class consist entirely or almost entirely of chloronemal tissue. The second class (*gam⁻*) are blocked in bud production, but produce both chloronemal and caulonemal filaments. The third class, *ove* mutants, has higher bud production than normal, resembling the wild type grown in the presence of added cytokinin.

12.4.1.1 ove Mutants Overproduce Cytokinins

Of the three classes, the physiological basis of *ove* mutants has been the most straightforward to establish. These mutants produce higher than wild-type levels of the cytokinins, isopentenyl adenine and zeatin. Complementation analysis shows that at least three genes can mutate to give this phenotype. The molecular basis of the *ove* phenotype has not yet been established. Almost all such mutants are recessive, and so the phenotype is likely to arise as a result of loss of gene function. Loss of gene function can result in the overproduction of a substance if the gene concerned acts as a negative regulator, coding, for example, for a repressor. There are however other possible explanations. Transfer RNAs contain purine residues which, in the free state, are cytokinins. The turnover of tRNAs is therefore a potential source of cytokinins. The rate of turnover of tRNAs is similar in wild-type and *ove* strains, and so an increase in tRNA turnover cannot account for the *ove* phenotype. However, the rate tRNA turnover is sufficient to generate higher levels of cytokinins than are present in the wild type. There must therefore be some mechanism which prevents this potential source of cytokinins flooding the cell. It may be this mechanism which is defective in *ove* strains. The pathway of cytokinin synthesis has not been established and, in theory, it would not be necessary to have a separate pathway for de novo synthesis of cytokinins, since this could be achieved by regulating their release during tRNA turnover.

12.4.1.2 Some cal⁻ Mutants Are Blocked in Hormone Synthesis, Others in Response

Mutants blocked early in development and unable to produce caulonemal filaments (*cal⁻* mutants), fall into three sub-classes with respect to their sensitivity to auxins and cytokinins. One class responds to the exogenous supply of auxin, by producing caulonemal filaments. Although it is difficult to restore this class of mutants to a completely wild-type phenotype by the addition of an auxin, the response is sufficient to suggest that these mutants are blocked in auxin production. Their phenotype therefore provides evidence that auxin is indeed necessary for the transition of primary chloronemal to caulonemal apical cells. Members of the second sub-class respond to the exogenous supply of cytokinin, by producing caulonemal filaments and buds. Again, these mutants cannot be restored to a wild-type phenotype by cytokinin treatment, but their phenotype is nevertheless consistent with their being impaired in cytokinin synthesis, indicating that cytokinin must also be required for the production of caulonemal apical cells.

The finding that normal development cannot be restored completely in either of these sub-classes of mutants by treatment with cytokinins or auxins, illustrates an important general point relating to developmental genetics. Since the synthesis of auxins and cytokinins is likely to be both temporally and spatially regulated, it is not surprising that their addition to the growth medium is a crude substitute for normal synthesis, and is unable to restore normal development.

The third of the mutant sub-classes blocked early in development, respond to neither cytokinins nor auxins. These mutants may therefore be altered in genes coding for the detection of cytokinin or auxin or in the transduction of these signals to bring about the production of caulonemal cells. No temperature-sensitive mutants of this class have yet been isolated but such mutants could be used to determine whether the genes involved are required only for the production of caulonemal apical cells or have a more prolonged role in caulonemal filament development.

12.4.1.3 Some gam⁻ Mutants Are Impaired in Hormone Synthesis, Others in Response

Mutants of the third class (*gam⁻* mutants) are blocked in bud production. Again members of this group may be divided into sub-classes. One sub-class responds to treatment with cytokinin, producing buds as copiously as the wild type. It is likely that members of this sub-class are also impaired in cytokinin synthesis, confirming that a cytokinin is required for the transition of side-branch initials to buds. Since members of this sub-class are able to produce caulonemal filaments, it seems that they are only partially blocked in cytokinin synthesis and this in turn tells us that the amount of cytokinin needed for caulonemal production is likely to be less than that required for bud production. The second mutant sub-class is insensitive to cytokinin treatment. Mutants of this sub-class are rarer and it may be that some are the result of more than one mutation. This group of mutants should include some which are unable to detect cytokinin and should also allow the identification of genes whose products are involved specifically in bud formation.

12.4.2 Continuous Medium Feeding with Auxins and Cytokinins

The principle conclusions for the roles of auxins and cytokinins drawn from the study of developmentally-abnormal mutants, have been confirmed using the technique of continuous medium feeding (see Sect. 12.2.1). When grown using this method and fed with minimal medium, a wild-type culture produces only chloronemal tissue, phenocopying *cal⁻* mutants. Caulonemal production can be restored to a continuously fed culture by including an auxin in the medium supply. Some bud production also occurs under these conditions. This confirms the role of auxin in caulonemal formation and shows that auxin must be leached out of tissue by continuous medium feeding. If the conclusion drawn from mutant studies, that cytokinin is needed for caulonemal production, is correct, then continuous medium feeding cannot leach out cytokinins so that their levels fall below that required for caulonemal production. The observation that some buds are produced on cultures fed with medium containing auxin suggests that cultures grown in this way retain even the higher level of cytokinins required for bud formation.

When a culture is fed with medium containing cytokinin, little effect is observed even when high levels (10 μM BA) are used; only chloronemal tissue is produced, no buds are formed, but there is some general inhibition of growth. This result is not surprising since caulonemal side-branch initials, the target cells required for bud production, are absent. The general inhibition of growth caused by high levels of cytokinins is also unsurprising since cytokinins are adenine analogues. If a culture is fed with medium containing both cytokinin and auxin, then a very large number of buds are produced, a result which is once again consistent with the conclusions drawn from mutant studies.

Somewhat different results are obtained when the technique of continuous medium feeding is applied to the closely related moss species, *Physcomitrium sphaericum*. When this species is fed with minimal medium, caulonemal filaments are still produced (Fig. 7A) and the addition of auxin to the medium has no effect (Fig. 7B). No response is observed in tissue fed medium to which only cytokinin has been added, even though side branch initials are present (Fig. 7C). Bud production can only be induced by feeding medium containing both auxin and cytokinin (Fig. 7D). This suggests that continuous medium feeding does not lower the auxin concentration within the tissue below that required for the production of caulonemal filaments. The requirement for auxin to be added to the medium for this species to show cytokinin-induced bud production, shows that auxin is also required for bud formation. A higher concentration of auxin must be required for bud formation than for the production of caulonemal apical cells, since continuous medium feeding must lower the auxin concentration enough to prevent bud production but not enough to impair caulonemal production.

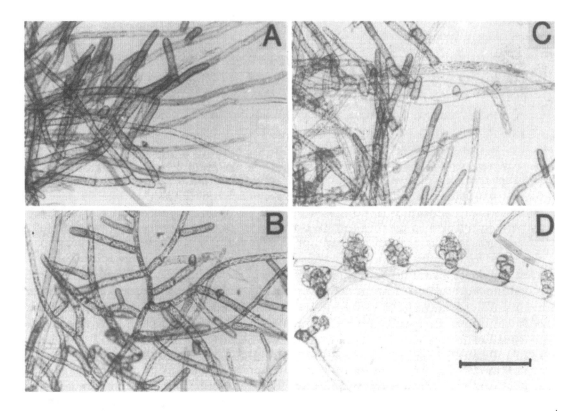

Fig. 7 A – D. Tissue from cultures of *Physcomitrium sphaericum* grown under a regime of continuous medium feeding. **A** Minimal medium. **B** Minimal medium + 1 μM naphthyl acetic acid (an auxin). **C** Minimal medium + 1 μM 6-benzyl adenine (a cytokinin). **D** Minimal medium + 1 μM naphthyl acetic acid + 1 μM 6-benzyl adenine. *Bar* 200 μm

12.5 Tropic Responses[4]

12.5.1 Phototropism

Most of the processes so far considered in this chapter involve the programming of cell fate. Processes which determine cell size and shape are also fundamental to development, and one such process, the tropic response of filament apical cells, has been studied extensively. The apical cells of primary chloronemal, secondary chloronemal and caulonemal filaments all grow by extension of the cell apex. The direction of tip extension is influenced by light direction (phototropism). The responses vary with the type of filament and with light quality. The tropic responses of apical cells may be summarised as follows:

1. At low red light levels, primary chloronemal apical cells grow towards light, but in high levels of red light, they grow perpendicular to the direction of the light source. The switch from the low to high intensity response occurs at different intensities depending on the wavelength of the light. In red and blue light only comparatively low intensities are required to trigger the high intensity response, in green and yellow light much higher intensities are required and for far-red light only the low level intensity response has been found even for the highest intensities tested.

2. In red light, caulonemal apical cells grow towards a low light intensity light source and perpendicular to a high intensity source.

3. Secondary chloronemal apical cells show positive phototropism in all intensities of light.

Although complex, it is not difficult to rationalise these phototropic responses. Spores germinating in low intensities produce primary chloronemal filaments which will need to grow towards the source of light. When germination occurs at higher light intensities, for example on the soil surface, or when a filament grows into light of a higher intensity, filament growth perpendicular to the direction of light is necessary.

Gametophore growth also shows phototropism, but appears to be simpler, being positive at all light intensities. *ptr* mutants have altered phototropic responses. All were originally isolated as a result of loss of gametophore phototropism, but their caulonemal apical cells have also lost their phototropic response. Mutations in at least three genes (*ptr*A, B and C) can lead to this phenotype and all mutants are recessive. The products of the genes concerned must therefore be required for both gametophore and caulonemal phototropism. However, investigation of the phototropic response of primary chloronemal apical cells in these mutants reveals that the phenotype is more complex. Such mutants have lost neither the low nor high intensity response of their primary chloronemal apical cells, but instead require a lower light intensity to switch from the low to the high intensity response. No explanation of this complex phenotype has so far been put forward.

12.5.2 Gravitropism

Caulonemal apical cells and gametophore stems are also sensitive to the direction of gravity, showing a negative gravitropic response, growing away from the gravity vector. These responses are only shown in darkness or in very low levels of light. It was thought originally that the lack of gravitropic response in high light intensities was simply a result of phototropic responses being morphogenetically stronger. However, *ptr* mutants, which respond to gravity in darkness in a way similar to the wild type, show no gravitropic response in light. It appears, therefore, that light actively switches off gravitropism.

Because of the requirement of gravitropic responses for darkness, their study requires special techniques. The response of individual filaments can be observed using infra-red video microscopy. Caulonemal filaments are intensitive to infra-red irradiation, growing as though in darkness, and so a television camera sensitive to infra-red can be used at these wavelengths to observe and record gravitational responses. Using these techniques it has been shown that the gravitational response of individual caulonemal apical cells is not constant. Each time the cell divides, not only is the negative gravitational response interrupted, but some growth towards the gravity vector occurs. This may be because the cell uses the nucleus as part of its graviperception mechanism and so is able to respond to gravity only while the nuclear membrane is intact. An alternative explanation is that some cell component, perhaps cytoskeletal, is required during alterations in the direction of apical extension and is also needed for nuclear and cell division. As a result, tropic responses cannot occur during the latter processes. Detailed studies of phototropic and polarotropic responses using similar techniques should reveal whether the effect of the cell cycle is specific to gravitropism. Neither of the proposed explanations accounts for the positive gravitropic response which occurs during cell division, but here mutant studies could be relevant.

At least 16 mutants showing abnormal gravitropic responses of caulonemal apical cells have been isolated. All of these mutants retain the normal negative gravitropic response of gametophore stems, indicating that the two responses must be at least partially distinct. Mutants showing an abnormal gravitropic response of caulonemal apical cells, fall into two classes, neither of which has lost completely its ability to respond to gravity. One class shows a much weaker response than the wild type but this is still negative. Mutations in at least two genes (*gtr* A and B) have this phenotype, and the mutant phenotype is recessive. The second class are mutant in the *gtr*C gene and respond strongly to gravity but in completely the opposite direction to the wild type, showing a positive gravitropic response. Most *gtr*C mutants are recessive and so loss of gene function can affect caulonemal gravitropism in two ways, neither leading to total loss of response. The positive gravitropism of *gtr*C mutants suggests that the *gtr*C gene product is required to interpret the graviperception signal to give a negative response, the default condition being to grow towards the gravity vector. This is consistent with the observation that a positive response occurs during cell division, since this too may result from an interruption of the transduction mechanism. The reason for the absence of a class of mutant which has completely lost the gravitropic response of caulonemal apical cells is not yet known.

12.6 Outlook

In the introduction to this chapter, it was argued that to study development, it was desirable to choose an organism which was accessible to the techniques of cell biology, and of traditional and molecular genetics. *P. patens* has already proved to be suitable on the first two counts, and molecular genetic studies now have a high priority. Once initial progress has been made, and the recent success in genetic transformation is a vital first stage, further progress should be rapid, since so much is already known about the developmental physiology of this organism. The next step must be the isolation of genes known to mutate to affect development. There are a number of strategies, already used in other organisms, which may achieve this. If transformation rates can be increased considerably, self-complementation, as is used in many microorganism species (see, e.g. Chap. 7, this Vol.), might be possible. However, the large genome size of *P. patens* (6×10^8 bp) makes it unlikely that this will be a successful method. More probably, a gene-tagging strategy will be successful, using intertional mutagenesis with either transposons or T DNA (see Chap. 15, this Vol.). If gene isolation is successful, there seems to be no limit to the work which could then be done since there is a large number of genes whose developmental roles are already characterised. The prospect of molecular analysis should encourage studies of the cellular events in the processes which have been described in this chapter.

12.7 Summary

Mosses, and particularly *P. patens*, are especially suitable for the study of development. The gametophyte, the dominant phase of the life cycle, is haploid, and this makes mutant isolation and genetic analysis straightforward. Culture is possible on a simple defined medium and in controlled environmental conditions. Many mutants which are affected in their development have been isolated. These include mutants which appear to be altered in the synthesis of the phytohormones, auxin and cytokinin. Analysis of such mutants has shown that these phytohormones are required at two crucial developmental stages, in a transition between cell types in the early filamentous stage of gametophyte development, and in bud formation. Other mutants are abnormal in their response to these phytohormones and are likely to be blocked in the signal transduction pathways involved in these processes.

P. patens shows a complex pattern of tropic responses to both light and gravity. Mutants abnormal in both types of response have been isolated, but those abnormal in their response to gravity, respond normally to light, and vice versa. The responses therefore appear to involve independent mechanisms.

Techniques for molecular genetic analysis are being devised, and genetic transformation is now available. Progress is being made in devising a technique of gene tagging using transposons, and this should soon make it possible to isolate developmentally important genes, which in turn should allow progress to be made in understanding mechanisms at the molecular level.

References

[1] Ashton NW, Boyd PJ, Cove DJ, Knight CD (1988) Genetic analysis in *Physcomitrella patens*. In: Glime JM (ed) Methods in bryology. Hattori Botanical Laboratory, Nichinan, Japan, pp 59–72

Batra A, Abel WO (1981) Development of moss plants from isolated and regenerated protoplasts. Plant Sci Lett 20:183–189

Boyd PJ, Hall J, Cove DJ (1988) An airlift fermenter for the culture of the moss *Physcomitrella patens*. In: Glime JM (ed) Methods in bryology. Hattori Botanical Laboratory, Nichinan, Japan, pp 41–45

Cove DJ (1983) Genetics of Bryophyta. In: Schuster RM (ed) New manual of bryology. Hattori Botanical Laboratory, Tokyo, Japan, pp 222–231

Knight CD, Cove DJ (1988) Time-lapse microscopy of gravitropism in the moss *Physcomitrella patens*. In: Glime JM (ed) Methods in bryology. The Hattori Botanical Laboratory, Nichinan, Miyazaki, Japan, pp 127–129

Knight CD, Cove DJ, Boyd PJ, Ashton NW (1988) The isolation of biochemical and developmental mutants in *Physcomitrella patens*. In: Glime JM (ed) Methods in bryology. The Hattori Botanical Laboratory, Nichinan, Miyazaki, Japan, pp 47–58

Schaefer D, Zryd J-P, Knight CD, Cove DJ (1990) Stable transformation of the moss *Physcomitrella patens*. Mol Gen Genet 226:418–424

[2] Bopp M (1952) Entwicklungsphysiologische Untersuchungen an Laubmoosprotonomen. Z Bot 40:119–152

Bopp M (1980) The hormonal regulation of morphogenesis in mosses. In: Skoog F (ed) Plant growth substances. Springer, New York Berlin Heidelberg, pp 351–361

Brandes H, Kende H (1968) Studies on cytokinin-controlled bud formation in moss protonemata. Plant Physiol 43:827–837

Cove DJ, Ashton NW (1984) The hormonal regulation of gametophytic development in bryophytes. In: Dyer AF, Duckett JG (eds) The experimental biology of bryophytes. Academic Press, London, England, pp 177–201

Cove DJ, Schild A, Ashton NW, Hartmann E (1978) Genetic and physiological studies of the effect of light on the development of the moss, *Physcomitrella patens*. Photochem Photobiol 27:249–254

Gorton BS, Eakin RE (1957) Development of the gametophyte of the moss *Tortella caespitosa*. Bot Gaz (Chicago) 119:31–38

Johri MM, Desai S (1973) Auxin regulation of caulonema formation in moss protonema. Nat New Biol 245:223–224

Reski R, Abel WO (1985) Induction of budding on chloronemata and caulonemata of the moss, *Physcomitrella patens*, using isopentenyladenine. Planta 165:354–358

[3] Ashton NW, Cove DJ, Featherstone DR (1979) The isolation and physiological analysis of mutants of the moss, *Physcomitrella patens*, which over-produce gametophores. Planta 144:437–442

Ashton NW, Grimsley NH, Cove DJ (1979) Analysis of gametophytic development in the moss, *Physcomitrella patens*, using auxin and cytokinin resistant mutants. Planta 144:427–435

Cove DJ (1984) The role of cytokinin and auxin in protonemal development in *Physcomitrella patens* and *Physcomitrium sphaericum*. J Hattori Bot Lab 55:79–86

Perry KC, Cove DJ (1986) Transfer RNA pool sizes and half lives in wild-type and cytokinin over-producing strains of the moss, *Physcomitrella patens*. Physiol Plant 67:680–684

[4] Cove DJ, Knight CD (1987) Gravitropism and phototropism in the moss, *Physcomitrella patens*. In: Thomas H, Grierson D (eds) Developmental mutants of higher plants. Cambridge University Press, London, pp 181–196

Jenkins GI, Cove DJ (1983a) Phototropism and polarotropism of primary chloronemata of the moss, *Physcomitrella patens*: responses of the wild type. Planta 158:357–364

Jenkins GI, Cove DJ (1983b) Phototropism and polarotropism of primary chloronemata of the moss, *Physcomitrella patens*: responses of mutant strains. Planta 158:432–438

Chapter 13   Plant Photoperception: the Phytochrome System

ROBERT A. SHARROCK

13.1 Introduction

Light plays an indispensable role in plant growth as the energy source for photosynthesis. However, in addition to driving metabolic processes, light also functions as a regulator of the pattern of plant development. Plants are sessile organisms which are unable to relocate to more favorable growth conditions once they have begun to develop, and they manifest a developmental strategy which allows a high degree of morphological adaptation within each individual and considerable responsiveness to environmental cues. Regulation of this developmental program by light cues, photomorphogenesis, requires the capacity to accurately gauge the light environment, a function of the activities of specific plant photoreceptor molecules. The biology of receptors in plants is an emerging science and the molecular properties and modes of action of plant receptors are poorly understood as compared to such systems in animals. Nonetheless, it is the importance of photoregulatory pathways in control of plant development, the possibility of elucidating novel photodetection and signal transduction mechanisms, and the potential to understand and modify agriculturally significant aspects of light-controlled plant physiology which invite research in this field.

13.2 Plant Growth and Development Are Responsive to Light Cues

The presence or absence of light has profound effects on plant morphology. The seeds of many species of flowering plant can be germinated in complete darkness and will grow until food storage reserves are exhausted. Such seedlings are referred to as etiolated (Fig. 1) and represent a plant growth pattern dedicated in large part to the search for light. In the natural environment, etiolated or partially etiolated growth occurs following seed germination under the soil. The plant's shoot is highly elongated, its leaves have not expanded, and it is white or yellow because it lacks chlorophyll. Moreover, the chloroplasts in etiolated plants have not fully developed but are present as etioplasts, precursor plastids poised for differentiation to mature photosynthetic organelles. In contrast to this, the morphology of a seedling grown in the presence of broad-spectrum white light shows true leaf development and expansion, inhibition of stem elongation, and the green color associated with mature chloroplasts (Fig. 1). This example illustrates only a few of the most obvious effects on plant morphology of two highly contrasting light environments, dark versus light. Indeed, the complex and often subtle role of light as an environmental regulator of development extends throughout the entire life of the plant.

13.3 Plant Regulatory Photoreceptors: the UV-B, UV-A/Blue, and Red/Far Red Responsive Systems [1]

Visual perception in many animals is mediated by specialized photoreceptor cells which absorb light and generate neural signals. These cells contain a dense stack of membranes in which the visual pigment rhodopsin is embedded. The photosensory apparatus in plants is fundamentally different from visual systems in that light is absorbed by photosensitive molecules present in a wide variety of cell types that are not specifically differentiated for the purpose of light perception. In this chapter, the

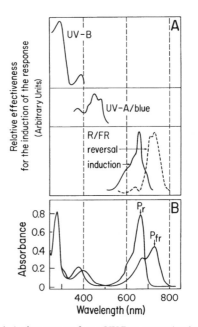

Fig. 2. A Action spectra for: a UV-B response (anthocyanin synthesis in sorghum), a UV-A/blue response (phototropism in oat), and a red/far red response (induction and reversal of hook opening in bean). B Absorption spectra of purified oat phytochrome in the red-absorbing P_r form and the far red-absorbing P_{fr} form. (Vierstra and Quail 1983, see [3])

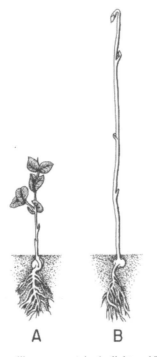

Fig. 1 A, B. Pea seedlings grown A in the light and B in complete darkness (etiolated)

term photoreceptor will be used not with reference to specialized cells but to the pigment molecules themselves which function in the initial absorption of informational light and transduction of intracellular signals. Like rhodopsin, the known plant photoreceptors are chromoproteins consisting of a light-absorbing prosthetic group or chromophore and an apoprotein. However, unlike rhodopsin, the one plant photoreceptor that has been purified and extensively characterized appears to have no defined physical association with membranes.

The spectral properties of the specific photoreceptor that controls a light-sensitive plant response can be assayed by determining an action spectrum, a plot of the effectiveness of different wavelengths of light for modulation of the response. Action spectra for a wide variety of plant photoresponses indicate that the photosynthetic pigments, the chlorophylls, do not play a major role in regulation of photomorphogenesis. Instead, there appear to be three major plant regulatory photoreceptor systems: one sensing UV-B light (280–320 nm), one

sensing UV-A and blue light (320–520 nm), and one sensing red (600–700 nm) and far red (700–780 nm) light. Figure 2A shows examples of action spectra for responses controlled by each of these receptor systems. The UV-B and UV-A/blue photoreceptors have not yet been isolated and their identities and molecular structures are subjects of current research. The reversible red/far red-sensing photoreceptor is phytochrome.

13.4 Phytochrome Biological Activity[2]

The spectral properties and biological activity of phytochrome are well illustrated by experimental observations first made in the 1940s and 1950s. Figure 3 shows the differential effects of red (R) and far red (FR) light in promoting and inhibiting two photomorphogenic events in flowering plants: light-sensitive seed germination and floral initiation in a plant species requiring long days to flower. Both of these events are promoted by a pulse of R but the inductive effect of R is reversed by subsequent exposure to a pulse of FR. Indeed, for both of these responses, plant tissue can be irradiated repeatedly with alternating doses of R and FR and the response is completely dependent upon the last stimulus given (Fig. 3). This successive cycle of induction and reversal results from the light-dependent interconversion of the two forms of phytochrome, an inactive red light-absorbing form, P_r, and an active far red light-absorbing form, P_{fr}. A very simple, generalized model for the action of these forms is:

$$P_r \xrightleftharpoons[\text{FAR RED}]{\text{RED}} P_{fr} \rightarrowtail \text{Response}$$

A wide variety of plant developmental responses are now known to be controlled by phytochrome. These include the triggering of major developmental events such as those shown in Fig. 3, environmentally controlled growth responses and morphological determination as seen in Fig. 1, the timing of stem cell differentiation leading to initiation of leaf and root primordia, and biosynthetic processes such as flavonoid pigment synthesis.

13.5 The Molecular Properties of Phytochrome[3]

Dark-grown (etiolated) plant tissue is a good source of phytochrome in that the tissue contains relatively high levels of the molecule but does not contain chlorophyll which interferes with the phytochrome spectral assay. Phytochrome has been purified to greater than 95% homogeneity from etiolated oat seedlings. It is a soluble homodimeric chromoprotein, each subunit consisting of a 124 kDa polypeptide and a light-absorbing prosthetic group or chromophore (Fig. 4A). The phytochrome chromophore is an open-chain tetrapyrrole which is covalently bound to a cysteine residue located approximately one third of the way from the amino-termi-

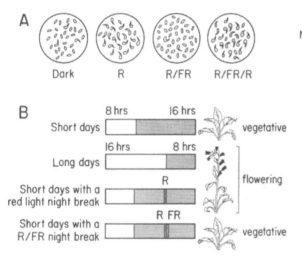

Fig. 3 A, B. Red/far red reversible physiological responses. **A** Lettuce seeds placed on wet paper in the dark require a brief pulse of red light to induce germination. This induction is cancelled if the red light pulse is followed by a far red pulse. **B** Initiation of floral development is dependent upon daylength for many plant species. In the case shown here, flowering requires a daylength longer than 8 h. The flowering response can be triggered under short-day conditions by a nightbreak pulse of red light and this is reversible with FR. The vegetative and flowering plants shown are the same age

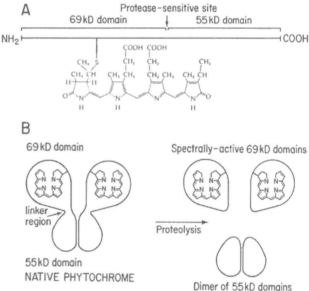

Fig. 4 A, B. The phytochrome molecule. **A** A linear representation of the 124 kDa oat phytochrome polypeptide and the attached linear tetrapyrrole chromophore. **B** Domain structure and proteolytic cleavage of phytochrome. The structure depicted is intended to illustrate the general properties of the phytochrome dimer and is not based upon a detailed understanding of phytochrome secondary or tertiary structure

nus of the apoprotein. Though the secondary structure of phytochrome has not been determined at the resolution of X-ray crystallographic studies, a combination of methods including biophysical analysis, proteolytic degradation studies, electron microscopy, and small angle X-ray scattering indicate that each phytochrome monomer is composed of two structural domains separated by an exposed and potentially flexible linker region (Fig. 4B). Following proteolytic cleavage in the linker region, the 69 kDa N-terminal domain is observed to be globular and to contain the chromophore and the polypeptide sequences necessary for correct interaction with the chromophore. The 55 kDa C-terminal domain is more elongated and contains the polypeptide sequences involved in dimerization.

We have seen that action spectra for a reversible R/FR light-controlled plant response indicate that phytochrome exists in either of two relatively stable, interconvertible conformations, one preferentially absorbing red light and the other preferentially absorbing far red light (Fig. 2A). The absorption spectra of phytochrome purified from etiolated oat seedlings (Fig. 2B) confirm this interpretation. The purified chromoprotein exists either as P_r with an absorption maximum at 666 nm or P_{fr} with a maximum at 730 nm (Fig. 2B). These two forms are reversibly interconvertible by light in vitro; irradiation with red light converts P_r to P_{fr} and, conversely, irradiation with far red light converts P_{fr} back to P_r. Conversion between these forms is rapid (less than 10 ms) and results in a conformational change in the recepter molecule. The chromophore is the site of R/FR light absorption and it is likely that it undergoes a Z,E *cis-trans* isomerization reaction as the primary phytochrome photoreaction. Light-induced changes in the secondary structure of the apoprotein are not precisely defined, though studies of the relative accessibility of regions of the polypeptide to chemical modification or proteolytic cleavage in the P_r versus the P_{fr} forms indicate that different polypeptide regions are exposed in the two conformations and that the N-terminal 6–10 kDa of the apoprotein are particularly important for maintaining the differential P_r and P_{fr} spectral properties.

Despite significant advances in characterization of the structure of the phytochrome molecule and the changes associated with photoconversion of the receptor, very little is known about the mechanism through which the active P_{fr} conformation modifies plant growth and development. As will be discussed below, phytochrome activation alters the expression of many plant genes and these alterations undoubtedly underlie many of the biochemical and physiological changes induced by red light. Nonetheless, no enzymatic activity or specific protein or nucleic acid binding activity has been assigned to the receptor itself. The complete amino acid sequences of phytochrome apoproteins from several plant species have been deduced from cDNA and genomic nucleic acid sequences. The apoprotein sequences from diverse flowering plant species are highly conserved (approximately 70% identical) but they exhibit no known homology to previously characterized proteins. Hence, the phytochrome receptor can be viewed as a unique regulatory molecule which presents a wide range of research challenges.

13.6 Photoequilibrium and Regulation of Phytochrome Abundance by Differential Stability [4]

Up to this point, we have dealt with phytochrome as a relatively static, binary biological switch. In fact, the biogenesis, photoconversion, and degradative turnover of this photoreceptor constitute a dynamic cycle, each component of which plays a role in the photosensory capacity of the plant. Etiolated plants contain high levels of phytochrome exclusively in the P_r conformation. This indicates that, in the absence of any activating light signal, the initial product of phytochrome biosynthesis and chromophore coupling is the inactive red-absorbing form. Phytochrome in the P_r conformation is very stable, with a half-life of about 100 h, and accumulates to approximately 0.5% of the soluble protein in dark-grown oat seedlings.

Absorption of R or FR light by phytochrome in the plant cell initiates at least two processes. First, phytochrome is photoconverted and a level of active P_{fr} is established. Since the interconversion of P_r and P_{fr} is rapid and reversible, the proportion of phytochrome in the inactive and active forms reflects a photoequilibrium based upon the proportion of R and FR light entering the cell. The higher the fluence (amount) of R relative to FR light at any time, the further the equilibrium is shifted to the active P_{fr} form. This equilibrium can change rapidly in response to changing light conditions and, as a result of this, the phytochrome system is sensitive not merely to the presence or absence of light but to the spectral composition of that radiation.

The second process that is initiated by conversion of P_r to P_{fr} is accelerated degradation of the phytochrome receptor itself. The active P_{fr} conformation of phytochrome is 50- to 100-fold less stable in the cell than the P_r form. Hence, conversion to P_{fr} results in simultaneous activation of the phytochrome system and greatly increased degradation of the receptor:

$$\left.\begin{array}{l}\text{Apoprotein} \\ \text{Chromophore}\end{array}\right\} P_r \underset{\text{FAR RED}}{\overset{\text{RED}}{\rightleftharpoons}} P_{fr} \rightarrow\!\!\rightarrow\!\!\rightarrow \text{Response}$$
$$\downarrow$$
$$\text{Degradation}$$

Reduction in number or "down-regulation" of a receptor molecule following exposure to its activating agent is observed in a wide variety of animal receptor systems such as those for steroid and adrenergic hormones, growth factors, and neurotransmitters. Its role is apparently modulation of the sensitivity of a receptor system from a sensitized state, where the threshold for activation is low, to a desensitized state. Such a desensitization of the phytochrome system may mark the transition from an etiolated plant growth strategy, in which a high concentration of receptor molecules increases the likelihood that a few photons of red light will be detected, to a photosynthetic growth strategy in which a lower concentration of photoreceptor molecules functions primarily in analysis of the spectral composition of relatively high levels of light.

13.7 Distribution of Phytochrome[5]

The phytochrome photoreceptor is found in almost all plant tissues and organs that have been examined. There is, however, variation in phytochrome abundance in different cell types and in different regions of the plant. Spectroscopic and immunochemical localization studies indicate that the photoreceptor is, in general, most abundant in cells recently derived from shoot and root meristematic regions. These are growing points of the plant where cells have only recently differentiated from perpetually embryonic meristem cells and are beginning rapid expansion. Immunodetectable phytochrome also shows qualitative specificity for certain cell types, illustrated by the strong phytochrome-associated immunofluorescence of cortical as compared to epidermal cells in sections of pea epicotyl tissue (Fig. 5A). It appears, then, that phytochrome localization reflects active control of receptor synthesis or stability that is responsive to the state of differentiation of specific cells and tissues.

13.8 Intracellular Localization of Phytochrome Is Dependent upon Its Conformation

The intracellular localization of a receptor is integral to its molecular mode of action. Many animal receptors fall into one of three general categories: membrane-bound receptors that function by initiating an intracellular signal transduction pathway, cytosolic receptors that translocate to the nucleus in their ligand-bound form and directly regulate gene expression, and nuclear-localized receptors which are also activated to function as transcription regulation factors. Phytochrome does not readily fit into any of these groups. Immunofluorescence visualization of phytochrome in the P_r form indicates a diffuse, cytoplasmic distribution (Fig. 5A). More detailed immunoelectronmicroscopic visualization reveals no preferential association of the receptor with subcellular compartments such as the nucleus or plastids or with membranes. This is consistent with the water-soluble nature of purified phytochrome. Conversion of phytochrome to the active P_{fr} form by red light causes a striking redistribution of the molecule (Fig. 5B). Within seconds of P_{fr} formation, the majority of phytochrome becomes associated with discrete, amorphous subcel-

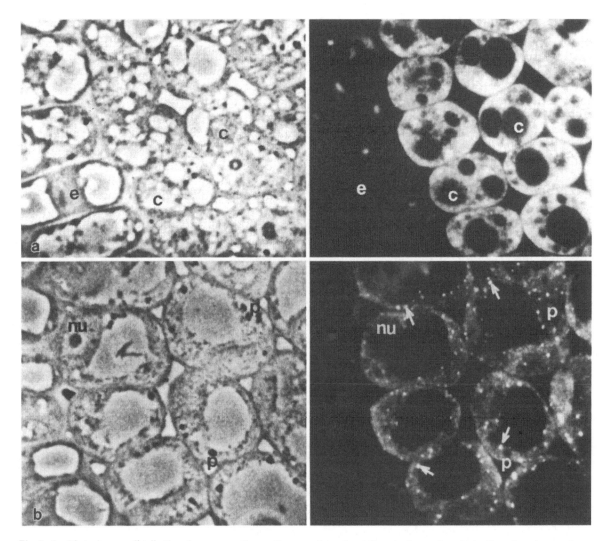

Fig. 5a, b. Phytochrome distribution in cross-sections of pea epicotyl tissue as visualized by immunofluorescence. **a** Phase contrast (*left*) and fluorescence (*right*) micrographs of etiolated tissue sections. The phase contrast image shows the cellular structure of the epidermal (*e*) and cortical (*c*) cells; the fluorescence micrograph shows diffuse phytochrome-associated fluorescence throughout the cytoplasm of cortical cells and weak or undetectable fluorescence in the epidermal cells. **b** Analogous micrographs of etiolated tissue sections that were treated with 20 s of red light and 10 min dark incubation prior to being fixed. The bright fluorescent aggregates of phytochrome are indicated with *arrows*; nuclei (*nu*) and plastids (*p*) are labeled. (Saunders et al. 1983)

lular bodies within the cytoplasm. This aggregation of the active form of phytochrome may correspond to association with defined intracellular sites involved in its molecular mode of action or, alternatively, with sites where the P_{fr}-specific destruction of phytochrome occurs. Once again, there is no indication that these receptor aggregates are preferentially associated with membranes or with the nucleus. Hence, though phytochrome undergoes a cycle of subcellular redistribution in association with its activation, this cycle may not be integral to its signaling mechanism.

13.9 There Are Multiple Types of Phytochrome in Higher Plants[6]

Almost all of the large and growing body of knowledge concerning the structure, spectral properties, localization, and regulation of phytochrome has been derived from studies of the relatively abundant phytochrome which is purified from etiolated plant tissue. However, it has never been clear that this is the only form of the red light photoreceptor present in plants. Analysis of tissue extracts from light-grown, green plants indicates that at least one additional low-abundance phytochrome is present and that this "green-tissue" phytochrome is spectrally and immunochemically distinct from the etiolated-tissue form. Furthermore, a gene family encoding several phytochrome-related proteins (the *phy* genes) has been described in *Arabidopsis thaliana*. Comparison of three of these *phy* polypeptide sequences shows that they are similar in overall structure and contain a conserved chromophore attachment site but that they are divergent (approximately 50% amino acid sequence identity), distantly related molecules. If phytochrome is indeed a receptor family, analogous to families of hormone and growth factor receptors in animals, it will be important to determine in what ways the different phytochromes vary in their physical properties, location in the plant, and regulation and what the specific roles of the individual photoreceptors are in the control of photomorphogenesis.

13.10 Phytochrome Regulates Plant Gene Expression[7]

One mechanism through which light controls complex plant developmental responses is modulation of gene expression and the patterns of expression of several plant nuclear genes have been shown to be specifically regulated by R and FR light. These include genes that encode abundant proteins involved in photosynthesis (the chlorophyll a/b binding proteins encoded by the *cab* genes and ferredoxin) and carbon fixation (the small subunit of ribulose-1,5-bisphosphate carboxylase-oxygenase encoded by the *rbcS* genes) and the enzymes NADPH-protochlorophyllide reductase and asparagine synthetase. Phytochrome control of the expression of a gene can be demonstrated by comparing the level of an mRNA in dark-grown seedlings with the level in seedlings which have received brief pulses of R, FR, or R followed by FR light (Fig. 6A). Several patterns of gene regulation can be observed. For a gene showing classical phytochrome regulation, the mRNA level is significantly altered shortly after a pulse of R but this change is prevented if the pulse of R is followed immediately by a pulse of FR (Fig. 6B). This pattern of mRNA regulation is referred to as a low-fluence (LF) phytochrome response and its induction requires exposure to an amount of red light that converts the bulk of phytochrome in the cell to the active P_{fr} form. In contrast to this, there are genes which show very low-fluence (VLF) phytochrome responses. Changes in expression of these genes are induced by conversion of less than 0.1% of phytochrome to P_{fr}. The absorption spectrum of P_r phytochrome extends very slightly into the far red region of the spectrum (Fig. 2B); hence, a saturating pulse of FR light converts 1–3% of P_r to P_{fr}. This level of P_{fr} exceeds that required to induce VLF responses so VLF responsive genes are induced by FR light and are not photoreversible (Fig. 6B). The example of phytochrome-regulated gene activity shown in Fig. 6B, *cab* mRNA levels in tomato, exhibits elements of both an LF and a VLF response.

Phytochrome-mediated control of gene expression can occur at the transcriptional or post-transcriptional level and can produce up- or down-effects. In addition, the phytochrome responses of different genes, even different members of gene families such as *cab* and *rbcS*, vary in the fluence of R light required for their induction, the kinetics of their induction, their extent of reversibility by FR light, and their dependence on other factors such as the developmental state of the cell or co-action of other receptor systems. It is clear from these considerations that the final pattern of response to phytochrome activation by a given gene is complex and may involve higher-order interactions with multiple regulatory circuits in the cell.

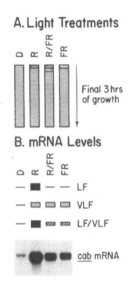

A. Light Treatments

Final 3 hrs of growth

B. mRNA Levels

LF

VLF

LF/VLF

cab mRNA

Fig. 6A, B. Regulation of mRNA levels by phytochrome. **A** Five to six day-old dark-grown plant seedlings are treated with pulses of R and FR light and placed back in the dark for 3 h to allow expression of the mRNAs. **B** The mRNA fraction is purified from the plants and transcripts showing low-fluence (*LF*), very low-fluence (*VLF*), and LF/VLF responses are distinguished by Northern blot analysis. The autoradiogram shows the LF/VLF response of the chlorophyll a/b binding protein (*cab*) mRNA in tomato seedlings

13.11 Autoregulation of the Phytochrome mRNA [8]

Photoregulation of the abundance of the phytochrome receptor itself through differential stability of the P_r and P_{fr} conformations has been discussed above. In some plant species, there is an additional level of cellular control of phytochrome biogenesis in that activation of the phytochrome system also strongly down-regulates transcription of the genes encoding the phytochrome apoprotein (the *phy* genes). This down-regulation serves as a good example of direct, rapid, and highly sensitive phytochrome-mediated control of gene expression.

Figure 7A shows the steady-state level of *phy* mRNA in etiolated oat seedlings and seedlings treated with R and FR light. Figure 7B shows a time course for transcriptional activity of the oat *phy* genes under the same conditions. Photorepression of phytochrome mRNA abundance has both low-

fluence (LF) and very low-fluence (VLF) components; a pulse of saturating FR light alone (1% P_{fr}) causes a significant reduction in the mRNA while a saturating R pulse (84% P_{fr}) causes a more severe reduction that is FR reversible (compare Fig. 7A to the LF/VLF response in Fig. 6). Nuclear run-on transcription assays show that transcription of the *phy* genes is reduced greater than 90% within 10 min of an activating light pulse and that R and FR light are equally effective in triggering the initial repression of *phy* gene transcription (Fig. 7B). Moreover, it is the *duration* of the transcriptional repression that depends upon the initial size of the P_{fr} pool. These characteristics of *phy* gene transcription explain both the VLF (reflecting the initial inhibition of transcription) and LF (reflecting the duration of inhibition) components of

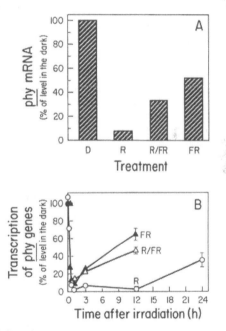

Fig. 7A, B. Phytochrome-mediated down-regulation of *phy* mRNA in etiolated oat seedlings. **A** Steady-state levels of *phy* mRNA as determined by blot analysis in oat seedlings grown in the dark (*D*) and 3 h following treatment with 5-s pulses of R, R/FR, or FR as shown in Fig. 6 (Colbert et al. 1985). **B** Time courses of the effects of R, R/FR, or FR light pulses on transcription of the *phy* genes. Tissue was harvested at the indicated times after the light treatments, and nuclei were isolated. Nascent transcripts were allowed to "run off" in the presence of a radioactive nucleotide and the amount of *phy* mRNA labeled was assayed by filter hybridization (Lissemore and Quail 1988)

the response. Further experiments have shown that the transcriptional photoregulation of the *phy* genes in oat is unaffected by the presence of protein synthesis inhibitors, indicating that the signal transduction pathway for phytochrome gene auto-regulation pre-exists in the uninduced cell. The time course of light-induced inhibition of *phy* gene transcription, which is similar to that of ligand-mediated regulation of hormone-responsive genes in animal systems, and its insensitivity to protein synthesis inhibitors suggest that the signalling pathway between the activated phytochrome receptor in the cytoplasm and the transcription machinery in the cell nucleus is both rapid and direct.

13.12 Approaches to the Analysis of Phytochrome Signal Transduction[9]

Little is known about phytochrome signal transduction following photoconversion of P_r to P_{fr}. The diversity of molecular and developmental responses affected by phytochrome suggests that multiple regulatory circuits are involved, many of which are probably removed by several steps from the initial absorption of light by the receptor. The high degree of sensitivity of some phytochrome responses may reflect amplification of the signal by a step in the transduction pathway involving a regulatory enzymatic activity such as a protein kinase or induction of the release or formation of a second messenger. Several approaches to defining the phytochrome intracellular signaling mechanism are currently being pursued. The first involves analyzing the receptor itself, an approach that we have seen has yet to yield much insight concerning its mode of action.

Another approach is that of choosing a defined cellular response which is accessible to analysis and tracing back through the sequential molecular steps of activation of that response to arrive back at the receptor. Control of gene expression at the level of transcription is a central function of phytochrome that is readily accessible to study through the techniques of molecular biology and a great deal of research is now directed at identifying and isolating the regulatory factors that mediate these responses. Short DNA sequences in the promoter regions of light-responsive plant genes, such as the *cab*, *rbcS*, and *phy* genes, have been identified and shown to function as *cis*-acting regulatory elements in determining the basal levels, tissue specificity, developmental timing, and photocontrol of transcription of these genes. As an example, the *cis*-acting elements that have been identified in the pea *rbcS*-3A promoter are illustrated in Fig. 8. The sequences involved in photoregulation are often called light-responsive elements (LREs) by analogy to the hormone-responsive elements (HREs) that bind regulatory factors in animal hormone-responsive promoters.

The regulatory roles of the LREs are deduced largely by testing the ability of these DNA sequence elements to confer light-responsive control on normally nonphotoregulated promoters (often truncated viral promoters that are missing their natural upstream activation sequences) and by assaying the effects of in vitro-induced mutations within the elements on promoter function in transgenic plants. Important experimental tools in these efforts are techniques that have been developed to introduce DNA sequences, such as chimeric promoters and gene constructs, into plant cells using *Agrobacterium*-mediated transformation and methods for regeneration of transformed cells into complete, fer-

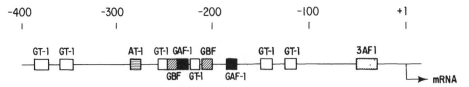

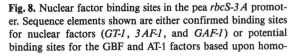

Fig. 8. Nuclear factor binding sites in the pea *rbcS-3A* promoter. Sequence elements shown are either confirmed binding sites for nuclear factors (*GT-1*, *3AF-1*, and *GAF-1*) or potential binding sites for the GBF and AT-1 factors based upon homo-logy to DNA sequence elements in related light-responsive promoters (Gilmartin et al. 1990). *Number +1* indicates the start site for transcription of the *rbcS-3A* gene

tile transgenic plants (see Chaps. 14 and 15, this Vol.). Using these approaches, LREs from the *cab*, *rbcS*, and *phy* promoter regions have been identified and shown to play roles in mediating phytochrome-responsive transcriptional activation or repression. Sequence-specific DNA binding proteins that interact with the LREs are currently being identified and clones corresponding to these proteins are being isolated. It appears from these studies that the level of transcriptional activity and the photoresponsive and tissue-specific characteristics of light-regulated plant promoters are determined by the interaction of a complex array of positive and negative *cis*-acting DNA elements with numerous soluble factors.

A third approach to identifying components of the phytochrome signaling mechanism is isolation of mutants that are defective in red light-mediated physiological responses. Genetic analysis has the potential to identify components of this complex regulatory network that will be difficult to resolve using molecular and biochemical methods. Though classical genetic analysis of phytochrome responses has been pursued to only a limited extent, it appears that the phytochrome system is tolerant of severe alterations and is, therefore, amenable to genetic analysis. Mutants of tomato (the *aurea* mutant) and *Arabidopsis thaliana* (some alleles of the long-hypocotyl or *hy* mutants) have been isolated that contain less than 5% of the wild-type level of assayable phytochrome. These mutant strains are viable and fertile but exhibit a partially-etiolated morphology – elongated internodes and yellow-green color – even in the presence of light and are deficient in phytochrome-mediated regulation of gene expression. Fifty-one independent *hy* mutants of *A. thaliana* showing this phenotype have been isolated and mapped to six loci. Mutations at three of these loci cause reduced levels of phytochrome and may affect expression or function of the receptor itself. Mutations at the other three *hy* loci retain apparently wild-type photoreceptor levels and may be located in genes for components of the signaling pathway. Methods developed for molecular mapping, chromosome walking, and gene complementation in *A. thaliana* should make possible the isolation of the genes that correspond to these important photoregulatory loci and characterization of their roles in the phytochrome pathway.

13.13 Outlook

Photoperception in plants and the mechanism of plant acclimation to the light environment are important links between the energy of the sun and life on this planet. Currently, there are many unresolved questions concerning the molecular genetics of plant photoperception. A longstanding question, and perhaps one of increasing urgency due to the loss of UV filtering in the earth's atmosphere, is the molecular structure of the UV-sensing pigments, the UV-B and UV-A/blue photoreceptors. The chemical natures of these vital molecules has eluded investigators for many years. By comparison, the phytochrome photoreceptor system is well characterized. Several properties of the phytochrome regulatory pathway make it particularly attractive as an experimental system for the study of environmental control of plant development. First, the amounts and relative proportions of red and far red light are environmental factors that influence many basic aspects of plant growth and development in a wide range of plant species. Second, the receptor for the light stimulus, phytochrome, has been identified and extensively characterized at the molecular level. The identities of most plant receptors, including those for shorter wavelength UV and blue light and for the plant hormones, are not yet known. Third, light is a stimulus whose effects are both very rapid, requiring no diffusion or transport to reach target cells or the receptor itself, and reversible. Because of this, activation of the photoreceptor can be accurately controlled in the laboratory and very rapidly changed. Finally, the patterns of expression of several very important plant gene products, proteins that play central roles in the processes of photosynthesis and carbon fixation, are controlled by phytochrome.

Currently, the most pressing problem in phytochrome research is to elucidate the molecular mechanism of action of the receptor and of the components of its signaling pathway. Initial progress in this will most likely be made in the area of photoregulation of gene expression. Nevertheless, when viewed in the larger perspective of control of photomorphogenesis, the question of the mode of action of phytochrome is seen to encompass diverse topics in the very broad areas of regulation of plant growth and development.

13.14 Summary

In addition to supplying the energy for photosynthesis, light plays very important and diverse roles as an environmental regulator of plant development. Photons of light that act as regulatory signals are absorbed by plant photoreceptor molecules, the most highly characterized of which is the cytoplasmically localized, dimeric chromoprotein phytochrome. The phytochrome molecule exists in either of two interconvertible conformations which differentially absorb red and far red light signals. The red-absorbing conformation is thought to be an inactive form of the receptor and photoconversion to the far red-absorbing conformation is the initial step in the triggering mechanism for light-mediated development or photomorphogenesis. Due to the rapid, reversible activation of the receptor, the phytochrome system functions as a sensor for the presence or absence, spectral quality, and photoperiodic cycling of long wavelength visible light. Though significant progress has been made in analyzing some of the molecular consequences of phytochrome activation, notably effects on the expression of several plant genes, the molecular mechanisms through which phytochrome intracellular signals are transmitted are not known.

References

General

Kendrick RE, Kronenberg GHM (1986) Photomorphogenesis in plants. Martinus Nijhoff Publishers, Dordrecht, The Netherlands

Kuhlemeier C, Green PJ, Chua N-H (1987) Regulation of gene expression in higher plants. Annu Rev Plant Physiol 38:221–257

Shropshire W, Mohr H (1983) Photomorphogenesis. Encyclopedia of plant physiology, new series, vols 16A, 16B. Springer, Berlin Heidelberg New York

Walbot V (1985) On the life strategies of plants and animals. Trends Genet 1:165–169

[1] Shropshire W, Withrow RB (1958) Action spectrum of phototropic tip curvature of *Avena*. Plant Physiol 33:360–-365

Withrow RB, Klein WH, Elstad V (1957) Action spectra of photomorphogenic induction and its inactivation. Plant Physiol 32:453–462

Yatsuhashi H, Hashimoto T, Shimizu S (1982) Ultraviolet action spectrum for anthocyanin formation in broom sorghum first internodes. Plant Physiol 70:735–741

[2] Borthwick HA, Hendricks SB, Parker MW, Toole EH, Toole VK (1952) A reversible photoreaction controlling seed germination. Proc Natl Acad Sci USA 38:662–666

Borthwick HA, Hendricks SB, Parker MW (1952) The reaction controlling floral initiation. Proc Natl Acad Sci USA 38:929–934

[3] Jones AM, Erickson HP (1989) Domain structure of phytochrome from *Avena sativa* visualized by electron microscopy. Photochem Photobiol 49:479–483

Jones AM, Vierstra RD, Daniels SM, Quail P (1985) The role of separate molecular domains in the structure of phytochrome from etiolated *Avena sativa* L. Planta 164:501–506

Rudiger W, Scheer H (1983) Chromophores in photomorphogenesis. In: Shropshire W, Mohr H (eds) Photomorphogenesis. Encyclopedia of Plant Physiology, new series, vol 16A. Springer, Berlin Heidelberg New York, pp 119–151

Tokutomi S, Nakasako M, Sakai J, Kataoka M, Yamamoto KT, Wada M, Tokunaga F, Furuya M (1989) A model for the dimeric molecular structure of phytochrome based on small-angle X-ray scattering. FEBS Lett 247:139–142

Vierstra RD, Quail PH (1982) Proteolysis alters the spectral properties of 124 kdalton phytochrome from *Avena*. Planta 156:158–165

Vierstra RD, Quail PH (1983) Purification and initial characterization of 124-kilodalton phytochrome from *Avena*. Biochemistry 22:2498–2504

Vierstra RD, Quail PH (1986) The protein. In: Kendrick RE, Kronenberg GHM (eds) Photomorphogenesis in plants. Martinus Nijhoff Publishers, Dordrecht, The Netherlands, pp 35–60

[4] Shanklin J, Jabben M, Vierstra RD (1987) Red light-induced formation of ubiquitin-phytochrome conjugates: identification of possible intermediates of phytochrome degradation. Proc Natl Acad Sci USA 84:359–363

Shimazaki Y, Cordonnier M-M, Pratt LH (1983) Phytochrome quantitation in crude extracts of *Avena* by enzyme-linked immunosorbent assay with monoclonal antibodies. Planta 159:534–544

[5] McCurdy DW, Pratt LH (1986) Immunogold electron microscopy of phytochrome in *Avena*: identification of intracellular sites responsible for phytochrome sequestering and enhanced pelletability. J Cell Biol 103:2541–2550

Pratt LH (1986) Localization within the plant. In: Kendrick RE, Kronenberg GHM (eds) Photomorphogenesis in plants. Martinus Nijhoff Publishers, Dordrecht, The Netherlands, pp 61–81

Saunders MJ, Cordonnier M-M, Palevitz BA, Pratt LH (1983) Immunofluorescence visualization of phytochrome in *Pisum sativum* L. epicotyls using monoclonal antibodies. Planta 159:545–553

[6] Sharrock RA, Quail PH (1989) Novel phytochrome sequences in *Arabidopsis thaliana*: structure, evolution, and differential expression of a plant regulatory photoreceptor family. Genes Dev 3:1745–1757

Tokuhisa JG, Daniels SM, Quail PH (1985) Phytochrome in green tissue: spectral and immunochemical evidence for two distinct molecular species of phytochrome in light-grown *Avena sativa* L. Planta 164:321–332

[7] Kaufman LS, Thompson WF, Briggs WR (1984) Different red light requirements for phytochrome-induced accumulation of *cab* RNA and *rbcS* RNA. Science 226:1447–1449

Nagy F, Kay SA, Chua N-H (1988) Gene regulation by phytochrome. Trends Genet 4:37–42

Tobin EM, Silverthorne J (1985) Light regulation of gene expression in higher plants. Annu Rev Plant Physiol 36:569–593

[8] Colbert JT, Hershey HP, Quail PH (1985) Phytochrome regulation of phytochrome mRNA abundance. Plant Mol Biol 5:91–101

Lissemore JL, Quail PH (1988) Rapid transcriptional regulation by phytochrome of the genes for phytochrome and chlorophyll a/b binding protein in *Avena sativa*. Mol Cell Biol 8:4840–4850

[9] Castresana C, Garcia-Luque I, Alonso E, Malik VS, Cashmore AR (1988) Both positive and negative regulatory elements mediate expression of a photoregulated CAB gene from *Nicotiana plumbaginifolia*. EMBO J 7:1929–1936

Gilmartin PM, Sarokin L, Memelink J, Chua N-H (1990) Molecular light switches for plant genes. Plant Cell 2:369–378

Green PJ, Yong M-H, Cuozzo M, Kano-Murakami Y, Silverstein P, Chua N-H (1988) Binding site requirements for pea nuclear protein factor GT-1 correlate with sequences required for light-dependent transcriptional activation of the *rbcS-3A* gene. EMBO J 7:4035–4044

Parks BM, Jones AM, Adamse P, Koornneef M, Kendrick RE, Quail PH (1987) The *aurea* mutant of tomato is deficient in spectrophotometrically and immunochemically detectable phytochrome. Plant Mol Biol 9:97–107

Sharrock RA, Parks BM, Koornneef M, Quail PH (1988) Molecular analysis of the phytochrome deficiency in an *aurea* mutant of tomato. Mol Gen Genet 213:19–41

Chapter 14 Exploration of *Agrobacterium tumefaciens*

Barbara Hohn

14.1 Introduction[1]

Bacteria of the genus *Agrobacterium* are soil-borne plant pathogens. Their unique contribution to biology consists of their natural system for delivering DNA into host plants, where it becomes integrated and thereby stabilized in the genome. As a consequence of expression of bacterial DNA in the plants, the normal developmental pattern is changed. *Agrobacterium tumefaciens* causes tumors (crown gall tumors), whereas *Agrobacterium rhizogenes* infection results in proliferation of roots (Fig. 1). The study of this system is interesting for the following reasons. (1) Since it is the only inter-kingdom DNA transfer system known, the understanding of the mechanism of this gene delivery is important. (2) The basis for the developmental transformation of plants has to be understood. (3) Detailed knowledge of the system as such allows its exploitation for plant transformation. The integration of the bacterial DNA into the plant chromosomes allows its use in marking particular plant genes, thereby mutating the host.

Oncogenic bacteria were shown to harbour large (~200 kb) plasmids, called Ti (for tumor-inducing), or Ri (for root-inducing). Transposon mutagenesis and other mapping experiments have revealed several important regions on the Ti-plasmid (Fig. 2). The T-DNA (transferred DNA) is the segment found integrated in the plant genome. It is flanked and thereby defined by two imperfect direct repeats of 25 bp called border sequences. The T-DNA, which expresses its genes only in the plant, codes for proteins that are responsible for the over-production of or increased sensitivity to plant hormones leading to tumor formation or root proliferation, respectively, and of enzymes responsible for the synthesis of opines. These are unique compounds associated with tumors which can be catabolized as carbon, nitrogen, and energy source by the inciting bacterium. Genes for this catabolism are located on the Ti plasmid as well.

Transposon mutagenesis has also revealed a stretch of DNA on the Ti-plasmid which is responsible for virulence. Mutants in this region are avirulent or exhibit severely attenuated virulence. Most of the virulence genes are induced by small phenolic compounds excreted from wounded plants. As a consequence of this induction the T-DNA undergoes processing, enters the plant nucleus and becomes integrated into the host's DNA.

In this chapter we will try to summarize molecular events leading to transfer, integration, and expression of T-DNA.

14.2 Exploration of *Agrobacterium tumefaciens*

14.2.1 Virulence, Function, and Induction[2]

The virulence (vir) region encompasses about 35 kb and consists of seven operons with one to 11 genes each. Figure 3 depicts the "induction cascade" which we can divide arbitrarily into eight steps. (1) The bacteria are chemotactically attracted towards a gradient of compounds produced by wounded plants, (2) bacteria attach to the plant cell wall and (3) the virulence region is induced. Induction may actually precede attachment. (4) The T-DNA is processed and turned into a mobilizable intermediate which is (5) transferred to the plant nucleus. (6) The T-DNA is integrated into the plant nuclear genome. (7) Genes localized on the T-DNA are expressed. (8) As a consequence, the cells are transformed to tumor cells (or root cells, in the case of *Agrobacterium rhizogenes*).

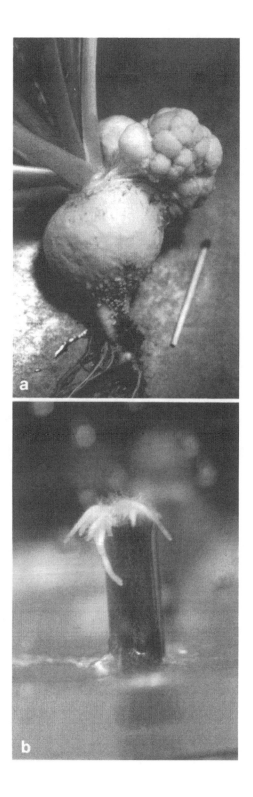

Fig. 1a, b. Phenotypes of plants infected with *Agrobacterium*. **a** *Brassica campestris*, infected with *Agrobacterium tumefaciens* C58 at the crown. (Courtesy G. Bakkeren). **b** *Brassica napus* petioles infected with *Agrobacterium rhizogenes*. (Courtesy of S. Gal)

Several virulence genes localized on the bacterial chromosome are responsible for attachment of *Agrobacterium* to plant cells. These genes are constitutively expressed and effect polysaccharide metabolism and secretion and synthesis of particular proteins. The component(s) in the plant cell wall to which these bacterial compounds attach is not known. Cellulose fibrils, produced under the direction of nuclear genes, aid in the attachment of bacterial aggregates to plant cells but are not essential for tumorigenesis.

In contrast to these chromosomally localized virulence genes, expression of the Ti plasmid-located genes is induced by plant signal molecules that are synthesized in wounded plant tissue. Acetosyringone (see formula in Fig. 3) and related phenolic products have been characterized as the active compounds. Virulence induction by these substances is mediated through the action of the VirA and VirG proteins. Mutations in their genes eliminate induction of all other virulence loci. The VirA protein has transmembrane topology and is suggested to act as environmental sensor and transmitter of the signal to the VirG protein. This transmission, manifested as a phosphorylation cascade (at least in

Ti-PLASMID

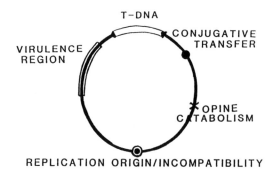

Fig. 2. Simplified map of Ti-plasmid

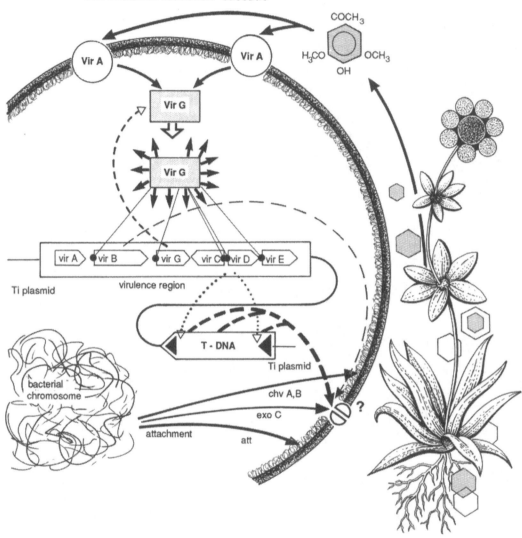

The virulence induction "cascade"

Fig. 3. Virulence gene induction cascade. Not all virulence loci are shown. *Triangles* bracketing the T-DNA symbolize border sequences. (Updated from Koukolíková-Nicola et al. 1987, see General References)

vitro), activates the VirG protein molecules, which leads to transcriptional activation of all other virulence loci. Special "vir-boxes", hexanucleotide motives in the promoter regions of inducible vir-operons, are involved in this activation. VirA and VirG are part of a two-component system common in bacteria, which combines a sensor for a specific environmental stimulus with a positive activator of transcription of genes that have to respond to it. VirG shares homology to other activators, VirA to other members of the sensor family.

The requirement for virulence A and G gene functions and their dependence on phenolic compounds excreted by plants is shared by a chemotactic response. However, bacteria are attracted by concentrations of acetosyringone which are two orders of magnitude below that required for virulence gene induction.

As will be discussed in greater detail in the next section, virulence proteins D1 and D2 are involved in processing the T-DNA. The VirC1 protein increases the efficiency of T-DNA processing. The functions of VirC2, VirD3, and VirD4 proteins are not known.

The virulence E locus, like VirC, codes only for auxiliary functions, as in its absence virulence is not abolished but only diminished. Virulence E2 protein is a sequence-unspecific single-stranded DNA binding protein, the stability of which seems to be guaranteed with the help of the small E1 protein.

The large *vir* B operon contains 11 genes. As deduced from their sequences, most proteins encoded by them are likely to be exported. Since these VirB proteins are absolutely essential for virulence, they may be involved in forming a membrane-associated device for directing the T-DNA out of the bacterial cell. The recent finding that VirB11 autophosphorylates and possesses ATPase activity is consistent with such a hypothesis.

In addition to these genes found in virulence regions of all Ti plasmids analyzed in detail, there are some strain-specific genes. For instance, nopaline plasmids contain the gene *tzs* (for transzeatin secretion) which is responsible for plant inducer-dependent production and excretion of the cytokinin transzeatin. Octopine strains, on the other hand, contain a *vir* F as well as a *pin* F (*pin* = plant-inducible) gene. The functions of these strain-specific genes are not clearly defined. In general, it should be pointed out that many effects may not be apparent under laboratory conditions and that natural conditions would have to be studied in order to understand fully the fine tuning of virulence and the adaptability of the organism.

14.2.2 T-DNA Transfer[3]

In this section we will discuss events following the induction of virulence gene expression, which lead to the production of transferable T-DNA, the actual transfer event, and the T-DNA in the plant, prior to integration.

14.2.2.1 T-DNA Processing

Two sequences are involved in *cis* in processing of T-DNA: the border sequences and the *overdrive*. Upon induction of the virulence region, specific nicks are introduced in the lower strand of border sequences (Fig. 4). Single-stranded molecules of lower strand polarity, composed exclusively of T-DNA, are formed as a result, possibly by displacement by a newly synthesized lower strand. Other T-DNA molecules found in induced *Agrobacterium* cells consist of double-stranded linear and a small amount of covalently closed circular molecules. Processing at the borders thus explains why border sequences are absolutely required for transformation, as shown by genetic experiments. The endonucleolytic activity resides in the combination of the proteins VirD1 and the N-terminal half of VirD2, as experiments in *Agrobacterium* and in *Escherichia coli* have shown. VirD1 exhibits topoisomerase activity whereas sequence specificity seems to reside in VirD2. Inspection of many natu-

T-DNA BOUNDARY SEQUENCES

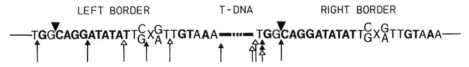

Fig. 4. The T-DNA in bacterium and in the plant. (↑) T-DNA/plant DNA junctions, (↑) T-DNA/T-DNA-junctions, (▲) nick in the lower strand of processed T-DNA. Compilation of data for nopaline- and actopine-Ti plasmids from the literature.

The *boxed parts* of the border sequences are conserved between all analyzed border sequences. Only the sequence of the upper strand is shown

ral border sequences reveals that the borders seem to consist of two blocks of absolutely conserved base pairs, bracketing a constant number of non-conserved nucleotides (Fig. 4). Mutation analysis has confirmed the requirement for the two blocks of sequences and an exact spacing in between. VirD2 protein, after having cleaved the T-DNA at the border sequences, remains bound to the 5′ end of the lower T-DNA strand. The bond seems to be of covalent nature and is able, at least in vitro, to protect the 5′ end of the right T-DNA terminus from exonucleolytic degradation. The binding of the D2 protein to processed T-DNA may be the main reason for the functional difference between left and right border, which has been detected by genetic experiments.

The *overdrive* sequence, a 24 bp DNA sequence (5′TAAPuTPyNCTGTPuTNTGTTTTGTTTG3′), found close to the right borders of octopine Ti plasmids, seems to enhance transfer efficiency. The virulence C1 protein, by binding to the *overdrive* sequence activates border cleavage and/or T-strand production by D proteins.

Another protein that possibly is directly involved in T-DNA transfer is VirE2. In its function as a single-strand-specific, sequence-unspecific DNA binding protein, it may well be required for the production of single-stranded T-DNA in the bacterium and/or its protection on its travel to the plant, although additional functions of this protein in, for example, recombination cannot be excluded. Since in *vir*E mutants virulence is reduced rather than abolished, one may postulate that a chromosomal single-strand binding protein would take over. Since *vir*E mutants can be complemented extracellularly, it can be imagined that this protein may be involved in the actual transfer.

14.2.2.2 Actual T-DNA Transfer

One requirement for T-DNA transfer is the attachment of bacteria to plant cells. Chromosomally located genes, conserved amongst at least some plant pathogenic and symbiotic bacteria, as well as virulence genes, specify some kind of transduction device that connects the two cells and allows the T-DNA to travel. The VirB and VirD4 proteins are

good candidates for participating in DNA export. Some of them have been shown to be associated with the bacterial membrane, and their predicted sequences contain putative secretion signals. VirB11, since it has been demonstrated to possess ATPase activity, may help to provide the necessary energy.

In what form the DNA travels to the plant is not known, although an involvement of single-stranded T-DNA at some stage is very likely. It may be completed to a double-stranded form in the bacterium, in the plant, or only upon integration. Adaptation to preexisting bacterial mechanisms seems to be a possible route for the evolution of this unique inter-kingdom DNA transfer. Several principles may have been combined: perception and transduction of environmental stimuli, followed by activation of genes responding to the stimuli, induction of pro-viruses, and conjugation. T-DNA may therefore be mobilized like a conjugating plasmid or like a virus. The need for attachment is not an argument against a virus theory, as only closely juxtaposed cells may represent the correct trigger for a virus transmission. Tumor-inducing virus particles have been looked for, but so far not found. The single-stranded nature of a large part of processed T-DNA

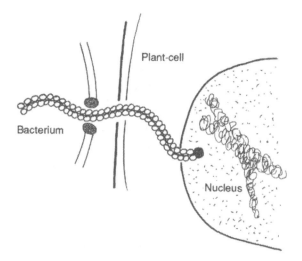

Fig. 5. Schematic representation of T-DNA transfer inder-mediate. The transfer complex consists of either single-stranded DNA, VirD2 protein (●), VirE2 protein (○) and possibly others, or of double-stranded DNA, VirD2 protein (●) and other Vir protein(s) (○). (◖) Pore in bacterial membrane

molecules may suggest their direct involvement in the transfer, along a relatively safe "mating bridge". Alternatively, accumulation of single-stranded T-DNA molecules following artificial induction in the absence of plants could be interpreted as deposition of molecules not active in transfer. Efficient transformation of plant cells in DNA-mediated transfer experiments does not prove involvement of single-stranded T-DNA in the actual transfer process. However, they clearly demonstrate the possibility. Figure 5 depicts a transfer model.

DNA transfer from bacterium to bacterium, together with the acquisition of genes responsible for attachment to plant cells could have evolved into bacterial-plant conjugation. Indeed, plasmid-mobilization functions in combination with virulence proteins have been shown to transfer genes to plants. This may mimic such an intermediate in evolution. Today, T-DNA transfer to plants and Ti plasmid transfer to *Agrobacterium* are separate processes, with border origins and virulence functions evolved for the former, and origin of transfer and transfer functions for the latter.

The recipient plant cell also appears to have a role in several stages of the T-DNA transfer, in addition to the initial induction of virulence genes. These possible roles will be discussed under T-DNA integration and host range.

14.2.2.3 *The T-DNA in the Plant*

Once inside the plant nucleus, by whatever route, T-DNA may not immediately integrate. A genetic approach to the analysis of T-DNA molecules in the plant, presumably before/independently of their insertion in the nuclear DNA, has been taken with the use of agroinfection. In these experiments T-DNA was constructed to consist exclusively of a plant viral genome, cloned between border sequences. Once inside a plant, the T-DNA's life as a plant virus can begin as soon as it is circularized. Analysis of circle joints, conveniently cloned and amplified by the virus, revealed the following: sequences at the right border are relatively well conserved, with 40% of all rescued molecules including the third base of the border corresponding to the nicking site in the bacterium (Fig. 4). This con-

trasts with a more ragged representation of sequences at the left border. However, even here the rule is followed that no base left of the left nick site is transferred (except in the case of a left border mutation).

These results imply that the right T-DNA end is protected from degradation during transfer. Since the VirD2 protein is found covalently attached to the 5′ end of the right T-DNA terminus in *Agrobacterium* (see Sect. 14.2.2.1), these results suggest the conservation of this bond in the plant. The bound D2 protein may perform further functions apart from inhibiting religation of T-DNA intermediates into the Ti plasmid in the bacterium and protecting T-DNA from exonucleases. It may serve to pilot the T-DNA into the nucleus, it may prime DNA replication, and/or it may aid in integrating T-DNA into plant-DNA. The latter would require the bound protein to retain an enzymatic function or to attract a cellular protein that actually performs the integration step.

14.2.3 Integration of T-DNA [4]

14.2.3.1 *The Plant Side of Integration*

T-DNA integrates into nuclear DNA of the host plant. Experiments employing in situ hybridization, linkage to known genetic markers or to restriction fragment length polymorphism markers have established that there seems to be no preferential integration target. The fact that many different integration points could be observed in individual plant cells also indicates that insertion is random. Also the wide host range of *Agrobacterium*, at least among dicot plants, indicates a certain target site independence.

Integration does not, however, seem to be random at the level of a gene. As will be described in more detail in the next chapter (see Chap. 15, this Vol.), transcriptional and translational fusions of a T-DNA gene with adjacent cellular sequences were found at similar frequencies in transformants of *Arabidopsis thaliana* and *Nicotiana tabacum*. Since these two species have a markedly differing density of transcribed sequences, these results cannot be interpreted as random insertions. It is suggested instead that T-DNA preferentially integrates into potentially transcribed areas of both genomes. Although direct proof for this interpretation is missing, it is a plausible hypothesis, since mammalian DNA in the process of being transcribed may more easily be accessible to invading viral DNA or, more specifically and possibly more relevant, may contain more nicks that can serve as entry points for integrating DNA.

14.2.3.2 The T-DNA Side of Integration

T-DNA insertion units consist of one to several copies of T-DNA. Independent locations of several inserts are also frequently found and have been shown to segregate in subsequent generations. Tandem arrangements of T-DNA have been found to be composed of direct and inverted repeats. Both of these have been found in single T-DNA arrays. The T-DNA may be complete (see below) or truncated, more infrequently containing internal deletions. T-DNA transfer is generally a more faithful transformation method compared to direct DNA transformation, in which rearrangement of transforming DNA tends to be the rule.

Several junctions between T-DNA elements and between T-DNA and plant DNA have been mapped and some of the joints sequenced. The finding that integration occurred close to the border repeats originally pointed to their importance in the transfer and integration process. A maximum of three bases of the right border repeat are retained and of 24 of the left border repeat are retained (Fig. 5). Junctions at the right border are more precise, on the T-DNA side at least, than at the left border. This is similar to what was found for the rescued T-DNA molecules described above, which presumably had not interacted with chromosomal DNA (although they did have to circularize). The integration event per se may therefore not represent a major distortion for the T-DNA termini. Internal junctions also seem to follow these rules. Interestingly, the breakpoints of each pair of T-DNA elements involved in an inverted repeat were always identical (at least at the resolution of restriction analysis). This may most easily be explained by replication of a double-stranded T-DNA molecule, with template strand switching of the replication machinery at the end of a particular element. Since such structures have never been detected in the bacterium, they most likely arose in the plant. This would imply that the T-DNA contains a replication origin. A double-stranded T-DNA molecule, the suggested template, either arrived from the bacterium as such or is a single-stranded molecule converted to the double-stranded form. Whether replication of T-DNA before integration is a prerequisite in general is not known. A different kind of mechanism has to be invoked to explain the head-to-head link found in the plant of two *different* T-DNA molecules originating from one *Agrobacterium.* In this case ligation of two double-stranded T-DNA elements must have occurred, since 5' ends of single-stranded DNA molecules cannot ligate to each other. Thus it seems that several mechanisms may operate within plants to replicate, assemble, and integrate T-DNA.

T-DNA integration can be explained as an event of illegitimate recombination. Short stretches of homology between the left T-DNA end and target DNA seem to be used, whereas the right T-DNA/plant DNA junction may (in some transformants) be explained as virulence protein aided recombination. A small stretch of target DNA was found deleted in all analyzed cases. Some minor rearrangements have been detected as well. Integration is certainly not as clean as transposon insertions with their duplication of short target sequences. This principal difference points to a rather different enzymology involved in integration. The tumor line which has been studied has been kept in culture for many years so that tissue culture-induced rearrangements cannot be excluded. Indeed, some clonal variation affecting integrated T-DNA has been reported. More sequences therefore will have to be analyzed before an integration mechanism can be suggested.

Once integrated, T-DNA is generally stable and usually transmitted to the offspring of transformants according to Mendelian laws. However, rearrangements have been observed after prolonged cultivation of tumor lines in tissue culture (see above) and in the offspring of plants containing inverted T-DNA repeats.

14.2.4 T-DNA Expression[5]

T-DNAs of different Ti or Ri plasmids may be composed of a single unit or may consist of two units separated by a stretch of non-T-DNA. Most importantly, however, they invariably contain genes coding for opine synthesis and genes responsible for the phenotypic change of the plant (Fig. 6). These genes carry typical eukaryotic expression signals and are active only inside the plant nucleus.

14.2.4.1 Opine Synthesis Genes

Opines, compounds made from an amino acid and a ketoacid or sugar, are found as a constant feature of tumors and other *Agrobacterium*-induced plant proliferations. *Agrobacterium tumefaciens* strains are also able to degrade specifically the opines of the tumors they have induced. Synthesis and catabolism of opines are coded for by Ti plasmid genes. Some opines have also been found to induce Ti plasmid transfer from bacterium to bacterium via conjugation. The tumors elicited by pathogenic strains of *Agrobacterium* may create ecological niches in which a favorable environment is provided for the pathogen. Direct experimental proof for this opine concept or genetic colonization theory is missing, however. On the contrary, most microorganisms found to catabolize opines in nature cannot themselves induce their synthesis. Thus they constitute competition for *Agrobacterium*. Moreover, a nonpathogenic *Agrobacterium radiobacter*

has been isolated that specifically kills *Agrobacterium tumefaciens* strains but carries genes for the catabolism of the opine produced by the victim.

Nevertheless, communication via opines between free-living bacteria and their fellows engaged in a symbiotic or pathogenic interaction may still be a general concept. Nitrogen-fixing nodules on a legume produce an opine-like compound, and the genes coding for the synthesis and for the catabolism of this "rhizopine" were shown to be closely linked on the plasmid carrying most of the other genes important for symbiosis.

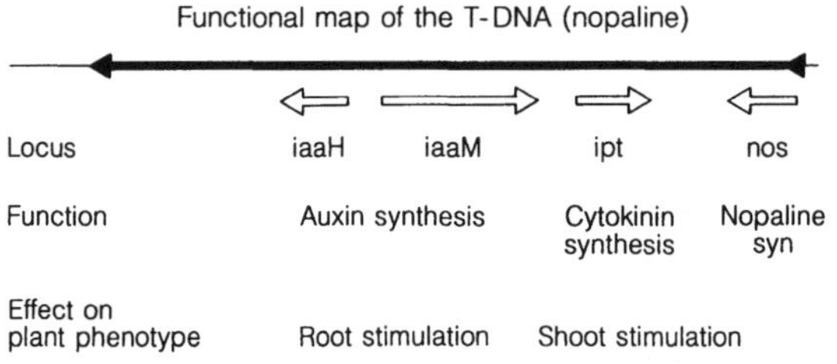

14.2.4.2 Oncogenes

Proliferation of infected plant tissue seems to be effected by the employment (and abuse) of the phytohormone pathways. Different strains achieve this by an overproduction of plant hormones or an improved use of the signal perception (and perhaps transduction) pathway.

Three genes coding for phytohormone production are responsible for oncogenicity of *Agrobacterium tumefaciens* T-DNA. Two code for enzymes which convert tryptophan to indole-3-acetamide and further to the auxin, indole-3-acetic acid. The other gene codes for an isopentenyl transferase which synthesizes an active cytokinin. The cooperation of these genes leads to the unregulated tumor growth which precludes regeneration of plants.

The origin of these genes seems to be prokaryotic. Similar genes are active in bacteria which induce tumorous proliferations on plants due to bacterial phytohormone production. Sequence comparison reveals a close similarity in the coding sequences. Similarly, the prokaryotic *tzs* gene of the virulence region of *Agrobacterium* (see Sect. 14.2.1, on virulence) closely resembles the eukaryotic T-DNA cytokinin gene in its structural part. The fact that in untransformed plants auxin biosynthesis probably follows a different pathway from that in plants transformed by *Agrobacterium* also argues against the bacterium having picked up the hormone genes in a plant. T-DNA genes have so far not been found to contain introns; this may be taken as another argument for the prokaryotic nature of T-DNA.

Fig. 6. Simplified functional map of nopaline T-DNA. *iaaM*, *iaaH* genes code for tryptophane mono-oxygenase and indole-3-acetamide hydrolase, respectively, the *ipt* gene for isopentenyl transferase, *nos* for nopaline synthase. *Triangles* symbolize the border repeats

In contrast to *Agrobacterium tumefaciens*, the root-inducing genes of *Agrobacterium rhizogenes* do not code for phytohormones. Proliferation of roots is mainly due to the activity of three *ro*ot *l*oci, *rol* A, B, and C. Genes for auxin biosynthesis, present in some *Agrobacterium rhizogenes* strains on a separate T-DNA, may expand the host range but are not a prerequisite for root formation. In contrast to plant cells transgenic for Ti plasmid oncogenes, Ri-transformed roots are capable of regeneration. The phenotype of the resulting plants (reduced apical dominance, wrinkled leaves, reduced gravitropism, abundant roots) are the result of the synergistic action of *rol* genes. Plants transgenic for single *rol* genes exhibit different phenotypes. These phenotypes seem to be the result of different gene functions and tissue-specific regulation of their activity. Overexpression of *rol* genes in transgenic plants identified a wrinkled phenotype (*rol* B) and a juvenile but male sterile *rol* C plant.

The root phenotype does not appear to involve the synthesis of plant hormones. Steps in the hypothetical signal transduction pathway of auxins have therefore been analyzed. Indeed, hairy root cells have been found to be several orders of magnitude more sensitive to auxin than normal cells, and this increase in sensitivity seems to be largely due to the activity of the *rol* B gene.

Rol gene-like sequences have been found in the genomes of *Nicotiana glauca* and other species of *Nicotiana*, although apparently in a transcriptionally inactive form. It has been shown recently that genetic tumors which can arise as a result of interspecies hybridization involving *Nicotiana glauca* exhibit a low but significant level of *rol* B and C transcription. The activation mechanism in still unknown.

14.2.5 Host Range[6]

Host-parasite interactions are extremely complex and involve specific steps from both partners. We will analyze the contribution of bacterium and plant separately, although the interrelationship always has to be kept in mind.

14.2.5.1 Host Range – the Bacterium

Although it may be imagined that specific differences in one of the bacterial genes responsible for attachment could yield a host range phenotype, no strains have yet been described that exhibit a completely altered host range as a result of such a mutation. This may mean that attachment is rather unspecific, at least as far as the bacterium is concerned. This may, at least in part, explain the wide host range of many analyzed strains. Small changes in host specificity are observed, however, due to mutations in the attachment genes *chvA,B, exo C*.

Two different steps in the induction cascade leading to general virulence have been found to be affected by or to affect different plant hosts differently: general induction via the VirA/VirG receptor/stimulator, and specific virulence genes. Certain limited host range (LHR) strains seem, due to the presence of a different *virA* gene, to be able to respond to inducers produced by only a small subset of plants. Certain "supervirulent" strains seem, due to overproduction of VirG protein, to be able to transform a wider range of plants. The phenotypes described have been shown directly to be linked to genes *A* and *G*, respectively.

virE and *virC* have also been termed host range genes. Effects of mutations in these genes, however, seem to be due to a general decrease in efficiency of transfer or possibly integration of T-DNA. Different hosts may simply require different minimal transfer efficiencies in order to respond and produce a tumor. Some plants may degrade or otherwise inactivate the transfer intermediate more easily. Other strains answer to invading *Agrobacterium* with a necrotic hypersensitive response unless the efficiency of transfer is reduced by introduction of a *virC* mutation. The octopine strain-specific *vir*F gene was found to be responsible for the block of T-DNA transfer from these strains to maize.

Different sets of T-DNA genes from different bacteria also influence tumor formation. As discussed in the last section, a tumor is the result of a hormonal imbalance in a particular infected plant. Due to differences in the endogenous hormone composition, a particular set of Ti plasmid-encoded genes can influence different plants differently. Thus, LHR strains which are not able to elicit tumors on particular plants can be complemented by the addition of a cytokinin gene from a wide host range *Agrobacterium* strain into its T-DNA.

14.2.5.2 Host Range – the Plant

As has been described above, many different factors, at many different levels of interaction, are needed from both the bacterium and the plant to establish the final result, a tumor. As outlined in a previous section, many steps are needed for this pathogenic relationship, and if one step is missing the infection may abort. Thus, it may be that plants avoid their pathogens by not permitting attachment, or they could "simply" not furnish the inducer. They could respond with a hypersensitive reaction to their invader (in the form of either bacterium or infecting DNA). Plants could block transport of T-DNA to the nucleus, inhibit it from integrating or from expressing their genes. Finally, as mentioned above, they could be insensitive to overproduction of T-DNA-specific phytohormones. Already in "generous" hosts there exists a window of competence, a particular time after wounding during which a cell has to heal before invasion can be successful. Production of an inducer during this period seems to be one process important for the host-parasite interaction, but induction of mitotic activity may be more important, since this is possibly connected with the invasion of T-DNA into the nucleus and its integration.

It is surprising that many dicot plants have not managed to avoid the pathogen, by blocking one of the many steps. This may mean that these steps represent processes which are vital for the plant and which therefore can be (ab)used by the bacterium. However, repression of gene activity by methylation has been observed as a frequent means to silence T-DNA genes. Most monocots do not respond to infection of *Agrobacterium* by tumor formation. Only some members of the orders Liliales and Arales show some transformation whereas all others have been classified as nonhosts. Transfer of T-DNA, however, has been shown to occur to cereals, using the sensitive technique of agroinfection. Plant viral genomes inserted into the bacterial T-DNA allow virus-symptom formation of plants inoculated with bacteria carrying the construct. This therefore constitutes an assay for T-DNA transfer which is independent of tumor formation and probably also integration. Maize and other cereals were shown to produce substances active as virulence inducers. Since the dicot-specific wound response and therefore the wound-induced mitotic activity is lacking in these cereals, integration may be a less frequent event than in dicots.

Since monocots and dicots can be infected by *Agrobacterium* (although with a different resulting phenotype) this particular plant-pathogen relationship may actually predate the monocot-dicot divergence. New evidence supports the hypothesis that angiosperms existed and underwent diversification some 300 million years ago. Plant-*Agrobacterium* interactions may be older still, because some gymnosperms were found to be susceptible. It seems likely that *Agrobacterium* coevolved with ancestral plants. The discovery of T-DNA sequences related to presentday *Agrobacterium rhizogenes* T-DNA in several species of the genus *Nicotiana* points also to the actual existence of these phytopathogens early in evolution.

14.3 Outlook and Summary

The fascinating aspects of the biology of *Agrobacterium* are only partially understood. It remains an intriguing question how the natural gene transfer system of *Agrobacterium* evolved and what is the origin of T-DNA-encoded genes. Other interesting questions, open for future experimentation, include: how is the cell-cell contact between a T-DNA delivering *Agrobacterium* and a plant cell established? what is the structure of a real T-DNA intermediate – will it be possible at all to capture an "active" T-DNA molecule and how should "activity" be defined and tested? is there an active involvement of the T-DNA transfer complex in targeting to the plant nucleus and in integration? By analyzing the rules of T-DNA integration it may be possible to target a specific T-DNA to a specific locus on the chromosome. Further understanding of various host range aspects may have an impact in improvement of plant transformation of recalcitrant plants, especially the monocots.

References

General

Binns AN, Thomashow MF (1988) Cell biology of *Agrobacterium* infection and transformation of plants. Annu Rev Microbiol 42:575–606

Koukolíková-Nicola Z, Albright L, Hohn B (1987) The mechanism of T-DNA transfer from *Agrobacterium tumefaciens* to the plant cell. In: Hohn Th, Schell J (eds) Plant DNA Infectious Agents. Springer, Berlin Heidelberg New York, pp 109–148

Melchers LS, Hooykaas PJJ (1987) Virulence of *Agrobacterium*. Oxford Surv Plant Mol Cell Biol 4:167–220

Ream W (1989) *Agrobacterium tumefaciens* and interkingdom genetic exchange. Annu Rev Phytopathol 27:583–618

Zambryski P (1988) Basic processes underlying *Agrobacterium*-mediated DNA transfer to plant cells. Annu Rev Genet 22:1–30

Zambryski P, Tempé J, Schell J (1989) Transfer and function of T-DNA genes form *Agrobacterium* Ti and Ri plasmids in plants. Cell 56:193–201

[1] Elsacker S, Zaenen I, Schilperoort RA, Schell J (1974) Large plasmid in *Agrobacterium tumefaciens* essential for crown gall inducing ability. Nature 252:169–170

Smith EF, Townsend CO (1907) A plant tumor of bacterial origin. Science 25:671–673

[2] Christie PJ, Ward Jr JE, Gordon MP, Nester EW (1989) A gene required for transfer of T-DNA to plants encodes an ATPase with autophosphorylating activity. Proc Natl Acad Sci USA 86:9677–9681

Gross R, Aricò B, Rappuoli R (1989) Families of bacterial signal-transducing proteins. Mol Microbiol 3:1661–1667

Melchers LS, Regensburg-Tuïnk TJG, Bourret RB, Sedee NJA, Schilperoort RA, Hooykaas PJJ (1989) Membrane topology and functional analysis of the sensory protein *virA* of *Agrobacterium tumefaciens*. EMBO J 8:1919–1925

Peters NK, Verma DPS (1990) Phenolic compounds as regulators of gene expression in plant-microbe interactions. Mol Plant-Microbe Interactions 3:4–8

[3] Bakkeren G, Koukolíková-Nicola Z, Grimsley N, Hohn B (1989) Recovery of *Agrobacterium tumefaciens* T-DNA molecules from whole plants early after transfer. Cell 57:847–857

Buchanan-Wollaston V, Passiatore JE, Cannon F (1987) The *mob* and *oriT* mobilization functions of a bacterial plasmid promote its transfer to plants. Nature 328:172–174

Dürrenberger F, Crameri A, Hohn B, Koukolíková-Nicola Z (1989) Covalently bound *virD2* protein of *Agrobacterium tumefaciens* protects the T-DNA from exonucleolytic degradation. Proc Natl Acad Sci USA 86:9154–9158

Rodenburg KW, de Groot MJA, Schilperoort RA, Hooykaas PJJ (1989) Single-stranded DNA used as an efficient new vehicle for transformation of plant protoplasts. Plant Mol Biol 13:711–719

van Haaren MJJ, Sedee NJA, de Boer HA, Schilperoort RA, Hooykaas PJJ (1989) Mutational analysis of the conserved domains of a T-region border repeat of *Agrobacterium tumefaciens*. Plant Mol Biol 13:523–531

[4] Jouanin L, Bouchez D, Drong RF, Tepfer D, Slightom JL (1989) Analysis of TR-DNA/plant junctions in the genome of a *Convolvulus arvensis* clone transformed by *Agrobacterium rhizogenes* strains A4. Plant Mol Biol 12:75–85

Koncz C, Martini N, Mayerhofer R, Koncz-Kalman Z, Körber H, Redei GP, Schell J (1989) High frequency T-DNA-mediated gene tagging in plants. Proc Natl Acad Sci USA 86:8467–8471

Mayerhofer R, Koncz-Kalman Z, Nawrath C, Bakkeren G, Crameri A, Angelis K, Redei GP, Schell J, Hohn B, Koncz C (1991) T-DNA integration: a mode of illegitimate recombination in plants. EMBO J 10:697–704

Gheysen G, Villarroel R, Van Montagu M (1991) Illegitimate recombination in plants: a model for T-DNA integration. Genes Dev 5:287–297

[5] Ichikawa T, Ozeki Y, Syono K (1990) Evidence for the expression of the *rol* genes of *Nicotiana glauca* in genetic tumors of *N. glauca × N. langsdorffii*. Mol Gen Genet 220:177–180

Shen WH, Petit A, Guern J, Tempé J (1988) Hairy roots are more sensitive to auxin than normal roots. Proc Natl Acad Sci USA 85:3417–3421

[6] De Cleene M (1985) The susceptibility of monocotyledons to *Agrobacterium tumefaciens*. Phytopathol Z 113:81–89

Furner IJ, Huffman GA, Amasino RM, Garfinkel DJ, Gordon MP, Nester EW (1986) An *Agrobacterium* transformation in the evolution of the genus *Nicotiana*. Nature 319:422–427

Grimsley N, Hohn T, Davies JW, Hohn B (1987) *Agrobacterium*-mediated delivery of infectious maize streak virus into maize plants. Nature 325:177–179

Hohn B, Koukolíková-Nicola Z, Bakkeren G, Grimsley N (1989) *Agrobacterium*-mediated gene transfer to monocots and dicots. Genome 31:987–993

Jarchow E, Grimsley N, Hohn B (1991) The host range determining virulence gene *virF* of *Agrobacterium tumefaciens* affects T-DNA transfer to *Zea mays*. Proc Natl Acad Sci USA 88:10426–10430

Martin W, Gierl A, Saedler H (1989) Molecular evidence for pre-Cretaceous angiosperm origins. Nature 339:46–48

Wolfe KH, Gouy M, Yang Y-W, Sharp PM, Li W-H (1989) Date of the monocot-dicot divergence estimated from chloroplast DNA sequence data. Proc Natl Acad Sci USA 86:6201–6205

Chapter 15 Exploitation of *Agrobacterium tumefaciens*

Csaba Koncz and Jozef Schell

15.1 Introduction[1]

During invasion of wounded plants, soil agrobacteria transfer a defined segment of their Ti and Ri plasmids into the plants. The transferred DNA, termed T-DNA, is integrated into the plant nuclear genome. Genes encoded by Ti and Ri plasmid T-DNAs are expressed in plants and confer the synthesis of plant growth factors as well as sugar and amino acid derivatives, called opines. Expression of T-DNA genes *iaa*M, *iaa*H, and *ipt* (see Hohn, Chap. 15, this Vol.) leads to production of phytohormones, auxin and cytokinin that induce proliferation of transformed cells to form undifferentiated tumors, crown galls. In contrast, cells transformed by *rol* A, B and C genes of Ri plasmid T-DNAs differentiate to hairy roots. While genetic analysis of the function and expression of these T-DNA genes provided a key for better understanding of various aspects of hormonal regulation and cell differentiation, studies of the T-DNA transfer and integration contributed directly to the development of T-DNA-based transformation vectors and transgenic plant technology. How T-DNA gene vectors are exploited to gain more insight to molecular biology of plants is the focus of this chapter.

15.2 T-DNA, a Universal Tool of Plant Molecular Biology[2]

Development of modern plant gene vectors derived from the T-DNA was based on the observations that:

- Foreign DNAs inserted in the T-DNA are accurately transferred from *Agrobacterium* to plants.

- T-DNA-encoded genes are not required for the transfer of the T-DNA.
- *vir*ABCDE and G operons of Ti and Ri plasmids encode an inducible DNA processing system that mediates the recognition of direct 25 bp repeats located at the borders of T-DNA segments of Ti and Ri plasmids.
- The function of these 25 bp repeats is analogous to that of conjugational transfer origins of bacterial plasmids, thus:
- Any DNA sequence flanked by these 25 bp end-repeats can be transferred from *Agrobacterium* into plants and:
- Separation of the T-DNA and virulence genes does not influence the transformation process.

Certain requirements are essential for T-DNA vector designs. Since *Escherichia coli* is used as a host for construction of vectors with transferable T-DNA segments, T-DNA vectors should contain selectable markers for both *E. coli* and *Agrobacterium* as well as a replicative or integrative maintainance function. Vectors made in *E. coli* can be transferred to *Agrobacterium* by transformation, electroporation, or conjugation. In the latter case, DNA sequences recognized by a suitable plasmid mobilization system have to be added to the constructs. Finally, for selection of transformed plant cells, a marker gene selectable in plants should also be inserted in the T-DNA. Notwithstanding its relatively short history, a great number of different T-DNA-based vectors have already been constructed along the following two lines.

15.2.1 Recombination-Based Ti Plasmid Vectors[3]

Recombination-based vectors are Ti or Ri plasmid derivatives from which some or all T-DNA oncogenes have been removed. Foreign DNAs are inserted into the T-DNA by homologous recombination using a target DNA sequence that is homologous to commonly used bacterial antibiotic resistance genes or to diverse *E. coli* cloning vectors and located within the T-DNA borders of Ti or Ri plasmids. The same target DNA is also part of a second plasmid, referred to as "intermediate vector." Foreign genes are subcloned into the target DNA of intermediate vectors and transferred from *E. coli* to *Agrobacterium* by plasmid conjugation or mobilization. A variety of genetic methods have been developed to select or screen for the integration of foreign DNA into the T-DNA using single or double cross-over recombination and diverse replicative or nonreplicative vectors in agrobacteria. To identify T-DNA-transformed cells, selectable and/or screenable marker genes are provided either within the T-DNA or in the intermediate vector carrying the cloned foreign DNA. Recombination-based vectors are commonly referred to as "armed" or "disarmed" plasmids depending whether or not their T-DNAs contain oncogenes. Apart from special applications, the "armed" Ti plasmids are no longer used as vectors. In contrast, "armed" Ri plasmid are frequently employed to obtain transgenic plants from hairy roots. Disarmed Ti plasmid vectors, such as pGV3850 or the SEV system, for example, are still in use today. These vectors are based on single cross-over recombination within a target site located inside the T-DNA borders that results in an intermediate vector-Ti plasmid cointegrate (Fig. 1a).

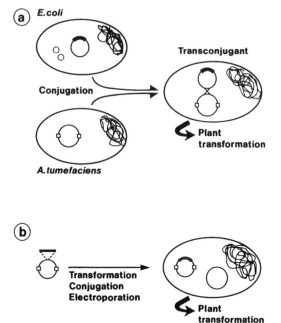

Fig. 1a, b. T-DNA plant gene vectors. **a** Insertion of cloned DNA into recombination-based vectors by single crossover. Foreign DNA is cloned into an intermediate vector and transferred into *E. coli* strain containing helper plasmids that promote conjugational transfer into *Agrobacterium*. The intermediate vector is unable to replicate in *Agrobacterium* but maintained by recombination with the T-DNA of Ti plasmid vector. *Open boxes* refer the T-DNA borders and the *thick line* for foreign DNA. **b** Use of binary vectors. Foreign DNA is cloned into the T-DNA of binary vectors and transferred into *Agrobacterium* that contains a T-DNA-less virulence helper Ti or Ri plasmid

15.2.2 Binary Vectors[4]

Binary vector systems consist of two elements: a helper Ti or Ri plasmid providing virulence functions, and a cloning vector containing bacterial and plant selectable marker genes and cloning sites flanked by T-DNA end-repeats (Fig. 1b). Most binary cloning vectors were derived from the well-characterized wide-host-range plasmid RK2 that can efficiently be mobilized between *E. coli* and *Agrobacterium*. An advanced binary system, referred to as PCV (*P*lant *C*loning *V*ector) system, is

based on a plant vector cassette (Fig. 2a) that carries only the ori_V (replication origin) and ori_T (origin of conjugational plasmid transfer) regions of plasmid RK2 in combination with diverse T-DNAs. ori_V and ori_T are active only when *trans*-acting RK2 functions *trf*a (replication function) and *tra* (plasmid transfer) are expressed in the same cell. To provide helper functions for replication and conjugation of PCV vectors, defective RK2 plasmid derivatives were inserted into the chromosome of *E. coli* and *Agrobacterium* hosts or into a T-DNA-less Ti plasmid pMP90RK. In *Agrobacterium* hosts carrying chromosomal insertions of *trf*a and *tra* RK2 genes, any Ti or Ri plasmids can be used as virulence helper. The presence of RK2 *tra* func-

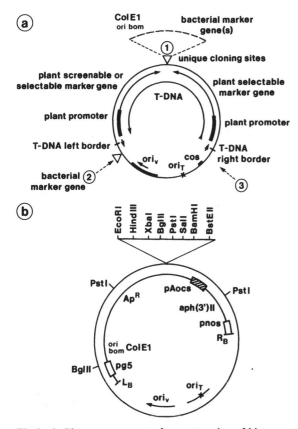

Fig. 2a, b. Plant vector cassette for construction of binary vectors. **a** Schematic design of a plant vector cassette that contains cloning sites, plant selectable and screenable marker genes within the T-DNA carried by an $ori_V - ori_T$ basic RK 2 replicon. *Arrows 1, 2,* and *3* are unique cleavage sites for introduction of additional elements into the cassette. **b** Plasmid pPCV 002, a prototype of simple binary plant gene vectors. p*nos* nopaline synthase promoter; *aph(3')II* coding sequence of kanamycin resistance gene of transposon Tn5; *pAocs* polyadenylation signal of octopine synthase gene; *Ap*^{R} ampicillin resistance gene; *ori* and *bom* of *ColE1* replication and conjugational transfer origins of plasmid ColE1; *pg 5* promoter of T-DNA gene 5; L_B and R_B left and right ends of the T-DNA

15.3 Gene Transfer and Transgenic Plant Technology[5]

One of the first observations during the pioneering experiments with wild-type T-DNAs showed that although T-DNA genes of agrobacteria were active in plants, other bacterial, yeast and animal genes inserted via T-DNA into the plant genome were not transcribed. It thus became evident that T-DNA-encoded genes must contain all signals necessary for transcription in plants. To achieve the expression of bacterial genes in plants, transcription promoter and terminator sequences of the nopaline synthase (*nos*) gene of the T-DNA were used first to construct chimeric genes with the coding sequence of neomycin phosphotransferase [*aph*(3')II] and chloramphenicol acetyltransferase (*cat*) genes of transposons Tn5 and Tn9. Expression of these chimeric antibiotic resistance genes could be followed by simple enzyme assays and permitted the selection of transformed plant cells. This opened the way for the development of transformation techniques using co-cultivation of agrobacteria with plant protoplasts, leaf-disks, stem and root explants; direct DNA uptake, protoplast fusion with charged liposomes, macro- and microinjection or bombardement with microprojectile-bound DNA. From cells transformed with disarmed T-DNA vectors fertile transgenic plants were regenerated that transmitted the introduced genes to their offsprings in a Mendelian fashion. Since little was known about gene expression in plants, most studies focused initially on the regulation of the expression of plant genes in foreign genetic background. Exploration of maize alcohol dehydrogenase and sucrose synthase, pea Rubisco and Cab, soybean lectin, and leghemoglobin genes, etc. in tobacco indicated that regulation of the transcription in response to environmental-, hormonal-, tissue-specific and developmental stimuli, is similar in diverse plant species. These studies, together with the analysis of the expression of plant virus genomes, contributed basic information and useful plant promoters to achieve regulated expression of foreign genes in plants.

tions in both hosts results in a ping-pong conjugation of binary vectors that helps to test the stability of T-DNA constructs before plant transformation. Since ori_V plasmids are maintained at a low copy number, most PCV vectors contain also a multi-copy ColE1 replicon to facilitate the cloning in *E. coli* (Fig. 2b).

15.4 Gene Expression in Plants[6]

Expression cassettes consisting of characterized plant promoter and polyadenylation signal sequences separated by suitable cloning sites were constructed using transcriptional regulatory elements of T-DNA genes (octopine, nopaline, mannopine synthase genes, etc.), 35S and 19S RNA genes of Cauliflower Mosaic Virus (CaMV) and plant genes, such as the light-regulated SSU or Cab. These cassettes were inserted into the T-DNA of binary vectors in linkage with diverse chimeric antibiotic resistance genes, to clone and express various foreign genes in plants. Certain empirical rules established during these studies indicated that:

- The active core of plant promoters contains a TATA-box followed by a transcription initiation site at 40 to 100 bp 3'-downstream.
- 5'-Upstream of the core ("minimal") promoter region are *cis*-regulatory elements located that modulate the level of transcription in a quantitative (i.e., SV 40 type positive enhancers) or qualitative (i.e., tissue-specific enhancers and silencers) fashion by interaction with *trans*-acting regulatory proteins (transcription factors).
- Increasing the copy number of certain enhancers results in a proportional increase in the level of transcription.
- Many promoters contain AT-rich DNA sequences (e.g., AT-boxes in heat shock, lectin, leghemoglobin, etc. promoters) in the 5'-upstream region that probably mediate interaction with common nuclear matrix proteins (e.g., HMG class) regulating conformational changes of active chromatin.
- From genes transcribed by RNA polymerase II polyadenylated transcripts are synthesized. Signals for polyadenylation (i.e., AATTAA/T) are located 3'-downstream of the translational stop codon. The distance between the stop codon and the polyadenylation site influences the steady-state level of transcripts.
- Most plant genes contain introns, therefore the derived primary transcripts undergo splicing. It is apparent that viroids that cause serious plant diseases affect splicing. The length of introns (especially that of the first intron) may regulate the level of gene expression.
- Ribosomes bind to the 5' leader sequence of mRNAs and initiate translation at the first ATG codon. The presence of nonframe ATG codons in chimeric gene constructs therefore greatly reduces the translation of foreign transcripts. Kozak's rule for efficient initiation of translation at ATG/G starts can also be applied to plant genes. A consensus sequence for ribosome-binding sites is not yet established for plant genes. However, it was observed that leader sequences of certain viral RNAs (i.e., tobacco or alfalfa mosaic virus) can be used as translational enhancers. Alteration of the coding sequence of foreign genes according to plant codon usage may also increase the efficiency of translation.
- Ribosomes do not necessarily dissociate at the stop codon during scanning the transcript. In the case of a dicistronic transcript this may lead to initiation of the translation of the second coding region. Since the translation of the first coding region is more efficient, chimeric constructs carrying a foreign gene as first, and a selectable marker gene as a second cistron can facilitate an increased production of foreign proteins in plants.
- N- and C-terminal sequences of plant proteins may be recognized by diverse processing mechanisms that mediate the targeting of the proteins to cellular compartments, such as chloroplast, mitochondrium, peroxisome, or endoplasmic reticulum. A fusion of corresponding DNA sequences to foreign genes can successfully be applied for targeting foreign proteins to plant organelles.

The use of T-DNA-based expression vectors resulted in a burst of applications. New selectable markers, such as hygromycin, bleomycin, gentamycin, streptomycin, and methotrexate resistance genes were constructed by expression of diverse bacterial genes and a dihydrofolate reductase gene from mouse. Reporter enzyme systems providing sensitive assays for monitoring gene expression in vitro, in vivo, and by histological methods were developed by expression of β-glucuronidase (*gus*), β-galactosidase (*lacZ*) and light-producing luciferase (*luc* and *lux*) enzymes from fireflies and *Vibrio harveyi*. To engineer plants resistant to or tolerant of insects, viruses, or herbicides, the crystal toxin protein of *Bacillus thüringiensis*, virus coat proteins, antisense virus transcripts and herbicide-resistant or inactivating enzymes, such as EPSP-synthase, acetolactate synthase, and phosphinotricine acetyltransferase, were expressed in transgenic plants with the help of T-DNA vectors.

15.5 Studying Gene Regulation by Chimeric Plant Promoters[6,7]

To understand how environmental, hormonal, or developmental signals are transmitted through various signal transduction pathways to individual genes, the interaction of *cis*-regulatory elements and transcription factors is studied using promoter and enhancer test T-DNA vectors. These vectors contain a reporter gene, the activity of which can easily be followed in transformed protoplasts (transient expression assays), or in diverse tissues of transgenic plants (transformation assays). Promoter derivatives carrying successive deletions in the 5'-upstream region are linked to a promoterless reporter gene. Alternatively, the reporter gene is linked to core promoter region of a known gene (i.e., minimal promoter region of *nos* or CaMV 35S RNA promoters), upstream of which suitable cloning sites are provided for insertion of putative *cis*-regulatory DNA sequences. This approach is used for functional analysis of individual promoter elements, as well as for studying possible interactions between promoter elements and transcription factors. Promoter analysis of light-regulated, heat shock, hormonally or anaerobically regulated genes revealed various DNA sequences that are targets for binding of transcription factors and/or for methylation in vivo. A complex picture emerging from these studies suggests that similar regulatory DNA sequences may be part of different promoters, thus an interaction between transcriptional factors and diverse promoter elements may mediate various and specific responses in transcription. Since promoters of all T-DNA genes are recognized by plant transcription factors that modulate their activity in a hormone-regulated and tissue-specific fashion, the T-DNA itself provides an excellent tool to gain more insight into hormone signal transduction and transcriptional regulation of plants.

15.6 Insertional Mutagenesis: a Link Between Classical and Molecular Plant Genetics[8]

From a genetic point of view, T-DNA is a unique insertion element that is integrated into the plant nuclear genome after transfer from agrobacteria, and therefore may cause insertional inactivation of plant genes. To identify T-DNA insertions in functional plant genes, a gene fusion approach was developed. A promoterless reporter gene was linked to the right border of a T-DNA, which also carried a bacterial plasmid replicon and a plant selectable marker gene. The ATG start codon of the reporter gene was either retained or deleted in order to generate either transcriptional or translational gene fusions (Fig. 3). Following selection of transformants using the selectable marker, the frequency of T-DNA-induced gene fusions was determined. In tobacco and *Arabidopsis* about 40% of all insertions resulted in transcriptional fusions, while 15 to 20% of T-DNA inserts induced translational gene fusions. Differences in the complexity and distribution of transcribed DNA sequences between tobacco and *Arabidopsis* excluded the possibility that a similar frequency of gene fusions in both plant species resulted from random T-DNA insertions. The data rather indicated that T-DNA is preferentially integrated in plant chromosomal loci that are potentially transcribed.

The gene fusion technique has the advantage that the expression of T-DNA-tagged plant genes can be followed in vitro and in vivo throughout the life cycle of plants or under influence of various external stimuli. Both transcriptional regulatory elements and coding sequences of T-DNA-tagged genes can be rescued from the nuclear DNA of transgenic plants with the help of a bacterial plasmid replicon carried by the T-DNA. Plant DNAs are digested with a restriction endonuclease that

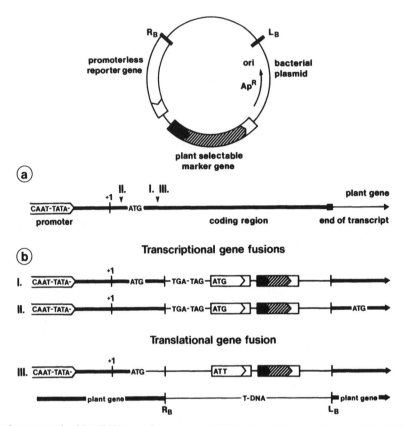

Fig. 3a, b. Insertional mutagenesis with T-DNA gene fusion vectors. **a** Schematic structure of T-DNA gene fusion vectors. A promoterless reporter gene is linked to the right border (R_B) of the T-DNA that also contains a bacterial plasmid replicon (Ap^R and *ori*) and a plant-selectable marker gene joined to the left T-DNA border (L_B). Underneath the T-DNA vector a hypothetical plant gene is depicted with promoter region (CAAT-TATA), transcription start (+1), translation start (ATG) and transcription termination site. *Arrows I, II,* and *III* indicate T-DNA integration sites. **b** Principle of T-DNA-induced transcriptional and translational fusions. *Line I* shows a T-DNA insert in the coding region of a plant gene, while in *line II* a T-DNA insert is depicted after integration in the transcribed but untranslated leader region of a plant gene (located between +1 and ATG). Due to the presence of stop codons in all reading frames upstream of the ATG codon of the reporter gene, *T-DNA insert in I* results in a dicistronic transcript. The *first coding region* of this transcript encodes a truncated plant protein, while the *second one* encodes the reporter enzyme. *T-DNA insert in II* leads to a monocistronic transcript starting at +1 position of the plant gene and terminating at the polyadenylation site of the reporter gene within the T-DNA. From both *I* and *II* transcriptional gene fusions an intact reporter protein is synthesized. *III* shows a T-DNA insert with a promoterless reporter gene that does not contain ATG translational start codon. T-DNA insertion in the coding region of a plant gene may lead to in-frame fusion between plant and reporter genes. This results in the translation of a fusion protein that consists of an N-terminal plant protein domain and a C-terminal reporter enzyme domain

has no recognition site within the T-DNA, circularized by self-ligation and transformed into *E. coli*, where the T-DNA and flanking plant DNA is recovered as a plasmid. Plant DNA sequences rescued in linkage with the promoterless reporter gene of the T-DNA are dissected and inserted into promoter and enhancer test vectors for further studies of the regulation of the identified plant promoters, or used as probes to isolate wild-type alleles of the tagged genes from genomic and cDNA libraries.

Arabidopsis thaliana, a plant with excellent genetics, became a model for plant molecular biology in general and for insertional mutagenesis studies in particular. A search for T-DNA-induced mutations in *Arabidopsis* showed that while insertions in diverse genes can be obtained at a fairly high frequency by gene fusions, only a portion of these mutations result in morphological alterations or other visible mutant phenotypes. Selectable antibiotic resistance genes carried by T-DNA gene-tagging vectors provide suitable markers for mapping of the induced mutations even in the absence of visible mutant phenotypes. A large number of mutations induced by irradiation or by chemical mutagenesis (e.g., EMS) is also available to test allelism with the T-DNA-induced mutations. Wild-type alleles cloned on T-DNA plant gene vectors (or corresponding cDNAs cloned in T-DNA-based expression vectors) are used for complementation of the induced mutations. Insertional mutagenesis in plants thus offers nearly as much flexibility as similar approaches in bacteria or in yeast. Recent isolation and characterization of *chlorata* (*ch*-42), *agamous* (*ag*), *apetala* (*ap*-2), and *glabrous* (*gl*-1) genes of *Arabidopsis* demonstrated that T-DNA-tagging is an efficient approach to identify genes regulating basic processes, such as photosynthesis or differentiation in plants.

15.7 Outlook

Application of T-DNA gene vectors in basic and applied plant science is virtually unlimited. As in the past, studies of the mechanism of T-DNA transfer are expected to facilitate further improvement of T-DNA vectors and plant transformation systems. Some rules established for *Agrobacterium*-plant interaction may lead to the discovery of new forms of interspecies gene transfer. Exploiting the growing knowledge on regulation of plant gene expression may help to achieve cell type-, tissue-, developmental stage-specific; hormone-, light-, heat-, gravity- or drought-induced; osmotically or chemically regulated expression of foreign or modified plant genes in a great variety of plants. Studies of transcription factors and corresponding genes will give more insight to the regulation of gene expression during plant development. T-DNA genes, exploited to alter plant morphology and development

genetically, may also be applied as experimental tools in studies of hormone signal transduction, cell division, or organ differentiation. Further development of insertional mutagenesis techniques involving the use of plant transposable elements may facilitate simple identification of new plant genes. T-DNA-induced gene fusions to suicide marker genes, that cause cell lethality, might help the isolation of genes that are expressed only in certain cell types or during certain stages of development. Alternatively, T-DNA-mediated insertion of strong promoters into the plant genome may be used for dominant activation of life defense genes involved in tolerance to drought, heat, salinity, or toxic chemicals. Transformation with T-DNAs carrying segments of the plant genome may result in recombination with homologous chromosomal loci that can be exploited in development of site-specific mutagenesis techniques for plant. Last but not least, combination of T-DNAs with telomeric elements of plant chromosomes might facilitate the identification of centromeric and autonomously replicating (ARS) DNA sequences to achieve chromosome engineering in plants.

15.8 Summary

The properties of T-DNA transfer and integration into the plant genome make it the system of choice for engineering of stably transformed plants. The versatility of T-DNA vectors that exploit the natural gene transfer process between agrobacteria and plants is such that T-DNA can be used for a variety of purposes other than simple gene transfer into plants. Over the years the unique ability of *Agrobacterium* to transfer the T-DNA to the plant cell has provided us with means to study plant-bacterial interaction, gene transfer and control of gene expression, differentiation, and development of plants. With the T-DNA vectors and marker genes currently available it is predictable that the collection of various approaches using T-DNA as a tool to investigate plant biology is far from being exhausted.

References

[1] Zambryski P (1988) Basic processes underlying *Agrobacterium*-mediated DNA transfer to plant cells. Annu Rev Genet 22:1–30

Zambryski P, Tempe J, Schell J (1989) Transfer and function of T-DNA genes from *Agrobacterium* Ti and Ri plasmids in plants. Cell 56:193–201

[2] Klee HJ, Horsch R, Rogers S (1987) *Agrobacterium*-mediated plant transformation and its further applications to plant molecular biology. Annu Rev Genet 20:467–486

Weising K, Schell J, Kahl G (1988) Foreign genes in plants: transfer, structure and expression. Annu Rev Genet 22:421–477

[3] Fraley RT, Rogers SG, Horsch RB, Eichholtz DA, Flick CA, Hoffman NL, Saunders PR (1985) The SEV system: A new disarmed Ti plasmid vector system for plant transformation. Bio/Technology 3:629–635

Koncz C, Kreuzaler F, Kalman ZS, Schell J (1984) A simple method to transfer, integrate and study expression of foreign genes, such as chicken ovalbumin and α-actin in plant tumours. EMBO J 3:1029–1037

Stougaard J, Abildstein D, Macker KA (1987) The *Agrobacterium rhizogenes* pRi TL segment as a gene vector system for the transformation of plants. Mol Gen Genet 207:251–255

Zambryski P, Joos H, Genetello C, Leemans J, Van Montagu M, Schell J (1983) Ti plasmid vector for the introduction of DNA into plant cells without alteration of their normal regeneration capacity. EMBO J 2:2143–2154

[4] An G, Watson BD, Stachel S, Gordon MP, Nester EW (1985) New cloning vectors for transformation of higher plants. EMBO J 4:277–284

Bevan M (1984) Binary *Agrobacterium* vectors for plant transformation. Nucl Acids Res 12:8711–8721

Hoekema A, Hirsch PR, Hooykaas PJ, Schilperoort RA (1983) A binary plant vector strategy based on the separation of the *vir* and T-region of agrobacteria. Nature 303:179–181

Klee HJ, Yanofsky MF, Nester EW (1985) Vectors for transformation of higher plants. Bio/Technology 3:637–642

Koncz C, Schell J (1986) The promoter of T_L-DNA gene 5 controls the tissue-specific expression of chimaeric genes carried by a novel type of *Agrobacterium* binary vector. Mol Gen Genet 204:383–396

[5] Schell J, Vasil IK (1989) Cell culture and somatic cell genetics of plants, molecular biology of plant nuclear genes. Academic Press, New York

Walden R, Koncz C, Schell J (1990) The use of gene vectors in plant molecular biology. Methods Mol Cell Biol 1:175–194

[6] Kuhlemeier C, Green PJ, Chua N-H (1987) Regulation of gene expression in higher plants. Annu Rev Plant Physiol 38:222–257

Schell J (1987) Transgenic plants as a tool to study the molecular organization of plant genes. Science 237:1176–1183

[7] Benfey DN, Chua N-H (1989) Regulated genes in transgenic plants. Science 244:174–181

[8] Feldmann KA, Marks MD, Christianson ML, Quatrano RS (1989) A dwarf mutant of *Arabidopsis* generated by T-DNA insertion mutagenesis. Science 243:1351–1354

Koncz C, Martini N, Mayerhofer R, Koncz-Kalman ZS, Körber H, Redei GP, Schell J (1989) High-frequency T-DNA-mediated gene tagging in plants. Proc Natl Acad Sci USA 86:8467–8471

Koncz C, Mayerhofer R, Koncz-Kalman ZS, Nawrath C, Reiss B, Redei GP, Schell J (1990) Isolation of a gene encoding a novel chloroplast protein by T-DNA tagging in *Arabidopsis thaliana*. EMBO J 9:1337–1346

Meyerowitz EM (1987) *Arabidopsis thaliana*. Annu Rev Genet 21:93–111

Redei GP (1975) *Arabidopsis* as a genetic tool. Annu Rev Genet 9:11–127

Yanofsky MF, Ma H, Bowman JL, Drews GN, Feldmann KA, Meyerowitz EM (1990) The protein encoded by the *Arabidopsis* homeotic gene *agamous* resembles transcription factors. Nature 346:35–39

Chapter 16 Development Genetics of *Arabidopsis*

Douglas Bradley and Robert E. Pruitt

16.1 Introduction

16.1.1 Biology of *Arabidopsis*[1]

Arabidopsis thaliana is a flowering annual belonging to the mustard (Brassicaceae) family. While it is itself of no agricultural importance, other members of the family, such as cabbage and oil seed rape, have great economic value as farm crops. *Arabidopsis* has a wide distribution in Europe, North America, Asia, and Africa, and has also been reported in Australia. The geographic origin of the species is thought to be in the Central Asian, Western Himalayan region.

Arabidopsis was recognized many years ago to have many features that made it an excellent experimental organism. It is a small plant (Fig. 1), and so

Fig. 1. *Arabidopsis thaliana.* (Hitchcock et al. 1964)

large numbers of individuals can be grown in a small space (ten plants per cm^2). Large populations of mutagenized plants can therefore be grown and screened for interesting mutant phenotypes. The life cycle of *Arabidopsis* is short, a generation can be completed in less than 6 weeks. *Arabidopsis* also has high fecundity, producing more than 10000 seeds per plant.

The *Arabidopsis* life cycle and developmental program is that of a typical dicotyledonous plant. Therefore, information derived from the study of *Arabidopsis* will find application to a wide variety of physiological, developmental, and biochemical questions that are common to all plants.

16.2 Genetics in *Arabidopsis*[2]

Arabidopsis is a self-fertile plant that outcrosses under natural conditions at a low frequency. This makes it easy to maintain genetically stable lines and to convert newly isolated mutant alleles to a homozygous state. Outcrossing can be performed by manual emasculation of a flower and then transfer of the desired pollen, allowing genetic mapping experiments and the construction of multiple mutant lines to be carried out with relative ease. A wide variety of genetically distinct ecotypes of *Arabidopsis* have been collected from natural populations around the world. These genetic resources are maintained at the *Arabidopsis* seed stock center in Frankfurt, Germany. The high level of genetic polymorphisms between the different ecotypes has proven to be very useful in the construction of restriction fragment length polymorphism (RFLP) genetic maps (see below).

A wide variety of mutagenic agents have been used to induce mutations in *Arabidopsis*. The most common mutagenesis method is to soak seeds in ethyl methane sulfonate (EMS). The treated seeds (called the M1 generation) are then planted, allowed to mature and produce seed through self-fertilization. The resulting mixture of progeny seeds (M2), which will include segregants that are heterozygous and homozygous for new mutant alleles, are then screened for the desired mutant phenotype. The dose of mutagen used must be carefully determined to effect efficient induction of mutations with out too much seed lethality. Optimal seed mutagenesis results in the ability to identify recessive mutants for many loci by looking at 2000 to 3000 M_2 plants.

16.2.1 Classical Genetics[3]

A large number of *Arabidopsis* mutants have been isolated from mutagenized plant populations. Genetic complementation and mapping experiments have allowed a genetic map to be constructed, consisting of 93 different genes located on five chromosome pairs (Fig. 2). The centromeres have been located on the genetic map through the use of lines that are trisomic for one arm of a chromosome. Mutant phenotypes range from obvious morphological abnormalities that are very useful as visual markers in genetic mapping, to specific enzymatic defects that can only be identified by biochemical analysis.

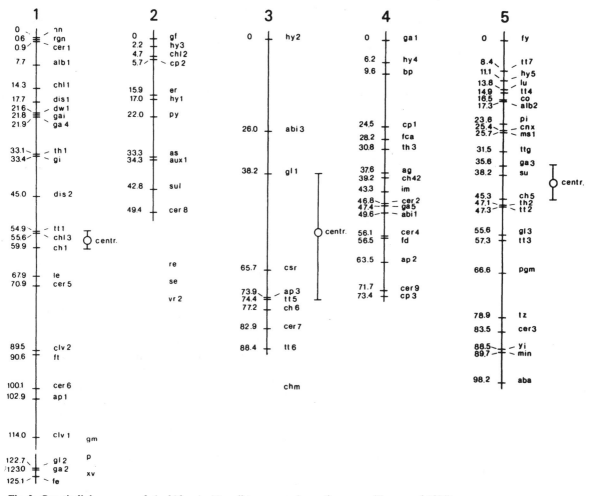

Fig. 2. Genetic linkage map of *Arabidopsis*. Map distances are in centimorgans (Koornneef 1990)

Many mutants have already been isolated which have phenotypes that are interesting to developmental biologists. To gain further knowledge of the role of these genes and gene products in plant growth and development, molecular methods are being utilized to facilitate the isolation of the genes defined by these mutations.

16.2.2 Physical and RFLP Maps of *Arabidopsis*[4]

The most general methods of gene isolation begin with the use of integrated genetic and RFLP (restriction fragment length polymorphism) or physical maps of the *Arabidopsis* genome. These types of maps help investigators clone mutationally defined genes in several ways. RFLP maps are essential when isolating genes through the use of "chromosome walking" methods. A complete physical map of the genome (i.e., the isolation and mapping of overlapping cosmid or yeast artificial chromosome clones encompassing the entire nuclear genome) allows immediate isolation of clones spanning the region to which an interesting mutant has been mapped.

To date, two independent RFLP maps of the *Arabidopsis* genome have been constructed that are roughly integrated with the classical genetic map. The two maps consist of over 200 DNA fragments, which, when combined into a single RFLP map, have an average spacing between markers of about 350 kilobase pairs. This density of RFLP markers makes it feasible to begin gene isolation efforts using chromosome walking techniques.

Several laboratories have begun efforts to construct a physical map of the *Arabidopsis* genome by isolating large numbers of yeast artificial chromosome (YAC) clones or cosmid clones. These clones are then mapped for restriction sites and pieced together into overlapping sets of clones called contigs. A complete physical map would consist of five different contigs representing each of the five *Arabidopsis* chromosomes. In practice, it is too difficult to obtain all of the clones necessary to arrive at a final map. As the established contigs (contiguous stretches) become larger, the probability of isolating a clone that is not already on the map becomes much smaller. Also, some regions of eukaryotic chromosomes are difficult to clone and maintain in cloning vectors and end up being under-represented in DNA libraries. Even so, a partial physical map is still very helpful. An incomplete physical map of *C. elegans* has been available for many years and has proven to be very useful in the isolation of genes that had been first identified by a mutationally induced phenotype (Edgar, Chap. 19, this Vol.). RFLP probes from the RFLP map will allow most of the contigs to be aligned with the RFLP and genetic maps. This is a large task, which for the *Arabidopsis* genome will take many years to complete. In time, though, it will be possible to rapidly isolate genes that have been defined by a mutant phenotype by first mapping the mutation on the genetic map, obtaining the YAC or cosmid clones covering that region and then testing each clone via plant transformation (Sect. 16.3.2) for complementation of the mutant phenotype. If the desired locus maps to a region not covered by the physical map, then chromosome walking methods have to be used to identify and clone that gene. In time, the gaps between the mapped contigs will be filled in by the accumulated effects of such chromosome-walking experiments. Finally, a long-range plan has been formulated to mount a coordinated, multinational research project on the *Arabidopsis* genome. One of the long-term goals of this project is to determine the complete DNA sequence of the *Arabidopsis* genome by the year 2000.

16.3 Molecular Biology of *Arabidopsis*

16.3.1 Genome Size and Organization[5]

Arabidopsis has the smallest genome size of any flowering plant species currently used in research or agriculture (Table 1). The best estimate of the haploid nuclear genome size of *Arabidopsis* is 10^8 base pairs. This is very similar to the genome size of *Caenorhabditis elegans* (Edgar, Chap. 19, this Vol.) and about seven times the size of the yeast *Saccharomyces cerevisiae*. The small size of the *Arabidopsis* genome greatly simplifies several aspects of molecular biological research and makes it

Table 1. Haploid genome size of various flowering plants

Plant	(in kilobase pairs)	clones for complete library[a]
Arabidopsis	100000	16000
Mung bean	470000	110000
Cotton	780000	180000
Tobacco	1600000	370000
Soybean	1800000	440000
Pea	4500000	1000000
Wheat	5900000	1400000

[a] The number of lambda clones that must be screened to have a 99% chance of isolating a single-copy sequence from these genomes. The library sizes are calculated assuming a random nuclear DNA library with an average clone insert length of 20 kb; 4.6 genome equivalents must be screened for a 99% probability of isolating any individual unique sequence (Meyerowitz and Pruitt 1985)

easy to screen an *Arabidopsis*-genomic DNA library to isolate specific gene clones.

A variety of experiments have been done to examine the fraction of the genome that is composed of repetitive DNA sequences. This is an important consideration when chromosome-walking methods are being performed to isolate a gene of interest. DNA reassociation analysis of *Arabidopsis* DNA shows that 10 to 14% of genome is composed of highly repeated or foldback sequences that reanneal rapidly. The middle repetitive and single-copy DNA sequences are present at 23 to 27% and 50 to 55%, respectively, of the whole plant DNA. A majority of the middle-repetitive sequence class have been shown to represent the chloroplast genome that is present at a high copy number in leaf mesophyll cells. Analysis of individual lambda phage clones containing *Arabidopsis* DNA inserts established that the average distance between repeated DNA sequences is generally large.

16.3.2 Gene Transfer in *Arabidopsis*[6]

The main goal of most DNA manipulations is to investigate the biological function of the cloned gene, and its encoded protein in the growth and development of the organism. To fully exploit recombinant DNA techniques it is essential that methods

exist to introduce the natural or modified genes back into the experimental organism. In the case of dicotyledonous plants, the bacteria *Agrobacterium tumefaciens* provides a natural gene-transfer system that can be manipulated experimentally to create transformed plants, containing any desired DNA sequence (see Hohn, Chap. 14 and Koncz and Schell, Chap. 15, this Vol.).

Several methods have been devised that use *Agrobacterium tumefaciens* to transform *Arabidopsis*. They all depend on the unique ability of *Agrobacterium* to recognize wounded plant tissues, then mobilize a specific sequence of DNA (called the transfer or T-DNA) from a resident plasmid (Ti plasmid), and finally, transfer the T-DNA to plant cells where it is inserted into the plant nuclear genome. This gene-transfer system has been adapted to genetic engineering by developing *Agrobacterium* plasmids that contain modified T-DNA regions that are still competent for plant transformation. The key modifications to the T-DNA are the inclusion of an antibiotic resistance gene that allows transformed plant cells to be selected by growth on the antibiotic and unique restriction enzyme sites for insertion of the DNA that is targeted for transformation into the plant genome.

The most efficient method of *Agrobacterium*-mediated transformation of *Arabidopsis* which has so far been developed utilizes root cuttings from plants grown under sterile culture conditions. The root explants are incubated on a medium containing both auxin and cytokinin to induce callus (undifferentiated plant cells) formation. The preincubated roots are then cocultivated for 48 h with *Agrobacterium* cells harboring the desired plant transformation vector. The *Agrobacterium* cells are then washed off from the root explants and the explants transferred to shoot induction medium which contains a high level of cytokinin to induce shoot formation. This medium also contains an antibiotic (e.g., carbenicillin) that prevents *Agrobacterium* growth but does not affect plant cells, and the appropriate antibiotic that will prevent the growth of untransformed plant cells but allow the proliferation of cells transformed with the T-DNA region containing the antibiotic resistance gene. In about 6 days, green calli appear, which over the next 2 to 3 weeks grow and produce shoots. The small plantlets have the capacity to flower and form seeds in culture. The F_1 seed produced from the primary transformant can then be collected and germinated on the selective antibiotic to identify transgenic plants.

16.4 Combining Genetics and Molecular Biology to Study Development in *Arabidopsis*[7]

A number of different strategies are emerging for *Arabidopsis* which allow the isolation of a gene that can be identified by its mutant phenotype. These methods are not unique to *Arabidopsis*, although the short generation time and small genome size make *Arabidopsis* an attractive system for using them.

16.4.1 Insertional Mutagenesis Using *Agrobacterium*[8]

Transgenic *Arabidopsis* plants can be produced using *Agrobacterium*-mediated transformation, as has been discussed in Section 16.3.3. When the T-DNA inserts into the *Arabidopsis* genome it will produce a genetic lesion at a more or less random location. By performing many transformation experiments it is possible to produce a population of plants which have been subjected to insertional mutagenesis. These plants can be screened for the mutant class of interest and the mutant gene identified using a T-DNA region as a DNA hybridization probe (see Chaps. 14 and 16, this Vol.).

There are a number of potential pitfalls which can cause problems in the use of this technique. First, the size of the mutagenized population is usually very small compared to what can be produced by conventional mutagenesis. This makes it difficult to search for a mutation in a single gene that has been identified previously by some other method. In flowering plants, regenerants from tissue culture (and transformants from the seed transformation method) often exhibit mutant phenotypes; therefore, not all new mutations which arise following T-DNA transformation are the result of

the T-DNA insertion. Careful genetic mapping of the T-DNA and mutation are necessary before attempting to clone the gene. These disadvantages may be more than outweighed by the rapidity with which a cloned gene can be obtained.

16.4.1.1 T-DNA Insertion Mutants with Dramatic Phenotypes

The first application of T-DNA insertional mutagenesis in *Arabidopsis* was done using a novel, nontissue culture method for plant transformation. This method works by germinating *Arabidopsis* seeds in the presence of *Agrobacterium* cells which contain a kanamycin-resistant plant transformation vector. The *Agrobacterium*-treated seeds are planted and plants grown to maturity. Seeds from these plants are then screened for resistance to kanamycin. These resistant plants are then examined for morphological mutants that cosegregate with the kanamycin-resistance trait. In one screen of 1800 T-DNA-induced mutants the most abundant phenotypes observed were seedling-lethals (10%), size variants (7%) and embryo-lethals (5%). About 1% of the T-DNA insertion mutants had dramatic effects on plant morphology. For example, T-DNA mutants have been isolated that alter trichome development (Sect. 16.5.3) and floral morphogenesis (Sect. 16.5.4). The mutant genes responsible for both of these phenotypes have been isolated by using the T-DNA insert to identify the desired clones from lambda libraries. Another interesting class of dramatic mutant was isolated which has a dwarf growth habit. The mutant plant is similar to wild type in its seed to seed cycle except that its morphological features are all reduced in size. Normal plants are 30 to 40 cm in height while the dwarf is 7 to 8 cm. This dwarf phenotype is different from the gibberellic acid dwarf mutants (discussed in Sect. 16.5.2) in that the phenotype cannot be suppressed by the addition of any growth hormones in the medium or when sprayed on the foliage. Genetic studies have been done to confirm that the dwarf phenotype cosegregates with the kanamycin-resistance phenotype, making it feasible to use the T-DNA tag to isolate the wild-type dwarf gene for further investigation.

16.4.2 Confirmation of Gene Identity by Complementation[9]

Use of any of the above methods is limited to identifying the region of DNA within which the gene of interest resides. The region identified might be quite well defined, as in the case of a "tagged" gene, or it might be very large, as in the case of the RFLP mapping/chromosome walking approach. In any event, some technique is required to determine what region precisely defines the gene being cloned. The best method to localize the gene in this way is to test the genomic DNA which has been cloned for its ability to functionally complement the original mutation. By creating a series of transgenic plants which bear different segments of genomic DNA, it is possible to determine the position of sequences required for gene function within the cloned chromosomal region (for an example see Sect. 16.5.3). Even in the case of the large region identified in a chromosomal walk, the use of this procedure in an iterative manner with binary cosmid transformation vectors will allow the location of the gene to be determined relatively quickly. Once this has been established, molecular techniques can be used to characterize the gene and its product in order to learn more about the mecha-

nism by which the gene acts in the biological process under study.

16.5 Developmental Systems Presently Under Study in *Arabidopsis*[10]

Given the generality of the approaches discussed above, many different interesting processes could be genetically dissected using *Arabidopsis*. In fact, although many different systems are being studied, the remainder of this chapter will deal with only a few major systems which are important to plant development. These systems are offered as examples of the type of work being done in *Arabidopsis*, as opposed to an exhaustive list of interesting systems.

16.5.1 Genetics of *Arabidopsis* Embryo Development[11]

Embryo development in flowering plants is a complex process that is commonly divided into several components: embryo morphogenesis and the establishment of the root and shoot apices; seed desiccation and dormancy; and preparation for germina-

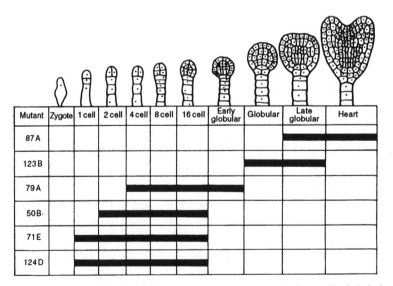

Fig. 3. Developmental arrest of mutant embryos from six embryo-lethal mutants of *Arabidopsis*. The lethal phase for each mutant is represented by a solid line beneath the corresponding stages of normal development (Meinke 1982)

tion. Descriptive and biochemical studies have established that the root and shoot apices are determined early in embryo development, while preparation of the seed for dormancy and the ability to germinate under the proper environmental conditions occur late in embryo maturation. Little is known about the genes, or their protein products, which regulate embryo development, though a large number of *Arabidopsis* mutants that are defective or altered in embryogenesis have been isolated.

The largest class of embryo mutants that have been isolated and studied are recessive embryonic lethal mutants. These mutants are easy to isolate and characterize because the wild-type embryos of *Arabidopsis* follow a consistent pattern of development. Over 75 recessive mutants have been isolated which arrest embryo development at a wide variety of stages (Fig. 3). A large number of the mutants also arrested with abnormal embryo morphologies such as: altered embryo size and coloring; distorted or fused cotyledons; and reduced hypocotyls. One problem with this approach is that many of the mutants that are isolated probably arrest embryogenesis because of defects in genes that are required for normal metabolism; the so-called housekeeping genes. Some of these mutants can be phenotypically rescued by culturing the mutant embryo on medium supplemented with nutrients. One example of this class of embryo-lethal mutants is a mutant deficient in biotin biosynthesis. Embryos from this mutant can be rescued by addition of biotin to the tissue culture medium.

A primary objective of future research on these embryo-lethal mutants will be to identify those mutants that define genes which play a regulatory role in *Arabidopsis* embryogenesis. It is this class of developmental mutants that will be the most interesting for gene isolation and further molecular biology studies. Several embryo-lethal T-DNA insertion mutants have been isolated (see Sect. 16.4.1), but it remains to be determined if the mutant phenotypes are due to lesions in housekeeping genes, or in genes encoding a regulatory function.

Another class of mutants that will be of great importance in understanding how embryogenesis proceeds is embryonic pattern mutants. Members of this mutant class exhibit an altered embryo body structure yet, do not arrest embryo growth and development. This class of mutants is very similar to some of the homeotic mutants of *Drosophila* (see Lobe and Gruss, Chap. 26, and Tautz, Chap. 21, this Vol.), and it is hoped that the study of the genes defined by these mutations will yield new information about how the pattern of a plant embryo is determined.

16.5.2 *Arabidopsis* Phytohormone Mutants[12]

Plant phytohormones play an important role in the development of higher plants, even so, almost nothing is known about the molecular basis of hormone action. There are five major classes of plant hormones: abscisic acid (ABA), gibberellins (GA), auxins, cytokinins, and ethylene. Each of these hormones has many effects on plant physiology. There is little information about which cell types synthesize plant hormones or how these sites of biosynthesis might shift through development.

Many mutations have been isolated in a variety of plant species that alter phytohormone biosynthesis and/or utilization (see Cove, Chap. 12, this Vol.). A large number of *Arabidopsis* mutants have been identified that exhibit phenotypes associated with four of the five of the major classes of plant hormones (Table 2). Efforts are under way in a large number of laboratories to isolate and study the genes defined by some of the *Arabidopsis* hormone mutations. These genes will be valuable tools in the study of the signal transduction pathways through which the hormones evoke a physiological response and how and where hormone biosynthesis occurs.

For example, a dominant mutation, *etr* (*e*thylene *r*esistant), has been isolated that is insensitive to gaseous ethylene. In wild-type plants, exposure to exogenous ethylene inhibits hypocotyl elongation in dark-grown seedlings. The *etr* mutant was identified by its ability to form an elongated hypocotyl in the presence of ethylene (Fig. 4). The *etr* mutant plants were found to be impaired in their ability to bind ethylene, suggesting that the *etr* mutation directly affects the receptor for ethylene. The *etr* mutation has been mapped on the RFLP map of *Arabidopsis*, and using chromosome walking methods,

Table 2. *Arabidopsis* hormone mutants

Hormone	Gene name	Gene symbol	Phenotype	Chromosome location
Abscisic acid (ABA)	ABA-deficient	*aba1*	Little or no endogeneous ABA, plants are wilted, fresh seeds germinate at high frequency (unlike wild type)	5 – 98.2
	ABA-insensitive	*abi1*	Germinates in presence of 10 µM ABA (unlike wild type), young plants wilted, fresh seeds germinate at high frequency	4 – 49.6
		abi2	Same as *abi1*	Not mapped
		abi3	Germinates in presence of 10 µM ABA, plants not wilted	3 – 26.0
Auxin	Auxin-resistant (2,4-D)	*aux1*	Agravitropic seedlng roots, root growth resistant to 2,4-D (50% inhibition at 14-fold higher level than wild type)	2 – 34.3
	Auxin-resistant	*axr1*	Decreased growth habit, agravitropic roots, reduction in apical dominance, roots and leaves resistant to inhibition of growth 2,4-D	1 – 2.6 cM from RFLP
Cytokinin	No cytokinin mutants have been isolated			488 (11.3)
Ethylene	Intensive to ethylene	*etr*	Dominant mutant, fivefold reduction in ethylene binding compared to wild type	1 – ND[a]
Gibberellin (GA)	Gibberellin requiring	*ga1*	GA responsive dwarf, some alleles require GA for germination	4 – 0.0
		ga2	Same as *ga1*	1 – 23.0
		ga3	Same as *ga1*	5 – 35.6
		ga4	GA responsive dwarf	1 – 21.9
		ga5	Same as *ga4*	4 – 47.4

[a] ND, not determined.

Fig. 4. Comparison of wild-type seedlings and one *etr* mutant grown in the presence of ethylene. Ethylene reduces the elongation and thickens the hypocotyls, and causes a pronounced apical hook. The *etr* hypocotyl looks the same as a wild-type seedling would if it were grown in the absence of ethylene (From the cover of Science 26 August 1988, vol. 241. Photo Kurt Stepnitz)

overlapping cosmid clones were isolated from an *etr* cosmid library. These cosmids were transformed into wild-type *Arabidopsis* plants and two overlapping clones were identified that conferred the dominant *etr* phenotype to the transgenic plants. Once the *etr* gene is located on the DNA region shared by these two cosmids (determined by subcloning and plant transformation) it will be possible to characterize the *ETR* structural gene, manipulate its expression in vivo and study its product in vitro. This will lead to a better understanding of the role *ETR* plays in ethylene binding and how ethylene induces changes in plant growth and development.

Auxins affect a wide variety of plant growth processes, such as root gravitropism, cell expansion, vascular bundle development, and apical dominance. Auxins have a powerful effect on plant physiology, and high levels of auxin cause plant death. This fact has led to the use of auxin analogs as herbicides.

Two nonallelic *Arabidopsis* mutants have been selected whose root growth is about tenfold more resistant to exogenous auxin than wild-type plants. Plants homozygous for recessive mutations in the *AUX1* gene are resistant to auxin, yet are morphologically normal, except for a defect in root gravitropism. However, five independent auxin-resistant mutants (all recessive) in the *AXR1* gene have been isolated that all exhibit morphological alterations throughout the plant (Fig. 5). The *axr1* mutants exhibit a decrease in plant height, hypocotyl elongation, root gravitropism, and fertility. The *AXR1* gene has been genetically mapped to chromosome 1 of *Arabidopsis* and chromosome-walking experiments are underway to isolate the *AXR1* gene. Biochemical studies on a similar auxin-resistant mutant in tomato (*diageotropica*), have shown that mutant *diageotropica* plants are missing an auxin-binding protein. Similar studies are underway with the *Arabidopsis axr1* mutant.

A large body of data has shown that abscisic acid (ABA) plays an important role in seed desiccation and dormancy, while gibberellins (GA) are required for germination. Five different genes (*GA1-5*) have been identified by mutation that are defective in GA metabolism. All of the *ga* mutants have a dwarf phenotype which can be suppressed by addition of exogenesis GA. None of the *GA* genes have been cloned, though chromosome walks have been initi-

Fig. 5. Comparison of wild-type (*left*) and *axr1–12* (*right*). Plants are 7 weeks old (From cover of Plant Cell. November 1990, vol. 2. Photo C. Lincoln and M. Estelle)

ated in a number of laboratories for those *GA* loci that are located near mapped RFLP markers.

Some mutant alleles of *ga1, ga2,* and *ga3* also require the addition of GA to overcome ABA-induced seed dormancy and germinate. Revertants of a nongerminating *ga1-1* mutant were selected that can germinate in the absence of GA. These mutants were all found to have lower ABA levels. Most of the second-site mutants were allelic, defining a gene called *aba1*. In the absence of the *ga1* mutation, *aba1* mutants have a pleotropic phenotype characterized by a brownish yellow leaf color, reduced ability to induce seed dormancy, wilting due to increased loss of water by transpiration rates (reversable if the plants are sprayed with ABA), and lowered ABA levels in seeds and leaves.

Arabidopsis mutants have also been isolated that are insensitive to the presence of exogeneous ABA. The ABA-insensitive mutants were genetically characterized and placed into three complementation groups (*abi1, abi2,* and *abi3*). Plants homozygous for mutant alleles belonging to each of the three complementation groups have phenotypes similar to the *aba1* mutants except that their ABA levels are slightly higher than in wild-type and much higher than in *aba1* mutants. Seed dormancy in each of the *aba1, abi1,* and *abi3* mutants is blocked, giving rise to fresh seeds that will germinate at a high frequency. Analysis of *aba* and *abi* double mutants suggests that two different ABA response pathways may exist. ABA induction of seed dormancy requires the *ABA1, ABI1,* and *ABI3* gene products, while some seed maturation processes (e.g., water loss and induction of seed storage proteins) require only that either the ABA*1* and ABI*3* protein be functional. Efforts are underway in several laboratories to isolate the *ABA* and *ABI* genes by chromosome walking methods. Perhaps then the role of these genes in the ABA signal transduction pathway and seed development will become clear.

16.5.3 Mutants of *Arabidopsis* That Alter Trichome Development[13]

Trichomes (plant hairs) are simple structures (Fig. 6A and B) that are readily visible on leaves and stems. Trichome morphogenesis is a good de-

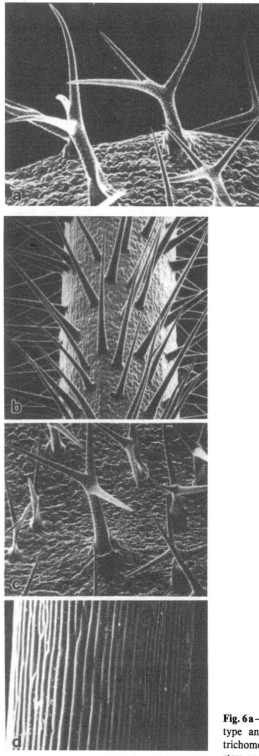

velopmental pathway to study because trichomes are nonessential, and mutant phenotypes are easy to score. They are formed from single protodermal cells at regular intervals on the surface of leaves and stems. Six nonallelic trichome mutants have been genetically characterized in *Arabidopsis*. They fall into three phenotypic classes: *glabrous1* (*gl1*) and *transparent testa, glabrous* (*ttg*) mutants lack almost all trichomes on both leaves and stems, mutants called *glabrous2* (*gl2*) and *glabrous3* (*gl3*) exhibit a reduction in trichome number and in the branch number of leaf trichomes, while *distorted1* (*dis1*) and *distorted2 (dis2)* mutants have enlarged trichomes. Efforts are underway to understand what role these genes play in the initiation and control of trichome morphogenesis.

As mentioned in Section 16.4.1.1, many T-DNA insertion mutants have been isolated that have dramatic morphological phenotypes. One mutant was found that had a new trichome phenotype; the leaf trichomes were normal, but the stem trichomes were totally missing (Fig. 6C and D). Genetic studies showed that the mutant trichome phenotype was linked to the T-DNA insert and that the insertion mutant was allelic to the previously mentioned *gl1* mutant. The *wildtype GL1* gene has been cloned through the use of the T-DNA insertion in *GL1* (*gl1−43*). The identity of the cloned *GL1* gene was then confirmed by plant transformation experiments in which a *gl1* mutant was complemented by the cloned gene. DNA sequencing of this clone has revealed that the *GL1* gene product has strong similarity to a previously identified family of transcriptional activators. It is interesting that the insertion mutation (*gl1−43*) affects only stem trichomes, while the *gl1* mutant is lacking both leaf and stem trichomes. In addition, genetic experiments have shown that the *gl1−43* trichome phenotype is dominant over the *gl1* phenotype. The molecular explanation of these phenotypic differences might be found when the DNA sequences from several mutant alleles are determined and compared to the wild-type *GL1* gene, and when in situ RNA expression studies have been completed.

Fig. 6a−d. Scanning electron micrographs of leaf and stem trichomes on wild-type and *gl1−43* plants (Marks and Feldmann 1989). (**a**) Wild-type leaf trichomes. (**b**) Wild-type stem trichomes. (**c**) *gl1−43* leaf trichomes. (**d**) *gl1−43* stem surface

16.5.4 Genetic Analysis of Floral Development in *Arabidopsis*[14]

16.5.4.1 Floral Induction

It is clear from classical work on plant physiology that the switch from vegetative growth to floral development is under many different types of regulation. A large number of plant species have some type of photoperiodic control in induction of floral development. These plants require days of certain lengths in order to be induced to flower. In *Arabidopsis* this requirement is not absolute: *Arabidopsis* will flower under any day length, although the plants will flower at an earlier developmental stage when grown under long day lengths. Another environmental influence which commonly affects the induction of flowering in a wide variety of plants is vernalization. These plant species require exposure to cold temperatures for a prolonged period of time before they can be induced to flower. In *Arabidopsis*, some wild-type strains have an absolute requirement for vernalization but most do not.

To apply the techniques described in Section 16.4, to the study of floral induction, it is first necessary to obtain mutations in the genes that are essential to this process. The most obvious and perhaps most common mutant phenotype might be an inability to flower. Other possible mutant phenotypes might be an altered response to photoperiod or perhaps a requirement for vernalization. To date, no mutants have been recovered in *Arabidopsis* which are unable to switch from vegetative shoot growth to a floral mode of development. A large number of mutants have been obtained which have been called *late flowering*. The mutant alleles fall into 12 different complementation groups with similar but not identical phenotypes. All mutations of the *late flowering* class take longer to produce flowers than the wild-type strains. Members of some complementation groups show a continued sensitivity to day-length and light quality, while others seem insensitive to these environmental variables. Additional experiments performed on these mutants indicate that in some, but not all, complementation groups the delay in flowering can be shortened by cold treatment. Inability to isolate a mutation which prevents the induction of flowering implies that there are multiple independent pathways which are capable of inducing a flowering response. The phenotypes of double and triple mutant combinations may reveal the existence of additional pathways which also are involved in the induction of flowering. None of the late flowering genes has been cloned to date, but a number of them are genetically mapped and research is under way in a number of laboratories to isolate them.

16.5.4.2 Floral Morphogenesis

After a plant has made the switch from vegetative growth to floral development, a new pattern of cell division emerges on the meristem surface, leading to the production of the right numbers of floral primordia in the right locations to produce a flower. This process is also under genetic investigation in *Arabidopsis*. Several classes of mutations have been recovered that result in plants which produce aberrant numbers of floral structures. The genes defined by these mutations are presumably essential to either the conversion of the meristem to its new pattern of cell division, or to the maintenance of that pattern once it is established.

The most extreme mutants of this class are those of the two genes *pin-formed* and *pinoid*. Plants homozygous for mutant alleles produce either a small number of flowers or none at all followed by an undifferentiated outgrowth of the apical meristem. In the case of the *pinoid* mutations which produce some flowers, these flowers are abnormal and contain fewer than normal sepals and stamens, an excess of petals, and a pin-shaped gynoecium, which is usually sterile. The effects of *pinoid* mutants are not limited to floral development but also cause other developmental defects, most notably in the embryo, which frequently contains extra cotyledons. Mutation of the gene *terminal flower* produces plants which have several normal flowers per stem, but this is followed by premature termination of the floral stem in a compound flower. Another gene of this class is *leafy*, mutation of which produces plants with highly branched inflorescences bearing abnormal flowers with few petals and stamens. Multiple mutant alleles of all of these loci have been isolated and attempts are underway to clone these genes. Characterization of these genes and their products may provide insight into the mechanism which establishes the pattern of floral apex development.

The mutants described above seem to alter the ability of the floral meristem to produce the correct number or type of primordia and, in some cases, alter the structure of the flower as well. Another class of mutants have floral meristems which produce flower primordia in the normal pattern, but where some of the organs are partially or completely transformed into organs that are normally located elsewhere in the flower (Fig. 7). These types of mutations have been referred to as homeotic and are reminiscent of homeotic mutations found in *Drosophila melanogaster*. A growing number of genes have been identified that appear to be involved in the process of determining the identity of the floral primordia.

Four loci, defined by mutations, have been shown to play major roles in the basic control of floral morphogenesis. These are *agamous* (*AG*), *apetala2* (*AP2*), *apetala3* (*AP3*), and *pistillata* (*PI*). Mutations in any of these genes produce plants with flowers in which one or more whorls of floral organs have been transformed into floral organs of a different type. The phenotypic changes seen in these four mutants are summarized in Table 3. Examination of single and double mutant phenotypes has led to a model for how the products of these four genes are directly involved in the determination of the identity of the individual floral organs. In this model, summarized in Fig. 8, the wild-type product of the *AP2* gene is required for the production of sepals, the *AP2*, *AP3*, and *PI* products are required for the production of petals, the *AG*, *AP3*, and *PI* products are required for the production of stamens, and the product of the *AG* gene is required for the production of carpels. In addition, it is hypothesized that the products of the *AP2* and *AG* genes are antagonistic and serve to limit the expression of each other to the perianth and reproductive organ primordia respectively. This model is similar to that proposed to account for the phenotypes of homeotic mutants affected in floral development in *Antirrhinum* (see Schwarz-Sommer et al., Chap. 17, this Vol.).

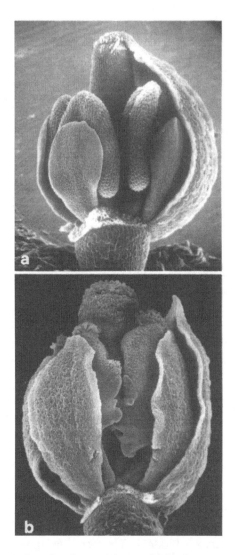

Fig. 7a, b. Scanning electron micrographs showing a representative *Arabidopsis* floral homeotic mutation. (Unpublished photographs J. Bowman). (**a**) Wild-type flower which has been dissected open to reveal normal petals and sepals developing within the flower bud. (**b**) Flower of the mutant *apetala3*. The second whorl organs are transformed into sepals, as evidenced by the nature of the epidermal surface, while the third whorl organs are transformed to carpelloid stamens (note the appearance of stigmatic papillae at the top of each stamen-like structure)

By taking advantage of temperature-sensitive alleles of two of these genes (*AP2* and *AP3*), it has been possible to determine the period during which the gene product is required. Interestingly, the tem-

Table 3. Organ transformations in *Arabidopsis* floral homeotic mutants

Genotype	1st Whorl	2nd Whorl	3rd Whorl	4th Whorl
Wild type	Sepals	Petals	Stamens	Carpels
agamous	Sepals	Petals	Petals	Buds
apetala 1	Bracts	Buds	Stamens	Carpels
apetala 2-1	Stigmoid leaves	Staminoid petals	Stamens	Carpels
apetala 3	Sepals	Sepals	Carpelloid stamens	Carpels
flower-10	Sepals	Petals	Stamens	Extra stamens
pistillata-1	Sepals	Sepals	Absent	Extra carpels

perature-sensitive period for these genes is quite distinct, with the *AP2* product required at or before the appearance of the floral organ primordia in which it acts, and the product of the *AP3* gene not required until the primordia in which it is active begin their differentiation. The *AG* gene has been cloned by T-DNA insertion (Sect. 16.4.1) and shown to have homology with known transcription factors from yeast and humans. The *AG* gene is also homologous with the recently cloned floral homeotic gene *defA* of *Antirrhinum*. The phenotype of the *defA* mutant is similar to that of the *ap3* and *pi* mutants, which may therefore belong to the same family of putative transcription factors. RNA homologous to the *AG* gene is found only in the primordia which will become stamens and carpels, consistent with the idea that it is directly involved in the determination of identity in those organs.

Two other genes which have phenotypes consistent with an important role in floral morphogenesis are *apetala 1* (*AP1*) and *flower development 10* (*FLO 10*). Plants that bear mutations of the *AP1* gene produce flowers which are lacking petals, but in which some of the petals are replaced with whole new flower buds. Mutations in *FLO 10* result in

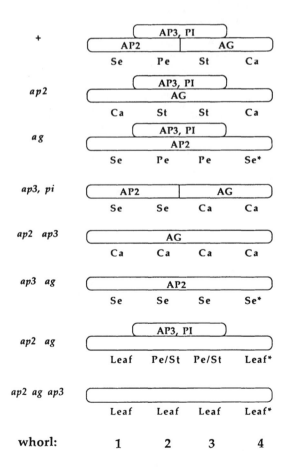

Fig. 8. Schematic representation of the model for floral development in *Arabidopsis*. A section through one-half of a floral primordium is represented as a *pair of boxes*, with the regions representing each whorl *labeled at the bottom*. The genotype under consideration is listed at the *left*, and in the boxes are the predicted pattern of activity of the products of the three classes of homeotic genes. The gene products are represented in *upper case*. *Under each diagram* is the predicted phenotype of the organs in each whorl. These predictions have all been confirmed by experiment. +, Wildtype; *Se* sepal; *Pe* petal; *St* stamen; *Ca* carpel. *Pe/St* represents the intermediate organs, with characteristics of both stamens and petals, found in *ag ap2* double mutants. *Leaf* either leaf or carpelloid leaf. Either can be found in these positions, depending on the *ap2* allele involved. The *asterisk* is a reminder that all *ag* combinations have additional whorls inside whorl four (Meyerowitz et al. 1991)

plants bearing flowers which have an increase in the number of stamens, but lack carpels. Genetic experiments to determine how these genes interact with the four genes detailed in the previous paragraph are underway.

16.5.5 Development of Roots in *Arabidopsis*[15]

Research is also in progress to examine the genetic control of root growth and development. One system that has been studied is the development of root hairs. Root hairs grow as roughly cylindrical outgrowths of root epidermal cells (Fig. 9). Root hairs are produced in a region behind the growing tip of a root, and elongate by a process of tip growth from only a limited subset of epidermal cells. Recent work has identified four genes which are essential to root hair development (*RHD1-RHD4*). Mutant alleles of any of these four genes produce root hairs with visible alterations in morphology. Plants homozygous for these mutations are healthy, fertile, and, with one exception, show no alterations except for the root hair phenotype. One of the genes (*RHD1*) appears to be involved in root hair initiation, while the other three genes appear to be essential for normal root hair elongation. Mutation of the *RHD2* gene results in extremely short root hairs, while lesions of the *RHD3* or *RHD4* genes results in wavy, irregular root hairs (Fig. 10). Analysis of double mutant combinations indicates that *RHD2* is epistatic to *RHD3* and *RHD4*, while all of the other interactions are additive.

Other recent work on root development concerns the mechanism by which roots grow, specifically the manner in which they avoid obstructions in their path. This analysis is based on an assay in which a root is made to grow in a wave pattern by growing *Arabidopsis* seedlings on a solid support

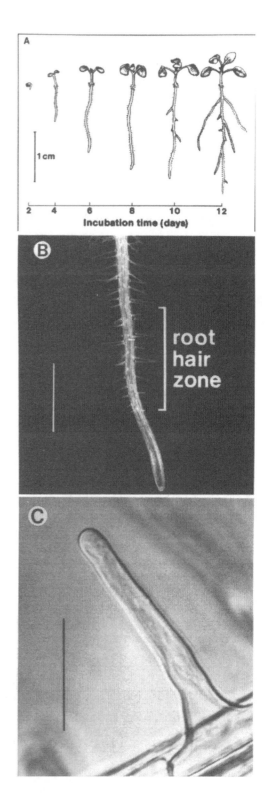

Fig. 9A–C. Growth of wild-type *Arabidopsis* roots and root hairs (Schiefelbein and Somerville 1990). **(A)** Drawing of *Arabidopsis* seedling growth from 2 to 12 days after germination. **(B)** Apex of 4-day-old root with the zone of root hair development indicated. *Bar* 1 mm. **(C)** Immature, elongating root hair examined with Normarski optics. *Bar* 50 μm

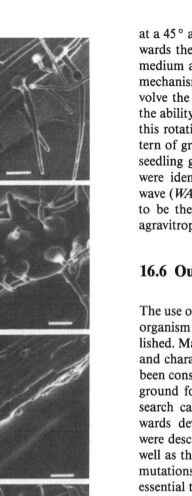

at a 45° angle. The root repeatedly grows towards the medium and then grows away from the medium after failing to penetrate the surface. The mechanism of this growth pattern appears to involve the rotation of the root tip as it grows, and the ability of the root tip to change the direction of this rotation. Mutations affecting this regular pattern of growth were selected by examination of M_2 seedling grown on this type of support. Six genes were identified which altered the pattern of the wave (*WAV1–WAV6*); two of these genes turn out to be the same as genes previously identified by agravitropic mutants (*aux1* and *agr1*).

16.6 Outlook

The use of *Arabidopsis thaliana* as an experimental organism for the study of genetics is well established. Many interesting mutants have been isolated and characterized, and a genetic/physical map has been constructed. These provide the essential background for present day research. Much of the research carried out recently has been devoted towards developing the tools and techniques that were described at the beginning of this chapter, as well as the isolation of many interesting classes of mutations affecting the different types of processes essential to the life of the plant. In the next decade we will see the application of these techniques to isolate interesting genes and the examination of the mechanisms that control plant processes at the molecular level.

16.7 Summary

Arabidopsis thaliana is a small flowering plant with a rapid life cycle. The haploid genome is only 10^8 bp and composed of five chromosomes. The genome has very few repetitive DNA sequences, which simplifies many aspects of both genetic and

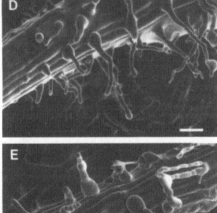

Fig. 10A–E. Scanning electron micrographs of wild-type and mutant *Arabidopsis* root hairs. *Bars* 50 µm. (Schiefelbein and Somerville 1990). (**A**) Wild type. (**B**) *rhd1*. (**C**) *rhd2*. (**D**) *rhd3*. (**E**) *rhd4*

molecular research. Hundreds of mutants have been isolated, with over 90 mutants being genetically mapped. These numbers are increasing rapidly each year. A physical map of *Arabidopsis* is being constructed which will be integrated with the existing genetic and RFLP maps. The resulting collection of mapped cosmid clones and RFLP markers will greatly simplify the isolation of interesting genes. Many different mutant phenotypes have been isolated that affect plant growth and development. T-DNA insertion mutagenesis and chromosome walking from defined RFLPs have proven to be powerful tools in the isolation of genes important for trichome development (*GL1*), ethylene binding (*ETR*), and floral morphogenesis (*AG*). The ability to transform *Arabidopsis* using *Agrobacterium* has allowed mutant complementation experiments to be performed in order to prove that these genes had been isolated. Mutants are also being isolated in *Arabidopsis* to aid in the study of complex biochemical pathways, such as lipid biosynthesis and photorespiratory metabolism, and to investigate host-pathogen interaction. Thus, while *Arabidopsis* may be a "useless" weed, the powerful experimental methods of genetics and molecular biology make this plant an attractive model system for the study of how plants grow and develop.

References

[1] Hitchcock CL, Cronquist A, Ownbey M, Thompson JW (1964) Vascular plants of the Pacific Northwest, part 2. University of Washington Press, Seattle, WA, pp 439–440

Redei GP (1970) *Arabidopsis thaliana* (L.) Heynh. A review of the genetics and biology. Bibliogr Genet 20:1–151

[2] Estelle MA, Somerville CR (1986) The mutants of *Arabidopsis*. Trends Genet 2:89–93

Meyerowitz E, Pruitt RE (1984) Genetic variation of *Arabidopsis thaliana*. California Institute of Technology, Pasadena, California

Redei GP (1975) *Arabidopsis* as a genetic tool. Annu Rev Genet 9:111–127

[3] Koornneef M, Van Eden J, Hanhart CJ, Stam P, Braaksma FJ, Feenstra WJ (1983) Linkage map of *Arabidopsis thaliana*. J Hered 74:265–272

Koornneef M (1990) Linkage map of *Arabidopsis thaliana* (2 n = 10). In: O'Brien SJ (ed) Genetic Maps; locus maps of complex genomes. Cold Spring Harbor Laboratory, Cold Spring Harborm, NY, pp 694–697

[4] Chang C, Bowman JL, DeJohn AW, Lander ES, Meyerowitz E (1988) Restriction fragment length polymorphism linkage map for *Arabidopsis thaliana*. Proc Natl Acad Sci USA 85:6856–6860

Nam H-G, Giraudat J, end Boer B, Moonan F, Loos WDB, Hauge BM, Goodman HM (1989) Restriction fragment length polymorphism linkage map of *Arabidopsis thaliana*. Plant Cell 1:699–705

[5] Leutweiler LS, Hough-Evans BR, Meyerowitz EM (1984) The DNA of *Arabidopsis thaliana*. Mol Gen Genet 194:15–23

Meyerowitz EM, Pruitt RE (1985) *Arabidopsis thaliana* and plant molecular genetics. Science 229:1214–1218

Pruitt RE, Meyerowitz EM (1986) Characterization of the genome of *Arabidopsis thaliana*. J Mol Biol 187:169–183

[6] Feldmann KA, Marks MD (1987) *Agrobacterium*-mediated transformation of germinating seeds of *Arabidopsis thaliana*: a non-tissue culture approach. Mol Gen Genet 208:1–9

Valvekens D, Van Montagu M, Van Lijsebettens M (1988) *Agrobacterium tumefaciens*-mediated transformation of *Arabidopsis thaliana* root explants by using kanamycin selection. Proc Natl Acad Sci USA 85:5536–5540

[7] Pruitt RE, Chang C, Pang PP-Y, Meyerowitz EM (1987) Molecular genetics and development of *Arabidopsis*. In: Loomis W (ed) Genetic regulation of development, 45th Symp Soc Dev Biol. Liss, New York, pp 327–338

[8] Errampalli D, Patton D, Castle L, Mickelson L, Hansen K, Schnall J, Feldmann K, Meinke D (1991) Embryonic lethals and T-DNA insertional mutagenesis in *Arabidopsis*. Plant Cell 3:149–157

Feldmann KA, Marks MD, Christianson ML, Quatrano RS (1989) A dwarf mutant of *Arabidopsis* generated by T-DNA insertion mutagenesis. Science 243:1351–1354

[9] Herman PL, Marks MD (1989) Trichome development in *Arabidopsis thaliana*. II. Isolation and complementation of the *Glabrous1* gene. Plant Cell 1:1051–1055

[10] Finkelstein R, Estelle M, Martinez-Zapater J, Somerville CR (1987) *Arabidopsis* as a tool for the identification of genes involved in plant development. In: Verma DPS, Goldberg R (eds) Plant gene research: temporal and spatial regulation of plant genes, vol 5. Springer, Berlin Heidelberg New York, pp 1–25

Somerville C (1989) *Arabidopsis* blooms. Plant Cell 1:1131–1135

[11] Jürgens G, Mayer U, Torres Ruiz RA, Berleth T, Miséra S (1991) Genetic analysis of pattern formation in the *Arabidopsis* embro. Development Supplement 1:27–38

Meinke DM (1982) Embryo-lethal mutants of *Arabidopsis thaliana*: evidence for gametophytic expression of the mutant genes. Theor Appl Genet 62:381–386

Meinke DM (1985) Embryo-lethal mutants of *Arabidopsis thaliana*: analysis of mutants with a wide range of lethal phases. Theor Appl Genet 69:543–552

Meinke DM (1986) Embryo-lethal mutants and the study of plant embryo development. In: Miflin BJ (ed) Oxford surveys of plant molecular and cell biology, vol 3. Oxford University Press, Oxford, pp 122–165

Schneider T, Dinkins R, Robinson K, Shellhammer J, Meinke

DW (1989) An embryo-lethal mutant of *Arabidopsis thaliana* is a biotin auxotroph. Dev Biol 131:161–167

[12] Bleecker AB, Estelle MA, Somerville C, Kende H (1988) Insensitivity to ethylene conferred by a dominant mutation in *Arabidopsis thaliana*. Science 241:1086–1089

Koornneef M, van der Veen JH (1980) Induction and analysis of gibberellin-sensitive mutants in *Arabidopsis thaliana* (L.) Heynh. Theor Appl Genet 58:257–263

Koornneef M, Jorna ML, Brinkhorst-van der Swan DLC, Karssen CM (1982) The isolation abscisic acid (ABA) deficient mutants by selection of induced revertants in nongerminating gibberellin sensitive lines of *Arabidopsis thaliana* (L.) Heynh. Theor Appl Genet 61:385–393

Koornneef M, Reuling G, Karssen CM (1984) The isolation and characterization of abscisic acid-insensitive mutants of *Arabidopsis thaliana*. Physiol Plant 61:377–383

Koornneef M, Hanhart CJ, Hilhorst HWM, Karssen CM (1989) In vivo inhibition of seed development and reserve protein accumulation in recombinants of abscisic acid biosynthesis and responsiveness mutants in *Arabidopsis thaliana*. Plant Physiol 90:463–469

Lincoln C, Britton JH, Estelle M (1990) Growth and development of the *axr1* mutants of *Arabidopsis*. Plant Cell 2:1071–1080

Mirza JI, Olsen GM, Iversen T-H, Maher EP (1984) The growth and gravitropic responses of wild-type and auxin-resistant mutants of *Arabidopsis thaliana*. Physiol Plant 60:516–522

[13] Herman PL, Marks MD (1989) Trichome development in *Arabidopsis thaliana*. II. Isolation and complementation of the *Glabrous1* gene. Plant Cell 1:1051–1055

Koornneef M (1981) The complex syndrome of *ttg* mutants. Arabidopsis Inf Serv 18:45–51

Marks MD, Feldmann KA (1989) Trichome development in *Arabidopsis thaliana*. I. T-DNA tagging of the *Glabrous1* gene. Plant Cell 1:1043–1050

[14] Bowman JL, Smyth DR, Meyerowitz EM (1989) Genes directing flower development in *Arabidopsis*. Plant Cell 1:37–52

Bowman JL, Smyth DR, Meyerowitz DM (1991) Genetic interactions among floral homeotic genes of *Arabidopsis*. Development 112:1–20

Irish VF, Sussex IM (1990) Function of the *apetala1* gene during *Arabidopsis* floral development. Plant Cell 2:741–753

Kunst L, Klentz JE, Martinez-Zapater J, Haughn GW (1989) *AP2* gene determines the identity of perianth organs in flowers of *Arabidopsis thaliana*. Plant Cell 1:1195–1208

Meyerowitz EM, Bowman JL, Brockman LL, Drews GN, Jack T, Sieburth LE, Weigel D (1991) A genetic and molecular model for flower development in *Arabidopsis thaliana*. Development (Suppl) 1:157–167

Yanofsky MF, Ma H, Bowman JL, Drews GN, Feldmann KA, Meyerowitz EM (1990) The protein encoded by the *Arabidopsis* homeotic gene *agamous* resembles transcription factors. Nature 346:35–39

[15] Okada K, Shimura Y (1990) Reversible root tip rotation in *Arabidopsis* seedlings induced by obstacle-touching stimulus. Science 250:274–276

Schiefelbein JW, Somerville C (1990) Genetic control of root hair development in *Arabidopsis thaliana*. Plant Cell 2:235–243

Chapter 17 Homeotic Genes in the Genetic Control of Flower Morphogenesis in *Antirrhinum majus*

Zsuzsanna Schwarz-Sommer, Heinz Saedler, and Hans Sommer

17.1 Introduction [1]

"Anyone who pays a little attention to the growth of plants will readily observe that certain of their external members are sometimes transformed, so that they assume − either wholly or in some lesser degree − the form of the members nearest the series".

J. W. von Goethe (1790)

During the development of a plant, differentiation of both vegetative and generative organs takes place. Vegetative growth of the shoot apex gives rise to leaves and other appendages. Floral organogenesis leads to the development of reproductive organs at sites at which shoots with leaves would normally develop during vegetative growth. For centuries observers of nature and scientists were puzzled by the close relation between these two types of differentiation processes. Goethe, for example, based his assumptions and conclusions on comparative morphological studies of normally and abnormally developing plants. According to him, the diverse appendages of a plant (thus also the organs of a flower) are due to metamorphosis (that is transformation) of a basic organ, the "ideal leaf". Goethe regarded the process of metamorphosis as being governed by a "sap" originating from basal parts of the plant, and becoming morphogenically distinct as it rises higher in the plant. Although at present we can formulate the problem in more precise terms of modern science, we are still far from being able to describe or understand the morphogenetic events in molecular terms.

This chapter deals with morphological, genetic and molecular aspects of flower development. At every level of observation we shall draw attention to possible conclusions and hypotheses which will become refined and substantiated as we proceed with our analysis of the problem of morphogenesis. Since molecular data are now available on homeotic genes controlling organogenesis in *Antirrhinum majus,* this chapter will emphasize this species in particular.

17.2 Normal Flower Development

17.2.1 General Rules of Normal Flower Development in Higher Plants

In higher plants, flower development follows a heritable pattern. Although the result of differentiation may appear different comparing, for example, the flower of a snapdragon with that of a rose or a tulip, certain universal rules are observed during the development process.

First of all, flower induction requires a stimulus (endogenous or exogenous) that triggers an irreversible transition in the vegetative shoot meristem to become reproductive tissue. This process, called floral evocation, is not the subject of this chapter.

Secondly, flowering always terminates vegetative and indeterminate growth. The slowly dividing meristematic cells at the center of the apex become consumed due to accelerated divisions during organ differentiation.

Thirdly, organogenesis is sequential and follows a precise spatial and temporal pattern in the initiation of organ primordia. Organs with identical functions are located in specific whorls. The sepal primordia in the first whorl are initiated first, followed by the primordia of petals, stamina, and carpels in their respective whorls. The temporal and spatial pattern of organogenesis and the number of individual organs within a whorl is species specific, and these differences contribute to differences in

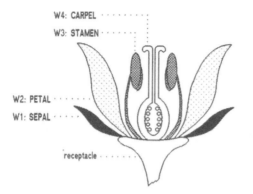

Fig. 1. Structure of the idealized flower of a higher plant. *W* whorl

the phenotypic appearance of the mature flower (shown as an idealized structure in Fig. 1).

Fourthly, the spatial and temporal developmental pattern (that is the plane and number of cell divisions and the pattern of cell elongation) of different organs differ from each other. This results in morphologically and functionally distinct structures. Thus stamen primordia, for example, differentiate into filaments carrying anthers in which pollen is produced, whereas carpel primordia differentiate into female organs, consisting of ovary, style and stigma, that are destined to produce ovules.

17.2.2 Determination of Organ Identity During Sequential Organ Initiation

Morphogenesis in plants occurs during the whole, occassionally decades long, lifetime of the organism. Thus it seems that organogenesis cannot be related to maternally determined positional information and, since plant cells do not change their position with respect to each other, cell migration may also not be involved in the developmental process. Therefore questions concerning the constancy of sequential developmental events are particularly intriguing. What is the mechanism by which cells in adjacent whorls interpret their position and differentiate precisely into the organ they are supposed to? In other words, what determines organ identity? And since the overall organization of flowers is

evolutionarily conserved in many species, are the genes controlling and governing it also conserved?

Several theoretical models have been proposed to answer these questions. One type of hypothesis assumes communication between organ primordia governed by the products of genes in one whorl which influence specific gene expression in the next. Communication could alternatively be governed by biophysical forces due to organ-specific cell divisions which produce specific stress patterns around developing primordia, and serve as the induction parameter for organs in the next whorl.

Recently, Holder suggested a different model, which is not necessarily incompatible with those outlined above. He assumes that a concentric field is established by gradients of morphogens (hormones or nutrients). This field has a specific information content and allows cells to define their position with respect to each other. The positional information is apt to change dynamically as morphogenesis proceeds. Organ-specific developmental patterns can be established as a result of the interpretation of the positional information at every stage of development, by specifically turning on genes in a particular type of organ.

In summary, there is little doubt that floral morphogenesis is a complex process involving the concerted action of environmental and genetic factors. The models may describe this complexity more or less satisfactorily, but the molecular mechanisms underlying the control of morphogenesis still remain a mystery.

17.2.3 Flower Morphogenesis in *Antirrhinum majus*

Floral morphogenesis can be followed easily in *Antirrhinum majus*. The lateral flower buds of the plant develop on an inflorescence from the base to the tip in a spiral order (Fig. 2). A single inflorescence carries up to 20 flowers at different stages of development, and on one plant several inflorescences can develop. The fully developed flower displays in zygomorphic (mirror-image) symmetry. This is best reflected by the structure of the (usually colored) corolla at the base of which petals are fused to form the corolla tube. The upper parts of

the petals compose the lobes, two of which form the upper lip, the three others the lower lip.

Floral induction is promoted by long irradiation periods combined with high light intensities, but flower formation only quantitatively depends on external factors. The flowering stimulus causes the vegetative shoot axis to become the inflorescence axis. On this, lateral bracts replace leaves and in the axil of the bracts, flower primordia develop instead of axillary shoots. The flower primordium displays an upper and a lower side due to its lateral position with respect to the inflorescence axis. The first visible morphological difference between vegetative and reproductive buds is the appearance of bilateral symmetry of the latter.

Initiation of sepal primordia in the first whorl is sequential, and the growth of individual sepals occurs according to different temporal patterns. The two lower sepals appear first and outgrow the three upper ones (Fig. 3a).

In the second whorl, five petal primordia develop at alternate positions with respect to the sepal primordia. Early in their development fusion of the basal parts of the young petals occurs (Fig. 3c–e), thus forming the corolla tube of the mature flower. Subsequently, the zygomorphic symmetry of the flower is initiated (Fig. 3c, d) due to differences in the individual development of the upper two and lower three petal primordia.

The five stamen primordia in the third whorl are initiated almost simultaneously with the primordia in the second whorl, at alternate positions with respect to these (Fig. 3b, c). Initially stamen primordia grow faster than petal primordia, with the exception of the fifth primordium, which is smaller than the others and retarded in its development, later becoming the degenerated staminodium (Fig. 2). The remaining stamen primordia, each of which becomes individually twisted during growth, differentiate into male organs. These consist of

Fig. 2. The inflorescence and the flower of *Antirrhinum majus*. The mature wild-type inflorescence at the *top* carries several zygomorphic flowers, which develop from the base to the tip along the inflorescence axis. The cross-section of a 5 mm-long immature flower bud shows the organization of organs in whorls. *Bar* 5 mm. *b* bract; *s* sepal; *p* petal; *st* stamen; *std* stamenoid; *g* gynoecium

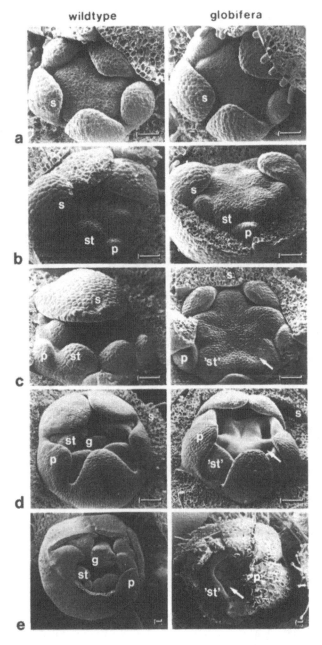

wildtype   globifera

a
b
c
d
e

Fig. 3a–e. Scanning electron micrographs of developing wild-type (*left panel*) and *globifera* mutant (*right panel*) flower buds from young *Antirrhinum majus* inflorescences. The photographs in each row (**a–e**) represent individual buds at a similar stage of development. The maturity of the buds increases *from top to bottom* (see Fig. 2). Organs in the first (**b–e**) and second (**e**) whorl were removed to detect inner whorl organs. Notice that the first visible change in development of globifera buds is revealed by sloping of the stamen primordia which then develop in a carpelloid manner. The *arrow* in row **d** points to developing locules which become filled with ovules as shown in row **e**. *Bar* 50 µm in the rows **a–c** and 100 µm in rows **d** and **e**. *o* ovules; for further abbreviations see Fig. 2 (Sommer et al. 1990, see [4])

floral meristem is consumed, and due to the initially accelerated growth of the carpels, an indentation in the center of the flower becomes visible. The developing carpels fuse at their lateral edges, which then develop toward the center and build a septum (Fig. 3d, e), later giving rise to a dimeric placenta. Still later ovules develop at this margin. The basal and outer parts of the fused carpels contribute to the ovary. The upper part differentiates into a style and a stigma.

Different organs differ not only in their overall morphology but also in that they are composed of different types of cells contributing to functionally different tissues. For example, only the female organ carries pipillate stigmatic cells at its tip that become functionally important during pollination. The epidermal layer of maturing organs also reveals some characteristic features, e.g., different patterns of hairs (trichomes). Whereas young petals possess an almost smooth outer surface, sepals are hairy as are also the basal parts of the gynoecium (e.g., see Fig. 6). In contrast, filaments of the male organs carry only few hairs on their outer surface.

17.3 Abnormal Flower Development[2,3]

17.3.1 Analysis of Homeosis Can Help to Unravel the Rules Underlying Morphogenesis

The reproducible sequence of morphogenetic events during flower organogenesis seems to be primarily based on temporal and spatial activity of genes. Altered or abolished gene action results in disruption of the sequence of events and leads to

long filaments carrying the anther composed of two dorsal and two ventral pollen sacks (see wild-type flower in Fig. 6).

The primordia of two carpels are initiated comparatively late during floral morphogenesis (Fig. 3c). During carpel development the rest of the

abnormal development. But such alterations can also be induced by manipulation of external and internal factors. A great deal of information on physiological factors interfering with flower development is available. Virtually any environmental factors (light, temperature, etc.) and many substances (hormones, nutrients, etc.) can cause some type of developmental abnormality often revealed as a homeotic alteration. In these the positional information is misinterpreted such that organs, or homologs of them, develop in the "wrong" place. But the experimental results are often contradictory and confusing, and hence do not contribute much to the understanding of the rules.

The understanding of animal development was greatly facilitated by the analysis of inherited abnormalities due to homeotic mutations. In plants as well, homeotic mutants exist in which the organs appear at the wrong place, mosaic organs develop, or the number and shape of organs is heritably altered. In these mutants cells may misinterpret their position and thus their developmental fate may be changed.

Observations on abnormal flowers have a long tradition, perhaps because deviations of such beautiful and regular structures are eye-catching. During the past centuries many scientists were aware of the value of such deviations for the understanding of normal development.

Antirrhinum majus was among the first plants in which floral abnormalities were shown to be heritable. As early as the beginning of this century, mutants with homeotic features were isolated and morphologically characterized. The discovery of such mutants was facilitated by several favourable properties of *Antirrhinum*. One advantage is the high mutation rate in this organism, which allows mutants to be isolated from relatively small populations. Today we know that several active transposable elements are responsible for the high rate of mutational events observed in *Antirrhinum*. In addition, it is possible to propagate *Antirrhinum* vegetatively. Because homeotic floral mutations frequently affect reproductive organs, vegetative propagation permits such sterile mutants to be maintained as homozygotes.

In *Antirrhinum* many genes are known which interfere with cell differentiation of different stages in flower development. In some mutants, the initiation and development of flowers is arrested at an early stage; in others, the symmetry or the form of floral organs is altered. In this chapter, however, we shall focus only on those homeotic genes which control organogenesis in the four whorls.

17.3.2 Three Types of Homeotic Genes Control the Establishment of Flower Organ Identity

In *Antirrhinum majus* all homeotic mutants which display incorrect organs can be assigned to three different types: type 1, in which the first and second whorls (the perianth of the flower) are affected; type 2, in which the third and fourth whorls (the reproductive organs) are altered with a concomitant increase in the number of organs; and finally type 3, in which the second and the third whorl is transformed (Figs. 4 and 5). These three types of mutants are also found in *Arabidopsis* (Fig. 4), and some of them in other species as well. Common features of these mutants provide the basis for some general conclusions.

Mutations in Homeotic Genes Affect Organs of Two Adjacent Whorls. To alter the fate of organs in two adjacent whorls, expression of homeotic genes would have to be established in the meristematic progenitor cells of the future organs, prior to the development of the organ primordia. Thus homeotic genes could be involved in both the establishment of positional information and in its interpretation. However, in homeotic mutants observed so far, organ primordia are formed at the correct position with respect to wild type, but these primordia differentiate into incorrect organs. Development of these transformed organs follows the time course of the genuine wild-type organs in the affected whorl. Further, the kinds of concomitant homeotic transformations of organs is limited. For example, sepal-like development of petals (sepalody) is always accompanied by feminization of stamens (carpellody), although stamens have the potential to undergo petal-like development (petalody) as well. This may indicate that positional information, at least in part, is established before homeotic genes act.

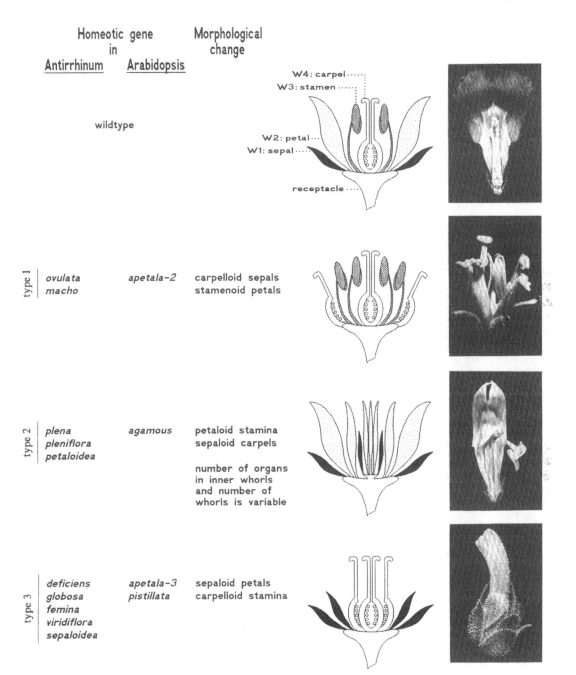

	Homeotic gene in Antirrhinum	Arabidopsis	Morphological change
wildtype			
type 1	ovulata macho	apetala-2	carpelloid sepals stamenoid petals
type 2	plena pleniflora petaloidea	agamous	petaloid stamina sepaloid carpels number of organs in inner whorls and number of whorls is variable
type 3	deficiens globosa femina viridiflora sepaloidea	apetala-3 pistillata	sepaloid petals carpelloid stamina

W4: carpel
W3: stamen
W2: petal
W1: sepal
receptacle

Fig. 4. Three types of homeotic mutants of genes controlling the identity of floral organs in *Antirrhinum* (Stubbe 1966, see [2]; Carpenter et al. 1990; Schwarz-Sommer et al. 1990) and *Arabidopsis* (Bowman et al. 1989). The flowers of representative *Antirrhinum* mutants are shown at the right. The lower corolla lobe of the wild-type and *plena* flower was removed (Schwarz-Sommer et al. 1990)

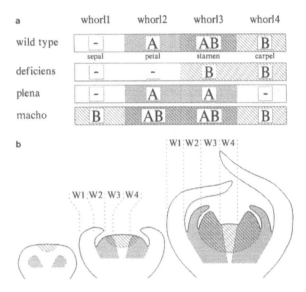

Fig. 5a, b. Theoretical scheme for determination of wild-type and homeotic mutant organ identity. **a** The two developmental pathways *A* and *B* include several homeotic genes whose hierarchical relation is not known. The pathways are assigned according to the pattern of expression of known homeotic genes. For example, pathway *A*, which is affected by mutation in *deficiens*, is induced in the second and third whorls, where *deficiens* is expressed. The scheme indicates the pathways that are induced, and the corresponding organs that develop in wild-type and in mutant flowers. Note that sepals are considered floral organs. The lack of induction of either of the two pathways is not meant to suggest that sepal development is a continuation of vegetative growth. **b** Schematic longitudinal sections of developing flowers at the stage of establishment of whorls. Two *different shaded areas* show the distribution of the two hypothetical morphogens when their concentration is above the threshold levels of their respective receptors. To induce pathway *A*, an eccentric gradient of a morphogen is proposed, while a concentric gradient of a different morphogen could induce pathway *B*. Notice that for stamen development the two gradients overlap in the third whorl only (After Schwarz-Sommer et al. 1990, courtesy P. Huijser)

Several Nonallelic Mutants Display Similar Phenotypes. The existence of several genes producing the same homeotic alteration upon mutation indicates that these genes may control each other or that they interact with each other in the control of other (target) genes.

Mutants of Two Different Plant Species Display Similar Homeotic Alterations. A comparison of homeotic mutants of *Antirrhinum* and *Arabidopsis* shows that at least some of the genes may have homologous functions in the determination of organ identity and thus may also be conserved in structure. The general rules mentioned to characterize the sequential differentiation process during floral organogenesis can perhaps be related to a structural and functional conservation of these genes.

But *Antirrhinum* and *Arabidopsis* differ in the number of genes which display a particular type of homeotic alteration upon mutation. Furthermore, certain types of organ transformations such as leaflike development (phyllody) have not been observed in *Antirrhinum*, but were frequently obtained in mutant alleles of the *apetala-2* gene in *Arabidopsis* (Fig. 4). This may reflect differences in the complexity of mechanisms directing organogenesis in these species.

Homeotic Phenotypes Are Influenced by Environmental Factors. The phenotypes of mutants listed in Fig. 4 are variable, because they are influenced by environmental conditions and also by the position of individual flowers on the inflorescence. This perhaps reflects the known but not well understood influence of external and internal factors on flower development.

17.3.3 Differential Induction of Two Pathways Determine Four Organs: a Model

Genetic analysis and morphological inspection of the three types of floral homeotic genes permit to propose a scheme (Fig. 5) which describes how homeotic genes could direct floral organogenesis. A model of this kind should explain how four different organs are generated in the four whorls and how known single gene mutations confer homeotic alterations on organs in two adjacent whorls.

The model postulates that after floral evocation, induction of only two developmental pathways is sufficient to define four different whorls of organs. According to the scheme in Fig. 5, stamen development, for instance, is initiated by the early and simultanous induction of the two pathways in the third whorl, and only there. If this is true, mutations in early-acting genes in either pathways will result in homeotically altered stamen development.

In fact, staminal carpellody or petalody is frequently observed in nature. In contrast, staminal sepalody is a rare event, but, as predicted by the scheme, can be induced by double mutations.

Due to recessive mutations in homeotic genes belonging to the respective pathways, two basic types of homeotic mutations, type 2 and type 3, can be generated. Type 1 mutations may be due to the mutationally established expression of pathway B in the first and second whorls. A possible epistatic relation between pairs of the three types of homeotic genes, that is, the mutual control of their expression, can be studied by generating double mutants with pairs of the three types of homeotic genes. For example, in a *globifera/macho* double mutant carpels are formed in all four whorls indicating that these two genes are expressed independently. Such a simple combinatorial phenotype was observed in the *agamous/apetala-3* double mutant of *Arabidopsis* which displays sepals only. These observations suggest that expression of a particular homeotic gene in one development pathway may be independent of the expression of homeotic genes in the other pathway.

In summary, the scheme in Fig. 5 proposes that homeotic genes are regulatory members of developmental pathways whose induction in different whorls results in differentiation of four different types of floral organs. However, it is still not clear which mechanism is responsible for the induction of the pathways.

17.4 The Role of the *deficiens* Gene in Floral Organogenesis[4]

17.4.1 Morphological Observations on Three Morphoalleles

Mutants of the *deficiens* gene were already isolated and genetically and morphologically characterized at the beginning of this century. It is a homeotic gene whose mutants display sepaloid petals and carpelloid stamens. Three classical mutant alleles are known which differ morphologically in the extent of organ transformation (Fig. 6). These so-called morphoalleles can be used to correlate mor-

phological features with molecular functions (see below).

deficiens[globifera] exhibits the most extreme phenotype. The petals in the second whorl are almost indistinguishable from sepals. The stamens in the third whorl are feminized in that they develop as carpels and bear fertile ovules in the mature flower. The genuine female organ either does not develop at all, or it fuses with the third whorl organs. Analysis of young *globifera* flower buds with the scanning electron microscope indicates that the initiation of organ primordia is not affected by the mutation (Fig. 3).

The flowers of morphoalleles like *nicotianoides* and *chlorantha* display chimeric features. Their second whorl organs are sepaloid; that is, they are reduced in size and greenish in color compared to wild-type petals. Feminization of stamens is reflected by various types of chimeric structures. For example, stamens of *chlorantha* strongly resemble wild-type organs, but bear ovules. On the other hand, stamens of *nicotianoides* display carpellody, as indicated by many morphological characters (broadened and shortened filaments, stigmatic tissue, hairs), but no ovules are formed. Furthermore, the grade of feminization of upper stamens can be stronger than that of lower stamens. These diverse morphological alterations in different organs and also in different parts of organs indicate that the *deficiens* function is diversified and specified in different tissues by modifying factors. Interestingly, the morphological characteristics of the morphoalleles can be influenced by changing environmental conditions and by introducing the alleles into different genetic backgrounds.

17.4.2 Instability of *deficiens*[globifera]

Transposable elements are genetic entities which can change their location within the genome. They can be inserted into a gene, and they can also excise and subsequently insert again somewhere else. The insertion of a transposon generates a mutation in the affected gene. Subsequent excision may restore gene activity (reversion). Genes whose expression is altered by the insertion of a transposon are thus genetically unstable. Transposons can be utilized to

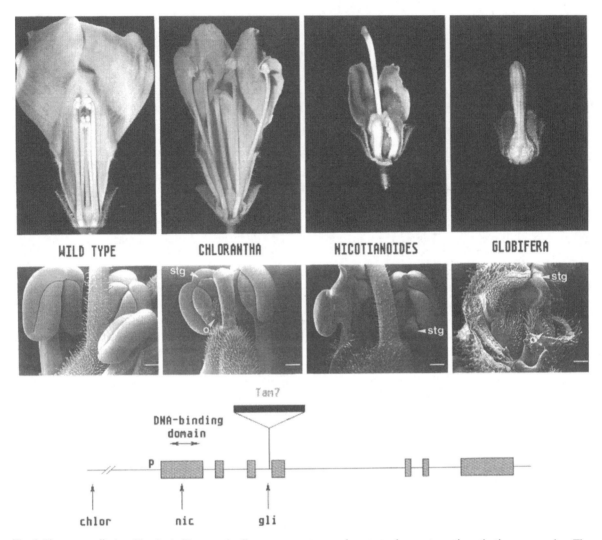

Fig. 6. Phenotypes displayed by *Antirrhinum majus* flowers carrying mutant alleles of the *deficiens* locus. The genotype of each flower is shown *below the photographs*. To reveal the structure of the inner whorls, some sepals and the lower lobe of the corolla were removed. The photographs show the real differences in the size of wild-type and mutant flowers. The scanning electron micrographs *in the middle* show the structure of inner whorl organs (stamens and gynoecium) of young flower buds of wild-type and mutant plants. *stg* = stigmatic tissue; *o* ovules. The structure of the *deficiens* transcription unit is shown at the *bottom*. *Shaded boxes* represent exons and *P* is the promoter of the gene. The first exon contains a conserved DNA-binding domain shown in detail in Fig. 8. *Arrows* point to the site of mutation determined in the *deficiens* alleles. Tam7 is a transposable element whose excision is responsible for the genetic instability of the *globifera* allele (Fig. 7)

isolate genes by transposon tagging and also to generate new mutants and mutant alleles. Unstable alleles and their germinal revertants are particularly useful for the cloning and identification of genes.

Excision may occur at virtually any time during development and is revealed by somatic sectors of revertant tissue in the mutant (Fig. 7). The mosaic character of mutant and revertant sectors provide some insight into the mode of action of the unstable gene. The analysis of somatic excision events in the unstable *deficiens*globifera mutant led to the following results: (1) Excision events that occur very late in development of the sepaloid petals restore petaloid features in a clonal manner indicating that

Fig. 7. Genetic instability of the *globifera* mutant. The plant depicted *at the left* displays somatic reversion events resulting in three different types of inflorescence. The inflorescence *at the right* includes flowers with wild-type morphology whose progeny also will be (wild-type) revertant. The inflorescence *at the left* carries one such revertant flower and *globifera* buds as well. The *globifera* inflorescence *in the middle* shows late somatic events resulting in petaloid structures. The *middle panel* shows a *globifera* flower bud with a chimeric organ in the second whorl. The plant *in the right panel* exhibits a revertant flower with almost perfect restoration of petals and petaloid organs in the second whorl. The chimney-like structure in place of stamens indicates that the third whorl is hardly affected by this reversion event

deficiens acts cell-autonomously. (2) Due to early, but still not heritable, somatic excision, single second whorl organs may display a sepaloid and a petaloid sector separated by a sharp boundary which extends from the base of the chimeric organ to its tip (Fig. 7). This may indicate that cell groups within a primordium are autonomously differentiating into a given part of the organ. (3) Early excisions may almost completely restore petals in the second whorl without simultaneously affecting the developmental fate of the feminized third whorl organs (Fig. 7). Furthermore, stamens or stamenoid characters in the third whorl are never restored without concomitant restoration of organs in the second whorl. This indicates that stamen development, in contrast to petals, requires early *deficiens* gene function.

In conclusion, early and late *deficiens* gene functions are dissectable and are different in the second and third whorls. During early petal development *deficiens* perhaps controls cell divisions, and thus contributes to the form of mature petals. Loss of this early function has a significant effect on the form of petals. Loss of late *deficiens* function, on the other hand, is reparable, because petaloid characters in the sepaloid second whorl reappear when *deficiens* expression is restored. In contrast, stamen differentiation in the third whorl depends on early *deficiens* function which cannot be compensated later in development. These observations support the model (Fig. 5) which predicts early and concomitant induction of two pathways for stamen development and induction of a single pathway for petal development.

17.4.3 Homeotic Genes Encode Transcription Factors: the MADS-Box

Recently two of the homeotic genes involved in the control of floral organogenesis have been cloned. Remarkably, DEF A, the protein encoded by *deficiens* in *Antirrhinum* and also AG, the protein product of the *agamous* gene in *Arabidopsis*, reveal a high degree of homology to the conserved DNA-binding and dimerization domain of two known transcription factors, SRF and MCM1 (Fig. 8). The serum response factor (SRF) is essential for the serum-inducible transcriptional control of the *c-fos* protooncogene in mammals. In yeast the MCM1 protein is essential for the control of a- and alpha-cell-type specific genes. Thus it seems that these four proteins participate in different organisms in the determination of the fate of cells during differentiation. In analogy to the homeobox domain, common to several proteins controlling developmental processes in animals, the conserved domain of MCM1, AG, DEF A, and SRF was designated MADS-box.

Whether the DEF A protein can bind to DNA is not known. But structural analysis of the *deficiens*[nicotianoides] allele, in which a single amino acid exchange causes the mutant phenotype, suggests that this domain is essential for the wild-type DEF A function. The observation that DEF A is a nuclear protein also supports the notion that it functions as a transcription factor.

17.4.4 Homology Between Floral Homeotic Genes

Both *deficiens* and *agamous* are involved in the determination of floral organ identity, although mutations in these genes result in two different types of homeotic alterations (Fig. 4). Since several other homeotic genes are known in *Antirrhinum*, the question arose as to whether some of them might also possess a MADS-box.

Ten additional MADS-box genes were detected in *Antirrhinum*, when the conserved domain of DEF A was used as a molecular probe to screen a cDNA library. All are expressed in floral organs and some of them are expressed in vegetative tissues as well. Two of the cDNAs could be assigned to known floral homeotic genes, *squamosa* and *globosa*. While *squamosa* is involved in the establishment of flower primordia after evocation (that is, it controls an early event in flower morphogenesis), the phenotype of the *globosa* mutant (Fig. 4) suggests that it participates in the control of petal and stamen development, similar to the *deficiens* gene.

The analysis of the MADS-box homologs is not completed yet, but one can assume that more of them will turn out to represent floral or vegetative morphogenic genes. Families of MADS-box genes were detected in *Arabidopsis*, tomato, yeast, mammals, frogs and flies. These findings may be significant for the question of the evolution of regulatory

```
        - - -DNA-binding───────────────────────────────────────
                        ───dimerization───────────────

  SRF     RVKIKMEFIDNKLRRYTTFSKRKTGIMKKAYELSTLTGTQCLLLVASETGHVYTFATRK
  MCM1    RRKIEIKFIENKTRRHVTFSKRKHGIMKKAFELSVLTGTQVLLLVVSETGLVYTFSTPK

                                    *

  DEF A   RGKIQIKRIENQTNRQVTYSKRRNGLFKKAHELSVLCDAKVSIIMISSTQKLHEYISPT
  AG      RGKIEIKRIENTTNRQVTFCKRRNGLLKKAYELSVLCDAEVALIVFSSRGRLYEYSNNS

  cons       R KI I   IEN   R   TF KRK GI KKA ELS L  A    LIV S    L   F

  cons^plant  RGKI IKRIEN TNRQVTY KRRNGL KKA ELSVLCDA ·VSLIV SS  KL EY    S
```

Fig. 8. Conservation of amino acids in the putative DNA-binding and dimerization domains (MADS-box) of proteins involved in the control of differentiation in mammals (*SRF*), yeast (*MCM1*), and plants (*DEF A* and *AG*). Conserved positions are typed in *bold face letters* and homologous exchanges by *light letters*. Two consensus sequences derived from all four proteins (*cons*) and from the two plant proteins only (*cons*[plant]) are shown *at the bottom*. The *asterisk* indicates a point mutation $(G-D)$ detected in the *deficiens*[nicotianoides] allele. The recognition site for calmodulin-dependent protein kinases is *underlined*

mechanisms for differentiation processes in diverse organisms.

17.4.5 Homeotic Genes Are Upregulated in Two Whorls: Transcriptional Control

Temporal and Spatial Expression Patterns. The *deficiens* gene is flower-specific, since it is expressed only in the flower and repressed or not induced in vegetative tissues. In situ hybridization experiments have revealed that *deficiens* is expressed most strongly in petals and stamens, in the organs that are homeotically transformed in *deficiens* mutants (Fig. 9). Similarly, mutation in the *agamous* gene in *Arabidopsis* results in petaloid stamens and sepaloid carpels and, accordingly, the wild-type gene is expressed at highly elevated levels in stamens and carpels. But the expression of homeotic genes in the wild type is not strictly confined to the organs affected in the mutant. Northern blot analysis of dissected *Antirrhinum* organs, for example, revealed that *deficiens* is weakly expressed also in sepals and carpels.

Deficiens gene expression is induced early in flower primordia and is virtually constant during flower development. This indicates that DEF A function is not only required as an early developmental switch, but that it is also essential late in differentiation. The function of the DEF A protein in the control of late developmental events is not clear. As already mentioned, sepaloid development of organs in the second whorl of the *globifera* mutant can be changed to petal-like development by late somatic reversion of the *deficiens* gene. The morphological variability of third whorl organs in the *deficiens* morphoalleles suggests that *deficiens* functions are also required late in stamen differentiation, subsequent to initiation of stamen development due to early *deficiens* function.

Genetic Control of deficiens Gene Expression. Structural analysis of the *chlorantha* allele of *deficiens* (Fig. 6) indicated that *deficiens* is under genetic control. Due to a small rearrangement in the promoter region expression of *deficiens* in *chlorantha* is reduced to 10–15% of that of wild type. This

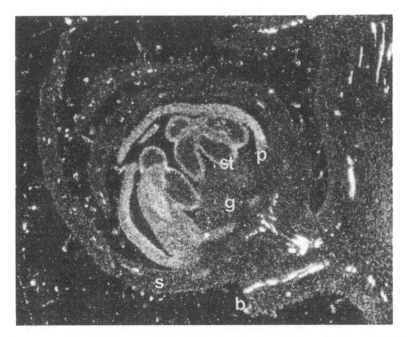

Fig. 9. In situ expression pattern of the *deficiens* gene. A longitudinal section through a young bud was hybridized with a molecular probe of the *deficiens* gene. For abbreviations see Fig. 2 (Courtesy P. Huijser)

reduction of expression of the gene is restricted to petals and stamens. Thus the deletion seems to affect the *cis*-acting binding site of a transcription factor which upregulates *deficiens* expression specifically in the two organs.

SRE (serum response element), the DNA-binding site of SRF, is functionally and structurally related to the binding sites of the MCM1 protein. Interestingly, SRE-homologous sequences have been found upstream of the promoters of *deficiens* and *agamous*. This could indicate that some of the other MADS-box proteins, for example those encoded by the *squamosa* or *globosa* genes, might control the spatial and temporal expression of *deficiens*. Alternatively, or in addition, *deficiens* and *agamous* gene expression might be autoregulated.

At this point one should recall that *deficiens* function is essential for the interpretation of positional information during floral organogenesis, but that the function is not necessarily required for the establishment of positional information. Thus the question is: what is the mechanism of genetic control of *deficiens* gene expression and that of other homeotic genes as well? In terms of the scheme developed to explain the role of homeotic genes in floral organogenesis (Fig. 5), one must ask what mechanism causes the induction of the two pathways which include the homeotic genes as regulatory switches.

At present no experimental data are available that might provide an answer to this question. But one may adopt Holder's field theory, mentioned above, to hypothesize that gradients of diffusible morphogenes may establish positional information (Fig. 5). This might then cause differential induction of two developmental pathways. Differential induction of each pathway could be accomplished by interaction of receptors with morphogenes in the respective fields, in response to the concentration of a particular morphogen. Below certain threshold levels the pathways are not induced (or remain repressed), and above specific threshold levels they become induced (or their repression abolished). For the concomitant induction of both pathways, leading to stamen development, the morphognes in the two gradients must simultaneously attain threshold levels of their corresponding receptors.

17.4.6 Posttranslational Processes Specify Homeotic Gene Functions

The simultaneous expression of homeotic genes in two different organs obviously contradicts the observation of distinct functions in each organ. Consequently, one has to postulate that the organ-specificity of homeotic genes is the result of posttranscriptional and posttranslational processes that determine their function in different organs.

The floral homeotic genes known so far are transcription factors and hence exert their function in the nucleus. Little is known about mechanisms which determine the nuclear transport of proteins subsequent to their synthesis in the cytoplasm, but any alteration in the rate of this putative transport process would influence the function of a homeotic gene.

Posttranslational events such as cooperative interactions with other proteins or chemical modification could also modulate the temporal and spatial pattern of homeotic gene function.

Interactions with Other Proteins. One striking feature of the DEF A protein is its small size (26.2 kDa) and its lack of structural properties (for example an acidic domain) necessary for functioning in activation of transcription. It is conceivable that this function is provided by other proteins which interact with DEF A and whose expression may be subject to temporal and spatial control. Cooperative interactions, for example, could modify the DNA-binding specificity of DEF A and modulate the stability of the complex. Experimental evidence for the existence and the nature of such accessory proteins is not available yet. However, some morphological and genetic data discussed in this chapter point to the existence of such modifying factors that determine the specificity of the pattern of *deficiens* gene functions.

Phosphorylation. DEF A is a phosphorylated nuclear protein, but the functional significance of its phosphorylation for DNA-binding is not known. Preliminary results have shown that the protein occurs in different states of phosphorylation in wild-type flowers. The relative abundance of the different forms seems to be altered in the *deficiens* morphoalleles *chlorantha* and *nicotianoides*. These results, once verified, could become very interesting for the following reasons. Firstly, the recognition sequence of calmodulin-dependent protein kinases is found in the DNA-binding domain of DEF A (Fig. 8), indicating that DEF A activity may be controlled by phosphorylation. Calmodulin-dependent protein kinases could mediate environmental influences on development via changes in Ca^{2+} concentration as a result of environmental and hormonal changes. Secondly, the alteration of the DEF A phosphorylation pattern in *deficiens* morphoalleles may indicate that DEF A is involved in the transcriptional control of a protein kinase. This protein kinase may not only phosphorylate DEF A but also other proteins as well.

Interestingly, the functions of SRF and MCM1 are also specified by interactions with other proteins and by phosphorylation. Hence one may speculate that the conserved structural homology between DEF A and SRF/MCM1 is not the only common feature of MADS-box proteins. Possibly also the mechanisms which regulate and establish their cell-specific functions are similar.

17.5 Outlook

One can expect rapid progress in the study of homeotic genes, their interactions with each other, and the regulation of their expression. This will contribute to our understanding of the basic mechanisms controlling flower development, but many questions will still remain open.

Nothing is known, for example, about genes (so-called targets) whose expression and function is regulated by homeotic genes. The same is true for genes that control expression of homeotic genes. To make progress in this field molecular studies must be complemented by genetic analyses. Such genes, for instance, can be identified, and perhaps isolated, if a search for modifiers and suppressors of homeotic mutations is carried out by genetic methods.

We often pointed to the lack of information about the establishment of positional information for subsequent organogenesis. The postulation of diffusible morphogens is not yet supported by experimental data. Not only the chemical nature of such factors is unknown; it is also not clear which part of a developing flower primordium can serve as the source of such morphogens and how morphologically distinguishable structures contribute to the establishment of the hypothetical gradients. Thus studies on the physiology of flowering and more detailed analysis of the morphology of developing flowers must complement molecular and genetic studies.

Similarities in processes of flower development among different plant species has often been emphasized. The homology between homeotic genes controlling organogenesis in *Antirrhinum* and *Arabidopsis* supports the current view that studying one plant species enhances our understanding of development of others as well. This may be true for the very basic molecular mechanisms which govern developmental processes by transcriptional control. But distinctive differences may also exist, such as the number and behavior of genes involved in a particular process, or the absence of certain types of homeotic transformations in some plant species. Therefore, it is possible that the complexity of processes directing flower development is different when comparing different plant species.

17.6 Summary

After evocation, floral development begins with the sequential appearance of flower and floral organ primordia and ends with the mature flower composed of structurally and functionally distinct organs. In this developmental process, first positional information is established, which allows cells to define their position with respect to each other. Subsequently, this positional information is interpreted by differential induction of a set of genes. Mutations in some of these genes lead to incorrect interpretation of the positional information and result in abnormal development. The function of such homeotic genes is revealed by the development of homeotically transformed organs in their mutants.

The known homeotic genes seem to be members of two developmental pathways whose differential induction during morphogenesis leads to the development of four different types of organs. How the positional information that controls the induction of these two pathways is established remains a matter of speculation.

Some floral homeotic genes belong to a family of genes encoding transcription factors with a conserved domain called the MADS-box. MADS-box proteins in plants, mammals, and yeast participate in the control of developmental processes, and exhibit several other common properties in addition to conserved structural features. The genes for these proteins are expressed constantly in the cells whose developmental fate they control and the proteins themselves all require posttranslational modification (such as chemical modification and intractions with accessory proteins) to determine the specificity of their spatial and temporal function. Whether these similarities reflect evolutionary conservation of the control of differentiation remains an open question.

References

[1] Goethe JW von (1790) Versuch die Metamorphose der Pflanze zu erklären. Translation: Arber A (1946) Goethe's botany. Chron Bot 10:67–124

[2] Steeves TA, Sussex IM (1989) Determinate shoots: thorns and flowers. In: Patterns in plant development. Cambridge University Press, Cambridge, pp 176–202

Stubbe H (1966) Genetik und Zytologie von *Antirrhinum majus* L. sect. *Antirrhinum*. VEB Gustav Fischer Verlag, Jena

Walbot V (1985) On the life strategies of plants and animals. Trends Genet 1:165–169

Wardlaw CW (1967) The inflorescence and the flower. In: Morphogenesis in plants. Methuen and Co Ltd, London, pp 307–355

Weberling F (1989) Morphology of flowers and inflorescences. Cambridge University Press, Cambridge

[3] Bowman JL, Smyth DR, Meyerowitz EM (1991) Genetic interactions among floral homeotic genes of *Arabidopsis*. Development 112:1–20

Carpenter R, Luo D, Doyle S, Coen ES (1990) Floral homeotic mutations produced by transposon-mutagenesis in *Antirrhinum majus*. Genes and Dev 4:1483–1493

Coen ES (1991) The role of homeotic genes in flower development and evolution. Ann Rev Plant Physiol Plant Mol Biol 42:241–279

Coen ES, Carpenter R (1986) Transposable elements in *Antirrhinum majus*: generators of genetic diversity. Trends Genet 2:292–296

Green P (1989) Shoot morphogenesis, vegetative through floral, from a biophysical perspective. In: Lord E, Bernier G (eds) Plant reproduction: from floral induction to pollination. The American Society of Plant Physiologists Symposium Series, vol I, pp 58–75

Haughn GW, Sommerville C (1988) Genetic control of morphogenesis in *Arabidopsis*. Dev Genet 9:73–89

Heslop-Harrison J (1963) Sex expression in flowering plants. Brookhaven Symp Biol 16:109–125

Hoder N (1979) Positional information and pattern formation in plant morphogenesis and a mechanism for the involvement of plant hormones. J Theor Biol 77:195–212

Kinet J-M, Sachs RM, Bernier G (1985) The development of flowers. In: The physiology of flowering, vol III. CRC Press, Boca Raton, pp 1–274

Meyer V (1965) Flower abnormalities. Bot Rev 32:165–195

Meyerowitz EM, Smyth DR, Bowman JL (1989) Abnormal flowers and pattern formation in floral development. Development 106:209–217

Sattler W (1988) Homeosis in plants. Am J Bot 75:1606–1617

Schwarz-Sommer Zs, Huijser P, Nacken W, Saedler H, Sommer H (1990) Genetic control of flower development by homeotic genes in *Antirrhinum majus*. Science 250:931–936

[4] Cohen P (1988) Protein phosphorylation and hormone action. Proc R Soc Lond B 234:115–144

Herskowitz I (1989) A regulatory hierarchy for cell specialization in yeast. Nature 342:749–757

Hunter T (1987) A thousand and one protein kinases. Cell 50:823–829

Mitchell PJ, Tjian R (1989) Transcriptional regulation in mammalian cells by sequence-specific DNA binding proteins. Science 245:371–378

Norman C, Runswick M, Pollock R, Treisman R (1988) Isolation and properties of cDNA clones encoding SRF, a transcription factor that binds to the *c-fos* serum response element. Cell 55:989–1003

Passmore S, Elble R, Tye B-K (1989) A protein involved in yeast a transcriptional enhancer conserved in eukaryotes. Genes Dev 3:921–935

Sommer H, Hehl R, Krebbers E, Piotrowiak R, Lönnig W-E, Saedler H (1988) Transposable elements of *Antirrhinum majus*. In: Nelson O (ed) Proc Int Symp on Plant transposable elements. Plenum, New York, pp 227–235

Sommer H, Beltran J-P, Huijser P, Pape H, Lönnig W-E, Saedler H, Schwarz-Sommer Zs (1990) *deficiens*, a homeotic gene involved in the control of flower morphogenesis in *Antirrhinum majus*: the protein shows homology to transcription factors. EMBO J 9:605–613

Yanofsky MF, Hong M, Bowman JL, Drews GN, Feldmann KA, Meyerowitz EM (1990) *agamous*, and *Arabidopsis* homeotic gene whose product resembles transcription factors. Nature 346:35–39

Chapter 18   The *Rhizobium*-Legume Symbiosis

ALLAN DOWNIE and NICHOLAS BREWIN

18.1 Introduction[1]

Nitrogen is an essential nutrient. As N_2 gas it is a major constituent of the atmosphere, but N_2 is chemically inert and therefore unavailable as a source of nitrogen for use by most living organisms. However, some bacteria have the ability to reduce N_2 and thereby "fix" atmospheric nitrogen using the enzyme nitrogenase. Many leguminous plants have capitalised on this special bacterial asset by going into partnership with nitrogen-fixing bacteria called rhizobia. In return for supplying nutrients to the bacteria, the plants receive a supply of reduced nitrogen. In essence, the legumes create a highly specialised environment within which the bacteria fix nitrogen. These specialised plant structures are called nodules; usually they are found on roots, but they also occur on the stems of some legumes.

The importance of nodules on legume plants has been recognised for a considerable time (Fig. 1); over 100 years ago Wilfarth and Hellriegel showed that legumes could grow well on nitrogen-free soil because they could fix atmospheric nitrogen and incorporate that nitrogen into plant tissue. More recently, the formation of nodules has been studied in great detail, giving insights into interactions between the plants and bacteria during the development of these plant organs.

18.1.1 Why Study Nodules?

Quite apart from the obvious agronomic importance of the *Rhizobium*-legume symbiosis in systems of sustainable agricultural production, what are its advantages as a model system for the study of the molecular genetics of cell differentiation? Are root nodules amenable to experimentation in

ways that would allow some of the important issues of developmental genetics to be addressed? A major virtue of the legume nodule is that the combined techniques of cytology, biochemistry and genetics can all be brought powerfully to bear on this microcosm of differentiation. Moreover, this important but inessential plant organ is the product of two totally different organisms, either of which can be genetically manipulated in order to understand the development of the other. Before describing nodules in detail it is important to recognise the breadth of potential systems that can be studied from the point of view of both the plant and the bacterium.

18.2 The Two Symbiotic Partners

18.2.1 Rhizobia[2]

Bacteria were originally placed in the genus *Rhizobium* (*rhizo* = root, *bios* = living) because they were found in root nodules. More recently, the bacteria have been found to fall into several distinct taxonomic groups and are currently placed in the genera *Rhizobium*, *Bradyrhizobium* and *Azorhizobium* (Fig. 2). Members of the genus *Rhizobium* are relatively fast-growing (forming colonies within about 2 days). They are closely related to *Agrobacterium* spp., the plant pathogens that form galls on plants by transferring bacterial DNA to the plant. There are several species of *Rhizobium*, and in general they have a fairly limited host range: each species can nodulate only a limited range of legumes. As indicated in Fig. 2, the species *R. leguminosarum* has been subdivided into three "biovars" which are defined on the basis of the specific range of legumes that they nodulate.

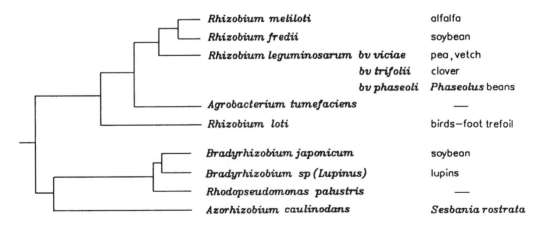

Fig. 1. The role of legumes in agriculture has long been recognised as important because legume crops can grow well in nitrogen-deficient soils. Crop plants grown in rotation with legumes also show much higher yields and orderly systems of crop rotations were recorded in the writing of Roman times. The earliest existing written description of nodules may be seen in the archaic Chinese pictographic character "shu" for soybean. (Courtesy of Hymowitz, Economic Botany). This character (top) can be traced to the 11th century B.C. and is thought to represent a shoot and a root with attached nodules represented as short lines below the root. The other part of the character is thought to represent a hand accepting the benefits of the soybean crop. Pictures of nodules are also found in early botanical drawings and the etching by Malpighi (*bottom*) drawn around 1675 shows nodules on *Phaseolus vulgaris*, the common bean (Courtesy of the Trustees of the John Innes Foundation)

The genus *Bradyrhizobium* is so called because the bacteria grow slowly (*brady* = slow), forming colonies after .4–5 days. In general, *Bradyrhizobium* spp. nodulate a broad range of tropical legumes whereas most *Rhizobium* spp. nodulate a narrow range of temperate legumes. Although *Rhizobium* and *Bradyrhizobium* spp. reduce N_2 in nodules, they cannot grow in culture using N_2 as nitrogen source. Some *Bradyrhizobium* spp. (unlike *Rhizobium* spp.) can reduce N_2 under certain laboratory growth conditions outside the plant, but paradoxically they appear to be unable to utilise that reduced N_2 for growth. A third genus, *Azorhizobium*, was recognised recently and these bacteria can utilise N_2 as a nitrogen source during growth in free-living culture. These bacteria were first

Rhizobium meliloti		alfalfa
Rhizobium fredii		soybean
Rhizobium leguminosarum bv viciae		pea, vetch
bv trifolii		clover
bv phaseoli		*Phaseolus* beans
Agrobacterium tumefaciens		—
Rhizobium loti		birds–foot trefoil
Bradyrhizobium japonicum		soybean
Bradyrhizobium sp (Lupinus)		lupins
Rhodopseudomonas palustris		—
Azorhizobium caulinodans		*Sesbania rostrata*

Fig. 2. Phylogenetic relationships of rhizobia and some related bacteria, with typical examples of legumes nodulated

isolated from stem nodules on the tropical legume *Sesbania rostrata* and the name *Azorhizobium caulonodans* (*azo* = nitrogen, *caulo* = stem) for the type species reflects its unusual characteristics.

It is important to recognise the potential advantages of studying bacteria from the different genera. For example, it has proven very useful in *Rhizobium* species that many of the nodulation genes (determining the type of legume nodulated) occur naturally on plasmids that can be transferred from strain to strain, thereby changing the range of legume hosts which can be nodulated. Additionally, *Bradyrhizobium* spp. have been used to identify mutants that can reduce N_2 in vitro but not in nodules, and it is possible that some of these mutants could be altered in the maturation of bacteroids within the plant rather than simply being mutated in genes required for nitrogen reduction. (In fact the reduction of N_2 is seldom measured; instead it is possible to measure the reduction of acetylene to ethylene because the nitrogenase enzyme can also catalyse this reaction.)

18.2.2 *Rhizobium* Genetics[3]

Genetic techniques that have been developed for rhizobia include conjugal transfer, chromosome mobilisation and phage-mediated transduction. Using such techniques, circular genetic linkage maps have been established for *R. meliloti* and *R. leguminosarum*. Efficient transformation of rhizobia has not been established, and usually gene cloning is carried out in *E. coli* using mobilisable broad host range vectors which are efficiently transferred to rhizobia by conjugation. Often the plasmids lack transfer functions and the transfer genes are supplied in trans by an additional plasmid that cannot replicate in rhizobia. There is now a wide variety of broad host range mobilisable plasmid vectors, some carrying reporter genes such as the *E. coli lacZ* (β-galactosidase) gene which can be placed under the control of the promoter of a cloned gene. Such plasmids have been important in understanding the regulation of nodulation and nitrogen fixation genes (see below). For work on symbiosis, a major constraint is usually plasmid stability because bacteria in nodules must go through many genera-

tions without any selection for retention of the plasmid. However, some plasmids are relatively stable and can be used to express genes within nodules.

Transposon mutagenesis has played an important role in defining rhizobial genes involved in the symbiosis. The transposon Tn5 has been most widely used because it has a reasonably high frequency of transposition and is random in its sites of insertion. Derivatives of Tn5 carrying reporter genes such as *lacZ* (β-galactosidase), *phoA* (alkaline phosphatase) or *gus* (β-glucuronidase) have been particularly useful because they can be used to monitor the regulation of the gene into which they were inserted. Thus there are many genetic techniques that can be used with rhizobia, an asset which has contributed significantly to their use in the study of the symbiosis.

18.2.3 Legumes[4]

The family Leguminosae contains a very large variety of plants. The great majority of species are nodulated by rhizobia although there are some non-nodulating genera, especially among the more primitive legumes. The taxonomic distribution suggests that the symbiotic characteristics arose well after the ancestral Leguminosae became isolated from the other families. Studies on a wide variety of nodulated plants have given insights into general principles of nodule formation. However, the most detailed experiments have been carried out with a limited range of legumes, in particular the crop and forage legumes of commercial importance such as soybean, pea, clover and alfalfa.

Specific legumes have been chosen to optimise the analysis of specific stages in the formation of nitrogen-fixing nodules. For example, small-seeded legumes such as clover, vetch and alfalfa are widely used to follow the early steps in the infection process. The roots of these legumes can be mounted intact on a microscope slide and the early events observed in situ. However, the prospect of picking large numbers of alfalfa nodules to carry out biochemical studies would be very daunting given the size of the nodules (which are about the size of a pin head). For that reason, large legumes such as soybean are chosen, making it possible to pick a kilogram of nodules for use in biochemical experiments.

18.2.4 Legume Genetics[5]

Thanks to the pioneering work of Mendel, the legume with the longest history of classical genetics is quite clearly the pea plant. Various mutant lines of peas have been identified and their phenotypes include complete or partial blocks in nodulation, increased nodulation, altered specificity for *Rhizobium* strains, inability to fix nitrogen and the ability to nodulate even in the presence of soil nitrate (which normally represses nodulation). Similar mutants have also been obtained with soybean, and many of the mutations have been chromosomally mapped. However, relatively little work has been done on the classical genetics of nitrogen fixation in other legumes.

Many attempts have been made to transform legume species with *Agrobacterium* and regenerate transgenic plants. Paradoxically, this has proven very difficult with peas and soybeans, whereas significantly better progress has been made with *Lotus* and clover both of which have been routinely transformed and regenerated. It is for this reason that the expression of nodule-specific genes cloned from soybeans has been studied in transgenic *Lotus* plants. As an alternative to regenerating complete transgenic plants it is possible to transform root cultures and to test the transgenic roots for nodulation or nodule-specific expression. Although this is a relatively rapid alternative to plant regeneration, it has some distinct limitations, since it is not possible to propagate the transgenic line through genetic crosses.

The nodules formed on legume plants fall into two very broad classes called "determinate" and "indeterminate" which reflect the nature of the respective meristems. Figure 3 illustrates an example of determinate nodules from *Phaseolus* beans, and contrasts them with the indeterminate type of nodule found on pea or clover roots. The spherical *Phaseolus* nodules are characterised by having a transient peripheral meristem, in which some of the cells are infected by rhizobia that will ultimately

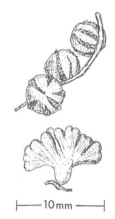

Fig. 3. Root nodules of *Phaseolus* (*above*) and clover (*below*) illustrating the morphology that arises from determinate and indeterminate meristems (Courtesy of H.D.L. Corby)

differentiate into nitrogen-fixing endosymbionts. By contrast, the cylindrical pea nodules are characterised by having an indeterminate apical meristem. The progeny of these cells are not invaded by rhizobia until the cells have moved out of the meristematic zone and ceased dividing, at which point they frequently become tetraploid or octaploid. The type of nodule formed is defined by the plant — there are examples of legumes that form determinate nodules after infection by either *Rhizobium* or *Bradyrhizobium* species. Conversely, one bacterial strain can induce spherical determinate nodules on one legume host and cylindrical indeterminate nodules on another. In fact, it is something of a paradox that a system which has very specific host-bacterial recognition can also be so promiscuous. The answer to this paradox must lie in the nature of the recognition events that occur during the infection and developmental process.

18.3 Nodule Development

Cell to cell signalling between rhizobia and legume root cells is required for the initiation and formation of nodules. Somehow this communication must establish that the only bacteria that will succeed in the infection process are those which will ultimately yield nitrogen to the plant. Clearly there must be a high degree of specificity in the interaction, and unravelling the signal exchange between plant and bacterium has relied heavily on bacterial genetics to identify early steps in this developmental process.

18.3.1 Preinfection Events[6]

In the early stages of the symbiotic interaction, rhizobia attach to root hair cells and induce a marked response in the plants – deformation and curling of root hairs – a prerequisite for infection

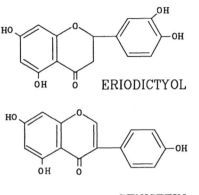

ERIODICTYOL

GENISTEIN

Fig. 5. The chemical structure of the plant-derived flavanone eriodictyol and the isoflavone genistein which induce *nod* gene expression in *R. leguminosarum* bv. *viciae* and *B. japonicum*, respectively

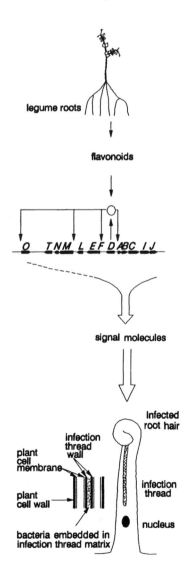

legume roots

flavonoids

Q T N M L E F D A B C I J

signal molecules

Infected root hair

infection thread wall

plant cell membrane

plant cell wall

infection thread

nucleus

bacteria embedded in infection thread matrix

to occur. This phenomenon may represent the initial stages in the redirection of root hair tip growth that occurs during infection thread formation (Fig. 4). These "pre-infection events" have been investigated by analysis of the nodulation (*nod*) genes of the bacterial partner. It was recognised that in *Rhizobium* species, genes essential for nodulation and host recognition are located on plasmids. Transfer of these so-called "symbiotic" (or *sym*) plasmids between rhizobia could alter the host range of the recipient strain. These nodulation genes (Fig. 4) are not normally expressed unless the bacteria are close to legume roots. It has been established that the roots exude flavonoids or isoflavonoids (Figs. 4 and 5) that act as potent inducers of *nod* gene expression. These inducers function via the products of the *nodD* gene, a con-

Fig. 4. Schematic representation of early stages of nodulation. Flavonoids secreted from legume roots induce *nod* gene expression after interacting with the regulatory protein NodD. The *nod* gene map is from *R. leguminosarum* biovar *viciae* showing induction of the *nodABCIJ*, *nodFEL*, *nodMNT* and *nodO* operons. Several of the *nod* gene products are involved in the formation of a signal molecule that induces root hair curling and nodule meristems. In the presence of the appropriate rhizobia, an infection thread develops following a redirection of the root hair cell wall growth, and the bacteria grow along this intracellular tunnel which elongates towards the developing nodule meristem

stitutively expressed regulatory gene whose product appears to interact with flavonoids and bind to promoters of other *nod* genes, thereby allowing their transcription. Because different legumes secrete different flavonoids or isoflavonoids, rhizobia have evolved a variety of closely related NodD proteins that recognise different flavonoids. In fact some strains, such as *R. meliloti,* contain multiple *nodD* genes which extend the range of legumes to which they can respond. Therefore the first step in the recognition process is a plant-made chemical which the bacteria must recognise via the activity of their *nodD* gene.

Once the bacterial *nod* genes are induced, the bacteria make a low molecular weight phytohormone that is secreted into the growth medium and functions at very low concentrations ($10^{-7}-10^{-11}$ M). This molecule (Figs. 4 and 6) induces deformation and curling of root hairs and in alfalfa can even induce nodule morphogenesis in the absence of bacteria. The *nodABC* genes are essential for the synthesis of the basic structure of the signal molecule, while some of the other nodulation genes are involved in modifying the compound to make it specific for its interaction with different legume roots. Thus, for example, *R. meliloti,* which normally nodulates alfalfa, does not induce root hair curling on vetch (normally nodulated by *R. leguminosarum*), but mutation of the *R. meliloti nodH* gene (absent from *R. leguminosarum*) allows it to induce root hair curling and to nodulate vetch whilst blocking alfalfa nodulation. Conversely, the *nodH* and *nodQ* genes were transferred from *R.*

meliloti to *R. leguminosarum*, which then acquired the ability to nodulate alfalfa, indicating that the *nodH* and *nodQ* genes can convert vetch-specific signals into alfalfa-specific ones. Apparently this occurs via an attachment of the sulphate group onto the molecule that is responsible for root hair curling (Fig. 6). Evidently, different rhizobia synthesise signal molecules that are related but quite specific for the appropriate legume host.

18.3.2 Infection[7]

Infection events have been analysed using the technique known as "spot inoculation" in which a very small volume of a bacterial suspension is applied on a resin bead at the site of emergence of new root hairs − the infectible zone on the seedling root (Fig. 7). This has allowed precise localisation of the time and place of infection, enabling the initial stages of *Rhizobium*-induced plant gene expression to be followed by in situ hybridisation techniques.

The bacteria gain access to the plant cortex via an infection thread which is made by the plant in response to the appropriate bacteria. This infection thread (Fig. 4) grows between and through root cells and bacteria grow and multiply within the lumen. The growing point of the infection thread appears to be closely controlled by the position of the plant nucleus which precedes its movement along

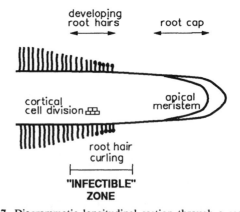

Fig. 7. Diagrammatic longitudinal section through a seedling root showing the infectible zone for *Rhizobium* invasion at the site of emergence of root hairs. This zone expresses the two early plant responses to *Rhizobium* infection, namely root hair curling and cortical cell division

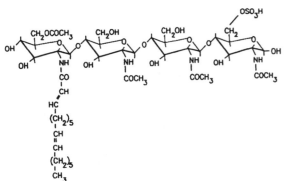

Fig. 6. The proposed chemical structure of the host-specific root hair curling glycolipid made by *R. meliloti*

the length of the root hair cell and appears to be connected to it by cytoskeletal structures. Similarly, the nuclei in other cells are also found close to the infection thread. The wall of the infection thread is like a plant cell wall that has invaginated and grown inward though the cell. An extension of the plasma membrane surrounds the infection thread. Therefore the bacteria within the infection thread can still be considered to be topologically outside the plant cell although they are embedded in an extracellular matrix that contains plant-made glycoproteins.

It is thought that the continued growth of the infection thread requires a constant supply of bacterially made root hair curling factor. The mechanism by which the plant recognises host-specific signals from the bacteria is unknown. However, there is strong evidence for involvement of a plant lectin in the recognition process. When a gene encoding pea lectin was transferred to clover plants they acquired the ability to be nodulated by a strain of *R. leguminosarum* bv. *viciae* (which normally nodulates pea and not clover).

While the infection thread is growing towards the inner cortical cells, a new meristem begins to form. The cell divisions that constitute its formation occur well in advance of the arrival of the infection thread carrying the bacteria. Therefore, there must be some diffusible signals that initiate the formation of the so-called nodule primordium. It has not been established whether this diffusible signal is made by the bacteria or depends on a plant-made secondary message. However, there is an induction of plant genes in the tissue surrounding the new meristem and the newly expressed protein products have been called early nodulins. Those that have been identified encode proline- or hydroxyproline-rich proteins similar to other plant structural cell wall proteins.

Experimentally, tissue invasion by rhizobia can be uncoupled from the initiation of the meristem. Mutants of *R. meliloti* that lack exopolysaccharide or have a modified exopolysaccharide cannot induce a normal infection process on alfalfa, but can nevertheless induce nodule structures that lack any bacteria or infection threads within the plant cells. This implies that: (a) the normal exopolysaccharide plays an important role in the infection process;

and (b) the inducer of the nodule meristem can act at a distance. It has been possible to identify at least one early nodulin gene that can be induced and expressed in empty nodules in the absence of bacteria.

18.3.3 Bacterial Release[8]

As the infection thread approaches the region of newly dividing nodule meristematic cells, it branches and bifurcates, carrying the bacteria into many cells in the developing nodule. Vesicles are then blebbed off into the cytoplasm each containing one bacterium surrounded by a plant membrane, the peribacteroid membrane (Fig. 8). Although similar to the plasma membrane, it appears to contain some characteristic proteins not found on the plas-

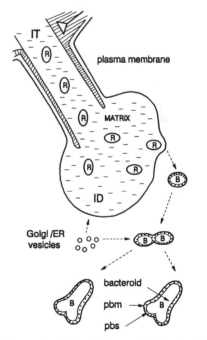

Fig. 8. Endocytosis and the development of bacteroids in the host plant cell. Rhizobia (*R*) are released from the walled infection thread (*IT*) into the unwalled infection droplet (*ID*) which contains plant-derived matrix glycoprotein. Individual bacteria are engulfed by the naked plant membrane surrounding this droplet. Within the plant cytoplasm, the bacteria grow, divide and differentiate into nitrogen-fixing bacteroids (*B*) enclosed by a plant-derived peribacteroid membrane (*pbm*) and peribacteroid space (*pbs*)

ma membrane. Little is known about why the infection thread branches or what signal causes the bacteria to be released at this stage rather than en route through other cortical cells. Bacterial mutants (altered in lipopolysaccharides) have been identified that are not properly released from the infection thread, implying that the lipopolysaccharide may confer some recognition function at this stage. When the bacteria have been released, they alter shape and increase in size (Fig. 8). The form of the bacteria within the peribacteroid membrane varies from legume to legume: in some legumes such as peas, single bacteria enlarge into Y-shaped pleiomorphic forms, whereas in other symbioses, the bacteria go through one or two doublings without concomitant division of the peribacteroid membrane envelope.

It is possible to use cytological techniques to examine the sequence of events associated with meristem organisation, plant cell differentiation, cell invasion by *Rhizobium*, and the differentiation of nitrogen-fixing bacteroids. A variety of molecular probes has been used, as illustrated by the following four examples: (1) antisense cDNA was used to localise expression of an early nodulin cell wall protein (Fig. 9) to a particular cell layer in the uninfected regions of the nodule; (2) antiserum to a purified nodule uricase was used to localise an enzyme of nitrogen assimilation to the peroxisome bodies of uninfected interstitial cells that adjoin the infected cells in the centre of the soybean nodule; (3) a monoclonal antibody (Fig. 9) was used to identify a hitherto unknown plant glycoprotein as

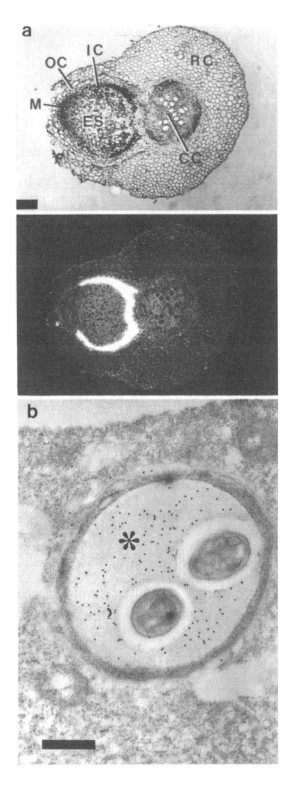

Fig. 9 a, b. In situ localisation with cDNA or antibody probes. **a** Expression of the early nodulin gene ENOD2 in uninfected cells of the pea nodule parenchyma, as detected using an antisense cDNA probe. Transection through a root with a 10 day old nodule visualised by bright field (upper) and dark field microscopy (lower), in which the autoradiographic signal is visible as white silver grains over the inner cortex (*IC*). *CC* central cylinder of the root; *RC* root cortex; *M* apical meristem of the nodule; *ES* early symbiotic growth zone of the nodule central tissue; *OC* nodule outer cortex. *Bar* 250 µm. (Photographs courtesy of C van de Wiel et al. EMBO Journal 9, 1–7 1990). **b** Transverse section of an infection thread from a pea nodule cell, showing the localisation of a plant glycoprotein (*asterisk*) in the luminal matrix using colloidal gold particles coupled to monoclonal antibody MAC265. *Bar* 0.5 µ. (Courtesy of K A VandenBosch)

a component of the *Rhizobium*-induced infection thread matrix; (4) fusions of cloned *nod* promoter sequences with alkaline phosphatase or β-galactosidase reporter genes have been used to demonstrate that *Rhizobium* genes involved in the synthesis of extracellular polysaccharides are still transcriptionally active within the infection thread. The level of sophistication of cytological analysis can be increased when combined with the use of genetics. For example, it is possible to arrest nodule development using particular bacterial mutants that stop at different stages in nodule development and ask what plant genes continue to be expressed in this mutant system.

18.3.4 Nodule Maturation[9]

The basic raison d'être for nodules is clearly to provide a suitable environment to allow biological nitrogen fixation to take place. This requires a microaerobic niche because *Rhizobium* is an obligate aerobe, but the nitrogenase enzymes are very sensitive to damage by atmospheric oxygen concentrations. A facilitated diffusion system for oxygen is needed to sustain oxidative phosphorylation by *Rhizobium* under microaerobic conditions. This is provided by leghaemoglobin, the most abundant nodule protein, which ensures a rapid oxygen flux while keeping the free oxygen concentration very low (Fig. 10). In parallel, the bacteroids have a very high affinity cytochrome oxidase that functions at very low oxygen concentrations and mutations affecting its functions can block nitrogen fixation. Furthermore, because nitrogen fixation consumes considerable amounts of energy, bacteroids require the supply of a good respiratory substrate provided by the plant through the peribacteroid membrane. Bacterial mutants impaired in their ability to transport dicarboxylic acids (succinate and malate) cannot fix nitrogen, indicating that these are major substrates for the bacteroids.

Many bacterial genes are required for the regulation, synthesis, and assembly of the bacteroid nitrogenase, the enzyme which reduces dinitrogen to ammonia. This enzyme requires a supply of electrons at a low redox potential and there are specialised electron transport enzymes which fulfil

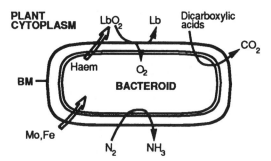

Fig. 10. Diagram illustrating bacteroid metabolism and interactions with the host plant cytoplasm. The bacteroid is enclosed within the peribacteroid membrane (*PBM*) forming an organelle-like structure, "the symbiosome". Bacteroids "fix" nitrogen (N_2) using the enzyme nitrogenase which requires iron (*Fe*) and molybdenum (*Mo*) assembled into a complex inorganic cofactor. Ammonia (NH_3), which is the immediate product of N_2 fixation, is excreted by bacteroids and assimilated through the activity of glutamine synthetase in the host plant cell. Dicarboxylic acids, supplied through the peribacteroid membrane, are the substrates for oxidative phosphorylation. In the microaerobic conditions prevailing in the centre of the nodule, leghaemoglobin (*Lb*) serves as a facilitated carrier of oxygen across the plant cytoplasm. The haem moiety of leghaemoglobin is thought to be synthesised by *Rhizobium* bacteroids while the plant genome encodes the leghaemoglobin genes which have been extensively characterised by molecular genetics

that task. The genes encoding all of the enzymes required for nitrogen fixation are known as *nif* and *fix* genes. (Rhizobial "*nif*" genes are homologous to *nif* genes found in *Klebsiella pneumoniae*, whereas "*fix*" genes are additional nitrogen genes that have been found in rhizobia.) The *nif* and *fix* genes are under the control of a cascade regulatory system controlled by the FixLJ proteins (Fig. 11). These two proteins are representatives of a class of prokaryotic regulatory systems known as the "sensor-kinase/response-regulator elements" which control many developmental processes in bacteria. They depend on a sensor protein, which responds to an environmental stimulus and then phosphorylates a second protein, the response-regulator, thereby modulating the biological activity of the regulator. The sensor-kinase proteins all share homology around their C-terminal domains which are involved in the kinase activity, but are somewhat different with regard to the N-terminal domains which are involved in recognising the environmental stimuli. The response-regulator proteins

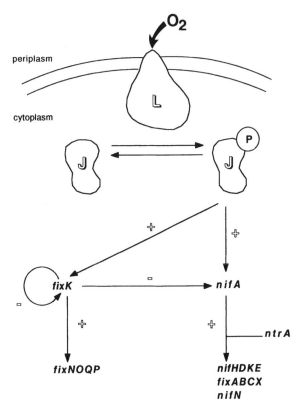

Fig. 11. Model for the regulation of the *nif* and *fix* genes in *R. meliloti*. (Courtesy of D. Kahn)

all show homologies around the N-terminal domain, the region which is phosphorylated, allowing bacteria to respond to a variety of stimuli such as the presence of available nitrogen, osmolarity, toxins, nutrient deprivation, phosphate limitation, plant exudates and dicarboxylic acids.

In the case of the *fixLJ* system, the principal effector appears to be the low concentration of oxygen which is sensed by a haemoprotein, the transmembrane sensor-kinase FixL (Fig. 11). This protein then phosphorylates the response-regulatory protein FixJ, such that it now activates the transcription of *fixK* and *nifA*. The *nifA* gene product then activates transcription of several *nif* and *fix* operons via an interaction with an upstream activator site about 100 nucleotides upstream of the transcription start site. The NifA protein appears to interact with a specialised RNA polymerase sigma factor encoded by *ntrA* (also known as *rpoN*) to stimulate the transcription of *nif* genes such as

nifHDK. In parallel the FixK protein can also activate transcription of another *fix* operon *fixNOQP* whilst it can also influence *nifA* expression. This complex control system ensures that the bacteria do not make the commitment to the process of nitrogen fixation unless the appropriate conditions prevail and such an environment is found within the nodule.

Ammonia, the product of the nitrogenase reaction, is excreted by the bacteroids and assimilated by the plant enzyme glutamine synthetase, which is encoded by genes that are specifically induced within nodules. Significantly, the induction of such late nodulin genes can occur even when the nodules are induced by mutant rhizobia that are unable to fix nitrogen. Therefore these plant genes must be under a developmental programme of control and not simply activated by a supply of fixed nitrogen from the bacteroid.

Our appreciation of the biochemistry of nitrogen fixation and nodule function has been extremely helpful in identifying and characterising the major tissue-specific gene products of mature nodules and their underlying genetic regulation during nodule development. However, during the first 10 days of nodule development, absolutely no nitrogen fixation can take place (because only the mature nodule structure provides the microaerobic niche that can permit this process to occur). Therefore all the early stages of plant-microbe interaction must proceed at a cost to the plant but without any obvious reference to the biochemistry of nitrogen fixation. We do not understand the underlying biochemistry in this area of cell-to-cell signalling which leads to the induction of a new plant meristem and its differentiation into several specialised cell types, but surely future research into the early stages of nodule development will contribute to a more general understanding of the processes of plant cell growth and differentiation and to the mechanisms of plant-microbe interactions.

18.3.5 Promoter Structure and Tissue-Specific Gene Expression[10]

The analysis of plant gene promoter structure has progressed significantly, particularly in relation to the leghaemoglobin gene family. This information has been obtained using transgenic plants containing various parts of leghaemoglobin promoters fused to a reporter gene such as chloramphenicol acetyltransferase. In parallel, DNA "footprinting" studies have identified protein-binding domains in the promoter regions of the leghaemoglobin gene. The sequences of the promoter regions of several nodulin genes have been determined and a comparison of these sequences indicates that there are conserved domains, parts of which may specify that the gene should be highly expressed within the nodule. However, it is far from clear how many plant nodulin genes are co-regulated, what DNA sequences determine their regulation, and what regulatory proteins are involved in their activation. Within nodules there are many different cell types − some of the cells are uninfected and these cells express some nodule-specific genes such as uricase that are involved in nitrogen assimilation. The picture that emerges points to a hierarchy of genetic control systems having different levels of specificity and controlling expression at the level of the individual gene, the tissue and the plant organ. Although not enough information is currently available for useful generalisations to be made, the study of nodulin gene promoters has obviously reached a very exciting stage.

18.4 Cell to Cell Communication and Gene Expression in Development[11]

Three general classes of signal can be recognised which trigger development through the activation of particular sets of genes: surface interactions, diffusible morphogens and physiological triggers. With few exceptions, we know very little about the transduction pathway that leads from the perception of a signal conveying developmental information to the genetic response observed as new patterns of gene expression. Therefore, we shall have to be content for the moment to discuss some examples of the three classes of developmental signals.

18.4.1 Cell to Cell Contact

Cell surface interactions imply that some kind of direct cell to cell contact is necessary for organised morphogenesis to occur. For example, the initiation of infection thread growth (Fig. 4) cannot occur without the presence of bacteria in contact with the root hair surface. Bacterial extracellular polysaccharides, perhaps in conjunction with some secreted plant and bacterial proteins, may be important components of this surface interaction. However, the mechanism and specificity of recognition are still unclear, as is the nature of the morphological and genetic response by the root hair cell. At a later stage in the tissue invasion by *Rhizobium,* the bacteria come into contact with the naked plant cell plasma membrane and are taken into the plant cell by endocytosis (an extremely rare phenomenon for plant cells which are normally ensheathed by a cell wall that denies access to invading bacteria). The surface interactions that occur at this stage probably involve the plant membrane (Fig. 8) in conjunction with bacterial lipopolysaccharide or some other component of the bacterial outer membrane, but again the nature and specificities of these interactions are still undefined.

Other aspects of cell-cell surface interactions involve communication between neighbouring plant cells during nodule morphogenesis (Fig. 12). How are the planes of cell division orientated in and around the nodule meristem? How is the cytoskeleton organised from cell to cell across the root cortex so as to channel the invading infection thread into the central tissues of the root? How is the fate map defined during differentiation of cells arising from the nodule meristem; is it dependent on cell position or on cell lineage? In most of these cases we are not even sure what kinds of interactions might be involved. Perhaps we have to consider cell growth and differentiation in relation to a combination of factors: the lines of physical stresses and strains within the growing tissue, molecular recognition between cell surfaces, intercellular communication through plasmodesmata, and positional effects relating to diffusion gradients and morphogenetic fields. Although these are interesting and fundamental questions, it will not be easy to obtain satisfactory answers in this or any other

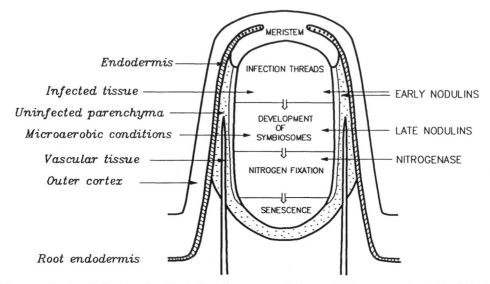

Fig. 12. Diagrammatic longitudinal section illustrating the organisation and development of an indeterminate nodule meristem. The pea nodule has an apical meristem which is uninfected and gives rise both to specialised uninfected tissues and to a central mass of cells which are invaded by rhizobia. Intracellular bacteria differentiate within organelle-like structures termed symbiosomes. They develop the capacity for nitrogen fixation and then ultimately senesce

system of plant differentiation. However, it could be argued that the nodule meristem provides a discrete well-organised and relatively simple system of organogenesis in which to approach these problems (Fig. 12).

18.4.2 Diffusible Signals

In addition to cell-cell interactions involving surface contact, other signals are diffusible, initiating cell differentiation at a distance. One example is the group of flavonoids which induces the transcriptional activation of *Rhizobium nod* genes (Fig. 5). Another example is the *Rhizobium*-made glycolipid (Fig. 6) which promotes root hair deformation and cortical cell division in the tissues of the legume host. Other more conventional plant growth regulators have also been implicated in the processes of nodule development. Cytokinins, auxins, and even oligosaccharides have been variously represented as possible promoters of nodule meristematic activity, and gibberellins, abscissins and ethylene have been implicated as possible inhibitors. Unfortunately, none of these plant systems has yet been analysed

to a satisfactory level of detail, but this situation could now change with the advent of more specific molecular probes with which to examine the onset of cell-specific gene expression.

18.4.3 Physiological Effects

A third class of morphogenetic signal affecting tissue-specific gene expression relates to the change in physiological status of the differentiating cells within the nodule. Examples of such physiological triggers are low oxygen concentrations, sucrose availability, nitrogen status, water and turgour relations, calcium status and perhaps various forms of physiological stress consequent on *Rhizobium* infection. However, the physiological triggers that might be anticipated are not always those actually found to operate during nodule development. For example, expression of the leghaemoglobin genes is not triggered by low oxygen concentrations, as might be expected, nor is transcription of the genes for ammonia assimilation triggered by the supply of ammonia from bacteroids. Perhaps these observations point to the fact that the integration of gene expression during the development of the *Rhizobium*-legume symbiosis is a highly evolved and coordinated process.

18.5 Outlook

In recent years, the molecular genetics of symbiosis has progressed faster with the bacterial partner than with the higher plant because the basic methodology has been much simpler to establish in prokaryotic systems. However, the molecular genetics of higher plants is now equipped with a formidable repertoire of techniques which should allow the genetic dissection of the *Rhizobium*-legume symbiosis to make rapid progress. Genes concerned with symbiosis can be identified either as tissue-specific transcripts or as transposon-induced mutations. Cloned genes can be modified and reintroduced into transformed and regenerated plant tissues. The resulting probes for tissue-specific gene expression will provide a completely new framework in which to analyse the sequence of signals and responses which gives rise to nodule development. It is now clear that the morphogenesis of the legume nodule requires a developmental programme that involves many plant genes and therefore the concept of extending the rhizobial symbiosis to nonlegume crop plants now appears somewhat naïve.

18.6 Summary

The development of nitrogen-fixing nodules on legumes is dependent upon the exchange of low molecular weight signal molecules between the plant and the rhizobia. Initially, the bacteria recognise the plant by sensing flavonoid molecules secreted from the roots. Subsequently, the rhizobia induce a group of genes whose products biosynthesize highly-specific molecules that initiate a programme of morphogenesis in those legume roots that recognise the appropriate signals. Cell surface interactions play an important role in the subsequent infection events that allow the bacteria to reach the developing nodule. A series of plant genes are induced,

thereby creating the appropriate specialised environment required for the nitrogen fixation reactions carried out by the bacteria.

References

[1] Bergersen FJ (1982) Root nodules of legumes: structure and functions. Research Studies Press, John Wiley and Sons, Chichester

Dart PJ (1977) Infection and development of leguminous root nodules. In: Hardy RWF, Silver WS (eds) A treatise on dinitrogen fixation, III, Biology. Wiley, New Nork, pp 367–472

[2] Young JPW, Johnston AWB (1989) The evolution of specificity in the legume-*Rhizobium* symbiosis. Trends Ecol Evol 4:341–349

[3] Long SR (1989) *Rhizobium* genetics. Annu Rev Genet 23:483–506

[4] Sprent JI (1989) Which steps are essential for the formation of functional legume nodules? New Phytol 111:129–153

[5] Carroll BJ, Mathews A (1990) Nitrate inhibition of nodulation in legumes. In: Gresshoff PM (ed) Molecular biology of symbiotic nitrogen fixation. CRC Press, Boca Raton, pp 159–180

Weeden NF, Kneen BE, LaRue TA (1990) Genetic analysis of *sym* genes and other related genes in *Pisum sativum*. In: Gresshoff PM, Roth J, Stacey G, Newton W (eds) Nitrogen fixation: achievements and objectives. Chapman and Hall, New York, pp 323–330

[6] Lerouge P, Roche P, Faucher C, Maillet F, Truchet G, Promé JC, Dénarié J (1990) Symbiotic host-specificity of *Rhizobium meliloti* is determined by a sulphated and acylated glucosamine oligosaccharide signal. Nature 344:781–784

Long SR (1989) Life together in the underground. Cell 56:203–214

Rolfe B, Gresshoff P (1988) Genetic analysis of legume nodule initiation. Annu Rev Plant Physiol 39:297–319

Truchet G, Roche P, Lerouge P, Vasse J, Camut S, deBilly F, Promé J-C, Dénarié J (1991) Sulphated lipo-oligosaccharide signals of *Rhizobium meliloti* elicit root nodule organogenesis in alfalfa. Nature 351:670–673

[7] Calvert HE, Pence MK, Pierce M, Malik NSA, Bauer DW (1984) Anatomical analysis of the development and distribution of *Rhizobium* infections in soybean roots. Can J Bot 62:2375–2384

Diaz CL, Melchers LS, Hooykaas, PJJ, Lugtenberg BJJ, Kijne JW (1989) Root lectin as a determinant of host-plant specificity in the *Rhizobium*-legume symbiosis. Nature 338:579–581

Scheres B, van de Wiel C, Zalensky A, Horvath B, Spaink H, van Eck H, Zwartkruis F, Wolters AM, Gloudemans T, Van Kammen A, Bisseling T (1990) The ENOD12 gene product is involved in the infection process during the pea-*Rhizobium* interaction. Cell 60:281–294

[8] Robertson JG, Wells B, Brewin NJ, Wood EA, Knight CD, Downie JA (1985) The legume-*Rhizobium* symbiosis: a cell surface interaction. J Cell Sci (Suppl) 2:317–331

Sharma SB, Signer ER (1990) Temporal and spatial regulation of the symbiotic genes of *Rhizobium meliloti* in planta revealed by transposon Tn5-*gusA*. Genes Dev 4:344–356

van de Wiel C, Scheres B, Franssen H, van Lierop M-J, van Lammeren A, van Kammen A, Bisseling T (1990) The early nodulin transcript ENOD2 is located in the nodule parenchyma (inner cortex) of pea and soybean root nodules. EMBO J 9:1–7

[9] Appleby CA (1984) Leghemoglobin and *Rhizobium* respiration. Annu Rev Plant Physiol 35:443–478

Batut J, Daveran-Mingot M-L, David M, Jacobs J, Garnerone AM, Kahn D (1989) *fixK*, a gene homologus with *frn* and *crp* from *Escherichia coli*, regulates nitrogen fixation genes both positively and negatively in *Rhizobium meliloti*. EMBO J 8:1279–1286

Mellor RB (1990) Bacteroids in the *Rhizobium*-legume symbiosis inhabit a plant lytic compartment: implications for other microbial endosymbioses. J Exp Bot 40:831–839

Saier MH, Wu L-F, Reizer J (1990) Regulation of bacterial physiological processes by three types of protein phosphorylating systems. TIBS 15:391–395

[10] Verma DPS, Delauney AJ (1988) Root nodule symbiosis: Nodulins and nodulin genes. In: Verma DPS, Goldberg RB (eds) Plant gene research. Springer, Berlin Heidelberg New York, pp 169–199

Forde BG, Cullimore JV (1989) The molecular biology of glutamine synthetase in higher plants. Oxford Surv Plant Mol Cell Biol 6:247–296

[11] Nap JP, Bisseling T (1990) Developmental biology of a plant-prokaryote symbiosis: the legume root nodule. Science 250:948–954

Verma DPS, Long S (1983) The molecular biology of the *Rhizobium*-legume symbiosis. In: Jeon K (ed) International review of cytology. Academic Press, New Nork, pp 211–245 (Suppl 14)

Section 3 **Animals**

Chapter 19 Embryogenesis in *Caenorhabditis elegans*

Lois Edgar

19.1 Introduction

Caenorhabditis elegans is a small soil nematode which is currently being extensively studied to discern general principles of how genes control development. The short life cycle, ability to culture in quantities sufficient for biochemical work, well-developed genetics, small cell number for a rather sophisticated animal, and rapidly increasing possibilities for molecular genetics are features that make this species a very productive system for study.

19.1.1 The *C. elegans* Life Cycle Is Rapid and Laboratory Culture Is Simple[1]

As a laboratory animal, *C. elegans* is cultured on the bacteria *E. coli* on Petri plates. Due to the small size of the worm (adults are about 1 mm in length), almost all handling of worms is done using a dissecting microscope. The life cycle of about 3 days (at 20 °C) consists of embryogenesis, which occurs inside eggs laid with an impermeable eggshell, four superficially similar larval stages (L1 through L4) separated by cuticle molts, and sexually mature adulthood (Fig. 1). An alternative nonfeeding Dauer larva stage replaces the L3 larva under conditions of starvation. Adults come in two sexes, self-fertilizing hermaphrodites (with two X chromosomes) and XO males which arise through occasional chromosomal nondisjunction (approximately one in 700 animals). Males will mate with hermaphrodites to produce equal numbers of male and hermaphrodite progeny. Strains may be frozen for preservation, a useful laboratory feature.

19.1.2 Each Cell of a Worm Can Be Identified Microscopically[2]

At hatching, the *C. elegans* larva (L1) has 558 nuclei; the adult hermaphrodite, exclusive of the germ line, has only 959 nuclei. Cell numbers, types, and position are almost invariant from one animal to the next. This small cell number, and the transparency of the cuticle, allow direct observation of individual cells in living animals with a compound microscope. Using differential interference contrast (Nomarski) optics, the complete cell lineage has been followed through both embryogenesis and postembryonic development, and the fate of each cell determined by a combination of direct observation and electron microscopy of sectioned embryos or worms. The existence of this complete "map" of normal development provides an extremely important framework for analysing either experimental manipulations or mutations that cause abnormal development in the worm at the level of individual cells.

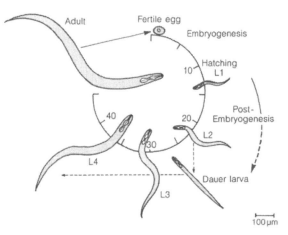

Fig. 1. The *C. elegans* life cycle at 25 °C, showing approximate duration of embryogenesis and developmental stages in hours

19.1.3 A Large Number of *C. elegans* Genes Have Been Identified and Mapped[3]

The genetics of *C. elegans* has been developing very rapidly in the 20 years since Brenner began to characterize the genetic system, and is at this point extensive. There are six linkage groups, including five autosomes and the X chromosome. Of an estimated 3000–5000 genes, about 900 have been identified by mutation and placed on the genetic map. Visible morphological mutants (such as *unc*oordinated, *dumpy*, or *long* worms) are generally used in standard mapping procedures. Deficiencies and duplications for many chromosomal regions have been identified, and are valuable tools for mapping, as well as generating molecular polymorphisms. For some linkage groups, balancer chromosomes which both suppress recombination and carry visible, sterile, or lethal mutations to identify heterozygotes allow the maintenance of lethal or sick mutations.

Gene function can be studied by several genetic methods. Analysis of somatic mosaics, created by the mitotic loss of free duplications, can identify subsets of cells which require a particular gene. Suppression of a mutation by amber suppressors is one of several criteria to identify a probable null allele of a gene, an important aspect of defining gene functions. Another type of suppressor, extragenic suppressors of mutations in specific genes, are extremely useful tools for identifying two types of interacting genes: those which function in a common regulatory pathway; or those which interact with the gene defined by the initial mutation at the gene product level, for example in a multimeric protein complex.

19.1.4 Rapidly Developing Molecular Genetics Provides Tools for Molecular Analysis[4]

C. elegans has a small genome, 8×10^7 base pairs, or about half that of *Drosophila*. This has speeded the development of molecular genetics and makes possible the goal of an entire genome sequence for this species within perhaps 10 years. Currently, a generally available physical map of the worm genome is being constructed by the group working at the Medical Research Council, Cambridge, England. A cosmid genomic library has been linked into contiguous stretches (contigs) that cover more than half the genome, and quite a few genetically identified loci have been placed on the physical map. Thus almost any newly cloned DNA sequence of interest can be hybridized against the cosmid or YAC (*y*east *a*rtificial *c*hromosome) libraries available from Cambridge, and placed on the physical map.

A number of methods are generally useful for cloning particular genes. One approach utilizes the transposable element Tc1 to create a mutation in the gene of interest. Another is to genetically map polymorphic Tc1 elements near the gene through repeated outcrossing to a strain with very few Tc1 inserts, clone the Tc1-associated sequences, and initiate a walk towards the gene, using the contigs available through the MRC group. A third method is to test the appropriate set of cosmids indicated by the correlation of genetic and physical map, and identify a fragment with either a mutational polymorphism, or an appropriately expressed transcript. Finally, transformation with a fragment that rescues the mutant phenotype provides good evidence that the right sequence has been identified.

Transformation by injection of cloned DNA sequences is also a valuable method for investigating gene function and regulation. In *C. elegans*, transformants can be recovered after injection of DNA into the adult gonad, and identified as progeny rescued for a coinjected visible marker. Most transformants contain large tandem arrays of nonintegrated DNA, and can be used for functional studies using modified sequences. Stable germline integrants arise at a lower frequency. Germline integration is not homologous, but random as in mammalian transformation systems.

19.2 An Overview of Embryogenesis[5]

Several central questions in the study of development are particularly appropriate to study in *C. elegans* because of the possibilities of combining the cellular, genetic, and molecular approaches. First, how does a cell become committed to a particular fate during early stages of embryogenesis? Second, how are patterning and morphogenesis initiated and controlled? Detailed answers to these questions require identification of the genes involved, of the molecular nature and functions of their products, and of the interactions and regulative hierarchies used to control the precise expression of a set of genes at the proper time and in the right tissue. At the cellular level, the developmental pattern depends on both cell-autonomous gene expression and on inductive responses dependent on cell interactions: what are their relative roles, and their genetic controls?

A third area of interest is defining the roles of the maternal and the embryonic genome in providing information and control. *C. elegans* appears typical of most invertebrates in that early embryogenesis runs on maternal transcripts and proteins; nevertheless, there is necessarily a transition to embryonic self-sufficiency. How and when does this occur?

19.2.1 Gametogenesis in *C. elegans* Occurs in a Linear Spatial Progression

In both sexes of *C. elegans*, germ cells mature sequentially as they progress down a tubelike gonad. The distal portion of the reflexed gonad (single in males, bilobed in hermaphrodites; see Fig. 2) consists of syncytial nuclei in mitotic proliferation; germ cells are formed and differentiate in the proximal arm. The male gonad has a single arm, which produces only sperm, and opens through a fan-like tail structure used in mating. In hermaphrodites, the first group of about 40 germ cells in each gonad arm differentiate as sperm, which are stored (as are male sperm from matings) in a spermatheca just distal to the uterus, and subsequently oocytes are produced. As immature oocytes separate from the syncytium, they begin to enlarge, incorporate yolk proteins transported in from the gut, and enter meiosis, arresting at first metaphase.

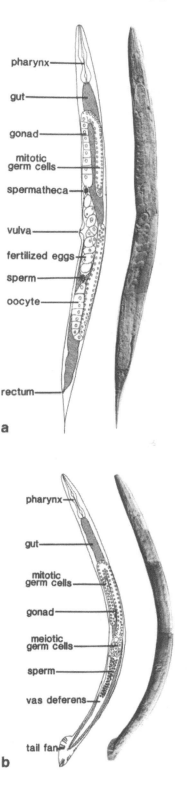

Fig. 2a, b. *C. elegans* adult hermaphrodite (a) and male (b): schematic diagram showing gonads, and Nomarski photomicrographs

19.2.2 Fertilization Events Designate Anterior-Posterior Polarity

As the oocyte matures, the female pronucleus is localized distally at the future anterior end of the embryo. Nuclear breakdown occurs as oocytes pass one by one into the spermatheca. Sperm enters at the opposite end (the future posterior), and an inner vitelline membrane and an ovoid eggshell rapidly form, the long axis indicating the future anterior-posterior axis of the embryo. As the zygote leaves the spermatheca, the female pronucleus resumes meiosis, producing two polar bodies which are extruded and remain inside the shell at the anterior end, a useful marker of polarity. At about 30 min after sperm entry, the male and female pronuclei appear at opposite ends of the egg, and the female pronucleus begins to move towards the male pronucleus, meeting it at the posterior end (Fig. 3 A, B). Concurrently, the cytoplasm undergoes vigorous movement, and a temporary constriction appears at the middle of the egg, which resembles a cleavage furrow and is termed pseudocleavage. After the pronuclei meet, they move together back to the center of the egg, and with a characteristic rocking, fuse and form a spindle along the anterior-posterior axis. This spindle shifts posteriorly, so that cytokinesis produces a

large anterior cell, the AB cell, and a smaller posterior daughter, P.

19.2.3 Early Cleavages Create Six Founder Cells

Early cleavages in *C. elegans* are rapid (about 15 min per cell cycle), and invariant in their pattern (Fig. 3). The P daughters continue to undergo asymmetric, longitudinally oriented divisions, while the AB daughters divide more rapidly, with spindles at 90° to the anterior-posterior axis. The asymmetric cleavages of the first four cell cycles produce six founder cells, AB, E, MS, P_4, C, and D, defined on the basis of their subsequent distinc-

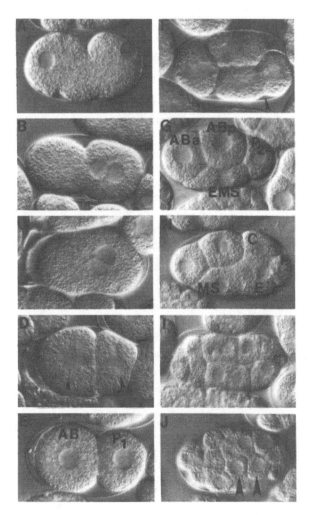

Fig 3 A–J. Fertilization and early cleavage stages, Nomarski photographs. (Anterior is to the *left*, ventral is *down*.) The egg is approximately 40×60 μm in size. (Photo F. Storfer). **A** Pronuclear formation and anterior membrane contraction, about 30 min after fertilization. The hermaphrodite pronucleus is at the *left* (anterior) end, the sperm pronucleus at the posterior end. **B** Pseudocleavage, about 45 min after fertilization. The hermaphrodite pronucleus has migrated posteriorly through the pseudocleavage furrow to meet the male pronucleus. **C** The two pronuclei move back toward the center of the egg and rotate 90° before nuclear breakdown and fusion. **D** First cleavage, producing a large AB (anterior) and a smaller P_1 (posterior) cell. *Arrowheads* indicate the different aster structures in the two cells. **E** Two-cell embryo. **F** Second cleavage. AB enters mitosis slightly before P_1. Here the AB cell has completed cytokinesis; P_1 is in telophase and the P_1 cleavage furrow of cytokinesis is appearing (*arrowhead*). **G** Four-cell embryo. EMS characteristically extends a "tail" around ABa. **H** Eight-cell embryo. Two AB cells are underneath and not visible in this focal plane. **I** 16-cell embryo. **J** 24-cell embryo, at the beginning of gastrulation. The two E cells (*arrowheads*) are beginning to move into the interior of the embryo

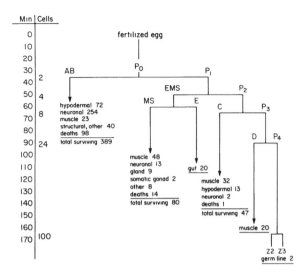

Fig. 4. Early lineages of *C. elegans*. The first four cleavages establish the six founder cells AB, E, MS, C, D, and P_4. The numbers and types of larval cells generated by each founder cell are indicated (Wood 1988)

19.2.4 Further Embryogenesis Consists of Gastrulation, Cell Proliferation, and Morphogenesis

Gastrulation occurs on the scale of movements of individual cells, beginning at the 28-cell stage (about 100 min after first cleavage at 20°C) when the two E cells move into the interior of the embryo. The MS and P_4 blastomeres follow, and cell movements and cell division continue concurrently to the 400-cell stage (300 min) (see Fig. 5). By this

tive patterns of cell division and cell fates (Fig. 4). E and P_4 are clonal progenitors of the gut and germ line, respectively; all of their descendants, and only their descendants, make up these tissues. D produces only body wall muscle, but only 20 of the total 81 body muscle cells. The other founder cells each produce several cell types; conversely, a particular differentiated cell type may arise from more than one founder cell. For example, MS, C, D, and AB descendants all contribute to the body wall muscle; MS, C, and AB also generate neurons, and C and AB contribute to the hypodermis. Thus mesoderm and ectoderm do not arise as clonal groups of cells. Cells are named on the basis of lineage and invariant directions of their division: e.g., ABalp is the posterior greatgranddaughter of the left granddaughter of the anterior daughter of the AB cell.

The timing of division is also very specific for each cell lineage. During pre-gastrulation cleavages, all AB descendants divide synchronously, whereas P_1 descendants divide more slowly, and each founder cell line follows an individual timetable. These characteristic patterns of division have been considered to indicate commitment to a given cell fate, and are useful clues for the analysis of mutants.

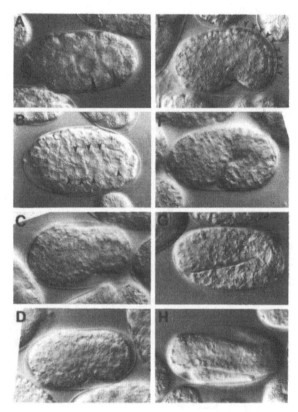

Fig. 5 A–H. Later embryogenesis of *C. elegans*; Nomarski photos (anterior at *left*, ventral at *bottom*) (Photo F. Storfer). **A** Mid-gastrulation (about 32 cells). Two E cells (*arrowheads*) are now in the interior of the embryo, and about to divide. **B** Mid-proliferation, about 100 cells. Two lines of interior E cells (*arrowheads*) are recognizable by their large nuclei. **C** Late "lima bean", about 300–400 cells. A pharyngeal primordium and gut have formed. **D** Comma stage, about 500 cells, at the onset of morphogenesis. **E** Morphogenesis. Hypodermal cells (*arrowheads*) are interdigitating and elongating from the dorsal surface to surround the embryo. **F** 1.5-fold embryo (500+ cells), beginning to elongate at the tail. **G** Twofold embryo. **H** Threefold "pretzel" (550 cells). The embryo has secreted a cuticle, is moving within the shell, and is ready to hatch

time, the embryo, although still a ball of cells, has organized the cell groups and locations for its major tissues (e.g., hypodermis, gut, pharynx, muscle, nerves) and begun some differentiation, as evidenced by immunological markers. The earliest markers of differentiation appear in the gut, as birefringent gut granules seen under polarized illumination at about 100 cells (200 min). By 350 min,

cell division is essentially complete. During the second half of embryogenesis, the embryo rapidly elongates into a worm, secretes a cuticle, and completes morphogenesis and cell differentiation, which includes 113 specific programmed cell deaths. Hatching occurs at about 15 h after fertilization.

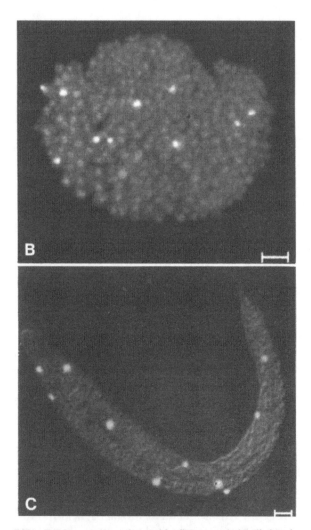

Fig. 6 A–C. Tracing parental chromosomes in embryos. **A** Experimental outline: labelled parental strands are represented by *thick lines*, newly synthesized (unlabelled) DNA strands by *thin lines*; possible experimental outcomes are shown at the *bottom right*. Nullo-X sperm shown; X-bearing sperm introduce 12 labelled parental strands. **B** Detection of parental DNA strands in the developing embryo. 200+ cell embryo labelled in sperm

DNA: BrDU strands are detected by fluorescently labelled antibody (10 *bright spots*) and all nuclei are stained with DAPI (*dimmer spots*). Parental chromatids show no continuing association. **C** Pretzel stage embryo labelled in sperm DNA. Parental labelled strands are scattered randomly through the worm, and appear in different locations in different worms (Ito and McGhee 1987)

19.3 Determination of Cell Fate: Manipulation of Normal Embryos

From the early *C. elegans* cell lineages, we know that specific founder cells invariably give rise to particular differentiated cell types. There is evidence that different mechanisms, under both maternal and embryonic control, operate in different lineages to initiate specific cell types. Direct observation and manipulation of embryos have defined some of the ground rules for various lineages. Experiments perturbing normal embryonic development have utilized manipulation, laser ablation and cell fusion, drugs blocking specific cellular processes, and antibodies developed to identify distinctive cytoskeletal components or differentiation markers for various cell types. These experiments have addressed the general questions of (1) the existence of nuclear or maternal cytoplasmic factors which are segregated to particular blastomeres and play a role in determining cell fate; (2) the cytoskeletal mechanisms for such cytoplasmic partitioning; and (3) the importance of cell-autonomous lineage-based factors versus cell interactions in cell fate decisions, that is, the roles of "mosaic" and regulative control of development.

19.3.1 Chromosomal Imprinting Does Not Appear to Affect Cell Fate Determination in *C. elegans*[6]

In mouse embryos the proper development of embryonic and extra-embryonic tissues requires the presence of both maternal and paternal chromosomes, which presumably contain equivalent sets of genes but may have different epigenetic factors associated with the male and female pronuclei. Whether segregated nuclear information other than the genes themselves plays a role in *C. elegans* embryogenesis was addressed by tracing the paths of maternal and paternal chromatids into the embryonic cells. Males or hermaphrodites were fed the thymidine analog 5-bromodeoxyuridine, mated to an unlabelled worm, and progeny embryos were stained with fluorescent antibodies to track the labelled maternal or paternal chromatids (Fig. 6).

The labelled chromatids persisted intact, and in each case were found to segregate randomly with respect to cell lineage. These experiments argue against chromosomal "imprinting" by sperm or oocyte-derived nuclear factors tightly associated with specifically segregated chromosomes.

19.3.2 Cytoskeletal Mechanisms Orient the Spindles Precisely in the Determinative Cleavages of the P Lineage[7]

Cytoskeletal organization during the early cleavages has been studied using fluorescently labelled antibodies to tubulin and F-actin, and drugs that perturb cytoskeletal organization. The movement of the centrosomes, which form the foci of spindle microtubules, is particularly informative as to the mechanism, although not the primary cause, of the precisely oriented P lineage divisions (Fig. 7). Dur-

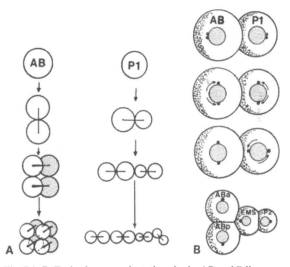

Fig. 7A, B. Early cleavage orientations in the AB and P lineages. **A** AB cells undergo proliferative divisions, cleaving synchronously and at 90° to the previous division (orthogonally). P_1 descendants cleave asynchronously on a longitudinal axis. P_1, P_2, and P_3 are stem cells, and undergo determinative divisions to generate the founder cells. **B** Migration and rotation of the centrosomes at second cleavage. In the AB cell, centrosomes migrate 90° around the nucleus and the cleavage spindle is oriented transverse to the anterior-posterior axis of the embryo. In P_1, the centrosomes migrate and then the nucleus and centrosomes rotate 90°; the ensuing spindle is oriented along the AP axis (After Hyman 1989)

ing each interphase, the centrosome, initially located at the pole of the previous division, divides and the two daughters migrate halfway around the nucleus, to lie on a transverse axis in the embryo. In the AB cell, division occurs on this axis; subsequent AB divisions follow the same pattern, with the centrosome migration offset by 90° at each successive division (producing an orthogonal pattern of division similar to the spiral cleavage of mollusk embryos). In the P_1 blastomere, however, the initial migration is followed by a rotation of the entire mitotic assembly of nucleus and centrosomes, so that the centrosomes come to lie on the anterior-posterior axis of the embryo. The spindle forms along this axis, and is also displaced to the posterior end of the cell. Treatment with the drugs nocodazole or colcemid, which inhibit microtubule polymerization, stops the migration of the centrosomes if applied during migration and, at considerably lower doses, prevents rotation and spindle displacement in P_0 and P_1 if applied before rotation, indicating that microtubules function in both processes. A model proposed for the rotation, supported by laser irradiation of various regions of the P_1 cytoplasm, suggests that microtubules connect one centrosome to the anterior part of the cytoplasmic cortex in the P cells, to provide both the torque for rotation and the precise alignment of the spindle.

19.3.3 Maternal Cytoplasmic Factors Play a Role in the Determination of Cell Fate[8]

The initial cleavage asymmetries just described imply that regional cytoplasmic differences exist even before first cleavage, and that P_1 cytoplasm is not the same as AB cytoplasm. Maternal cytoplasmic components of P lineage cells appear to be implicated in determining both germline and gut cell fates, and in setting the early cell cycle rhythms.

19.3.3.1 Cytoplasmic Particles Called P Granules Segregate with the Germ Line[9]

Studies of the mechanism and timing of the segregation of P granules, which are germ-line specific cytoplasmic particles identified by antibody stain-

ing, have illuminated the process of cytoplasmic localization during the early cell cycles. Before each of the first four cleavages in the early P lineage divisions, P granules are moved to the end of the cell which becomes the daughter leading to the germ line (Fig. 8). When embryos in first cleavage are stained with fluorescent phalloidin, a specific binding protein for polymerized actin, extensive foci of actin filaments are observed in the anterior portion of the zygote after pseudocleavage. Treatment with cytochalasin D, which prevents microfilament polymerization, both disrupts this actin network and abolishes P granule localization, implicating actin microfilaments in the mechanism localizing P granules.

Brief cytochalasin pulses during the first cell cycle show that microfilaments are crucial to P granule segregation only during an 8-min period shortly before pseudocleavage. Pulses during this period abolish cytoplasmic movements and the pseudocleavage furrow; the pronuclei meet in the middle of the egg, rather than at the posterior end; and the first cleavage spindle is centered rather than displaced posteriorly (Fig. 9). After such treatment, embryos will recover and continue to divide, but P granules are not segregated, and a normal embryo does not develop. Whether P granules are indeed direct determinants for the germ line is not yet clear.

19.3.3.2 The Timing of Early Cell Cycles Is Cytoplasmically Controlled[10]

When a laser beam was used to break down cell membranes during early cleavages, fusing AB- and P-derived cells, all nuclei in a common cytoplasm divided synchronously. In a second experiment, enucleated AB or P cytoplasts were fused to P or AB blastomeres, respectively. After the fusion, the period of the nuclear cell cycle changed to reflect that of the added cytoplasm. Perturbing the cell cycle also affects the expression of gut markers (see Fig. 11), but the relation of lineage-specific cell cycle periods to the specification of cell fates is not at all clear. Correct timing, as studies on the gene *cib-1* suggest, may be a necessary but not sufficient condition for correct determination in both the germ line and the gut.

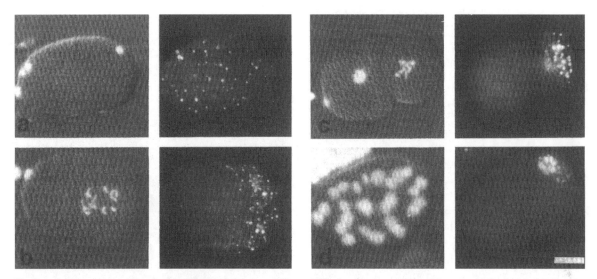

Fig. 8a–d. Segregation of P granules during early cleavage. The *left panels* show embryos with nuclei stained by the fluorescent dye DAPI, the *right panels* show P granules labeled with immunofluorescent antibody in fixed embryos of the same stages. Anterior is to the left (Photo S. Strome). **a** One-cell embryo during pronuclear formation prior to pseudocleavage. P granules are not localized. **b** One-cell embryo just before first cleavage.

The chromosomes area condensed, and P granules are localized to the posterior, to be partitioned into P_1 at division. **c** Two-cell embryo late in the cell cycle. P granules are again localized at the posterior and partitioned into P_2 at division. **d** Embryo at approximately 30 cells. P granules are localized in a single cell, P_4, the germ-line progenitor

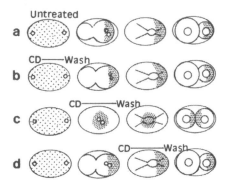

Fig. 9a–d. Effects of 10-min cytochalasin D pulses on localization of P granules. Anterior to the *left*; *stipples* represent P granule distribution; ✕ represents the spindle position. *Lines* represent duration of the cytochalasin D exposure (After Hill and Strome 1990). **a** Untreated embryo, showing normal localization of P granules. **b** Early cytochalasin D pulse, ending before pronuclear movement. Localization is normal. **c** Pulse during pronuclear movement and the normal period of pseudocleavage. Pseudocleavage does not occur and P granules are not localized to the P_1 cell. **d** Pulse beginning after pronuclear meeting and pseudocleavage. Localization occurs normally

19.3.3.3 Gut Determination Also Requires Cytoplasmic Segregation[11]

The gut (E cell) lineage is clonally established even before the germ line, at the third cleavage. Expression of gut cell fate has been followed by the expression of autofluorescent "gut granules" and a gut-specific esterase which can be cytochemically stained. Both of these markers appear early, at about the 100-cell stage. Their expression is blocked in embryos treated with alpha-amanitin before the 16–20-cell stage, suggesting that the genes involved are transcribed even earlier, shortly after the E lineage arises.

Cytochalasin D treatment of embryos before first cleavage abolishes expression of the gut markers, implicating the microfilament system in the initial segregation of potential for gut development as well as that of P granules. Experiments blocking subsequent cleavages show that the capacity to produce the gut markers is under cytoplasmic control,

Fig. 10a−e. Staining for gut esterase in embryos blocked with cytochalasin D and incubated for 18 h. As division proceeds, the capacity to make gut-specific markers is passed from P_1 to EMS to the E cell, and thereafter is present in all E descendants. However, one-cell embryos treated with cytochalasin D do not express the markers (Edgar and McGhee 1986). **a** Cytochalasin D treated one-cell embryo. **b** Two-cell embryo, with P_1 expressing esterase. **c** Four-cell embryo, with EMS expressing esterase. **d** Eight-cell embryo, with E expressing esterase. **e** 16-cell embryo, with esterase expression in both E cells

and is passed into the E-cell progenitor at each of the first three cleavages.

Permeabilized cytochalasin-treated two-cell embryos, after overnight incubation, will produce gut granules and esterase only in the P_1 cell; four-cell embryos in the EMS, eight-cell embryos in the E cell, 16-cell embryos in the two E cells, and so on (Fig. 10). Blastomeres isolated at any of these stages will produce the markers without interactions with other cells. In another type of experiment, the P_1 nucleus of a two-cell embryo was extruded through a laser-cut hole in the eggshell, and the remaining P cytoplasm was fused to an AB cell (Fig. 11). Some of the descendants from the AB nucleus subsequently produced gut granules, strengthening the idea that a localized cytoplasmic component is instrumental in gut determination.

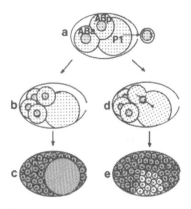

Fig. 11a−e. Cytoplasmic localization of gut cell potential. **a** The P_1 nucleus of a two-cell embryo was extruded through a laser-cut hole in the eggshell, leaving an enucleate cytoplast. **b, c** Control embryos receiving no further treatment. **d** When AB had divided to four cells, one of these was laser-fused to the P_1 cytoplast. The AB nucleus entered the P_1 cytoplasm and continued division. **e** After overnight incubation, cells in the region derived from the P_1 cytoplasm showed gut granule expression (light area) (After Schierenberg 1985)

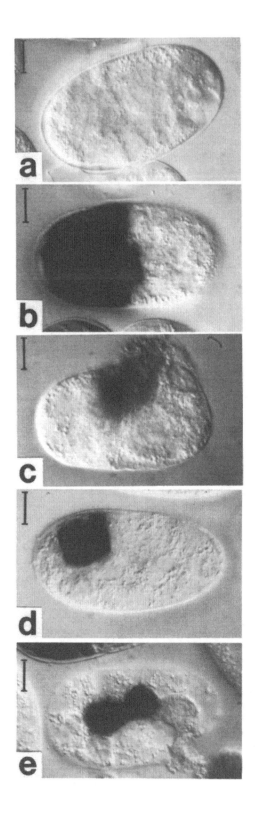

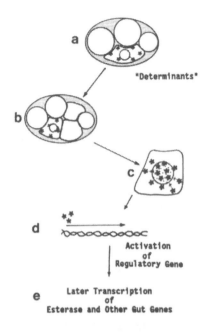

Fig. 12a – e. Model for gut-cell determination and activation. In this model, *stars* represent a cytoplasmic determinant segregated into the E cell, which then enters the nucleus and activates a nuclear regulatory gene during the first E-cell cycle of DNA replication

These experiments generate the model that some cytoplasmic factor is segregated into the E cell, which then must initiate transcription of a cascade of gut-specific genes (Fig. 12). If permeabilized early embryos are treated at various times with the drug aphidicolin to block DNA synthesis, the eventual expression of gut granules and esterase is completely inhibited until the first round of DNA synthesis in the E cell is in progress, but is no longer inhibited after the end of this cell cycle. This time frame for nuclear activation suggests that the cytoplasmic determinant may be a binding factor which gains access to newly synthesized DNA during this particular round of DNA synthesis.

19.3.3.4 Cloning the Esterase Gene May Identify Its Molecular Controls[12]

In the effort to identify such regulatory proteins, the gut esterase gene has been cloned. Binding factors are indeed indicated by gel retardation assay, in which the cloned DNA runs more slowly in gel electrophoresis if it has first been incubated with nuclear extracts from early embryos, although specific factors have not yet been isolated. With respect to controls built into the gene itself, short nucleotide sequences in the 1300 nucleotides upstream of the first exon appear to confer specificity for gut expression. Deletions of different parts of this region cause ectopic expression of esterase in the developing pharynx, muscle, or hypodermal cells when transient expression is assayed following injection of the deleted sequences. These experiments suggest both that an activator initiates E lineage expression, and that repressors prevent esterase expression in the non-gut tissues derived from the P lineage.

19.3.3.5 Pluripotent Blastomeres Blocked in Cleavage Express Only One Differentiation Pathway[13]

The P_0, P_1, and EMS cells are pluripotent cells, and give rise to other cell types besides gut and germ line. In further cleavage-block experiments on two-cell embryos, fluorescently labelled antibodies were used to assay muscle myosin and an embryonic hypodermal antigen as well as gut granules in the blocked P_1 cell (Fig. 13). It was found that, while the P_1 cell could differentiate muscle, gut or hypodermal markers after incubation, it never produced more than one marker in a particular embryo. This exclusivity suggests an all-or-none aspect of gene activation as cell fates are narrowed down.

19.3.4 Most Lineages Are Established by the 51-Cell Stage, but Some Cell-Fate Decisions Depend on Early Cell Interactions[14]

Because *C. elegans* embryos display an invariant set of cell movements as well as divisions, it is not clear whether cell fate is always specified by intrinsic cell-autonomous factors, i.e., determined by lineage, or in addition requires cell interactions which are masked because cell contacts are invariant. Extensive laser ablations after the 51-cell stage show little or no regulation by neighboring or related cells,

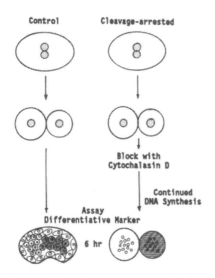

Fig. 13. Diagram of the cleavage-arrest assay for differentiation potential. Two-cell embryos are blocked in cleavage with cytochalasin D. DNA synthesis continues at a normal rate for many cell cycles. After incubation, differentiative markers for hypodermis, muscle, and gut are assayed by staining with fluorescent antibodies. In the illustration, *dark stippling* represents gut granules, found only in P_1 in the blocked embryo, and in the gut in normal embryos. This result indicates that the P_1 blastomere has the potential to initiate a specific differentiation program without further cleavage (After Cowan and McIntosh 1985)

suggesting that most cell fates, at least with respect to differentiation, are set before this stage. The P-derived germ line and gut lineages appear cell-autonomous with respect to cell fate from the beginning, as described above. However, there is increasing evidence that early cell interactions control the choice of cell fate in AB-derived cell lineages.

19.3.4.1 Pharyngeal Muscle Differentiation in ABa Cells Is Induced by P Lineage Cells

Both ABa and P descendants contribute muscle cells to the pharynx. Following laser removal of the three ABa-derived precursors at the 28-cell stage, the remaining P-derived cells make a posterior half-pharynx; in the reciprocal ablation of the two P-derived pharyngeal muscle precursors, the remaining ABa precursors make an anterior half-pharynx. However, if cell lines are isolated much earlier, by removing either the AB blastomere at two cells, or EMS at four cells with a glass microneedle, and the

partial embryo is assayed for differentiation with an antibody specific to early pharyngeal muscle cells, surviving EMS derivatives will produce the antigen but ABa derivatives will not. Thus the EMS line appears to induce muscle development in the ABa cells entering the pharynx, and acts before the 28-cell stage.

19.3.4.2 Establishment of Left-Right Polarity and Bilateral Symmetry Depends on Cell Contacts

At four cells, the worm embryo has established its dorsal-ventral axis, with EMS located ventrally (see Fig. 14). The ABa and ABp cells then make a left-right division on a diagonal axis, so that the left pair of daughter cells, ABal and ABpl, lies slightly anterior to their respective sisters ABar and ABpr. Thus their descendants establish somewhat different cell contacts; they also follow different, though similar, lineage patterns. From this initial left-right asymmetry, the embryo eventually achieves a high degree of bilateral symmetry, by a rather surprising method: cells from different lineages in similar contralateral positions achieve similar differentiative cell fates. Left-right differences remain, however, in at least 20 cells. How and when are the left-right axis and its polarity differences established?

Because very early ablations are technically difficult to perform without killing the entire embryo, the possibility of important cell interactions before the 51-cell stage has been approached by direct manipulation. If the AB cell is pressed with a glass needle during cleavage, changing the relative positions of ABa and ABp in the embryo, a normal worm, with normal left-right polarity, develops. Thus ABa and ABp are still equivalent cells, and the dorsal AB cell makes a normally skewed left-right division. However, shifting the skew by pressure at the next cleavage, so that ABal and ABpl become the posterior set and ABar and ABpr lie slightly anterior, produces a mirror-image worm with reversed left-right asymmetries (Fig. 14).

Thus the two AB cell pairs are still equivalent in the 15-cell embryo but the "forward" pair will follow the left-side lineage even though it is physically on the right side, to give the mirror image. Instructions from cell contacts after the six-cell stage must therefore determine left-right polarity. Since the laser ablation data indicate that asymmetries are well established by 51 cells, significant cell interactions must occur within this early window of development.

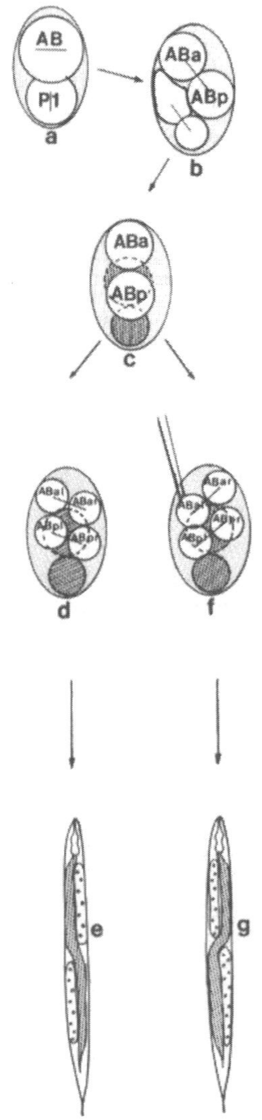

19.4 Mutational Analysis of Embryogenesis in *C. elegans*

Genetics provide a powerful approach to the investigation of development, in that individual gene functions can be identified and studied as mutations in one gene at a time, and the interactions between mutations can identify interactions of gene products and regulatory pathways. At this point, genetic studies of embryogenesis in *C. elegans* have included a number of general mutational screens attempting to identify both maternally and embryonically required genes important in aspects of cell fate determination, patterning, and morphogenesis. So far, only a few of these genes have been cloned and analyzed at a molecular level. A complete molecular understanding will require both classical genetics, to identify and analyze the functions of critical genes in the organism; and molecular genetics, to characterize the molecular nature and the interactions of the gene products. Current strategies for study are moving in both directions, either from the genetic identification of mutations in genes of interest to their molecular cloning and analysis, or in reverse, from the molecular characterization of a cloned gene to some understanding of its role in the organism through isolating or creating mutations or disrupting gene function through transformation or other means.

19.4.1 Maternal Gene Products Control and Support Much of Early Embryogenesis[15]

In many organisms, the genes affecting the earliest embryonic stages are transcribed maternally, and many of these transcripts are translated in the embryo after fertilization. Gene products made maternally and required for embryogenesis can be identified by means of maternal-effect lethal mutations,

Fig. 14a–g. Left-right asymmetry in six-cell normal and reversed embryos (After Wood, 1991). **a** Side view of a two-cell embryo (anterior at top). **b** Side view of a four-cell embryo. The dorsal-ventral axis is established when ABp moves over the P_1 cell as the AB cell divides at second cleavage. **c** Dorsal view of the four-cell embryo, showing upper (dorsal) plane, anterior at top. EMS lies under ABp. **d** Dorsal view of a six-cell embryo, after the left-right divisions of ABa and ABp. These divisions are skewed, producing a left-right asymmetry, and lineage-equivalent cells on the left and right later follow different lineage patterns. **e** Dorsal view of a normal adult worm showing left-right asymmetries of the gut and gonad. **f** Dorsal view of a reversed six-cell embryo, produced by pressure with a microneedle during the ABa and ABp divisions. **g** Dorsal view of the reversed adult worm arising from this operation

by the criteria that heterozygous mothers produce all viable progeny, while homozygous mutant mothers are viable but produce only dead offspring on self-fertilization (see Fig. 15). Essential genes which must be transcribed in the embryo are identified through nonmaternal embryonic lethal mutations, by the criteria that heterozygous mothers will produce 1/4 dead progeny.

A great majority of the nonmaternal, nonconditional lethal mutants identified to date in *C. elegans* can survive through embryogenesis, and die during larval stages. Because screens isolating mutations causing death anywhere in the life cycle might be expected to identify an unselected spectrum of essential genes, these findings further support the idea that most of embryogenesis is supported by the activity of the maternal genome. Conversely, in *C. elegans* screens for temperature-sensitive *ts* (conditional) embryonic lethal mutations, a majority of the genes defined are required both maternally and at other stages of the life cycle, and thus probably participate in generally required functions.

One strategy to search for a subset of genes whose normal products are maternally supplied and required only for embryogenesis is the isolation of nonconditional, strict maternal-effect mutants. Such a screen covering linkage group II identified 17 loci with strict maternal effects, but only four of these loci mutated at the frequencies typical

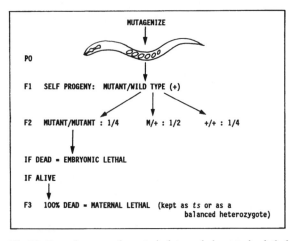

Fig. 15. Screening procedures to isolate and characterize lethal mutations

of null mutations in *C. elegans*, the others much less frequently. Thus it is likely that the majority of these mutants represent rare alleles of genes also used during other times of the life cycle, rather than genes used only during embryogenesis. From this study, the total number of strictly maternally active genes in the genome is estimated as no more than 12. Similar conclusions were drawn from a study of X-linked strict maternal-effect mutations in *Drosophila*, suggesting that the number of genes used strictly in embryogenesis in flies is also small.

19.4.1.1 Some Developmentally Important Genes Have Been Identified by Mutant Phenotypes[16]

If most of the maternally active genes involved in key steps in embryogenesis are also required embryonically, general maternal screens may not identify all genes which have important maternal roles in embryogenesis, because most homozygous mutations in such genes would give lethal embryonic, rather than sterile maternal, phenotypes. An alternative approach is to look for mutations with particular phenotypes that reflect disturbances in a specific process. This approach has been highly successful in *Drosophila*, where scoring cuticle markers on dead embryos has revealed underlying pattern defects. Although phenotypic assays for *C. elegans*, where mutant embryos display a limited range of death phenotypes, are not as well developed, specific screens have been conducted for maternal effect mutants defective in early cleavage patterns, that affect cell fate specification; for dominant maternal effect mutations, that should identify members of multi-gene families; and for "grandchildless", or maternal-effect sterile mutants, that may affect genes involved in germ line determination. Of the genes identified in these screens, many affect early cleavage patterns and probably represent components of the cytoskeletal mechanisms for precise early cell divisions, which thus govern the distribution of maternal information specifying cell fate (see Table 1). Among these, the five *par* genes have been most extensively analysed. In *par-1*, *par-2*, and *par-4* mutations, both cells often show AB-like cleavage orientation at the second cleavage; conversely, in *par-3* mutants, both

Table 1. Genes acting maternally in *C. elegans* development

Gene	Mutant defects	Terminal mutant phenotype	P granule segregation	Probable lesion	Cloned?/homologies
I. Genes involved in mechanisms of early cleavage					
par-1 *par-2* *par-4* *par-5*	Abnormal pseudo-cleavage and early cleavages; second cleavage often synchronous and transverse (both cells follow AB division pattern)	Several hundred cells; no morphogenesis; abnormal numbers and localization of differentiated cells	Abnormal; no localization	Cytoskeletal system of early mitotic spindles; abolished spindle rotation in P_1	No Yes/none No No
par-3	Abnormal pseudo-cleavage and early cleavages; second cleavage often synchronous and longitudinal in both cells (both cells follow P division pattern)	Several hundred cells; no morphogenesis; abnormal numbers and localization of differentiated cells	Abnormal; no localization	Cytoskeletal system of early mitotic spindles; adds rotation in AB	No
spe-11	*par*-like early cleavages	Several hundred cells; no morphogenesis; abnormal numbers and localization of differentiated cells	Abnormal; no localization	Cytoskeletal system of early mitotic spindles; abolishes spindle rotation in P_1	No
mei-1 *mei-2* *mel-26* *zyg-9*	Aberrant meiosis and early cleavages	Several hundred cells; no morphogenesis; abnormal numbers and localization of differentiated cells	Abnormal; no localization	Cytoskeletal system of early mitotic and meiotic spindles	No
II. Genes affecting germ line determination					
mes-1	Grandchildless: progeny of homozygous mutants lack a germline	Sterile progeny from homozygous mothers	Abnormal; no localization at fourth cleavage	Segregation mechanism for germ line determinants	No
mes-2	Same	Same	Normal	Germ line determinants?	No
mes-3	Same	Same	Normal		No
mes-4	Same	Same	Normal	Same	No
mes-5	Same	Same	Normal	Same	No
III. Genes affecting cell fate					
cib-1	Abnormal early cleavage, P divisions slow	Several hundred cells, little morphogenesis; after delayed cleavage, P_1 generates gut, muscle, but no germ line (acts like daughter cell EMS)	Abnormal; no localization	Mitotic apparatus of P lineage	No
glp-1	Maternal alleles: no anterior pharynx	Late morphogenesis	Normal	Receptor/effector for cell communication: P to AB, before 28 cells	Yes/*lin-12* (*C. elegans*) Notch (*Drosophila*)
	Nonmaternal alleles; no germ line proliferation	Sterile adults	Normal	Receptor/effector for cell communication: somatic distal tip cell to germ line, postembryonic	EGF-like motifs Ankyrin motif transmembrane protein

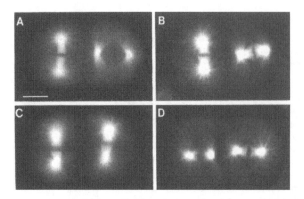

Fig. 16 A–D. Spindle orientations in two-cell embryos of wild-type and *par* mutants. Embryos in the process of mitosis have been fixed and stained with fluorescently labeled antibodies against tubulin. Anterior (the AB cell) is at the left (Photo K. Kemphues). **A** Wild-type embryo, showing a transverse spindle in the AB cell and a longitudinal spindle in the P cell. Note asynchrony of mitosis in the two cells; P_1 is slower than AB. **B** *par-1* mutant embryo with normal spindle orientation but synchronous mitoses in AB and P. **C** *par-2* mutant embryo with abnormal transverse orientation of the P_1 spindle. **D** *par-3* mutant embryo with abnormal longitudinal orientation of the AB spindle. *Bar* 10 μm

cells cleave with the longitudinal P lineage orientation (Fig. 16). Somewhat surprisingly, in the double mutant *par-2, par-3* the P-like cleavage prevails, suggesting that *par-3* normally represses spindle rotation, but *par-2* prevents it from acting in P cells.

Mutants in the gene *spe-11* also show aberrant early cleavage, but the genetic tests localize the defect to sperm. Because the sperm contributes the centrosome to the zygote, the *spe-11* gene product is most likely a centrosome-associated molecule, implying that the centrosome is also involved in the control of the important asymmetric early cleavages in the zygote.

Other genes being studied, the *mes* genes, *cib-1*, and *glp-1*, appear to be more directly involved in the mechanisms determining particular cell fates. Current data on these genes is summarized in Table 1.

19.4.1.2 glp-1 Is Involved in Induction and Cell Communication in Two Distinct Sets of Cells[17]

Maternal effect mutants in the *glp-1* (germline proliferation) gene, isolated in screens for maternal effect lethals, produce embryos missing the anterior portion of the pharynx. As described above, this part of the pharynx is derived from descendants of the ABa blastomere, and requires the presence of P-derived cells in order to express the proper fate. The pharyngeal (embryonic) temperature-sensitive period occurs very early, between the 4-cell and 28-cell stage. Another set of *glp-1* alleles are nonmaternal; in these mutants the germ line fails to proliferate and homozygotes are sterile. Previously, laser ablation experiments killing the distal tip cell of the somatic gonad produced the same phenotype, indicating that communication between the distal tip cell and the germ line represses meiosis.

Both of these mutant phenotypes suggest that the *glp-1* gene is involved in cell interactions. Mosaic analysis of the germ line phenotype indicates the *glp-1* protein is a receptor: normal gene product must be present in the germ line and is not necessary in the distal tip cell (Fig. 17).

Molecular cloning of the *glp-1* gene has produced interesting results. The sequence is highly homologous to the closely-linked *C. elegans* gene *lin-12* and to the *Drosophila* gene *Notch*, both of which appear to function in cell communications. The sequences of the three genes all predict transmembrane proteins. All contain several copies of an epidermal growth factor-like sequence and three copies of a unique cysteine-rich motif on the probable extracellular sequences, and six copies of a motif homologous to ankyrin, also found in the yeast cell cycle genes *cdc 10* and *swi 6*, on the putative intracellular domains. Presumably the three gene products function similarly to receive a signal at the cell surface and activate the next step inside the cell. Extragenic suppressors of *glp-1*, which should identify interacting gene products, include two known collagen genes, suggesting that the basement membrane may be involved in signalling. *glp-1* and *lin-12* most likely arose from gene duplication, and their differences may provide insight into how specificity is controlled for the multiple cell interactions involved in development.

MOSAIC ANALYSIS OF *glp-1*

A STRAIN MUTANT FOR *glp-1* BUT CARRYING A FREE DUPLICATION CONTAINING *glp-1*⁺
IS ANALYZED FOR ANIMALS LOSING THE DUPLICATION IN PARTICULAR TISSUES.

OCCASIONALLY THE DUPLICATION IS LOST DURING MITOSIS

WILD TYPE CELL LINE   MUTANT CELL LINE

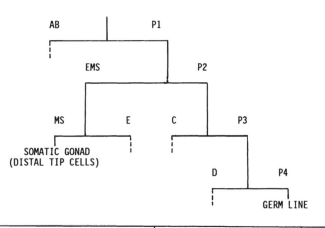

DUPLICATION IN SOMATIC GONAD	DUPLICATION IN GERM LINE	PHENOTYPE
I. LOST	PRESENT	WILD TYPE
II. PRESENT	LOST	*glp-1*
FROM THESE TWO CASES, *glp-1* ACTIVITY IS NEEDED IN THE GERMLINE, BUT NOT IN THE DISTAL TIP CELLS. BECAUSE LASER ABLATION OF THE DISTAL TIP CELLS ALSO PRODUCES THE *glp-1* PHENOTYPE, THE DISTAL TIP CELL MUST BE SENDING A SIGNAL, AND THE *glp-1* GENE PRODUCT MUST BE RECEIVING IT IN THE GERM LINE.		

Fig. 17. Mosaic analysis of the *glp-1* sterile (nonmaternal) phenotype, indicating that germ-line cells require *glp-1* activity. (Data from Austin and Kimble 1987)

19.4.2 Embryonic Gene Activity also Plays a Role in Cell Fate Determination and Morphogenesis

The maternal activity of the genes described in the previous section indicates that the mechanisms for segregating cytoplasmic components that specify cell fate, some of the determinants themselves, and at least one of the later inductive cell interactions, are primarily under maternal control. At some point in development, however, an embryo must switch to using its own transcripts. Does *C. elegans* utilize embryonically acting genes that function like the pair-rule and homeotic genes of *Drosophila* to refine the developing embryonic pattern?

The early perceptions of *C. elegans* development as highly mosaic and maternally controlled assumed little postmaternal influence on cell fates.

However, the evidence discussed in Sections 19.3.4 and 19.4.2.1 demonstrates that cell interactions are important early in embryogenesis and that transcription starts earlier than previously believed. A more extensive screening for nonmaternal embryonic lethal mutations is now in progress, and should provide a clearer understanding of the role of embryonic gene activity in cell fate determination and morphogenesis.

19.4.2.1 Embryonic Transcription Begins Early, and Reaches a High Level Before 100 Cells[18]

An assay for the appearance of nuclear poly-A containing message by in situ hybridization to a poly-U probe showed that a high level of embryonic transcription is operating by the 90- to 125-cell stage, or at the end of gastrulation. More recently, direct radioactive labelling of permeabilized embryos indicates that transcription begins considerably earlier, at eight to 16 cells, and is appreciable by 20 to 30 cells. Run-on transcription per nucleus is also as high using RNA extracts from embryos averaging less than 30 cells as that obtained from extracts of 100- to 200-cell embryos. Clones isolated by using these extracts to probe genomic or cDNA libraries have identified several genes transcribed at high levels in embryos at about the 30-cell stage, but not in later embryos. Thus it is quite likely that some of the key genes in cell fate determination and morphogenesis are activated very early in embryonic development, just as in *Drosophila* the pattern genes are expressed one or two cell cycles before general transcription begins.

19.4.2.2 Deficiency Analysis Suggests Maternal Control of Early Cleavage Patterns and Initial Cell Proliferation[19]

The phenotype of embryos homozygous for a given chromosomal deficiency should be equivalent to that resulting from null mutations in the earliest embryonically required gene in that region. As a rapid way to assess the functions of a large number of essential genes, the pregastrulation lineages and terminal phenotypes of homozygous deficiency embryos, which segregate as a quarter of the progeny from heterozygous mothers, were analyzed for a set of deficiencies covering approximately half of the known genes. Almost all of the lethal deficiencies produced embryos with normal pregastrulation cleavages, which typically arrested as "monsters" with 200 to 500 cells and showed markers for muscle, gut, and hypodermal differentiation. In most cases, however, defined organs such as the gut or pharynx were absent or rudimentary.

From this study, it seems likely that crucial embryonically active genes will affect aspects of morphogenesis, which requires precise spatial and temporal patterns of gene expression, and is likely to involve cell interactions. The capacity to differentiate at the cellular level seems to display much greater cell autonomy. However, if all transcription is blocked with alpha-amanitin before 16 cells, embryos arrest at about 100 cells, but do not show any markers of differentiation. Even if maternal provisions can support cell division and maintenance to this stage, early embryonic transcription appears necessary for key aspects of differentiation.

19.4.2.3 Screens for Non maternal Embryonic Lethal Mutations Have Identified Embryonically Transcribed Genes[20]

Several mutagenesis screens are in progress with the goal of saturating the genome for mutations in embryonically required genes, and identifying among them the most developmentally interesting: those contributing to fate determination, differentiation of particular cell types, or formation of morphological structures as opposed to genes required for general maintenance function.

By Nomarski microscopic observation, most embryonic lethal phenotypes can be sorted into about ten classes on the basis of degree of elongation, hypodermal integrity or rupture, paralysis indicating muscle involvement, and degree of morphogenesis. This limited number of easily identified terminal phenotypes makes analysis of which system a gene affects, or its implied functions, rather difficult to approach. One very promising method is analysis of mutant patterns of antibody markers for specific differentiated cell products, identifying

Table 2. Genes acting embryonically in *C. elegans* development

Gene	System affected	Terminal mutant phenotype	Probable lesion	Cloned?/homologies
pha-1	Pharynx	Elongated larva without a pharynx; muscle, marginal, and gland cells missing	Cell interactions determining cell fate	No
emb-29	All cells	Embryos arrest at 140 cells; cells differentiate	Cell cycle component	No
zen-1	Ectodermal tissues	Embryos arrest at 500 cells; no morphogenesis; hypodermis fails to enclose embryo	Cell fate specification; reduced hypodermal, muscle cell numbers; extra neurons	No
hlh-1	Muscle	No mutants yet identified	Expressed only in body wall muscle precursor cells from the 28-cell stage	Yes; homologous to vertebrate MyoD

early or late markers for such cells as neurons, hypodermal cells, or muscle. A number of such antibodies, both monoclonal and polyclonal, have been made. As the repertoire of markers is expanded, more specific mutant phenotypes can be defined. Mutations have been isolated that affect cell division, hypodermal, muscle, and neuronal development, and morphogenesis of the pharynx, identifying genes that appear to be involved in cell fate determination and morphogenesis in these ectodermal and mesodermal tissues. Table 2 summarizes information on several of the most fully defined genes.

19.4.2.4 Molecular Screening Has Identified Homologs to Developmentally Important Genes in Other Organisms[21]

The continuing expansion of the physical genomic map for *C. elegans* makes it realistic to use molecular approaches to identify genes homologous to those important in developmental processes in other organisms. For functional analysis, genes localized on the *C. elegans* physical map must also be identified on the genetic map, either by correlation to mutants that already exist or by creating mutants with DNA polymorphisms in that region. Mutant phenotypes can then be used to determine how the gene under study functions in normal embryos.

A Large Number of Homeobox Genes Have Been Identified. The molecular search for homeobox sequences in *C. elegans* has identified a large number throughout the genome. One group forms a cluster of four genes with sequence homologies to genes in the same relative order in the HOM-C (*Antennipedia* homeobox complex) in *Drosophila* and vertebrates, suggesting that this complex, involved in antero-posterior pattern formation, may have evolved before segmentation. Few of the homeobox genes are as yet associated with known genes, except for *mec-3*, *unc-86*, *pal-1*, and *mab-5*, and even fewer with genes known to affect embryogenesis. At least one homeobox gene, the *engrailed* homolog identified by cross-reaction with an antibody to the *Drosophila* gene product, is expressed in developing embryos at the 70-cell stage, and only in future hypodermal cells, although its function is not yet known.

hlh-1, a Homolog of Vertebrate Myo-D, Is Expressed in Muscle Precursors at the 28-Cell Stage. MyoD is a mammalian gene which appears to act as a "master switch" for myogenesis. Its DNA sequence contains a conserved helix-loop-helix (HLH) region found in the oncogene *myc* and in various developmental regulatory genes such as *achaete-scute* of *Drosophila*. A *C. elegans* homolog of the MyoD gene, *hlh-1*, has been identified, which closely resembles the vertebrate MyoDs in the HLH region, although flanking sequences are apparently not related. Like its vertebrate counterparts, the CeMyoD protein, as demonstrated by antibody staining or following transformation with a β-galactosidase fusion gene, is localized to nuclei and is expressed only in muscle cells or muscle progenitor cells. The earliest expression occurs in the four MS cells of the 28-cell embryo; later the fusion product appears in posterior embryonic cells, presumed to be C and D descendants. Interestingly, since the four MS representatives in the 28-cell embryo generate other cell types in addition to muscle, the CeMyoD signal must later disappear or be overridden in the nonmuscle lineages.

19.5 Outlook

The identification and cloning of genes important in *C. elegans* embryogenesis is moving rapidly at this point. Genes affecting cell fate specification and morphogenesis in neurons, hypodermal cells, muscle, gut, germ line, and the pharynx are now under investigation. As the molecular identities, expression patterns, and genetic interactions of more and more such genes are analyzed, the hierarchies of genetic regulation of worm development should become increasingly understood at the molecular level. Questions of importance will focus on how maternal and embryonic regulation interact in determination of cell fate and differentiation, and on what cell interactions are required for differentiation and successful morphogenesis. How do cellular and molecular interactions set up proper spatial and temporal patterns and initiate new gene activities? Is negative regulation of gene interactions as important in *C. elegans* embryonic development as in the well-defined genetic pathway of sex determination?

Continuing pursuit and investigation of *C. elegans* homologs of the homeobox genes, growth-factor-related peptides, steroid hormones and their receptors, and the tyrosine kinase gene family, all active in other embryonic systems, should identify genes also important in *C. elegans* development. The cloning of stage-specific genes expressed maternally and in early embryos, using both the run-on transcript system and polymerase chain reaction (PCR) amplification, is also being initiated, and may identify additional genes functioning at the earliest steps of cell fate specification. In a complementary fashion, the current pursuit of saturation

mutagenesis screens should define more of the genes important in embryogenic development, and allow genetic definition of their functions and interactions.

The growing number of *C. elegans* homologs to genes important in other developmental systems underscores that the genes used in development are probably quite a universal set. An understanding of *C. elegans* development should extend our knowledge of how genes are modified to control different patterns of embryogenesis in widely diverse organisms.

19.6 Summary

Studies of *C. elegans* development have combined cellular, genetic, and molecular approaches to approach general principles of how genes control development. Questions particularly appropriate to explore in this organism include how a cell initially becomes committed to its fate, and how patterning and morphogenesis are initiated and controlled. Maternal information controls early steps of cell fate determination for the germ line and gut lineages; this information is segregated by cytoskeletal mechanisms involving microfilaments during the early cleavages. However, cell interactions before the 51-cell stage establish left-right polarity and bilateral symmetry in the early embryo and determine cell fate in lineages contributing to the pharynx.

Mutational analysis of embryogenesis indicates that the early cleavage mechanisms defining cytoplasmic localization, gut and germ line determination, and cell division up to the premorphogenesis stage, i.e., much of early embryogenesis, are supported by maternally supplied gene products. However, only a few genes are required strictly maternally; thus many maternal-effect mutations may identify "housekeeping" genes. Embryonic gene activity begins early, as evidenced both by transcriptional analysis and the isolation of nonmaternal embryonic lethal mutations which show phenotypes before and during morphogenesis. Homologs of homeobox genes, vertebrate MyoD, and epidermal growth factor play roles in *C. elegans* embryogenesis. Molecular analysis of both maternally and embryonically active genes is progressing at a rapid rate, and should illuminate mechanisms and interactions of key genes in control of embryogenesis in this organism.

References

[1] Emmons SW (1987) Mechanisms of *C. elegans* development. Cell 51:881–883

Kenyon C (1988) The nematode *Caenorhabditis elegans*. Science 240:1448–1453

White J (1988) The Anatomy. In: Wood WB (ed) The nematode *Caenorhabditis elegans*, chapter 4. Cold Spring Harbor Laboratory, New York, pp 81–122

Wood WB (ed) (1988) The nematode *Caenorhabditis elegans*. Cold Spring Harbor Laboratory, New York

Wood WB (1988) Introduction to *C. elegans* biology. In: Wood WB (ed) The nematode *Carnorhabditis elegans*, chapter 1. Cold Spring Harbor Laboratory, New York, pp 1–17

[2] Sulston J (1988) Cell Lineage. In: Wood WB (ed) The nematode *Caenorhabditis elegans*, chapter 5. Cold Spring Harbor Laboratory, New York, pp 123–156

Sulston J, Schierenberg E, White J, Thomson N (1983) The embryonic cell lineage of the nematode *C. elegans*. Dev Biol 100:64–119

[3] Herman RK (1988) Genetics. In: Wood WB (ed) The nematode *Caenorhabditis elegans*, chapter 2. Cold Spring Harbor Laboratory, New York, pp 17–46

Herman R (1988) Mosaic analysis in the nematode *Caenorhabditis elegans*. J Neurogenet 5:1–24

Hodgkin J, Kondo K, Waterston R (1987) Suppression in the nematode *Caenorhabditis elegans*. Trends Genet 3:325–329

[4] Coulson A, Sulston J, Brenner S, Karn J (1976) Toward a physical map of the genome of the nematode *C. elegans*. Proc Natl Acad Sci USA 83:7821–7825

Emmons SW (1988) The genome. In: Wood WB (ed) The nematode *Caenorhabditis elegans*, chapter 3. Cold Spring Harbor Laboratory, New York, pp 47–80

Fire A (1986) Integrative transformation of *Caenorhabditis elegans*. EMBO J 5:2673–2680

[5] Kimble J, Ward S (1988) Germ-line development and fertilization. In: Wood WB (ed) The nematode *Caenorhabditis elegans*, chapter 7. Cold Spring Harbor Laboratory, New York, pp 191–214

Wood WB (1988) Embryology. In: Wood WB (ed) The nematode *Caenorhabditis elegans*, chapter 8. Cold Spring Harbor Laboratory, New York, pp 215–242

[6] Ito K, McGhee J (1987) Parental DNA strands segregate randomly during embryonic development of *Caenorhabditis elegans*. Cell 49:329–336

[7] Hyman A (1989) Centrosome movement in the early divisions of *Caenorhabditis elegans*: a cortical site determining centrosome position. J Cell Biol 109:1185–1193

Hyman A, White J (1987) Determination of cell division axes in the early embryogenesis of *Caenorhabditis elegans*. J Cell Biol 105:2123–2135

[8] Kemphues K (1989) *Caenorhabditis*. In: Grover DM, Hames BD (eds) Genes and embryos. IRL Press at Oxford University Press, Oxford, pp 95–126

Schierenberg E (1989) Cytoplasmic determination and distribution of developmental potential in the embryo of *Caenorhabditis elegans*. Bio Essays 10:99–104

Strome S (1989) Generation of cell diversity during early embryogenesis in the nematode *Caenorhabditis elegans*. Int Rev Cytol 114:81–123

[9] Hill D, Strome S (1988) An analysis of the role of microfilaments in the establishment and maintenance of asymmetry in *Caenorhabditis elegans* zygotes. Dev Biol 125:75–84

Hill DP, Strome S (1990) Brief cytochalasin-induced disruption of microfilaments during a critical interval in 1-cell *C. elegans* embryos alters the partitioning of developmental instructions to the 2-cell embryo. Development 108:159–172

Strome S, Wood W (1983) Generation of asymmetry and segregation of germ-line granules in early *C. elegans* embryos. Cell 35:15–25

[10] Schierenberg E, Wood W (1985) Control of cell-cycle timing in early embryos of *Caenorhabditis elegans*. Dev Biol 107:337–354

[11] Edgar L, McGhee J (1986) Embryonic expression of a gut-specific esterase in *Caenorhabditis elegans*. Dev Biol 114:109–118

Edgar L, McGhee J (1988) DNA synthesis and the control of embryonic gene expression in *C. elegans*. Cell 53:589–599

Laufer J, Bazzicalupo P, Wood W (1980) Segregation of developmental potential in early embryos of *Caenorhabditis elegans*. Cell 19:569–577

Schierenberg E (1985) Cell determination during early embryogenesis of the nematode *Caenorhabditis elegans*. Cold Spring Harbor Laboratory Symp Quant Biol 50:59–68

[12] Aamodt E, Chung M, McGhee J (1991) Spatial control of gut-specific gene expression during *C. elegans* development. Science 252:579–582

McGhee J (1991) Gut esterase expression in the nematode *Caenorhabditis elegans*. In: Wassarman PM (ed) Advances in Developmental Biochemistry. JAI Press, Conneticut

[13] Cowan AE, McIntosh JR (1985) Mapping the distribution of differentiation potential for intestine, muscle, and hypodermis during early development in *Caenorhabditis elegans*. Cell 41:923–932

[14] Priess J, Thomson J (1987) Cellular interactions in early *C. elegans* embryos. Cell 48:241–250

Sulston J, Schierenberg E, White J, Thomson N (1983) The embryonic cell lineage of the nematode *C. elegans*. Dev Biol 100:64–119

Wood WB (1991) Reversal of handedness in *C. elegans* embryos: new evidence for early cell interactions determining cell fates. Nature 349:536–538

[15] Herman R (1978) Crossover suppressors and balanced recessive lethals in *Caenorhabditis elegans*. Genetics 88:49–65

Kemphues K (1988) Genetic analysis of embryogenesis in *Caenorhabditis elegans*. In: Malacinski G (ed) Developmental genetics of higher organisms: a primer in developmental biology. MacMillan, New York, pp 193–220

Kemphues K (1989) *Caenorhabditis*. In: Grover DM, Hames BD (eds) Genes and embryos. IRL Press at Oxford University Press, Oxford, pp 95–126

Kemphues K, Kusch M, Wolf N (1988) Maternal-effect mutations on linkage group II of *Caenorhabditis elegans*. Genetics 120:977–986

Miwa J, Schierenberg E, Miwa S, von Ehrenstein G (1980) Genetics and mode of expression of temperature-sensitive mutations arresting embryonic development in *Caenorhabditis elegans*. Dev Biol 76:160–174

Sigurdson D, Spanier G, Herman R (1984) *Caenorhabditis elegans* deficiency mapping. Genetics 108:331–345

Wood WB (1988) Embryology. In: Wood WB (ed) The nematode *Caenorhabditis elegans*. Cold Spring Harbor Laboratory, New York, pp 215–241

Wood W, Hecht R, Carr S, Vanderslice R, Wolf N, Hirsh D (1980) Parental effects and phenotypic characterization of mutations that affect early development in *Caenorhabditis elegans*. Dev Biol 74:446–469

[16] Hill DP, Shakes DC, Ward S, Strome S (1989) A sperm-supplied product essential for initiation of normal embryogenesis in *Caenorhabditis elegans* is encoded by the paternal-effect embryonic-lethal gene, *spe-11*. Dev Biol 136:154–166

Kemphues K, Priess J, Morton D, Cheng N (1988) Identification of genes required for cytoplasmic localization in early *C. elegans* embryos. Cell 52:311–320

Kirby C, Kusch M, Kemphues K (1990) Mutations in the *par* genes of *Caenorhabditis elegans* affect cytoplasmic reorganization during the first cell cycle. Dev Biol 142:203–215

Mains P, Sulston I, Wood W (1990) Dominant maternal-effect mutations causing embryonic lethality in *Caenorhabditis elegans*. Genetics 125:351–369

Mains P, Kemphues K, Sprunger S, Sulston I (1990) Mutations affecting the meiotic and mitotic divisions of the early *Caenorhabditis elegans* embryo. Genetics 126:593–605

Schnabel R, Schnabel H (1990) Early determination in the *C. elegans* embryo: a gene, *cib-1*, required to specify a set of stem-cell-like blastomeres. Development 108:107–119

Capowski E, Martin P, Garves C, Strome S (1991) Identification of Grandchildless loci whose products are required for normal germ-line development in the nematode *Caenorhabditis elegans*. Genetics 129:1061–1072

[17] Austin J, Kimble J (1987) *glp-1* is required in the germ line for regulation of the decision between mitosis and meiosis in *C. elegans*. Cell 51:589–599

Austin J, Kimble J (1989) Transcript analysis of *glp-1* and *lin-12*, homologous genes required for cell interactions during development of *C. elegans*. Cell 58:565–571

Greenwald I (1989) Cell-cell interactions that specify certain cell fates in *C. elegans* development. Trends Genet 5:237–241

Maine E, Kimble J (1989) Identification of genes that interact with *glp-1*, a gene required for inductive cell interactions in *Caenorhabditis elegans*. Development 105:133–143

Priess J, Schnabel H, Schnabel R (1987) The *glp-1* locus and cellular interactions in early *C. elegans* embryos. Cell 51:601–611

Yochem J, Greenwald I (1989) *glp-1* and *lin-12*, genes implicated in distinct cell-cell interactions in *C. elegans* encode similar transmembrane proteins. Cell 58:553–563

[18] Edgar L (1991) Personal communication

Hecht R, Gossett L, Jeffery W (1981) Ontogeny of maternal and newly transcribed mRNA analyzed by in situ hybridization during development of *Caenorhabditis elegans*. Dev Biol 83:374–379

Schauer I, Wood WB (1990) Early *C. elegans* embryos are transcriptionally active. Development 110:1303–1317

[19] Storfer F (1990) Contributions of embryonic gene expression to early embryogenesis in the nematode *Caenorhabditis elegans*: a genetic analysis. Ph D Thesis, University of Colorado, Boulder, CO

[20] Hecht R, Berg-Zabelshansky M, Rao P, Davis M (1987) Conditional absence of mitosis-specific antigens in a temperature-sensitive embryonic-arrest mutant of *Caenorhabditis elegans*. J Cell Sci 87:305–314

Rothman J (1991) Personal communication

Schnabel R, Schnabel H (1990) An organ-specific differentiation gene, *pha-1*, from *Caenorhabditis elegans*. Science 250:686–688

[21] Burglin TR, Finney M, Coulson A, Ruvkun G (1989) *Caenorhabditis elegans* has scores of homeobox-containing genes. Nature 341:239–242

Kamb A, Weir M, Rudy B, Varmus H, Kenyon C (1989) Identification of genes from pattern formation, tyrosine kinase, and potassium channel families by DNA amplification. Proc Natl Acad Sci USA 86:4372–4376

Kenyon C, Wang B (1991) A cluster of *Antennapedia*-class homeobox genes in a non-segmented animal. Science 253:516–517

Krause M, Fire A, Harrison S, Priess J, Weintraub H (1990) CeMyoD accumulation defines the bodywall muscle cell fate during *C. elegans* embryogenesis. Cell 63:907–919

Rothman J (1990) Personal communication

Chapter 20 Generation of Temporal and Cell Lineage Asymmetry During *C. elegans* Development

GARY RUVKUN

20.1 Introduction

Genetic analysis of *C. elegans* development has focused on developmental events that take place after hatching, during postembryonic development. After hatching with 558 cells, about 10% of these are blast cells that undergo further cell divisions (Fig. 1) to generate a total of 959 neurons, muscles, intestinal and hypodermal cells in the hermaphrodite and 1031 cells in the male. Like embryonic development (see Edgar, Chap. 19 this Vol.), the pattern of cell division and differentiation during *C. elegans* postembryonic development is nearly invariant and has been completely described. The cell lineage of wild-type, mutant, or laser-ablated animals can be determined by direct observation of development using Nomarski optics. Because most cells during *C. elegans* development generate unique patterns of descendents (though symmetries in the lineage exist), the cell lineage produced by a particular blast cell during development is a signature of that cell's identity. Any changes in cell identity, induced, for example, by laser ablation of neighboring cells or by mutation, can be recognized

by a change in the lineage produced by that cell. By laser ablation, it has been shown that in many cases, the patterns of cell lineage executed by particular cells do not depend on their neighbors and instead reflect some intrinsic developmental program. On the other hand, the lineages of particular blast cells, for example, those that generate the hermaphrodite vulva, have been shown by laser ablation experiments to depend on interactions with their neighbors. Thus the pattern of cell divisions and differentiations that normally occur during *C. elegans* development depends on the ancestry of cells in some cases and on their neighbors or positional signals in other cases. Two major goals of developmental genetic analysis in *C. elegans* have been to explain how genes couple cell lineage information to cell identity and to explain how genes control and mediate cell-cell interactions. As described below, this analysis has revealed molecular mechanisms for the generation of lineage asymmetry and for intercellular signaling that are general to perhaps all metazoans.

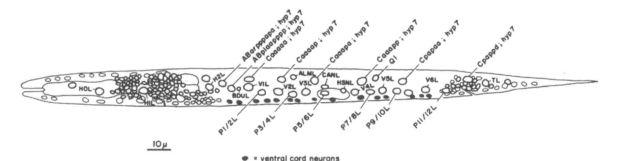

= ventral cord neurons

Fig. 1. The blast cells of a newly hatched larval stage 1 animal. Arrangement of neuronal and hypodermal nuclei on the *left side* of a newly hatched L1. Cells which do not divide postembryonically are named by their lineage, for example, Cpaapp. Postembryonic blast cells are renamed H1, H2, etc. Nuclei which were difficult to identify, i.e., the nerve ring, are unnamed in this figure. Pattern on the *right side* is very similar. (Sulston et al. 1983)

This chapter describes three genetic pathways that have been revealed by genetic analysis of *C. elegans* development. Each of these genetic pathways represents a distinct process that occurs during *C. elegans* development: (1) Analysis of the *unc*-86 gene has revealed how mother and daughter cells in a number of cell lineages are differentially specified – that is how lineage asymmetry is generated in the cell lineage. (2) Analyses of the genes *let*-23, *let*-60, *lin*-15, and *lin*-12 has revealed components of a cell-cell signaling mechanism. (3) The heterochronic genes have revealed how genes coordinate the overall temporal pattern of the postembryonic cell lineage. Not only have mutations identified these genes, the detailed analysis of particular mutations has revealed regulatory regions of the genes that are necessary for their proper function.

20.2 *unc-86* Couples Cell Lineage and Cell Identity[2]

Mutations in *unc*-86 cause particular neuroblasts to be transformed into their neuroblast parents, executing patterns of cell lineage more like those parent cells. For example, in *unc*-86 null mutants the lineage of a neuroblast called Q is altered such that after division, its posterior daughter, Q.p., acts as if it were Q (Fig. 2). In addition, the fates of some neurons are affected. Thus *unc*-86 gene function is necessary for particular neuroblast daughter cells to become distinct from their mother cells and to specify the identity of at least some of their daughter neurons as well.

wild type

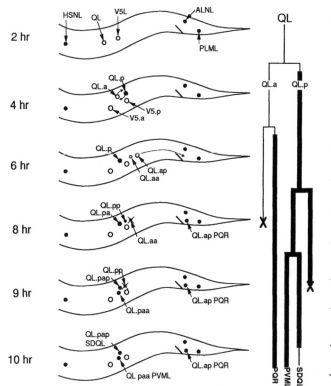

unc-86

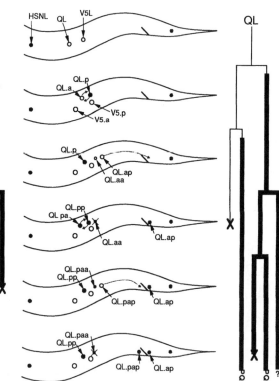

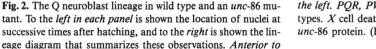

Fig. 2. The Q neuroblast lineage in wild type and an *unc*-86 mutant. To the *left in each panel* is shown the location of nuclei at successive times after hatching, and to the *right* is shown the lineage diagram that summarizes these observations. *Anterior to* *the left. PQR, PVML,* and *SDQL* correspond to neuronal cell types. *X* cell death. *Thickened lines* denonte nuclei containing *unc*-86 protein. (Finney and Ruvkun 1990)

Evidence that the *unc*-86 gene is activated by lineage controls rather than by extracellular signals has been indirect. First, the affected lineages occur at disparate times and locations during development. Second, *unc*-86 gene activity specifies proper neuroblast and neuronal cell identity even when some of those cells are misplaced in space or time by other mutations. These data suggest that either signals intrinsic to the lineage or a disseminated extracellular signal regulate *unc*-86. The gene was cloned to establish whether the gene is, in fact, regulated by lineage control and to dissect where within the lineage it many function to specify mother/daughter identity.

The *unc*-86 DNA sequence revealed a 160 amino acid region called the POU domain that has sequence similarity to three mammalian transcription factors, Pit-1, Oct-1, and Oct-2. The POU domain has two parts: an upstream domain that is specific to POU proteins, and a homeodomain of a new class. This sequence similarity suggested that *unc*-86 specifies the identities of neuroblasts and neurons by regulating their patterns of transcription.

From an *unc*-86 cDNA clone, the *unc*-86 protein was produced in *E. coli*, and used to immunize rabbits. The antisera stain the nuclei of 57 neurons and many of the neuroblasts that generate those neurons. All these neurons and neuroblasts were identified. In this way it was assessed if the *unc*-86 protein is expressed in the mother or daughter cells of the lineages affected by *unc*-86 mutations. For example in the Q lineage, *unc*-86 protein is never observed in the mother cell Q, but begins to be detectable in the nucleus of the daughter cell Q.p shortly after division (Fig. 2). Similar results were found for other neuroblast lineages and are summarized in Fig. 3. Thus the appearance of *unc*-86 protein must be sensitive to some factor segregated asymmetrically at division, and *unc*-86 must act in the daughters to make them distinct from their mothers. It is not yet known whether it is *unc*-86 transcription or translation that responds to the cell division asymmetry.

Besides its role in cell lineage, *unc*-86 is also required for the specification of particular neural identities. The protein is found in the nuclei of 57 neurons of 27 types, as defined by patterns of con-

nectivity, anatomy, and cell type-specific markers, and is required for aspects of the differentiated phenotypes of some or all of these neurons. Interestingly, these 27 neuron types share no known common features, suggesting that *unc*-86 does not direct any particular cellular phenotype, but carries some other form of information, such as distinguishing these neurons from their close lineal relatives. Thus in most of the lineages shown in Fig. 3, a dopaminergic neuron does not express *unc*-86, whereas one of the closely related neurons that expresses *unc*-86 also expresses a neural specification gene *mec*-3. Thus it is likely that a common lineage control mechanism activates *unc*-86 in all these lineages. Because distinct neuron types express *unc*-86, it may control different genes in different cells. One such gene that is activated by *unc*-86 in some cell lineages is *mec*-3, which is normally expressed in five of the 27 *unc*-86-expressing cell types, and whose expression depends on *unc*-86 function. However, expression of *unc*-86 is not sufficient to activate *mec*-3; it is expressed in only 10 of the 57 neurons and in none of the 22 neuroblasts that express *unc*-86. Presumably other genes must interact with the *unc*-86 protein to activate *mec*-3 specifically in those cells.

The *unc*-86 sequence similarity to other proteins containing the POU domain suggests mechanisms for combinatorial gene control: the ability of the mammalian POU protein Oct-1 to activate transcription of particular Herpes viral and host genes depends on its interaction with the Herpes viral protein VP16 via residues in the homeodomain. An *unc*-86 interaction with a transcription factor like VP16 could modify its ability to activate transcription of particular downstream genes, such a *mec*-3.

The rat POU domain gene, Brn-3, which is also expressed in the nervous system, is 87% identical to *unc*-86 in the upstream subdomain and 79% identical in the homeodomain. Thus Brn-3 is the best candidate for the *unc*-86 homologue is mammals. It is not unreasonable to expect that like *unc*-86, Brn-3 will control neuroblast as well as neuronal cell identities, and thus be necessary for the generation of multiple neuronal cell types. There is precedent for such genetic control by a mammalian POU protein: the Pit-1/GHF-1 POU protein transcription factor is pituitary-specific and activates tran-

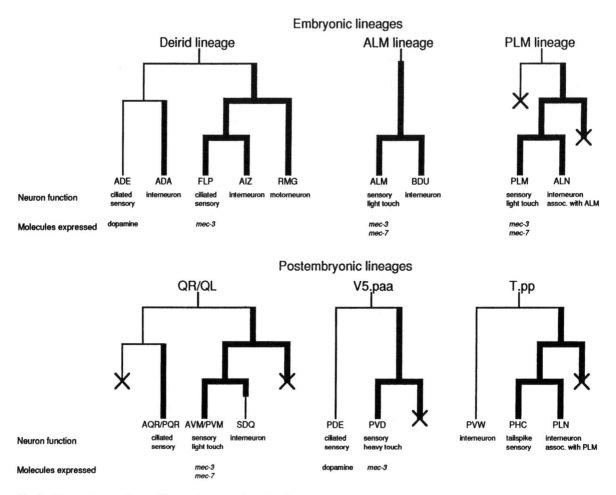

Fig. 3. The pattern of *unc*-86 protein expression in the neuroblast lineages, showing the modular nature of its expression pattern. Cells expressing *unc*-86 are indicated by thickened lines. Characteristics of the neurons are shown below. (Finney and Ruvkun 1990)

scription of the pituitary growth hormone and prolactin genes. Genetic data show that the Pit-1/GHF-1 gene is necessary for production of pituitary blast cells as well as differentiation of three pituitary cell types. Two dwarf mutations in mouse have been shown to be mutations in the Pit-1/GHF-1 gene. These mutations cause a decrease in pituitary gland size, and lead to complete loss of expression of the growth hormone and prolactin genes.

These experiments have revealed that lineage asymmetry in *C. elegans* is generated by production of a transcription factor in particular cells. It remains to be analyzed exactly how such asymmetry of *unc*-86 expression is generated. Is *unc*-86 transcription or translation activated in particular daughter cells? If it is *unc*-86 transcription, is there a particular transcription factor that is asymmetrically segregated or activated to activate *unc*-86 expression? At the level of cell specification, with what proteins does *unc*-86 interact to specify particular cell types? And what particular downstream genes are activated in the 27 distinct types of cells that express *unc*-86?

20.3 Genetic Analysis of Cell-Cell Signaling During *C. elegans* Development[3]

The *C. elegans* vulva is an organ through which eggs are laid. The organ is generated from hypodermal vulval precursor cells named P3.p, P4.p, P5.p, P6.p, P7.p, and P8.p that undergo specific cell lineages during postembryonic development (Fig. 4). The patterns of cell lineage generated by P3.p to P8.p fall into three classes, named 1 °, 2 °, and 3 °, and occur in the order shown in Fig. 4. The 1 ° and 2 ° lineages generate the cells of the vulva, and the 3 ° lineage generates hypodermal cells that fuses with a large hypodermal syntitial cell. The Pn.p cells depend on cell-cell signaling rather than lineage per se to specify their particular pattern of division: laser ablation of the gonadal anchor cell ac, which normally comes to lie directly adjacent to P6.p, causes all six Pn.p cells to undergo 3 °cell lineages (Fig. 5a, b). In addition, laser ablation of, for example, just P5.p, causes P4.p to move into the P5.p position and change from a 3 ° fate to a 2 ° fate. Thus signaling exists from the ac cell to the vulval precursor cells and between the vulval precursor cells themselves.

To identify genes and molecules that mediate this cell-cell signaling or respond to it, mutations that perturb vulval cell lineages have been identified. Many of these mutations cause the vulval precursor cells to divide as if they were sensing too much or too little signal from each other or from the ac cell: mutations that cause the vulval precursor cells to execute 3 ° cell lineages are what would be expected from lesions in genes that are necessary to produce or respond to the a signal from the ac cell, and mutations that cause more of the vulval precursor cells to produce 1 ° and 2 ° cell lineages are what would be expected from lesions in genes that negatively regulate the production or response to the ac signal (or signals between vulval precursor cells). By a combination of laser ablations in these mutants and observing the Pn.p cell lineages in double mutants, possible sites of action of these genes have been inferred.

20.3.1 *let*-23 Is Necessary for Intercellular Signaling and Is a Member of the EGF-Receptor Superfamily

The *let*-23 gene is necessary for 1 ° and 2 ° Pn.p cell identities. Recessive decrease of function mutations in *let*-23 cause cells P3.p to P8.p to all undergo simple 3 ° patterns of division like P3.p rather than the more complex 1 ° and 2 ° lineages necessary to generate the vulva (Fig. 5c). Because of these lineage changes, no vulva is generated. These mutations cause the vulval precursor cells to act like they receive no signal from the anchor cell. However, the gene could act either in the ac cell, or in the vulval precursor cells, or elsewhere. No genetic mosaic analysis has been done to address this issue.

The *let*-23 gene has been cloned and shown to be a membrane-bound tyrosine kinase related to the EGF receptor. This discovery of a member of the membrane bound tyrosine kinase receptor superfamily in a known signaling pathway is satisfying. The result is quite consistent with the gene acting in the Pn.p cells in response to the ac signal, though no expression data nor laser ablation data yet suggests this. *let*-23 may also act outside the Pn.p cells because null mutations in *let*-23 are lethal, whereas ablation of the ac cell or its precursor cell are viable and only cause defects in vulval development.

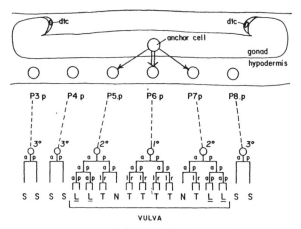

Fig. 4. Anatomy and cell lineage of the Pn.p cell lineages that generate the vulva, and the anchor cell that signals the Pn.p cells. (Sternberg and Horvitz 1984)

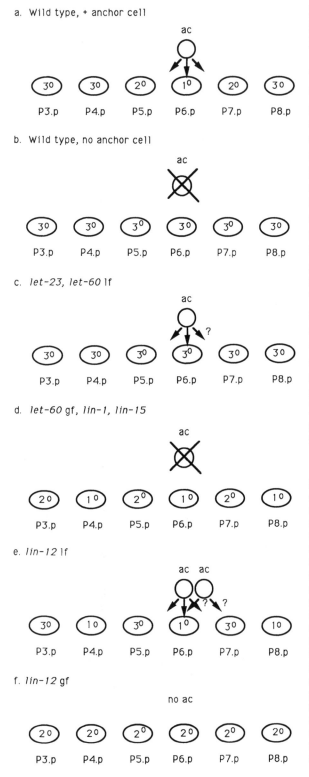

a. Wild type, + anchor cell

b. Wild type, no anchor cell

c. *let-23, let-60* lf

d. *let-60* gf, *lin-1, lin-15*

e. *lin-12* lf

f. *lin-12* gf

20.3.2 *let-60* Is Necessary for Intercellular Signaling and Encodes the *C. elegans ras* homolog

The *let-60* gene is also necessary for 1° and 2° Pn.p cell identities. Recessive mutations in the *let-60* gene cause cells P3.p to P8.p to undergo 3° cell lineages, as if they received no ac signal (Fig. 5c). In addition, some *let-60* alleles cause the same phenotype but are genetically dominant. These alleles appear to interfere with normal *let-60* gene function, either by forming a nonfunctional complex with the wild-type *let-60* gene product or with an interacting gene product. Other *let-60* alleles are genetically dominant but have the opposite phenotype: these mutations cause P3.p to P8.p to undergo 1° and 2° cell lineages and result in the production of multiple ectopic pseudovulvae (Fig. 5d). These gain-of-function *let-60* alleles appear to cause unregulated *let-60* gene activity. These alleles cause 1° and 2° cell lineages even in combination with *let-23* mutations, suggesting that *let-60* acts downstream of *let-23* (Fig. 6). Again genetic mosaic analysis to distinguish whether *let-60* functions in the ac cell or in the Pn.p cells has not been done. However, laser ablation of the ac cell in a *let-60* gain of function mutant that produces all 1° and 2° Pn.p cell lineages, has no effect on the mutant

Fig. 5a–f. The effects of mutations or laser ablations on Pn.p cell lineages. **a** In wild type, a signal or signals from the anchor cell in the gonad induces the nearest Pn.p cell to express a 1° lineage and flanking cells to express 2° lineages. The cells generated by P5–7.p form the vulva. Pn.p cells more distant from the ac cell express 3° lineages, which generate hypodermal cells that fuse with the hypodermal syntitium. In wild type but with no anchor cell (**b**) or in *let-60* (*lf*) and *let-23* mutants (**c**), all Pn.p cells express 3° lineages. In *let-60* gf, *lin-1*, or *lin-15* mutants (**d**), either with or without an ac cell, the Pn.p cells express 1° or 2° lineages and generate ectopic pseudovulvae. The Pn.p cells in *lin-12* (*lf*) mutants (**e**) express 1° or 3° lineages. Due to the *lin-12* effects on another lineage, two ac cells that are also generated by these mutants. Ablation of these two ac cells causes P3.p to P8.p to express 3° cell lineages. The Pn.p cells express only 2° cell lineages in *lin-12* gf mutants (**f**) even though no ac cell is produced in these mutants. An *arrow* denotes the putative signal(s) produced by the anchor cell. *Question marks* near the anchor cell denote that the signal from the anchor cell to the Pn.p cells may or may not be produced in these mutants. An *X* through the cell denotes a laser ablation of that cell. (Beitel et al. 1990)

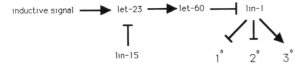

Fig. 6. The genetic pathway of *let*-23, *let*-60, and *lin*-15. *Arrows* indicate positive regulation and *T bars* indicate negative regulation. The pathway is derived by epistasis analysis of double mutants with both loss of function and gain of function alleles of *let*-60. (Han et al. 1990)

phenotype. Thus it is probable that this mutant *let*-60 gene, and by inference, the wild-type *let*-60 gene does not act in the ac cell. However, it is far from proven that *let*-60 acts in the Pn.p cells.

The *let*-60 gene has been cloned and shown to be the *C. elegans ras* homolog. In mammals and yeast, *ras* has been shown to act downstream of receptor proteins in the transduction of extracellular signals to unknown downstream genes that mediate cellular response to those signals. Thus it is reasonable that such a gene is involved in signal transduction between the ac and Pn.p cells.

The *let*-60 gain of function mutation that causes unregulated *let*-60 gene activity, is Gly to Glu mutation at amino acid 13. The same mutation in the mammalian *ras* gene has been shown to convert *c-ras* to an oncogene.

The genetic and molecular data on *let*-23 and *let*-60 suggest that *let*-23 acts as a receptor, perhaps for the signal from ac, and that *let*-60/*ras* acts to transduce that signal to downstream genes. Null mutations in *let*-60 are lethal, suggesting that the gene acts in other developmental processes in addition to specifying the P3.p to P8.p cell identities. Given that *let*-60 and *let*-23 have the same phenotype, including the lethal null phenotype, it is possible that the *let*-60 ras protein is necessary to transduce the signal produced by the *let*-23 receptor in these unknown other cells as well.

20.3.3 Other Genes Involved in Intercellular Signaling

lin-15. The *lin*-15 gene is necessary for 3° cell identities. Mutations in *lin*-15 cause the opposite phenotype from *let*-60 and *let*-23 loss of function mu-

tations — cells P3.p to P8.p execute 1° and 2° patterns of cell lineage normally only observed for cells P5.p, P6.p, and P7.p (Fig. 5d). This phenotype is not ameliorated by laser ablation of the ac cell, suggesting that the *lin*-15 mutation does not act in the ac cell (or more correctly does not act solely in the ac cell). These data alone suggest that the *lin*-15 mutation may inappropriately activate a signal reception molecule in the Pn.p cells. However, genetic mosaic analysis of *lin*-15 has revealed that the gene acts not in the cell lineages that generate the Pn.p cells, but rather in those lineages that generate the syntitial hypodermal cell hyp7 with which Pn.p cells with 3° fates must fuse. These data suggest that hyp7 may require *lin*-15 gene activity for Pn.p cells to express their 3° fate. *lin*-15 could simply be necessary for these particular cell fusions (but not all cell fusions, for hyp7 is generated normally in this mutant and depends on many cell fusions) or that hyp7 may produce a signal or substance that the Pn.p cell requires for 3° cell fates. This result complicates the simple model of ac/Pn.p cell signaling in that now a third variable must be included. *lin*-15 has not yet been cloned.

lin-12. The *lin*-12 gene appears to regulate signalling between Pn.p cells. There are two classes of *lin*-12 mutations. Loss of function mutations cause the P3.p to P8.p to undergo 1° or 3° but not 2° patterns of division (Fig. 5e), whereas gain of function *lin*-12 mutations causes all these cells to undergo 2° patterns of cell lineage (Fig. 5f). Ablation of the ac cell in *lin*-12 loss of function mutants causes P3.p to P8.p to undergo 3° patterns of division, whereas an ac cell in strains bearing *lin*-12 gain of function mutations induces a 1° cell lineages in the cell closest to ac, P6.p. These data argue that: (1) the *lin*-12 gene activity is necessary for 2° patterns of cell lineage but not for 1° or 3° lineages; (2) P3.p to P8.p can respond to the ac signal without *lin*-12 gene activity, so that *lin*-12 is not involved in reception of the ac signal; (3) the *lin*-12 activity must be low in order to express a 3° fate. *lin*-12 affects more cell lineages than just the Pn.p cells, including the germline

The *lin*-12 gene has been cloned and shown to encode a membrane bound receptor that is similar to the *Drosophila* Notch protein, also involved in cell-cell signaling. In addition *lin*-12 has sequence similarity to a motif present in a yeast cell cycle control gene *cdc*16. The gain of function *lin*-12 mutations map to the extracellular domain of this protein, suggesting that the *lin*-12 activity is normally negatively regulated. It is not known whether *lin*-12 functions in the Pn.p cells or in adjacent cells.

The combination of genetic and molecular data suggest that *lin*-12 is receptor involved in signaling between Pn.p cells. Within the P3.p to P8.p cells, the gene may normally be active only in the P5.p and P7.p that undergo 2° cell lineages under the influence of both the ac cell inductive signal and a signals from P6.p that it is 1°. In a *lin*-12 gf mutant, the hyperactive *lin*-12 receptor protein signals independently from these signals, causing 2° cell lineages. In *lin*-12 loss of function mutants, the cells cannot specify the 2° fate and either execute 1° or 3° cell lineages.

20.3.4 Summary of Genes Involved in Cell-Cell Signaling

Thus genetic and molecular analysis has revealed many of the genes involved in cell-cell signaling in the vulva. It is gratifying that three of the members of this set are recognizable members of the intercellular signaling hall of fame – *ras*, EGF receptor, and the Notch/*lin*-12 family. The analysis of regulatory mutations in these genes will identify regions of the proteins necessary for their up and down regulation in response to intercellular signaling. Most conspicuously absent from the set of genes so far analyzed in detail is the actual signal molecule from the ac cell. If the signaling molecule is not necessary for viability, it is probable that a mutation leading to loss of the ac signal should exist in the set of vulval mutants so far isolated. Genetic mosaic analysis and molecular analysis should reveal if the other members of the set identify the ac signal.

In the future it remains to dissect how these proteins interact to specify cell type. In which cells do they act? Do those that are expressed in the same cell interact directly? How does the interaction regulate the activity of the protein? And how does the activity of these proteins finally lead to changes in cell lineage and cell differentiation?

20.4 Genetic and Molecular Studies of Heterochronic Genes in *C. elegans*[4]

The heterochronic genes of *C. elegans* control the temporal pattern of the postembryonic cell lineage. Heterochronic mutations cause particular cells in a variety of cell lineages and tissues, for example, the intestinal (E lineage), the mesoblast (M lineage), and the hypodermal cell lineages (H0, H1, H2, V1 to V6, T and P lineages), to adopt fates normally associated with cells at earlier or later stages of development (Fig. 1 and 7).

The *lin*-14 heterochronic gene plays a central role in this regulation of the cell lineage. Loss-of-function (*lin*-14 lf) alleles cause the precocious execution of cell lineages normally observed in descendent cells one or two larval stages later. Gain-of-function (*lin*-14 gf) alleles affect the same cell lineages but cause the opposite transformations in cell fate: early cell lineages are normal, but later cells reiterate the early cell lineages, normally associated with their ancestor cells. For example, in the development of the lateral hypodermis, in *lin*-14 loss-of-function mutants, the blast cell T skips its characteristic larval stage 1 (L1) sublineage and instead expresses a sublineage normally associated with its granddaughter blast cell T.ap (Fig. 7). In *lin*-14 gain-of-function mutants, the blast cell T expresses its normal L1-specific sublineage, but its granddaughter T.ap reiterates this L1-specific sublineage normally associated with blast cell T (Fig. 7). The *lin*-14 gain-of-function mutations have been shown by genetic criteria to cause excess *lin*-14 gene activity, while the *lin*-14 loss-of-function mutations are due to loss of *lin*-14 gene activity. These data suggest that during normal development, a relatively high *lin*-14 gene activity during early larval stages, for example in cell T, is reduced later in development, for example in cell T.ap, to form a temporal developmental switch. In an analogous way the *lin*-14 gene coordinately controls the postembryonic fates of cells in a number of lineages: the E, M, H0, H1, V1 to V6, T, and P lineages.

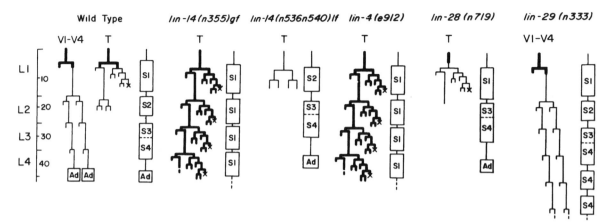

Fig. 7. Summary of the temporal and cellular expression profile of *lin*-14 protein in wild-type, *lin*-14 gain-of-function, *lin*-14 loss-of-function, *lin*-28, *lin*-4, and *lin*-29 heterochronic mutants. Shown are the lineage changes induced by these mutations and *superimposed in dark line* are those cells that also express the *lin*-14 protein. Wild type expresses *lin*-14 protein only during the L1 stage. *lin*-14 (n355)gf and *lin*-4(e912) express the *lin*-14 protein at all developmental stages and therefore reiterate L1-specific cell lineages. *lin*-14(n536n540) expresses no *lin*-14 protein so that precocious L2-specific lineages occur during the L1 stage.

lin-28(n719) expresses *lin*-14 protein before the L1 cell division, but not after it so that a normal L1-specific lineage occurs, but L3- and later specific lineages occur precociously during the L2 stage. *lin*-29(n333) expresses *lin*-14 normally durign the L1 stage of that L1-specific lineages and L2-specific lineages are normally expressed. This mutant fails to express adult specific traits regardless of whether *lin*-14 is mutant or not. The effects of heterochronic mutations on postembryonic cell lineages and expression of *lin*-14 in postembryonic lineages in these mutants

20.4.1 A Temporal Gradient in the *lin*-14 Nuclear Protein

The product of the *lin*-14 gene is a nuclear protein whose level forms a steep temporal gradient during *C. elegans* development: it is present at high levels in many but not all nuclei of embryos and larval stage 1 animals but is undetectable at later stages. For example, in the case of the T lineage, during normal development, the *lin*-14 nuclear protein is present at high levels in the T cell but is not observable in cell T.ap (Fig. 7). In *lin*-14 gain-of-function mutants, this temporal gradient is disrupted: the *lin*-14 protein is now observed at all stages of development, for example in both cells T and T.ap (as well as T.apap, etc.), and these cells reiterate L1-specific cell lineages (Fig. 7). In some *lin*-14 null mutants, no *lin*-14 protein is detectable, and cells express L2-specific lineages during the L1 stage. Thus the normally sharp decrease in *lin*-14 protein levels during the L1 stage causes cells to switch from L1-specific cell lineages to L2-specific cell lineages, and in *lin*-14 gain-of-function mutants, the inappropriate presence of the *lin*-14 nuclear protein late

in development prevents this temporal switch in cell fate.

The *lin*-14(*n355*) and *lin*-14(*n536*) gain-of-function mutations are located in 3′ untranslated region of the *lin*-14 mRNA. The *n355* mutation is an insertion or inversion of at least 10 kb of unknown DNA sequences 256 bases 3′ to the termination codon of the *lin*-14 protein coding region common to both *lin*-14 transcripts (Fig. 8). The other *lin*-14 gain-of-function mutation, *n536*, is a 607 bp deletion, 300 bases downstream from the 3′ end of the *lin*-14 open-reading frame (Fig. 8) and overlaps the region which is rearranged in the *n355* mutation.

The location of both *lin*-14 gain-of-function mutations in the 3′ untranslated region suggests that they do not affect the *lin*-14 protein. Immunoblot analysis of the *lin*-14 proteins from wild type and both the gain-of-function mutants confirmed this prediction; no difference in the size of the *lin*-14 proteins was observed in these mutants.

In staged wild-type worms, the amount of the 3.5 kb *lin*-14 mRNA relative to actin-1 mRNA is relatively high in eggs and L1, decreases about 3 fold by L2, and continues to decrease until it is about

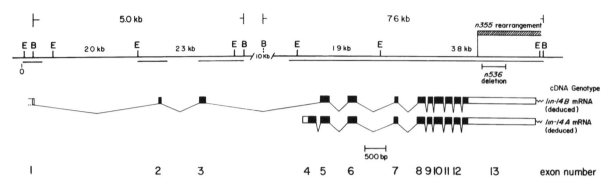

Fig. 8. A molecular map of the *lin*-14 region showing the organization of the gene with the two *lin*-14 transcripts inferred from cDNAs isolated from both wild type (N2) and *lin*-14 mutann strains. The cDNAs were isolated using either hte 3.8 b EcoRI fragment, the 7.6 kb BglII fragment, or the 5.0 kb BglII fragment as probe (B = BglII, E = EcoRI).

Exons containing open reading frames are shown in *black*, whereas those containing untranslated regions are shown in *white*. Exons are *numbered below*. Those genomic regions that have been sequenced are shown under the genomic restriction map and generally flank the exons detected in the cDNA clones. This analysis could have missed exons that are incorporated rarely into *lin*-14 transcripts. The location of the two *lin*-14 gain of function mutations in the 3' untranslated region are shown above and below the genomic restriction map

tenfold less by L4. The level of *lin*-14 proteins decreases at least ten times more precipitously after L1 than the *lin*-14 mRNA. RNAs isolated from staged strains bearing the gain-of-function *lin*-14 mutations *n355* or *n536* show that the temporal regulation of *lin*-14 mRNA level in these mutants is normal. Therefore the molecular basis of the *lin*-14 gain-of-function mutations is not increased *lin*-14 transcription or mRNA stability.

This normal transcriptional regulation of the *lin*-14 gene in the gain-of-function *lin*-14 mutants is discordant with the dramatically altered temporal pattern of *lin*-14 protein accumulation in these mutants. These data suggest that the gain-of-function mutations in the 3' UTR affect posttranscriptional regulation of the *lin*-14 gene.

Thus the 3' region of the *lin*-14 transcripts contains an element that encodes the posttranscriptional down regulation of *lin*-14 protein levels after larval stage 1. Posttranscriptional regulation of translation or transcript stability has been demonstrated in the 3' UTR of other eukaryotic genes and proteins have been identified that bind to those elements. In HIV, the *rev* transactivator protein has been shown to interact with an RNA structure in the 3'UTR of the viral transcript and to mediate export of the mRNA from the nucleus. Thus, sequences in the 3'UTR can regulate export of a tran-

script from the nucleus, the half-life of the transcript, or translation of that transcript.

The *lin*-14 gene encodes two protein products. The *lin*-14 DNA sequence revealed 13 exons in the *lin*-14 gene, with introns ranging in size from about 12 kb to 44 bp. Differential splicing of exons 1, 2, and 3 (the *lin*-14B transcript) or exon 4 (the lin-14A transcript) to the common exons 5 to 13 was observed in *lin*-14 cDNAs (Fig. 8). The longest open reading frame in the lin-14B transcript yields a protein of 539 amino acids. The *lin*-14A transcript encodes a 537 amino acid *lin*-14A protein. This protein has a 63 amino acid N terminal domain that is distinct from the 65 amino acid N-terminal domain of the *lin*-14B protein. No similarity between the two amino-termino of the *lin*-14 proteins could be detectd. No significant similarity to any proteins in databases could be detected. RNase protection experiments showed that the two transcripts are expressed in wild type at similar levels and show the same temporal regulation.

The *lin*-14 proteins also contain a high proportion of prolines, serines, threonines, and glutamates, or PEST sequences, as has been observed in various unstable proteins. Given that the protein appears to be quickly degraded late in larval stage 1 to control an L1/L2 switch in cell fates, these sequences could mediate that instability.

20.4.2 How Other Heterochronic Genes Interact with *lin*-14 to Generate the Temporal Gradient or Interpret It

Other heterochronic genes could control developmental timing of the *C. elegans* cell lineage by participating in the generation or reception of the *lin*-14 temporal gradient. A recessive mutation in the gene *lin*-4 causes the same retarded phenotype as *lin*-14 gain-of-function mutations. In addition, the *lin*-4 phenotype depends on a functional *lin*-14 gene: *lin*-14 loss-of-function mutations are epistatic to *lin*-4. These data show that the *lin*-4 gene negatively regulates the *lin*-14 gene, though they do not suggest whether this is direct or at what level the interaction occurs. The *lin*-4 gene is necessary for the down regulation of *lin*-14 protein levels: inappropriate *lin*-14 protein staining at late stages is observed in this mutant (Fig. 7). This inappropriate expression of *lin*-14 is not due to inappropriate transcriptional regulation of the gene, because *lin*-14 mRNA levels are normal in a *lin*-4 mutant. Thus, directly of indirectly, *lin*-4 negatively regulates *lin*-14 posttranscriptionally. It is possible that the *lin*-4 protein directly interacts with the *lin*-14 3' regulatory sequences.

Mutations in *lin*-28 lead to precocious expression of L3-specific hypodermal cell lineages during L2 the stage, and so are similar to *lin*-14 mutations in these hypodermal cell lineages only (Fig. 7). In these hypodermal lineages the level of *lin*-14 protein is decreased in *lin*-28 mutants, suggesting that the gene acts upstream of *lin*-14 in these cells. *lin*-28 could activate *lin*-14 transcription or translation, or stabilize the *lin*-14 protein in these hypodermal cells by, for example, a heterodimer interaction.

Loss-of-function mutations in the *lin*-29 gene only affect larval stage 4 cell fates in the hypodermal lineages: these cells continue to express L4 fates during adult stages in *lin*-29 mutants (Fig. 7). *lin*-29 mutants cause this phenotype in double mutant combinations with any of the other heterochronic genes, suggesting that it is farthest downstream in the pathway and directly regulates an L4/adult switch.

20.4.3 Summary: Generation of the *lin*-14 Temporal Gradient

The DNA sequence of the *lin*-14 gene from wild type and three *lin*-14 mutants, and its expression pattern in wild-type and various mutants has revealed aspects of how the *lin*-14 protein gradient is generated (Fig. 9). First, during normal development the down regulation of *lin*-14 mRNA and protein levels begins after a feeding signal initiates postembryonic development. Both the *lin*-14 and the *lin*-28 gene activities are necessary to maintain the level of *lin*-14 protein high before this food signal. After initiation of postembyronic development, posttranscriptional down regulation of *lin*-14 is triggered, perhaps by *lin*-4. This leads to an at least 25-fold decrease in *lin*-14 protein level from L1 to L2 that causes cells to switch to L2-specific fates. *lin*-14 gain-of-function mutations abrogate the negative posttranscriptional regulation of *lin*-14 protein levels. This failure to markedly reduce the *lin*-14 protein levels prevents or delays the normal L1 to L2 switch in cell fates in these mutants.

The stability of the *lin*-14 protein is also relevant to the formation of the *lin*-14 temporal gradient. At mid-larval stage 1, posttranscriptional control stops new *lin*-14 protein synthesis from the *lin*-14 mRNAs still present. The half-life of the previously synthesized *lin*-14 protein must be less than 1 h to account for this observed rate of disappearance. The *lin*-14 protein levels decrease at the same rate in both dividing cells and non-dividing cells, suggesting that, at least in the nondividing cells, breakdown of the nuclear membrane is not necessary for the degradation of the *lin*-14 protein and that nuclearly localized proteases must control this process. The presence of PEST sequences in the *lin*-14 protein supports the notion that the rate of *lin*-14 protein degradation is relevant to the formation of the *lin*-14 protein gradient.

While the DNA sequence of the *lin*-14 gene did not reveal any homology that would suggest its molecular mechanism, the nuclear localization of the *lin*-14 proteins suggests that they may regulate the pattern of gene expression of the cells that accumu-

A model for the *lin-14* heterochronic switch:

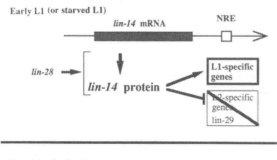

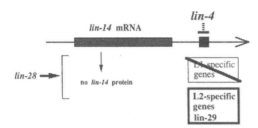

Fig. 9. A model for the generation of the *lin-14* protein gradient. In early L1 or starved L1 wild-type animals, *lin-14* proteins is actively translated and specifies L1-specific cell lineages, perhaps by activating L1-specific genes in each blast cell and by repressing L2-specific and later gene activities including *lin-29*. During normal development, *lin-14* transcript levels decrease 2–3 fold from the L1 to the L2 stage while *lin-14* protein levels decrease at least 25 fold. Both gain-of-function and loss-

of-function mutants which show inappropriately high and low levels of *lin-14* protein respectively, do not show any variation in the regulation of *lin-14* transcript levels compared to wild type. The normal temporal pattern of *lin-14* protein level is regulated posttranscriptionally by the heterochronic genes *lin-4* and *lin-28* which act antagonstically to define the *lin-14* protein gradient. *lin-28* is required either to positively activate *lin-14* expression or to stabilize the *lin-14* protein during the L1 stage of wild type development and during the L2 and later stages of development in *lin-4* and *lin-14gf* mutants. A decrease in *lin-14* protein levels during the mid-L1 stage is triggered by feeding and/or by post-embryonic developmental events. This triggering event may activate *lin-4* expression or activity to negatively regulates *lin-14* at a posttranscriptional level. This negative regulation of *lin-14* does not operate by down-regulating *lin-28* because the *lin-28* gene is still active at the L2 stage as shown by the fact that *lin-28* mutations are epistatic to *lin-14* gf mutations. Thus lin-4 negatively regulates *lin-14* independently of *lin-28*. The *lin-4* gene may directly or indirectly negatively regulate *lin-14* expression via the negative regulatory element (NRE) in the *lin-14* 3′ UTR (depicted as an *open box* in the L1 stage, when it is not acting and as a *black box* in the L2 stage, when it is necessary for post-transcriptional regulation). This negative regulatory element is inactivated in *lin-14* gain-of-function mutants allowing inappropriate expression of *lin-14* protein at L2 and later stages. Hence, this negative regulatory element functions at the same time as *lin-4*. The *lin-4* protein could bind to the *lin-14* NRE or another protein could act at the *lin-14* NRE while *lin-4* acts to regulate, for example, *lin-14* translation or protein stability. This disappearance of the *lin-14* protein prevents the expression of L1-specific cell lineage genes and allows the expression of L2-specific and subsequent cell lineage genes, including *lin-29*, that specifies the switch from larval to adult-stage cuticle formation. *Arrows* denote positive regulation; *bars* denote negative regulation

late them. The observation that the *lin*-14 protein is normally present only in embryos and larval stage 1 animals suggests that it either activates early genes or represses late genes so that the disappearance of *lin*-14 after larval stage 1 causes as transition from the expression of early cell lineage genes to late cell lineage genes.

20.5 Outlook

We can expect that molecular genetic analysis of the development events described above, as well as many others that space did not permit to be described, will continue to reveal how genes and their products conspire to generate patterned arrays of

cells during development. We must explain how these control genes form a regulatory network that decodes the genetic information to specify the spatial and temporal pattern of cells produced during ontogeny. Even though the cell lineage of *C. elegans* is a very different developmental strategy from other organisms, we now know that the DNA sequences of many of the genes that control it (e.g. *lin*-12, *unc*-86, *glp*-1, *mab*-5, *unc*-5) have been conserved throughout evolution. These genes must have been conserved because of their molecular interactions that mediate how they become asymmetrically activated in particular spatial, temporal, or lineal domains, and/or how these genes and gene products interact with other genes and their products to specify the pattern and types of cells generated during development. The principles for control of timing that have been revealed by these studies may also be general: the heterochrony observed in phylogeny is very reminiscent of that described here in *C. elegans* and suggests that heterochronic genes like *lin*-14 may be major players in evolutionary change and developmental control across phylogeny.

While many of the players in this complicated process have been identified and are under study, genes used in multiple developmental processes and genes that are members of redundant gene families will be systematically missed in genetic screens that depend on a later or more limited phenotype. This is especially true of genetic analysis that depends on viable mutants, as has been the rule for most *C. elegans* genetics to date. The *C. elegans* genome sequence and exhaustive searches for members of gene superfamilies as well as more biochemical experiments may reveal genes of this type.

While genetic studies to date have focused on genes that specify cell identity, and have revealed mostly genes involved in regulation of gene expression or signal transduction, the next frontier will be to study the genes downstream of these regulatory genes – the gene whose products actually mediate morphogenesis and the production of specialized cells. We will then begin to understand not just how multicellular animals are encoded but also how they are assembled. In one sense, the solution to the coding problem is clearly delineated – we now can see the outlines of how molecules like *unc*-86, *lin*-12, etc. specify cell type based on lineage or intercellular signaling. Next we must explain how the genes regulated by these cell identity genes actually cause changes in the behavior of cells.

References

[1] Horvitz HR (1988) The genetics of cell lineage in the nematode *Caenorhabditis elegans*. CSH Press, Cold Spring Harbor Laboratory, NY

[2] Chalfie M, Horvitz HR, Sulston JE (1981) Mutations that lead to reiterations in the cell lineages of *C. elegans*. Cell 24:59–69

Finney M, Ruvkun G (1990) The unc-86 gene product couples cell lineage and cell identity in *C. elegans*. Cell 63:895–905

Li S, Crenshaw EB, Rawson EJ, Simmons DM, Swanson LW, Rosenfeld MG (1990) Dwarf locus mutants lacking three pituitary cell types result from mutations in the POU domain gene pit-1. Nature 347:528–533

[3] Arroian RV, Koga M, Mendel JE, Ohshima Y, Sternberg PW (1990) The *let*-23 gene necessary for *Caenorhabditis elegans* vulval induction encodes a tyrosine kinase of the EGF receptor subfamily. Nature 348:693–699

Beitel GJ, Clark SG, Horvitz HR (1990) Caenorhabditis elegans ras gene let-60 acts as a switch in the pathway of vulval induction. Nature 348:503–509

Han M, Sternberg PS (1990) let-60, a gene that specifies cell fates during *C. elegans* vulval induction, encodes a ras protein. Cell 63:921–931

Han M, Aroian R, Sternberg PS (1960) The let-60 locus controls the switch between vulval and non vulval cell fates an *Caenorhabditis elegans*. Genetics 126:899–913

Herman RK, Hedgecock EM (1990) Limitation of the size of the vulval primordium of *C. elegans* by lin-15 expression in surrounding hypodermis. Nature 348:169–171

Ferguson E, Sternberg PS, Horvitz HR (1987) Genetic pathway for the specification of vulval cell lineages in *Caenorhabditis elegans*. Nature 326:259–267

Yochem J, Weston K, Greenwald LS (1988) The *Caenorhabditis elegans lin*-12 gene encodes a transmembrane protein with overall similarity to *Drosophila* Notch. Nature 335:547–550

Sulston JE, Schierenberg E, White JG, Thomson JN (1983) The embryonic cell lineage of the nematode *Caenorhabditis elegans*. Dev Biol 100:64–119

Sternberg PW, Horvitz HR (1984) The genetic control of cell lineage during nematode development. Annu Rec Genet 18:489–524

[4] Ambros VA, Horvitz HR (1984) Heterochronic mutations in *Caenorhabditis elegans*. Science 266:409–416

Ruvkun GB, Guisto J (1989) The *Caenorhabditis elegans* heterochronic gene *lin*-14 encodes a nuclear protein that forms a temporal developmental switch. Nature 338:313–319

Chapter 21 Genetic and Molecular Analysis of Early Pattern Formation in *Drosophila*

21.1 *Drosophila* as a Genetic System[1]

Drosophila is a small dipteran fly and belongs to the group of holometabolous insects (Fig. 1). Both the larvae and the adult flies live on rotting fruit, and feed mainly on the fungi and bacteria which grow on these fruit. It probably had its evolutionary origins in tropical Africa, but a large number of different species can now be found throughout the world. *D. melanogaster* was introduced into the genetic laboratory by T.H. Morgan around 1910, mainly because it was easy to grow (generation time two weeks with about 300 offspring per female) and because it showed a wealth of morphological markers, which could be used for genetic experiments. *D. melanogaster* quickly became one of the genetically best analyzed organisms. Morgan received the Nobel prize for medicine in 1933 for describing the principles of genetic recombination and for the discovery of the chromosomal sex determination. One of his coworkers, H.J. Muller, received the Nobel prize in 1946 for the discovery and the description of the mutagenicity of X-rays. This international recognition has boosted *Drosophila* research and *Drosophila* has since been continuously used for the analysis of a variety of basic biological questions, ranging from biochemistry over behavioral genetics to evolution. The exchange of information and stocks among the *Drosophila* researchers has traditionally been very good and provides us nowadays with an exceptionally wellfounded genetic system. The degree of sophistication of the genetic experiments that can be done with *Drosophila* is unparalleled and this is the real strength of the system.

Drosophilids are diploid organisms containing between three to six chromosome pairs. *D. melanogaster* has four chromosome pairs, three autosomal and the sex chromosomes. Chromo-

somes one to three are large metacentric chromosomes, while the fourth is a very small acrocentric chromosome. The first chromosome is the X-chromosome, which is involved in sex determination. The XX constitution results in female development, the XY constitution in male development.

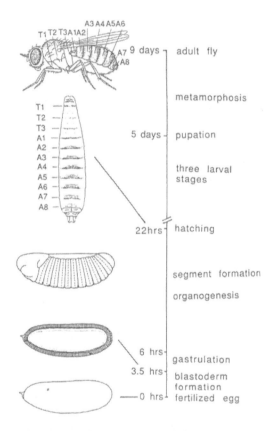

Fig. 1. Developmental stages of *Drosophila melanogaster.* The total generation time depends on the temperature and is about 10–14 days at 25 °C. One day is required to complete embryogenesis, 4 days to go through three larval stages and another 4 days to go through pupation. The adult fly hatches after 9 days and is fertile after another 1 or 2 days. The flies live about 4 weeks and a single female lays about 300 eggs during this time

The Y-chromosome contains no vital gene function apart from fertility genes. The genes can be grouped into a linear order on these chromosomes by recombination mapping. Several specialized cells, most notably the salivary gland cells of the larvae, replicate their chromosomes without undergoing mitotic divisions. This leads to the production of giant or polytene chromosomes, which are essentially composed of bundled threads of interphase DNA. The giant chromosomes show a characteristic pattern of densely and loosely packed chromatin regions, resulting in a defined pattern of bands and interbands. The interbands represent apparently transcriptionally competent chromatin regions, while the bands contain the genes which are not activated in the respective tissue. Early assumptions that one band would represent one gene could not be verified. Chromosomal deletions, inversions, and translocations can be directly mapped on these polytene chromosomes by comparing the banding pattern of the paired homologous chromosomes in heterozygous larvae. It is thus possible to relate the genetic map to the physical map on the chromosome.

Defined regions of giant chromosomes can be dissected under the microscope, the DNA can be extracted and can subsequently be cloned in a process called "microcloning". This yields valuable entry points into the genome in order to start chromosomal walks. In turn, cloned DNA segments can be mapped on the chromosomes by hybridizing labeled probes on chromosome squashes ("in situ hybridization") (Fig. 2). There is an ever increasing number of cloned and mapped DNA fragments from *Drosophila*, which are systematically listed in a "clone list" compiled by J. Merriam, UCLA. This provides entry points into almost every region of the *Drosophila* genome and cloning genes of interest has become a laboratory routine. A more systematical approach, namely cloning the whole *Drosophila* genome in overlapping DNA fragments, has also now been started.

A number of different mutagens can be used to induce mutations in *Drosophila*. The most common ones are X-rays, which induce chromosome breakage and result therefore often in deletions and translocations, and EMS which induces predominantly point mutations. Mutations of vital genes in

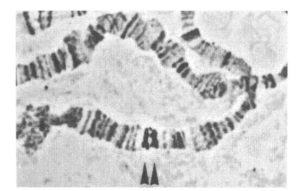

Fig. 2. Example for an in situ hybridization experiment on polytene chromosomes. Two probes which lie 250 kb apart were hybridized simultaneously using biotin-labeled DNA and peroxidase as a detection system. The two probes span the chromosome region 77E to 78E (indicated by *arrowheads*), which includes the map position of the gap gene *knirps*. (Photo U. Nauber, Göttingen). The conventions for the physical chromosome map are as follows: the whole chromosome complement is divided into 100 divisions, whereby divisions 1–20 cover the X-chromosome, 21–40 the left arm of chromosome 2, 40–60 the right arm of chromosome 2, 61–80 the left arm of chromosome three and 81–100 the right arm of chromosome 3. Each division is further divided into six subdivisions, which normally mark the most prominent bands and which are designated by the letters A–F

Drosophila can on the other hand be rescued in germline transformation experiments. This is possible by employing P-element vector constructs carrying the appropriate cloned DNA from the gene region. P-elements occur as natural transposable elements, which are specifically active in the germline of the flies. They encode an enzyme called transposase, which recognizes the ends of the element and causes it to "jump" to other chromosomal locations. P-element plasmid constructs which are injected into the pole plasm region of the embryo where the germ line cells will form, can also serve as a substrate for the initiation of the "jumping". The process is independent of the DNA which lies between the ends of the element, provided that a plasmid which encodes the transposase is coinjected. These features provide a perfect genetic vehicle for the introduction of cloned DNA into the germline. P-element-mediated transformations allow the functional analysis of any piece of DNA in the context of the whole organism. This has become an indispensable tool for the determination of molecular interactions and gene functions.

DNA or RNA for maternal, or early-expressed genes can also be injected directly into the embryo and can partially or fully rescue a mutant phenotype. The reverse, namely phenocopying a known mutation by injection of antisense RNA, is also possible. Injection and transplantation techniques, which include also the transplantation of cytoplasmic material, as well as the transplantation of germline cells (pole cells, see below), therefore play a major role in the functional analysis of the early-acting genes.

A very powerful technique for the analysis of the functional interactions among later acting genes is the induction of mitotic recombination by X-ray treatment during development. If the chromosomes are appropriately marked, one can analyze mutant cell "clones" and their interactions with neighboring cells. The details and the use of this technique are described in Chapter 22, this Volume by Garcia-Bellido.

21.2 Oogenesis and Embryogenesis[2]

Drosophila eggs are formed in meroistic ovarioles (Fig. 3). A characteristic of these ovarioles is that they consist of 16 germline-derived cells, only one of which will eventually form the oocyte. The other 15 cells act as nurse cells, which produce most of the material needed for the building of the oocyte. This includes the deployment of localized determinants, which eventually specify the anterior-posterior polarity. The yolk proteins are synthesized in the fat body of the female and are then imported into the oocyte. It is unclear whether any other proteins follow such a pathway, in particular, whether proteins or signals involved in pattern formation may be derived from nongermline cells. It now seems likely, however, that at least the somatic follicle cells, which surround the developing oocyte and which are mainly involved in producing the chorion (the "egg shell"), provide essential positional cues for the dorsoventral and the terminal system (see below).

Once the egg is fertilized and laid, it starts a series of 14 synchronous nuclear division cycles,

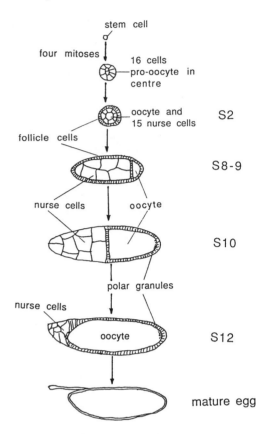

Fig. 3. Oogenesis in *Drosophila*. A germline derived stem cell divides into 16 cells, one of which will form the oocyte, while the other 15 function as nurse cells. This cluster is surrounded by somatic follicle cells. The development of the oocyte includes the deposition of the maternal determinants, which organize the anterior-posterior and the dorso-ventral axes. The surrounding follicle cells apparently play an important role in providing the appropriate signals (see text). The *numbers to the right* indicate the different stages during oogenesis. (After Manseau and Schüpbach 1989)

which are not accompanied by concomitant cell divisions. Beginning with the seventh nuclear division, the nuclei migrate to the surface of the egg and form eventually a homogeneous layer around the whole egg. This stage is commonly referred to as the syncytial blastoderm stage, though the correct biological expression would be plasmodium and not syncytium. Such a system with delayed cellularization is obviously open to any diffusible regulatory substances which may be involved in pattern formation, be it small molecules or whole proteins. The cell membranes start to form only after the 14th nuclear division, by growing in between the nuclei from the surface and eventually surrounding them along with some cytoplasmic material. A specialized set of cells, the pole cells, form after the tenth nuclear division at the posterior pole of the embryo and will give rise to the germline cells (cf. Figs. 1 and 4).

Gastrulation as well as morphogenetic movement begins, once cellularization is complete (Fig. 4). Gastrulation is characterized by the invagination of a ventral furrow, which results in mesoderm formation. At the same time, a posterior plate forms, shifting the pole cells towards the dorsal side of the embryo. Anteriorly, a cephalic fold appears, which marks the future head region of the embryo (Fig. 4). The cell layers, which are going to form the segmented region of the animal – the germ band – extends then towards the posterior. However, since the egg shell does not provide the space for such an extension, the germ band curves around the posterior end and elongates along the dorsal side (extended germ band stage, Fig. 4). The first morphological signs of segmentation appear at this stage as reiterated grooves in the ectoderm. These grooves reflect the borders between the functional segment units (parasegments), but are not identical with the later visible divisions, which separate the segments in the larvae and the adults. The latter are only defined on a morphological basis and are related to the attachment points of the longitudinal muscles. Various experiments and observations have, however, shown that the functional building blocks for the reiterated segmental units are those which are demarcated by the early ectodermal grooves. These units have therefore been called parasegments, in order to allow a distinction between these and the differentiated structures of the larvae and the adult. A parasegment constitutes roughly the region between the posterior half of one segment and the anterior half of the next segment (cf. Fig. 4).

The germ band starts to shorten again and retracts eventually completely from the dorsal side (retracted germband stage). It starts to stretch anteriorly now, causing the involution of a large part of the head segments, which follow the stomodaeal opening and thus come to lie inside the larva later on. Various other morphogenetic movements con-

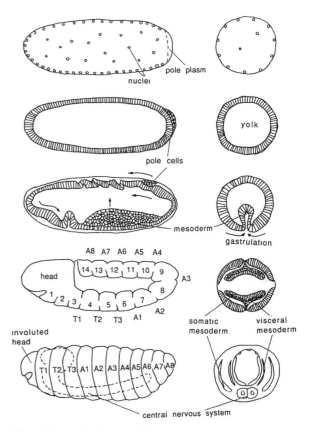

Fig. 4. Schematic diagram of the early stages during *Drosophila* embryogenesis. The nuclei in the freshly laid egg go through 13 division cycles without concomitant cell division. Starting with the eighth cycle, most of them move to the surface of the egg. The cell walls are then formed (cellular blastoderm) and gastrulation begins with the invagination of the mesoderm at the ventral side of the embryo. At the same time, the posterior plate forms and shifts towards anterior, a process which is called germband extension. The first signs of segmentation appear after germ band extension. The visible segment boundaries at this stage correspond to the parasegments (numbered *1–14*). The segments seen at the larval and adult stages become visible after germband retraction. See text for further details. (After Poulson 1950)

cern the formation of the gut and the central nervous system. The larval cuticle is then formed and the larva is ready to hatch (Fig. 4).

Analysis of pattern defects is usually done by examining the first instar larval cuticle. This can be prepared even from embryos which fail to hatch, since the cuticle is laid down at the end of embryogenesis. Analysis of cuticular preparations implies however that defects of the internal organs are usually missed, unless they have a specific effect on the cuticular pattern. Accordingly, apart from the nervous system (see Campos-Ortega Chap. 23, this Vol.) little is known about the genetic control of organogenesis so far. The following discussion focuses therefore exclusively on the genetic control of the early pattern formation events, as revealed by the analysis of the cuticular structures with the denticle belts, which are characteristic for the thoracic and abdominal segments.

21.3 Experimental Embryology

21.3.1 The Blastoderm Fate Map[3]

The essential pattern elements of the embryo are already determined at early stages of embryogenesis. This can be shown by using a laser beam to destroy small numbers of cells at defined positions at the blastoderm stage and by correlating these sites with cuticular defects in first instar larvae. These experiments allowed a detailed map for the cells at the blastoderm stage to be drawn. This map correlates the cells with the structures they would acquire at later stages (fate map) (Fig. 5). The fate map for *Drosophila* shows that only a region between 20–60% EL (EL = egg length; 0% = posterior pole) will give rise to the overtly segmented structures of the larva. The remaining cells specify the head and "tail" elements, which come to lie inside the larva.

The observation that the cells have already acquired their fate at the blastoderm stage could suggest that they respond to a complex pattern of prelocalized inductive factors which were deposited in the egg during oogenesis. Such a "mosaic" mode of development seemed a particularly attractive model for insect embryos, after it was shown that

there existed at least one prelocalized determinant, namely the posterior pole plasm which is necessary for the formation of the pole cells. However, after

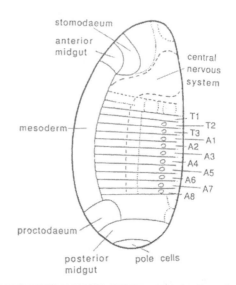

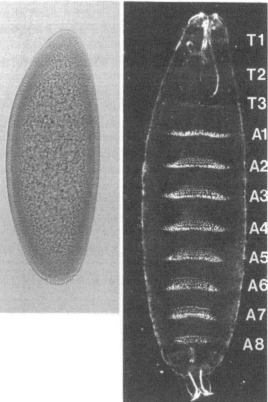

Fig. 5. Representation of the blastoderm fate map as it can be derived from laser ablation experiments (*top*). A blastoderm stage embryo is shown to the *left* (*bottom*) demonstrating the lack of any morphological signs of segmentation. A cuticle preparation of a wild type larvae is shown to the *right* (*bottom*), with the characteristic denticle belts for each segment. (After Nüsslein-Volhard et al. 1982)

the discovery of a number of mutations in *Drosophila*, which lead to mirror image duplications of abdominal structures (the maternal *Bicaudal* mutations, see below) a mosaic mode of development seemed to be unlikely. Such mutations are much easier to reconcile with the assumption of gradients of morphogenetic molecules, which are interpreted by the cells. Moreover, ligation experiments on other insect embryos, most notably of the leafhopper *Euscelis plebejus* had strongly suggested the existence of such gradient systems. *Drosophila* embryos did not easily lend themselves to similar types of experiments. However, the fact that mutations existed already in *Drosophila*, which had defined effects on pattern elements and the fact that *Drosophila* is by far the best manageable genetic system led to the idea that systematic mutagenesis screens should allow the detection of genes which are involved in the specification of the pattern elements. This analysis would enable a general description of the early embryonic processes at the molecular level.

21.3.2 Systematic Mutagenesis [4]

One can expect two modes of inheritance for genes involved in pattern formation, maternal and zygotic. Maternal inheritance would be expected for all processes which are involved in building the oocyte, including the deposition of RNA or proteins, which provide the positional cues for pattern formation. Zygotic inheritance would be expected for all processes which are involved in the interpretation of these positional cues, including all transcripts and proteins which are produced by the cells of the developing embryo. These two inheritance modes result in two different gross phenotypes of the mutants, namely either apparent female "sterility" for maternal inheritance (because all of the laid eggs of a homozygous mutant female would not develop normally) or "embryonic lethality" for zygotic inheritance (a quarter of the eggs does not develop normally, if the allele is recessive and both parents are heterozygous for the allele). These phenotypes can be fairly easily scored for a large number of mutagenized lines. Those lines which show such a phenotype are then analysed further to see

whether their embryos display specific cuticular pattern defects. A maternal inheritance mode can of course only be detected, if homozygous mutant females are viable. Thus, genes which are also involved in crucial functions during the life cycle of the fly will be missed. However, since the aim of such screens is to identify genes which are specifically involved in the pattern formation events, this risk is presumably small, but has to be taken into account.

The *Drosophila* genome encodes at least 5000 genes, mutations in less than 5% of which will result in female sterility and in a quarter of which will result in embryonic lethality. It turned out that only a minor fraction of those display specific pattern defects in the embryo, namely about 30–40 maternal genes and 50–60 zygotic. A screen is considered to reach saturation when no new genes are detected and additional mutations lead only to further alleles of already known genes. The most extensive screens have been performed for zygotic genes on the first, second and third chromosome and it has been estimated that at least a 95% saturation was reached. The fourth chromosome is very small and codes presumably for less than 50 genes. It probably contains only one gene involved in pattern formation.

Several inherent assumptions on the nature of the pattern formation processes have to be made if such systematic screens are to be successful. The first is that the information for building the three-dimensional pattern of the organism must be encoded in the linear structure of the DNA and more specifically in the nuclear DNA. Determination of basic body patterns could, for example, also be mediated by an epigenetic cytoskeletal framework, which could not be analyzed by DNA mutagenesis experiments. The second is that the genetic circuits which determine pattern formation are not at the same time involved in the normal "housekeeping" functions of the cell, since this would preclude the formation of scoreable patterns from the start. Finally, it is necessary to assume that different pattern element are not specified by redundant pathways, but by single genes or single regulatory circuits. Redundant gene functions would not normally be detected, since the chances of hitting two genes specifying the same pathway in a single mutagenesis experiments are very small. All three assumptions are clearly nontrivial and were even more so at the time when these screens were started. This may have been the reason why only a few laboratories have undertaken such screens. The screens, however, proved to be exceedingly successful. Mutations altering specific pattern elements were detected, classified, and analyzed in their effects. This work has led to an increasingly detailed insight into the principles of pattern formation in the *Drosophila* embryo and now allows old questions on the nature of the developmental processes to be answered.

The various mutations affecting the pattern formation process in the *Drosophila* embryo can be grouped into several different classes according to the phenotypes they produce. For a further functional characterization, it is, however, necessary to utilize additional techniques. These are either injection and transplantation techniques or cloning of the genes. The cloned probes can then be used for in situ hybridization experiments, or for in vitro experiments. Injection and transplantation techniques have proven very successful for the further analysis of maternal genes, while the use of cloned molecular markers has been most revealing in the analysis of the regulatory hierarchy and the interactions among the zygotic genes. In contrast, functional analysis of the molecular interactions in vitro is only now beginning to show results.

21.4 The Developmental Gene Hierarchy[5]

The basic polarity and the different pattern elements in the embryo are specified along two axes, the anterior-posterior one and the dorso-ventral one. Both axes are apparently specified independently of each other. Both require a set of maternal and a set of zygotic gene products, which interpret the maternal positional cues. The respective genes can be grouped into hierarchies according to the timepoints during which they are needed during development. There is a further class of genes, the homoeotic genes, which are active along the anterior-posterior axis. They are required to provide the

individual segments with their identity. If they are mutant, they do not lead to pattern defects, but to transformations of one segment identity into another one. Such transformations can have drastic phenotypes, such as legs instead of antennae, or a second pair of wings instead of the halteres. Many details are known about the genes involved and their mode of action. However, since this chapter focuses on the principles of pattern formation rather than on the principles of providing segment identity, they will not be further discussed in this context.

21.4.1 Anterior-Posterior Determination

The maternal genes for the anterior-posterior axis can be grouped into three classes, one which specifies anterior pattern elements, one which specifies posterior pattern elements, and one which specifies the pattern elements of the termini (Fig. 6). The existence of the latter class came as a surprise from the genetic screens and was not predicted from classical experiments.

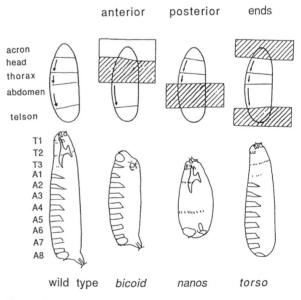

Fig. 6. The three maternal genetic systems which organize the anterior-posterior axis. The regions affected by mutants in each system are shown by the *hatched region* at the *top* of the figure. The resulting phenotypes are diagrammed at the bottom. (After Nüsslein-Volhard et al. 1987)

The zygotic genes for the anterior-posterior axis fall again into three classes, the gap genes, which specify large overlapping domains of the embryo, the pair rule genes, which act in double segmental units, and the segment polarity genes, which affect every segmental unit. These genes were grouped solely on the basis of their mutant phenotypes. However, the existence of these different groups im-

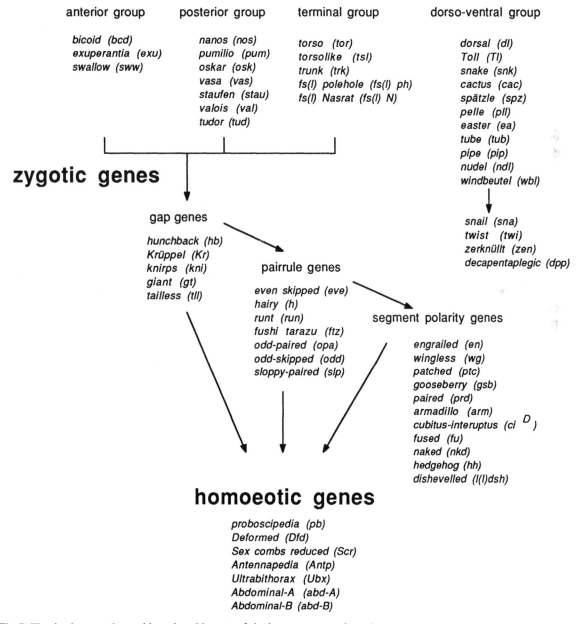

maternal genes

anterior group

bicoid (bcd)
exuperantia (exu)
swallow (sww)

posterior group

nanos (nos)
pumilio (pum)
oskar (osk)
vasa (vas)
staufen (stau)
valois (val)
tudor (tud)

terminal group

torso (tor)
torsolike (tsl)
trunk (trk)
fs(l) polehole (fs(l) ph)
fs(l) Nasrat (fs(l) N)

dorso-ventral group

dorsal (dl)
Toll (Tl)
snake (snk)
cactus (cac)
spätzle (spz)
pelle (pll)
easter (ea)
tube (tub)
pipe (pip)
nudel (ndl)
windbeutel (wbl)

snail (sna)
twist (twi)
zerknüllt (zen)
decapentaplegic (dpp)

zygotic genes

gap genes

hunchback (hb)
Krüppel (Kr)
knirps (kni)
giant (gt)
tailless (tll)

pairrule genes

even skipped (eve)
hairy (h)
runt (run)
fushi tarazu (ftz)
odd-paired (opa)
odd-skipped (odd)
sloppy-paired (slp)

segment polarity genes

engrailed (en)
wingless (wg)
patched (ptc)
gooseberry (gsb)
paired (prd)
armadillo (arm)
cubitus-interuptus (ci ^{D})
fused (fu)
naked (nkd)
hedgehog (hh)
dishevelled (l(l)dsh)

homoeotic genes

proboscipedia (pb)
Deformed (Dfd)
Sex combs reduced (Scr)
Antennapedia (Antp)
Ultrabithorax (Ubx)
Abdominal-A (abd-A)
Abdominal-B (abd-B)

Fig. 7. The developmental gene hierarchy with most of the important genes in each group

plies already that a hierarchy may exist which leads to an increasingly refined determination of the pattern elements during development (Fig. 7).

21.4.1.1 Maternal Genes

The Anterior System[6]. Anterior pattern elements are affected by several maternal mutations, the key gene is, however, believed to be *bicoid*. Embryos coming from mothers which are homozygous mutant for *bicoid* show a loss of head and thoracic structures. Such embryos can be rescued by transplanting cytoplasm from the anterior region of wild-type embryos into the anterior region of the mutant embryo. The rescuing activity is localized at the anterior tip of the wild-type embryos, indicating an asymmetric distribution of the *bicoid* product. The *bicoid* gene was cloned and molecular probes verified these inferences. The *bicoid* RNA is localized at the anterior tip of the early embryo. Its translation products spread towards the posterior region of the embryo and form an anterior-posterior concentration gradient. It seems that the relative concentration values which are produced by this gradient provide positional cues for subsequently acting genes. This can be shown by raising or lowering the gene dosage for *bicoid* by manipulating the number of *bicoid* genes in the maternal genome and monitoring the corresponding effects on the fate map of the embryo. Fate map changes can be inferred from the changed relative position of early morphological markers like for example the head fold, or by analysing the expression pattern of zygotic segmentation genes, as for example the pair rule gene *fushi tarazu*, which is normally expressed in seven stripes at defined positions along the embryo (see below). Raising the gene dose of *bicoid* causes both, the headfold and the *fushi tarazu* stripes to shift towards posterior in a coordinated manner, lowering the *bicoid* gene dose causes them to shift towards anterior (Fig. 8). Interestingly, though, these early fate map changes are in some way counterbalanced during the later development of the embryo, since they can not be seen in the larval cuticle pattern. This shows very clearly that it is not the absolute positions which have to be specified by a morphogenetic molecule, but only the po-

larity and the relative order of the individual pattern elements.

The morphogenetic gradient of the *bicoid* protein is produced by a source-diffusion mechanism, whereby the tightly localized RNA provides the source (Fig. 8a). Much attention focuses therefore now on the question of how this localization is achieved. Two genes of the anterior system, namely *exuperantia* and *swallow* appear to play a role in this process. The *bicoid* RNA is mislocalized in embryos from mothers mutant for these genes, and the phenotypic effects of these mutants on the cuticle pattern can be roughly correlated with this mislocalization. However, both mutants show more pleiotropic effects and the real mechanistic interactions are not yet understood.

The *bicoid* gene codes for a transcription factor with a homeo domain as the DNA-binding element. The protein is thus directly involved in activating other genes. One of these interactions, namely the activation of the primary zygotic expression of the gap gene *hunchback*, has been studied in detail using in vitro techniques. It is, in fact, the best studied example which convincingly demonstrates direct molecular interactions among segmentation genes. This is partly due to the fact that this interaction is straightforward and simple. The *hunchback* gene has a specialized promotor for the *bicoid* regulation, which is activated in a concentration dependent manner (Fig. 9). The *bicoid* protein binds upstream of this promotor at at least three sites, which are roughly 100, 200, and 300 bp away from the transcriptional start point. It appears to then act as a transcription activating factor, which utilizes the general RNA polymerase II machinery to initiate transcription. This was shown by constructing a hybrid gene containing the specific DNA binding part (the homeo domain) of the *bicoid* protein and the activating part of the GAL 4 transcription factor from yeast, or other heterologous activating sequences. The resultant fusion protein was able to rescue the *bicoid* mutant phenotype completely, showing that a very general, evolutionarily conserved part of the transcriptional machinery is used for this regulation. Thus no other specific interactions other than the binding of the *bicoid* protein to the *hunchback* promotor appear to be required for the spatial determination of the

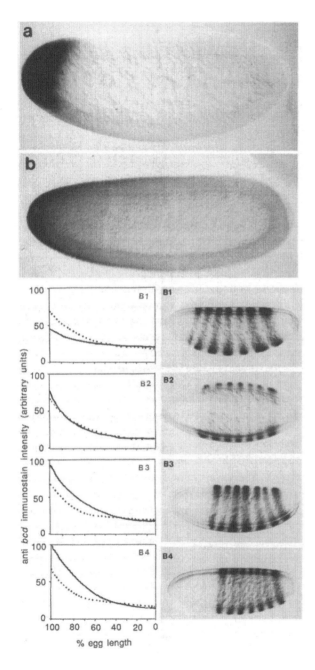

Fig. 8. Function of the *bicoid* gradient. *Top* the *bicoid* RNA (*a*) is localized within the first 10% of the embryo, while the *bicoid* protein (*b*) diffuses towards posterior and forms an anterior posterior gradient of protein distribution, which may cover the anterior 2/3 of the embryo. (St. Johnston et al. 1989). *Bottom* effect of 1–4 maternal doses of the *bicoid* gene on the expression of pair rule stripes in the developing embryo. Raising the *bicoid* gene dosage results in the production of more *bicoid* protein and thus equivalent relative concentrations occur at more posterior positions in the egg. This can be shown by densitometric tracings of embryos stained with the *bicoid* antibody (*left*). The *broken lines* represent the measurements of the wild-type gradient, which is equivalent to two copies of *bicoid*. The *full lines* represent the measurements from embryos with 1–4 copies. The pattern elements which depend on the positional values provided by the relative concentration of the *bicoid* protein are also shifted towards posterior (*right*). The figure shows staining of early embryos with an antibody against the pair rule gene *fushi tarazu*, which reveals *bicoid* dosage-dependent changes of the relative positions of the stripes. (Driever and Nüsslein-Volhard 1988)

The Posterior System[7]. Several genes of the posterior system produce similar or identical phenotypes, when maternally mutant. They all lead to a loss of abdominal segments. Several of them lead in addition to a loss of pole plasm, which is necessary for the formation of the pole cells. With the help of an elegant series of transplantation experiments, it has been possible to analyze these genes further. Their products are synthesized in the nurse cells and get localized during oogenesis at the posterior end of the oocyte (see below). Rescuing activity, as seen by the reappearance of abdominal denticle belts in the injected embryos from mutant mothers, can be obtained either from the nurse cells, or from the posterior pole of freshly laid eggs. However, in order to obtain a good rescue of embryos maternally mutant for one of the genes, it is necessary to inject this rescuing activity at about 20–50% egg length in the posterior region of the embryo, which corresponds roughly to the prospective abdominal region in the fate map. This indicates that the posterior activity has to be moved from the posterior end towards more anterior positions in wild-type embryos. This happens apparently not merely by diffusion, since one of the genes in the posterior group, *pumilio*, is specifically involved in this process. This has been shown by demonstrating that rescuing activity is present at the posterior pole in

primary *hunchback* expression domain. As shown below, most other regulatory interactions, including other *bicoid*-mediated regulatory events, appear to be far more complex, involving the binding of multiple factors which may have both activating and repressing effects.

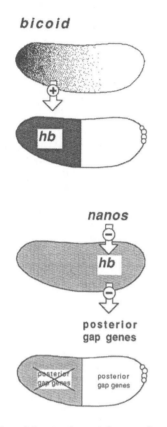

Fig. 9. Function of the anterior and the posterior systems. *Top: bicoid* forms an anterior-posterior gradient of protein distribution, which directly activates the transcription of its zygotic target gene *hunchback. Bottom: nanos* inhibits the translation of the homogenously distributed *hunchback* mRNA in the posterior region of the embryo. This allows the expression of the posterior gap genes, which would otherwise the repressed by *hunchback* protein. Both regulatory events lead to a similar subdivision of the embryo into two fields, which is the basis of the partial redundancy of the two systems (see text)

type, it seems likely that it is the key gene for the posterior group of maternal genes.

The function of the posterior group of genes, with its presumptive key gene *nanos* is radically different from the function of the anterior group of genes with its key gene *bicoid*. It was possible to show that *nanos* does not provide a morphogenetic molecule for the abdomen itself. Instead it is necessary for the repression of a molecule, which is in turn a repressor of abdomen formation. Intriguingly, this repressor of abdomen formation is the maternal product of the gap gene *hunchback*. Maternal *hunchback* RNA is provided homogeneously throughout the embryo, but its translation is inhibited in the posterior half of the embryo. This inhibition is mediated by the posterior maternal system and may be directly dependent on *nanos*. A perturbation of abdomen formation occurs, when this inhibition is artificially circumvented. On the other hand, a rescue of abdomen formation is achieved in embryos which are maternally mutant for both, *nanos* and *hunchback*. Thus, the sole function of *nanos* appears to be the translational repression of the maternal *hunchback* mRNA. Nonetheless, the posterior system clearly has a morphogenetic function. By causing the asymmetry of the maternal *hunchback* expression, it divides the embryo into two fields, the anterior one, which will repress formation of the abdomen, and the posterior one, which allows abdomen formation (Fig. 9). This probably represents the first functional subdivision of the embryo along its anterior-posterior axis and gives it at the same time a polarity which acts as a reference system for the action of the subsequent genes. Note that the same subdivision is also caused by the anterior system acting via *bicoid* (Fig. 9), though this occurs at a later time during development. Nonetheless, it can functionally replace the effect of the maternally caused subdivision and therefore represents a redundant pathway.

Several genes of the posterior system have been cloned so far, namely *vasa, nanos, oskar,* and *pumilio. vasa* codes for a protein which has homology to a whole family of genes which share similarity with the eukaryotic translation initiation factor-4A. This is compatible with its presumptive role in processing or regulating the RNA of the *nanos* gene. *vasa* product is, however, in addition

embryos maternally mutant for *pumilio* and that this activity can be used to rescue the same embryos, if it is mechanically transplanted towards more anterior positions. One of the genes of the posterior system, namely *nanos*, behaves somewhat differently. Rescuing activity for it cannot be recovered form the nurse cells, but only from the posterior region of freshly laid wild-type eggs. This indicates that it may be the last gene in the chain of posterior-acting genes. Furthermore, since it is one of the genes which does not show also a pole cell pheno-

required for normal oogenesis and for pole cell formation. *vasa* mRNA is produced during oogenesis, but is not localized in any defined way. Instead, the translation products of this RNA become localized during oogenesis at the posterior pole, enter the pole cells during early embryogenesis and stay in the germline cells throughout the life cycle. In contrast to *vasa*, the *nanos* RNA is tightly localized at the posterior pole of the embryo. This localization is apparently mediated by a specific receptor, since the RNA is homogeneously distributed during oogenesis. The receptor may be in excess and could act as a "sink" for the more or less freely diffusible RNA. The formation of the receptor and its localization would thus be the task of the other genes of the posterior system. This process may be tightly linked to the process of the formation of the pole plasm, which explains why mutations in these genes usually affect both processes at the same time.

The product of the *Bicaudal D* gene is apparently directly involved in localizing the posterior factors. It codes for a protein with homology to the myosin heavy chain and may thus be part of a cytoskeletal framework which is necessary for pole plasm formation. Certain dominant mutations of this gene cause the posterior factors to be localized at the anterior of the oocyte, which may result in perfectly mirror-image symmetric double abdomen structures.

The Terminal System. [8] The existence of the terminal system, which affects two separate regions of the embryo in a similar manner, was neither expected, nor predicted from any previous experiments. Only the systematic mutagenesis screen allowed this system to be defined. At least six genes belong to this class with *torso* being a presumptive key gene. Embryos from mothers mutant for *torso* lack the structures of the acron and the telson, which correspond to the anterior 10% and posterior 20% of the blastoderm fate map. A clue to the function of the terminal system came from the analysis of neomorphic (gain of function) alleles of *torso*, as well as from the cloning of the *torso* gene product. *torso* RNA and protein surprisingly are not localized, but are distributed throughout the whole embryo. The protein is found in the cell membranes and codes for a receptor tyrosine kinase. This suggests that the region-specific effect of the *torso* gene is mediated by a localized external signal. This signal most likely comes from specialized follicle cells at the termini of the oocyte. A gene which is involved in providing this signal is *torsolike*, since this maternal effect gene was shown not to be re-

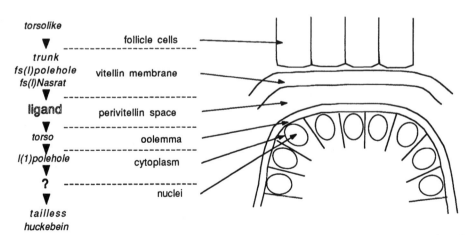

Fig. 10. Hypothetical function of the terminal system. The presumptive hierarchy of genes involved in generating the signal for the terminal system is shown to the *left*. A very schematic diagram of the terminal region of an embryo is shown to the *right*. Note that this diagram does not represent the true situation, since the follicle cells are only present during oogenesis, while the nuclei in the embryo develop only at blastoderm stage. The figure shows, however, that several membranes have to be transversed eventually to generate the signal. Most notable is the perivitellin space in which a freely diffusible ligand may be found, which is, however, immediately trapped by a receptor in the oolemma (see text for further details)

quired in the germline cells which form the nurse cells and the oocyte, but instead in the somatic cells of the female and thus most likely in the follicle cells which surround the oocyte. The process of how these cells obtain in turn their spatial determination is currently unknown. The signal which is provided by the follicle cells is apparently recognized by the *torso*-encoded receptor molecule, which in turn may start a phosphorylation cascade, which eventually results in the regionalized activation of at least one zygotic target gene, namely *tailless* (Fig. 10). This conclusion stems from the analysis of the neomorphic alleles of *torso* described above. These molecules have lost their receptor function and lead to the activation of the zygotic target gene throughout the embryo, since they can no longer recognize the region-specific signal. This leads to a phenotype with hugely expanded terminal structures and loss of the structures in between. This phenotype can however be partially rescued, if *tailless* is mutant at the same time, proving that *tailless* is at least one of the zygotic target genes for the abnormal activation. This interpretation suggests furthermore that the effects of the terminal system can override both the anterior and the posterior systems by repressing their functions. The terminal system may therefore be viewed as a system which sets the frame within which the segmentation genes of the anterior and the posterior system can act.

21.4.1.2 Zygotic Genes[9]

The zygotic genes interpret the maternally provided positional information and convert it into segmental subdivisions. Their products are therefore required in the developing embryo and this is the reason why they show a zygotic inheritance mode. This does not however preclude their also having a maternal expression phase, as is for example the case for *hunchback* (see above). It becomes increasingly clear that most of the segmentation genes have more than one function, and that these may be partly redundant. Redundancies can, of course, hamper the genetic analysis or may even preclude the detection of the respective genes in mutagenesis screens. This has been the case, for example, for a

gene called *caudal*, which was isolated in a molecular screen for genes containing homeoboxes. It was subsequently shown to be expressed maternally and zygotically and that both expressions can complement each other genetically. A segmentation phenotype can be obtained only if caudal is mutant for both maternal and zygotic expression. This gene cannot yet be ordered into the segmentation gene cascade and its primary function is still unclear. It is also unclear how many more of these genes may exist, since mutagenesis screens in which both maternal and zygotic functions are scored are very difficult to do. Preliminary attempts for such a screen suggest that up to 10% of the zygotically lethal mutations may in addition have a maternal component.

The genes discussed below were isolated because they showed a clear zygotic segmentation phenotype. From their analysis and from understanding the logic of their interaction, it can be inferred that the collection is reasonably complete. Certain other functions can, however, be seen, which are not yet represented by a known gene.

Gap Genes.[10] Gap genes were originally defined as those genes which when mutated, produce large contiguous gaps in the segment pattern of the embryo. The classic representatives of this class are *hunchback*, which leads to loss of head and thoracic structures, *Krüppel*, which leads to a loss of thoracic and abdominal structures and *knirps*, which leads to a loss of most of the abdominal structures. It is clear that gap genes stand at the beginning of the segmentation gene hierarchy and their definition has accordingly changed somewhat. Gap genes are now defined more functionally as those genes which are expected to be directly regulated by maternal signals and which are in turn directly involved in regulating the next level of the hierarchy. *tailles*, for example, is clearly a gap gene in this view, as discussed above.

Four gap genes, *hunchback*, *Krüppel*, *knirps*, and *tailles* have been cloned. All four are presumptive transcription factors, with Zn-finger domains as the DNA binding motifs. *Krüppel* and *hunchback* show the Cys-His motifs which were first characterized for the 5S rRNA transcription factor TFIIIA from *Xenopus*, while *knirps* and *tailless*

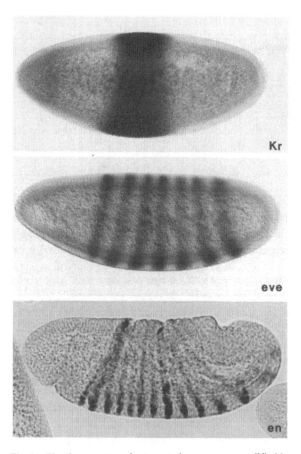

Fig. 11. The three groups of segmentation genes, exemplified by one representative from each group, namely *Krüppel* (*top*, *Kr*) as a gap gene, *even skipped* (*middle*, *eve*) as a pair rule gene and *engrailed* (*bottom*, *en*) as a segment polarity gene. The expression pattern of these genes is visualized by staining early embryos with antibodies against the respective gene products. Note that the full development of the expression pattern of the three groups of genes occurs at progressively later stages. All segmentation genes show in addition secondary expression patterns, either at blastoderm stage, or at later stages during neurogenesis.

show the Cys-Cys finger motifs which are known from hormone receptor genes.

The expression domains of the gap genes are roughly consistent with their phenotypic effects, *hunchback* is expressed primarily in an anterior domain, *Krüppel* in the middle of the embryo (Fig. 11) and *knirps* in a posterior domain. The RNA expression domains of these genes are smaller than the regions in the blastoderm fate map, which are affected by the mutations. This has led to con-

siderable speculation about the function of the gap genes. It is only now becoming clear that the regions in which the gap gene proteins act are in fact larger than was previously assumed and that they may indeed cover the whole phenotypic regions. The gap gene proteins may diffuse from their primary sites of synthesis and may form short range morphogenetic gradients by themselves. It has been shown that very low concentrations of the *hunchback* product are required for the full activation of the *Krüppel* domain and that, in turn, low concentrations of the *Krüppel* product are required for the full acivation of the *knirps* domain. Interestingly. both *Krüppel* and *knirps* are also activated by other factors, *Krüppel* by *bicoid*, and *knirps* by an as yet unknown factor. Both *hunchback* and *bicoid*, at high protein concentrations, may also act as repressors of *Krüppel*. This double effect – repression at high concentrations and activation at low concentrations – is responsible for determining the anterior and posterior boundaries of the *Krüppel* expression domain. This expression domain may in addition be influenced by other factors, it can for example be repressed by *tailless* (see above). These partially redundant interactions may be necessary to ensure proper development under adverse conditions. However, redundant regulatory pathways clearly complicate their analysis and in particular in vitro studies. Proving direct regulatory interactions at this level is therefore not an easy task.

The current model of the interactions among the gap genes is depicted in Fig. 12. The key features are that the regulatory cascade proceeds from anterior towards posterior and that the gap genes themselves provide short-range overlapping morphogenetic gradients, which are likely to be further involved in specifying the different stripes of the pair rule genes.

Pair rule Genes.[11] The existence of this class of genes also came as a surprise from the mutagenesis screens. The fact that there is a transient organization in double segmental units was not predicted from previous experiments. Mutations in pair rule genes affect alternating segments in a specific way. There is for example a loss of thoracic denticle belts T1 and T3 as well as of even numbered abdominal belts A2, A4, A6 and A8 in embryos mutant for

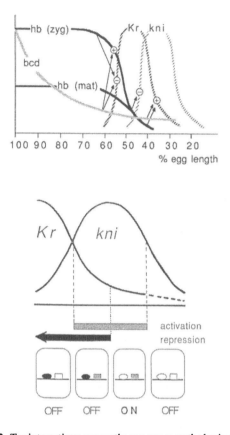

Fig. 12. *Top* interactions among the gap genes on the basis of the formation of short range gradients of protein distribution. The primary expression domains of *hunchback* (*hb*) *Krüppel* (*Kr*) and *knirps* (*kni*) arc indicated with respect to their relative position in the egg. The distribution of the maternal *bicoid* and the maternal *hunchback* protein are shown in addition. The interactions are represented by the "+" (activation) and "–" (repression) signs. *hunchback* is zygotically activated by *bicoid*. The resulting high concentrations of *hunchback* protein repress *Krüppel* transcription, while low concentrations cause an activation of *Krüppel* transcription. These two regulatory effects are also provided by *bicoid*. Medium levels of *hunchback* cause a repression of *knirps* transcription. *Krüppel* activation and *knirps* repression are also regulated by the maternal expression of *hunchback*. (After Hülskamp et al. 1990). *Bottom* once the basic gap gene pattern is set up, the resulting gradients may further serve to specify the primary pair rule stripes in a combinatorial manner. The scheme shows the specification of the spatial position of the sixth stripe of the pair rule gene *hairy*. Relatively high concentrations of *knirps* protein are necessary to activate the promotor element for this stripe, which delimits its potential expression to the *two oval squares in the middle* (*shaded bar*). Low concentrations of *Krüppel* protein act as a repressor on the same promotor element, thus ensuring that the stripe expression is limited to a subregion of only one square. (After Pankratz et al. 1990)

even skipped, or loss of T2 and odd-numbered belts in *fushi tarazu*. Cloning of several pair rule genes has allowed the use of molecular probes in order to determine their expression pattern, as well as to clarify the regulatory interactions which lead to the pattern. Pair rule genes are normally expressed in seven more or less equally spaced stripes throughout the embryo (Fig. 11), whereby the stripes of the different genes overlap in a characteristic way with each other. This pattern is dependent on the regulation by the gap genes, but also on interactions among the pair rule genes themselves. Accordingly, one can define a class of primary pair rule genes, which are thought to be directly regulated by the gap gene products and a class of secondary pair rule genes, which are necessary for the refinement of the pattern and which depend in their expression pattern on the primary pair rule genes. *even skipped*, *hairy*, and *runt* are currently believed to be primary pair rule genes. Interestingly, analysis of the promotor structures has shown at least for *hairy* and *even skipped* that the expression of each of their stripes may be regulated by a different "enhancer" element. One can therefore artificially separate almost each stripe and follow its regulation individually. In contrast, secondary pair rule genes like *fushi tarazu* appear to have only a single "enhancer" element, which specifies all its stripes simultaneously. It appears therefore that each of the primary pair rule stripes is set up independently, by responding to a different combination of spatial cues from the maternal genes and the gap genes (Fig. 12), while the secondary pair rule genes respond coordinately to these primary stripes and deliver eventually their information to the segment polarity genes. The detailed interaction among the pair rule genes is still not known. A number of them code again for potential transcription factors, or at least nuclear proteins; *even skipped* and *fushi tarazu* contain a homeobox as the DNA binding domain and *hairy* a helix-loop-helix motif. The interactions among them may therefore be direct and at the DNA level. Again, the fact that the embryo is still an open blastoderm at the time of activation of the pair rule genes may be significant, since this allows their products to diffuse to neighboring cells and thus provide the necessary positional information.

Segment Polarity Genes.[12] Mutations in this class of genes cause the loss of defined pattern elements within each segment, which is often accompanied by inverted duplications of the remaining structures. The segment primordia at blastoderm stage are about four cells wide and molecular probes for segment polarity genes show that the borders between the segments are already very tightly defined (Fig. 11), *engrailed*, for example, defines the anterior border of each parasegment, while *wingless* defines the posterior border. The expression regions of each of these genes abut each other and are interdependent. *engrailed* is not expressed in *wingless* mutants and vice versa. How this regulation is achieved is not completely clear. *engrailed* is a putative transcription factor, containing again a homeobox as the DNA binding element. Since cellularization is complete at the time of the first expression of the segment polarity genes, these genes must make use of cell communication pathways to transmit their positional information. The *wingless* product may, in fact, be directly involved in such a pathway. It is not nuclear, but shows a signal sequence, which destines it for secretion. The protein can, in fact, be found outside of the *wingless* expressing cells and even within the *engrailed* expressing cells, which never express *wingless* mRNA. The product of *patched*, another segment polarity gene, may also be involved in cell communication, since it is apparently located in the cell membranes. However, any detailed mechanism for how this communication occurs is not yet obvious and remains to be analyzed further. One significant feature of some segment polarity genes is, however, that their expression persists throughout embryonic development, indicating that they are required to maintain the cellular states they define. This is different from the gap and pair rule genes, whose expression fades earlier.

21.4.2 The Dorso-Ventral System[13]

The major pattern elements which are specified by the dorso-ventral system are the prospective mesodermal region at the most ventral side, the neurogenic ectoderm region at the ventro lateral sides, the dorsal ectodermal region at the dorso lateral sides, and the prospective amnioserosa cells at the most dorsal side (Fig. 13). Fate map studies have shown that these basic pattern elements are already specified during early blastoderm formation. In contrast to the determination of the anterior-posterior axis, there is only a single genetic sys-

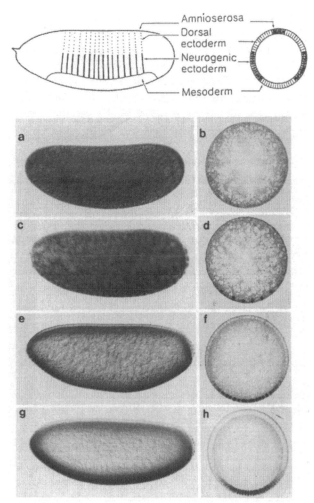

Fig. 13a–h. Determination of dorso-ventral polarity. *Top* the blastoderm fate map of dorso-ventral structures. (After Anderson 1987). *Bottom* formation of the gradient of *dorsal* protein during blastoderm formation. The *dorsal* protein is homogeneously distributed in the cytoplasm of the early embryo. As a result of the specification of dorso-ventral polarity, the protein is imported into the nuclei at the ventral side, but not at the dorsal side, resulting in a gradient of differential nuclear transport **a,c,e,g** lateral view; **b,d,f,h** cross-sections of progressively older blastoderm stage embryos. (Roth et al. 1989)

tem defining the various pattern elements. A number of maternal and zygotic gene functions are involved in this process. If mutant, they lead either to a loss of dorsal, or a loss of ventral pattern elements in the embryonic cuticle. Interestingly, most of the maternally defined loci belong to the latter class, the key gene of which is *dorsal* (Fig. 13). *dorsal* protein apparently provides a morphogen which directly determines the fate of the cells in a concentration dependent manner. Intriguingly, neither the *dorsal* RNA nor the *dorsal* protein is asymmetrically distributed in the embryo. Instead, the necessary asymmetry is achieved by the differential uptake of the protein into the nuclei. *dorsal* protein enters the nuclei at the ventral side in a tightly regulated way, but not those at the dorsal side of the embryo, establishing a ventro-dorsal gradient of nuclear uptake. The role of the other genes in the dorso-ventral cascade is to establish the polarity and the coordinates of the *dorsal* gradient. The *Toll* gene appears to play a key role in this process. It encodes a putative membrane-bound receptor that is apparently activated by a local signal coming from the follicle cells which surround the developing oocyte. As noted above, the follicle cells are not derivatives of the germline cells, but belong to the somatic cells of the animal. Using pole cell transplantation experiments (pole cells form the germline cells), it is therefore possible to decide whether a particular gene is required in the follicle cells only and whether it could therefore be involved in providing the signal for *Toll*. Using such experiments, it was possible to define three genes being necessary in the somatic follicle cells, namely, *nudel*, *pipe*, and *windbeutel*. These genes are involved in releasing a regional signal into the perivitelline fluid which surrounds the embryo. To act as a ligand for *Toll*, this signal, however, must be further processed. Two of the factors necessary for this processing have been cloned (*easter* and *snake*) and their genes code for secreted serine proteases, suggesting that they are involved in a signal amplification mechanism, similar to that observed for the blood clotting cascade. The activated *Toll* product then establishes the point where *dorsal* protein uptake into the nuclei will be highest, thus defining the ventral most point of the embryo. The most dorsal point of the embryo is defined by the exclusion of *dorsal* protein

from the nuclei (Fig. 13). This is apparently mediated by the product of the *cactus* gene, which is the only known maternal gene leading to a loss of dorsal structures when mutant. It thus formally acts as an antagonist of *dorsal*, though its exact role is not yet clear. The *dorsal* protein may subsequently act directly as an activator of the subordinate zygotic genes, though there is no known motif for a DNA binding element in its protein sequence. The most likely zygotic target genes would be *twist*, *snail*, *zerknüllt*, and *decapentaplegic*, *twist* and *snail* being activated while the other two would be repressed.

21.5 Outlook [14]

The genetic dissection of the pattern formation processes in *Drosophila* has revealed a number of principles of how positional information can be generated at the molecular level. It appears that there is no unique process underlying these mechanisms, but that virtually every regulatory pathway which is used within the cells can also be used for defining the spatial pattern between the cells. There is direct use of DNA binding transcription factors, as seen for *bicoid* and the gap gene products. There is regulation at the level of the RNA, as seen for the posterior system, which requires in addition molecules of intermediate filament types for the deployment of the localized factors. There is the use of cell surface receptor molecules acting as kinases, as seen for the terminal system and for the dorso-ventral system. The latter system further uses specific proteolytic enzymes in a regulatory cascade, as well as regulated nuclear uptake of a transcription factor. There is finally also intercellular communication, apparently by the secretion and uptake of molecules, as is seen for the segment-polarity gene *wingless*. This list may expand further once more and more gene products are cloned and analyzed.

The finding that so many different processes are involved in pattern formation is, however, somewhat discomforting. It raises the question whether crucial regulatory circuits may not be hidden among basic cellular functions and may thus have been missed in the mutagenesis screens. However, those functions which have been identified seem to suffice to explain the formation of all pattern elements or allow at least the definition of those parts of the process which cannot yet be correlated with known gene functions. Nonetheless, redundant pathways may exist and have, in fact, already been identified in some cases. It is therefore always necessary to evaluate critically the strength and weaknesses of the mutagenesis screens. They did not intend to find all pattern forming processes and they have purposely been restricted to those pattern elements which determine cuticular structures of the larvae. There is therefore plenty of room for the analysis of further processes. These include in particular the processes of oocyte formation and of organogenesis, formation of the nervous system, and the formation of adult structures, such as, for example, the eye. These fields have been taken on by several laboratories now and promise to become equally successful as the screens aimed at the detection of the processes of early pattern formation events. An increasingly, useful tool for such screens is the use of so-called enhancer trap lines. Such lines consist of flies which have been transformed with a P-element construct containing the bacterial β-galactosidase gene under a weak promotor. When integrated into the genome, this promotor is often regulated by the enhancer element near the site of integration. If this element is a tissue specific enhancer element, β-galactosidase activity is found in a subset of cells and tissues. Since this can be detected in a simple cytological staining procedure, it is possible to screen large numbers of lines. A bewildering array of different patterns has so far been identified in this way, and it was possible to clone several genes which may at least partly be involved in setting up these patterns. The full value of the technique, in particular for further genetic analysis, has still to be established, but it seems likely that it will lead to many new insights into the pattern formation events.

21.6 Summary

The systematic genetic approach to the analysis of the early pattern formation processes in *Drosophila* has yielded a very detailed insight into the principles which govern embryogenesis in insects. It is possible to define hierarchies of genetic interactions which are required to specify the progressive subdivision of the early embryo into segmental units along the anterior-posterior axis and into dorso-ventral fields. The earliest events are determined by morphogenetic gradients, which provide positional information by concentration-dependent interactions with their targets. Several classes of molecules can be involved in these interactions, ranging from transcription factors to cell surface receptor molecules. The genetic analysis has also shown that the regulatory events leading to pattern formation can be redundantly specified.

References

General

Akam M (1987) The molecular basis for metameric pattern in the *Drosophila* embryo. Development 101:1 – 22
Ashburner M (1989) *Drosophila* – a laboratory handbook. Cold Spring Harbor Laboratory Press, New York
Campos-Ortega JA, Hartenstein V (1985) The embryonic development of *Drosophila melanogaster*. Springer, Berlin Heidelberg New York
Ingham PW (1988) The molecular genetics of embryonic pattern formation in *Drosophila*. Nature 335:25 – 34
Roberts DB (1986) *Drosophila* – a practical approach. IRL Press, Oxford
Anderson KV, Nüsslein-Volhard C (1984) Information for the dorsal-ventral pattern of the *Drosophila* embryo is stored as maternal mRNA. Nature 311:223 – 227
Asburner M, Berendes HD (1978) Puffing of polytene chromosomes. In: Asburner M, Wright TRF (eds) The genetics and biology of *Drosophila*, vol 2b. Academic Press, London, pp 315 – 395
Pirotta VH, Jäckle H, Edström JE (1983) Microcloning of microdissected chromosome fragments. In: Hollaender A, Setlow JK (eds) Genetic engineering: principles and methods, vol 5. Plenum Press, New York, pp 1 – 15
Poulson DF (1950) In: Demerec M (ed) The biology of *Drosophila*. Wiley, New York, pp 168 – 274
Rosenberg UB, Preiss A, Seifert E, Jäckle H, Knipple DC (1985) Production of phenocopies by *Krüppel* antisense RNA injected into *Drosophila* embryos. Nature 313:703 – 706
Spradling AC, Rubin GM (1982) Transposition of cloned P-elements into the *Drosophila* germ line chromosomes. Science 218:341 – 347

[2] Manseau LJ, Schüpbach T (1989) The egg came first, of course! Anterior-posterior pattern formation in *Drosophila* embryogenesis and oogenesis. Trends Genet 5:400–405

Martinez-Arias A, Lawrence PA (1985) Parasegments and compartments in the *Drosophila* embryo. Nature 313:639–642

[3] Lohs-Schardin M, Cremer C, Nüsslein-Volhard C (1979) A fate map for the larval epidermis of *Drosophila melanogaster*: localized cuticle defects following irradiation of the blastoderm with an ultraviolet laser microbeam. Dev Biol 73:239–255

Sander K (1976) Specification of the basic body pattern in insect embryogenesis. Adv Insect Physiol 12:125–238

[4] Jürgens G, Wieschaus E, Nüsslein-Volhard C, Kluding H (1984) Mutations affecting the pattern of the larval cuticle in *Drosophila melanogaster* II. Zygotic loci on the third chromosome. Roux's Arch Dev Biol 193:283–295

Nüsslein-Volhard C, Wieschaus E, Kluding H (1984) Mutations affecting the pattern of the larval cuticle in *Drosophila melanogaster* I. Zygotic loci on the second chromosome. Wilhelm Roux's Arch Dev Biol 193:267–282

Wieschaus E, Nüsslein-Volhard C, Jürgens G (1984) Mutations affecting the pattern of the larval cuticle in *Drosophila melanogaster* III. Zygotic loci on the X-chromosome and the fourth chromosome. Roux's Arch Dev Biol 193:296–307

[5] Lewis EB (1978) A gene complex controlling segmentation in *Drosophila*. Nature 276:565–570

Nüsslein-Volhard C, Frohnhöfer HG, Lehmann R (1987) Determination of anteroposterior polarity in *Drosophila*. Science 238:1675–1681

Nüsslein-Volhard C, Wieschaus E (1980) Mutations affecting segment number and polarity in *Drosophila*. Nature 287:795–801

[6] Driever W, Nüsslein-Volhard C (1988) A gradient of *bicoid* protein in *Drosophila* embryos. Cell 54:83–93

Driever W, Nüsslein-Volhard C (1988) The *bicoid* protein determines position in the *Drosophila* embryo in a concentration dependent manner. Cell 54:95–104

Driever W, Ma J, Nüsslein-Volhard C, Ptashne M (1989) Rescue of *bicoid* mutant *Drosophila* embryos by *bicoid* fusion proteins containing heterologous activating sequences. Nature 342:149–154

Frohnhöfer HG, Nüsslein-Volhard C (1986) Organization of anterior pattern in the *Drosophila* embryo by the maternal gene *bicoid*. Nature 324:120–125

St Johnston D, Driever W, Berleth T, Richstein S, Nüsslein-Volhard C (1989) Multiple steps in the localization of *bicoid* RNA to the anterior pole of the *Drosophila* oocyte. Development 107:13–19 (Suppl)

Struhl G, Struhl K, Macdonald P (1989) The gradient morphogen *bicoid* is a concentration-dependent transcriptional activator. Cell 57:1259–1273

[7] Ephrussi A, Dickinson LK, Lehmann R (1991) *oskar* organizes the germ plasm and directs localization of the posterior determinant *nanos*. Cell 66:37–50

Hay B, Jan LY, Jan YN (1990) Localization of *vasa*, a component of *Drosophila* polar granules, in maternal-effect mutants that alter embryonic anteroposterior polarity. Development 109:425–433

Hülskamp M, Schröder C, Pfeifle C, Jäckle H, Tautz D (1989) Posterior segmentation of the *Drosophila* embryo in the absence of a maternal posterior organizer gene. Nature 338:629–632

Lehmann R, Nüsslein-Volhard C (1987) Involvement of the *pumilio* gene in the transport of an abdominal signal in the *Drosophila* embryo. Nature 329:167–170

Wang C, Lehmann R (1991) *nanos* is the localized posterior determinant in *Drosophila*. Cell 66:637–648

Wharton RP, Struhl G (1989) Structure of the *Drosophila Bicaudal D* protein and its role in localizing the posterior determinant *nanos*. Cell 59:881–892

Wharton RP, Struhl G (1991) RNA regulatory elements mediate control of *Drosophila* body pattern by the posterior morphogen *nanos*. Cell 67:955–967

[8] Casanova J, Struhl G (1989) Localized surface activity of *torso*, a receptor tyrosine kinase, specifies terminal body pattern in *Drosophila*. Genes Dev 3:2025–2038

Klingler M, Erdélyi M, Szabad J, Nüsslein-Volhard C (1988) Function of *torso* in determining the terminal anlagen of the *Drosophila* embryo. Nature 335:275–277

Schüpbach T, Wieschaus E (1986) Maternal effect mutations altering the anterior-posterior pattern of the *Drosophila* embryo. Roux's Arch Dev Biol 195:302–317

Stevens LM, Frohnhöfer HG, Klingler M, Nüsslein-Volhard C (1990) Localized requirement for *torso-like* expression in follicle cells for development of terminal anlagen of the *Drosophila* embryo. Nature 346:660–663

[9] Macdonald PM, Struhl G (1986) A molecular gradient in early *Drosophila* embryos and its role in specifying the body pattern. Nature 324:537–545

Perrimon N, Engstrom L, Mahowald AP (1989) Zygotic lethals with specific maternal effect phenotypes in *Drosophila melanogaster*. I. Loci on the X-chromosome. Genetics 121:333–352

[10] Hülskamp M, Tautz D (1991) Gap genes and gradients – the logic behind the gaps. BioEssays 13:261–268

Hülskamp M, Pfeifle C, Tautz D (1990) A morphogenetic gradient of *hunchback* organizes the expression of the gap genes *Krüppel* and *knirps* in the early *Drosophila* embryo. Nature 346:577–580

[11] Caroll SB (1990) Zebra patterns in fly embryos: activation of stripes or repression of interstripes? Cell 60:9–16

Gehring WJ (1985) Homeotic genes, the homeobox and the genetic control of development. CSHSQB 50:243–251

Hiromi Y, Kuroiwa A, Gehring WJ (1985) Control elements of the *Drosophila* segmentation gene *fushi tarazu*. Cell 43:603–613

Pankratz M, Seifert E, Gerwin N, Billi B, Nauber U, Jäckle H (1990) Gradients of *Krüppel* and *knirps* gene products direct pair-rule stripe patterning in the posterior region of the *Drosophila* embryo. Cell 61:309–317

[12] Martinez-Arias A, Baker NE, Ingham PW (1988) Role of segment polarity genes in the definition and maintenance of

cell states in the *Drosophila* embryo. Development 103:157–170

Nakano Y, Guerrero I, Hidalgo A, Taylor A, Whittle JRS, Ingham PW (1989) A protein with several possible membrane-spanning domains encoded by the *Drosophila* segment polarity gene *patched*. Nature 341:508–513

van den Heuvel M, Nusse R, Johston P, Lawrence PA (1989) Distribution of the *wingless* gene product in *Drosophila* embryos: a protein involved in cell-cell communication. Cell 59:739–749

[13] Anderson KV (1987) Dorsal-ventral embryonic pattern genes of *Drosophila*. Trends Genet 3:91–97

Hashimoto C, Hudson KL, Anderson KV (1988) The *toll* gene of *Drosophila*, required for dorsal-vental embryonic polarity, appears to encode a transmembrane protein. Cell 52:269–279

Roth S, Stein D, Nüsslein-Volhard C (1989) A gradient of nuclear localization of the *dorsal* protein determines dorsoventral pattern in the *Drosophila* embryo. Cell 59:1189–1202

Stein D, Roth S, Vogelsang E, Nüsslein-Volhard C (1991) The polarity of the dorsoventral axis in the *Drosophila* embryo is defined by an extracellular signal. Cell 65:725–735

[14] Grossniklaus U, Bellen HJ, Wilson C, Gehring WJ (1989) P-element-mediated enhancer detection applied to the study of oogenesis in *Drosophila*. Development 107:189–200

O'Kane CJ, Gehring WJ (1987) Detection in situ of genomic regulatory elements in *Drosophila*. Proc Natl Acad Sci USA 84:9123–9127

Tautz D (1992) Redundancies, development and the flow of information. BioEssays (in press)

Chapter 22   Generation of Pattern in *Drosophila melanogaster* Adult Flies

Antonio Garcia-Bellido and Manuel Mari-Beffa

22.1 Introduction[1,2]

Adult flies of *Drosophila melanogaster* have an epidermis which is produced from groups of cells which become distinct from those of the blastoderm, early in embryo development. Two types of cell groups are involved. The first, imaginal disks, give rise to most adult structures (imago is the technical term for the adult fly, hence imaginal disks). The second, histoblast nests, develop into the dorsal epidermis (tergites) and ventral epidermis (sternites) of the abdomen. There are a number of imaginal disks, each of which generates a characteristic adult structure. Each disk consists of a single layer of cells which forms a sac, connected to the larval epidermis by a stalk. The cells of the imaginal disks divide continuously during larval development. For example, the imaginal disk which in *Drosophila* gives rise to a wing, develops from a few cells until, at maturity, it contains about 50000 cells. At metamorphosis, the disks evaginate, expand, and substitute for the larval epidermis, which undergoes programmed death. In contrast, the histoblast nests, groups of cells within the larval epidermis, remain quiescent during larval development and do not start to proliferate until metamorphosis, giving rise to the about 4500 cells in each hemitergite or 3000 cells in each hemisternite. Like the imaginal disks, each histoblast nest only gives rise to structures on the left- or right-hand side of a segment, hence each tergite or sternite is produced in two halves (hermitergites and hemisternites).

Epidermal cells differentiate cuticular structures such as chaetae (bristles which act as sensory organs of the peripheral nervous system; see Sect. 22.4.3), trichomes and veins as well as fixed patterns of pigmentation. The pattern of these structures is characteristic for each segment of the adult. These patterns are also specific for a developmental stage and are associated with constant shapes. Two mechanisms might operate to generate pattern in the adult epidermis of *Drosophila*. Elements may be specified on a primordium (= anlage) which has already developed. Alternatively, intercalary cell proliferation leads to the successive specification of developmental territories and hence of differentiated cell types. The first of these mechanisms operates early in development at the blastoderm stage (see Tautz, Chap. 21, this vol.) and in some adult structures. e.g., the ommatidia in the eye disk. The second mechanism occurs in the formation of imaginal disks. More promitive insects (and annelids) generate segments by a different process, by budding from a terminal region of growth. In each of these three types of processes, cell-cell interactions occur and specific genes may be involved which define the actual size and shape of the patterns which characterize the species.

Experimental manipulations can lead to variations in size while retaining the normal pattern, and it has therefore been argued that cell proliferation is distinct from pattern formation. However, many observations in insects suggest that this conclusion is wrong. The analysis of pattern generation in these organisms uncovers a tight causal relationship between cell proliferation and pattern formation. This chapter presents an overview of what is at present known about the generation of some imaginal patterns in *Drosophila*.

22.2 Developmental Analysis[3]

The programming of the developmental fate of imaginal disks in *Drosophila* can be studied by transplanting disks into the abdominal cavity of either a larva or an adult. When transplanted into a larva, the cultured disks differentiate when the larva undergoes metamorphosis, and give rise to structures which are characteristic of their origin, e.g., a wing disk gives rise to structures which are found on the adult wing. The disks are therefore said to be determined. When cultured in an adult, the disks proliferate (see below) but do not differentiate. They can be isolated, divided and recultured in another adult. After several such subcultures, the disks can be transplanted into a larva and their developmental potential assessed. It is found that the original determination may break down upon culture and structures and patterns characteristic of other disks may appear, a phenomenon known as transdetermination. Transdetermination does not occur in a random manner, but in frequencies characteristic of the region of origin.

Experiments involving the culture of imaginal disks have also provided results which support the conclusion that the final size of an imaginal disk is determined by internal cues rather than the hormonal milieu in which it develops. Wing disks can be taken from newly hatched larvae and transplanted into the abdominal cavity of an adult, where they continue to grow in a very different hormonal milieu. The disks nevertheless grow to a maximum size similar to that which they would have normally achieved if left in situ. If disks are instead transplanted into younger larvae and cultured for several extra days before entering metamorphosis, the disks do not show any increase beyond their normal size. Further evidence for the autonomous development of imaginal disks comes from experiments in which larvae are induced to undergo metamorphosis prematurely, so that the imaginal disks have not reached their normal size. It is found that the imaginal disks give rise to adult organs that are smaller than normal and contain less cells. In extreme cases, organs are produced which lack some pattern elements such as chaetae. Results of this type show that patterns within the adult fly are dependent on size, and determined by the information within the cells which contribute to the adult structure.

Experiments in other insects reveal that epidermal cells express positional information. In *Rhod-*

nius, pieces of tergite can be cut out and transplanted. When a portion of tergite from one segment is transplanted to an equivalent position in another segment(= isotopic graft), healing takes place reconstructing the normal cuticular patterning, although the graft retains all features characteristic of its original segment. However, when a graft is made to a different position along the anterior-posterior axis, either within the same or in a different segment (= heterotopic graft), a reaction occurs at the border of the graft and host tissue. The final pattern is the result of the local connection of cells originally at the same position (Fig. 1a). If the graft is rotated through 90 °C before isotopic grafting, healing occurs in such a way as to connect cells in host and graft, which originally had the same position along the anterio-posterior axis (Fig. 1b). These observations suggest that cells have adhesion properties which are dependent on their position within a tergite. We may conclude that cells must contain information which specifies their position,

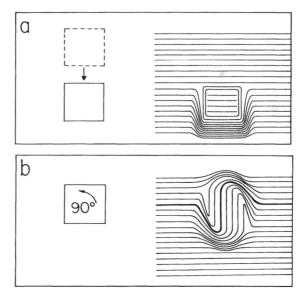

Fig. 1a, b. Transplantation of pieces of epidermis in abdominal tergites of *Rhodnius*. The transplant operation is shown schematically to the *left*, and to the *right*, the resultant pattern of regeneration as visualized by the formation of ripples on the epidermis. In **a**, the pattern of regeneration suggests that host epidermis cells with similar positional values become reconnected, leaving the heterotropic transplant isolated. In **b**, an isotopic transplant, following 90° rotation, results in a pattern in which tissue with similar positional values are again reconnected

i.e., each cell has a positional value. It also shows that the positional values of cells of tergites from different segments are equivalent.

Similar experiments have been carried out in *Leucophaea* and *Drosophila* using imaginal disks which give rise to appendages such as legs. Insect legs are composed of segments, each of which has a characteristic size and pattern. When a leg is amputated, cells at the stump divide to regenerate all the distal structures which have been cut off. It is also possible to cut off a leg and to graft it on to a stump. However, when leg and stump have been cut at the same level within the same segment healing occurs without cell proliferation. If the cuts are made at the same level in two different leg segments, healing again occurs without cell proliferation and intermediate segments are not regenerated (Fig. 2a, graft A−C). When the cuts are made at different levels, so that the stump/graft has regions missing, cell division leads to the intercalary regeneration of a complete segment (Fig. 2a, graft A−B). When stump and graft duplicate part of a

segment (Fig. 2a, graft A−D), cell division again leads to intercallary growth but here structures are generated in the reverse of normal polarity. In all cases, the regenerated structures are formed by the offspring of cells at the border of both the stump and graft, and are of normal dimensions. Regeneration therefore occurs when tissue from different positions along the proximo-distal axis of a leg segment are brought together. Leg segments must therefore contain positional information, all segments behaving as homologous units.

Other experiments reveal that positional information must exist in the circular axis of appendages (i.e., the dorso-ventral axis of the insect). Rotation of a leg graft relative to the stump leads, as in tergites, to a reaction. One or more extra legs develop, containing cells derived from both the stump and graft. When the graft is rotated 180° (see Fig. 2b), so that there is a maximum difference in positional values between stump and graft, two extra legs are usually produced. In insects, there is therefore positional information which drives cell

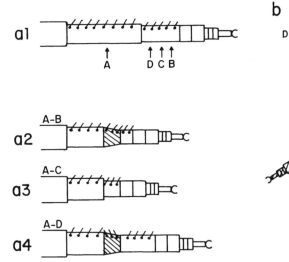

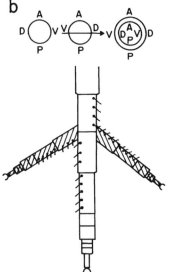

Fig. 2a, b. The results of grafts involving legs from larvae of *Leucophaea*. **a1** Schematic representation of the leg with the positions of the cuts (*arrows* in **a1**) made for the grafts leading to the regeneration patterns shown in **a2, 3** and **4**. **a2** A distal portion of leg cut at *B* was grafted to a stump cut at *A*. Intercalary regeneration (*shaded*) occurs in the normal polarity to restore the original positional values of the segment. **a3** The grafted portion was cut at *C*. *A* and *C* have equivalent positional values

within the segment and so healing occurs without regeneration. **a4** The grafted portion was cut at *D*. The host and graft have positional values which overlap, and intercalary regeneration again occurs but in this case the regenerated tissue shows a reverse of normal polarity (as shown by the orientation of chaetae). **b** The distal portion of a right leg was grafted to the stump of a left leg and regeneration has lead to the development of extra limbs. (*A* anterior; *P* posterior; *V* ventral; *D* dorsal)

division along the anterio-posterior and dorso-ventral body axis as well as the proximo-distal axis of appendages.

It is also possible to study the programming of regeneration by manipulating single cells. Cells can be isolated either from a blastoderm (see Tautz Chap. 21, this Vol.) or from an imaginal disk, and transplanted into another blastoderm where they develop to give rise to structures which are characteristic of their region of origin (see Fig. 3a). These results again indicate that adult patterns result from the properties of individual cells without immediate reference to the more general features of the cell population in which they lie.

Further information about these local properties can be obtained from experiments involving cell mixing. Cells derived from imaginal disks can be dissociated, allowed to re-aggregate and the resulting cell pellet can be cultured within the abdominal cavity of an adult. When induced to differentiate by transplantation into a larva, the cultured cells give rise to tissue with identifiable cuticular structures. These normally correspond to those produced by the disk from which they were derived,

showing that determination is maintained even when cells are dissociated and re-aggregated. It is possible to aggregate cells from two different disks (see Fig. 3b) or disk regions. The origin of the cells can be marked by using genetic stocks having different cell phenotypes (see Sect. 22.3). When mixed aggregates are derived from different disk types (e.g., wing and leg) or disk regions, the cells are able to sort out from one another, but when cells from the similar disks or disk regions are mixed, then inter-mingling can occur to form mosaics. Cells of the same origin generate more or less integrated territories, although these extend only to a few thousand cells. Individual cuticular elements are usually correctly located but sometimes their polarity is reversed (Fig. 3b). It is possible to alter the behaviour of cells in a number of ways both genetically and by manipulating the experimental procedure. Mutations in homoeotic genes (see Sect. 22.4.1) can lead to alterations in cell fate and cell recognition. Culture of cells before implantation into a host larva results in transdetermination (see above). Thus cells derived from a wing disk, which have been maintained in culture prior to mixing, can give rise,

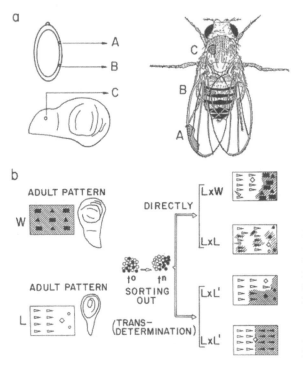

Fig. 3a, b. Cell explantation (**a**) and mixing (**b**) experiments. **a** Cells from different regions of the blastoderm (*A, B*) or from the presumptive notum region of wing imaginal disks (*C*) are transplanted into a host blastoderm. After growth and metamorphosis, they differentiate to give adult pattern fragments of normal dimensions and in a position characteristic of their origin (*A, B,* and *C*). **b** Dissociated cells from wing (*W*) and leg (*L*) imaginal disks can be mixed and allowed to aggregate. The aggregated pellet of cells can either be cultured by transplantation into the abdominal cavity of an adult or can be induced to differentiate by transplantation into a larva which undergoes metamorphosis. Upon differentiation, the cells give rise to tissue which has patterns characteristic of adult cuticular structures. The disk origin can be identified by the use of cell markers (*black* and *white*). The resultant patterns depend on the cell combinations and the time of culture after aggregation before inducing differentiation. If differentiation occurs directly following aggregation, *L × W* combinations show total sorting out of components from the two disks, and formation of partial pattern within tissue originating from a single disk, while *L × L* aggregates generate pattern in mosaics. Notice that some elements show a wrong orientation in an otherwise correct pattern. If the aggregates are maintained in culture prior to inducing differentiation, cells from wing disks can transdetermine to leg (*L'*) and mix to form either integrated patterns or mirror-image duplications upon the metamorphosis of the host larva, to structures characteristic of leg and mix with leg cells but sort from wing cells.

22.3 Clonal Analysis [4, 5]

To study the generation of pattern during normal development, it is desirable to follow individual cell lineages. A method is required so that the progeny of a single cell in a disk can be marked and the adult structures to which it gives rise be identified. This would allow the analysis of the dynamics of cell proliferation as well as demonstrating whether cell lineages existed which are associated with specific adult structures. In *Drosophila*, these analyses are carried out using genetic markers which lead to alteration of cellular phenotypes without affecting the processes of pattern formation. An example of such a marker is *mwh*, multiple wing hair, which leads to cells producing a tuft of small trichomes rather than the single trichome produced by single wild-type cells. The *mwh* allele is recessive, and can be used in the mixing experiments described in Section 22.2, by taking cells form homozygous *mwh* stock and from a wild-type stock.

Cell lineage can also be studied using such markers, by inducing a genetic change during development which leads to a lineage with an identifiable cell phenotype. The first such experiments used gynandromorphs, sexual mosaics which result from the loss of an X chromosome from a cell early in development. The starting point is a female zygote (XX) carrying a recessive cell marker on one X chromosome. If the X chromosome carrying the wild-type dominant allele is lost from an early cleavage nucleus, this will give rise to a lineage of male tissue (XO) which is phenotypically mutant. The territories which are male (mutant) and female (normal) can then be mapped. Such an analysis shows that there is no major mixing or migration during the subsequent proliferation of cells. It also shows that any adult structure can be produced as an integrated mosaic of female and male cells. Thus it can be argued that at the earliest stage of development at which X chromosome loss can occur, cell lineages are not determined to give rise to any particular adult structure.

It is also possible to label cell lineages at later stages of development using mitotic recombination which can be induced by x-irradiation in embryos, heterozygous for a recessive cell marker. If a mitotic recombination event involving the marked chromosome occurs at any stage at which cells are dividing, a cell clone will be produced which is homozygous for the marker, and therefore phenotypically distinct. Such a procedure allows the identification of all the descendants of a particular cell and gives useful information about the dynamics of cell proliferation, allowing estimation of cell cycles times, the number of cell divisions during the development of adult structures, and can be used to deduce the number of founder cells for any imaginal disk or histoblast nest (Fig. 4 and 8a).

22.3.1 Developmental Compartments Are Supra-Cellular Units, Polyclonal in Origin

When mitotic recombination involves a cell in the developing wing disk, the clone that arises occupies a compact territory, with a shape characteristic of the region occupied. When a series of clones are studied from different individuals, and therefore of independent origin, their territories can be seen to overlap. There are, however, restrictions to the territory a single clone may occupy. There exist virtual lines, lines which are not associated with any structure but which are established as result of the behavior of cell clones. These virtual lines subdivide the wing into "developmental compartments". Thus clones formed as a result of mitotic recombination early in development at the cellular blastoderm stage will occupy territories in any region of the anterior or posterior part of the wing but will not extend across the anterior/posterior boundary, even though the clones may abut the boundary for hundreds of cells (Fig. 4a). Other compartments can be identified in the development of the wing disk; one divides dorsal from ventral structures, the other notum from wing. Homologous compartments, at least for the anterior/posterior division, have been found in structures derived from other imaginal disks, including legs, proboscis, terminalia, and antenna. Adult structures do not therefore develop from founder cells with rigid de-

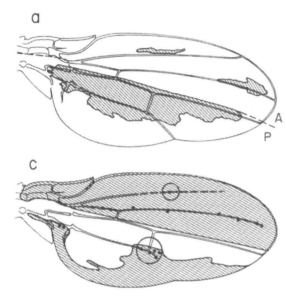

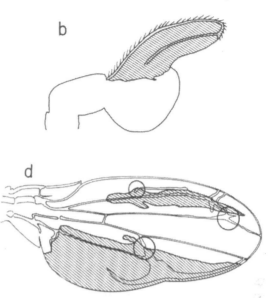

Fig. 4a–d. Different types of genetic mosaic in wing and haltere. **a** Two types of clone initiated at the same time in development in M/M$^+$ flies. In the anterior (A) compartment, marked clones (*fine shading*) have retained the M/M$^+$ genotype. Note the preferred line of clone borders along veins. In the posterior (P) compartment, an M$^+$/M$^+$ clone (*coarse shading*) is shown. Note the clone border along the A/P compartment boundary. **b** A Ubx^-/Ubx^- clone (*coarse shading*) in a Ubx^-/Ubx^+ haltere. The clone is labeled by a cell marker genetically linked to Ubx and has cell types and patterns characteristic of wing. **c** Two dorsal clones homozygous for ve, h, ri. The ve mutant allele leads to a lack of the distal part of all longitudinal veins. The h mutant allele leads to the production of extra chaetae. The ri mutant allele leads to the absence of the LII vein and suppresses the production of extra chaetae caused by h. *Circles* indicate regions of nonautonomy of vein and the differentiation of extra chaetae in the borders between mutant and normal territories. **d** A ventral (in the anterior compartment) and a dorsal (in the posterior compartment) clone of px^{72}/px^{72} cells showing partial and local nonautonomy (*circles*). The px^{72} mutant allele leads to the differentiation of extra veins. Note that surrounding wild-type cells (genotype px^{72}/px^+) also show extra veins connecting with the px veins of the px^{72}/px^{72} clone

velopmental fates, as in the nematode, *Caenorhabditis elegans*, but are instead organised into compartments, supra-cellular units of development. These are polyclonal in origin, a number of unrelated cells becoming committed at a stage in development together to give rise to a specific part of the adult, e.g., the anterior half of a wing. These commitments apparently arise by a series of sequential decisions made as the cells of the disk proliferate (see Sect. 22.4.2).

22.3.2 The Elements of the Epidermis Develop in Place

Cells belonging to a clone usually remain together, although they may give rise to different cuticular elements. Clones formed late, in the last three divison

cycles of the wing imaginal disk, appear to be restricted to a specific cuticular elements, i.e., chaetae or wing veins. The development of cuticular elements therefore occurs in situ and not by the migration of previously determined cells.

22.3.3 Size and Shape of Patterns Do Not Result from Mechanisms Which Count Cell Divisions

These analyses of cell lineage help us to define both qualitatively and quantitatively, events in the development of wild-type individuals. The studies can be extended to mutant individuals to ascertain how a particular mutation affects development. The study of homoeotic transformations (see Sect. 22.4.1), in which the developmental fate of one segment or compartment is apparently substituted by that of another, reveals that these transformations are associated with corresponding changes in the parameters governing cell proliferation, i.e., if an antenna is transformed to a leg, the cells now show the division characteristics of leg. In some cases, a recessive mutant allele of a gene which affects development can be linked in the same chromosome arm to a cell marker. When mitotic recombination involving this chromosome arm occurs, a marked clone is produced which is homozygous for both the cell marker and the morphogenetic mutant allele. This type of procedure has allowed the demonstration that the *Ubx* gene acts in a cell autonomous manner, since the *Ubx* phenotype is shown by all cells in the marked clone, even though adjacent cells have a wild-type phenotype (see Fig. 4b). Once again, development is governed by the genes which are active within a cell rather than its developmental context. The results also show that, for normal development to occur, wild-type *Ubx* function is required throughout the entire period during which cell proliferation occurs.

Mutants having different effects on cell behaviour allow different questions to be addressed. Minute mutations (M) lead to the production of abnormal ribosomal proteins. M^+/M heterozygous individuals develop more slowly than the wild-type, but eventually attain the same adult size. Mitotic recombination in a M^+/M individual can lead to the production of a M^+/M^+ cell clone, which shows a cell-autonomous increase in the rate of cell division. Normal-sized wings are produced but M^+/M^+ clones occupy much larger territories than M^+/M clones of the same age (Fig. 4a), suggesting that compensatory mechanisms exist which control cell proliferation. The clones also respect the previously formed compartment boundaries and form normal sized patterns. These experiments show that the control of cell proliferation and pattern formation does not utilize mechanisms which count the number of cell divisions nor are fixed cell lineages involved.

22.3.4 Cell-Cell Interactions Are Involved in Pattern Formation

Similar analyses can be carried out using recessive lethal alleles. About 10% of such alleles lead to the death of epidermal cells in homozygous clones. Even in these cases it is often possible to use alleles whose phenotypic effect is less extreme to gain information about the function of the gene during development. Some pattern mutants, e.g., vein mutants (see Sect. 22.4.2), show autonomy of phenotype in the centre of the cell clone but partial correction at its borders. This local nonautonomy is not due to the rescue of mutants cells by the surrounding wild-type cells because in many cases the surrounding wild-type cells themselves locally show mutant phenotypes (Fig. 4b, c). Local cell-cell interactions must therefore be occurring in the course of normal pattern formation (see Sects. 22.4.2 and 22.6).

22.4 Genetic and Molecular Analyses

22.4.1 Systemic Genes [5, 6]

Several types of gene have roles in pattern formation. Homoeotic genes are involved in the specification of segment identity, while proneural genes (see Sect. 22.4.3) are required in the determination of elements of the peripheral and central nervous system.

Cuticular patterns differ from segment to segment and also between different regions within a segment. These patterns are affected globally in homoeotic mutants. The *Ubx* gene is an example of a homoeotic gene (Fig. 5). *Ubx* gene function is essential for development; embryos lacking functional *Ubx* alleles die, having specific parasegments (see Tauts, Chap. 21, this Vol.) substituted for one another. Using cell clones produced by mitotic recombination, it can be shown that impairment of *Ubx* function leads to the transformation of all (or part) of the anterior (for *abx* and *bx* alleles) or posterior (*pbx* and *bxd* alleles) compartments of the third thoracic segment into the anatomical counterparts of the second thoracic segment (e.g., haltere to wing, see Fig. 5b and c). Alleles which result in enhanced *Ubx* expression (*Cbx*) cause the reciprocal transformation (e.g., wing to haltere, see Fig. 5d and e). The observed transformations caused by *Ubx* alleles are cell autonomous in clones or cell aggregates, suggesting that the function of this gene does not involve cell-cell signaling.

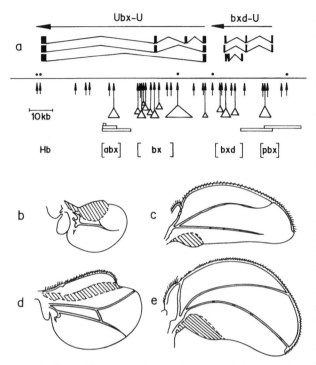

Fig. 5 a. Map of *Ubx* gene. *Ubx-U* unit of transcription which by alternative splicing gives rise to different mRNAs and proteins; *bxd-U* second unit of transcription giving rise to mRNAs which are untranslated. *cis*-acting regulatory elements map in this region; *Hb* exon-containing homeobox. abx, bx, bxd *and* pbx are *cis*-acting *regulatory regions* of UBx-*U* transcription, which may act as enhancers. *Vertical arrows* indicate positions of different mutations. *Horizontal empty bars* represent regions absent as a result of deletion mutations. *Dots* indicate positions of rearrangements associated with *Cbx* alleles. *Triangles* represent insertions of transposable elements affecting the normal transcription of *Ubx*. **b** Partial transformation of haltere into posterior wing as a result of *pbx*. *Shaded area* has retained haltere development. **c** As b for *bx*; anterior wing structures have developed. **d** Partial transformation of wing into haltere (*coarse shaded area*) as a result of *Cbx³*. **e** As **d**, but for *Cbx¹*

Molecular genetic analysis of the *Ubx* gene has revealed that alternative splicing can give rise to a number of different transcripts and hence proteins (Fig. 5a). Transcription appears to be driven by DNA sequences which act as *cis*-acting regulatory regions, possibly acting as enhancers. Each regulatory sequence results in a specific spatial pattern of transcription and *Ubx* alleles affecting particular regions are affected in the sequence needed for transcription in that region. The *Ubx* gene product is localized within the nucleus, a finding which cor-

relates with the probable role of the protein as a DNA binding protein. The protein contains a "homeodomain" which has affinity for specific DNA sequences including some found in the *Ubx* gene itself. The role of the *Ubx* gene is likely to be the regulation of genes which are further downstream in the regulatory hierarchy. Such downstream genes may include genes involved in wing-patterning (see below), but we still know very little about their molecular function and the way in which *Ubx* may modulate the developmental operations which these genes control.

The identification of genes under the control of regulatory genes is now being undertaken using several molecular approaches. One is based on the location of target genes to which a regulatory protein binds, by the immunostaining of polytene chromosomes using specific antibodies against regulatory gene products. In this way the dependence of *Ubx* expression on *Polycomb*, which can be inferred from their genetic interaction, can be confirmed. The chromosome band in which *Ubx* is located can be immunostained using anti-Pc antibodies. When regulatory proteins are not expressed in salivary glands, it may be possible to achieve ectopic expression by heatshocking a transformant fly which carries a hybrid gene composed of the transcribed region of the regulatory gene coupled to a temperature-inducible promoter.

22.4.2 Cell Communication and Proliferation Genes[7, 8]

Wing veins arise as hollows, left as a result of longitudinal swellings of epidermal cells, between the wing surfaces (see Figs. 4 and 7). Many genes are known in *Drosophila*, the mutant alleles of which affect the pattern of wing venation. There are two main classes; those in which mutation causes the absence of veins (e.g., genes of the *veins* (*vn*) group affect all veins, genes of the *radius incompletus* (*ri*) group affect specific veins) and those in which mutation leads to the appearance of extra veins (e.g. genes of the *plexus* (*px*) group). These mutants usually also have associated displacements of pattern towards or away from the anterior/posterior boundary (Fig. 6).

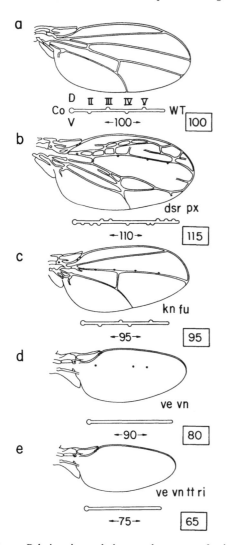

Fig. 6a–e. Relative size and shape and patterns of veins for wings of different genotypes. *Cross-sections below each wing represent wing width and positions of veins.* The *numbers in boxes* indicate the number of wing cells in mutants relative to the wild type. The *numbers between the arrows* represent the width relative to the wild type. *Small circles on veins* represent *sensilla campaniformia.* These are absent in *ve, vn, tt, ri* wings. *D* dorsal compartment; *V* ventral compartment; *Co* costal vein; *II–V* longitudinal veins L2–L5

Genes which can mutate to affect wing veins, include some, e.g., *Notch* (*N*), *Delta* (*Dl*), *decapentaplegic* (*dpp*) and *torpedo* (*top*), which are homologs of vertebrate proto-oncogenes and growth factor genes, suggesting that these genes have a role in regulating cell proliferation and dif-

ferentiation during wing vein development (Fig. 7). We will consider two of these in more detail, *torpedo*, which may be involved in both cell proliferation and differentiation, and *Notch*, which appears to be involved only in cell differentiation.

Evidence is accumulating to support the hypothesis that *torpedo*, a *Drosophila* homolog of the v-erb-V chicken oncogene and human epidermal growth factor receptor, is involved in both the differentiation and proliferation of cells. Mutations can occur at the *torpedo* locus which lead to a range of phenotypic effects, including embryo or larval lethality, as well as rudimentary imaginal disks. Some viable alleles lead to an absence of wing veins (Fig. 7e) while other alleles, e.g., *Ellipse* (*E*) lead to the development of extra wing veins as well as affecting the eye, notum, and chaetae. Genetic combination of viable *top* alleles with other mutants affecting wing venation, suggests synergistic interactions both with mutants leading to an absence of wing veins and with those leading to extra veins. Mosaics involving near null alleles of *top* provide further information about its role in imaginal patterning. The *top*⁻ clones contain only very few cells, do not differentiate wing veins and differentiate epidermal cells which are very small. Interestingly, if a *top*⁻ clone occupies a position in which a wing vein would normally develop, then extra wing veins may be produced in adjacent wild-type territories. This nonautonomous effect suggests that there are stripes of cells adjacent to the sites of normal vein development which are potentially able to form veins and that there must be lateral inhibition which normally prevents this. Other proto-oncogene homologs, such as *Notch*, are also involved in this process.

Three different types of alleles with pleiotropic effects map at the *Notch* locus. First, there are several recessive alleles with allele specific phenotypes which complement each other in heteroallelic combinations. Second, there are alleles which when heterozygous, lead to scalloped wings (breaks in the wing margin) and thick veins (Fig. 7b). Finally, alleles such as *Abruptex* (*Ax*), when heterozygous lead to an absence of veins (Fig. 7c). Molecular analyses have shown that *Notch* codes for a transmembrane protein with extracellular domains homologous to vertebrate epidermal growth factors

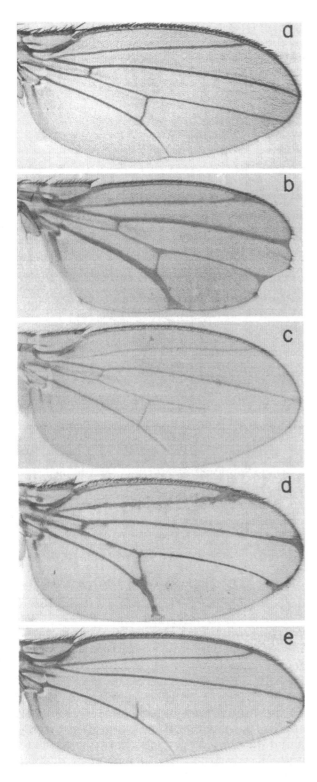

and with intracellular regions homologous to nucleotide binding and phosphorylation domains. The gene is transcribed ubiquitously in imaginal disks with higher rates of transcription in regions associated with cell proliferation. However, this pattern of gene expression cannot explain the patterned phenotype of the adult. In mosaics, both N and Ax clones show normal sized clones in adult structures, arguing against a role of N^+ in cell proliferation. Mutant clones also show cell autonomous phenotypes for wing vein development, Ax clones do not develop veins and N clones develop thicker veins in competent regions. Thus N^+ may act as receptor in the process of cell-cell communication which operates to bring about lateral inhibition of vein development. The wing scalloping phenotype of some N alleles may be due to interaction of its product(s) with cell adhesion molecules, since this phenotype is also associated with alleles of other genes known to be involved in cell adhesion. These are being studied using mutant combinations with alleles of other related genes such as *Delta* (*Dl*), (Fig. 7d) another gene with motifs homologous to epidermal growth factor, and *Enhancer of split* [*E(spl)*] which has several coding regions with homologies to *c-myc* and with the β subunit of G-proteins (see Campos Ortega, Chap. 23, this Vol.).

Finally, the competence of cells to differentiate vein may be defined by other genes coding for transcription factors and by proto-oncogene homologues. This process involves sequential compartmentation during the growth of wing imaginal disks. The A/P compartment boundary is generated during early segmentation of the blastoderm prior to disk singularization and evagination (see Fig. 8a, b). Cell interactions are involved in the establishment of this boundary. The *wingless* gene, a *Drosophila* homolog of the human *int-1* proto-oncogene, acts to produce a signal involved in both the setting up the A/P boundary and the early specifications necessary for normal wing development. The wing disk may then inherit spatial speci-

Fig. 7a−e. Wing vein patterns resulting from various mutant alleles of the genes coding for homologs of vertebrate proto-oncogenes. **a** wild-type; **b** Notch ($N^{264.39}$/+); **c** Abruptex (Ax^{M1}/+); **d** Delta (Dl^{M1}/+); **e** Torpedo (top^1/top^{4a})

fications, defined by segment polarity genes (Tautz, Chap. 21, this Vol.), from the blastoderm. During development, further regional specifications dependent on the A/P boundary occur. Genes like *decapentaplegic* (*dpp*), homologous to *Transforming Growth Factor-β*, are required later in the anterior margin of this boundary for normal patterning in both the A/P and proximo/distal axes. Still later, a new plane of symmetry subdivides each A and P compartment into dorsal and ventral compartments (Fig. 8a and c). Within these four compartments, new subdivisions, defined by late clonal restrictions along the veins, divide the wing into intervein regions. Some mutations, e.g., *shaggy* and

patched, a segment polarity gene coding for a transmembrane protein, have different cell behaviors in morphogenetic mosaics, depending of the region affected.

22.4.3 Terminal Differentiation Genes[9]

Within very specific morphogenetic coordinates, defined by pattern genes, cell differentiation of pattern elements occurs late in development. Cuticular sensory elements such as sensilla and chaetae are produced by epidermal cells in patterns which are segment specific. Their development is dependent on the function of the *achaete-scute* gene complex (AS-C).

Loss-of-function alleles of AS-C lead to the absence of particular sensory elements in allele-specific patterns, while gain-of-function alleles lead to the development of extra elements. The total deletion of AS-C is not lethal to cell clones and results in a cell-autonomous elimination of sensory elements. Molecular analysis of the complex shows that four transcripts are produced from within a region of 100 kb. Most mutations map to the intervals between coding regions and uncouple the coding regions from discrete *cis*-acting regulatory regions which are probably involved in the regulation of their expression in a position-dependent manner (Fig. 9a). Sequence analysis has shown that the four transcripts have the potential to code for proteins very similar to each other, each containing helix-loop-helix motifs. These are typical of DNA transcription factors which are associated with cell differentiation and like them also have potential phosphorylation domains.

The spatial expression of AS-C genes is dependent on other *trans*-acting regulatory genes, which operate in several steps. The earliest expression of AS-C genes detected by in situ hybridization techniques, is found in clusters of neighbouring cells (Fig. 9b1). The genes *extramacrochaetae* (*emc*) and *hairy* (*h*) act antagonistically to the AS-C genes, changing the number and spacing of sensory elements and probably affecting the early expression of the genes. The proteins coded by *emc* and *h* both have helix-loop-helix motifs (dimerization domains) suggesting that they may interact directly

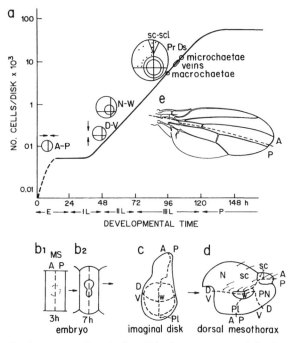

Fig. 8a, b. Dynamics of cell proliferation, compartmentalization and determination events in the development of the dorsal mesothorax. **a** Cell proliferation. Along the time axis (*abscissa*): *E* embryo; *I.L, II.L,* and *III.L* larval instars; *P* pupa. The time at which compartments are determined is shown diagrammatically above the cell proliferation curve. The time of chaetae mother cell and vein determination is also indicated on the proliferation curve. **b1, b2, c, d,** and **e** Representation of how compartments are successively determined in the second thoracic segment of the embryo (**b1** and **b2**), in the mesothoracic disk (**c**), in the dorsal mesothorax (**d**) and in the wing (**e**). *A* anterior; *P* posterior; *V* ventral; *D* dorsal; *Pr* proximal; *Ds* distal; *N* notum; *PN* postnotum; *W* wing; *Pl* pleura; *Sc* scutum; *Scl* scutelum

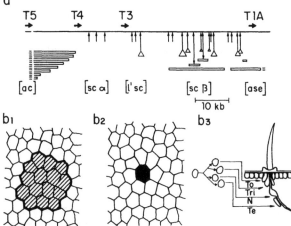

Fig. 9. a Map of AS-C. *Vertical arrows* indicate positions of different mutations. *Horizontal empty bars* represent regions absent as a result of deletion mutations. *Triangles* represent insertions of transposable elements. *T5*, *T4*, *T3*, and *T1A* are transcribed regions; the direction of transcription is shown by the *arrow*. The location of genetic regions is indicated within parentheses as follows: *ac*: *achaete*; *sc-a* and *sc-β*: *scute*, *l'sc*: *lethal of scute*; *ase*: *asense*. **b** Successive restriction of expression of the *scute* T4 protein in the epidermis from groups of cells (**b1**) to a single (mother of chaetae) cell (**b2**) which divides to produce four cells which do not express T4 (**b3**). *To* tormogen; *Tri* trichogen; *N* neuron; *Te* thecogen

with AS-C proteins. Other genes are involved in the cell interactions which lead to a single cell from among the cluster of cells expressing the AS-C genes, becoming established as the mother cell (Fig. 9 b 2), which gives rise to the sensory organ by two further divisions (Fig. 9 b, c). Just as the role of *top* in lateral inhibition of wing vein differentiation was established by the use of *top⁻* mosaics (see above), the inhibitory role of AS-C was established long ago by the use of genetic mosaics involving mutations in AS-C genes. If the territory of an AS-C mutant clone occupies a position in which a chaeta should normally develop, no chaeta develops, but a neighboring wild-type cell may differentiate to produce a chaeta nearby. Mutant *Dl* or *N* clones also show abnormalities in this inhibitory mechanism, developing several sensory organs from one cell cluster. The effects of both *Dl* and *N* are mediated by the repression of genes in AS-C in neighboring cells, since the phenotype of these mutants varies depending on the number of copies of

the AS-C genes present in the genome. There are, however, no AS-C alleles known which lead to one type of sensory element being transformed to another. The type of sensory organ is defined by other genes including *cut* and *lozenge*.

22.5 Comparative Analysis of Patterns[7, 10]

The understanding of the mechanisms which generate pattern requires the study of variation, not only within, but also between species. The mechanisms regulating development must not only be capable of bringing about the morphology of the species in which they have been investigated, but also the different morphology of related species. The comparison of patterns between phylogenetically related species may reveal processes which cannot be detected by the study of mutants or by physiological experiments. In fact, even within the same insect species there may be dramatic morphological differences between developmental stages, casts or sexes. We can therefore ask how specific are the genetic mechanisms involved in achieving a particular morphology. It may be that different developmental systems use essentially similar gene complexes to generate overtly different morphologies.

The existence of homologous mutations in related species has been known for a long time. With the introduction of molecular techniques, remarkable conservation of DNA sequences and of gene function has been shown to occur across different animal species. Genes concerned with developmental processes ranging from segmentation to cell communication and signal transduction are conserved from annelids to vertebrates. Chapter 26 (Lobe and Gruss, this Vol.) deals in detail with one example of this, the finding that mammals contain genes homologous to *Drosophila* homoeotic genes. Molecular comparisons between species are bound to be powerful tools for understanding many developmental processes. For example, the *engrailed* gene, which contains a homeodomain, is conserved in many organisms. In *Drosophila*, *engrailed* is expressed in the embryo and in imaginal disks exclusively in cells of the posterior compartment of all segments. Its transcription in the *Drosophila* syncytial blastoderm is the result of a hierarchy of both positive and negative *trans*-acting factors which is independent of cell proliferation (see Tautz, Chap. 21, this Vol.). However, in more primitive insects such as the locust, *engrailed* expression appears in the abdominal segments as they are generated sequentially. This suggests that in these species there is a progressive specification of posterior compartments which occurs alongside cell proliferation.

The conservation, not only of genes, but also of developmental operations, can be inferred from the analysis of patterns of sensory organs such as chaetae. The comparative analysis of eighty species of the genus *Drosophila* and of related genera, reveals a common pattern of chaetae in the mesonotum, which consists of a grid of a few regularly distributed large chaetae with many small chaetae in between. Chaetae patterns vary between species in that individual or a small set of large chaetae may be missing in particular positions, leaving the general pattern invariant. This suggests that species patterns may result from the local suppression of individual chaetae while leaving the remaining chaetae in their normal places, and thus that there is conservation of developmental genetic operations which generate patterns. Similar conclusions can be drawn from the comparative analysis of wing venation in insects.

22.6 Outlook and Summary [2, 11]

Classical experiments in a number of insect species, together with new molecular and genetic data on spatial gene expression and cell behavior in *Drosophila* have started uncovering morphogenetic properties that demand alternative interpretations to those provided by prevalent models of pattern formation. These existing models have been proposed to account for the size invariant properties of some developmental systems and are based on gradients of concentration of diffusible morphogens leading to the definition of position within borders. According to these models, cells become specified by global properties or their position in embryonic fields, independently of their neighbours.

Data from *Drosophila* systems, both embryonic and adult, provide evidence that the morphogenetic mechanisms investigated depend more on cell-cell interactions. Examples given in the previous sections include:

1. Wild-type imaginal disks sent to premature metamophosis show incomplete patterns and are not therefore size-invariant.
2. Patterns formed by the growth of isolated cells transplanted into a host blastoderm, correspond to fragments of perfectly dimensioned patterns (Fig. 3a).
3. Dissociated and mixed cells retain not only very precise positional values but also a fixed polarity. They are therefore able to reconstruct by cell recognition, fragments of patterns in mosaics (Fig. 3b).
4. Genetic mosaics of morphogenetic mutants are autonomous except at their borders where both mutant and wild-type clones show pattern accommodation (see Fig. 5c, d).
5. The pattern of regeneration following translocation of pieces of *Rhodnius* tergites (Fig. 1) and after the distal, intercalary and circular manipulation and grafting of appendages (Fig. 2) also emphasizes the role of cell-cell interactions.

Molecular genetic analysis has started to identify the role of genes which are involved in development. Thus in the embryo and in imaginal disks there are genes which are expressed in spatially reiterated patterns whereas others are expressed ubiquitously. The products of some, by homology, are likely to have DNA binding domains, while others are related to vertebrate genes coding for growth factors, receptors and intracellular signal transducers. The question is, how these different genes operate to control cell proliferation and spatial differentiation? How too, if patterns are size-invariant only within a very small range of size, is cell division controlled so that it ceases when the final dimensions characteristic of a species are achieved? The latter question rephrases the old Aristotelian problem of entelechy, the property of innate completion or perfection, in modern terms of genes and cells.

Any entelechy model will have to explain how cell proliferation is controlled by cells communicating with their nearest neighbours. We must explain how cell division ceases when differences in positional values between cells become minimal. Studies in wing development and vein patterning have suggested that there are genes whose activity may define these positional values (genes of the *vn* and *px* groups) and others which are involved in signal transduction (*torpedo*) that may define the increments of positional value between neighbouring cells. During the development of the adult *Drosophila*, the regional specification occurs sequentially as it does in the embryo (see Tautz, Chapt. 21, this Vol.). These regional specifications (e.g., compartments, vein regions, macrochaetae regions) are possibly defined by specific transcription factors which are activated as a result of cell-cell interactions. However, in order to achieve final completion, regional specification and cell proliferation must be coupled. Genes controlling cell proliferation and differentiation have to be coupled to genes specifying regions so that specific levels of gene expression are fixed at specific positions (e.g., a region border). Each region will achieve its final dimensions when the full spectrum of positional values (varying levels of gene function) are reached. In this way the level of expression of genes controlling proliferation can be coupled to dimensions in terms of the number of cells. Visible cell differentiation, e.g., of chaetae or veins, will take place at specific levels of positional values.

The entelechy problem obviously extends to the processes that occur in the development and regeneration of a variety of animals, including vertebrates. Our intellectual challenge is to determine whether these apparently diverse processes are controlled by common underlying genetic and developmental mechanisms.

References

[1] Akam M (1987) The molecular basis for metameric pattern in *Drosophila* embryo. Development 101:1–22

Anderson DT (1972) The development of hemimetabolous Insects. In: Counce SJ, Waddington CH (eds) Developmental systems: Insectis, vol 1. Academic Press, London, pp 95–163

Ursprung H, Nöthiger R (1972) Results and problems in cell differentiation, vol. 5. Springer, Berlin Heidelberg New York

[2] Wolpert L (1969) Positional information and the spatial pattern of cellular differentiation. J Theor Biol 25:1–47

[3] Bohn H (1970) Interkalare regeneration und segmentale Gradienten bei den Extremitäten von *Leucophaea*-Larven (Blattaria). Wilhelm Roux's Arch Dev Biol 165:303–341

French V, Bryant PJ, Bryant SV (1976) Pattern regulation in epimorphic fields. Science 193:969–981

García-Bellido A (1965) Larvalentwicklung transplantierter Organe von *Drosophila melanogaster* im Adultmilieu. J Insect Physiol 11:1071–1078

García-Bellido A (1972) Pattern formation in imaginal disk. In: Ursprung H, Nötiger R (eds) Results and problems in cell differentiation, vol 5. Springer, Berlin Heidelberg New York, pp 59–91

García-Bellido A, Nöthiger R (1976) Maintenance of determination by cells of imaginal disks of *Drosophila* after dissociation and culture in vivo. Wilhelm Roux's Arch Dev Biol 180:189–206

Hadorn E (1965) Problems of determination and transdetermination. In: Genetic control of differentiation. Brookhaven Symp Biol 18:148–161

Illmensee K (1978) *Drosophila* chimeras and the problem of determination. In: Gehring WJ (ed) Results and problems in cell differentiation, vol 9. Springer, Berlin Heidelberg New York, pp 51–69

Lawrence PA (1966) Gradients in an insect segment: the orientation of hairs in the milkweed bug *Oncopeltus*. J Exp Biol 44:607–620

Locke M (1967) The development of patterns in the integument of insect. Adv Morphog 6:33–38

Stumpf HF (1966) Mechanism by which cells estimate their location within the body. Nature (Lond) 212:430–431

[4] Crick FMC, Lawrence PA (1975) Compartments and polyclones in insect development. Science 189:340–347

García-Bellido A (1975) Genetic control of wing disk development in *Drosophila*. In: Cell patterning. Ciba Found Symp 29. Elsevier, Amsterdam, pp 161–182

García-Bellido A, Merriam JR (1971a) Parameters of the wing imaginal disks development of *Drosophila melanogaster*. Dev Biol 24:61–87

Janning W (1978) Gynandromorph fate maps in *Drosophila*. In: Gehring WJ (ed) Results and problems in cell differentiation, vol 9. Springer, Berlin Heidelberg New York, pp 1–28

Morata G, Ripoll P (1975) Minutes: Mutants of *Drosophila* autonomously affecting cell division rate. Dev Biol 42:211–221

[5] García-Bellido A (1977a) Inductive mechanisms in the process of the wing vein formation in *Drosophila*. Wilhelm Roux's Arch Dev Biol 182:93–106

Morata G, García-Bellido A (1976) Developmental analysis of some mutants of the *bithorax* system of *Drosophila*. Wilhelm Roux's Arch. Dev Biol 179:125–143

Ripoll P, García-Bellido A (1973) Cell autonomous lethals in *Drosophila melanogaster*. Nature New Biol 241:15–16

[6] Beachy PA (1990) A molecular view of the *Ultrabithorax* homeotic gene of *Drosophila*. TIG 2:46–51

Capdevila MP, García-Bellido A (1981) Genes involved in the activation of the *bithorax* complex of *Drosophila*. Wilhelm Roux's Arch Dev Biol 190:339–350

Ducan I, Lewis EB (1982) Genetic control of body segment differentiation in *Drosophila*. In: Subtelny S (ed) Developmental order: its origin and regulation. Alan R. Liss, New York, pp 533–554

Zink B, Paro R (1989) In vivo binding pattern of a *trans*-regulator of homeotic genes in *Drosophila melanogaster*. Nature 337:468–471

[7] Hoffman FM (1989) Roles of *Drosophila* proto-oncogene and growth factor homologs during development of the fly. In: Vogt PK (ed) Current topics in microbiology and immunology, vol 147. Springer, Berlin Heidelberg New York, pp 1–29

[8] Díaz-Benjumea FJ, González-Gaitán MA, García-Bellido A (1989) Developmental analysis of wing vein pattern formation. Genome 31:612–619

Baker NE (1988) Embryonic and imaginal requirements for *wingless*, a segment-polarity gene in *Drosophila*. Dev Biol 125:96–108

de Celis JF (1989) Análisis genético y de desarrollo de la formación del patrón de quetas en *Drosophila melanogaster*. Thesis, U.A.M., Madrid

de Celis JF, Marí-Beffa M, García-Bellido A (1990) Cell autonomous role of *Notch,* and epidermal growth factor homologue, in sensory organ differentiation in *Drosophila*. Proc Natl Acad Sci USA 88:632–636

Díaz-Benjumea FJ, García-Bellido A (1990a) Genetic analysis of the wing vein pattern of *Drosophila*. Wilhelm Roux's Arch Dev Biol 198:336–354

Díaz-Benjumea FJ, García-Bellido A (1990b) Behaviour of cells mutant for an EGF receptor homologue of *Drosophila* in genetic mosaics. Proc Soc Dev Biol 242:36–44

Gelbart WM (1989) The *decapentaplegic* gene: a *TGF-β* homologue controlling pattern formation in *Drosophila*. Development 104 (Suppl) 65–74

Johansen KM, Fehon RG, Artavanis-Tsakonas S (1989) The *Notch* gene product is a glycoprotein expressed on the cell surface of both epidermal and neuronal precursor cells during *Drosophila* development. J Cell Biol 109: 2427–2440

Markopoulou K, Artavanis-Tsakonas S (1989) The expression of the neurogenic locus *Notch* during the imaginal development of *Drosophila melanogaster* and its relationship to mitotic activity. J Neurogenet 6:11–26

Phillips RG, Roberts IJH, Ingham PW, Whittle JRS (1989) The segment-polarity gene *patched* is involved in a position-signalling mechanism in imaginal discs. Development 110 (1):105–114

Price JV, Clifford RJ, Schüpbach T (1989) The maternal ventralizing locus *torpedo* is allelic to *faint little ball,* an embryonic lethal, and encodes the *Drosophila* EGF receptor homolog. Cell 56:1085–1092

Simpson P, El Messal M, Moscoso del Prado J, Ripoll P (1988) Stripes of positional homologies across the wing blade of *Drosophila melanogaster*. Development 103:391–401

[9] Campuzano S, Carramolino L, Cabrera CV, Ruiz-Gómez M, Villares R, Boronat A, Modolell J (1985) Molecular genetics of the *achaete-scute* gene complex of *Drosophila melanogaster*. Cell 40:327–328

de Celis JF, Marí-Beffa M, García-Bellido A (1990) Function of *trans*-acting genes of the *achaete-scute* complex in sensory organ patterning in the mesonotum of *Drosophila*. Wilhelm Roux's Arch Dev Biol (in press)

García-Alonso L, García-Bellido A (1988) *Extramacrochaete,* a *trans*-acting gene of the *achaete-scute* complex of *Drosophila* involved in cell communication. Roux's Arch Dev Biol 197:328–338

García-Bellido A (1985) Cell lineages and genes. Philos Trans R Soc Lond B 312:101–128

García-Bellido A, Santamaría P (1978) Developmental analysis of the *achaete-scute* system of *Drosophila melanogaster*. Genetics 88:469–486

Garrell J, Modolell J (1990) The *Drosophila extramacrochaetae* locus, an antagonist of proneural genes that like these genes, encodes a helix-loop-helix protein. Cell 61:39–48

O'Kame CJ, Gehring WJ (1987) Detection in situ of genomic regulatory elements in *Drosophila*. Proc Natl Acad Sci USA 84:9123–9127

Romani S, Campuzano S, Macagno E, Modolell J (1989) Expression of *achaete scute* genes in *Drosophila* imaginal disks and their function in sensory organ development. Genes Dev. 7:997–1008

Ruiz-Gómez M, Modolell J (1987) Deletion analysis of the *achaete-scute* locus of *Drosophila melanogaster*. Genes Dev 1:1238–1246

Stern C (1954) Two or three bristles. Am Sci 43:213–247

Villares R, Cabrera CV (1987) The *achaete scute* gene complex of *Drosophila:* conserved domains in a subset of genes required for neurogenesis and their homology to *myc*. Cell 50:415–424

[10] Beeman RW (1987) A homeotic gene cluster in the red flour beetle. Nature 327:247–249

García-Bellido A (1977b) Homoeotic and atavic mutations in insects. Am Zool 17:613–629

García-Bellido A (1983) Comparative anatomy of cuticular patterns in the genus *Drosophila*. In: Goodwin BC, Holder N, Wylue CC (eds) Development and evolution. Cambridge University Press, Cambridge, pp 227–255

Gehring WJ (1987) The homeobox: structural and evolutionary aspects. In: Firtel RA, Davidson EH (eds) Molecular approaches to developmental biology. Alan R. Liss, New York, pp 115–129

Karr TL, Weir MP, Ali Z, Kornberg T (1989) Patterns of engrailed protein in early *Drosophila* embryos. Development 105:605–612

Patel NH, Kornberg TB, Goodman CS (1989) Expression of *engrailed* during segmentation in grasshopper and crayfish. Development 107:201–212

[11] Bryant PJ, Simpson P (1984) Intrinsic and extrinsic control of growth in developing organs. Q Rev Biol 59:387–415

Ingham PW (1988) The molecular genetics of embryonic pattern formation in *Drosophila*. Nature 335:25–34

Martínez-Arias A (1989) A cellular basis for pattern formation in the insect epidermis. Trends Genet 5:262–267

Chapter 23 Genetic Mechanisms in Early Neurogenesis of *Drosophila melanogaster*

JOSÉ A. CAMPOS-ORTEGA and ELISABETH KNUST

23.1 Introduction[1-5]

Insect neurons are generated by the proliferation of progenitor cells called neuroblasts. In *Drosophila melanogaster*, the neuroblasts develop from a special region of the ectoderm, the neurogenic region (NR), or neuroectoderm; in this region neighboring cells have to decide between one of two alternative fates and develop either as neuroblasts or as epidermoblasts (progenitor cells of the epidermis). Due to its apparent simplicity, the decision of neuroectodermal cells for the neural or the epidermal fate is a good example to investigate the mechanisms of origin of cell diversity in a multicellular organism.

The sum of processes that lead to the segregation of neuroblasts from epidermoblasts, and therefore to the formation of a neural primordium, is denominated early neurogenesis to distinguish it from late neurogenic events, as outgrowth of axons (axonogenesis, pathway finding), formation of synapsis (synaptogenesis), etc. Early neurogenesis is a complex problem that includes a large variety of different aspects. Most of the work done in the past 10 years on early neurogenesis in *Drosophila melanogaster* has been concerned with the mechanisms of segregation of neuroblasts and epidermoblasts to give rise to the neural and epidermal cell lineages, whereas very little is known about the formation of the neurogenic ectoderm as such, or about the mechanisms of origin of the different types of neuroblasts and epidermoblasts and the lineages to which they give rise. Therefore, in this chapter we shall be concerned with the process of separation of the progenitor cells for the epidermal and central neural lineages.

23.2 The Neuroectoderm Gives Rise to a Constant Array of Neuroblasts[1,6]

In the *Drosophila* embryo, the NR becomes morphologically manifest at about 3.5 h after egg laying (embryogenesis takes approximately 22 h at 25 °C). At this time, virtually all cells of the NR enlarge to become conspicuously different from the remaining, non-neurogenic ectodermal cells. In striking contrast to *D. melanogaster*, only single cells among groups of several cells of the NR seem to enlarge during early neurogenesis in various species of grasshoppers which have been investigated; the enlarged cells are the prospective neuroblasts themselves, which will segregate from the remaining ectodermal cells.

The segregation of the neuroblasts lasts in *D. melanogaster* for approximately 3 h and proceeds in three discrete pulses which give rise to three subpopulations of neuroblasts, called SI, SII (Fig. 1) and SIII neuroblasts. At these three pulses, single cells leave the NR to move internally and become

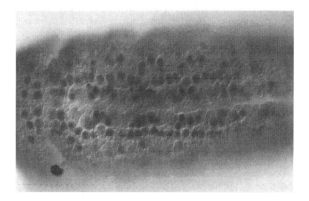

Fig. 1. An early stage 11 wild-type embryo to illustrate the regular pattern of neuroblasts, as stained with an antibody against the segmentation gene *hunchback* (kindly provided by D. Tautz, Munich). The neuroblasts are arranged in *three regular rows*, median, intermediate, and lateral, on either side of the midline

located between the mesoderm and the ectoderm, where they form the neural primordium (Fig. 1). This means, the operations which lead to the separation of the two cell lineages are actually repeated at three consecutive stages. Approximately 500 neuroectodermal cells will finally develop as neuroblasts, whereas the remaining cells will develop as epidermoblasts.

23.3 Cellular Interactions Lead to Cell Commitment [1, 7, 8]

In insects, the decision as to whether a cell of the NR will adopt the epidermal or the neural fate is mediated by cell-cell interactions. Two pieces of experimental evidence support this contention, one obtained by laser ablation of individual cells carried out in grasshoppers and the other by cell transplantations in *D. melanogaster* and several other Drosophilidae. If a neuroblast of the grasshopper is ablated with a laser beam, another cell of the neuroectoderm replaces the ablated one and takes neural fate, although this latter cell would have adopted epidermal fate under normal circumstances. This observation suggests that neuroblasts and epidermoblasts interact in normal development, such that the latter appear to be inhibited by the former from adopting the neural fate.

Laser ablations are difficult to carry out in *Drosophila* due to the small cell size. However, individual cells can be transplanted from one embryo (donor) into another (host). Cells are transplanted either to the same site in the host (homotopic transplantation) or to a different site (heterotopic transplantation). The way in which a cell differentiates after transplantation allows one to conclude as to whether its fate depends on its place of origin or on its environment. Upon their homotopical transplantation, *Drosophila* wild-type ectodermal cells behave in the host in the same way as they would have done in the donor. Single cells from the NR in the ventral ectoderm homotopically transplanted into the NR take on either neural or epidermal fate. Consequently, the transplanted cells give rise to three types of clones: neural cells, epidermal cells, and mixed clones of neural and epidermal cells.

Single cells from the dorsal epidermal anlage transplanted homotopically give rise only to epidermal clones upon homotopic transplantation. However, ectodermal cells exhibit a differential behavior following heterotopic transplantation. Ventral cells transplanted from the NR into the dorsal ectoderm develop according to their origin and differentiate either epidermal (more frequently) or neural histotypes. In contrast, dorsal ectodermal cells transplanted ventrally into the NR develop according to their new location, acquiring the capacity to differentiate into neural clones; thus, they give rise to either epidermal or neural clones.

This latter observation is striking, for under normal conditions dorsal cells do not develop as neuroblasts, either in situ or even upon homotopic transplantation. A possible interpretation of this result is that the cells in the dorsal epidermal anlage are normally prevented from developing as neuroblasts by an inhibitory process, and that they are relieved from this inhibition upon their transplantation into the NR. Another possibility, however, is that the transplanted dorsal cells are actively induced by their neighbors in the NR to adopt a neural fate. In order to see whether intercellular influences actively prevent neurogenesis within dorsal regions, ventral cells are transplanted dorsally. In this case, some of the cells develop neural and others develop epidermal progenies. Thus, no inhibition can be demonstrated experimentally in the dorsal region. Therefore, the results support the existence of neuralizing signals within the neuroectoderm.

Signals with epidermalizing character do also occur within the NR. This can be shown by, for example, the transplantation of neuroblasts back into the NR of younger animals; in some cases, the transplanted cells are induced by their neighbors to change their fate and develop as epidermoblasts, thus producing progenies with epidermal histotypes.

23.4 Genes Required for Neurogenesis [5, 9-11]

In *D. melanogaster*, the correct separation of neural and epidermal progenitor cells is controlled by two groups of genes, one composed of the so-called neurogenic genes and the other of the proneural genes (see Fig. 2). Poulson called *Notch* (the first neurogenic gene discovered) a "neurogenic" gene following the convention in *D. melanogaster* genetics of naming a gene according to the phenotype of

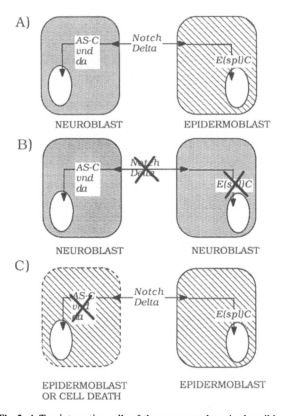

Fig. 2. A Two interacting cells of the neuroectoderm in the wild-type, the white ovals represent the cell nuclei. A large number of genes, i.e., the genes of the *achaete-scute* complex (AS-C), *ventral nervous system condensation defective* (*vnd*) and *daughterless* (*da*), and the neurogenic genes, including *Notch*, *Delta*, the genes of the E(SPL)-C, and several others not shown in this diagram, encode the proteins of a regulatory signal chain which allows the cells to develop either as neuroblasts or as epidermoblasts. The gene products of the AS-C, *vnd* and *daughterless* are required to regulate the genetic activity of the neuroblasts, those encoded by the various genes of the E(SPL)-C to regulate the genetic activity of the epidermoblasts. Regulatory signals are conveyed by the proteins encoded by *Notch* and *Delta*. **B** Mutation of one of the neurogenic genes results in the development of all neuroectodermal cells as neuroblasts. **C** Mutation in the genes of the AS-C, *vnd* or *da* results in either the development of additional epidermoblasts, at the expense of neuroblasts, or in cell death

the mutation that leads to its discovery. However, *Notch*, as well as the other genes of the group, exerts a proepidermal function in the process that we are considering. Loss of function of any of the neurogenic genes causes all the neuroectodermal cells to develop as neuroblasts, i.e., neurogenic mutants initiate neurogenesis with 2000 instead of the normal 500 neuroblasts. Consequently, since these 2000 neuroblasts proliferate at a normal rate, embryos which lack any of the neurogenic genes develop very large (hyperplasic) nervous systems; at the same time they lack the ventral and cephalic epidermis because the corresponding progenitor cells have developed as neuroblasts (Fig. 3). Thus, the wild-type functions of the neurogenic genes are apparently required to suppress the neural fate in the neuroectodermal cells and allow them to develop as epidermoblasts.

The "proneural genes" comprise the various members of the *achaete-scute* complex (AS-C), the gene *ventral nervous system condensation defective* (*vnd*), and the gene *daughterless* (*da*). The phenotype of loss-of-function mutations in the proneural genes is the opposite of that of neurogenic mutations: embryos lacking the genes of the AS-C and/or *vnd* initiate neurogenesis with less than the normal 500 neuroblasts – depending on the mutant, 20–25% of all neuroblasts are missing. In addition, during later stages, large numbers of cells degenerate within the neural primordium of all these mutants. Consequently, the fully differentiated mutant embryos die with a smaller central nervous system (Fig. 3). The genes of the *AS-C* and *vnd* are thus required for the commitment of the normal 500 neuroblasts. Since the populations of neuroblasts affected by mutations in different genes do not seem to overlap significantly, *AS-C, vnd,* and probably other as yet unidentified genes, may each control the development of particular sets of neuroblasts. The participation of the gene *da* in neuroblast commitment and segregation is not yet clear. The complement of neuroblasts in *da⁻* mutants is initially normal; however, most of these cells die early during embryonic development. The most remarkable trait of *da⁻* mutants is that they lack a peripheral nervous system, due to defective commitment of the progenitor cells of the sensory organs.

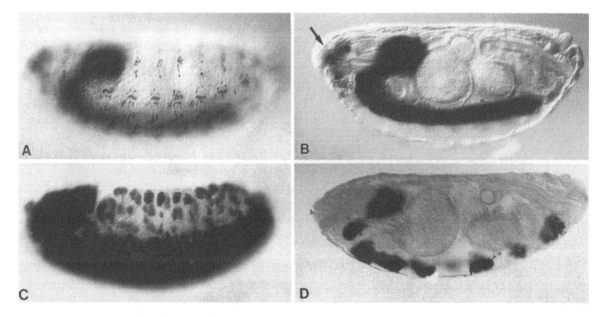

Fig. 3 A – D. Lateral (**A**) and median (**B**) planes of focus through the same wild-type embryo; **C** is a neurogenic mutant; **D** is a *daughterless* mutant. All three embryos have been stained with a neural specific antibody (44 c 11, kindly provided by Y. N. Jan, San Francisco) that recognizes an antigen present in all neuronal nuclei of the embryo; therefore only the nuclei of neurons are stained. **A** shows the exquisite pattern of some of the sensory neurons, **B** the central nervous system. The *arrow* points to one of several sensory organs that are visible on this plane of focus. Notice the conspicuous (central and peripheral) neural hyperplasia of the neurogenic mutant shown in **C** and the neural hypoplasia of the *daughterless* mutant shown in **D**. Notice that the cephalic sensory organs (*arrow* in **B**), as well as all other sensory organs, are missing

Neurogenic and proneural genes are currently referred to as two different groups, and such a distinction is justified by the phenotype of their mutations: loss-of-function of any of the neurogenic genes causes neural hyperplasia, and loss-of-function of any of the proneural genes causes the opposite phenotype, neural hypoplasia. In the following sections, we shall discuss that the products of all of these genes are apparently involved in a complex genetic network and together contribute to a single process – the separation of neural and epidermal cell progenitors.

23.5 The Neurogenic Genes Are Functionally Interrelated [1, 4, 7, 12–14]

There is abundant evidence from various genetic analyses to support the notion that the neurogenic and the proneural loci are functionally interrelated.

Since the products of these genes participate in different processes besides neurogenesis, a wide variety of phenotypic effects has been observed in the mutants and, consequently, in the various genotypic combinations that have been studied. Insofar as their participation in the segregation of neuroblasts from epidermoblasts is concerned, six of the neurogenic loci, i.e., *Notch* (*N*), *almondex* (*amx*), *master mind* (*mam*), *Delta* (*Dl*), *neuralized* (*neu*) and the genes of the *Enhancer of split* complex [E(SPL)-C], have been found to be functionally linked with some of the proneural genes (those of the AS-C and *da*; see below) in a chain of epistatic relationships; *big brain* (*bib*) was found to act independently of the others (see Fig. 4). Hence, the function of each of these genes appears to be dependent on that of another gene and, consequently, the function of the entire chain is perturbed if any of the links is missing.

Genetic mosaics support the notion that mutations in the different neurogenic genes affect specif-

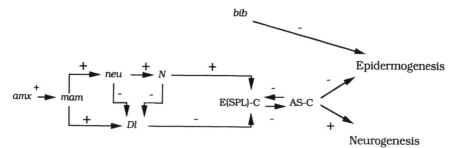

Fig. 4. The network of relationships of proneural (represented by the AS-C) and neurogenic genes. This diagram is based on formal arguments derived from transmission genetics and gives no indication concerning the molecular level at which the proposed interactions take place. However, there are indications that some interactions occur at the transcriptional level, whereas others reflect protein-protein interactions. Positive or negative signs reflect the kind of functional influences, i.e., repression or activation, assumed to be exerted by one gene product upon the next one. The data suggest that the regulation of the cell decision is mediated by the AS-C. E(SPL)-C provides an epidermalizing signal to the neuroectodermal cells. Although *bib* is one of the neurogenic genes and takes part in the process of neuroblast segregation, its function is apparently independent of the other six neurogenic genes

ic parts of a signal chain. Mutations in *N* express their phenotype in a cell autonomous manner, suggesting that *N* is required in the epidermoblasts to provide functions related to the reception of a signal. In addition, when neuroectodermal cells lacking the genes of the E(SPL)-C are transplanted into the NR of a wild-type embryo, the mutant cells express autonomously their mutant phenotypes; that is to say, the functions of the E(SPL)-C are also required to receive and/or process the signal. In contrast, cells lacking *amx, bib, mam, Dl* or *neu* develop as epidermoblasts with the same frequency as wild-type cells upon their transplantation into the wild-type NR. These genes therefore behave as if their products were required to produce and/or relay the regulatory signals that lead to epidermogenesis.

23.6 NOTCH and DELTA Encode Transmembrane Proteins with Similarity to the Epidermal Growth Factor, EGF [15]

The participation of NOTCH and DELTA in cell communication processes is strongly supported by the primary structure of their proteins. Both proteins show typical features of transmembrane proteins: two hydrophobic domains, an amino terminal signal peptide and a membrane spanning do-

main (Fig. 5). A striking feature of the extracellular domains of the NOTCH and DELTA proteins are tandem arrays of 36 and 9 cysteine-rich motifs. Each of these repeats consists of about 38 amino acids and has striking similarity to the epidermal growth factor (EGF) and other proteins of vertebrates and invertebrates (Fig. 5). The available data on the expression of these genes do not yet allow any firm conclusion to be drawn about their function. Preceding the segregation of the neuroblasts, both *N* and *Dl* are transcribed in all neuroectodermal cells; but after the segregation of neural and epidermal progenitor cells, *Dl* RNA can be detected mainly in neuroblasts.

23.7 Physical Interactions Between NOTCH and DELTA [11, 16, 19]

The EGF-like repeats in the extracellular domains of DELTA and NOTCH may well represent essential parts of the cell communication pathway by mediating protein-protein interactions. The participation of EGF-like repeats in specific protein-protein interactions has been demonstrated in several cases, for example for the binding of EGF and LAMININ to their corresponding receptors and, recently also for the NOTCH and DELTA proteins. Cells of a permanently growing cell line were

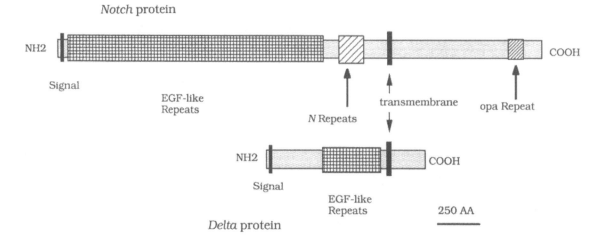

Fig. 5. A comparison of the primary structures of the proteins encoded by the genes *Notch* and *Delta*. The *Notch* protein comprises 36 EGF-like repeats. *N* and *opa* are other repetitive sequences which are not present in the *Delta* protein. The *Delta* protein exhibits a single hydrophobic transmembrane domain and a stretch of nine EGF-like repeats

transfected with DNA of either *N* or *Dl*, and the corresponding proteins were expressed after induction with an appropriate inducer. Thus, the cells, which normally do not express these proteins, now express them on their cell surfaces. Mixing NOTCH and DELTA expressing cells leads to an increase of the adhesivity of the cells and to the formation of aggregates of the otherwise nonadhesive cells, which can be explained by interactions between the two proteins.

These interactions are likely to be mediated by the EGF-like repeats and can help to understand phenotypic traits of some *N* and *Dl* point mutations. The proteins encoded by the *split* (*spl*) and several *Abruptex* (*Ax*) alleles (all these are mutations in the *N* locus) differ from the NOTCH wild-type protein by just single amino acid exchanges in particular EGF-like repeats, depending on the allele. On the other hand, two alleles of *Dl* are known which suppress the phenotype of the *spl* mutation, but do not modify that of any of the *Abruptex* mutations. Such allele specific interactions support direct physical relationships between DELTA and NOTCH.

Other data, e.g., the effects exerted by temperature-sensitive *N* alleles on compound eye development, suggest that the NOTCH protein acts as a kind of general cell adhesion molecule in various cellular interactions occurring at different developmental stages, thus playing a permissive rather than instructive role in these processes. So far, no conclusive statement on the mode of *Dl* function can be made, since no data are available on the distribution of the DELTA protein. If the DELTA protein would be present in all the neuroectodermal cells during lineage segregation, it would be difficult to think of it as serving instructive functions; such a distribution would rather be compatible with cell adhesive properties similar to those proposed for *N*, that would facilitate specific interactions exerted by other still unknown proteins.

23.8 The Enhancer of Split Gene Complex [E(SPL)-C] Comprises Several Related Functions [4, 13, 17]

The development of epidermoblasts is regulated by the genes of the E(SPL)-C, the composition of which (see Fig. 6) is not yet well understood. The E(SPL)-C received its name from a dominant mutation, called *E(spl)*[D], that enhances the phenotype of a *N* mutation called *split* (*spl*), in particular the effects of the *spl* mutation on compound eye development. *spl* flies have smaller and rougher eyes than the wild type; in combination with *E(spl)*[D], the eyes of *spl* flies are even smaller and very rough.

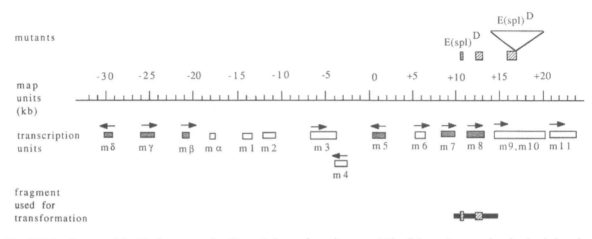

Fig. 6. Molecular map of the *E(spl)* gene complex. Transcription units within this region and direction of transcription are indicated beneath the physical map. Six transcription units are emphasized by shadowing because they encode conserved proteins with a b-HLH motif. *E(spl)*^D is a dominant allele that gives the locus its name. This allele carries several molecular lesions in transcription units m8 and m9/m10. The bar below represents the DNA fragment including the *m8* transcription unit of the *E(spl)*^D mutant that causes enhancement of the *spl* phenotype upon transformation in the germ line

The *E(spl)*^D mutation is a small deletion of the *m8* transcription unit, which leads to the truncation of the encoded protein at its carboxy-terminus. With the help of a transposable P-element, genetically transformed flies were created which carry one or two copies of the dominant *m8* gene, besides two copies of the wild-type *m8* gene. In combination with *spl*, the mutant *m8* gene inserted by the P-element into the genome brings about the same enhancement of the *spl* phenotype as the original *E(spl)*^D mutation.

Surprisingly, flies which carry a deletion of the *E(spl)* gene, and thus lack any *m8* transcript and protein, are perfectly viable. This, together with several other genetic data, suggests that the *m8* function is to some extent redundant, i.e., related functions can substitute the *E(spl)* gene. Indeed, the molecular characterization of the genes of the E(SPL)-C has revealed another five transcripts (*m7, m5, mβ, mγ, mδ*, Fig. 6) encoding conserved proteins. These proteins are characterized by a basic and a helix-loop-helix motif (b-HLH-motif; see below, and Fig. 7). Furthermore, the spatial expression pattern of these six transcripts as well as that of another one (named *m4* in Fig. 6) is highly suggestive of their participation in the process of epidermoblast commitment: it is nearly identical during the neurogenic period, and the regions where these RNAs are expressed match remarkably well those zones of the neuroectoderm in which the decision between the two developmental fates is made.

Another transcription unit comprising two RNAs (*m9/m10*), deserves mention. The deduced amino acid sequence of the putative protein contains a repeated motif which is similar to the *β*-subunit of TRANSDUCIN, a G-protein involved in phototransduction. Mutations in the *m9/m10* tran-

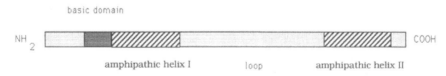

Fig. 7. Scheme of the basic and helix-loop-helix (b-HLH) motifs, found in many different DNA binding proteins of mammals and insects

scription unit are embryonic lethal in homozygosity, but the mutant embryos exhibit only a weak neurogenic phenotype. Recent genetic experiments indicate that this gene may interact with one or several of the other b-HLH proteins encoding genes of the complex.

Thus, the available evidence indicates that the genes of the E(SPL)-C regulate the development of the epidermoblasts. The results of transplanting cells lacking the E(SPL)-C into the wild-type NR, showing that these cells cannot develop as epidermoblasts, strongly support this hypothesis. Since six of the genes of the E(SPL)-C encode putative DNA-binding proteins, it is likely that this regulation is at the level of transcription of subordinate genes.

This motif consists of a basic domain and two amphipathic helices separated by a loop (Fig. 7). DNA-binding properties have been demonstrated for some of these proteins, which are thus likely to act as transcriptional regulators. Moreover, various members of the HLH protein family are able to form homo- and heterodimers, which opens the possibility of a high degree of complexity in the regulatory functions of the corresponding genes through the combination of different proteins. The recent finding that the protein encoded by *l'sc* and the DAUGHTERLESS protein can form heterodimers corroborates the observation of genetic interactions between some of the AS-C genes and *da*. Formation of heterodimers between the various proteins encoded by the E(SPL)-C would also explain the functional redundancy found in the genetic analysis of the gene complex.

23.9 The AS-C Gene, *daughterless* and the Genes of the E(SPL)-C Are Members of the Same Gene Family[3, 18]

The *AS-C* includes four genes *achaete* (*ac*), *scute* (*sc*), *lethal of scute* (*l'sc*), and *asense* (*as*), the names being derived from the phenotypic effects of their mutations on bristle development and viability. The gene complex comprises a number of transcripts, four of which, T5, T4, T3, and T1a (also named T8) have been identified as corresponding to *ac, sc, l'sc,* and *as* (see Fig. 8). Sequence analyses have shown that the proteins they encode are similar to each other and contain the same b-HLH motif (Fig. 7) found in six proteins encoded by the E(SPL)-C. The *daughterless* (*da*) locus has recently been cloned and found to encode a protein with the same conserved b-HLH motif.

The presence of the b-HLH motif in several proteins in vertebrates and invertebrates is striking.

23.10 The AS-C Genes Are Required for Neuroblast Commitment[9]

The spatial distribution of the T3, T4, and T5 transcripts of the AS-C shows a high degree of correlation with the processes of neuroblast segregation and development of sensory organs and the stomatogastric nervous system, i.e., those processes in which the functions of the genes are known to be required from the analysis of mutants. During early neurogenesis, the three transcripts of the AS-C are expressed in partially overlapping clusters of cells within the neuroectoderm (Fig. 9). Since their domains of expression overlap partially, some neuroblasts may contain all three RNAs, and possibly their products, whereas other neuroblasts contain only one or two of them. This pattern of transcription is suggestive of a role for the AS-C genes in neuroblast commitment.

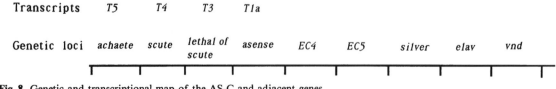

Fig. 8. Genetic and transcriptional map of the AS-C and adjacent genes

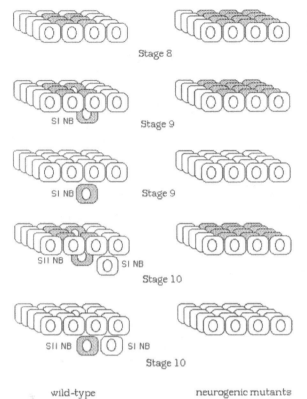

Stage 8

SI NB   Stage 9

SI NB   Stage 9

SII NB   SI NB
Stage 10

SII NB   SI NB
Stage 10

wild-type   neurogenic mutants

Fig. 9. Spatial distribution of T5 (*achaete*) transcripts in wild-type and neurogenic mutants. An array of neuroectodermal cells is shown at different stages in both cases. In the stage 8 wild-type embryo, five to six clustered cells transcribe the T5 RNA; in stage 9, one of the cells segregates as a neuroblast and continues transcribing T5 for some time. A the end of stage 9, transcription of T5 reappears in another cluster of similar size, one cell of which will segregate as a SII neuroblast and continue to transcribe T5. In the neurogenic mutants, the number of cells transcribing T5 is larger than in the wild-type preceding the stage at which SI neuroblasts normally segregate. Like in the wild-type, transcription of T5 is interrupted until the end of stage 9, reappearing in a large cluster of cells preceding the time of SII neuroblast segregation

We have already mentioned that approximately 20–25% of the normal neuroblast complement is lacking in AS-C$^-$ mutants: the corresponding neuroectodermal cells are not committed to their normal fates, but instead they have taken on epidermal fate, or died. Similar findings have been made with *vnd* mutants as well. In addition, deletion of particular AS-C genes correlates with defects in particular subsets of sensory organs, suggesting

specific roles for the AS-C genes during development of central and peripheral neural progenitor cells. Thus, it seems likely that the AS-C and *vnd* gene products provide the neuroectodermal cells with the properties necessary for their development as neuroblasts.

23.11 Interactions Between Neurogenic and Proneural Genes [19]

The AS-C genes (and *da*) are functionally interconnected with the neurogenic genes forming part of the same genetic network. On the one hand, changes in the pattern of transcription of the genes *l'sc* and *ac* have been observed in embryos carrying any of several neurogenic mutations. In these embryos, at the stage when the neuroblasts normally segregate, the corresponding RNAs are found in many more cells than in the wild-type (Fig. 9). Hence, the neurogenic genes define the domains of transcription of the AS-C genes. On the other hand, the phenotype caused by any of the neurogenic mutants is considerably reduced if the same embryo lacks the AS-C of the *da* functions.

23.12 Outlook and Summary

One important conclusion to be drawn from the data above is that the proteins encoded by the neurogenic genes, the genes of the AS-C and *vnd*, and *da* are functionally interrelated, forming a regulatory network that permits the cells to take on the neural or the epidermal fate. Evidence derived from cell transplantations and from genetic mosaics indicates that cellular interactions are mediated by direct contact between the cells of the NR and not by diffusing substances. Several pieces of evidence suggest that the decision to adopt the neural developmental fate is taken in a cell autonomous manner; cell communication is subsequently required to implement this decision – with the participation of neuralizing signals – as well as to permit epidermal development – with the participation of epidermalizing signals. In the neurogenic mutants, the cell communication process is impaired, the primary neurogenic fate of these cells cannot be suppressed and all cells of the neuroectoderm develop as neuroblasts.

The molecular structures of the various proteins encoded by the AS-C and *da*, together with the phenotypes of their mutants, strongly suggest that these proteins carry out the regulatory functions necessary for development of the neuroblasts. From the modifications of the pattern of transcription of AS-C genes found in neurogenic mutants, we assume that cells that normally develop as epidermoblasts are misrouted into neurogenesis because they continue to express the AS-C genes. With respect to epidermogenesis, several of the proteins encoded by the genes of the E(SPL)-C contain basic DNA binding and HLH dimerization motifs; this is compatible with their function as transcriptional regulators. Therefore, we assume that these proteins regulate the specific genetic activities of the neuroectodermal cells that enable them to develop as epidermoblasts. It seems improbable that proteins encoded by the E(SPL)-C activate directly transcription of the "realizators" of the epidermal pathway (i.e., the genes whose products eventually make epidermis), since the complete deletion of the E(SPL)-C still permits some epidermal development in the dorsalmost embryonic regions. This, and other genetic considerations, leads to propose that the regulation of epidermogenesis by the E(SPL)-C occurs indirectly, via the proneural genes. We assume a dual function for the proneural genes: to suppress the epidermal "realizator" genes and to activate the neural "realizator" genes. Support for this hypothesis is provided by the observation that the neurogenic phenotype of double mutants AS-C$^-$ and, for example, E(SPL)-C$^-$, is reduced as compared to that of the E(SPL)-C$^-$ mutants alone. If the AS-C, as well as the other proneural genes, suppress the epidermal "realizator" genes, the deletion of the AS-C would eliminate this suppressive effect and thus permit the development of epidermis.

We propose that, at the membrane of neuroectodermal cells, the cellular interactions are mediated by the DELTA and NOTCH proteins, probably by the EGF-like repeats present in the extracellular domains of both proteins. Since we postulate a signal chain, a very appealing possibility is that the interactions between both proteins represent relationships between ligand and receptor. Such a relationship would imply an asymmetrical distribution – or a somehow asymmetrical function – of the two proteins. The NOTCH protein is present in all neuroectodermal cells, whereas so far, no data are available on the distribution of the DELTA protein. Another possibility is that both DELTA and NOTCH are capable of passing signals between neighbouring cells in both directions, as opposed to the unidirectional flow of information implicit in ligand-receptor relationships. Hence, with the available data we cannot yet decide as to whether the complex assumed to be formed by both proteins plays an instructive or a permissive role in the cell communication process.

Thus, we envisage the segregation of neural and epidermal cell lineages as the result of a delicate balance between the activity of two groups of transcriptional regulators, encoded by the proneural genes (AS-C, *da*, perhaps *vnd*) for the neural fate, and by the genes of the E(SPL)-C for the epidermal fate. A possible sequence of events, which is compatible with the available data, is that all the cells in the neuroectoderm express a particular set of proneural genes and acquire initially the competence to develop as neuroblasts (i.e., a primary neurogenic fate). Individual cells would then become committed to the neural fate following stochastic fluctuations in the expression of these latter genes, probably reinforced by neuralizing signals mediated by DELTA and NOTCH coming from neighboring cells; lateral inhibition exerted by the neuroblast on the surrounding cells would lead to activation of the genes of the E(SPL)-C and to repression of the proneural genes in these cells, allowing their development as epidermoblasts.

References

[1] Campos-Ortega JA (1988) Cellular interactions during early neurogenesis of *Drosophila melanogaster*. Trends Neurosci 11:400–405

[2] Doe CQ, Hiromi Y, Gehring WJ, Goodman CS (1988) Expression and function of the segmentation gene *fushitarazu* during *Drosophila* neurogenesis. Science 239:170–175

Jiménez F (1988) Genetic control of neuronal determination in insects. Trends Neurosci 11:378–380

[3] Ghysen A, Dambly-Chaudière C (1988) From DNA to form: the *achaete-scute* complex. Genes Dev 2:495–501

[4] Campos-Ortega JA, Knust E (1990) Molecular analysis of a cellular decision during embryonic development of *Drosophila melanogaster*: epidermogenesis or neurogenesis. Eur J Biochem 190:1–10

Klämbt C, Knust E, Tietze K, Campos-Ortega JA (1989) Closely related transcripts encoded by the neurogenic gene complex *Enhancer of split* of *Drosophila melanogaster*. EMBO J 8:203–210

[5] Jan YN, Jan LY (1990) Genes required for specifying cell fates in *Drosophila* embryonic sensory nervous system. Trends Neurosci 13:493–498

[6] Campos-Ortega JA, Hartenstein V (1985) The embryonic development of *Drosophila melanogaster*. Springer, Berlin Heidelberg New York Tokyo, viii + 227 pp

Thomas JB, Bastiani MJ, Bate M, Goodman C (1984) From grasshopper to *Drosophila*: a common plan for neuronal development. Nature 310:203–207

[7] Technau GM, Campos-Ortega JA (1987) Cell autonomy of expression of neurogenic genes of *Drosophila melanogaster*. Proc Natl Acad Sci USA 84:4500–4504

[8] Becker T, Technau GM (1990) Single cell transplantation reveals interspecific cell communication in *Drosophila* chimeras. Development 109:821–832

Doe CQ, Goodman CS (1985) Early events in insect neurogenesis. II. The role of cell interactions and cell lineages in the determination of neuronal precursor cells. Dev Biol 111:206–219

Taghert PH, Doe CQ, Goodman CS (1984) Cell determination and regulation during development of neuroblasts and neurones in grasshopper embryos. Nature 307:163–165

[9] Jiménez F, Campos-Ortega JA (1990) Defective neuroblast commitment in mutants of the achaete-scute complex and adjacent genes of *Drosophila melanogaster*. Neuron 5:81–89

[10] Lehmann R, Jimenez F, Dietrich U, Campos-Ortega JA (1983) On the phenotype and development of mutants of early neurogenesis in *Drosophila melanogaster*. Wilhelm Roux's Arch Dev Biol 192:62–74

Poulson DF (1937) Chromosomal deficiencies and embryonic development of *Drosophila melanogaster*. Proc Natl Acad Sci USA 23:133–137

[11] Brand M, Campos-Ortega JA (1988) Two groups of interrelated genes regulate early neurogenesis in *Drosophila melanogaster*. Wilhelm Roux's Arch Dev Biol 197:457–470

[12] de la Concha A, Dietrich U, Weigel D, Campos-Ortega JA (1988) Functional interactions of neurogenic genes of *Drosophila melanogaster*. Genetics 118:499–508

[13] Campos-Ortega JA, Knust E (1990) Defective ommatidial cell assembly leads to defective morphogenesis: a phenotypic analysis of the *E(spl)*[D] mutation of *Drosophila melanogaster*. Wilhelm Roux's Arch Dev Biol 198:275–285

[14] Hoppe PE, Greenspan RJ (1990) The *Notch* locus of *Drosophila* is required in epidermal cells for epidermal development. Development 109:875–885

Simpson P (1990) *Notch* and the choice of cell fate in *Drosophila* neuroepithelium. Trends Genet 6:343–345

[15] Artavanis-Tsakonas S (1988) The molecular biology of the *Notch* locus and the fine tuning of differentiation in *Drosophila*. Trends Genet 4:95–100

Haenlin M, Kramatschek B, Campos-Ortega JA (1990) The pattern of transcription of the neurogenic gene *Delta* of *Drosophila melanogaster*. Development 110:905–914

Johansen KM, Fehon RG, Artavanis-Tsakonas S (1989) The *Notch* gene product is a glycoprotein expressed on the cell surface of both epidermal and neuronal precursor cells during *Drosophila* development. J Cell Biol 109: 2427–2440

Kidd S, Baylies MK, Gasic GP, Young MW (1989) Structure and distribution of the Notch protein in developing *Drosophila*. Genes Dev 3:1113–1129

Wharton KA, Johansen KM, Xu T, Artavanis-Tsakonas S (1985) Nucleotide sequence from the neurogenic locus *Notch* implies a gene product that shares homology with proteins containing EGF-like repeats. Cell 43:567–581

Vässin H, Bremer KA, Knust E, Campos-Ortega JA (1987) The neurogenic locus *Delta* of *Drosophila melanogaster* is expressed in neurogenic territories and encodes a putative transmembrane protein with EGF-like repeats. EMBO J 6:3431–3440

[16] Cagan RL, Ready DF (1989) *Notch* is required for successive cell decisions in the developing *Drosophila* eye. Genes Dev 3:1099–1112

Fehon RG, Kooh PJ, Rebay I, Regan CL, Xu T, Muskavitch MAT, Artavanis-Tsakonas S (1990) Molecular interactions between the protein products of the neurogenic loci Notch and Delta, two EGF-homologous genes in *Drosophila*. Cell 61:523–534

Greenspan RJ (1990) The *Notch* gene, adhesion, and developmental fate in the *Drosophila* embryo. New Biol 2:595—600

[17] Hartley DA, Preiss A, Artavanis-Tsakonas S (1988) A deduced gene product from the *Drosophila* neurogenic locus Enhancer of split shows homology to mammalian G-protein β-subunit. Cell 55:785–795

[18] Alonso MC, Cabrera CV (1988) The *achaete-scute* gene complex of *Drosophila melanogaster* comprises four homologous genes. EMBO J 7:2585–2591

Busch JS, Sassone-Corsi P (1990) Dimers, leucine zippers and DNA-binding domains. Trends Genet 6:36–40

Davis RL, Cheng PF, Lassar AB, Weintraub H (1990) The MyoD DNA binding domain contains a recognition code for muscle specific gene activation. Cell 60:733–746

Jones N (1990) Transcriptional regulation by dimerization: two sides to an incestuous relationship. Cell 61:9–11

Olson EN (1990) *MyoD* family: a paradigm for development? Genes Dev 4:1454–1461

Villares R, Cabrera CV (1987) The achaete-scute gene complex of *Drosophila melanogaster*: conserved domains in a subset of genes required for neurogenesis and their homology to myc. Cell 50:415–424

[19] Brand M, Campos-Ortega JA (1990) Second site modifiers of the split mutation of Notch define genes involved in neurogenesis in *Drosophila melanogaster*. Wilhelm Roux's Arch Dev Biol 198:275–285

Cabrera CV (1990) Lateral inhibition and cell fate during neurogenesis in *Drosophila*: the interactions between *scute*, *Notch* and *Delta*. Development 109:733–742

Chapter 24   *Xenopus* Embryogenesis

TOMAS PIELER

24.1 Introduction

Classical experimental embryology using amphibians as an experimental test system was established in the first half of this century, largely by the work of Spemann and colleagues. These studies, together with early classical experiments by Wilhelm Roux and Hans Driesch, identified two major developmental control mechanisms, inductive interactions and the cytoplasmic localization of information in the egg and early embryo. During the past four decades, amphibians, and in particular *Xenopus laevis*, have become one of the major systems to study the biochemistry and molecular biology of these two mechanisms which govern early vertebrate development. Use of *Xenopus* as an experimental system has several practical advantages, including the easy accessibility of embryonic material and the relatively large size of the embryo, which makes it available for direct physical manipulations. In particular, oocyte and embryo microinjections have proven to be a powerful, almost universal tool for the study of different aspects of the regulation of gene expression. The aim of the first part of this chapter is to describe the principle features of early *Xenopus* embryogenesis and to relate relevant phenomena to what has been discovered at the molecular level more recently. The second part will then be devoted to a more detailed review and discussion of the utility of *Xenopus* embryo microinjection as a system for the analysis of both the developmental regulation of gene expression and the function of individual genes in the regulation of development.

24.2 The Oocyte[1]

The developmental history of the frog is given in Fig. 1. Oocyte growth in *Xenopus* takes about 3 months. In this time, the oocyte increases enormously in size, which is primarily due to the storage function of this cell. Oocytes accumulate large amounts of materials for use in the early post fertilization stages of embryogenesis (Table 1). They have a darkly pigment half, the animal pole, and a bright half, the vegetal pole (Fig. 2). The vegetal region contains a stockpile of yolk platelets, a mixture of phospho- and lipoproteins, as well as of other substances used for embryonic nutrition in the first days of development. These internally stored materials allow the experimentator to grow embryos for days in simple saline solutions without extra supply of energy. The darkly pigmented animal region is highly enriched in ribosomes, mito-

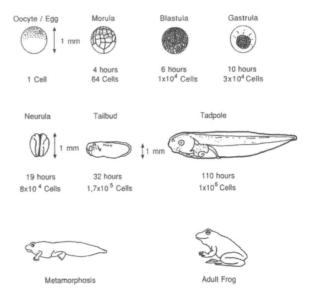

Fig. 1. Frog development from fertilized egg to adult frog. Diagramatic representation of successive stages in frog development. Stages up to and including the 110-h tadpole are drawn to the same scale. Metamorphosis stage and adult frog are drawn out of scale

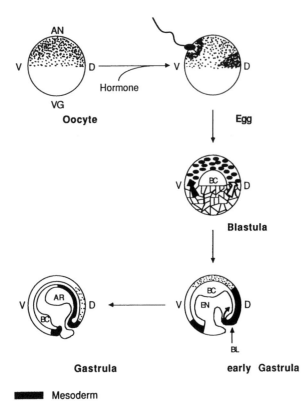

Mesoderm

Neuroectoderm

Fig. 2. Diagramatic representation of various stages in early *Xenopus* embryogenesis. The fully grown (*stage VI*) *Xenopus* oocyte has a darkly pigmented animal (*AN*) and a lightly pigmented, yolk-loaden vegetal (*VG*) half. Hormone-induced maturation from oocyte to egg allows fertilization and establishment of the ventral (*V*)/dorsal (*D*) polarity with a typical, asymetrical distribution of cytoplasmic components. Upon fertilization, a fast, synchronous series of cell divisions results in the blastula stage embryo, with the formation of a liquid containing inner cavity, the blastocoel (*BC*). At this stage, vegetal cells interact with the overlying responding animal cells, leading to the differentiation of the responding cells to form mesodermal tissue. The *open arrow* represents a diffusible dorso-vegetal signal, which is probably distinct from a diffusible ventral-vegetal signal (*filled arrow*). The inducing vegetal cells will form endoderm (*EN*) and the cells of the animal cap ectoderm. the beginning of gastrulation is marked by the invagination of mesodermal cells. The cells continue to move along the blastocoel roof and induce the overlying responding ectodermal cells to differentiate into neural tissue. Another consequence of these gastrulation movements is the formation of the primitive gut, the archenteron (*AR*)

Table 1. General properties and molecular features of a fully grown *Xenopus* oocyte[a]

Diameter volume	Total nucleus	1.25 mm
		1.00 µl
		0.04 µl
RNA content	28S and 18S ribosomal DNA	5 µg
	5S RNA	0.06 µg
	tRNA	0.06 µg
	polyA$^+$ mRNA	0.08 µg
Protein content	Yolk	250 µg
	Nonyolk	27.5 µg
	Histones	0.14 µg
	TFIIIA[b]	0.003 µg
DNA content	Chromosomal	12 pg
	Nucleolar	25 pg
	Mitochondrial	4000 pg
RNA synthesis	Per day	20 ng
Protein synthesis	Per day	400 ng

[a] After Gurdon and Wickens (1983).
[b] Data from Shastry et al. (1984).

chondria and other cell organelles and it harbors the large nuclear compartment, referred to as the germinal vesicle (GV). The size of the oocyte and its nucleus, as well as the high degree of biosynthetic activity make the *Xenopus* oocyte a suitable experimental system for the analysis of many aspects in the regulation of gene expression.

24.2.1 The *Xenopus* Oocyte as a Test Tube[2]

The utility of the *Xenopus* oocyte for the transcription of injected genes and for the translation of injected mRNA has been established by the work of Gurdon and coworkers. It has been demonstrated that genes injected into the oocyte nucleus are actively transcribed and faithfully processed. The RNA product then migrates into the cytoplasm and mRNA is efficiently translated into biologically active proteins. Thus, the oocyte is a suitable system for the experimental analysis of every single step in this chain of events.

In contrast to other eukaryotic experimental systems, the activities listed above proceed in single cells and in the absence of cell division and nuclear envelope breakdown, since the fully grown *Xenopus* oocyte is arrested in the diplotene stage of the first meiotic prophase. Thus, and also facilitated by other unique features such as the large size of these cells, and, more importantly, of their nuclei, this system allows nuclear and/or cytoplasmic microinjection to be employed for the direct analysis of nucleocytoplasmic transport of macromolecules by means of manual microdissection of the two cellular compartments following the microinjection procedure.

24.2.2 Nucleocytoplasmic Transport of RNA and Proteins[3]

Intracellular migration of small RNA molecules in *Xenopus* oocytes has been studied in detail. The basic experiment, which is shown in Fig. 3, was carried out by De Robertis and colleagues in the following way: mixtures of radiolabelled human tRNA, 5S RNA and snRNAs were injected into either cytoplasm or nucleus of the *Xenopus* oocyte. Both experiments gave the same results: tRNA was found to accumulate in the cytoplasm, 5S RNA distributed between cytoplasm and nucleoli, and snRNAs were able to concentrate in the nucleus. A similar picture had emerged from the study of proteins using the same experimental approach. Accumulation of proteins in the nucleus is highly selective and requires the presence of a signal sequence, which allows large proteins to passage through the nuclear pores. The early RNA injection experiments did not address the important question of whether or not specific proteins were critically involved in the observed migratory pathways of RNA. This problem was first analyzed for the snRNAs. After transcription, the monomethyl capped U2 snRNA leaves the nucleus and enters the cytoplasmic compartment. There, it binds to the Sm protein which is required for a second modification event in the cytoplasm, trimethylation. Thus, Sm binding and trimethylation are cytoplasmic events preceding nuclear migration of the U2 snRNP. Mutant U2 RNAs which do not bind the Sm protein

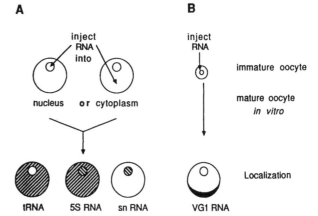

Fig. 3 A, B. Nucleocytoplasmic migration of small RNA and cytoplasmic localization of mRNA. Injection of mixtures of radiolabeled RNA into either nucleus or cytoplasm of the fully grown *Xenopus* oocyte (**A**) results in a differential nucleocytoplasmic distribution for tRNA (cytoplasm), 5S RNA (nucleus and cytoplasm) and snRNAs (nucleus). Microinjection of some specific RNA species (i.e., *VG1*, **B**) results in a discrete cytoplasmic localization in the process of oocyte maturation at the vegetal pole

fail to accumulate in the nucleus. Since the Sm protein by itself remains cytoplasmic, it has been proposed that a karophilic domain in the polypeptide may become exposed upon formation of the RNA/protein complex. Recently, it could be demonstrated that a secondary cytoplasmic event, which depends on Sm binding, trimethylation, is causally linked to nuclear migration. Since microinjection of the m_3G cap structure alone inhibits transport of U snRNAs to the nucleus, involvement of a cap binding protein in the transport mechanism appears likely.

The first, post-transcriptional step in nucleocytoplasmic migration of small RNAs has been analyzed for tRNA and 5S RNA. Point mutations introduced into tRNA often result in accumulation of unprocessed tRNA precursors in the nucleus, indicating that these two processes are interconnected. A pre-tRNA 5'-processing enzyme has been found to copurify with a 16S macromolecular complex, and it has been suggested that this complex processing nuclease might play a role in transport of tRNA from the nucleus.

Direct evidence for the involvement of specific RNA binding proteins in nucleocytoplasmic transport of 5S RNA in *Xenopus* oocytes has been ob-

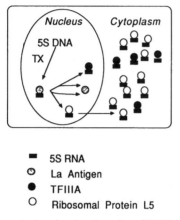

- ■ 5S RNA
- ◎ La Antigen
- ● TFIIIA
- ○ Ribosomal Protein L5

Fig. 4. Nucleocytoplasmic migration of small RNPs, containing 5S ribosomal RNA, in fully grown *Xenopus* oocytes. Newly transcribed 5S RNA forms a transient complex with the La protein. The La protein is an autoantigen probably serving as a general RNA polymerase III termination factor. Subsequently, La protein is replaced by either TFIIIA or ribosomal protein L5 and these complexes migrate to the cytoplasmic compartment, where they accumulate

tained most recently. It was found that two different proteins, the 5S gene specific transcription factor TFIIIA and ribosomal protein L5, promote the migration of newly transcribed 5S RNA from nucleus to cytoplasm (Fig. 4). A 5S RNA variant which was impaired in its ability to interact with L5 and TFIIIA, was retained in the nucleus. These findings suggest that nucleocytoplasmic transport of RNA might in other cases as well be mediated by specific RNA binding proteins.

24.2.3 Cytoplasmic Localization of mRNA[4]

RNA migration to specific sites within the *Xenopus* oocyte is also observed with certain mRNA species. This phenomenon is of particular interest in respect to early embryogenesis since, as mentioned in the introduction, maternally expressed cytoplasmic determinants are believed to distribute asymetrically in early cleavage stages and thereby to contribute to the determination of cell fate. On these grounds, Melton and coworkers went out to identify maternal RNAs from *Xenopus* oocytes, which are localized in either the animal or the vegetal pole region. Three animal and one vegetal cDNA clone were

isolated. Most interestingly, the vegetal clone (VG1) was found to be structurally related to transforming growth factor β (TGFβ), a factor which could be involved in mesoderm induction (as discussed below).

The VG1 mRNA is uniformly distributed in the cytoplasm of immature oocytes and becomes localized as a crescent at the vegetal pole of mature oocytes. Exogenous VG1 mRNA injected into immature oocytes becomes localized in a similar fashion to the endogenous message during oocyte growth (Fig. 3). As a first step toward the elucidation of the mechanism which mediates the translocation of specific RNAs like VG1 in the *Xenopus* oocyte, it was shown that localization is independent of the translatability of the injected RNA. It seems that specific structural features of the RNA are required and it is easy to imagine that specific RNA binding proteins might be involved, as described for nucleocytoplasmic migration of small RNAs in the previous paragraph.

24.3 Egg Maturation and Fertilization[5]

The fully grown oocyte undergoes a series of major morphological changes in the process of hormone-induced maturation and subsequent insemination. The outer and inner morphology of the egg is related to that which has been described for the oocyte maintaining the animal/vegetal polarity with the typical pattern of pigmentation. Maturation requires 3–6 h from the time of exposure to hormone and is accompanied by the disintegration of the nuclear membrane as the cell proceeds through the entire first meiotic cycle and arrests in metaphase of the second meiotic cycle. The plasma membrane becomes capable of fusing with the sperm. The point of sperm entry is usually located in the darkly pigmented animal pole region. The sperm entry point then defines (under normal conditions) the orientation of the dorso-ventral polarity which is established by a series of active cytoplasmic and surface movements. These events are completed when the pronuclei pair. At this time, a gray crescent, visible on the outer surface, forms in the animal half, opposite to the site of sperm entry. The area of the gray crescent defines the dorsal side of the fertilized egg (Fig. 2). Formation of the dorsal-ventral polarity involves internal, cytoplasmic rearrangements, such as the shift of a yolk-free region of cytoplasm from the center of the animal hemisphere (where the germinal vesicle was previously located) toward the gray crescent side.

In summary, the fertilized *Xenopus* egg is characterized by the asymetric distribution of materials, including information-carrying molecules such as specific mRNAs, which allow vegetal and animal halves (the two hemispheres) of the egg to be distinguished and which also establish the dorsal-ventral polarity (Fig. 2).

24.3.1 Regulation of Translation During *Xenopus* Oocyte Maturation[6]

As stated above, unfertilized eggs of *Xenopus laevis* contain large amounts of maternal mRNA. Regulation of expression of many of these stored RNAs takes place upon egg maturation and fertilization at the level of translation suggesting a functional role for these proteins in early development. It seems that translational regulation in response to maturation is achieved via selective polyadenylation or deadenylation at the 3'-end of individual mRNAs. Translation inactivation of mRNA coding for ribosomal proteins during oocyte maturation has been found to be accompanied by deadenylation, whereas translation activation of specific mRNAs requires poly(A) elongation. Both specific polyadenylation as well as specific deadenylation appear to be mediated with the help of RNA sequence elements located near the 3'-end of the transcripts.

In the case of selective adenylation, a short conserved oligonucleotide sequence (UUUUUAU), in close proximity 5' to the common polyadenylation signal (AAUAAA), has been identified and shown to regulate the maturation-specific poly(A) elongation of several different mRNA species in *Xenopus* eggs. It is not know what factors are involved in maturation specific polyadenylation. It is also not known if some of those activities which have been identified to mediate nuclear polyadenylation in HeLa cells have homologs in *Xenopus* and if these,

if they exist, participate in the events discussed. It is, however, easy to imagine that specific RNA binding proteins will be involved.

24.4 Mesoderm Induction[7]

The first 10–12 cell cycles following fertilization of the *Xenopus* egg occur synchronously and at high speed (about 35 min per cell cycle). There is no increase in the total volume of the embryo and this early phase of development up to midblastula stage occurs in the absence of transcription activity. During the same period, a liquid-containing cavity forms in the interior of the animal hemisphere, which is referred to as the blastocoel (Fig.2). The midblastula embryo maintains the pattern established in oogenesis in that it preserves animal/vegetal as well as dorsal/ventral polarity. Asymmetric distribution of maternal cytoplasmic materials is probably also involved in the process of induction. Induction is defined as the interaction between two different tissues, one inducing and the other responding. Mesoderm induction represents the interaction of inducing vegetal cells with the overlying responding animal cells. This leads the totipotent animal cells to form mesodermal tissues, including the different forms of muscle, notochord, and somites. Evidence for this comes from experiments in which regions of the blastula embryo were cultured either in isolation or in combination (Fig. 5A). If blastomeres from the animal pole region of a blastula-stage embryo are removed by microsurgery and cultured in isolation, they will form epidermis. If, however, they are brought into contact with cells from the vegetal pole, they will give rise to mesodermal tissues. It has also become clear that vegetal cells from the dorsal region of the embryo have mesoderm-inducing capacities different from those derived from the ventral region.

The molecular identity of the endogenous mesoderm-inducing factor(s) has not yet been firmly established. It is, however, becoming increasingly likely that members of the growth factor family of proteins are involved. It has been demonstrated by the animal cap explant technique that one class of mesoderm-inducing factors could be related to

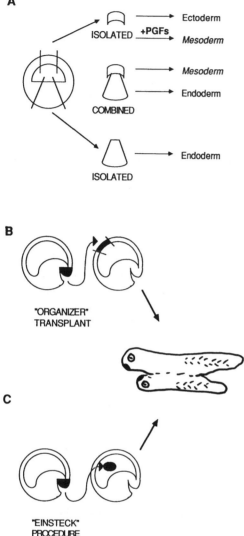

transforming growth factor β (TGFβ), and a second one to the group of heparin-binding growth factors such as fibroblast growth factor (FGF). Members of both growth factor families were found to be expressed in early *Xenopus* embryos. Mesoderm-inducing capacity was directly demonstrated for *Xenopus* FGF, making it almost certain that it has a role in mesoderm induction. A TGF-β-like mRNA was also identified in *Xenopus* oocytes, the vegetal clone VG1. Most interestingly, the mRNA encoding this protein becomes localized in the vegetal pole region of the oocyte making it an excellent candidate for a localized cytoplasmic determinant. However, although the circumstantial evidence is compelling, direct experimental proof for a function of VG1 in mesoderm induction has not yet been obtained. Finally, culture medium from a *Xenopus* cell line (XTC cells) contains a potent mesoderm-inducing activity which, after purification and sequence analysis, turned out to be another member of the peptide growth factor family, structurally related to activin. Other activin-related factors with mesoderm inducing capacity are indeed expressed at blastula stage and these may finally turn out to be the real inducers in vivo.

24.4.1 Early Molecular Responses to Mesoderm Induction and the Regulation of Muscle Gene Expression[8]

The identification of inducing agents discussed in the previous sections immediately leads to questioning how the extracellular signal in form of a growth factor is transmitted. It is, thus, a very reasonable approach in this context to search for genes which are activated as a direct response to mesoderm induction. Furthermore, it is of primary interest to identify possible regulators of gene expression, i.e., DNA binding transactivators. This has been achieved in two different ways.

Dawid and coworkers have developed methods which allow the construction of subtracted cDNA libraries, that is libraries which are enriched for regionally or temporally restricted sequences. More recently, this strategy has been employed for the isolation of cDNAs which respond to mesoderm induction. Most interestingly, one out of 22 clones

Fig. 5 A–C. Embryonic induction. Explants from the animal cap region of a blastula-stage embryo will form ectodermal tissue in isolation, corresponding to their normal fate in the developing embryo. Similarly, explants from the vegetal region will form endodermal tissues in isolation and in the embryo. If, however, both explants are cultivated in contact with each other, the vegetal cells will induce the animal cells to form mesodermal tissues. The same response in animal cap cells, namely to form mesodermal derivatives, can be obtained by use of peptide growth factors (*PGFs*) as inducing signal (**A**). Transplantation of the dorsal blastopore lip (the "organizer" region, **B**) of an early gastrula embryo to the ventral side of a second early gastrula embryo leads to the formation of a complete secondary embryonic axis. The same effect can be obtained by transplanting the dorsal lip into the blastocoel cavity of an early gastrula embryo (Einsteck method, **C**)

Table 2. Marker genes for early *Xenopus* embryogenesis

Name	Function	Structure	Expression
VG1	Signal molecule[a]	Peptide growth factor	Maternal (localized in oocyte vegetal pole)
Cardiac actin	Structural protein		Mesoderm (somites)
XlHbox1	Transcription regulator[a]	Homeo domain	Mesoderm (anterior)
Xhox3	Transcription regulator[a]	Homeo domain	Axial mesoderm (in posterior/anterior gradient)
Myo D	Transcription regulator[a]	Helix-loop-helix	Mesoderm (somites)
Xtwi	Transcription regulator[a]	Helix-loop-helix	Mesoderm (notochord, lateral plate)
Mix.1	Transcription regulator[a]	Homeo domain	Mesoderm (marginal zone) and presumptive endoderm
CG1	(Glue protein?)		Ectoderm (cement gland)
Epidermal keratin	Structural protein		Ectoderm (epidermis)
N-CAM	Cell adhesion		Neuroectoderm (general)
XlHbox6	Transcription regulator[a]	Homeo domain	Neuroectoderm (posterior)
XIF3	Structural protein	Intermediate filament type III	Neuroectoderm (anterior)

[a] Predicted from structural homology.

with differential expression was found to encode a homeobox protein (Mix. 1), suggesting that this protein may bind DNA and play a regulatory role in development.

A different approach is to isolate structural homologs of genes with a demonstrated role in development regulation and, more specifically, in mesoderm differentiation in other organisms. By use of this strategy, Gurdon and coworkers have isolated a cDNA related to a gene product from mouse cells, Myo D. Expression of the transfected Myo D cDNA is sufficient to convert fibroblasts to myoblasts. Mouse Myo D and its relative in *Xenopus* have significant sequence homology to the proto-oncogene *c-myc* and other DNA binding proteins, indicating a transcription regulatory function for Myo D in the process of mesoderm differentiation. Similarly, another member of the *c-myc*-related multi-gene family of DNA binding proteins (referred to as the Helix-Loop-Helix proteins), the *Drosophila twist* gene, which is expressed in mesoderm and critically involved in its formation in the fly, has been used as a probe for the isolation of the *twist* gene homolog in *Xenopus*, *Xtwi*. A specific feature of the *Xtwi* gene is that it is transcriptionally activated as a response to two different sequen-

tial early inductions in frog development: mesoderm and neural induction (see below).

In addition to these three marker genes for mesoderm induction, which all exhibit structural features identifying them as coding for DNA-binding and, therefore, as proteins with potential to regulate gene expression, earlier studies have described a marker for differentiation of a mesoderm-derived cell type, namely *a*-cardiac actin genes, expressed in the somites of amphibian embryos. A summary of cloned *Xenopus* marker genes is given in Table 2.

24.4.2 Regulatory Elements in *Xenopus* Muscle Gene Activation[9]

Gene elements involved in the acitvation of the *Xenopus* cardiac actin gene in response to mesoderm induction have been studied by use of the *Xenopus* embryo as a transient expression system. Cloned fragments of genomic DNA encoding cardiac actin were microinjected into fertilized eggs and transcription activity from the injected DNA analyzed. Injected genes were transcribed at the same time and in the same tissue as the endogenous muscle actin genes. A sequence element upstream of the muscle actin gene was identified, which is required for the response to mesoderm induction (the CArG box). The CArG box shares sequence similarity with the Serum Response Element (SRE) which mediates the transcription activation of several other genes in response to serum, i.e., the proto-oncogene *c-fos*. Moreover, it was demonstrated that CArG box and SRE are functionally interchangeable and that they serve as a binding site for a protein, the serum response factor (SRF). However, there is no simple relationship between the presence of SRF and the transcriptional responses mediated by SRE, since different SRE-regulated *Xenopus* genes have different temporal and spatial patterns of expression. Nevertheless, although the picture emerging for the molecular mechanism regulating the *Xenopus* cardiac actin gene as a response to mesoderm induction is far from being complete, it provides us with molecular details and partial insights into how an extracellular signal which is critically involved in the first events of mesoderm induction might be transmitted to the nucleus, in order to establish mesoderm specific gene expression.

24.5 Gastrulation and Neural Induction[10]

In *Xenopus laevis* the periods of gastrulation and neurulation require about 16 h, which is twice as long as the time from fertilization to gastrulation. With developmentally important characteristic morphogenetic movements, the body axis of the future tadpole is established. Gastrulation begins with the formation of the blastopore lip in the area corresponding to the site where the gray crescent had been visible in the egg (Fig. 2). Occurrence of the blastopore lip signifies the beginning of invagination of cells at the dorsal side of the embryo; these cells will then move on along the blastocoel roof and will form the presumptive dorsal mesoderm (chordamesoderm). This chordamesoderm will induce the overlying ectodermal cell layer to differentiate into neural plate which will develop into the central nervous system. This process defines

the second major inductive chain of events in early *Xenopus* development, neural induction.

The most direct early experimental evidence for the requirement of intercellular communication between the invaginating mesoderm and the overlying ectodermal cells in this first step of neural development has been provided by Holtfreter. He was able to demonstrate that the entire prospective neuroectoderm develops into epidermis if it grows isolated from the mesoderm. This was achieved by the removal of the embryo membrane envelope and growth of these embryos in high salt medium. Such a procedure prevents mesoderm invagination and leads to "exogastrulation" such that the mesoderm is excluded from the inner embryo with the ectoderm forming an empty sac connected to the mesoderm by a narrow stalk.

In their famous "organizer experiment", Spemann and Mangold transplanted the blastopore lip region from an early gastrula embryo to the ventral side of a different early gastrula host embryo (Fig. 5B). This operation can lead to the formation of an almost complete secondary body axis. Since most of the tissue in this secondary axis is derived from the host embryo and due to extraordinary transforming capacity of the blastopore lip, this region was termed the "organizer".

Induction of a secondary embryonic axis can also be achieved by use of a differnt experimental strategy, the so-called Einsteck-Methode. Here, the dorsal lip of an early gastrula embryo is grafted into the blastocoel of an early gastrula host embryo, also resulting in the induction of a secondary embryonic axis (Fig. 5C).

24.5.1 Differential Gene Expression in Neural Induction[11]

Several genes have been identified in the early *Xenopus* embryo which are either activated at the level of transcription in response to neural induction or which are neural-specific (Table 2). *Cis*-acting elements in neural-specific regulation of transcription have not yet been identified, but recent evidence that retinoic acid might be involved in neural induction indicates that nuclear receptors, in conjunction with appropriate responsive elements, could be mediating a transcription response to an inductive signal directly.

Other findings raise a second possibility, namely that growth factor-like substances act as neural inducers, in a way similar to the mesoderm induction described above. Experimental evidence for this is provided by increases in the concentration of both adenylate cyclase activity and cAMP concentration during neural induction. These observations support the concept that receptor-mediated signal transduction is of primary importance in mesoderm and neural induction.

24.6 Analysis of Gene Function in *Xenopus* Embryogenesis [12]

Microinjection techniques allow the study of gene function and gene regulation in the developing *Xenopus* oocyte, egg, or embryo. In the following section, achievements, prospects and limitations of the different existing experimental strategies which make use of this powerful tool will be reviewed. We have already learned in the example of mesoderm-specific transcription of *Xenopus* cardiac actin genes, that injected genes can be subject to the same regulatory mechanisms as the endogenous genes. This is indeed not an isolated example and many other genes have been characterized in this in vivo system for the identification of *cis*-acting elements which confer temporal and spatial specificity of transcription. All these studies agree in two important aspects: (1) injected genes were found to be transcribed in the correct temporal and spatial framework; (2) key regulatory elements reside within the first 1000 nucleotides of the 5'-flanking region in genes transcribed by RNA polymerase II. Thus, the *Xenopus* embryo has the prospect of serving as a suitable system to study molecular aspects of the developmental regulation of many genes in addition to the cardiac actin gene.

In many cases it is desirable to be able to analyze the biological function of a gene isolated on the basis of interesting structural features or a suggestive expression pattern, or a combination of both. It is immediately evident from the embryo microinjection experiments, utilizing specifically regulated promoters, that it should be possible to achieve targeted transcription of optional DNA sequences by assembling the appropriate regulatory modules together with the sequences to be transcribed, and injection of the resulting DNA constructs into the *Xenopus* embryo. Analysis of gene function in development by these and other means can be achieved in two fundamentally different ways, either by irregular (over-) expression of the gene to be studied (gain of function) or by specific suppression of gene activity (loss of function).

24.6.1 Inhibiton of Gene Expression by Antisense RNA/DNA Techniques [13]

DNA microinjection into *Xenopus* embryos was successfully used in order to inhibit expression of a skeleton protein gene. This was achieved by producing a partial antisense RNA constitutively transcribed under the control of a viral promoter. The phenotypic result was perturbation of normal cellular interactions in the retina. Direct injection of antisense RNA into *Xenopus* oocytes had previously been used to inhibit translation. In the embryo microinjection experiments described above, it was found that production of partial antisense RNA resulted in the reduction of the level of endogenous sense transcripts in the embryo from the blastula stage onwards. It should be noted that direct injection of anti-sense RNA into the fertilized *Xenopus* egg does *not* result in sense transcript ablation. This phenomenon can be attributed to the presence of an unwinding/modifying activity in the egg and early embryo which prevents formation of stable RNA double strands.

A different complication for experiments with anti-sense RNA in embryos produced from injected plasmid DNA comes from the observation that the injected DNA will not be propagated in all cells. At least for some DNA constructs tested its distribution becomes mosaic as development proceeds.

Targeted, sequence-specific destruction of RNA can also be achieved by a different class of (anti-sense) RNA, the so-called ribozymes. Ribozymes are small RNAs capable of catalyzing RNA cleavage reactions in a sequence specific manner. Ribozymes have a functionally essential core structure and catalytic specificity is obtained by base-pairing with a short stretch of target RNA. It has been shown that such molecules work in vivo, in *Xenopus* oocytes, when transcribed from a DNA construct inserted into the transcribed region of an RNA polymerase III gene. However, although ribozyme activity provides us with a potentially useful alternative for specific inhibition of gene expression in the embryo, it is hampered by at least one of the same limitations which apply to the antisense RNA approach, mosaicism.

Destruction of specific RNA molecules can be triggered by injection of synthetic oligonucleotides into the *Xenopus* oocyte or unfertilized egg. Complementary oligonucleotides efficiently promote RNA degradation by an endogenous RNase H activity. The advantage of using synthetic oligonucleotides is their easy availability and the efficiency of specific RNA cleavage induced by this technique in the oocyte. Furthermore, Wylie and coworkers have developed an experimental strategy to generate embryos from oocytes manipulated in this or others ways. It is achieved by in vitro maturation of oocytes followed by reimplantation of these eggs into a frog. It will package such eggs into a jelly coat which is required to make the eggs fertilizable in vitro or in vivo. Thus, in principle it is possible to generate embryos which are devoid of one or multiple specific maternal mRNAs; clearly, the utility of this approach is principally limited to the study of maternally transcribed genes which are subject to translational regulation upon or after maturation. Since these prerequisites per se define a group of genes with a potential role in developmental regulation, anti-sense oligonucleotide-mediated destruction of maternal RNA in the oocyte should be applicable to the analysis of gene function in early *Xenopus* development, and a function in oocyte maturation has indeed been demonstrated for the proto-oncogene c-*mos* by these means.

An alternative experimental strategy to interfere with the function of a specific protein can be applied if antibodies are available. Injection of an antibody directed against the *Xenopus* homeobox protein *Xl*Hbox1 into fertilized eggs has been found to interfere with the proper formation of the anterior spinal cord. Other examples for the successful application of this technique to proteins expressed in the early embryo are likely to follow.

24.6.2 Embryo Microinjection of Synthetic mRNA [14]

Based on the possibility of in vitro synthesis of translatable mRNA from cloned cDNA sequences, microinjection of such artificial RNAs into *Xenopus* embryos has been performed with the aim of increasing the concentration of specific regulatory proteins in early development. This approach was pioneered by Andrews and Brown, who were able to show that injection of synthetic mRNA coding for the 5S gene-specific transcription factor TFIIIA resulted in synthesis of increased amounts of the proteins and an elevated level of 5S gene transcription (as discussed Chap. 25, this Vol on 5S gene transcription by A. P. Wolffe). The same approach was then subsequently used by Melton and coworkers in the analysis of two different *Xenopus* homeobox proteins with unknown function. In one case, for Xhox-1A, overexpression of the proteins resulted in distortion of the segmented somite pattern, indicating parallels between mechanisms of segmentation in flies and vertebrates (see Tautz Chap. 21, this Vol., on *Drosophila* development). A different *Xenopus* homeobox gene, Xhox3, was found to be expressed in a graded fashion along the anterior-posterior axis in the axial mesoderm of a gastrula embryo with high concentration in posterior and low concentration in anterior regions. Injection of Xhox3 mRNA into the prospective anterior region of developing embryos results in axial defects with a loss of the ability to form anterior (head) structures. These results strongly suggested that the Xhox3 gene product is involved in establishing cell identity in the anterior-posterior axis of the *Xenopus* embryo. Bringing their analysis another step further, Ruiz i Altaba and Melton found that transcription of Xhox3 in animal cap explants is inducible by growth factors. Moreover, the in vivo situation for Xhox3 transcription in anterior mesoderm (low level Xhox3) can be generated in animal cap explants as a response to XTC-MIF, a *Xenopus* peptide growth factor-like substance of the TGF-β family; a high level expression of Xhox3 is the result of administering FGF. We have already learned that the organizer region when transplanted into the blastocoel of an early gastrula embryo will induce a nearly complete secondary embryonic axis (Fig. 5D, Einsteck-Methode). Now, using the animal cap explant in a similar way as the organizer in such an Einsteck-Experiment, Ruiz i Altaba and Melton have been able to demonstrate that transplantation of animal cap cells induced with XTC-MIF results in the formation of a second body axis with anterior (head) structures, whereas FGF-treated animal caps will induce secondary posterior (tail) structures (Fig. 6). Taken together, these experiments have several important implications; firstly, they argue in favor of a regulatory role for

the homeobox protein Xhox3 in the specification of antero-posterior position in mesoderm. Secondly, they give additional support for the view that a graded signal of one or more peptide growth factors is involved in mesoderm induction and in the determination of its antero-posterior character. These experiments also nicely illustrate the potential of a combination of experimental strategies from classical developmental embryology, on the one hand, and from modern molecular biology, on the other.

A final example of the successful application of the mRNA microinjection approach is the functional analysis of a mamalian proto oncogene, int-1. The *Drosphila* int-1 homologue (wingless) is a segment polarity gene encoding a signal molecule; ectopic expression of int-1 in the *Xenopus* embryo by means of mRNA injection leads to an anterior duplication of the embryonic axis which has its visible expression in the bifurcation of the neural tube. Thus, in contrast to the experiments described above, where a second embryonic axis was obtained only after transplantation of tissue with "organizer" activity, ectopic expression of a single gene was sufficient to give a similar effect. It has been proposed that int-1 interferes with the action of the organizer itself, resulting in an apparent "splitting" of the organizer.

In summary, we have seen that an alteration in the relative and absolute concentrations of functional DNA binding proteins by pseudogenetic techniques in *Xenopus* embryos can indeed have clear phenotypic effects which give a good indication for the functional context which is relevant for these proteins. The more examples we are able to collect, the more we will be able to begin to understand the complex network of regulatory molecules which control early vertebrate development.

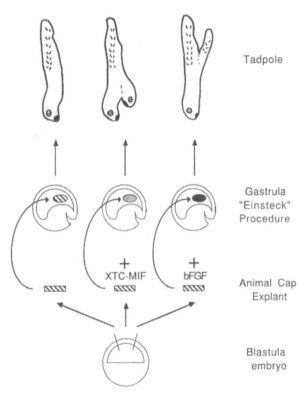

Tadpole

Gastrula
"Einsteck"
Procedure

Animal Cap
Explant

Blastula
embryo

Fig. 6. In vivo transplantation of animal cap explants induced with peptide growth factor into the blastocoel of early gastrulae. Isolated animal caps were incubated in the presence or absence of peptide growth factors. Mesoderm induced by XTC-MIF results in the formation of secondary anterior structures (i.e., head), whereas mesodern induced by bFGF leads to the formation of secondary posterior structures (i.e., tail). Noninduced mesoderm does not result in the genesis of secondary axial structures

24.6.3 Multiple Families of Nucleic Acid Binding Regulatory Proteins [15]

Many regulatory, DNA-binding proteins have been found to be composed of multiple, evolutionarily conserved, functionally distinct domains. Three major classes of DNA binding peptide motifs which are present in regulatory proteins of early

Xenopus development have been defined: (1) the helix-turn-helix motif, also found in prokaryotic DNA repressor molecules, (2) the helix-loop-helix motif which is functional not only in DNA binding but also in protein/protein interactions, and (3) in Zn-Finger motif which was first discovered in *Xenopus* TFIIIA (see Wolffe Chap. 25, this Vol.). Finger proteins are characterized by tandem repeats of sequence elements with a characteristic pattern of conserved cysteine and histidine residues, probably involved in the coordination of Zn. All three DNA binding motifs define multigene families of regulatory proteins expressed in early *Xenopus* development. In preceding paragraphs, *Xenopus* homeobox genes were discussed on several occasions and for a further review of the structure and function of vertebrate homeobox proteins, the reader is referred to Wright and coworkers. We have also mentioned two known examples of the newly discovered helix-loop-helix family of proteins (Myo D and X*twi*) in the section on mesoderm induction.

Except for the well-characterized TFIIIA, virtually nothing is known about Zn Finger protein function in early *Xenopus* development. However, first steps in addressing this question have been undertaken. mRNAs encoding Finger proteins have been isolated from staged *Xenopus* cDNA libraries, and it has become evident from these studies that the Zn Finger motif defines a large multigene family with probably more than 100 members transcribed in early *Xenopus* development. Spatial and temporal expression patterns have been analyzed for many of these RNAs encoding Finger proteins and the picture emerging is in contrast to what has been discovered for the *Xenopus* homeobox genes as a group. Most of the Zn Finger mRNAs are maternally transcribed (in the oocyte) and seem to disappear in blastula/gastrula stages. Only a small subgroup is zygotically transcribed (only after the mid-blastula transition) and, in most cases, no indication for localization has been obtained. In contrast, homeobox protein expression is mostly non-maternal and spatially restricted in the early *Xenopus* embryo. Another distinguishing property of Finger versus homeobox proteins is the potential RNA binding activity of Zn Finger proteins, as demonstrated for TFIIIA. As pointed out earlier, specific RNA binding proteins might be involved in

post-transcriptional regulatory events in the control of gene expression. *Xenopus* Zn Finger proteins display several interesting structural features. In a comparative analysis it has been found that for many of these proteins, groups of tandem finger repeats are organized in distinct higher-order structural units, with a pair of mutually distinct fingers being the most frequently observed second-order repeat unit. A different structural feature defines a further subgroup of *Xenopus* Finger proteins. About 50% of these Finger proteins share a region of extensive sequence homology (FAX domain), N-terminally associated with distinct sets of finger elements. The function of this evolutionarily conserved element is not yet understood. These many different Finger protein encoding cDNAs are now available for a functional analysis in vivo with experimental tools discussed in the previous sections. These studies should contribute to the understanding of molecular events that regulate early *Xenopus* development.

24.7 Outlook

In the field of experimental embryology, *Xenopus* has its place as a vertebrate system which is particularly suited to study first events in embryonic differentiation. Experimental strategies have been established which allow the study of gene function in the absense of the powerful tools provided by classical genetics. The pseudogenetic techniques discussed will hopefully lead to a molecular description of what has been ingeniously discovered by Spemann and Mangold early this century: the organizer principle. Starting out with the fertilized egg, we would like to understand on a molecular level how the regulated expression of specific structural genes, which can be viewed as markers for terminally differentiated tissue, is established; a number of transcription regulator genes as well as signal molecules that can determine differentiation events have been identified. We are now beginning to understand how such signals are transmitted and how they can control the expression and/or biological activity of transcription factors which, in turn, initiate a cascade of regulatory events. A complete molecular description of mesodermal markers in early *Xenopus* embryogenesis seems in reach.

24.8 Summary

Various aspects of gene regulation and gene function in early vertebrate development have been studied using *Xenopus* oocytes, eggs, and early embryos as a biological test system. The oocyte has been succesfully employed to analyze post-transcriptional regulatory mechanisms such as RNA transport and localization. Microinjection of cloned genes into fertilized eggs allows the identification of transcriptional, regulatory elements which respond to cell/cell communication in the process of mesoderm induction. Many genes which encode potential regulatory proteins have been isolated on the basis of specific structural properties, such as encoded DNA binding motifs and/or homology to heterologous regulators of development and transcription, or on the basis of indicative expression patterns, such as inducibility or localized expression. Several different experimental approaches which combine classical experimental biology with the powerful tools of DNA technology, have been established in an attempt to identify their biological function in early *Xenopus* embryogenesis. Functional analysis of homeobox proteins in the specification of the anterior/posterior axis and of a proto-oncogene in egg maturation are discussed.

References

General

Davidson EH (1986) Gene activity in early development. Academic Press, Orlando, Florida

Gilbert SF (1988) Developmental biology. Sinauer Associates, Sunderland, MA

Spemann H (1938) Embryonic development and induction. Yale University Press, New Haven

[1] Gurdon JB, Wickens MP (1983) The use of *Xenopus* oocytes for the expression of cloned genes. Methods Enzymol 101:370–386

Shastry BS, Honda BM, Roeder RG (1984) Altered levels of a 5S gene-specific transcription factor (TFIIIA) during

oogenesis and embryonic development of *Xenopus laevis*. J Biol Chem 259:11373–11382

[2] Gurdon JB, Melton DA (1981) Gene transfer in amphibian eggs and oocytes. Annu Rev Genet 15:189–218

[3] De Robertis EM, Lienhard S, Parisot RF (1982) Intracellular transport of microinjected 5S and small nuclear RNAs. Nature 295:572–577

Dingwall C, Laskey RA (1986) Protein import into the cell nucleus. Annu Rev Cell Biol 2:367–390

Fischer U, Lührmann R (1990) An essential signalling role for the m3G cap in the transport of U1 snRNP to the nucleus. Science 249:786–790

Guddat U, Bakken AH, Pieler T (1990) Protein-mediated nuclear export of RNA: 5S RNA containing small RNPs in *Xenopus* oocytes. Cell 60:619–628

Hamm J, Darzynkiewicz E, Tahara S, Mattaj IW (1990) The trimethylguanosine cap structure of U1 snRNA is a component of a bipartite nuclear targeting signal. Cell 62:569–577

Mattaj IW, De Robertis EM (1985) Nuclear segregation of U2 snRNA requires binding of specific RNP proteins. Cell 40:111–118

Tobian JA, Drinkard L, Zasloff M (1985) tRNA nuclear transport: defining the critical regions of human tRNA$_i^{met}$ by point mutagenesis. Cell 43:415–422

[4] Rebagliati MR, Weeks DL, Harvey RP, Melton DA (1985) Identification and cloning of localized maternal RNAs from *Xenopus* eggs. Cell 42:769–777

Yisraeli JK, Melton DA (1988) The maternal mRNA VG1 is correctly localized following injection into *Xenopus* oocytes. Nature 336:592–595

[5] Gerhart JC (1980) Mechanisms regulating pattern formation in the amphibian egg and early embryo. In: Goldberger RF (ed) Biological regulation and development, vol 2. Molecular organization and cell function. Plenum Press, New York, pp 133–316

[6] Dworkin MB, Shrutkowski A, Dworkin-Rastl E (1985) Mobilization of specific maternal RNA species into polysomes after fertilization in *Xenopus laevis*. Proc Natl Acad Sci USA 82:7636–7640

Fox CA, Sheets MD, Wickens MP (1989) Poly(A) addition during maturation of frog oocytes: distinct nuclear and cytoplasmic activities and regulation by synthetic UUUUUAU. Genes Dev 3:2151–2162

McGrew LL, Dworkin-Rastl E, Dworkin MB, Richter JD (1989) Poly(A) elongation during *Xenopus oocyte* maturation is required for translational recruitment and is mediated by a short sequence element. Genes Dev 3:803–815

[7] Kimelman D, Kirschner M (1987) Synergistic induction of mesoderm by FGF and TGF-β and the identification of a mRNA coding for FGF in the early *Xenopus* embryo. Cell 51:869–877

Nieuwkoop PD, Johnen AG, Albers B (1985) The epigenetic nature of early chordate development: inductive interaction and competence. Cambridge University Press, Cambridge

Rosa F, Roberts AB, Danielpur D, Dart LL, Sporn MB, Dawid JB (1988) Mesoderm induction in amphibians: the role of TGF-β2-like factors. Science 239:783–785

Slack JMW, Darlington BG, Heath JK, Godsave SF (1987) Mesoderm induction in early *Xenopus* embryos by heparin-binding growth factors. Nature 326:197–200

Smith JC (1989) Mesoderm induction and mesodern inducing factors in early amphibian development. Development 105:665–677

Smith JC, Price BMJ, Van Nimmen K, Huylebroeck D (1990) Identification of a potent *Xenopus* mesoderm inducing factor as a homologue of activin A. Nature 345:729–731

Sokol S, Wong GG, Melton D (1990) A mouse macrophage factor induces head structures and organizes a body axis in *Xenopus*. Science 249:561–564

Weeks DL, Melton DA (1987) A maternal mRNA localized to the vegetal hemisphere in *Xenopus* eggs codes for a growth factor related to TGF-β. Cell 51:861–867

[8] Davis RL, Weintraub H, Lassar AB (1987) Expression of a single transfected cDNA converts fibroblasts to myoblasts. Cell 51:987–1000

Hopwood ND, Pluck A, Gurdon JB (1989a) Myo D expression in the forming somites is an early response to mesoderm induction in *Xenopus* embryos. EMBO J 8:3409–3417

Hopwood NP, Pluck A, Gurdon JB (1989b) A *Xenopus* mRNA related to *Drosophila twist* is expressed in response to induction in the mesoderm and the neural crest. Cell 59:893–903

Rosa FM (1989) Mix. 1, a homeobox mRNA inducible by mesodern inducers, is expressed mostly in the presumptive endodermal cells of *Xenopus* embryos. Cell 57:965–974

Sargent TD, Dawid JB (1983) Differential gene expression in the gastrula of *Xenopus laevis*. Science 222:135–139

[9] Gurdon JB, Mohun TJ, Sharpe CR, Taylor MV (1989) Embryonic induction and muscle gene activation. TIG 5:51–56

Mohun TJ, Brennan S, Dathan N, Fairman S, Gurdon JB (1984) Cell type specific activation of actin genes in the early amphibian embryo. Nature 311:716–721

Taylor M, Treisman R, Garrett N, Mohun T (1989) Muscle-specific (CArG) and serum-responsive (SRE) promoter elements are functionally interchangeable in *Xenopus* embryos and mouse fibroblasts. Development 106:67–78

Wilson C, Cross GS, Woodland HR (1986) Tissue-specific expression of actin genes injected into *Xenopus* embryos. Cell 47:589–599

[10] Holtfreter J (1983) Die totale Exogastrulation, eine Selbstablösung des Ektoderms vom Entomesoderm. Roux' Arch J Entwicklungsmech 129:669–793

Mangold O (1933) Über die Induktionsfähigkeit der verschiedenen Bezirke der Neurula von Urodelen. Naturwissenschaften 21:761–766

Spemann H, Mangold H (1924) Über Induktion von Embryonalanlagen durch Implantation artfremder Organisatoren. Roux' Arch J Entwicklungsmech 100:599–638

[11] Durston AJ, Timmermans JPM, Hage WJ, Hendricks HFJ, de Vries NJ, Heideveld M, Hienarkoop PD (1989) Retinoic acid causes and anteroposterior transformation in the developing central nervous system. Nature 340:140–144

Kintner CR, Melton DA (1987) Expression of *Xenopus* N-CAM RNA in an early response of ectoderm to induction. Development 99:311–325

Otte AP, van Run P, Heideveld M, van Driel R, Durston AJ (1989) Neural induction is mediated by cross-talk between the protein kinase C and cyclic AMP pathways. Cell 58:641–648

Sharpe CR, Fritz A, De Robertis EM, Gurdon JB (1987) A homeobox-containing marker of posterior neural differentiation shows the importance of predetermination in neural induction. Cell 50:749–758

[12] Brown DD, Schlissel MS (1985) A positive transcription factor controls the differential expression of two 5S RNA genes. Cell 42:759–767

Busby SJ, Reeder RH (1983) Spacer sequences regulate transcription of ribosomal gene plasmids injected into *Xenopus* embryos. Cell 34:989–996

Jonas EA, Snape AM, Sargent TD (1989) Transcriptional regulation of a *Xenopus* embryonic epidermal keratin gene. Development 106:399–405

Krieg PA, Melton DA (1985) Developmental regulation of a gastrula specific gene injected into fertilized *Xenopus* eggs. EMBO J 4:3463–3471

Krieg PA, Melton DA (1987) An enhancer responsible for activating transcription at the midblastula transition in *Xenopus* development. Proc Natl Acad Sci USA 84:2331–2335

[13] Bass LB, Weintraub H (1988) An unwinding activity that covalently modifies its double-stranded RNA substrate. Cell 55:1089–1098

Cotten M, Birnstiel ML (1989) Ribozyme-mediated destruction of RNA in vivo. EMBO J 8:3861–3866

Giebelhaus DH, Eib DW, Moon RT (1988) Antisense RNA inhibits expression of membrane skeleton protein 4.1 during embryonic development of *Xenopus*. Cell 53:601–615

Harland R, Weintraub H (1985) Translation of mRNA injected into *Xenopus* oocytes is specifically inhibited by antisense RNA. J Cell Biol 101:1094–1099

Haseloff J, Gerlach WL (1988) Simple RNA enzymes with new and highly specific endoribonuclease activities. Nature 334:585–591

Kloc M, Miller M, Carrasco AE, Eastman E, Etkin L (1989) The maternal store of Xlgv7 mRNA in full-grown oocytes is not required for normal development in *Xenopus*. Development 107:899–907

Melton DA (1985) Injected anti-sense RNAs specifically block messenger RNA translation in vivo. Proc Natl Acad Sci USA 82:144–148

Sagata N, Oskarrsson M, Copeland T, Brumbaugh J, Van de Woude GF (1988) Function of c-*mos* proto-oncogene product in meiotic maturation in *Xenopus* oocytes. Nature 335:519–525

Shuttleworth J, Colman A (1988) antisense oligonucleotide-directed cleavage of mRNA in *Xenopus* oocytes and eggs. EMBO J 7:427–434

Wormington MW (1986) Stable repression of ribosomal protein L1 synthesis in *Xenopus* oocytes by microinjection of antisense RNA. Proc Natl Acad Sci USA 83:8639–8643

Wright CVE, Cho KWY, Hardwicke J, Collins RH, De

Robertis EM (1989) Interference with function of a homeobox gene in *Xenopus* embryos produces malformations of the anterior spinal cord. Cell 59:81–93

[14] Andrews MT, Brown DD (1987) Transient activation of oocyte 5S RNA genes in *Xenopus* embryos by raising the level of the trans-acting factor TFIIIA. Cell 51:445–453

Harvey RP, Melton DA (1988) Microinjection of synthetic Xhox-1A homeobox mRNA disrupts somite formation in developing *Xenopus* embryos. Cell 53:687–697

McMahon AP, Moon RT (1989) Ectopic expression of the proto-oncogene int-1 in *Xenopus* embryos leads to duplication of the embryonic axis. Cell 58:1075–1084

Ruiz i Altaba A, Melton DA (1989a) Involvement of the *Xenopus* homeobox gene Xhox3 in pattern formation along the anterior-posterior axis. Cell 57:317–326

Ruiz i Altaba A, Melton DA (1989b) Interaction between peptide growth factors and homeobox genes in the establishment of antero-posterior polarity in frog embryos. Nature 341:33–38

[15] Köster M, Pieler T, Pöting A, Knöchel W (1988) The finger motif defines a multigene family represented in the maternal mRNA of *Xenopus laevis* oocytes. EMBO J 7:1735–1741

Knöchel W, Pöting A, Köster M, El-Baradi T, Nietfeld W, Bouwmeester T, Pieler T (1989) Evolutionary conserved modules associated with Zn fingers in *Xenopus laevis*. Proc Natl Acad Sci USA 86:6097–6100

Miller J, McLachlan AD, Klug A (1985) Repetitive Zinc-binding domains in the protein transcription factor IIIA from *Xenopus* oocytes. EMBO J 4:1609–1614

Nietfeld W, El-Baradi T, Mentzel H, Pieler T, Köster M, Pöting A, Knöchel W (1989) Second order repeats in *Xenopus laevis* finger proteins. J Mol Biol 208:639–659

Ruiz i Altaba A, Perry-O'Keefe H, Melton DA (1987) Xfin: an embryonic gene encoding a multifingered protein in *Xenopus*. EMBO J 6:3065–3070

Wright CVE, Cho KWY, Oliver G, De Robertis EM (1989) Vertebrate homeodomain proteins: families of region-specific transcription facotrs. TIBS 14:52–56

Chapter 25 The Developmental Regulation of the Genes Coding for 5S Ribosomal RNA in *Xenopus laevis*

ALAN P. WOLFFE

25.1 Introduction[1]

Xenopus laevis, the South African clawed toad, is a popular animal for research into oogenesis and embryogenesis. The large oocytes and eggs of *Xenopus* can be microinjected easily and many early experiments focussed on introducing first nuclei, and later DNA, RNA, and protein into these living 'test tubes'. The results of these experiments, many of them initiated in the laboratories of J. B. Gurdon and D. D. Brown, have vastly increased our understanding of replication, transcription, and translation of eukaryotes. These studies have led to the discovery of DNA amplification, the isolation and sequencing of the first eukaryotic gene and transcription factor, in vitro transcription, chromatin assembly, and replication. The rapid development of *Xenopus* embryos, following fertilization in vitro, also provides a simple vertebrate system in which the molecular mechanisms responsible for causing the differentiation of the egg into an animal can be determined.

25.2 Ribosome Biogenesis[2]

Stored within the egg of the frog *Xenopus laevis* are most of the cellular components required to create an embryo. This rich storehouse contains over 200000 times the normal cellular complement of ribosomes. The commitment of the *Xenopus* oocyte and embryo to ribosome biogenesis made this a system of choice for early research into the regulation of eukaryotic gene transcription.

The molecular mechanisms facilitating high rates of ribosome synthesis in *Xenopus* oocytes differ for each component. The genes encoding 18S and 28S ribosomal RNA are amplified extrachromosomally in oocytes. In contrast, many copies of the genes encoding 5S ribosomal RNA are present within the genome. Most copies of these genes belong to a large oocyte type family that is transcribed in oocytes, but not in somatic cells. A smaller somatic-type family is transcribed in both oocytes and somatic cells. This is an example of a very simple type of developmental control in which two similar gene families encoding the same product are both initially transcriptionally active, but during development, transcription of one family is switched off. The molecular mechanisms responsible for this selective gene inactivation are the focus of this chapter. There is no known developmental significance to having two types of 5S RNA in ribosomes. Oocyte and somatic ribosomes are believed to be equally active in translation.

25.3 DNA

25.3.1 The 5S RNA Gene Families[3]

The abundance of the 5S RNA genes facilitated the purification, cloning, and sequencing of both the oocyte and somatic gene families. The large oocyte gene family, consisting of 20000 copies per haploid genome, is organized into clusters of 1000 copies at the telomeres of most *Xenopus* chromosomes. Each copy varies in length between 650 and 850 bp, but invariably contains a gene and a pseudogene separated by an A-T rich spacer. Most of the somatic gene family, which consists of 400 copies per haploid genome, are found in a single cluster on one chromosome. Each copy of the somatic type is uniquely 880 bp in length and consists of a single gene and a G-C rich spacer. Oocyte and somatic

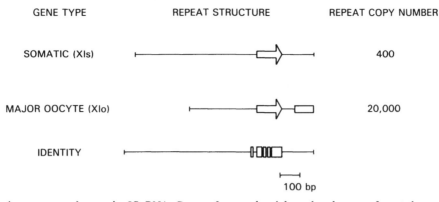

Fig. 1. The major oocyte and somatic 5S RNA Genes of *Xenopus laevis*. Each class is organized in clusters of simple tandem repeats. One repeating unit of each is drawn to scale. Each repeating unit of the somatic family (*Xls*) is 880 bp in length. There are approximately 400 copies of Xls, most of which are in a single cluster on one chromosome. The intergenic spacers of Xls are very GC-rich. The major oocyte gene family (*Xlo*) is lo-cated mainly at the telomeres of most chromosomes in clusters of over 1000 repeats. Each repeating unit of 650 to 850 bp consists of a gene and a pseudogene separated by an AT-rich spacer. Regions of sequence identity are indicated beneath the genes (*open arrows*) by the boxed region. A pseudogene sequence next to the oocyte 5S RNA gene is also indicated by an open box. Only the gene region and a short region 5' to the gene are homologous

Xenopus laevis 5S RNA genes are identical in length (120 bp) and differ at only 5 bp in their sequences. The spacer regions share little or no homology. The structure of oocyte and somatic 5S RNA gene repeats (5S DNA) is summarized in Fig. 1. As will be discussed below, all of the sequence differences between the genes that are important for developmental regulation appear to lie within the genes. A role for the spacer regions or chromosomal position in regulating differential gene activity has yet to be found. Having examined the large-scale organization of 5S DNA, further analysis required the development of the means to manipulate the sequence of 5S DNA and simple functional assays for gene activity.

25.3.2 The 5S RNA Gene[4]

Each repeat within the 5S DNA was shown to contain all of the sequences required to synthesize 5S RNA by in vivo procedures in which purified 5S DNA repeats or cloned single copies of each gene family were introduced by microinjection into *Xenopus* oocyte nuclei. These results were confirmed in vitro using extracts of isolated oocyte nuclei. Deletion and point mutational analysis of cloned single repeats defined the essential promoter elements (Fig. 2). The 5S RNA gene promoter (internal control region) actually lies within the gene itself. This internal control region extends over 50 bp and in isolation is sufficient to direct the assembly of a transcription complex and initiation of transcription by RNA polymerase III (see below). Four of the five differences between oocyte and somatic genes lie within this sequence. A region 5' to the gene is required for the accurate initiation of transcription by RNA polymerase III. A short run of thymines (>3) flanked by G-C base pairs is sufficient to direct RNA polymerase III itself to terminate transcription. There are therefore three functional regions of a 5S RNA gene as defined by these simple in vitro transcription assays: the internal control region, a short region 5' to the gene, and the termination signal (Fig. 2). The proteins interacting with these regions have been fractionated and the significance of their binding for gene transcription defined.

25.4 Transcription: the Active State

25.4.1 The Transcription Complex[5]

Eukaryotic RNA polymerases do not initiate specific transcription on naked DNA. In order to have

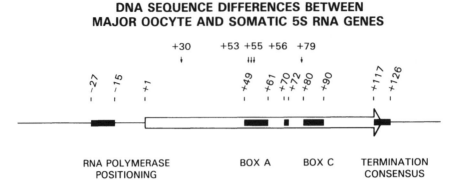

**DNA SEQUENCE DIFFERENCES BETWEEN
MAJOR OOCYTE AND SOMATIC 5S RNA GENES**

Fig. 2. The 5S RNA gene. A typical 5S RNA gene (*open arrow*) is shown including sequence features of defined significance (*boxed areas*). A region just 5' to the gene (−27 to −15) is required for accurate initiation of RNA polymerase. The regions inside the gene constitute the binding sites of TF III A and TF III C. A consensus termination signal at the end of the 5S RNA gene (+120) is also shown. *Vertical arrows* indicate the five sequence differences (single base pair) between *Xenopus laevis* major oocyte and somatic 5S RNA genes

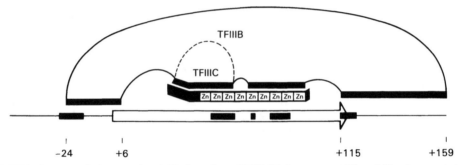

Fig. 3. The 5S RNA gene transcription complex. Aside from the involvement of at least three separate activities in the assembly of a transcription complex on a *Xenopus* somatic 5S RNA gene little is known about the composition of the complex. TF III A appears to be an integral component (see Fig. 4) although its interaction with the ICR is stabilized during complex formation. TF III C is known to rapidly stabilize the association of TF III A with a somatic 5S RNA gene via the A box region (Fig. 2) although no change in protein-DNA interaction can be detected. In the final assembled complex, protein-DNA interactions involving TF III A and TF III B extend over 180 bp of DNA as indicated

accurate initiation of transcription, proteins have to be associated with a gene. Purified transcription factor III A (TF III A) (see next section) and RNA polymerase III are necessary, but not sufficient, for transcription of 5S DNA. Two crude fractions eluted from phosphocellulose columns (TF III B and TF III C) are also required for transcription. TF III A, B and C associate with the 5S RNA gene in the absence of RNA polymerase. This complex of proteins is described as a transcription complex, since only when it is assembled can RNA polymerase initiate and synthesize 5S RNA. TF III A must bind to the internal control region before TF III C and TF III B can associate with the gene. The incor-

poration of TF III C and TF III B into the transcription complex greatly extends protein-DNA interactions on the 5S RNA gene (Fig. 3) and alters those of TF III A with the internal control region. These poorly understood protein-protein and protein-DNA interactions have important consequences for gene regulation.

25.4.2 Transcription Factor (TF) III A [6]

Growing *Xenopus* oocytes store 5S RNA in a ribonucleoprotein particle which contains a single molecule of 5S RNA and a single molecule of a protein. This protein (TF III A) also binds to the internal control region of a 5S RNA gene and is required for transcription of the gene. This suggested a potential autoregulatory role of 5S RNA in transcription, a possibility which will be discussed later. The gene encoding TF III A was cloned and sequenced, revealing a repetitive structure within the protein. This repeating unit of the protein has been called a zinc finger since each of the repeats might associate with a zinc atom.

The structure of TF III A is believed to consist of nine zinc fingers and a carboxyl terminal domain which does not bind to DNA. A single molecule of TF III A binds to each 5S RNA gene. The zinc fingers are arrayed in a linear fashion along the internal control region, such that the carboxyl terminus of the protein projects towards the 5' end of the 5S RNA gene. Our current understanding of this protein-DNA interaction is summarized in Fig. 4. The precise structure of these zinc fingers and how TF III A actually contacts the 5S RNA gene is unknown. The association of TF III A with the gene is necessary to initiate the binding of the other transcription factors required to assemble a transcription complex. Transcription only occurs when RNA polymerase III recognizes a complete transcription complex.

25.4.3 Transcription Complex Stability [7]

The stable association of multiple transcription factors with eukaryotic genes has been described in vitro and in vivo. The significance of such stable interactions is that in many cells, stable patterns of gene activity are maintained for long periods of time, and in the case of a terminally differentiated cell, until cell death. However, for genes requiring

transcriptional activity to be modulated the transient association and dissociation of transcription factors is advantageous. Transcription complexes assembled on somatic and oocyte 5S RNA genes typify these two modes of interaction between transcription factors and promoter elements.

First, the stable association of transcription factors with a somatic 5S RNA gene will be considered. The binding of TF III A alone to a 5S RNA gene is relatively weak ($K_D = 10^{-9}$). TF III C interacts with the specific TF III A-5S DNA complex via protein-protein contacts with TF III A and protein-DNA contacts with the internal control region. The transcription factors TF III A and C together bind so tightly to a somatic 5S RNA gene that they do not dissociate even if free factor molecules are removed from solution or if excess competitor 5S RNA is used to challenge the complex. The stabilization mechanism is unknown, but does not appear to involve a closed to open complex transition analogous to RNA polymerase transcription initiation in prokaryotes. Possible mechanisms include a cooperative binding of TF III A and TF III C, a conformational change in TF III A structure perhaps catalyzed by TF III C or a covalent modification of TF III A, for example a dephosphorylation or phosphorylation mediated by TF III C. The stability of the entire transcription complex, including TF III B, reflects that of the TF III A/C/5S DNA complex.

In contrast to the stabilization following the association of TF III C with TF III A bound to a somatic 5S RNA gene, little or no change in the binding of TF III A to an oocyte 5S RNA gene follows the binding of TF III C. TF III A alone binds with equal affinity to the somatic and oocyte type 5S RNA genes. Therefore the changes in the internal control region between oocyte and somatic 5S

Fig. 4. Alignment of TF III A along the internal control region of a 5S RNA gene. TF III A, a protein of 38 500 Da, binds to the internal control region (ICR) of a 5S RNA gene. The structure of TF III A is believed to consist of nine zinc fingers (boxes) and linkers between the fingers. Each finger is believed to contain a single zinc atom. Regions of the protein not involved in binding DNA are tilted away from the gene

RNA genes have a much greater effect on the sequestration (or action) of TFIIIC than on TFIIIA binding (Fig. 5). The stability of the whole transcription complex on the oocyte 5S RNA gene again reflects the stability of the TFIIIA/C/5S DNA complex.

A major consequence of the differences in transcription complex stability on somatic and oocyte 5S RNA genes is that conditions that limit transcription factor (TFIIIA and TFIIIC) activity, concentration, or availability will selectively restrict transcription from the unstable oocyte 5S RNA gene transcription complex. One obvious candidate for sequestering TFIIIA is 5S RNA, hence the possible autoregulatory role of the TFIIIA/5S RNA complex. However, the selective removal of transcription factors from oocyte 5S RNA genes by purified 5S RNA has yet to be demonstrated. Differential transcription complex stability provides the most probable basis for the developmental control of differential 5S RNA gene transcription.

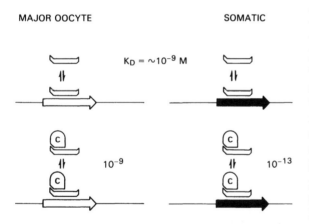

Fig. 5. TFIIIC discriminates between major oocyte and somatic 5S RNA genes. TFIIIA binds with equal affinity to major oocyte (*open arrow*) and somatic (*closed arrow*) 5S RNA genes. This interaction is relatively weak ($K_D = 10^{-9}$ M). In the presence of TFIIIC, the interaction of TFIIIA with a somatic 5S RNA gene is greatly stabilized. The affinity of the complex of TFIIIA and TFIIIC for the 5S RNA gene is over 4 orders of magnitude greater than that of TFIIIA alone. A comparable stabilization does not occur on the major oocyte 5S RNA gene. Although the molecular basis of stabilization is not yet defined, this phenomenon appears to be the basis of differential transcription complex stability

25.5 Regulation – Establishing Differential Transcriptional Activity

25.5.1 The in Vivo Regulation of Differential 5S RNA Gene Transcription[8]

In vitro studies of differential 5S RNA gene transcription only demonstrate possible molecular mechanisms for the developmental regulation of 5S RNA gene expression. In order to test the validity of these mechanisms, experimental strategies have to be developed to test them in vivo. The improvement in methodologies for accurately quantitating levels of specific RNA synthesis led to the demonstration that although the inactivation of oocyte type 5S RNA genes begins at the end of oogenesis, it is not complete until gastrulation.

Xenopus embryogenesis is characterized by 11 rapid cycles of cell division following fertilization. This cleavage of the egg occurs in the absence of transcription. RNA synthesis ceases following the breakdown of the oocyte nucleus during maturation of the oocyte into an egg. The developmental stage at which gene activity begins again is called the mid-blastula transition (MBT); the embryo at this time has 4000 cells. At the MBT, equivalent low levels of somatic and oocyte type 5S RNA are synthesized. Because of the large excess of oocyte over somatic type 5S RNA genes, this represents a 50 times higher rate of transcription for somatic than for oocyte 5S RNA genes. Two or three cell divisions later, near the end of gastrulation (10 – 20000 cells), the final state of differential gene expression is established (summarized in Fig. 6). The vast majority of oocyte 5S RNA genes are inactivated, such that the ratio of somatic to oocyte type 5S RNA gene transcription is over 1000:1.

Various attempts have been made to examine both the transcription factor and primary sequence requirements necessary to reproduce this developmental regulation using cloned 5S DNA injected into *Xenopus* eggs, and by perturbing transcription factor levels in the embryo. The injection of cloned repeats of oocyte and somatic type 5S RNA genes demonstrated that sequences within the internal control region were responsible for at least a 50-fold discrimination between oocyte and somatic

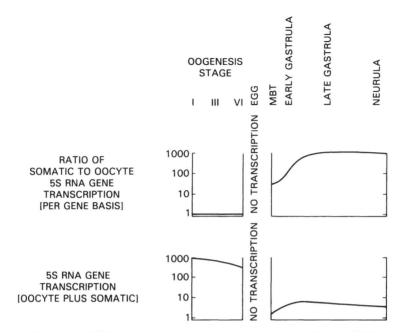

Fig. 6. Developmental regulation of 5S RNA gene transcription. Both the ratio of somatic to oocyte 5S RNA gene transcription on a transcription rate per gene basis (*upper panel*) and the total amount of 5S RNA gene transcripts on a per nucleus basis (*lower panel*) are shown. Although both oocyte and somatic 5S RNA genes are maximally transcribed during the earlier stages of oogenesis, the total amount of 5S RNA transcription falls towards the end of oogenesis. Following oocyte maturation into an egg (EGG), there is no transcription of 5S DNA during fertiliza-tion and cleavage until the mid blastula transition (MBT) at which stage the embryo has 4000 cells. At this time 5S RNA gene transcription gradually increases, while the proportion of transcription from the oocyte 5S RNA genes gradually decreases. Towards the end of gastrulation the final ratio of 1000 transcripts of a somatic 5S RNA gene to every transcript from an oocyte 5S RNA gene is established and maintained. Overall 5S RNA gene transcription falls during later embryogenesis and in some tissues somatic 5S RNA genes may become partially repressed

5S RNA genes in *Xenopus* eggs. Elevating the level of TFIIIA in the *Xenopus* egg, either by injecting the purified protein or synthetic messenger RNA, enhanced oocyte type 5S RNA gene transcription relative to that of somatic 5S RNA genes. In spite of the continued presence of elevated TFIIIA, this increase in oocyte 5S RNA transcription was only transient. Towards gastrulation oocyte 5S RNA genes were repressed. This result implies that either a transcription factor other than TFIIIA becomes limiting or that a dominant repressive effect begins to operate on oocyte-type 5S DNA. The eventual repression of oocyte-type 5S RNA genes is attribut-ed to the assembly of chromatin onto 5S DNA.

25.5.2 The Assembly of Chromatin and 5S RNA Gene Transcription[9]

The assembly of the chromatin fiber in response to histone H1 addition appears responsible for the repression of class III genes. This evidence for a role for chromatin structure in the developmental regulation of 5S RNA gene expression has led to numerous attempts to reconstruct a chromatin environment on cloned 5S DNA. These attempts have so far failed to demonstrate any influence on differential 5S RNA gene transcription that can be construed as having developmental significance. It may be that small plasmid DNA, even when assembled into chromatin, is not large enough to fold into a higher-order structure such as the chromatin fiber. This possibility could be examined by using larger DNA templates for chromatin reconstitution, such as those cloned in bacteriophages M13 and λ.

25.5.3 A Model for Differential 5S RNA Gene Expression During Early Development[10]

Transcription factors interact differentially with oocyte and somatic 5S RNA genes. Although TFIIIA binds with equivalent affinity to oocyte and somatic 5S RNA genes, discrimination between the two gene types occurs during the binding of TFIIIC. A consequence of this discrimination is that stable transcription complexes are assembled on somatic type 5S RNA genes, whereas unstable complexes are assembled on oocyte-type 5S RNA genes. When transcription factors are in large excess, as in the growing oocyte, both oocyte and somatic 5S RNA genes will be actively transcribed. Following fertilization of the *Xenopus* egg, the excess of transcription factors relative to 5S DNA will fall as cell division increases the number of nuclei (and therefore 5S RNA genes). Limiting transcription factors will associate with the somatic 5S RNA genes on which they can be stably seques-

tered. The oocyte 5S RNA genes will lose their associated transcription factors and chromatin will assemble on the DNA. Chromatin structural changes will occur during *Xenopus* embryogenesis as the embryonic nuclei change from an adaptation for a state of rapid DNA synthesis to the normal somatic nuclear structure in a cell possessing a normal cell cycle. The compaction of chromatin concomitant with the lengthening of the cell cycle and the association of histone H1 is proposed to directly influence differential 5S RNA gene transcription. This might occur by providing a competing protein-DNA interaction that drives the dissociation of the weakly associated oocyte 5S RNA gene transcription complexes (Fig. 7).

25.6 Regulation – Maintaining the Differentiated State

25.6.1 Maintenance of Differential Gene Expression[11]

Once a state of differential 5S RNA gene activity is established in a somatic cells, this must be main-

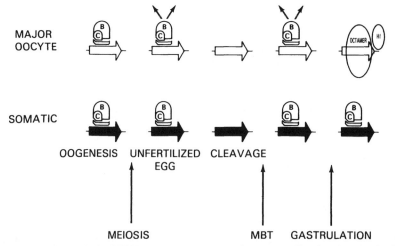

Fig. 7. Model for developmental control of 5S RNA gene transcription. This diagram summarizes the occupancy of major oocyte (*open arrow*) and somatic (*closed arrow*) 5S RNA genes with transcription complexes or chromatin during oogenesis and embryogenesis. A complex is represented on each gene. The *absence of arrows* implies that the complex is stable; the presence of arrows that the complex is unstable. The end result is that by late gastrulation, stable transcription complexes are assembled on somatic 5S RNA genes and on oocyte 5S RNA genes adopt a repressed chromatin structure

tained through the processes of DNA replication and cell division. We have already discussed how a stable transcription complex assembled on a somatic 5S RNA gene might maintain that gene in an active state indefinitely in a terminally differentiated nondividing cell. The more interesting question is what happens when DNA has to be replicated and the pattern of gene activity or inactivity propagated to daughter cells.

REPLICATION

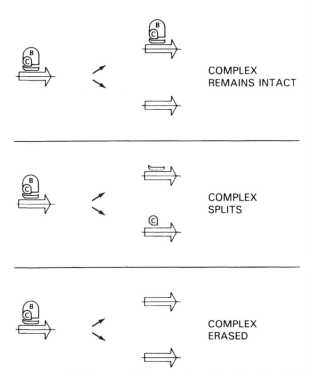

COMPLEX REMAINS INTACT

COMPLEX SPLITS

COMPLEX ERASED

Fig. 8. A role for transcription complexes in maintaining 5S RNA gene activity. A 5S RNA gene transcription complex is represented before (*left*) and after (*right*) replication of the 5S RNA gene. Three possible results are shown: the complex might segregate intact to one daughter duplex. This would maintain gene activity in 50% of the cells. The complex may be split, which would inactivate both daughter genes if transcription factors were not available. If they were available, then cooperative interaction between transcription factors might establish active transcription complexes on both daughter duplexes. Finally the complex might be completely disrupted and transcription factors displaced from both daughter DNA duplexes by replication. The latter appears to be the case under certain in vitro conditions

DNA replication will perturb a transcription complex, simply because the duplex DNA that is part of the complex is duplicated. The transcription complex must also be duplicated if the gene is to remain active. Three possibilities exist (Fig. 8): the complex could remain intact on one DNA duplex following replication and a complex assemble de novo on the other daughter duplex. Alternatively, the complex could be disrupted such that transcription factors segregate to both DNA duplexes. New transcription complexes would then be established on both daughter genes perhaps making use of cooperative protein-protein interactions. Finally, the transcription complex could be erased entirely by replication. Transcription complexes would then have to form de novo after every replication fork passage. Experimental evidence supports the latter possibility. If we assume that similar events operate on repressed chromatin, then after every round of DNA synthesis active and repressed states must be reestablished on somatic and oocyte-type 5S RNA genes. It is possible that limitation of transcription factors and the maturation of chromatin following replication fork progression will lead to repression of the oocyte type 5S RNA genes much as appears to occur during development. However, other processes may serve to facilitate this repression such as the differential replication timing of oocyte and somatic 5S DNA.

Differential replication timing is known to occur in many eukaryotic systems. Perhaps the clearest example concerns the two X chromosomes in female mammals. The active X-chromosome replicates early in S phase whereas the inactive X-chromosome replicates late in S phase. This type of phenomenon led to the proposal that if somatic type 5S RNA genes are replicated prior to the replication of oocyte type 5S DNA, then limiting transcription factors might be sequestered onto the somatic 5S DNA. One part of this model has experimental support: somatic 5S RNA genes are replicated earlier than oocyte 5S RNA genes during S phase in *Xenopus* somatic cell lines. The basis of this differential replication timing is unknown. It could be related to gene activity, sequence, or chromosomal position. Although this phenomenon may be important in adult somatic cells, it should be emphasized that oocyte 5S RNA genes are repressed in the embryo with transcription factors in considerable excess. We presume, therefore, that transcription factors would be available even for late replicating genes.

25.6.2 Developmental Regulation of TFIIIA Concentration [12]

Early models for differential 5S RNA gene transcription focused on TFIIIA concentration and a possible differential binding of TFIIIA to oocyte and somatic 5S RNA genes. A reduction in TFIIIA concentration was proposed to exaggerate differential 5S RNA gene transcription. The fact that oocyte and somatic 5S RNA genes bind TFIIIA equivalently, and can be differentially transcribed when TFIIIA is in excess, reduces the significance of the regulation of TFIIIA abundance for 5S RNA gene regulation. However, high TFIIIA levels are a prerequisite for high rates of oocyte 5S RNA gene transcription and levels of TFIIIA are much reduced in somatic cells. This will facilitate oocyte 5S RNA gene inactivation and repression. The level of TFIIIA protein per cell falls by 6 orders of magnitude between the immature oocyte and an adult cell. This fall is a reflection of a comparable drop if TFIIIA mRNA levels. This suggests that TFIIIA concentration is regulated primarily at the transcriptional level. TFIIIA apparently play no role in the transcriptional activation or repression of its own gene. The promoter of the TFIIIA gene contains a number of positive and negative regulatory elements whose developmental significance remains to be defined.

25.6.3 Multiple TFIIIA Protein Variants [13]

Although there is only a single gene encoding TFIIIA, several proteins resolved on denaturing polyacrylamide gels are bound by polyclonal antibodies against TFIIIA. Different proteins are detected at different developmental stages by different antibodies. Although post-translational modifications of TFIIIA have not been detected, it remains possible that a modified TFIIIA or a protein se-

quence variant may also play a role in differential 5S RNA gene expression.

25.7 Outlook [14]

Over the past two decades the major 'players' in the regulation of *Xenopus* 5S RNA genes have been defined through a combination of in vivo and in vitro analysis. Although we know that transcription fractions, histones, specific regions of DNA sequence, and RNA polymerase III are involved in regulation, we still know little about how these components fit together. We anticipate that the further cloning and mutagenesis of the transcription factors will at least elucidate the mechanistic basis of transcription and transcription complex stability. Progress in the fractionation of the chromatin assembly apparatus should allow the construction of a biochemically defined chromatin such that the function of transcription factors and RNA polymerase can be examined in a chromosomal context. The further dissection of differential 5S RNA gene expression is likely to devolve into an analysis of the biophysical nature of protein-protein and protein-DNA interactions.

There is, however, a larger question, that is likely to be of continuing interest to biologists: why does the elevated content of TFIIIA protein occur only in oocytes? In other words, we know that high levels of TFIIIA contribute to transcription of the oocyte 5S RNA genes in oocytes, but what is responsible for the oocyte-specific transcription of the TFIIIA gene? The problem of the transition from the oocyte to the somatic program of gene expression has been moved back a level and yet is still not solved. Recently transcription factors that maybe involved in the regulation of oocyte-selective gene expression have been cloned.

25.8 Summary

In *Xenopus laevis* two genes, each present in multiple copies (oocyte and somatic type) encode 5S ribosomal RNA. Both require the association of similar transcription factors with similar promoter elements before they can be transcribed by RNA polymerase III. The complexes of transcription factors differ between the two genes. One complex is stable and binds transcription factors with high affinity, while the other complex is unstable. Transcription factors readily dissociate from unstable transcription complexes. A reduction in transcription factor concentration during *Xenopus* embryogenesis contributes to the selective inactivation of oocyte 5S RNA genes as a consequence of the dissociation of transcription factors from the unstable transcription complexes assembled on the oocyte 5S DNA. Alterations in chromatin structure, such as the compaction of chromatin associated with the binding of histone H1 may contribute to the dissociation of weakly bound transcription factors from the oocyte 5S RNA genes. A chromatin structure dependent on the presence of histone H1 is responsible for maintaining inactive oocyte 5S RNA genes in the repressed state. The differential expression of oocyte and somatic 5S RNA genes in somatic cells may well be maintained by different mechanisms from those responsible for establishing differential transcription during development.

References

[1] Dawid IB, Sargent TD (1988) *Xenopus laevis* in developmental and molecular biology. Science 240:1443–1448

Gurdon JB (1974) The control of gene expression in animal development. Oxford University Press, Oxford

Slack JMW (1983) From egg to embryo. Cambridge University Press, Cambridge

[2] Brown DD, Littna E (1964) DNA synthesis during the development of *Xenopus laevis*, the South African clawed toad. J Mol Biol 20:95–112

Brown DD, Littna E (1966) Synthesis and accumulation of low molecular weight RNA during embryogenesis of *Xenopus laevis*. J Mol Biol 20:95–112

Gall JG (1968) Differential synthesis of the genes for ribosomal RNA during amphibian oogenesis. Proc Natl Acad Sci USA 60:553–560

Wegnez M, Monier R, Denis H (1972) Sequence heterogeneity of 5S RNA in *Xenopus laevis*. FEBS Lett 25:13–20

[3] Brown DD, Wensink PC, Jordan E (1971) Purification and some characteristics of 5S RNA from *Xenopus laevis*. Proc Natl Acad Sci USA 68:3175–3179

Brown DD, Carroll D, Brown RD (1977) The isolation and

characterization of a second oocyte 5S DNA from *Xenopus laevis*. Cell 12:1045–1056

Callan HG, Gall JG, Berg CA (1987) The lampbrush chromosomes of *Xenopus laevis*: preparation, identification and distribution of 5S DNA sequences. Chromosoma (Berl) 95:235–250

Fedoroff NV, Brown DD (1978) The nucleotide sequence of oocyte 5S DNA in *Xenopus laevis* I. Tha AT-rich spacer. Cell 13:701–716

Miller JR, Cartwright EM, Brownlee GG, Fedoroff NV, Brown DD (1978) The nucleotide sequence of oocyte 5S DNA in *Xenopus laevis*. II. The GC-rich region. Cell 13:717–725

Peterson RC, Doering JL, Brown DD (1980) Characterization of two *Xenopus* somatic 5S DNA and one minor oocyte-specific DNA. Cell 20:131–141

[4] Birkenmeier EH, Brown DD, Jordan E (1978) A nuclear extract of *Xenopus laevis* oocytes that accurately transcribes 5S RNA genes. Cell 15:1077–1086

Bogenhagen DF, Sakonju S, Brown DD (1980) A control region in the center of the 5S RNA gene directs specific initiation of transcription: II. The 3′ border of the region. Cell 19:27–35

Bogenhagen DF, Brown DD (1981) Nucleotide sequences in *Xenopus* 5S DNA required for transcription termination. Cell 24:261–270

Brown DD, Gurdon JB (1977) High-fidelity transcription of 5S DNA injected into *Xenopus* oocytes. Proc Natl Acad Sci USA 74:2064–2068

Cozzarelli NR, Gerrard SP, Schlissel M, Brown DD, Bogenhagen DF (1983) Purified RNA polymerase III accurately and efficiently terminates transcription of tS RNA genes. Cell 34:829–835

Gurdon JB, Brown DD (1978) The transcription of 5S DNA injected into *Xenopus* oocytes. Dev Biol 67:346–356

Korn LJ, Brown DD (1978) Nucleotide sequence of *Xenopus borealis* oocyte 5S DNA: comparison of sequences that flank several related eucaryotic genes. Cell 15:1145–1156

Pieler T, Hamm J, Roeder RG (1987) The 5S gene internal control region is composed of three distinct sequence elements, organized as two functional domains with variable spacing. Cell 48:91–100

Sakonju S, Bogenhagen DF, Brown DD (1980) A control region in the center of the 5S RNA gene directs specific initiation of transcription. 1. The 5′ border of the region. Cell 19:13–25

[5] Bieker JJ, Martin PL, Roeder RG (1985) Formation of a rate limiting intermediate in 5S gene transcription. Cell 40:119–127

Bogenhagen DF, Wormington WM, Brown DD (1982) Stable transcription complexes of *Xenopus* 5S RNA genes: a means to maintain the differentiated state. Cell 28:413–421

Lassar AB, Martin PL, Roeder RG (1983) Transcription of class III genes: formation of preinitiation complexes. Science 222:740–748

Parker CS, Roeder RG (1977) Selective and accurate transcription of the *Xenopus laevis* 5S RNA genes in isolated chromatin by purified RNA polymerase III. Proc Natl Acad Sci USA 74:44–48

Segall J, Matsui T, Roeder RG (1980) Multiple factors are required for the accurate transcription of purified genes by RNA polymerase III. J Biol Chem 255:11986–11991

Setzer DR, Brown DD (1985) Formation and stability of the 5S RNA transcription complex. J Biol Chem 260:2483–2492

Shastry BS, Ng S-Y, Roeder RG (1982) Multiple factors involved in the transcription of class III genes in *Xenopus laevis*. J Biol Chem 257:12979–12986

Wolffe AP, Morse RH (1990) The transcription complex of the *Xenopus* somatic 5S RNA gene: a functional analysis of protein-DNA interactions outside of the internal control region. J Biol Chem 265:4592–4599

Wolffe AP, Jordan E, Brown DD (1986) A bacteriophage RNA polymerase transcribes through a *Xenopus* 5S RNA gene transcription complex without disrupting it. Cell 44:381–389

[6] Brown RS, Sander C, Argos S (1985) The primary structure of transcription factor TFIIIA has 12 consecutive repeats. FEBS Lett 186:271–274

Engelke DR, NG S-Y, Shastry BS, Roeder RG (1980) Specific interaction of a purified transcription factor with an internal control region of 5S RNA genes. Cell 19:717–728

Ginsberg AM, King BO, Roeder RG (1984) *Xenopus* 5S gene transcription factor TFIIIA: characterization of a cDNA clone and measurements of RNA levels through development. Cell 39:479–489

Hanas JS, Hazuda J, Bogenhagen DF, Wu FH-Y, Wu C-W (1983) *Xenopus* trancription factor A requires zinc for binding to the 5S RNA gene. J Biol Chem 258:14120–14125

Miller J, McLachlan AD, Klug A (1985) Repetitive zinc-binding domains in the protein transcription factor IIIA from *Xenopus*. EMBO J 4:1609–1614

Pelham HRB, Brown DD (1980) A specific transcription factor that can bind either the 5S RNA gene or 5S RNA. Proc Natl Acad Sci USA 77:4170–4174

Pelham HRB, Wormington WM, Brown DD (1981) Similar 5S RNA transcription factors in *Xenopus* oocytes and somatic cells. Proc Natl Acad Sci USA 78:1760–1764

Smith DR, Jackson IJ, Brown DD (1984) Domains of the positive transcription factor specific for the *Xenopus* 5S RNA gene. Cell 37:645–652

Taylor W, Jackson IJ, Siegel N, Kumar A, Brown DD (1986) The developmental expression of the gene for TFIIIA in *Xenopus laevis*. Nucl Acids Res 15:6185–6195

Vrana KE, Churchill MEA, Tullius TD, Brown DD (1988) Mapping functional regions of transcription factor TFIIIA. Mol Cell Biol 8:1684–1696

[7] Bogenhagen DF, Wormington WM, Brown DD (1982) Stable transcription complexes of *Xenopus* 5S RNA genes: a means to maintain the differentiated state. Cell 28:413–421

Darby MK, Andrews MT, Brown DD (1988) Transcription complexes that program *Xenopus* 5S RNA genes are stable in vivo. Proc Natl Acad Sci USA 85:5516–5520

Hayes J, Tullius TD, Wolffe AP (1989) A protein-protein interaction is essential for stable complex formation on a 5S RNA gene. J Biol Chem 264:6009–6012

Lassar AB, Martin PL, Roeder RG (1983) Transcription of class III genes: formation of preinitiation complexes. Science 222:740–748

Majowski K, Mentzel H, Pieler T (1987) A split binding site for TFIIIC on the *Xenopus* 5S gene. EMBO J 6:3057–3063

McConkey GA, Bogenhagen DF (1988) TFIIIA binds with equal affinity to somatic and major oocyte 5S RNA genes. Genes Dev 2:205–214

Millstein L, Eversole-Cire P, Blanco JM (1987) Differential transcription of *Xenopus* oocyte and somatic type 5S genes in a *Xenopus* oocyte extract. J Biol Chem 262:1–11

Peck LJ, Millstein L, Eversole-Cire P, Gottesfeld JM, Varshavsky A (1987) Transcriptionally inactive oocyte-type 5S RNA genes of *Xenopus laevis* are complexed with TFIIIA in vitro. Mol Cell Biol 7:3503–3510

Wolffe AP (1988) Transcription fraction TFIIIC can regulate differential *Xenopus* 5S RNA gene transcription in vitro. EMBO J 4:1071–1079

Wolffe AP, Brown DD (1987) Differential 5S RNA expression in vitro. Cell 51:733–740

[8] Andrews MT, Brown DD (1987) Transient activation of oocyte 5S RNA genes in *Xenopus* embryos by raising the level of the transacting factor TFIIIA. Cell 51:445–453

Brown DD, Schlissel MS (1985) A positive transcription factor controls the differential expression of two 5S RNA genes. Cell 42:759–767

Wakefield L, Gurdon JB (1983) Cytoplasmic regulation of 5S RNA genes in nuclear-transplant embryos. EMBO J 2:1613–1619

Wormington WM, Brown DD (1983) Onset of 5S gene regulation during *Xenopus* embryogenesis. Development 99:248–257

[9] Schlissel MS, Brown DD (1984) The transcriptional regulation of *Xenopus* 5S RNA genes in chromatin: the roles of active stable transcription complexes and histone H1. Cell 37:903–913

Wolffe AP (1989a) Transcriptional activation of *Xenopus* class III genes in chromatin isolated from sperm and somatic nuclei. Nucl Acids Res 17:767–780

Wolffe AP (1989b) Dominant and specific repression of *Xenopus* oocyte 5S RNA genes and satellite I DNA histone H1. EMBO J 8:527–537

[10] Wolffe AP (1988) Transcription fraction TFIIIC can regulate differential *Xenopus* 5S RNA gene transcription in vitro. EMBO J 4:1071–1079

Wolffe AP, Brown DD (1987) Differential 5S RNA expression in vitro. Cell 51:733–740

[11] Andrews MT, Brown DD (1987) Transient activation of oocyte 5S RNA genes in *Xenopus* embryos by raising the level of the transacting factor TFIIIA. Cell 51:445–453

Brown DD (1984) The role of stable complexes that repress and activate eucaryotic genes. Cell 37:359–365

Gilbert DM (1986) Temporal order of replication of *Xenopus laevis* 5S ribosomal RNA genes in somatic cells. Proc Natl Acad Sci USA 83:2924–2928

Goldman MA, Holmquist GP, Gray MC, Caston LA, Nag A (1984) Replication timing of genes and middle repetitive sequences. Science 224:686–692

Gottesfeld J, Bloomer LS (1982) Assembly of transcriptionally active 5S RNA gene chromatin in vitro. Cell 28:781–791

Guinta DR, Korn LJ (1986) Differential order of replication of *Xenopus laevis* 5S RNA genes. Mol Cell Biol 6:2536–2542

Guinta DR, Tso JY, Narayanswami S, Hamkalo A, Korn IJ (1986) Early replication and expression of oocyte type 5S RNA genes in a *Xenopus* somatic cell line carrying a translocation. Proc Natl Acad Sci USA 83:5150–5154

Wolffe AP, Brown DD (1986) DNA replication in vitro erases a *Xenopus* 5S RNA gene transcription complex. Cell 47:217–227

Wormington WM, Schlissel M, Brown DD (1983) Developmental regulation of the *Xenopus* 5S RNA genes. Cold Spring Harbor Symp Quant Biol 47:879–884

[12] Brown DD (1982) How a simple animal gene works. Harvey Lect 76:27–44

Brown DD (1984) The role of stable complexes that repress and activate eucaryotic genes. Cell 37:359–365

Brown DD, Schlissel MS (1985) A positive transcription factor controls the differential expression of two 5S RNA genes. Cell 42:759–767

Ginsberg AM, King BO, Roeder RG (1984) *Xenopus* 5S gene transcription factor TFIIIA: characterization of a cDNA clone and measurements of RNA levels through development. Cell 39:479–489

Pelham HRB, Wormington WM, Brown DD (1981) Similar 5S RNA transcription factors in *Xenopus* oocytes and somatic cells. Proc Natl Acad Sci USA 78:1760–1764

Scotto KW, Kaulen H, Roeder RG (1989) Positive and negative regulation of the gene for transcription factor IIIA in *Xenopus laevis* oocytes. Genes Dev 3:651–662

[13] Blanco J, Millstein L, Razik MA, Dilworth S, Cote C, Gottesfeld J (1989) Two TFIIIA activities regulate expression of the *Xenopus* 5S RNA gene families. Genes Dev 3:1602–1612

Pelham HRB, Wormington WM, Brown DD (1981) Similar 5S RNA transcription factors in *Xenopus* oocytes and somatic cells. Proc Natl Acad Sci USA 78:1760–1764

[14] Tafuri S, Wolffe AP (1990) *Xenopus* Y box transcription factors: molecular cloning, functional analysis and developmental regulation. Proc Natl Acad Sci USA 87:9028–9032

Chapter 26 From *Drosophila* to Mouse

C. G. LOBE and P. GRUSS

26.1 Introduction

The study of development in mammals is of particular interest, not only because of important relevant applications to medicine and agriculture, but also because there is a natural curiosity to learn about the process of embryology in humans. However, compared to other systems, the mammal is not simple to work with. The eggs and embryos are small, therefore not as available to manipulation as frog or chick, and mammalian embryos develop in utero, again putting a barrier against manipulation and observation. Nevertheless, technical advances in the past years have made mammalian development more amenable to study and, in fact, afford exciting new possibilities. This is particularly true for the mouse, which is most often used as the representative mammalian species since its genetics are best characterized, inbred and many mutant lines are available, and it has a relatively short gestation period. In recent years it has become possible to introduce foreign genetic information into a fertilized egg to generate transgenic animals, pre- or post-implantation embryos can be incubated for short periods of time in vitro for study, in situ detection methods have been refined to enable an analysis of endogenous or transgene expression in early development, embryo stem cell lines are available, and homologous recombination of DNA and subsequent generation of chimeric mice using these stem cells has been achieved. At the molecular level, many genes which are implicated to play a role in embryogenesis have been identified and cloned, based in large part on homologies to developmental genes of lower species. These genes and their analysis using the new technologies will be discussed here.

26.2 Embryogenesis of the Mouse[1]

The developing mouse embryo undergoes three stages common to most animals; cleavage gastrulation, and organogenesis (Fig. 1).

Shortly after fertilization, the pronuclei form followed by the first round of DNA replication, so that the first cleavage can occur between 17 and 20 h. The two-cell embryo then undergoes the second round of DNA replication and it is not until this point that mRNA synthesis (expression of zygotic genes) begins. Cleavage, mitotic cell divisions without growth, continues, albeit at a much slower rate in mammals than in other systems. Compaction begins at the eight-cell stage, that is the cells become a compact ball with tight junctions between the cells. The first differentiation occurs in the 16-cell embryo, the morula, so that the external cells will become the trophoblast cells which produce the chorion (outer cells of the placenta). It is interesting that the first differentiation is toward placenta, which will provide food and oxygen, since mammals do not have large quantities of yolk. The trophoblast cells secrete fluid into the center of the morula causing a central cavity, the blastocoel, to form. Thus, the cleavage stage generates a blastocyst consisting of an inner cell mass, which will make up the embryo, surrounded by trophoblast cells, which will contribute to extraembryonic tissue. The inner cell mass subsequently generates the epiblast and the primitive endoderm, the latter of which lines the blastocoel cavity. The epiblast goes on to differentiate into the primitive ectoderm, from which the three germ layers of ectoderm, mesoderm, and endoderm are derived.

At gastrulation, cells migrate through the primitive streak, an invagination which forms along the anterior-posterior (A-P) axis of the primitive ec-

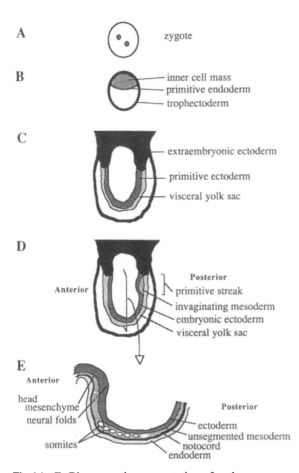

A   zygote

B   — inner cell mass
     — primitive endoderm
     — trophectoderm

C   — extraembryonic ectoderm
     — primitive ectoderm
     — visceral yolk sac

D   Posterior
     Anterior   primitive streak
                 invaginating mesoderm
                 embryonic ectoderm
                 visceral yolk sac

E   Anterior
     head
     mesenchyme
     neural folds   Posterior
                 ectoderm
                 unsegmented mesoderm
     somites   notocord
                 endoderm

Fig. 1A–E. Diagrammatic representation of early mouse embryo development. **A** Zygote with two pronuclei. **B** Blastocyst stage. **C** Egg cylinder stage. **D** Early gastrulation. Mesodermal cells begin to migrate through the primitive streak to form a layer between the ectoderm and viscernal yolk sac. **E** Somitogenesis. Progressing from anterior to posterior, somites form by condensation of mesodermal tissue and the ectoderm begins to fold to form the neural tube

toderm. They then move anteriorly and laterally to form the mesoderm and endoderm tissues, including the notochord along the dorsal midline. At the same time, two longitudinal neural folds form in the ectoderm along the A-P axis. In birds and amphibia, it has been shown that the notochord induces the formation of these folds in a process termed neural induction. The neural folds fuse together to form the neural tube, destined to become the brain and spinal cord. Cells which lie at the junction between the newly formed neural tube and surface ec-

toderm, the neural crest cells, will migrate to make up the peripheral nervous system, sympathetic and parasympathetic systems, pigment cells, and certain areas of head cartilage. At the same time that the neural folds are closing to form the neural tube and the primitive streak is regressing, the paraxial mesoderm on either side of the notochord condenses giving rise to pairs of somites at day 8 post coitum (p. c.). These condensations represent the first obvious signs of segmentation. Both the neural tube and somites are established progressively from the anterior to the posterior of the developing embryo. Later, cells at the ventral side of the somites undergo mitosis and become mesenchymal cells, the sclerotome. These differentiate to form chondrocytes, which will form the axial skeleton. The remainder of the somites is dermomyotome, which will partition into dorsal dermatome and myotome, destined to become mesenchymal connective tissue of the skin and striated muscles of back and limbs, respectively. The mesoderm which flanks the somites, the intermediate mesoderm, will form the segmented nephrotomes of the pronephros, mesonephros, and metanephric kidney, and parts of organs such as the lung, liver, and gut. Further away from the midline is the lateral plate mesoderm which will form heart, blood vessels, and blood cells, and limb tissues apart from the muscles. The endoderm will form the primitive gut, which gives rise to digestive and respiratory tubes.

26.3 Developmental Control Genes Derived from *Drosophila*

The starting point for identification of putative developmental genes in the mouse was provided by *Drosophila*, for which developmental patterns of the embryo have been extensively characterized (see Chaps. 21, 22, and 23, this vol.). In brief, the zygote undergoes superficial cleavage in which a series of rapid nuclear divisions takes place without cellularization. The nuclei migrate to the egg periphery and are surrounded by membranes to form the "cellular blastoderm". Gastrulation then proceeds with the invagination of cells along the ventral midline to generate a mesodermal layer within the ventral ectoderm. The double layer of ectoderm and mesoderm produces the characteristic segmented pattern of the *Drosophila* embryo by a series of condensations along the mesoderm and corresponding indentations of the ectoderm. These segments will give rise to the head, thoracic, and abdominal adult segments.

Development of the fertilized egg has a similar basic pattern for *Drosophila* and mouse in that there is an initial cleavage period, formation of an ectoderm layer, migration of cells through that layer to form the mesoderm and endoderm, and segmentation of the germ layers which make up the embryo. However, some important differences exist: for example, in *Drosophila* the unfertilized egg is polarized, whereas in the mouse there is no obvious polarization until the cephalic furrow and primitive streak form; for *Drosophila*, nuclear cleavage without cellularization allows substances to diffuse more readily through the early blastoderm than for the mouse; segmentation of the mouse embryo is less defined and extensive than in *Drosophila* and the mouse embryo has a later contact with the maternal system.

The question of how embryogenesis occurs in *Drosophila* was addressed by generating numerous developmental mutants for study. Classical genetics in concert with molecular biology allowed the identification of genes corresponding to the mutations. Subsequently, conserved domains of these genes were used to screen for homologous counterparts in vertebrates in the hopes of identifying related developmental control genes. With this approach, many genes have been isolated (Table 1), but the extent to which their functional role has been conserved has yet to be determined.

26.3.1 Homeobox Genes[2]

The fly developmental genes which were the best characterized initially were the homeotic genes, contained in two gene complexes: the Antennepedia complex (ANT-C) and the Bithorax complex (BX-C). Each member of the homeotic gene family contains a DNA sequence homology of 180 base pairs (bp), designated the homeobox. Using

Table 1. Murine genes which have been cloned using conserved sequences from *Drosophila* developmental genes

Drosophila gene	DNA-binding domain	Murine homologs
Antennapedia	Homeobox	Hox 1, 2, 3, 5 clusters
Engrailed	Homeobox	En1, En2
Paired	Paired-box	Pax 1, 2, 3, 4, 5, 6, 7, 8
Even-skipped	Homeobox	Evx1, Evx2
Krüppel	Finger	Zfp 1, Krox20
Caudal	Homeobox	Cdx1

The first column lists the gene from which the conserved sequence was used, the second column lists the conserved DNA-binding domain from each gene, and the last column lists the murine genes which have been cloned using the conserved sequence.

the prototype homeobox from the Antp gene as a probe, homeobox genes were identified in *Xenopus*, mouse, and human genomic libraries. In the mouse, the genes (Hox genes) reside as clusters on chromosomes 2, 6, 11, and 15, as depicted in Fig. 2. The existence of the two *Drosophila* gene clusters and multiple representations of that array in mouse and other higher vertebrates probably arose initially by local gene duplication and diversification, followed by duplication of entire homeobox gene clusters. That these genes not only represent conserved sequences, but might also perform conserved functions, was first indicated by Northern blot and in situ data since they were found to be expressed with spatial and temporal specificity during embryogenesis.

At early gastrulation, Hox 1.5 and 1.6 RNAs have been detected in the posterior presomitic mesoderm and overlying ectoderm by in situ analysis of 7.5 day p.c. embryos. Hox 2.1 and Hox 1.3 RNAs have also been detected at this time, using Northern and RNase analysis. Later during gastrulation, at day 8 to 8.5, as the notochord forms and the somites start to condense, many of the homeobox genes are expressed. Generally, transcripts are detected in the ectoderm and pre-somitic mesoderm. The anterior and posterior boundaries of expression vary, however, between the genes. Further on in development, at day 12 p.c., when the level of most of the homeobox genes transcripts has

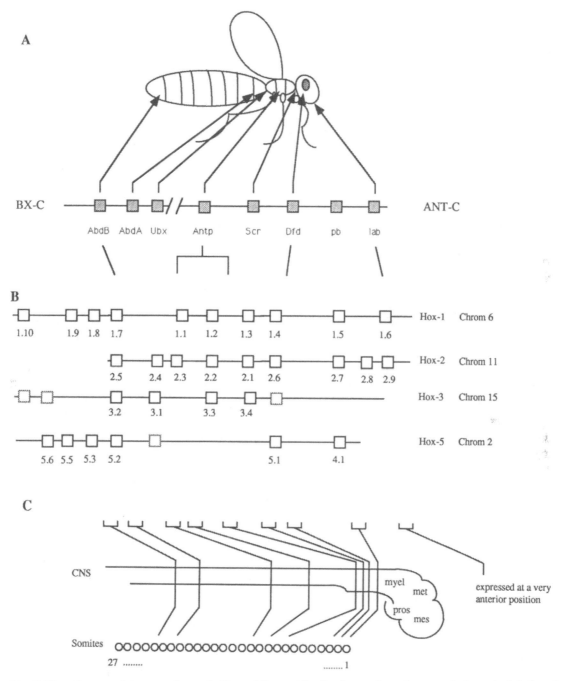

Fig. 2 A – C. Homeobox-containing gene clusters in *Drosophila* and mouse. In the *center* of the figure the gene clusters are schematically represented, aligned vertically to indicate genes that belong to the same subfamily. *Above*, the anterior border of expression of each of the *Drosophila* genes during embryogenesis is indicated. *Below*, the anterior border of expression for each subfamily of mouse homeobox genes is shown, both in the ectoderm-derived central nervous system (*CNS*) and the mesoderm-derived prevertebrae. *pros*, *mes*, *met*, and *myel* indicate prosencephalon, mesencephalon, metencephalon and myelencephalon, respectively, which are the compartments of the developing brain. (After Duboule and Dolle 1989)

peaked, a similar pattern is seen. That is, the ectoderm-derived tissues (brain and spinal cord) continue to express the homeobox genes with a variation in the anterior boundary, ranging from the myelencephalon for Hox 1.6 through to the hindbrain and cervical region of the spinal cord for other genes. Likewise, transcripts are detected in the mesoderm compartment, in the sclerotome (prevertebrae), myotome, and in gut tissue derived from mesoderm flanking the somites, but for each gene this expression has anterior and posterior limits which correspond, but are slightly offset posteriorly, to their limits of expression in the ectoderm-derived tissue.

An interesting feature of this variation in the anterior limit of expression is that there is a direct correlation between the anterior border of expression of a given gene and its position in the gene cluster (Table 1). For example, considering genes within the Hox 1 cluster, transcripts of Hox 1.6 extend to the presumptive myelencephalon of the neural tube whereas, proceeding upstream in the cluster, for Hox 1.5, 1.4, 1.3 and 1.2 the limits of expression lie more posteriorly in the hindbrain. Hox 1.6 expression begins anterior to the first somatic mesoderm, Hox 1.5 expression in pre-vertebra 1, Hox 1.4 expression in pre-vertebra 2, Hox 1.3 in pre-vertebra 3 to 4, and Hox 1.2 in pre-vertebra 8. A similar situation has been reported for the Hox 2 cluster. The genes are also turned on over time from the 3'-most to the 5'-most in the cluster, since development proceeds from anterior to posterior in the embryo. This may be of significance to the mechanism of coordinate gene regulation.

A correlation between gene position in the cluster and expression pattern was also found for the Hox 4 cluster in the developing limb. In this case, the more 5' a gene's position, the later and more distal is the expression. In the chick and amphibia limb bud, grafting experiments have demonstrated a "zone of polarizing activity" (ZPA) at the posterior (distal and caudal) edge of the limb, which establishes the correct limb polarity. A gradient of retinoic acid exists in the limb with highest concentration posteriorly, and the ZPA can be mimicked or disturbed by localized application of retinoic acid, suggesting that it acts as a type of morphogen. That retinoic acid may establish the gradient of

Hox gene expression has been demonstrated using human teratocarcinoma cells, which can be differentiated with retinoic acid. The genes of the human Hox 2 cluster were found to respond differentially to retinoic acid, with the genes at the 3' end being expressed more quickly and at lower concentrations of retinoic acid than more 5'-located genes. Since both Hensen's node (at the anterior of the primitive streak) and the neural plate also possess ZPA activity, it is tempting to speculate that a morphogen such as retinoic acid exists which establishes a gradient through concentration or time of exposure and this gradient is represented in the Hox gene pattern of expression.

A similar relationship, in which the order of genes within the cluster corresponds to the order of anterior boundaries of expression of each gene, exists among the *Drosophila* homeotic genes (see Fig. 2). However, in *Drosophila*, the A-P axis of the embryo is represented once in the two complexes of ANT-C and BX-C, whereas in vertebrate multiple copies apparently exist, one per Hox cluster. This may reflect the more complex body pattern of the vertebrates and the need for more detailed information. The single representation of the A-P axis of the embryo in *Drosophila* and the multiple representation of this information in higher vertebrates is also in accord with the idea of duplication of a primordial gene cluster and the observation that greater homology is shown between genes of different clusters than between genes within the same cluster.

In order to determine the function of the mouse homeobox genes, two approaches have been taken to generate mouse mutants, aimed either at achieving a dominant gain of function or a dominant loss of function. Using the "gain of function" approach, the coding sequence of the Hox 1.1 gene was put under the control of the chicken β-actin promoter to give ectopic expression of Hox 1.1, and this construct was used to generate transgenic mice. Expression of the transgene in newborn mice corresponded to morphologic abnormalities in the head region, which does not normally express the Hox 1.1 gene (Fig. 3). These deformities included cleft palate, ear malformations, open-eyes at birth, and lack of fusion of the pinae. This mutant phenotype is reminiscent of the deformities caused by expo-

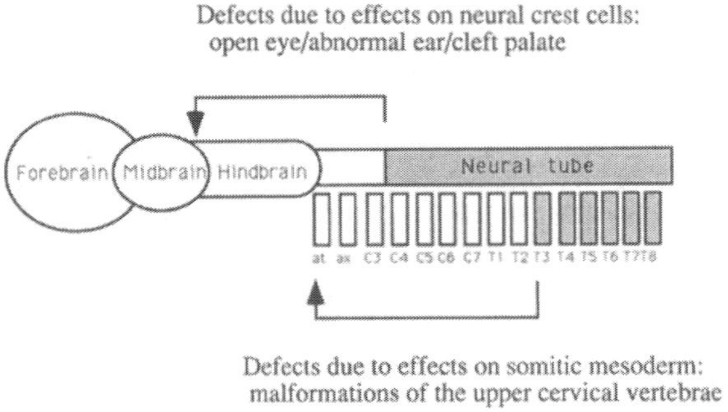

Defects due to effects on neural crest cells:
open eye/abnormal ear/cleft palate

Forebrain Midbrain Hindbrain   Neural tube

at ax C3 C4 C5 C6 C7 T1 T2 T3 T4 T5 T6 T7T8

Defects due to effects on somitic mesoderm:
malformations of the upper cervical vertebrae

Fig. 3. Anterior expression boundaries and sites of malformations due to Hox 1.1 ectopic expression in transgenic mice. The anterior neuroectoderm (CNS) and the prevertebrae as metameric units are schematically represented. The expression domains of Hox 1.1 are shown as *stippled areas* in the CNS (up to the fourth cervical ganglion) and in the prevertebrae (up to T 3). The *arrowheads* point out on the anteroposterior axis the sites of malformation occurring in transgenic mice: at the level of the rostral hindbrain (first visceral arch) and at the craniovertebral transition (*at* atlas; *ax* axis)

sure of developing embryos to retinoic acid, a powerful teratogen, and may result from a disturbance in the migration of the cranial neural crest cells. Later, it was also noted that more severely affected animals also showed malformations of the upper cervical vertebral column, in which both rostral vertebrae had characteristics of more posterior vertebrae, reflecting an effect on somite differentiation into sclerotome and dermomyotome. Similarly, overexpression of the homeobox gene, Hox 1.4, was achieved by inserting extra gene copies, which resulted in an adult megacolon phenotype. Mis-expression of a homeobox gene clearly has some deleterious effect on development and, in the case of Hox 1.1 ectopic expression, is suggestive of the homeotic type of mutation seen in *Drosophila* wherein parts of one segment are transformed into the corresponding structures of another segment. A more obvious homeotic transformation probably does not occur because segmentation in vertebrates has become a much more complicated or fundamentally different process, or because vertebrates have evolved many more genes which are overlapping and complementary in function, so that aberrant expression of one of the genes is not so deleterious.

Using the "dominant loss of function" approach, one allele of the Hox 1.1 gene and of the En-2 gene has been inactivated in embryonal stem cells by a targeted recombination event. In order to produce chimeric mice, the stem cells have been introduced into mouse blastocysts, with the ultimate goal of breeding these chimeras in order to establish a homozygous mouse mutant strain. When achieved, this homozygous mutant should lack the function of the Hox 1.1 or En-2 gene and thus indicate the roles of these genes during murine embryogenesis.

26.3.2 Engrailed-Type Homeobox Genes [3]

Engrailed is a gene of the segment polarity class, expressed in a band of cells just posterior to the A-P boundary of each segment in *Drosophila*, and is crucial for the development of the posterior half of each segment. The gene contains a homeobox, distantly related to the Antp homeobox, which was used to isolate two mouse genes, En-1 and En-2. An in situ analysis of expression detected transcripts of both genes during the early somite stage in a band restricted to the anterior part of the neural folds. At day 12 p.c., both genes are still expressed in a ring at the midbrain-hindbrain junction of the neural tube but En-1 transcripts are also observed extending down the spinal cord. Additionally, En-1 is expressed in the mesoderm compartment, specifically in sclerotome-derived prevertebrae and in dermatome-derived cells under the skin throughout the length of the embryo, and in the limb and tail buds. The pre-vertebrae expression displays a periodic pattern, present in the pre-vertebrae but not the intervertebral disks. Later, at 17.5 days p. c., En-1 and En-2 are expressed in different but overlapping areas of the mid- and hindbrain. This expression pattern is consistent with a role, together with the Antp-type homeobox genes, in compartmentalization of the early central nervous system (CNS) and later in neural differentiation. The periodic expression of En-1 in the pre-vertebrae is suggestive of the *en* pattern in *Drosophila*, but apparently occurs too late for establishment of segmentation and may instead serve to extend segmentation by playing a role in vertebral column formation (see below).

26.3.3 Paired-Box Genes[4]

In two other classes of *Drosophila* developmental genes, another conserved sequence is found. This is the paired-box domain, a 128 amino acid sequence shared by the pair-rule gene, paired (*prd*), and the segment-polarity genes, gooseberry-proximal (*gsb-p*) and gooseberry-distal (*gsb-d*). A family of murine genes, designated Pax for paired-box, was identified using the *Drosophila* paired-box sequence (Fig. 4). One member, Pax 1, was detected in situ starting at 9 days p.c. in sclerotome cells and therefore, like murine En-1, appears to be induced after the primary segmentation of the mesoderm. At 12 days p.c. this pattern has shifted so that expression is seen only in periodic bands of the sclerotome which correspond to the intervertebral disk anlagen (Fig. 5). Interestingly, in the process of vertebral column formation the sclerotome is believed to re-segment, with the fusion of the posterior half of one sclerotome to the anterior half of the next, creating a single vertebra. Pax 1 might therefore play a role in directing the formation of the vertebrae, dividing anterior and posterior halves of the somite, similar to the *Drosophila* segment polarity genes which determine anterior and posterior halves of segments. In accord with such a role for Pax 1 was the finding that a mouse mutant, *undulated*, has a mutated Pax 1 sequence. Mice homozygous for *undulated* exhibit a reduction of the posterior part of the vertebrae, abnormally large intervertebral disks and small vertebrae centers. Analysis of the Pax 1 gene from *undulated* mice revealed that the sequence contained a point mutation which would cause a Gly-Ser replacement in a highly conserved part of the paired-box. Later it was found by genetic analysis that another related mutant which exhibits a more severe phenotype, *undulated short tail*, has a deletion of the Pax 1 gene. This marks the first correspondence of one of the mouse genes identified through *Drosophila* homology to a mouse developmental mutant. Further proof that the *undulated* phenotype is caused by a Pax 1 mutation awaits a rescue experiment, in

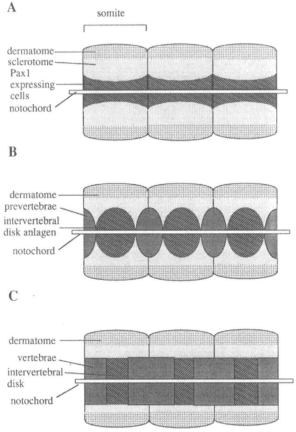

Fig. 4A–C. Frontal section showing Pax 1 expression during vertebral column development. **A** 10 day p.c.: expression is in part of the sclerotome. **B** 12 day p.c.: expression coincides with intervertebral disk anlagen. **C** 14 day p.c.: expression occurs in intervertebral disks

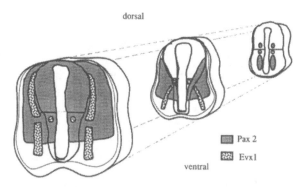

Fig. 5. Schematic representation showing Evx 1 and Pax 2 expression in sections of the neural tube at different stages of development, from back to front (*right to left*) day 10, 11 and 12

which the wild-type gene introduced into a mutant mouse restores a normal phenotype.

Other members of the Pax gene family are expressed throughout the length of the neural tube, at least to the rhombencephalic isthmus. This is also true of Evx1, a murine gene which was cloned via homology to the *Drosophila eve* gene, which contains a homeobox distantly related to the Antp and En homeoboxes. In cross-sections of the neural tube, these genes are expressed in specific patterns of the neural epithelium and mantle layer and may specify the fate of subsets of neurons in the spinal cord. In addition, some developing organs have been shown to contain Pax RNAs, such as Pax 2 in the developing kidney.

26.3.4 Finger-Containing Genes[5]

Murine homologs to *Krüppel* (*Kr*), a member of the gap gene class, have also been isolated. The *Drosophila Kr* gene is expressed in a broad central band in the blastoderm and later in the developing nervous system, where it interacts with other segmentation genes to regulate the spatial pattern of segmentation and homeotic genes. *Krüppel* and two other members of the gap gene class, *Hunchback* (*Hb*) and *knirps* (*kni*), contain the finger-like DNA-binding domains first recognized in the *Xenopus* transcription factor TF III A. This conserved domain from *Kr* was used to isolate murine homologues, including murine Krüppel 2 (mKr 2, now designated Zfp 2 for Zinc-finger protein). Transcripts from the Zfp 2 gene were detected beginning at day 10 p. c., with expression limited to the developing nervous system throughout the length of the spinal cord and brain. Another Zfp gene, Krox 20, which was first identified as a serum-responsive gene, was subsequently detected in neural crest cells early in mouse embryogenesis by in situ hybridization. This suggests that both Zfps may be required for the differentiation of neurons. However, the possibility that they also take part in laying down the homeobox expression pattern, analogous to the situation in *Drosophila*, must await more sensitive assays for the detection of early expression.

26.4 Octamer-Binding Proteins and Genes[6]

The search for molecules which may play a role in embryogenesis initially focused on identification of genes via their conserved sequences. More recently, molecules which may be involved in the initial regulation of other developmental control genes have been discovered, not at the DNA but at the protein level. These factors were detected by virtue of their ability to bind the octamer motif (ATTGCAT). The transcription factors previously described that bind this motif and thereby regulate gene expression include a ubiquitously expressed factor, Oct 1, a B cell-specific factor, Oct 2, and the pituitary protein, Pit-1 (GHF-1). These proteins, together with another octamer-binding protein, *unc-86* from *C. elegans* (the POU family), share a conserved 150 to 160 amino-acid long region containing a homeobox-related subdomain which binds the octamer, and a POU-specific subdomain. A family of octamer binding proteins, OBPs, were identified, which are differentially expressed during early stages of development by using the immunoglobulin heavy chain octamer as a DNA target. Two of these, Oct 4 and 5, are present in oocytes and embryonal stem cells but are not present in differentiated cells, as determined by gel shift assays. To see if transcription might be activated early in mouse development by these OBPs, the octamer motif was inserted into a tk promoter/β galactosidase gene construct and this construct used for injection of fertilized mouse oocytes. Expression was detected using β-galactosidase staining and showed that this construct was expressed in the blastocyst and, more interestingly, expression was restricted to the inner cell mass. It is possible that Oct 4 and 5 represent early factors involved in cell determination in the zygote and/or initiation of the cascade of gene expression which directs embryogenesis. Oct 4 may also be a marker of totipotent cells which carry the flow of genetic information since it is also present in primordial germ cells and maternally in oocytes but not in sperm.

Using, once again, the approach of identifying genes by their homology to conserved sequences, four new mammalian POU-domain genes have been identified. Two were isolated from a human brain cDNA library, one from a rat brain cDNA library, and one from a rat testes cDNA library. Characterization by in situ analysis showed that these four genes, as well as Oct 1, Oct 2, and Pit 1, are all expressed in the developing embryonic nervous system. These seven members of the Pou family are therefore expressed in various patterns in the neural tube of the embryo and then with different tissue specific patterns in adult.

26.5 The Protein Products as Transcription Factors [7]

If embryogenesis is indeed orchestrated by a hierarchy of gene regulation, then the protein products of the genes in the cascade must be able to regulate other genes. The conserved domains discussed above were proposed to be DNA-binding domains based on their homology to prokaryotic and yeast gene regulatory proteins, in the case of the homeobox and to the *Xenopus* transcription factor, TFIIIA, for the finger structures. A consensus binding site for several *Drosophila* homeobox proteins was identified based on in vitro binding to 5' flanking sequences of homeobox genes themselves and later in vivo using cotransfection assays in culture cells and a yeast expression system. Other proteins which were shown to bind sequences in their own promoters are the murine Hox 1.5 and Hox 1.3 proteins, produced in bacterial and insect expression systems respectively. The ability of the proteins to regulate gene transcription upon binding to DNA promoter elements is exemplified by the POU proteins and steroid hormone receptors. These proteins were known to bind to specific DNA sequences and thereby activate transcription, and when biochemically characterized, were found to contain homeobox and finger domains, respectively. Another demonstration that the *Drosophila* homeobox proteins can regulate transcription via their DNA-binding activity has been provided by cotransfection experiments. Multiple copies of the *Drosophila* homeobox-binding consensus sequence were linked to heterologous promoters and reporter

genes and these constructs were cotransfected with expression vectors which produced various homeobox proteins. Cotransfection of some homeobox genes (*ftz*, *prd*, *zen*) led to stimulation of transcription of the receptor gene construct, whereas others (*en*, *eve*) had no effect. The proteins which stimulated transcription through the homeobox binding sequence could also act synergistically, and the proteins which did not activate transcription inhibited the activity of those which did.

The experiments described above suggest that the homeobox proteins recognize common binding sites but are able to impart different effects on promoter activity. It may be that various proteins compete for the same sites and a range of protein affinities is provided through deviations from the consensus binding site. Since more than one consensus binding element is usually present in eukaryotic gene promoters, binding of more than one homeobox protein species would allow for interaction, synergistic or antagonistic. The final outcome of whether genes containing the promoter response element(s) are transcriptionally active or silent in that cell would be determined through competition and interaction between the set of homeobox proteins present in a cell.

26.6 Outlook

There are two questions which the future research on developmental genes will address: one is how the regulation of expression of these genes is achieved, and the other is what the functions of the protein products are. Both of these questions will be more easily answered once the exact expression patterns of all of the genes is known, particularly in comparison to one another, since the gene products probably interact to regulate both their own expression and expression of other genes, thereby determining cell fate.

To determine how the genes are regulated, genomic DNA sequences are linked to reporter genes such as CAT, Luciferase, or β-galactosidase, and these constructs are then introduced into tissue culture cells by transfection or used to generate transgenic mice. In this way, DNA sequences which are able to reproduce the correct endogenous expression pattern can be identified. A finer analysis can then be done using gel retardation assays or DNase footprinting, in which protein extracts are mixed with the DNA fragment and their binding sites mapped.

Several approaches to determine the function of the gene products are now being taken. On a molecular level, expression vectors can be utilized to synthesize large amounts of the protein which the gene encodes. This protein can then be used in DNA binding studies to find out what the correct transcriptional element is which the protein normally recognizes in other genes. By introducing deletions in the expressed protein, the functional domains of the protein can also be deduced.

On a more biological approach, the genes can be mis-expressed to see what the resultant phenotype is (as discussed above for Hox 1.1, 1.4 and En 2). One way to do this is to knock out the gene under study using homologous recombination. Another way is to generate transgenic mice with the gene of interest expressed under a heterologous promoter which gives incorrect expression. The mice which are generated can then be examined for a changed morphology as well as an alteration in expression of other genes. Similarly, when the genes are mapped, they may turn out to correlate with an existing mouse or human mutation, which would also give clues to what the function of the wild-type gene is.

As we accumulate information on the regulation and exact function of the genes expressed during development, a picture of how cell division and differentiation are directed during embryogenesis will emerge.

26.7 Summary

Although their definitive function has yet to be established, the expression pattern and biochemical nature of the murine homeobox, paired box, and finger-containing genes suggests they play a role in development. The Hox genes seem to be involved in regional specification, Pax 1 in developmental of the vertebral column, and some Zfp genes in neu-

ron differentiation. The protein products of these genes probably function as transcriptional regulators, in accord with a model which proposes that the information for the body plan of a mouse is contained in control genes, expressed during development to regulate other genes. An intriguing but difficult task is now to find promoter targets outside of the homeobox gene family.

The exact function of these genes should be clarified through further analysis of their pattern of expression and the effects of altering that pattern. If indeed they share functional as well as structural characteristics with the *Drosophila* developmental control genes, then establishing their time and mode of actions will at last illuminate how development and cell determination occurs in vertebrates.

References

[1] Hogan BLM, Holland PWH, Schofield PN (1985) How is the mouse segmented? Trends Genet 1:67–74

Hogan BLM, Costantini F, Lacy E (1986) Manipulating the mouse embryo: a laboratory manual. Cold Spring Harbor Laboratory, Cold Spring Harbor, New York

[2] Akam M (1989) Hox and HOM: homologous gene clusters in insects and vertebrates. Cell 57:347–349

Balling R, Mutter G, Gruss P, Kessel M (1989) Craniofacial abnormalities induced by ectopic expression of the homeobox gene Hox 1.1 in transgenic mice. Cell 58:337–347

Bastian H, Gruss P (1990) A murine *even-skipped* homologue, *Evx-1*, is expressed during early embryogenesis and neurogenesis in a biphasic manner. EMBO J 9:1839–1852

Brockes JP (1989) Retinoids, homeobox genes, and limb morphogenesis. Neuron 2:1285–1294

Carrasco AE, McGinnis W, Gehring WJ, DeRobertis EM (1984) Cloning of an *X. laevis* gene expressed during early embryogenesis coding for a peptide region homologous to *Drosophila* homeotic genes. Cell 37:409–414

Colberg-Poley AM, Voss SD, Chowdhury K, Gruss P (1985) Structural analysis of murine genes containing homeobox sequences and their expression in embryonal carcinoma cells. Nature 314:713–718

Dolle P, Iypisua-Belmonte J-C, Falkenstein H, Renucci A, Dubould D (1989) Coordinate expression of the murine Hox-5 complex homoeobox-containing genes during limb pattern formation. Nature 342:767–772

Dressler GR, Gruss P (1989) Anterior boundaries of Hox gene expression in mesoderm-derived structures correlate with the linear gene order along the chromosome. Differentiation 41:193–201

Duboule D, Dolle P (1989) The structural and functional organization of the murine HOX gene family resembles that of *Drosophila* homeotic genes. EMBO J 8:1497–1505

Fainsod A, Bogarad LD, Ruusala T, Lubin M, Crothers DM, Ruddle FH (1986) The homeo domain of a murine protein binds 5' to its own homeo box. Proc Natl Acad Sci USA 83:9532–9536

Gaunt SJ, Sharpe PT, Duboule D (1988) Spatially restricted domains of homeo-gene transcripts in mouse embryos: relation to a segmented body plan. Development 104 (Suppl):169–179

Graham A, Papalopulu N, Krumlauf R (1989) The murine and *Drosophila* homeobox gene complexes have common features of organization and expression. Cell 57:367–378

Holland PWH, Hogan BLM (1988) Expression of homeo box genes during mouse development: a review. Genes Dev 2:773–782

Jackson IJ, Schofield P, Hogan BLM (1985) A mouse homeobox gene is expressed during embryogenesis and in adult kidney. Nature 317:745–748

Joyner AL, Kornberg T, Coleman KG, Cox DR, Martin GR (1985) Expression during embryogenesis of a mouse gene with sequence homology to the *Drosophila* engrailed gene. Cell 43:29–37

Joyner AL, Skarnes WC, Rossant J (1989) Production of a mutation in mouse En-2 gene by homologous recombination in embryonic stem cells. Nature 338:153–156

Kappen C, Schughart K, Ruddle FH (1989) Two steps in the evolution of Antennapedia-class vertebrate homeobox genes. Proc Natl Acad Sci USA 86:5459–5463

Kessel M, Balling R, Gruss P (1990) Variations of cervical vertebrae after expression of a Hox-1.1 transgene in mice. Cell 61:301–308

Levine M, Rubin GM, Tjian R (1984) Human DNA sequences homologous to a protein coding region conserved between homeotic genes of *Drosophila*. Cell 38:667–673

Lewis EB (1978) A gene complex controlling segmentation in *Drosophila*. Nature 276:565–570

Mahaffey JW, Kaufman TC (1987) The homeotic genes of the Antennepedia complex and the Bithorax complex of *Drosophila*. In: Malacinski GM (ed) Developmental genetics of higher organisms. Macmillan, New York, pp 329–359

McGinnis W, Levine MS, Hafen E, Kuroiwa A, Gehring WJ (1984a) A conserved DNA sequence in homoetic genes of the *Drosophila* Antennapedia and bithorax complexes. Nature 308:428–433

McGinnis W, Hart CP, Gehring WJ, Ruddle FH (1984b) Molecular cloning and chromosome mapping of a mouse DNA sequence homologous to homeotic genes of *Drosophila*. Cell 38:675–680

Muller MM, Carrasco AE, DeRobertis EM (1984) A homeobox-containing gene expressed during oogenesis in *Xenopus*. Cell 39:157–162

Odenwald WF, Taylor CF, Palmer-Hill FJ, Friedrich V, Tani M, Lazzarini RA (1987) Expression of a homeodomain protein in non contact-inhibited cultural cells and post mitotic neurons. Genes Dev 1:482–496

Schneuwly S, Klemeny R, Gehring WJ (1987) Redesigning the body plan of *Drosophila* by ectopic expression of the homoeotic gene Antennapedia. Nature 325:816–818

Scott MP, Weiner AJ (1984) Structural relationships among genes that control development: sequence homology between the Antennapedia, Ultrabithorax and fushi tarazu loci of *Drosophila*. Proc Natl Acad Sci USA 81:4115–4119

Struhl G (1985) Near-reciprocal phenotypes caused by inactivation or indiscriminate expression of the *Drosophila* segmentation gene Antennapedia. Nature 318:677–680

Sundin OH, Busse HG, Rogers MB, Gudas LJ, Eichele G (1990) Region-specific expression in early chick and mouse embryos of Ghox-lab and Hox 1.6, vertebrate homeobox-containing genes related to *Drosophila* labial. Development 108:47–58

Wolgemuth DJ, Behringer RR, Mosteller MP, Brinster RL, Palmiter RD (1989) Transgenic mice overexpressing the mouse homeobox-containing gene Hox 1.4 exhibit abnormal gut development. Nature 337:464–467

Zimmer A, Gruss P (1989) Production of chimaeric mice containing embryonic stem (ES) cells carrying a homeobox Hox 1.1 allele mutated by homologous recombination. Nature 338:150–153

[3] Davidson D, Graham E, Sime C, Hill R (1988) A gene with sequence similarity to *Drosophila* engrailed is expressed during the development of the neural tube and vertebrate in the mouse. Development 104:315–316

Davis CA, Joyner AL (1988) Expression patterns of the homeo box-containing genes En-1 and En-2 and the proto-oncogene int-1 diverge during mouse development. Genes Dev 2:1736–1744

Joyner AL, Martin GR (1987) En-1 and En-2, two mouse genes with sequence homology to the *Drosophila* engrailed gene expression during embryogenesis. Genes Dev 1:29–38

[4] Balling R, Deutsch U, Gruss P (1988) *Undulated*, a mutation affecting the development of the mouse skeleton, has a point mutation in the paired box of Pax 1. Cell 55:531–535

Bopp D, Burri M, Baumgartner S, Frigerio G, Noll M (1986) Conservation of a large protein domain in the segmentation gene paired and in functionally related genes of *Drosophila*. Cell 47:1033–1040

Deutsch U, Dressler GR, Gruss P (1988) Pax 1, a member of a paired box homologous murine gene family, is expressed in segmented structures during development. Cell 53:617–625

[5] Chowdhury K, Dressler G, Breier G, Deutsch U, Gruss P (1988) The primary structure of the murine multifinger gene mKr2 and its specific expression in developing and adult neurons. EMBO J 7:1345–1353

Gaul U, Swifert E, Schuh R, Jäckle H (1987) Analysis of *Krüppel* protein distribution during early *Drosophila* development reveals post-transcriptional regulation. Cell 50:639–647

Ollo R, Maniatis T (1987) *Drosophila Krüppel* gene product in a baculovirus expression system is a nuclear phosphoprotein that binds to DNA. Proc Natl Acad Sci USA 84:5700–5704

Schuh R, Aicher W, Gaul U, Cote S, Preiss A, Maier D, Seifert E, Nauber U, Schroder C, Kemler R, Jäckle H (1986) A conserved family of nuclear proteins containing structural elements of the finger protein encoded by *Krüppel*, a *Drosophila* segmentation gene. Cell 47:1025–1032

Wilkinson DG, Bhatt S, Chavrier P, Bravo R, Charnay P (1989) Segment-specific expression of a Zinc-finger gene in the developing nervous system of the mouse. Nature 337:461–464

[6] Bodner M, Castrillo J-L, Theill LE, Deerinck T, Ellisman M, Karin M (1988) The pituitary-specific transcription factor GHF-1 is a homeobox-containing protein. Cell 55:505–518

He X, Treacy N, Simmons DM, Ingraham HA, Swanson LW, Rosenfeld MG (1989) Expression of a large family of POU-domain regulatory genes in mammalian brain development. Nature 340:35–42

Herr W, Sturm RA, Clerc RG, Corcoran LM, Baltimore D, Sharp PA, Ingraham HA, Rosenfeld MG, Finney M, Ruvkun G, Horvitz HR (1988) The POU domain: a large conserved region in the mammalian pit-1, oct-1, oct-2, and *Caenorhabditis elegans* unc-86 gene products. Genes Dev 2:1513–1516

Schöler HR, Balling R, Hatzopoulos AK, Suzuki N, Gruss P (1989) Octamer-binding proteins confer transcriptional activity in early mouse embryogenesis. EMBO J 9:2551–2558

Schöler HR, Hatzopolous AK, Balling R, Suzuki N, Gruss P (1989) A family of octamer-specific proteins present during mouse embryogenesis: evidence for germline-specific expression of an Oct-factor. EMBO J 9:2583–2556

[7] Desplan C, Theis J, O'Farrell PH (1988) The sequence specificity of homeodomain-DNA interaction. Cell 54:1081–1090

Green S, Chambon P (1987) Oestradiol induction of a glucocorticoid-responsive gene by a chimaeric receptor. Nature 325:75–78

Han K, Levine MS, Manley JL (1989) Synergistic activation and repression of transcription by *Drosophila* homeobox proteins. Cell 56:573–583

Hoey T, Levine M (1988) Divergent homeobox proteins recognize similar DNA sequences in *Drosophila*. Nature 332:858–861

Hollenberg SM, Giguere V, Segui P, Evans RM (1987) Colocalization of DNA-binding and transcriptional activation functions in the human glucocorticoid receptor. Cell 49:39–46

Jaynes JB, O'Farrell PH (1988) Activation and repression of transcription by homoeodomain-containing proteins that bind a common site. Nature 336:744–749

Krasnow MA, Saffman EE, Kornfeld K, Hogness DS (1989) Transcriptional activation and repression by Ultrabithorax proteins in cultured *Drosophila* cells. Cell 57:1031–1043

Laughon A, Scott MP (1984) Sequence of a *Drosophila* segmentation gene: protein structure homology with DNA-binding proteins. Nature 310:25–31

Miller J, McLachlan AD, Klug A (1985) Repetitive zinc-binding domains in the protein transcription factor III A from *Xenopus* oocytes. EMBO J 4:1609–1614

Odenwald WF, Garbern J, Arnheiter H, Tournier-Lasserve E, Lazzarini RA (1989) The Hox 1.3 homeo box protein is a sequence-specific DNA-binding phosphoprotein. Genes Dev 3:158–172

Samson M-L, Jackson-Grusby L, Brent R (1989) Gene activation and DNA binding by *Drosophila* Ubx and Abd-A proteins. Cell 57:1045–1052

Shephard JCW, McGinnis W, Carrasco AE, DeRobertis EM, Gehring WJ (1984) Fly and frog homoeodomains show homologies with yeast mating type regulatory proteins. Nature 310:70–71

Winslow GM, Hayashi S, Krasnow M, Hogness DS, Scott MP (1989) Transcriptional activation by the Antennapedia and fushi tarazu proteins in culture *Drosophila* cells. Cell 57:1017–1030

Chapter 27 Cloning Development Mutants from the Mouse t Complex

DENISE P. BARLOW

27.1 Introduction

27.1.1 Why Clone Mouse Developmental Mutants?[1]

An understanding of the molecular basis of development depends on the identification and functional analysis of key regulatory molecules. One way in which this can be done is by cloning the genes responsible for developmental mutations. In organisms such as *Drosophila* and *Caenorhabditis* this approach has been successful in identifying a large number of molecular pathways that regulate early decisions in development. A similar approach in the mouse to study mammalian development is made more difficult by the long generation time, the genome size, and the inaccessibility of the early embryo. Despite these difficulties, more than 700 developmental mutations have been identified in the mouse, of which 100 are embryonic lethals, and the possibility of using new techniques in molecular genetics to isolate the genes responsible for these mutations has stimulated fresh interest in classical mouse genetics. One area of the mouse genome where a large number of developmental mutations have been identified is the t complex, and the purpose of this chapter is to summarize what is now known of the structure and organization of the t complex and to describe how the new techniques in mammalian molecular genetics can be specifically applied to clone mutant genes from this region.

27.1.2 The Mouse t Complex[2]

The name "t complex" has been given to a region that includes most of the proximal part of mouse chromosome 17. The extent of this region is now known, from genetic mapping and in situ-hybridization, to be in the order of 15–20 centiMorgans (cM), or 30–40 Megabases (Mb). A diagram of mouse chromosome 17 showing the position and size of the t complex and the location of the reference loci used to define its limits is shown in Fig. 1. Forty-two developmental mutants, including thirty-two that are lethal at various stages of embryogenesis, have been mapped into this region. With a few exceptions, these mutants are defined only by their phenotype and map position, very little is known of the tissue and embryonic stage affected. The chromosomal map, which positions the mutant gene relative to cloned chromosomal markers, offers at present the best opportunity for molecular access to these genes.

27.1.3 Historical Perspective[3]

Interest in the t complex began more than 60 years ago with the discovery of a dominant short-tailed mutant in the offspring of an X-ray-irradiated male mouse. This mutation was mapped to the linkage group later identified as chromosome 17, and named *Brachyury* (locus symbol *T* for tail phenotype, with a capital letter because the effect on tail length is dominant). The existence of the t complex itself was later identified when crosses between the *T* mutant and wild mice generated an additional tailless phenotype. Approximately 20% of wild mice when mated to *Brachyury* mutant mice produced offspring that had no tails. These mice were initially thought to carry only a recessive mutant *T* allele and the mutation was named "*t*" (t for tail phenotype and a lower case letter for a recessive mutation). While *T* produced a short tail in heterozygous animals, the recessive allele did not. The chro-

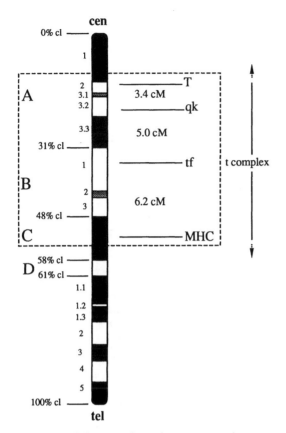

Fig. 1. Extent of the t complex region on mouse chromosome 17. The standard chromosome idiogram after Giemsa banding is shown. The labels on the *left hand side* of the idiogram refer to the numbered Giemsa bands, the percentage chromosome length (*cl*) is also shown. The *right hand side* shows the reference loci *T, qk, tf,* and *MHC* (see Table 1) that define the chromosomal region involved in the t complex, with distances given in centiMorgans (cM). tf is a recessive gene affecting hair development, mutant homozygotes show repeated waves of hair loss. The approximate proximal and distal borders of the t complex have been determined by chromosomal in situ hybridization. *cen* centromere; *tel* telomere

complex, also located on chromosome 17, and the name *T/t* complex was then applied to indicate the existence of a series of functionally related genes linked to *T.* The name "t complex" is in use today but it is now an anachronism, since although experiments have confirmed the existence of a large number of embryonic-lethal genes on chromosome 17, there is no evidence to suggest any functional or structural relationship to each other.

27.2 Current Understanding of t Complex Genetics [4]

Genetic and molecular studies by many laboratories throughout the world have in the last decade revolutionized our understanding of the mouse t complex. The first breakthrough came with the demonstration that the *T/t* locus and the different embryonic-lethal genes were nonallelic, and that the embryonic lethal genes could be separated into three clusters. These data suggested that the size of the chromosomal region associated with the t complex spanned almost the entire proximal half of chromosome 17. Furthermore, the recessive embryonic-lethals could be separated into different complementation groups (14 in total, see Table 1) such that the homozygotes from two different complementation groups were viable. However, because of the population structure of *Mus musculus* these viable homozygotes are not found in the wild. *Mus musculus* is the name for the common house mouse, from which most laboratory strains are derived. The second breakthrough came with the understanding that two different structural forms of the t complex region co-exist in *Mus musculus* populations. One form of the t complex is now considered the "wild type", and chromosomes containing this form are called wild-type chromosomes and given the symbol " + ". The second form which contains the recessive "*t*" allele linked to an embryonic lethal gene, is called the "t haplotype". Chromosomes containing this form are called t haplotype chromosomes and give the symbol "t". Twenty percent of wild mice carry the t haplotype form of the t complex. Wild mice can exist as homozygotes for the wild-type form of the t complex

mosomes carrying the recessive "*t*" allele were later shown also to carry a recessive embryonic-lethal gene. Several different recessive embryonic-lethal genes were identified on chromosomes carrying the recessive "*t*" allele. Initial mapping studies and the analysis of the developmental lethal phenotypes were interpreted as suggesting that "*t*" associated embryonic lethals were a group of tightly linked genes that acted in a coordinated manner in early development. An analogy was made with the *H-2*

Table 1. Embryonic lethal developmental mutants mapped to the t complex

Locus	Phenotype and biology	Locus	Phenotype and biology
T	*brachyury,* six alleles known Semi-dominant gene: heterozygotes have short tails, homozygotes are lethal at E10 and show absence of the notochord. The gene has been cloned		plasma triglyceride and cholesterol levels due to lypolytic enzyme deficiency and die perinatally
		qk	*quaking,* three alleles known. Recessive gene with pleiotropic effects. qk homozygotes show neonatal onset mylelin deficiency and whole body tremors, male mice are sterile. qk^L (see above) a lethal allele of unknown etiology
Fu	*Fused,* three alleles known Semi-dominant gene: heterozygotes have kinked and shortened tails, homozygotes vary in severity; *Fused* is viable but with more severe tail defects, but Fu^{ki} and Fu^{kb} are lethal at E7 with neuroectoderm hyperplasia and twinning of embryonic axis	tcl-0	Complementation group includes t^0, t^6. Recessive: death at E6, failure of embryonic ectoderm differentiation, possibly a cell lethal
Tme	*T-associated maternal effect.* The *Tme* gene is expressed only from the maternal chromosome and heterozygotes that delete the maternal allele (T^{hp} and t^{wlub2}) are lethal at E15, generalized edema is the only pathology observed. The defect is nuclear encoded, and the *Tme* locus is thought to be differentially modified during maternal and paternal gametogenesis leading to differential expression during embryogenesis. An imprinted gene has been identified at this locus	tcl-4	Complementation group includes t^4, t^9, t^{w18}. Recessive: death at E10, abnormal at gastrulation, defect in survival of mesoderm precursors. Homozygous embryonic stem cell line established
		tcl-12a, 12b,	Complementation group includes t^{12}, t^{w32}. Recessive: death at E2 late morula stage, failure of compaction and blastocysts do not form. May not be rescued by aggregation chimeras, suggesting generalized cell lethal
L(17)-1 to 12	are ethylnitrosourea (ENU)-induced mutation, with the exception of L(17)-1 only approximate map positions within the t complex have been determined	tcl-wl	Complementation group includes t^{wl}, t^{wl2}. Recessive: death between E9 and birth, with degeneration of ventral half of neural tube and brain. Homozygous fibroblast cells are viable
L(17)-1	Embryonic lethal, shown to be an allele of quaking	tcl-w5	Complementation group contains t^{w5}. Recessive: death at E6.5, embryonic ectoderm degenerates. Homozygous embryonic stem cell line has been established
L(17)-2	Unknown embryonic lethal maps distal to T and to D17Leh66D		
L(17)-3	Unknown embryonic lethal maps close to Fu	tcl-w73	Complementation group includes t^{w73} recessive: death at E6, failure of implantation, probably at the level of trophoblast outgrowth. Contained within the T^{hp} and t^{wlub2} deletions
L(17)-4	Unknown embryonic lethal maps between tf and the MHC		
L(17)-5	Unknown embryonic lethal maps between tf and the MHC	tcl-wlubl	Complementation group contains t^{wlubl}. Recessive: unknown presumed embryonic lethal. The partial t haplotypes t^{wlub2} and Tt^{Orl} were derived from unequal recombination between t^{wlubl} and a wild-type chromosome and these have been used to map the *Tme* and t^{w73} loci
L(17)-6	Unknown perinatal lethal maps close to Fu		
L(17)-7	Unknown embryonic lethal maps in proximal t complex		
L(17)-8	Unknown perinatal lethal maps between Fu and tf	tcl-wPa[1]	twPa-1 Unknown embryonic lethal[a] (extinct)
L(17)-9	Unknown embryonic lethal maps distal to MHC	tcl-tTuw20	tTuw20 Unknown embryonic lethal
		tcl-tTuw2	tTuw2 Unknown embryonic lethal
L(17)-10	Unknown embryonic lethal maps maps in proximal t complex	tcl-tTuw25	tTuw25 Unknown embryonic lethal
		tcl-tTuw27	tTuw27 Unknown embryonic lethal
L(17)-11	Unknown embryonic lethal maps distal of tf	tcl-tTuw24	tTuw24 Unknown embryonic lethal
L(17)-12	Unknown embryonic lethal maps close to tf	tcl-tTuw28	tTuw28 Unknown embryonic lethal
cld	*combined lipase deficiency* Recessive gene, homozygotes show abnormal	tcl-tTuw12	tTuw12 Unknown embryonic lethal
		tcl-tTuw11	tTuw11 Unknown embryonic lethal

[a] Lethal phenotype not yet described but an embryonic lethal gene is presumed to exist in all complete t haplotypes (see text)

(+/+), or, as heterozygotes for the wild type and t haplotype forms (+/t), but not as t/t homozygotes (because of the recessive embryonic-lethal gene). t/+ mice are morphologically indistinguishable from +/+ mice but they can be distinguished genetically by the phenomena listed below. Evolutionary studies of the t complex in *Mus musculus* and the closely related *Mus spretus* species have led to the suggestion that the t haplotype chromosome is more ancient than the form considered nowadays as the wild type.

27.3 t Complex-Associated Phenomena

27.3.1 Interaction with *Brachyury (T)*[5]

The *Brachyury (T)* locus marks the proximal end of the t complex and is one of the reference loci of this region (see Fig. 1 and Table 1). So far, more than six alleles have been identified, and a candidate for the wild-type gene has been cloned. The original mutant *Brachyury* allele was identified in the wild-type t complex and has a dominant effect on tail length when placed opposite a chromosome also carrying the wild-type form of the t complex. Thus T/+ mice are short-tailed. However, the mutant effect of *T* is enhanced when it is opposite the t haplotype

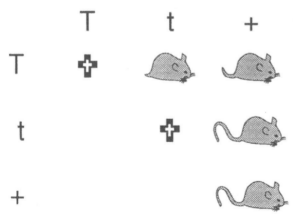

Fig. 2. Interaction of wild type and t chromosomes with T. The tail phenotypes of genetic crosses obtained from T/t (no tail), T/+ (short tail), t/+ (normal tail) or +/+ (normal tail) are illustrated diagrammatically, cross-symbol embryonic-lethal

form of the t complex, such that T/t mice are tailless. This interaction in T/t mice identified a recessive allele present on all t haplotype chromosomes, and is now used as an assay for the presence of the t haplotype form of the t complex (see Fig. 2).

27.3.2 Embryonic Lethality[6]

Sixteen recessive embryonic-lethal genes, that define 16 complementation groups, have been identified in the t haplotype form of the t complex. The intensive interest in the t complex region has led over the years to the identification of additional embryonic-lethal genes in the wild-type form of the t complex. In total, 32 recessive embryonic-lethal developmental mutants have now been mapped to the 30–40 Mb t complex region. The identification of such a high density of essential embryonic genes on chromosome 17 is attributable to the intensive analysis this system has received over the past 60 years, and does not reflect a concentration of specific embryonic genes on this chromosome. The t complex embryonic-lethal genes were identified in both the wild-type and t haplotype forms of the t complex, in natural and laboratory-bred populations, and also following chemical mutagenesis. These mutant loci are listed and briefly described in Table 1, and their mapped position within the t complex shown in Fig. 3. The past 3 years have seen the identification, using the methods described in this chapter, of the *T* gene that is essential for mesoderm formation, of an imprinted gene at the *Tme* locus, of several genes near the t^{w5} mutation, and of three candidates genes involved in segregation distortion (see below). Several other genes, including t^{w18}, *Fused*, and *quaking* are now the subject of intensive cloning efforts.

27.3.3 Male Sterility and Altered Gamete Transmission Ratios[7]

In addition to the embryonic lethal developmental mutants, many genes affecting male germ cell development and function have been mapped within the t complex region (see Table 2 and Fig. 4). The

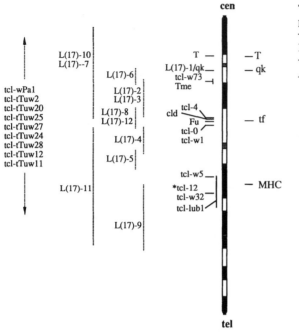

Fig. 3. Thirty-two embryonic-lethal developmental mutants map to mouse chromosome 17. The genetically mapped position, relative to the reference loci, of 32 embryonic-lethal developmental mutants is indicated. The mutants listed on the *left hand side* of the diagram, whose position is indicated by *dotted lines,* are mapped with a low level of resolution, but they are known to be nonallelic to the other chromosome 17 mutants. *t cl-12 and tcl-w32 each have two separate but closely linked loci

Table 2. Male germ-cell function mutants mapped to the t complex

Locus	Phenotype and biology
Hst-1	Hybrid sterility-1. Three alleles exist and males heterozygous for these are sterile. Spermatogenesis is abnormal. Females unaffected
Tas	T-associated sex reversal. A locus affecting primary sex determination, linked to the Thp deletion, but now may not map to chromosome 17
Tcd	t-Complex ratio distorter. A series of four loci distributed throughout the t complex region, which act cumulatively to increase the transmission ratio in males of the chromosome carrying the t form of the Tcr locus
Tcs	t-Complex sterility. A series of four loci which may be identical with the Tcd loci, but are recessive. Tcs loci act cumulatively to cause sterility in males in the presence of wild-type and t forms of the Tcr locus. A candidate gene for Tcs-1 has been cloned
Tcr	t-Complex ratio responder. A single locus regulated by the Tcd and Tcs series of loci. A candidate gene for this locus has been cloned

most interesting of these are the series of loci that affect the transmission ratio of gametes in t/+ heterozygote male mice. These mutant alleles, in contrast to the embryonic-lethal genes, are common to all t haplotype complementation groups, and are a key factor in the propagation of the t haplotype chromosome in wild mouse populations. Male mice, which are heterozygous for the wild type and t haplotype form of the t complex, transmit the t chromosome-bearing gamete to up to 99% of their progeny because the gametes containing the wild-type form become dysfunctional during meiosis. The mechanism is unknown, the disadvantaged wild-type gametes appear morphologically normal with normal motility, but it has been suggested that in some cases they can show a premature acrosome reaction which would block fertilization. Genetically two types of loci have been identified, multiple distorter loci and a single responder

locus, and a model has been proposed in which the products of the distorter loci act on the responder locus producing transmission distortion in favor of the chromosome containing the t form of the responder locus. It is not known if the interaction between the distorter loci and the responder locus takes place at the RNA or protein level, but the recent cloning of a candidate gene for the responder locus and for one of the distorter loci will allow insights into the molecular basis of this phenomenon. It is possible to breed mice that are homozygous for the t form of the t complex if they are derived from different embryonic-lethal complementation groups (e.g., t^{w5}/t^{w73}, see Table 1); in this case female mice are normal but the males are sterile. The loci involved in this sterility are genetically coincident with those affecting transmission distortion and it has been suggested that the sterility and distortion genes are identical.

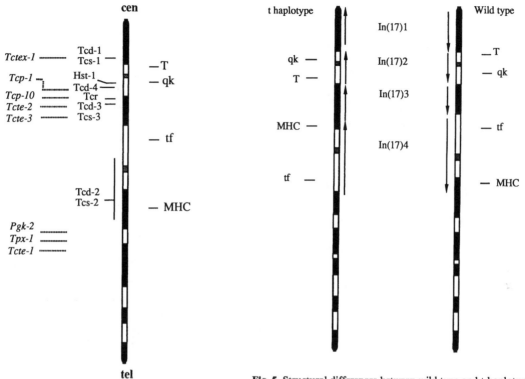

Fig. 4. Male germ-cell mutants and testes-specific genes mapped to chromosome 17. The genetically mapped position, relative to the reference loci, of male germ-cell mutants and cloned testes-expressed genes is indicated. The mutants are listed in *plain type* and the genes, which show either testes-specific or increased expression, are in *italics*. *qk*, although a neurological mutant, also has effects on male fertility, see Table 1

Fig. 5. Structural differences between wild type and t haplotype forms of chromosome 17. The t haplotype chromosome contains four neighboring inversions In(17)1, In(17)2, In(17)3 and In(17)4. The extent of these inverted regions and the equivalent regions on the wild-type chromosome is indicated by the *arrows*. The positions are shown of four reference loci *T, qk, tf*, and the *MHC*, that are inverted in the t haplotype chromosome with respect to the wild-type chromosome

27.3.4 Recombination Suppression[8]

Identification of the existence of the t haplotype form of the t complex soon led to the observation that recombination was suppressed along the length of the t complex region in t/+ heterozygotes. Recombination occurred freely outside the t complex, and also within the t complex region in +/+ and t^x/t^y homozygotes (where x and y are embryonic-lethal genes from different complementation groups). The phenomenon of recombination suppression was a source of continual frustration in the early work on the t complex; genetic analysis had identified the phenomenon but could provide few if any answers and the puzzle was not completely solved until molecular markers were available to

analyze the structure of the t and + forms of the chromosome. This molecular analysis showed that the t haplotype chromosome contains four neighboring inversions when compared to the wild-type form (see Fig. 5). Once these structural differences were identified many features of the t complex became interpretable and potentially accessible using current molecular biology techniques.

27.4 Cloning t Complex Developmental Mutants

The developmental mutant genes identified in the t complex are not readily accessible by standard cloning strategies either because the predicted time of expression occurs too early in development to allow access to tissue (e.g., t^{w5}, see Table 1) or because the phenotype cannot be used to predict which tissue would express the gene (e.g., *Tme*, see Table 1). The cloning strategy described here is that of cloning the genomic region that has been defined by high-resolution mapping to contain the gene, and then to search this genomic DNA for the target gene. This strategy of moving from the phenotype to the gene has also been called reverse genetic cloning and is particularly applicable to the t complex because of the high density of target genes and chromosomal markers. Four steps are involved.

Step 1. Mapping the mutant gene relative to cloned flanking chromosomal markers.

Step 2. Cloning the genomic DNA spanned by the flanking chromosomal markers.

Step 3. Identifying coding sequences in this genomic DNA.

Step 4. Correlation of a candidate gene with the developmental mutant.

27.4.1 Step 1: Mapping the Mutant Gene with Respect to Flanking Chromosomal Markers [9]

The goal here is to position the mutant gene with respect to the nearest cloned chromosomal markers, in order to determine the maximum region containing the gene. This is done initially by genetic mapping which gives the relative order of the gene and chromosomal markers, and secondarily by physical mapping to give an accurate measure of the distance separating chromosomal markers that flank the gene. These flanking markers can then be used as molecular entry points into the genomic region.

27.4.1.1 t Complex Chromosomal Markers

More than 80 chromosomal markers (excluding 50 markers from the relatively small region occupied by the Major Histocompatability Complex, MHC)

have now been genetically mapped to mouse chromosome 17, of these 60 map to the t complex. These markers have been obtained from several sources including the random mapping of genes and the directed cloning of chromosome 17-specific DNA sequences, obtained by microdissection of metaphase chromosomes and from somatic-cell hybrids. In addition comparative mapping between the human and mouse genome has identified human chromosome 6 as showing homology to mouse chromosome 17, and many genes mapped to the t complex were identified because they were first mapped to the human chromosome.

27.4.1.2 Genetic Mapping of the t Complex [10]

Figure 6 shows a genetic map of 45 chromosomal markers using data compiled from several genetic mapping systems. Genetic mapping crosses are generated by crossing two different inbred strains of mice that differ (i.e., are polymorphic) at a large number of gene loci, and then crossing the first generation (F_1) back to one of the parental strains. Meiotic recombination that occurred in the F_1 generation will then result in chromosomes in the second generation (F_2) that are composites from both parents. The recombination breakpoints will occur at random along the length of the chromosome. Chromosomal markers are mapped in such crosses by firstly identifying that the marker can detect a difference between the two parental chromosomes, and secondly by asking if two markers always segregate together if a large number of meioses are examined, or if they are frequently separated by recombination. Closely linked markers will not be separated by recombination. The distance-separating markers analyzed in a genetic cross is given in centiMorgans (a centiMorgan is defined as a distance in which the probability of a crossover event in 100 meiosis is 1%). Many genetic mapping systems have been used for the t complex including Recombinant Inbred (RI) strains and t/+ heterozygotes. A recent addition useful because of the high probability of gene polymorphisms, is the use of an interspecies cross using *Mus spretus* and *Mus musculus*. An extra mapping system is the use of a panel of deletion chromosomes that cover most of the t complex (see Fig. 7), which have been

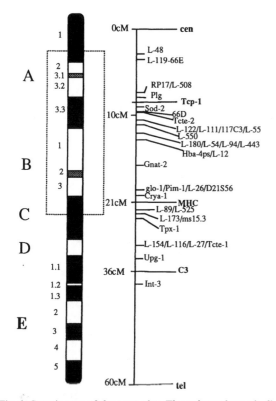

Fig. 6. Genetic map of the t complex. The estimated genetic distance separating 40 chromosomal markers, spanning a region of approximately 38 cM from the wild-type t complex is shown. The position of Tcp-1 and the MHC was determined by chromosomal in situ hybridization. The alignment of chromosomal markers with Giemsa banding shown here is speculative, and may change when more markers are available. It is derived from three sources of information, in situ hybridization of two markers, an estimate of total chromosome length as 115 Mb and an estimate of the recombination length as approximately 60 cM (1 cM is approximately equal to 2 Mb in the mouse). For reasons of space the complete locus name has been abbreviated for the random chromosomal markers, e.g., *L-48 = D17Leh-48*

essential in defining the physical limits of the genomic region that contained genes such as *T*, *Tme*, and *Fu*.

27.4.1.3 Physical Mapping of the t Complex[11]

Genetic maps are able to order markers and to provide a measure of distance in terms of recombination frequency. One cM can normally be considered equal to 2000 kb in the mouse. However, the control of recombination is not understood and this number can vary dramatically due to factors that suppress or enhance recombination such as the proximity of the centromere and the sex of the animal. It is essential to have an accurate measure of genomic distance for the strategy described here because it is difficult and time-consuming to clone and analyze for coding sequences, long regions of the mammalian genome. A number of factors such as the presence of repetitive DNA and low or restricted levels of expression of potential genes restrict this type of analysis to regions of between 500 and 1000 kb. Pulsed-field gel long-range restriction mapping is able to measure genomic distance in kilobases and thus provide the accuracy essential for reverse genetic cloning strategies (see Fig. 8). Pulsed-field gel (PFG) electrophoresis allows the separation of very large DNA fragments by subjecting DNA to two field pulses which are applied sequentially at an angle of from $120-180°$ to each other; the ability of these large fragments to reorientate themselves to each new field pulse is inversely dependent on the size of the molecule. Large DNA fragments of between 0.5 to 4 Megabase pairs (Mb) can be generated by digestion with enzymes that cut rarely in the mammalian genome. These enzymes contain one or two CpG dinucleotides in their recognition sequence and are "rare cutters" because CpG is under-represented in the mammalian genome, and also because cleavage is inhibited if the cytosine of the CpG dinucleotide is methylated. The original PFG apparatus has been improved by several modifications, and detailed protocols have made this a routine laboratory technique. The physical map is constructed by testing it two chromosomal markers, known to be close from genetic mapping, recognize the same large DNA fragment using standard DNA blot and hybridization techniques. Attempts to construct a full length long-range restriction map encompassing the entire t complex region have met with partial success. Currently six long-range restriction maps that total 19 Mb have been constructed using 23 chromosomal markers (Fig. 9). Completion of the t complex restriction map will be delayed until more chromosomal markers can be obtained or until methods can be developed to extend the size of the map that can be generated around any single

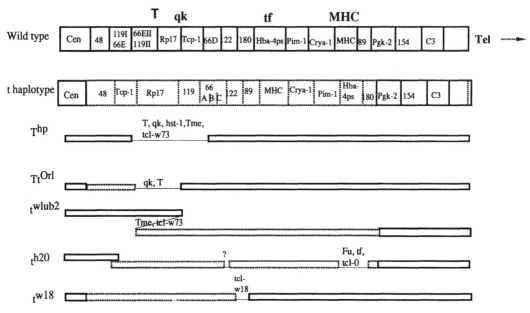

Fig. 7. Gene mapping with deletion chromosomes. Five deletion chromosomes that have been used to define the limits of t complex developmental mutants are shown. The proximal half of chromosome 17 is shown and this has been divided into 17 segments each defined by a DNA marker. The *top diagram* shows the order of markers in the wild-type chromosome with the approximate position of the reference loci shown above (*solid lines*), underneath the order of markers in the t haplotype is given (*dotted lines*). The names for random chromosomal markers have been abbreviated to the number, e.g., 48 = D17Leh48. The *thin solid line* in each of the five chromosomes indicates the region that is deleted for each chromosome. The developmental mutants that have been correlated with marker loss in these deletions are *listed above this line*. Some chromosomes, that resulted from recombination across In(17)2 (t^{wlub2}) or In(17)4(t^{w18}) inversion, have been shown to contain *duplicated regions*

DNA marker. However, despite the current limitations, PFG maps are being applied to analyze specific sub-regions in the t complex, and maps have been constructed that determine the limits of the *Brachyury*, *Tme* and t^{w73}, and *Tcr* genes.

27.4.2 Step 2: Cloning the Genomic DNA Containing the Mutant Gene Locus[12]

The approach taken in this step will depend on the results of the genetic and physical mapping. If the only information available is one of direction, i.e., that a marker is proximal or distal to the defined position of the mutant gene, then this stage will require movement along the chromosome in the direction of the gene until the point is reached when the gene is identified or when a marker is isolated that is now on the distal side of the mutant gene. This strategy has been successfully used to clone

the *T* gene, which had been mapped to a position distal to the *D17Leh119II* locus but proximal to the *D17Rp17* locus (see Fig. 6). Since these markers are separated by 3 cM and unlinked by PFG mapping, it was not feasible to analyze all the DNA spanned by these markers, and the search for the *T* gene relied on tight genetic linkage following analysis of 380 meioses, which positioned the mutant very close to the *D17Leh119II* locus. If the target gene can be identified within closely spaced flanking markers then both of the markers can be used as molecular entry points into the genomic region containing the mutant gene. Several new technical advances have made cloning large regions of the mammalian genome less labor intensive. There are three choices for cloning contiguous stretches of genomic DNA: phage vectors (mean insert size of 8 – 15 kb), cosmid vectors (mean insert size of 38 – 42 kb) and yeast artificial chromosomes (mean insert size 200 – 800 kb). Two additional methods

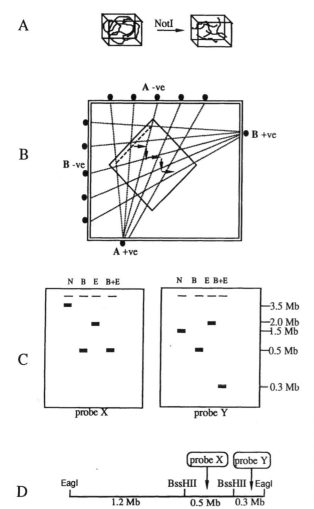

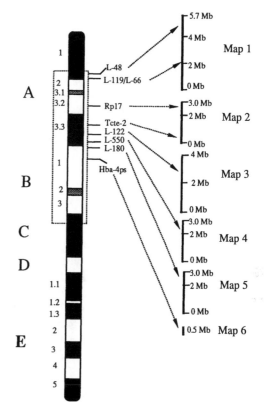

Fig. 8A–D. Physical mapping by pulsed field gels. **A** Genomic DNA prepared by lysing cells, embedded in agarose in order to prevent mechanical shearing of the DNA, is digested with "rare cutter" enzymes such as Not I that generates fragments in the 1–3 Megabase size range. **B** The agarose-embedded DNA is loaded into a standard agarose gel and subject to pulsed-field gel electrophoresis using sequential field pulses (Fields *A* and *B*) applied at an angle of 120°. **C** After separation the DNA is transferred to a nylon membrane and hybridized, sequentially with radiolabeled chromosomal markers X and Y (N, Not I, B, BssH II, E, Eag I). **D** The sizes of the bands that are co-recognized by probes X and Y are used to construct a long-range restriction map

Fig. 9A–E. Not I restriction map of the proximal part of chromosome 17. Long-range restriction maps of Not I DNA fragments spanning approximately 19 Mb from the wild-type t complex of the BALB/c mouse (Barlow and Lehrach, unpublished data). The six noncontiguous maps span the region from *D17Leh-48*, the most proximal t complex marker, to *HBa-4ps*, a marker from the In(17)3 (see Fig. 5). *Map 1* includes a additional inverted duplication of the 119-66 region.[34,55] *Map 2* is a contiguous restriction map including *D17Rpl7, D17Leh-508, Plg, Igfr-2, Tcp-1, Sod-2, D17Leh66D, Tcte-2*. This map contains four mutant loci of interest, *Tme, t^{w73}, Tcr,* and *Tcd-3/Tcs-3* (described in Tables 1 and 2), and crosses the distal boundary of In(17)2 (see Fig. 5). *Map 3* contains *D17Leh-122, D17Leh111, 117C3* (renamed *Tcte-3*), *D17Leh-55;* the orientation of these four loci is unknown, but they are all in the middle region and In(17)3. *Map 4* is defined by only one locus, *D17Leh-550*, it is not yet clear from genetic mapping if this marker is in In(17)3 or in In(17)4. *Map 5* contains two loci from In(17)4, *D17Leh-180* and *D17Leh-54*, that are located within the *t^{w18}* deletion. *Map 6* contains two loci *Hba-4ps* and *D17Leh-12* in In(17)4 that are located within the *t^{h20}* deletion

for spanning large genomic regions are "band cloning" where large DNA fragments separated by preparative PFGs can be cloned, and chromosome jumping or hopping, a cloning technique that allows the cloning of both ends, but not the intervening region, of DNA molecules up to several hundred kilobase pairs long. Because genomic libraries are not fully representative it is possible that all these types of library would be used at some time.

27.4.3 Step 3: Identifying Coding Sequences in Genomic DNA

The intergenic distance in mammalian DNA is in the order of 50–100 kb, thus a typical phage or cosmid clone may not contain coding sequences. What are the signals that distinguish coding from noncoding sequences in genomic DNA? Described below are three approaches towards identifying genes in cloned DNA. The first two have the advantage that long regions of up to 42 kb of genomic DNA can be analyzed at one time.

27.4.3.1 CpG Islands and Genes[13]

Clusters of the dinucleotide CpG, in which the CpG/GpC ratio is equal and which are free of cytosine methylation, have been found to be associated with the 5' end of many genes. These are known as CpG islands. The exact proportion of genes that are associated with this signal is not clear, but one estimate is that there are 30000 CpG islands and 60000 genes in the mammalian genome. Thus a simple assay is to ask if the cloned genomic DNA contains such an island. This is done by restriction of DNA using enzymes that cleave at CpG rich sequences (the same as are used for PFG analysis), and subsequent confirmation that these sequences are unmethylated in mammalian DNA. This technique was used to successfully isolate a testis-specific transcript from the In(17)3 region of the t complex and the candidate T gene, and has the advantage that genes can be identified when the tissue and time of expression are unknown.

27.4.3.2 RNA and cDNA Analysis[14]

One problem with using RNA blots to assay for genes is that frequently the tissue and time of expression are unknown. However, it may be possible to infer the time of expression. If, for example, the mutant gene is lethal in early development, it can be assumed that the gene is expressed before the lethal period. mRNA can be made from nearly all the early embryonic stages, but generally it is not feasible to consider preparing large amounts from stages earlier than E8 (day 8 of embryonic development). A problem with later stages is that the embryo contains a wide range of tissues and organs, and a tissue-specific gene may not be detected in preparations made from whole embryos. Again, the work involved makes it unfeasible to dissect embryos into organs for large-scale mRNA preparations before E15. mRNA preparation and analysis is expensive in terms of time and materials and a possible alternative is to analyze cDNA libraries by preparing DNA from pools of up to 10000 clones and assaying for the presence of a gene by DNA blot analysis. One advantage of using either RNA or cDNA blots to search for genes is that large genomic regions can be scanned at one time, for example; a candidate gene for the Tcr locus was detected by RNA blots using radiolabeled cosmid-cloned genomic DNA.

27.4.3.3 Conserved Sequences and Genes[15]

Coding sequences, or parts of coding sequences can be conserved among closely related species. Hybridization probes made from mouse genes for example, will frequently detect sequences in rats and hamsters under normal stringency or slightly reduced stringency conditions. Thus it is possible to identify coding sequences in genomic DNA by asking if the sequence is conserved in the DNA from closely related species. This type of DNA blot has been called a "zoo blot", and one of the first successful uses of this approach was the identification of the human Dystrophin gene.

27.4.4 Step 4: Correlation of a Candidate Gene with the Developmental Mutant

It is possible that a successful gene search will identify more than one gene in the chromosomal region that has been determined as containing the mutant gene. How can the candidate gene be distinguished? The final correlation will rely on a gene inactivation experiment, with the demonstration that inactivation of the candidate gene recreates the mutant phenotype (see Wagner and Keller, Chap. 30, this Vol.). However, other data could also be used to pre-select the right candidate. Information from the mutant phenotype could be used make predictions, which could be tested, about the developmental onset and the tissue-specific pattern of expression of a correct candidate gene. Further evidence would come from a demonstration of a difference at the RNA or DNA level between the mutant and wild-type mouse, this evidence would be further strengthened if more than one mutant allele showed a difference in the expression of the same candidate transcript, as was shown for the analysis of the *Brachyury* gene.

27.5 Outlook

One question that needs to be asked is whether mouse geneticists can reach similar goals in understanding mammalian development, as achieved by those working on development problems using *Drosophila* and *Caenorhabditis*. Is a molecular analysis of developmental pathways possible starting from a comparatively limited collection of mouse mutants? The answer is positive. This is because the strategy adopted with other organisms that have simpler genetics, of collecting all of the mutants that act in one developmental cascade, is now not necessary because molecular biology has advanced over the past ten years. The cloning of a single mouse mutant gene such as the *Brachyury* gene, an essential gene for the developmental pathway that results in the formation of mesoderm and mesoderm derivatives, can now be the starting point to isolate other genes 'upstream' and 'downstream'. Transcription factors that regulate expres-

Table 3. Mouse mutants classified by their phenotype

Biological system	Number of mutants
Coat color and white spotting	66
Skin and hair texture	77
Skeleton	116
Tail and other appendages	84
Eye	62
Inner ear and circling behavior	42
Neurological and neuromuscular	97
Behavioral	15
Hematological	24
Endocrine defects, growth, obesity	41
Reproductive system, sterility	30
Internal defects of viscera	44
Immune defects	28
Viral, disease, and tumor resistance	43
Homozygous lethality	102

sion and other proteins that interact with a cloned gene, can now be isolated directly, without prior identification by a mutant screen. One extra bonus for mouse developmental geneticists is the use of Embryonal Stem cell techniques which provide the opportunity to use homologous recombination to specifically inactivate genes, allowing the function of any gene to be analyzed. Thus the prospects for a molecular analysis of mouse development, using many of the overlapping approaches described in this book, are both realistic and exciting, giving us much to look forward to over the next 5 years.

27.6 Summary [16]

The strategy of cloning genes from their mapped chromosomal position offers a new approach to the study of mammalian development because it allows the possibility of cloning any mouse mutant gene of interest. Although a large effort in terms of time is still required to isolate a developmental mutant, it is expected this could be reduced as resources, skills, and techniques improve. One factor that would have a key impact on this strategy would be an improvement of the mouse map in terms of DNA marker density and accuracy, and efforts are being made now in this direction to co-ordinate genetic and physical mapping strategies in the mouse.

For example, a Committee for the Mouse Chromosome 17 has been formed to coordinate the mapping information for this chromosome. The number of available mouse mutants in diverse biological systems is impressively large (more than 700, see Table 3), and is being continually expanded by monitoring activities at the major mouse breeding laboratories. The t complex, because it has the highest density of mapped developmental mutant genes and chromosomal markers, offers the best chance of cloning key embryonic genes by the reverse genetics strategy outlined above. In addition, the number of embryonic mutants mapped to the t complex is being continually increased by chemical mutagenesis experiments using ethylnitrosourea which aim to identify all the genes in this region that are essential for embryonic development. The t complex since its identification in 1927 can now be considered one of the best understood and accessible mouse genetic systems. The identification of the *T* gene, of an imprinted gene at the *Tme* locus and of genes from the segregation distortion system, will illuminate phenomena that have not been accessible through other cloning strategies. The information locked in the mouse t complex is now accessible because of advances in mammalian molecular genetic techniques and will provide data that is essential for understanding the molecular regulation of mammalian development and differentiation.

References

[1] Alberts B, Bray D, Lewis J, Raff M, Watson JD (1989) Molecular biology of the cell, 2nd edn. Garland Pub Co, New York

[2] Evans EP (1985) Standard normal chromosomes. In: Lyon M, Searle AG (eds) Genetic strains and variants of the laboratory mouse, 2nd edn. OUP, UK, pp 576−577

Frischauf A-M (1983) The T/t complex of the mouse. TIG 1:100−103

Silver LM (1985) Mouse t haplotypes. Annu Rev Genet 19:179−208

[3] Dobrovolskaia-Zavadskaia N (1927) Sur la mortification spontanée de la queue chez la souris nouveau-nee et sur l'existence d'un caractère (facteur) hereditaire, non viable. CR Soc Biol 97:114−116

[4] Artzt K (1984) Gene mapping within the T/t complex of the mouse. III. t lethal genes are arranged in three clusters on chromosome 17. Cell 39:565−572

Committee for the mouse chromosome 17 (1991) Maps of mouse chromosome 17: First Report. Mamm Genome 1:5−29

Hammer MF, Schimenti J, Silber LM (1989) Evolution of mouse chromosome 17 and the origin of inversions associated with t haplotypes. Proc Natl Acad Sci USA 86:3261−3265

Klein JK, Sipos P, Figueroa F (1984) Polymorphism of t complex genes in European wild mice. Genet Res Camb 44:30−46

[5] Herrmann BG, Labeit S, Poustka A-M, King TR, Lehrach H (1990) Cloning of T, a gene required in mesoderm formation in the mouse. Nature 343:617−622

[6] Abe K, Weil J-F, Wei F-S, Hsu YC, Uehara H, Artzt K, Bennet D (1988) Searching for coding sequences in the mammalian genome: the H-K region of the mouse MHC is replete with genes expressed in embryos. EMBO J 7:3441−3449

Alton AK (1982) The histological, cellular and molecular correlates of the T (hairpin) maternal defect. Ph D Thesis, Cornell University Medical College, USA

Axelrod HR (1985) Altered trophoblast functions in implantation-defective mouse embryos. Dev Biol 108:185−190

Barlow DB, Stöger R, Herrmann BG, Saito K, Schweifer N (1991) The mouse Igf2 receptor is imprinted and closely linked to the Tme locus on chromosome 17. Nature 349:87

Bennett D (1975) The T-locus of the mouse. Cell 6:441−454

Bennett WI, Gall AM, Southard JL, Sidman RL (1971) Abnormal spermiogenesis in quaking, a myelin deficient mutant mouse. Biol Reprod 5:30−58

Bucan M, Herrmann BG, Frischauf A-M, Bautch V, Bode V, Silver LS, Martin GR, Lehrach H (1987) Deletion and duplication of DNA sequences is associated with the embryonic lethal phenotype of the t9 complementation group of the mouse t complex. Genes Dev 1:367−385

Green MC (1989) Catalogue of mutant genes and polymorphic loci. In: Lyon M, Searle AG (eds) Genetic strains and variants of the laboratory mouse, 2nd edn. OUP, UK, pp 12−404

Greenspan RJ, O'Brien MC (1986) Genetic analysis of mutations at the Fused locus in the mouse. Proc Natl Acad Sci USA 83:4413−4417

Frail DE, Braun PE (1985) Abnormal expression of myelin associated glycoprotein in the central nervous system of dysmyelinatin mutant mouse. J Neurochem 45:1071−1075

Johnson DR (1975) Further observations on the hair-pin tail (T^{hp}) mutation in the mouse. Genet Res Camb 24:207−213

Lyon MF, Jarvis SE, Sayers I (1979) Complementation reactions of a lethal t haplotype believed to include a deletion. Genet Res Camb 33:153−161

Martin GR, Silver LM, Fox HS, Joyner AL (1987) Establishment of embryonic stem cell lines from preimplantation mouse embryos homozygous for lethal mutations in the t complex. Dev Biol 120:20−28

McGrath J, Soltor D (1984) Maternal Thp lethality in the mouse is a nuclear, not cytoplasmic, defect. Nature 308:550−551

Paterniti JR Jr, Brown WV, Ginsberg HN, Artzt K (1983) Combined lipase deficiency (*cld*): a lethal mutation on chromosome 17 of the mouse. Science 221:167–169

Shedlovsky A, King TR, Dove WF (1988) Saturation germ line mutagenesis of the murine t region including a lethal allele at the quaking locus. Proc Natl Acad Sci USA 85:180–184

Shin H-S (1989) The T/t complex and the genetic control of mouse development. In: Litwin SD (ed) Human immunogenetics: basic principles and clinical relevance. Marcel Dekker Inc, New York, pp 443–471

Shin H-S, Bennett D, Artzt K (1984) Gene mapping within the T/t complex of the mouse. IV: the inverted MHC is intermingled with several t-lethal genes. Cell 39:573–578

Wiking H, Silver LM (1984) Characterization of a recombinant mouse t haplotype that expresses a dominant lethal maternal effect. Genetics 108:1013–1020

7 Brown J, Cebra-Thomas JA, Bleil JD, Wasserman PM, Silver LM (1989) A premature acrosome reaction is programmed by mouse t haplotypes during sperm differentiation and could play a role in transmission distortion. Development 106:769–773

Eicher E, Cherry M, Flaherry L (1978) Autosomal phosphoglycerate linked to mouse major histocompatability complex. MGG 158:225–228

Forejit J, Ivani P (1975) Genetic studies on male sterility of hybrids between laboratory and wild mice (*Mus musculus* L.). Genet Res 24:189–206

Lader E, Ha H-S, O'Neil L, Artzt K, Bennet D (1986) Tctex-1: a candidate gene family for a mouse t complex sterility locus. Cell 44:357–363

Lyon MF (1984) Transmission ratio distortion in mouse t haplotypes is due to multiple distorter genes acting on a responder locus. Cell 37:621–628

Lyon MF (1986) Male sterility of the mouse t complex is due to homozygosity of the distorter genes. Cell 44:357–363

KAsahara M, Passmore HC, Klein J (1989) A testes-specific gene maps between *Pgk*-2 and *Mep*-1 on mouse chromosome 17. Immunogenetics 29:61–63

Olds-Clark P, Pietz B (1986) Fertility of sperm from t/+ mice: evidence that + bearing sperm are dysfunctional. Genet Res 47:49–52

Sarvetnick N, Tsai J-Y, Fox H, Pilder S, Silver LM (1989) A mouse chromosome 17 gene encodes a testes-specific transcript with unusual properties. Immunogenetic 30:34–41

Schimenti J, Cebra-Thomas JA, Decker CL, Islam SD, Pilder SH, Silver LM (1988) A candidate gene for the mouse t complex responder (Tcr) locus responsible for haploid effects on sperm function. Cell 55:71–78

Silver LM, Kleene KC, Distel RJ, Hecht NB (1987) Synthesis of mouse t complex proteins during haploid stages of spermatogenesis. Dev Biol 119:605–608

Washburn LL, Lee BK, Eicher EM (1990) Inheritance of T-associated sex reversal in mice. Genet Res 56:185–191

Willison K, Dudley K, Potter J (1986) Molecular cloning and sequence analysis of a haploid expressed gene encoding t complex polypeptide 1. Cell 44:727–738

8 Committee for the mouse chromosome 17 (1991) Maps of mouse chromosome 17: first Report. Mamm Genome 1:5–29

Fox HS, Martin GR, Lyon MF, Herrmann B, Frischauf A-M, Lehrach H, Silver LM (1985) Molecular probes define different regions of the mouse t complex. Cell 40:63–69

Rogers JH (1986) The mouse t complex is composed of two separate inversions. TIG 2:145–146

9 Committee for the mouse chromosome 17 (1991) Maps of mouse chromosome 17: first report. Mamm Genome 1:5–29

Nadeau JH, Reiner AH (1989) Linkage and synteny homologies in mouse and man. In: Lyon M, Searle AG (eds) Genetic strains and variants of the laboratory mouse, 2nd edn. OUP, UK, pp 506–536

Nadeau JH, Herrmann BG, Bucan M, Burkart D, Crosby JL, Erhardt MA, Kowosky M, Kraus JP, Michiels F, Schnattinger A, Tchetgen M-B, Varnum D, Willison K, Lehrach H, Barlow DP (1991) Genetic maps of mouse chromosome 17 loci including 12 new anonymous DNA loci. Genomics 9:78–89

Vincek V, Kawaguchi H, Mizuno K, Zaleska-Rutczynska Z, Kasahara M, Forejit J, Figueroa F, Klein J (1989) Linkage map of mouse chromosome 17: localization of 27 new DNA markers. Genomics 5:773–786

10 Babiarz BS (1983) Deletion mapping of the T/t complex: evidence for a second region of critical embryonic genes. Dev Biol 95:342–351

Committee for the mouse chromosome 17 (1991) Maps of mouse chromosome 17: first report. Mamm Genome 1:5–29

Erickson RP, Lewis SE, Slusser KS (1978) Deletion mapping of the t complex of chromosome 17 of the mouse. Nature 274:163–164

Fox HS, Silver LM, Martin GR (1984) An alpha globin pseudogene is located within the mouse t complex. Immunogenetics 19:125–130

Guenet J-L, Simon-Chazottes D, Avner PR (1986) The contribution of wild derived mouse inbred strains to gene mapping methodology. Curr Top Microbiol Immunol 127:109–113

Mann EA, Silver LM, Elliott RW (1986) Genetic analysis of a mouse t complex clone that is homologous to a kidney cDNA clone. Genetics 114:993–1006

Silver LM, Lukralle D, Garrells JI (1983) TOrl is a novel, variant form of mouse chromosome 17 with a deletion in a partial t haplotype. Nature 310:422–424

Taylor B (1981) Recombinant inbred strains. In: Lyon M, Searle AG (eds) Genetic strains and variants of the laboratory mouse, 2nd edn. OUP, UK, pp 773–796

11 Barlow DP, Lehrach H (1987) Genetics by gel electrophoresis: the impact of pulsed-field gel electrophoresis on mammalian genetics. TIG 3:167–171

Carle GF, Olson MV (1987) Orthogonal-field-alternation gel electrophoresis. In: Wu R (ed) Methods in enzymology, vol 155. Recombinant DNA Part F. Academic Press, New York, pp 468–480

Committee for the mouse chromosome 17 (1991) Maps of mouse chromosome 17: first Report. Mamm Genome 1:5–29

Razin A, Riggs AD (1980) DNA methylation and gene function. Science 210:601–610

Sambrook J, Fritsch EF, Maniatis T (1989) Molecular cloning. A laboratory manual, 2nd edn. CSH Press, Cold Spring Harbor, New York

Schwartz DV, Cantor CR (1984) Separation of yeast chromosome-sized DNAs by pulsed-field gradient gel electrophoresis. Cell 37:67–75

Smith CL, Klco SR, Cantor CR (1989) Pulsed-field gel electrophoresis and the technology of large DNA molecules. In: Davies K (ed) Genome analysis. A practical approach. IRL Press, Oxford, pp 41–72

[12] Berger SL, Kimmel AR (1987) I. Guide to molecular cloning techniques: Section VI. Genomic cloning, pp 173–214. In: Berger SL, Kimmel AR (eds) Methods in Enzymology Vol 152. Acad Press, London, UK

Collins ES (1986) Chromosome Jumping. In: Davies K (ed) Genome analysis, a practical approach. IRL Press, UK, pp 73–93

Michiels F, Burmeister M, Lehrach H (1987) Derivation of clones close to met by preparative field inversion gel electrophoresis. Science 236:1305–1308

Poustka A, Lehrach H (1988) Chromosome jumping – a long range technique. In: Setlow J (ed) Genetic engineering 10. Plenum, New York, pp 169–193

Schlessinger D (1990) Yeast artificial chromosomes: tools for mapping and analysis of complex genomes. TIG 6:248–254

[13] Bird AP (1986) CpG-rich islands and the function of DNA methylation. Nature 321:209–213

Lindsay S, Bird AP (1987) Use of restriction enzymes to detect potential gene sequences in mammalian DNA. Nature 327:336–338

Ohno S (1986) The total number of genes in the mammalian genome. TIG 2:8

Rappold GA, Stubbs L, Labeit S, Crkvenjakov R, Lehrach H (1987) Identification of a testes specific gene from the mouse t complex next to a CpG island. EMBO J 6:1975–1980

[14] Abe K, Wei J-F, Wei F-S, Hsu YC, Uehara H, Artzt K, Bennett D (1988) Searching for coding sequences in the mammalian genome: the H-2K region of the mouse MHC is replete with gene expressed in embryos. EMBO J 7:3441–3449

Hogan BLM, Constantini F, Lacy E (1986) Manipulating the mouse embryo. A laboratory manual. CSH Press, CSH New York

[15] Monaco AP, Neve RL, Colletti-Feneer C, Kunkel LM (1986) Isolation of candidate cDNAs for portions of the Duchenne Muscular Dystrophy gene. Nature 323:646–650

[16] Green MC (1989) Catalogue of mutant genes and polymorphic loci. In: Lyon M, Searle AG (eds) Genetic strains and variants of the laboratory mouse, 2nd edn. OUP, UK, pp 12–404

Shedlovsky A, Guenet J-L, Johnson LL, Dove W (1986) Induction of recessive lethal mutations in the T/t-H-2 region of the mouse genome by a point mutation. Genet Res Camb 47:135–142

Chapter 28   The Use of in Situ Hybridisation
to Study the Molecular Genetics of Mouse Development

David G. Wilkinson

28.1 Introduction

The study of mammalian development is a challenging problem, as even the most amenable system, the mouse, has a number of technical limitations, in particular the difficulty of systematic developmental genetics and of microsurgical manipulations of the embryo. Nevertheless, significant advances towards understanding molecular mechanisms of mouse development have been made through the cloning of many genes with potential roles in embryogenesis. An important step in the analysis of these genes is to determine the spatial and temporal regulation of their expression during development. When combined with other lines of evidence, expression patterns can provide preliminary clues to the developmental function of genes and can also provide insight into mechanisms of development. These lines of evidence include homologies with previously characterised genes (see Lobe and Gruss, Chap. 26, this Vol.), analogies with other developmental systems and the study of mutants. In addition, knowledge of expression patterns is essential for the design and interpretation of reverse genetic manipulations in which genes are inactivated by homologous recombination or misregulated in transgenic mice.

Patterns of gene expression can be analysed at the level of mRNA or protein. The detection of cellular RNA by in situ hybridisation is an amenable method, as specific probes can be rapidly obtained from cloned genes. However, the detection of protein by immunocytochemistry is also important, as patterns of protein and RNA distribution can differ, and knowledge of the cellular location of protein is essential for the analysis of its function. The visualisation of protein distribution is particularly important for secreted proteins, as this can reveal

their sites of action. Nevertheless, the analysis of RNA distribution is frequently an accurate guide to the sites of gene action in embryos, and in situ hybridisation has found widespread use in studies of many developmental systems. This chapter will emphasise the application of in situ hybridisation to analyse gene expression during mouse development. In addition, studies of other developmental systems that illustrate the value, limitations and future applications of the technique will be discussed.

28.2 The Technique of in Situ Hybridisation [1]

The principle behind in situ hybridisation is the specific annealing of labelled nucleic acid probes to mRNA in sections of fixed tissue (Fig. 1). The procedure involves a series of steps: the preparation of probe, the preparation of sections of fixed tissue, hybridisation and washing, and the detection of annealed probe (see references for detailed protocols and discussion of important parameters and controls).

Nucleic acids can be labelled with radioisotopes, or with modified analogues of nucleotides that can be detected by immunological or affinity techniques. Many variations on the type of probe and method of labelling are possible, each with different advantages and drawbacks (Table 1). Single-stranded RNA probes, synthesised by in vitro transcription and labelled with ^{35}S-UTP, have found widespread use due to their high sensitivity. As an alternative, oligonucleotides can be readily designed as specific hybridisation probes, and end-labelled with radioisotopes. However, most radiolabelled probes have a limited shelf-life, give signals

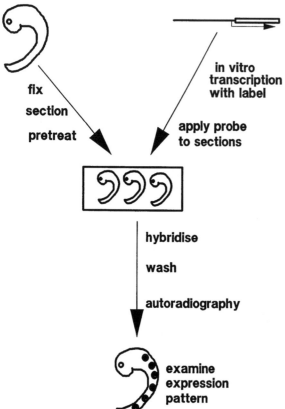

fix

section

pretreat

in vitro
transcription
with label

apply probe
to sections

hybridise

wash

autoradiography

examine
expression
pattern

Fig. 1. The method of in situ hybridisation. The preparation of tissue sections of embryos and of radiolabelled single-stranded RNA probes and their use for in situ hybridisation are depicted. Many variations on this technique are possible, including the use of oligonucleotide probes, modified nucleotides that can be detected with immunochemical reagents, and the whole-mount staining of embryos

Table 1. Advantages and limitations of different methods of in situ hybridisation

	Advantages	Limitations
Labelling method		
^{32}P		Low resolution
^{35}S	Rapid results	
	Good resolution	
^{3}H	Excellent resolution	Long exposure needed
Digoxigenin/ biotin	Very rapid results	
	Stable probe	
	Excellent resolution	
	Whole mount staining	
Type of probe		
Oligonucleotide	Ease of design	Limited sensitivity
dsDNA		Low sensitivity
ssRNA	High sensitivity	
	Reduce background with ribonuclease	

embryos can be precisely orientated to obtain appropriate planes of section. In general, crosslinking fixatives, such as paraformaldehyde, are used, as these have been found to be optimal for the retention of RNA in the section during hybridisation.

Tissue sections are treated with acetic anhydride in order to reduce nonspecific binding of the probe and are digested with protease to increase the accessibility of the RNA. Hybridisation is carried out under coverslips, and then nonspecifically bound probe is removed by washing under moderate or high stringency conditions and, if RNA probes are used, by digestion with ribonuclease A. Finally, the localisation of hybridised probe is revealed, either by autoradiography or by immunological detection methods.

that are not at a cellular resolution, and may require exposure times of 1 week or longer. These limitations can be overcome by nonradioactive methods, for example by the use of nucleotides coupled to digoxigenin and an antibody against this modified base.

Two methods are available for the preparation of tissue sections. Embryos can be frozen, sections cut on a cryostat and then mounted and fixed on slides. Alternatively, embryos can be fixed and then embedded in paraffin wax prior to cutting sections on a microtome. The latter method has a number of advantages, including a superior histological preparation of the tissues and the ease with which the

28.3 The Value of Analysing Expression Patterns

The spatial distribution of transcripts can be determined by the in situ hybridisation analysis of a complete series of longitudinal and transverse sections, and by analysing embryos at different developmental stages, a picture can be built up of how the expression pattern changes through time. How can this information be used to advance our understanding of the function of the gene in development? In isolation, the expression pattern says nothing about gene function, and thus the interpretation of in situ hybridisation data is dependent on other lines of evidence.

28.3.1 Correlation of Gene Expression with Developmental Events[2]

A number of genes identified by their biological effects on adult tissues or cell culture systems are found to be expressed in embryos, and it is frequently assumed that they exert similar effects in these different contexts. Similarly, newly identified members of multigene families are often presumed to have analogous developmental roles to their better characterised relatives. The analysis of developmental expression patterns provides important circumstantial evidence that can argue for or against these proposals. Studies of proto-oncogene expressions are a good example of how in situ hybridisation analysis can yield evidence contrary to ideas based on the initial context of their discovery. A simple view of the association of these genes with uncontrolled cell growth in tumours is that they have a normal role in cell proliferation. However, in situ hybridisation studies of embryos have, almost without exception, suggested a more complex situation (Table 2). Some proto-oncogenes are expressed in post-mitotic cells, for example neurons, arguing strongly that the gene normally functions in this tissue in processes other than proliferation. Other proto-oncogenes are expressed in complex patterns that correlate with proliferation in some tissues, but not others, suggestive of multiple, tissue-specific functions. Indeed, there is increasing evidence that certain proto-oncogenes are components of signal transduction pathways that are utilised for many different purposes during development, and that others have more specialised roles in, for example, pattern formation and cell differentiation.

The correlation of gene expression with developmental events can also provide positive evidence re-

Table 2. In situ hybridisation used to examine relationship between proto-oncogene expression and cell proliferation in embryos

Proto-oncogene	Expression in proliferating tissue Correlates with proliferation?	Expression in post-mitotic cells
c-*myc*	Yes/no	Yes
c-*src*	No	Yes
c-*mos*	Not applicable	Yes
int-1	No	No
int-2	No	Yes
c-*fos*	Yes	Yes
c-*jun*	Yes	Yes

Proto-oncogenes are widely assumed to have roles in the control of cell proliferation. In situ hybridisation studies reveal that the developmental expression of some proto-oncogenes correlates with proliferation in certain tissues, but not others (c-*myc*). For other proto-oncogenes, this correlation does not hold in any proliferating tissues (e.g., *int*-2). Moreover, many of these genes are expressed in post-mitotic cells. These data indicate that the roles of proto-oncogenes in development are not solely associated with cell proliferation.

garding potential gene function, and focus attention on particular lines of enquiry. Examples of this are provided by studies of the developmental expression of members of the TGFβ (transforming growth factor β) gene family. TGFβ has been shown to evoke tissue-specific responses in the proliferation, migration and differentiation of cells, and thus this and related genes seem likely to have important roles as cell signalling molecules during development. Thus, it is pertinent to examine whether developmental expression correlates with changes in cell phenotype, or with known sites of cell-cell interactions. Indeed, in situ hybridisation studies have provided several examples of such correlations. In developing skin, TGFβ2 expression first occurs in the mesenchyme, then shifts to the epithelium, and thus could have a role in the progressive interactions that occur between these two cell populations. Similarly, TGFβ1, -β2 and -β3 are each expressed in distinct patterns that shift between the different, interacting cell populations of developing teeth. Studies of TGFβ1 are particularly good examples of the value of studying gene expression both at the level of RNA and protein. In situ hybridisation studies revealed TGFβ1 RNA in a number of epithelia, while immunocytochemistry revealed TGFβ1 protein in adjacent meseenchymal cells, indicating that this gene product acts as a paracrine signal during development. Since a number of epithelial-mesenchymal interactions can be reproduced in the in vitro culture of explanted tissues, for example teeth, these in situ hybridisation studies should now lead to tests of the functions of these genes in appropriate systems.

28.3.2 Conserved and Divergent Expression Patterns in Different Organisms[3]

An increasing number of genes are being found in the mouse that are structurally related to genes identified in more amenable developmental systems, in particular the fruit fly *Drosophila melanogaster* (see Lobe and Gruss, Chap. 26, this Vol.). In certain cases, in situ hybridisation studies have revealed similar expression patterns in the mouse as in *Drosophila*, suggesting that these genes have a similar developmental role in these different systems. In addition, these studies indicate that conserved genes have also been recruited to serve new functions in vertebrates.

The Antennapedia-like homeobox genes provide the clearest example of a conserved gene family that may have similar roles in *Drosophila* and mice. In *Drosophila*, the Antp-like homeotic genes are organised in two clusters, and genetic analysis shows that successive genes (with respect to the direction of transcription) specify the phenotype of progressively more anterior segments. The expression patterns correlate with this function, as high levels of transcripts accumulate only in the segments whose phenotype they influence. The Antp-like *Hox* genes of the mouse are organised in clusters with extensive structural similarity to the *Drosophila* complexes, indicating that they are derived from a common progenitor. Moreover, in situ hybridisation studies show that successive Hox genes are expressed in increasingly more anterior regions of the embryo, the same relationship as exists in *Drosophila* (Fig. 2). This similarity in expression patterns suggests that the Antp-like genes have

the same role in patterning along the A-P axis in the mouse as in *Drosophila*, a function that may predate the divergence of the arthropod and chordate lineages. Other genes implicated in processes fundamental to animal development, such as mesoderm formation, have also been found to have homologues with similar expression patterns in insects and vertebrates.

However, extrapolations between different systems may not always be valid, as although similarities can reflect derivation from a common ancestor (homology), they can also arise by convergent evolution (analogy). A possible example is provided by studies of the vertebrate hindbrain. The hindbrain is a segmented structure and Hox genes are expressed in this tissue in segment-restricted patterns similar to those of their counterparts in the *Drosophila* embryo (Fig. 2). This similarity could indicate that the Antp-like genes specified segment phenotype in the common ancestor of arthropods and chordates, and that this role has been conserved during subsequent evolution. However, several lines of evidence suggest that the segments of insects and vertebrates are not homologous, but rather the segmentation of tissues arose independently in the evolution of these phyla. Thus, it is likely that although the role of Antp-like genes in specifying A-P pattern is ancient and conserved, the coupling of their expression to segments occurred subsequent to the divergence of the arthropod and chordate lineages and is an example of convergent evolution.

A further complication to extrapolations between developmental systems is the recruitment of conserved genes to serve functions in multiple developmental events. Several *Drosophila* genes, including *even-skipped*, *fushi tarazu*, *wingless* (*wg*) and *engrailed* (*en*) are expressed in stage-specific patterns, initially in stripes in each epidermal segment and later in specific cells in the nervous system. Genetic evidence reveals that these patterns reflect stage-specific roles, in segmentation and neuronal specification, respectively, and this raises the question as to which of these functions is ancestral. It is therefore significant that the vertebrate homologues of *en* (*En*-1 and -2) and *wg* (*int*-1) are not expressed in patterns that correlate with the formation of segments, but rather are expressed in specif-

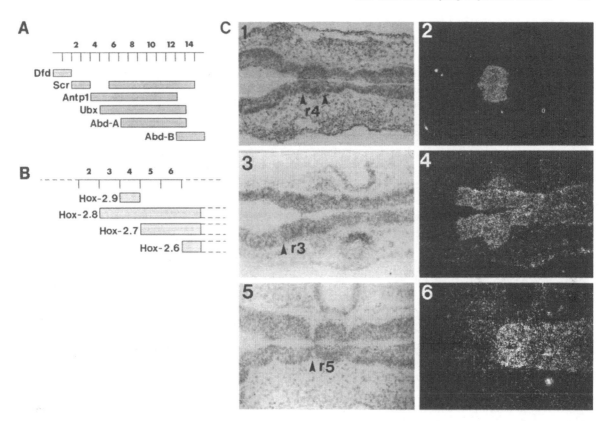

Fig. 2 A–C. Similar expression patterns of homeobox genes in *Drosophila* and mouse embryos. The *Antp*-like genes of *Drosophila* and mouse are organised in structurally homologous clusters, and there is a similar correlation between the position of each gene and its A-P domain of expression. They are also expressed in segment-restricted expression patterns in these diverse organisms. **A** The expression domains of *Antp*-like homeobox genes in the early *Drosophila* embryo are depicted. The domains have borders coincident with the boundaries of parasegments (*numbered*). **B** The expression domains of Hox-2 genes (Hox-2.6, -2.7, -2.8, -2.9) in the mouse hindbrain also have limits at segment boundaries. **C** Some examples of Hox-2 expression in the mouse hindbrain. *1, 2* Hox-2.9; *3, 4* Hox-2.8; *5, 6* Hox-2.7. Hindbrain segments (rhombomeres; *r*) are *numbered* and expression boundaries indicated with *arrows*. The *left* and *right panels* are bright- and dark-field photographs, respectively. (**A** after Akam 1987; **B** and **C** after Wilkinson et al. 1989b)

ic regions of the nervous system. These data suggest that the ancient role of these genes is in neural specification, and that the *en* and *wg* genes were later recruited to have roles in segmentation in arthropods (but not in vertebrates). As many structurally conserved genes are expressed in vertebrate tissues that have no homologous counterpart in insects (for example, Hox gene expression in the vertebrate limb), it is highly likely that they too have been reutilised in development.

28.3.3 Genetics and Expression Studies Are Powerful in Combination[4]

Studies of gene function in vertebrate embryos have in nearly all cases relied on indirect, circumstantial evidence, including the interpretation of expression patterns. However, the rigorous testing of gene function, and the elucidation of regulatory cascades requires the application of genetics. This is exemplified by the use of developmental mutants for the cloning and analysis of genes involved in patterning the early *Drosophila* embryo. In situ hybridisation and immunocytochemical studies have revealed that in many cases expression of these genes correlates with the tissues affected in mutants. Exceptions to this include genes involved in cell-cell interactions, where mutations are found to affect the development of nonexpressing cells responsive to the gene product. Moreover, the combination of expression studies with genetics have given insight into two aspects of developmental gene function not apparent from either technique alone. First, by studying patterns of gene expression in embryos mutant for another gene(s), it is possible to infer the existence of positive and negative regulatory interactions between genes. These studies have led to insight into the cascade of interactions during early development (see Tautz, Chap. 21, this Vol.). Second, expression studies have suggested roles unsuspected from the analysis of mutants. A good example, mentioned in the previous section, is the observation that certain "segmentation" genes are expressed at later stages in distinctive patterns in the nervous system of *Drosophila*. These findings led to the devising of experiments that revealed stage-specific roles of these genes in the specification of neuronal phenotype.

Can a combination of genetics and in situ hybridisation be used to study development in the mouse? Recent technical advances in reverse genetics to alter or eliminate the expression of genes makes this a likely prospect (see Wagner and Keller, Chap. 30, this Vol.). In addition, the genes for a number of developmental mutants have been mapped and several have been cloned (see Chaps. 26, 27, 29, this Vol.). The *undulated* gene is an excellent example of a correlation between expression pattern and mutant phenotype. Mutants of this gene have disrupted formation of the vertebrae due to defects in the differentiation of sclerotome cells. Sequence analysis of the *Pax*-1 gene revealed a point mutation in *undulated* mutants, suggesting that these genes are allelic. Consistent with this, in situ hybridisation studies revealed expression of *Pax*-1 in ventral sclerotome cells, and later in the intervertebral discs. Another developmental mutant of the mouse for which the gene has been cloned is the *T* gene. Homozygous mutants for this gene have a reduced amount of mesoderm and severely disrupted morphogenesis of a specific mesodermal derivative, the notochord.

Expression of the *T* gene correlates with this phenotype, as transcripts are found restricted to early mesoderm and its progenitor, and to the notochord (Fig. 3). Together, these expression studies and genetic data provide good evidence for a direct role of the *T* gene in mesoderm development.

28.3.4 Expression Patterns Can Provide Insight into Developmental Mechanisms[5]

An important issue in studies of embryogenesis is the elucidation of when and where groups of cells make committments to particular pathways of development. The expression patterns of certain genes provide new insight into this issue, since their expression is detected in advance of morphological changes in cells or tissues. One example is the expression of the *T* gene both in primitive ectoderm next to the primitive streak and in early mesoderm (Fig. 3); this finding indicates that the former cells have made some committment towards forming mesoderm before migrating into the primitive streak.

Genes with spatially restricted (rather than cell-type-restricted) expression are particularly interesting as they may indicate the existence of committed groups of cells important in pattern formation that are not discernable on the basis of morphology. Examples of this are provided by studies of the zinc-finger gene *Krox*-20 and the homeobox gene *En*-2, both of which have restricted domains of expression in the early neural plate. *Krox*-20 is expressed in the early neural epithelium, first in one stripe, and then in two stripes in the hindbrain (Fig. 4). Subsequently, *Krox*-20 is found to be expressed in two alternating rhombomeres in the hindbrain of 9.5-day mouse embryos. Studies of the chick hindbrain revealing rhombomeric patterns of neuronal development in the chick suggested that rhombomeres are segments. An important property of segments in insects is that they are domains of gene action, and thus the expression pattern of *Krox*-20 provides supportive evidence for segmentation in the hindbrain. The generation of the stripes of *Krox*-20 expression in the early neural plate suggests that segments are forming at this stage, prior to the morphological appearance of rhombomeres.

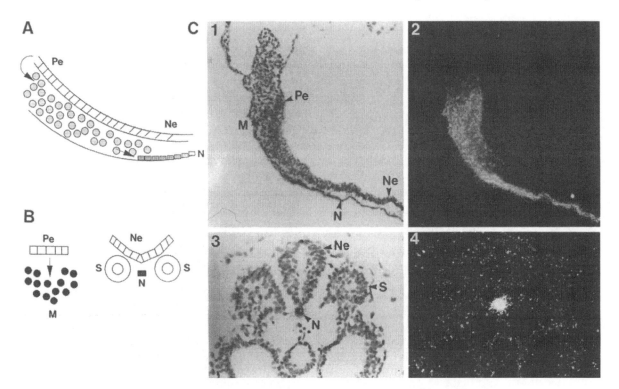

Fig. 3 A–C. In situ hybridisation and genetics are powerful in combination. In mouse embryos mutant for the *T* gene, formation of mesoderm and one of its derivatives, the notochord, are severely disrupted. **A** The formation and differentiation of mesoderm in the primitive streak of a mouse embryo is depicted. The epithelial primitive ectoderm (*Pe*) gives rise to mesenchymal mesoderm cells. In more anterior regions, older mesoderm becomes organised into tissues, including the notochord (*N*) and somites. *NE* neural epithelium. **B** Diagrams of transverse sections cutting through the indicated regions shown in **A**. The tis-

sues affected in *T* mutants are indicated by *shading. M* Mesoderm; *S* somite. **C** Expression of the *T* gene revealed in a longitudinal section of an 8.5-day embryo (*1, 2* compare with **A**) and a transverse section of a 9.5-day embryo (*3, 4* compare with **B**). The *left* and *right panels* are bright- and dark-field photographs, respectively. *T* transcripts are restricted to early mesoderm and the notochord, the tissues that are severely disrupted in mutants, and to primitive ectoderm destined to form mesoderm. (**C** after Wilkinson et al. 1990)

Similarly, the expression of *En-2* in the prospective midbrain and anterior hindbrain may reflect a committment towards forming these specific regions of the CNS. This idea has been substantiated by experiments showing that when this tissue is transplanted to a new location it not only continues to express *En-2*, but also forms midbrain/hindbrain structures.

28.4 Some Pitfalls in the Interpretation of Expression Patterns[6]

Although the analysis and interpretation of expression patterns is clearly of value, it must be em-

phasised that in situ hybridisation studies provide only indirect evidence regarding gene function and developmental mechanisms. Therefore, there are a number of potential complications and caveats that must be taken into account.

In situ hybridisation analysis gives a static picture of expression patterns, and in some cases sites of expression at one stage of development may not correspond to those at another. Thus, gene expression is not always a reliable marker of cell lineage, and it is important to examine expression patterns at frequent intervals during development. Indeed changes in expression patterns may reflect stage-specific roles, as for the *Drosophila en* gene. In addition, inferences based on previously characterised

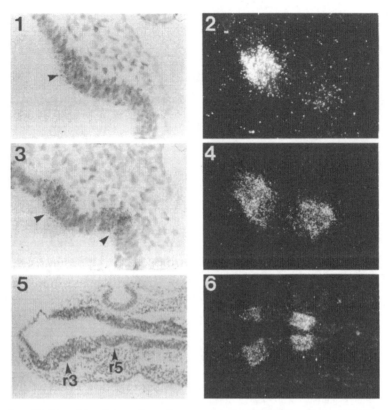

Fig. 4 1–6. Expression domains detected in advance of morphological changes. The expression pattern of the *Krox*-20 gene reveals committed groups of cells in the early mouse embryo. One domain of *Krox*-20 expression is found in the neural epithelium of an 8-day embryo (**1, 2**), and two domains in an 8.5 day embryo (**3, 4**). At 9.5 days (**5, 6**) the two domains are found to correspond to hindbrain segments (rhombomeres; r3 and r5). The *left* and *right panels* are bright- and dark-field photographs, respectively. These data indicate that hindbrain segmentation is occurring in the 8-day embryo, in advance of the formation of rhombomeres at the morphological level. (After Wilkinson et al. 1989a, b)

functions of genes should be treated with caution given the diverse functions of some gene products (e.g., growth factors). Likewise, the extrapolation of roles between diverse organisms may not always be valid. In some cases functionally distinct gene products can arise by differential splicing, and thus it is important to use in situ hybridisation probes that distinguish different transcripts. Furthermore, the analysis of transcripts provides no insight into the regulation of gene function by translational or post-translational mechanisms; examples of this in development include the degradation of c-*mos* protein following fertilisation, and the targeting of *dorsal* protein to specific nuclei in the *Drosophila* embryo.

Finally, genetic studies in *Drosophila* have revealed that simplistic interpretations of expression patterns can be misleading in two respects. First, correlations with particular developmental events may not be meaningful given the simultaneous occurence of many different processes: expression of the *Drosophila* homologue of the *abl* proto-oncogene correlates with proliferation in the early embryo, yet in *abl* mutants tissue morphogenesis is disrupted, not proliferation. Second, some sites of gene expression may be functionally redundant: *Scr* transcripts are first expressed in alternating segments, as are pair-rule segmentation genes, yet *Scr* has no function in segmentation. Rather, *Scr* has a homeotic function in certain anterior segments, and *Scr* protein accumulates only in these segments. Other *Drosophila* homeobox genes also have domains of expression that extend beyond the segments whose phenotype they influence, and this could also be true in vertebrate embryos. The absence of selective pressure on the expression of genes in redundant sites may in some cases lead to variable patterns in related organisms. Thus, one test of the functional relevance of specific sites of expression in the mouse is the in situ hybridisation analysis of other vertebrate species.

28.5 Outlook

28.5.1 Technical Advances in in Situ Hybridisation

Some technical advances that overcome limitations in current methods of in situ hybridisation are likely to find wide application. Sensitive, nonradioactive methods have been successfully used for studies of *Drosophila* embryos and these offer the advantages of safety, rapid results and the resolution of signal at the cellular level. In addition, since the hybridisation and immunocytochemical detection of probe can be carried in whole mount, spatial patterns of gene expression can be directly visualised without the need for reconstruction from serial sections. Since an important aspect of development is the differential or coordinate regulation of different genes, it would be very useful to analyse the spatial distribution of multiple gene products in the same tissue section. One way in which this can be achieved is by establishing conditions that allow the immunocytochemical detection of protein, followed by in situ hybridisation to RNA. However, a more universally applicable method may be offered by nonradioactive in situ hybridisation techniques; the simultaneous detection of multiple transcripts could be achieved by labelling probes with different haptens and using immunocytochemical reagents that yield different coloured products for each of these.

The sensitivity of in situ hybridisation, although high, can be a limiting factor. For example, genetic studies of *Drosophila* have revealed sites of specific gene function in which the levels of transcripts and protein are below the limits of current detection methods. While improvements in sensitivity are likely to occur, an attractive alternative is the use of *lac*-Z constructs in transgenic mice. By linking *lac*-Z coding sequences to the promoter of the gene under analysis, β-galactosidase will be expressed in the pattern of the latter, and can be then detected with great sensitivity by using a chromogenic substrate. An additional virtue of this method is that the staining can readily be carried out in whole mount preparations. However, it is clearly important that all of the appropriate regulatory elements are present in the construct in order to reproduce the actual expression pattern of the normal gene. This may prove to be technically difficult for genes with large and complex regulatory regions, and thus the patterns of β-galactosidase staining should be correlated with the results of in situ hybridisation as far as possible. Another drawback of this transgenic approach is that it is not quantitative, yet the relative levels of transcripts accumulation in different regions of the embryo can be highly significant. An example is the expression of *Drosophila* homeotic genes at higher levels in the segments whose phenotype they influence; the other, lower level sites of expression may be functionally redundant.

28.5.2 Insights into Gene Function and Mechanisms of Development[7]

Without doubt the establishment of methods for the manipulation of gene expression in embryos are the most significant technical advances for future studies of mouse development. In combination with studies of expression patterns, the developmental effects of inactivating and altering gene expression will provide insight into gene function. Particularly significant will be the analysis of regulatory interactions by testing whether mutations in one gene affects the expression of others. Given the combinatorial and tissue-specific actions of many gene products, essential background information for the unravelling of such interactions will be comparative studies of the spatial and temporal expression patterns of all of the relevant genes. Comparative studies of different members of gene families may be especially important, as genetic studies in *Drosophila* indicate that related genes can duplicate each others function in regions of overlapping expression. The results of inactivating the wnt-1 gene by homologous recombination in the mouse are a possible example of this phenomenon. This inactivation lead to the deletion of midbrain and anterior hindbrain structures, yet the gene is also expressed in other regions of the neural tube that are apparently normal in these mutants. As wnt-1 is a member of a gene family, it is clearly important to ascertain whether the expression domains of related genes overlap that of wnt-1, and potentially duplicate its function.

There are already a number of examples of gene expression marking domains of cells prior to morphological differentiation, thus providing new insight into temporal and spatial aspects of cell committment. It is likely that further studies will reveal useful information regarding the development of some poorly understood regions of the vertebrate embryo. One such region is the early forebrain. Recent work has shown certain homeobox genes to be expressed in neuromeres of the diencephalon, and this should now allow study of mechanisms of regional specification in the forebrain.

The detection of transcripts can also be used in manipulative experiments to test gene function and analyse the signals that induce and pattern tissues. Thus far, inductive signals have largely been analysed by the use of tissue morphology and the products of differentiated cell types as markers of induction. However, it is likely that a series of regulatory events has occurred between the initial signal and the detection of these late markers. The cloning of early molecular markers of induced tissues (for example, mesoderm and the neural plate) and their detection by in situ hybridisation or immunocytochemistry should allow the dissection of this chain of events. The promise of this approach is shown by experiments using homeobox genes as markers of neural ectoderm in Xenopus which have led to new insight into the timing of committment to neural differentiation and the role of growth factors in patterning the early embryo. Although mouse embryos are less amenable than frog embryos, recent technical advances will aid the interpretation and analysis of manipulative experiments. The production of transgenic mice that constitutively express β-galactosidase is particularly significant,

as embryonic tissues from these mice can be unambiguously identified after transplantation by histochemical staining. By combining this technique with in situ hybridisation, it will be possible to address a number of important questions; for example, the in situ hybridisation analysis of Hox gene transcripts after appropriate transplantations will give insight into the timing and distribution of signals that establish the spatial patterns of Hox gene expression.

28.6 Summary

The analysis of patterns of gene expression by in situ hybridisation is an important technique in the molecular analysis of development. Although expression patterns do not always correlate with sites of gene function, in situ hybridisation data can provide evidence for or against specific roles of genes in development. This approach is particularly powerful in combination with other clues, such as the conservation of gene structure and expression between diverse systems and the use of genetics. In addition, the analysis of expression patterns provides an essential background for the design and interpretation of direct approaches towards studying gene function by reverse genetics and the use of appropriate cell or organ culture systems. Gene expression can be used as a marker of cell phenotype in the absence of morphological changes, and thus the in situ hybridisation analysis of certain genes provides insight into when and where cells become committed to particular pathways of development. Further understanding of the steps leading to this committment will come by combining the analysis of gene expression patterns with classical embryological manipulations, such as tissue grafting and excision.

References

[1] Cox KH, DeLeon DV, Angerer LM, Angerer RC (1984) Detection of mRNAs in sea urchin embryos in in situ hybridisation using asymmetric probes. Dev Biol 101:485–502

Hemmati-Brivanlou A, Frank D, Bolce ME, Brown BD, Sive HL, Harland RM (1990) Localisation of specific mRNAs in Xenopus embryos by whole-mount in situ hybridisation. Development 110:325–330

Tautz D, Pfeifle C (1989) A non-radioactive in situ hybridisation method for the localisation of specific RNAs in *Drosophila* embryos reveals translational control of the segmentation gene hunchback. Chromosoma 98:81–85

Wilkinson DG, Green J (1990) *In situ* hybridisation and the three-dimensional reconstruction of serial sections. In: Copp AJ, Cockroft D (eds) Postimplantation mammalian embryos. IRL Press, Oxford, pp 155–171

[2] Heine UI, Munoz EF, Flanders KC, Ellingsworth LR, Lam H-YP, Thompson NL, Roberts AB, Sporn MB (1987) Role of transforming growth factor-β in the development of the mouse embryo. J Cell Biol 105:2861–2876

Lehnert SA, Akhurst RJ (1988) Embryonic expression pattern of TGF beta type-1 RNA suggests both paracrine and autocrine mechanisms of action. Development 104:263–273

Pelton RW, Nomura S, Moses HL, Hogan BLM (1989) Expression of transforming growth factor $\beta 2$ RNA during murine embryogenesis. Development 106:759–767

Pelton RW, Dickinson ME, Moses HL, Hogan BLM (1990) In situ hybridisation analysis of TGF$\beta 3$ expression during mouse development: comparative studies with TGF$\beta 1$ and $\beta 2$. Development 110:609–620

[3] Akam ME (1987) The molecular basis for metameric pattern in *Drosophila* embryos. Development 101:1–22

Doe CQ, Scott MP (1988) Segmentation and homeotic action in the developing nervous system of *Drosophila*. Trends Neurosci 11:101–106

Dolle P, Izpisua-Belmonte J-C, Falkenstein H, Renucci A, Duboule D (1989) Coordinate expression of the murine *Hox-5* complex homeobox-containing genes during limb pattern formation. Nature 342:767–772

Duboule D, Dolle P (1989) The murine *Hox* gene network: its structural and functional organisation resembles that of *Drosophila* homeotic genes. EMBO J 8:1497–1505

Graham A, Papalopulu N, Krumlauf R (1989) The murine and *Drosophila* homeobox gene complexes have common features of organisation and expression. Cell 57:367–378

Holland PWH (1990) Homeobox genes and segmentation: co-option, co-evolution and convergence. Sem Dev Biol 1:135–145

Hopwood ND, Pluck A, Gurdon JB (1989) A *Xenopus* mRNA related to *Drosophila twist* is expressed in response to induction in the mesoderm and the neural crest. Cell 59:893–903

Murphy P, Davidson D, Hill RE (1989) Segment-specific expression of a homeobox-containing gene in the mouse hindbrain. Nature 341:156–159

Patel NH, Martin-Blanco E, Coleman KG, Poole SJ, Ellis MC, Kornberg TB, Goodman CS (1989) Expression of *engrailed* proteins in arthropods, annelids, and chordates. Cell 58:955–968

Wilkinson DG, Bailes JA, McMahon AP (1987) Expression of the proto-oncogene *int*-1 is restricted to specific neural cells in the developing mouse embryo. Cell 50:79–88

Wilkinson DG, Bhatt S, Cook M, Boncinelli E, Krumlauf R (1989b) Segmental expression of *Hox*-2 homoeobox-containing genes in the developing mouse hindbrain. Nature 341:405–409

[4] Balling R, Deutsch U, Gruss P (1988) *undulated*, a mutation affecting the development of the mouse skeleton, has a point mutation in the paired box of *Pax*-1. Cell 55:531–535

Herrmann BG, Labeit S, Poustka A, King TR, Lehrach H (1990) Cloning of the *T* gene required in mesoderm formation in the mouse. Nature 343:617–622

Ingham PW (1988) The molecular genetics of embryonic pattern formation in *Drosophila*. Nature 335:25–34

Wilkinson DG, Bhatt S, Herrmann BG (1990) Expression pattern of the mouse *T* gene and its role in mesoderm formation. Nature 343:657–659

[5] Davis CA, Noble-Topham SE, Rossant J, Joyner AL (1988) Expression of the homeo box-containing gene *En*-2 delineates a specific region of the developing mouse brain. Genes Dev 2:361–371

Martinez S, Wassef M, Alvarado-Mallart RM (1991) Induction of a mesencephalic phenotype in the 2-day-old chick prosenecephalon is preceded by the early expression of the homeobox gene *en*. Neuron 6:971–988

Wilkinson DG, Bhatt S, Chavrier P, Bravo R, Charnay P (1989a) Segment-specific expression of a zinc-finger gene in the developing nervous system of the mouse. Nature 337:461–464

[6] Gertler FB, Bennett RL, Clark MJ, Hoffmann FM (1989) *Drosophila abl* tyrosine kinase in embryonic CNS axons: a role in axonogenesis is revealed through dosage-sensitive interactions with *disabled*. Cell 58:103–113

Ingham PW, Hidalgo A, Taylor AM (1990) Advantages and limitations of in situ hybridisation as exemplified by the molecular genetic analysis of *Drosophila* development. In: Harris N, Wilkinson DG (eds) *In situ* hybridisation: application to developmental biology and medicine. Cambridge University Press, Cambridge, pp 97–114

Roth S, Stein D, Nusslein-Volhard C (1989) A gradient of nuclear localisation of the *dorsal* protein determines dorso-ventral pattern in the *Drosophila* embryo. Cell 59:1189–1202

Watanabe N, Vande Woude GF, Ikawa Y, Sagata N (1989) Specific proteolysis of the c-*mos* proto-oncogene product by calpain on fertilisation of *Xenopus* eggs. Nature 342:505–511

[7] McMahon AP, Bradley A (1990) The *Wnt*-1 (*int*-1) gene is required for development of a large region of the mouse brain. Cell 62:1073–1085

Price M, Lemaistre M, Pischetola M, Di Lauro R, Duboule D (1991) A mouse gene related to *Distal-less* shows a restricted expression in the developing forebrain. Nature 351:748–751

Ruiz i Altaba A, Melton DA (1989) Interaction between peptide growth factors and homeobox genes in the establishment of antero-posterior polarity in frog embryos. Nature 341:33–38

Sharpe CR, Fritz A, De Robertis EM, Gurdon JB (1987) A homeobox-containing marker of posterior neural differentiation shows the importance of predetermination in neural induction. Cell 50:749–758

Thomas KR, Cappechi MR (1990) Targeted disruption of the murine *int*-1 proto-oncogene resulting in severe abnormalities in midbrain and cerebellar development. Nature 346:847–850

Chapter 29   Insertional Mutagenesis and Mouse Development

MICHAEL R. KUEHN and JONATHAN P. STOYE

29.1 Introduction

Classical genetic analysis of spontaneous or induced mutations has identified a number of candidate developmental genes. Further analysis of mutant phenotypes has helped define possible functions for many of these genes. However, to complete our understanding of their roles in development, an extensive molecular genetic analysis will ultimately be required. To accomplish this goal, these genes must be cloned. This can be a daunting prospect when not much is known except a mutant phenotype or genetic map location.

There is one type of mutation, however, which does allow an affected gene to be cloned readily. Insertional mutations result from the integration of new DNA into a gene. These mutations can be as simple as the addition of a single base pair causing frame shifting within the coding region of a gene, or as complex as the integration of a large block of DNA into a regulatory region. While all mutations can be useful, insertional mutations arising from the integration of defined DNA sequences are especially valuable. Here the integrated sequence serves a dual purpose. It not only creates the mutation but also acts as a molecular marker "tagging" the locus. By recloning the integrated sequence the affected gene can be easily recovered as well. Insertional mutations, therefore, can bridge the gap between classical and molecular genetics.

In developmental studies, the term insertional mutagenesis covers a variety of approaches to identify and isolate genes. One exploits the presence of mobile genetic elements in the mouse. These are discrete DNA sequences that are capable of moving to new chromosomal locations. In recent years it has become clear that certain mouse developmental mutations, well known from classical genetic analysis, are due to insertions of these mobile elements.

The ability to recognize the association of defined sequences with previously studied mutations has been an exciting development. Complementing these studies are approaches to create new insertional mutations. One of the most powerful utilizes transgenic mice, animals that have acquired new DNA sequences in their germ line through experimental manipulation. In the generation of a transgenic mouse, an insertional mutation can arise due to the integration of foreign DNA into normal mouse genes. Although the technique is relatively new, experimentally induced insertional mutations have already identified a number of genes with developmental effects.

In this chapter, we examine how insertional mutations have made it possible finally to clone and characterize genes which have intrigued developmental biologists for many years. In addition, transgenic approaches for inducing new mutations will be described and some of the genes that have been isolated will be discussed. We examine the mechanisms by which insertions of mobile DNA can lead to both dominant and recessive mutations, how induced mutations are shown to be due to insertion and how the affected gene ultimately can be cloned and analyzed. Finally, we discuss future prospects for utilizing insertional mutagenesis to further our understanding of mammalian development. We begin with an introduction to mobile genetic elements, the endogenous agents of insertional mutation in many organisms including the mouse.

29.2 Mobile Genetic Elements [1]

The extreme instability of certain mutations in maize led McClintock to postulate the existence of mobile genetic elements over 40 years ago. However, the elucidation of the molecular nature of these sequences did not take place until the discovery of insertion sequences and transposons in bacteria. Since that time, mobile DNA elements have been shown to represent an important cause of genetic variation in nature, having been implicated in spontaneous mutations in virtually every organism studied experimentally (Table 1). Even one of the traits Mendel originally studied, wrinkled seed, is caused by a mobile element. Different types of mobile elements move by different mechanisms, but one consequence of their movement is shared: the potential to disrupt gene function by transposition and insertion.

The insertion elements found in bacteria were the first to be characterized. Their transposition probably occurs by direct DNA to DNA recombination, mediated by a "transposase" enzyme encoded by the element itself. Transposition to a new site is often accompanied by complete excision of the transposon from its original location. The *D. melanogaster* P elements, as well as the Ac/Ds elements in *Z. mays* and Tc1 elements in *C. elegans* are structurally and mechanistically similar to bacterial transposons. While all of these are now exploited for insertional mutagenesis studies in their species of origin, similar sorts of elements have not been observed in vertebrates. The mobile DNA elements which have been found, retrotransposons and retroposons, transpose by very different mechanisms but are equally effective at causing insertional mutations.

As their names imply, retrotransposons and retroposons acquire new sites of integration by a mechanism involving reverse transcription of an RNA intermediate. The element at the original site of integration is maintained, while copies do the actual "transposing" to new locations. It is at this step where a new mutation of a gene can occur. The best characterized members of the retrotransposon group are retroviruses. These have been the focus of considerable attention mainly due to their ability to cause tumors in vertebrates. The extensive analysis done over the past decades has led to a thorough understanding of the retrovirus replication cycle, as illustrated in Fig. 1. During the conversion of the RNA to DNA, regions at the ends of the viral RNA

Table 1. Examples of spontaneous insertional mutations in nature

Organism	Mobile element	Mode of transposition[a]	Mutant allele
Yeast (*S. cerevisiae*)	Ty1	RNA	*his4-912*
Pea (*P. sativum*)	lps-r	DNA	*wrinkled-seed*
Maize (*Z. mays*)	Ac/Ds	DNA	*bronze*
Nematode (*C. elegans*)	Tc1	DNA	*unc-22*
Fruitfly (*D. melanogaster*)	copia	RNA	*white-apricot*
Mouse (*M. musculus*)	MLV	RNA	*dilute*
Human (*H. sapiens*)	Line L1	RNA	*hemaphilia A*

[a] RNA indicates an RNA intermediate in the transposition process, DNA implies a direct DNA-mediated mode of transposition.

are duplicated to form long terminal repeat (LTR) sequences which delimit the proviral DNA (see Fig. 2). The LTR regions contain the sequences that regulate viral gene expression whereas the genes for the structural proteins that make up the virion and the gene for reverse transcriptase are located between the LTRs. Examination of the copia elements of *D. melanogaster* and Ty elements in *S. cerevisiae*, other members of the retrotransposon family, has revealed proviral-like structures similar to that of retroviruses. Although it has been demonstrated only for Ty elements, transposition of all these elements probably involves an RNA intermediate. However, the extracellular phase is probably unique to retroviruses.

Retroposons are a collection of elements with structures quite different from retrotransposons. In mammals, most are members of two families of interspersed repeated DNA, the short and long interspersed nuclear element families (SINE and LINE families). Intact LINE elements are greater than 5 kilobase pairs (kbp). They appear to encode a reverse transcriptase-like enzyme, but lack the LTR sequences distinctive of retrotransposons. There is growing evidence that intact LINEs have the capacity to actively transpose. SINE sequences are around 500 base pairs (bp) in length and are present in extremely high copy number, sometimes in excess of 100000. The Alu repeats in humans and the B1 family in rodents are SINEs. These and other retroposons, including the processed pseudogenes found in mammals, probably also have been inserted into novel chromosomal positions by a mechanism dependant on reverse transcription. However, these elements do not encode their own enzymatic function.

As can be seen from this brief description, a variety of elements are present in the genomes of animals that have the potential to insert into and disrupt gene activity. We focus now on those elements that have been found in the laboratory mouse and discuss their potential for insertional mutagenesis.

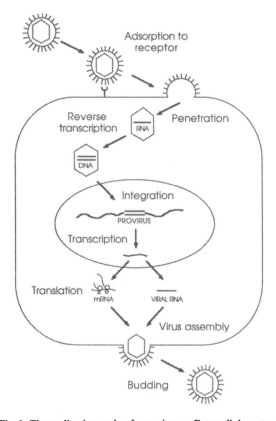

Fig. 1. The replication cycle of retroviruses. Extracellular retroviral particles bind to specific cell membrane receptors and enter the cell by receptor-mediated endocytosis. Once in the cell, reverse transcription of the viral RNA genome takes place within the remnant of the viral particle to generate a linear double-stranded DNA molecule. The DNA-protein complex moves to the nucleus and the DNA integrates into the host chromosome. The integrated viral DNA is called a provirus. For every virus that enters a cell only a single provirus is formed. The proviral genome is transcribed by host RNA Polymerase II and used both as messenger RNA for the translation of viral proteins and as genome RNA for progeny virus. New viral particles assemble and bud out through the plasma membrane, without causing host cell death

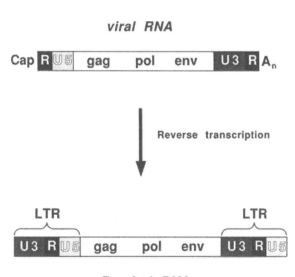

Fig. 2. RNA and DNA forms of a retrovirus. The only differences between the RNA and DNA forms of a retrovirus are at the ends. At the extremities of the viral RNA are modified sequences typical of most eukaryotes namely a methylated capping group at the 5'end and a string of around 200 A residues at the 3'end. Within these lie viral direct repeats (R). The long terminal repeats (LTRs) at the ends of the provirus are formed during reverse transcription. The left LTR is composed of the unique sequences derived by duplication from the 3'end of the viral RNA (U3), the R segment and unique sequences from the 5'end of the RNA (U5). The right LTR is identical, containing a U3, R, and a duplicated U5 region. The LTRs contain the regulatory sequences controlling the transcription of the viral genes. These include the gag and env genes which encode proteins required for viral particles, and the pol gene which encodes the enzymes involved in reverse transcription and integration of the provirus

29.3 Endogenous Retroviruses and Insertional Mutagenesis of the Mouse Germ Line [2]

As mentioned above, infectious retroviruses have been the focus of considerable study in the past decade because they cause cancer in mice. As part of efforts to understand the molecular basis of oncogenesis, the integrated forms of exogenous proviruses have been cloned and characterized. Some have even been completely sequenced. The molecular probes made available by these studies have revealed the vast number of endogenous retroviral elements that exist in the germ line of mice. In fact, it has been estimated that endogenous retroviral and retroviral-like sequences may constitute up to 1% of the mouse genome.

To date, four families of endogenous retroviral sequences have been well characterized. Two are very closely related to exogenous retroviruses and are probably the result of recent germ line infection. Murine leukemia virus (MLV) sequences are found at a copy number of around 50 in most inbred strains. Mouse mammary tumor virus (MMTV) sequences are also present in most strains, but at a lower copy number than MLV. The other two families of endogenous retrovirus-like elements are not closely related to any known infectious virus. These families are present in very high copy numbers. Sequences coding for the virus-like 30S RNAs (VL30 elements) are present in as many as several hundred copies in the germ line of some strains, while the copy number of intracisternal A particle sequences (IAPs) in the germ line is probably as great as 1000. Taken together, all of these endogenous elements represent a huge potential for insertional mutations.

In principle, any of the integration events leading to the present complement of endogenous MLVs in various strains could have been mutagenic. However, with so many endogenous proviruses, recognizing the association of a particular provirus with a specific mutation can be technically difficult. Fortunately, the many MLVs can be classified into subgroups that each contain far fewer members. Ecotropic and nonecotropic MLVs differ in the types of cells that they can productively infect. For ecotropic viruses this is limited to mouse cells, whereas nonecotropic viruses have a much broader host range. This cell type specificity is conferred by the *env* gene product. Probes specific for ecotropic or nonecotropic MLV sequences have been developed based on sequence differences that have been found in the *env* gene in these two families.

29.3.1 Ecotropic Retroviruses and Insertional Mutation of the Dilute Gene [3]

Probes specific for the endogenous ecotropic proviruses were the first to be developed and have revealed at least 18 different endogenous proviruses in the various mouse strains studied so far. Most strains possess two or fewer of these genetic loci, named *emv-1* to *emv-18* (for *e*cotropic *m*urine *v*irus). From this analysis has come the finding that the recessive coat-color mutation at the dilute locus (*d*) on chromosome 9 is due to insertion of an ecotropic MLV.

The original *d* mutation is an old mutation of mouse fanciers and was one of the first to be studied by mouse geneticists in the 1930s. It causes a lightening of coat color when homozygous. The dilution effect is not caused by reduced pigment, but is due to clumping of pigment granules in the hair shaft. This phenotype is apparently the result of abnormal melanocyte function. An additional 200 mutant alleles have been found at the *d* locus, mostly from radiation mutagenesis experiments carried out since the 1960s. Unlike the original *d* allele, these others are associated with neurological disorders and postnatal lethality in addition to the effect on coat color.

Using the ecotropic specific probe, a perfect correlation was found to exist between the genetic segregation of one endogenous ecotropic locus, *emv-3*, and the *d* mutation. However, even when two loci show such tight genetic linkage, they can still be physically quite far apart on the chromosome. Additional proof, therefore, had to be obtained that the *d* mutation is actually caused by the insertion, rather than simply being linked to it. The extensive genetic analysis already done on the *d* mutation was extremely helpful in this regard.

Over the years, 11 animals that were spontaneous revertants to a normal coat color phenotype have been identified in a survey of over a million mice. Although a rare event, the differences in intensity of hair pigmentation made revertants to d^+ relatively easy to identify. These mice have proven to be extremely informative. Molecular analysis of each revertant revealed the loss of sequences corresponding to *emv-3*. The absolute concordance of loss of the provirus and reversion of phenotype provided firm but indirect evidence that *emv-3* causes the *d* mutation. In addition, this finding provided the impetus to clone *emv-3* so that the *d* gene could be isolated as well.

Once it has been established that a mutation is due to insertion, how is the affected gene ultimately isolated and its normal function studied? First, the region surrounding the integration site must be cloned. Second, sequences must be found that are unique to the region surrounding the insertion, and devoid of repeated sequences that can frustrate molecular analysis. The unique flanking sequence probes can then be used to analyze RNA extracted from tissues likely to express the gene of interest. Although the basic approach is straightforward, problems can arise. Perhaps the sequences immediately flanking the inserted DNA do not detect expression. If the provirus is in an intron, the closest coding sequences could be many kbp from the insertion site. Even if probes can be isolated which contain exon sequences, the gene may be expressed at undetectably low levels or at developmental stages that are inaccessible.

The cloning of the *d* locus is a good illustration of this basic approach and some of the inherent problems, but also of how additional alleles made available by classical genetic analysis can greatly assist molecular studies. Using the *emv-3*-specific probe to screen a library of genomic DNA from mice homozygous for *d*, a molecular clone containing the provirus and flanking sequences was obtained. The clone was further characterized in order to identify sequences unique to this chromosomal region. An immediate use for the flanking probe was in a more detailed examination of the structure of the d^+ revertants. Analysis of two revertants revealed the presence of a single LTR at the *emv-3* integration site (Fig. 3). DNA sequence

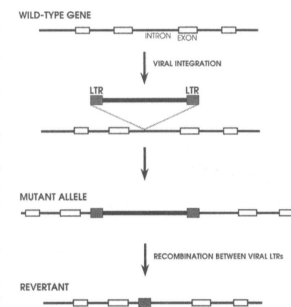

Fig. 3. Retroviral insertional mutagenesis and reversion. A schematic representation of the structures of wild-type, mutant, and revertant alleles of a gene mutated by virus insertion, similar to what has been seen for both the dilute and the hairless mutations. Proviral integration into an intron of a gene creates a mutant allele. Intramolecular recombination between the two LTRs results in most of the viral sequences being lost except a single LTR. This leads to reversion to normal phenotype

analysis confirmed that this single LTR is the remnant of homologous recombination between the two LTRs, an event that excised the provirus.

The fact that the remaining LTR does not disrupt *d* gene function is suggestive of two points. First, the *emv-3* integration site is probably within a noncoding region, perhaps an intron. A single LTR would be equally disruptive as the whole provirus if the integration was in a coding region. Further analysis confirmed the lack of any open reading frames (ORFs) around the integration site. Second, it suggests that the insertional inactivation of cellular genes need not involve simple disruption of coding sequences. In *Drosophila* and yeast, insertions of retrotransposons that cause mutations often occur outside of the coding regions. Presumably, proviruses in all these cases interfere with some aspect of RNA transcription or processing.

Although transcriptionally active sequences were not found close to the integration site of the *emv-3* provirus, molecular characterization of another *d* allele was informative. This allele was shown to be a deletion of about 3.5 kbp of DNA, located 2 kbp away from the site of proviral integration in the original *d* mutant. The fact that this deletion produces the dilute coat color and neurological phenotype, suggested that the deleted region is integral to *d* gene function. In fact, sequences from the deleted region were shown to be transcribed in cell lines of melanocyte origin and provided the starting point for molecular analysis of the structure of the *d* gene. Preliminary charcterization of the gene suggests that it encodes a cytoskeleton component. It an apparently spans around 150 kbp, and several large transcripts are expressed. Studies have shown that when the provirus is present, expression of only some of the RNA species within melanocytes are affected. The pattern of expression of the RNA presumed necessary for neurological function remains unaffected, explaining why the original dilute allele (now called d^v for *d*ilute-viral) affects only coat color. Hopefully, in the near future, molecular analysis of *d* will provide valuable insights into melanocyte function and neurological development.

29.3.2 Nonecotropic Retroviruses and Insertional Mutation of the Hairless Gene[4]

Whereas the number of ecotropic proviruses is fairly low, most inbred strains of mice have between 40 and 60 nonecotropic proviruses in the germ line. Recently, probes have been developed which identify subfamilies of nonecotropic viruses: the xenotropic viruses that infect nonmurine cells, and the polytropic and modified polytropic viruses with host ranges overlapping those of ecotropic and xenotropic viruses. About 15 endogenous copies of each of these three types of nonecotropic viruses are present in most inbred strains. The availability of these probes has allowed a genetic analysis of each small subgroub of nonecotropic loci, similar to that done with the ecotropic loci.

The genetic location of over 100 nonecotropic proviral loci have been determined. Genetic linkage between nonecotropic proviruses and a number of known mutations has been observed and in one case, convincing genetic evidence has been obtained implicating proviral insertional mutagenesis. The hairless mutation (*hr*) is closely linked to the *pmv-43* nonecotropic locus. Again, the examination of a revertant was decisive in proving that the provirus actually causes the mutation. As is the case for the d^+ revertants, the hr^+ revertant also contains a solo LTR (see Fig. 3). The strategy used for the isolation of the *hr* gene has been similar to that for *d*. The *pmv-43* provirus and sequences surrounding it were cloned and analyzed for expressed sequences. The approach used illustrates a general strategy also being applied in a variety of molecular studies where genes are being searched for within defined chromosomal regions. The method takes advantage of new insights into the structure of the genome.

Interspersed throughout the genome are small regions containing an unusually high concentration of the dinucleotide CG in unmethylated form. It is now becoming clear that these small stretches, called CpG islands, are usually found in the vicinity of transcribed DNA and, therefore, can be considered markers for genes. Restriction enzymes having recognition sites containing 1 or 2 CG doublets, cut DNA predominantly in CpG islands. These enzymes can be used to locate CpG islands in genomic or cloned DNA. Examination of sequences cloned from *hr* revealed a CpG island 8 kbp upstream of the *pmv-43* proviral integration site. Subsequent analysis demonstrated that a transcript initiates within the *hr* CpG island and is expressed in normal skin. This is just one illustration of how examining the structure of the cloned flanking DNA for clues to gene activity can help focus further efforts.

Analysis of the *hr* gene should provide information about hair formation and skin development. Since the hairless mutation also apparently causes a mild perturbation of the immune system, analysis of normal gene function might shed some light on the relationship between skin and thymus. Mutations in several other genes also affect both tissues to a greater or lesser degree.

29.3.3 Spontaneous Reintegrations and Insertional Mutations in Hybrid Mouse Strains[5]

All of the endogenous elements discussed above probably entered the germ line some time ago. What about the potential for new integration events? Certain strains of mice, such as AKR, have high levels of endogenous ecotropic viral expression. At a low frequency, these mice can acquire new germ-line proviruses. Theoretically, this should provide a simple and straightforward approach for inducing new insertional mutations in the mouse. Unfortunately, the amplification frequency in most strains is far too low to provide new insertional mutants within a reasonable time scale (at best only about one new germ-line integration in 15 years). However, with the recent observation that certain hybrid strains of mice have a much higher rate of spontaneous germ-line proviral acquisition, this strategy can now be put into place. As many as one in every ten offspring of these hybrid mice gain one or more new proviruses. With such a high rate, it is conceivable that these mice may provide a research tool analogous to the strains of *Drosophila* that undergo hybrid dysgenesis.

The frequency of reinsertion depends both on the genotype of the mice and the virus involved. The phenomenon has been studied in greatest detail in a hybrid mouse strain derived by crossing SWR/J by RF/J mice. The RF/J strain carries two closely linked ecotropic proviruses, *Emv-16* and *Emv-17*. Expression of these endogenous viruses in hybrid females eventually results in infection of their offspring. New integrations have been found at a rate of 0.35 per mouse. These new proviral integration sites have been named *Srev* loci (for *S*WR/J-*R*F/J *e*cotropic *v*irus). The mechanism of amplification has been shown to be due to the susceptibility of the oocyte to infection. The source of virus for oocyte infection in these animals probably is the surrounding ovarian cells. In fact, viral expression has been detected within the ovary recently.

The frequencies of reinsertion observed should make it possible to derive many new developmental mutations. The first results are certainly encouraging. Of the first 18 *Srev* loci studied, one was found to be associated with a recessive embryonic lethal mutation. Although not much is yet known about the *Srev-5* mutation, only that homozygous embryos die around midgestation, the method that was used to identify it has been applied in many insertional mutagenesis studies. For recessive mutations that are viable and result in visible phenotypes, homozygotes can be easily distinguished. However, if a recessive mutation is lethal during embryonic or fetal development, homozygotes will not be found among the offspring. To prove that a prenatal mutation existed, matings between animals heterozygous for *Srev-5* were carried out. A significant number of live-born progeny were analyzed to determine if any were homozygous for the provirus. In practice, this meant measuring the hybridization intensity of bands diagnostic for the *Srev-5* provirus on Southern blots. Homozygotes, which would have had twice the copy number and therefore twice the hybridization intensity compared to heterozygotes, were never found. This provided indirect but strong evidence for a recessive prenatal lethal mutation. As will be discussed later, similar approaches have been used to identify other recessive prenatal lethal insertional mutations.

29.4 Exogenous Retroviruses and Insertional Activation of Oncogenes in Somatic Cells[6]

It has long been known that retrovirus infection can lead to cellular transformation and tumor formation. Another bonus of the intense molecular study of retroviruses has been the demonstration that acutely transforming retroviruses, which cause tumors with very short latency periods, encode novel transforming genes. These viral oncogenes are modified versions of, and ultimately derived from, normal cellular genes. The normal function of these so-called proto-oncogenes is under intense study. Several are presumably involved in the regulation of cell differentiation and growth.

Other retroviruses lead to tumor formation only slowly and these have been found to contain nothing beyond the usual viral genes. How do these retroviruses cause cancer if they lack oncogenes? The answer lies once again in the ability of retroviruses to insertionally mutate cellular genes. Insertions close to proto-oncogenes have been found to affect their expression in a positive manner. Several mechanisms have been proposed to account for insertional activation, some of which are illustrated in Fig. 4. Whatever the mechanism, these results show that retroviral insertion not only can lead to recessive loss of gene activity but also can cause dominant activation of gene expression.

Insertional activation of a cellular proto-oncogene was first demonstrated for the avian c-*myc* gene. The association of this gene with tumor formation was already well known through the characterization of acute viruses containing the modified oncogenic form (v-*myc*). Avian leukosis virus (ALV), a slowly transforming virus that induces tumors of the chicken bursa, does not carry v-*myc*. However, over 90% of bursal tumors do contain ALV proviral insertions a few kbp upstream from the c-*myc* gene. Transcripts initiating in the viral 3'LTR and containing the coding exons of c-*myc* have also been found. The finding that viral insertion can modify the expression of the c-*myc* oncogene suggested that additional transforming genes could be isolated by searching for common proviral integration sites in other tumors. Such searches were first rewarded in virally induced mammary tumors and subsequently in a number of other tumor types.

29.4.1 Identification and Isolation of *int-1*[7]

The strategy for cloning transforming genes can be illustrated best by describing the first of several genes found to be activated in mammary tumors. Southern analysis of DNA from a number of independently derived tumors of C3H mice revealed many additional proviruses in addition to the expected endogenous MMTVs. One of the new proviruses was cloned and unique sequence probes were identified in the flanking DNA. These probes were hybridized to the original panel of tumor DNA samples to see if integration in the same region had occurred in other tumors. Out of 26 tumors examined, 18 (70%) contained integrations within this 19 kbp segment of mouse DNA. In ad-

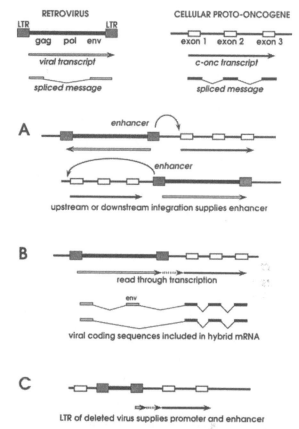

Fig. 4A–C. Mechanisms of retroviral insertional activation. Three different ways that insertion of a retrovirus can affect expression of a cellular proto-oncogene (c-onc) are shown. In **A**, integration of a retrovirus either upstream or downstream of a gene provides an enhancer that works at a distance to activate the gene. The proto-oncogenes *int-1* and *Pim* have both been activated by downstream enhancement. Almost invariably, the transcriptional orientation of the integrated provirus is away from the activated gene. In **B**, transcription of an integrated retrovirus extends into the flanking cellular gene. Splicing of the hybrid transcript produces a mature mRNA that contains exons from both the virus and from the flanking gene. Transcripts including sequences from the viral *gag* gene have led to c-*myb* activation. Transcripts that also include the viral *env* sequences, as has been seen for c-*erb-B*, can produce altered function at the protein level. In **C**, the 3'LTR of a deleted retrovirus integrated upstream of a gene provides a new set of regulatory sequences for transcriptional initiation. This arrangement has been observed often at the avian c-*myc* gene

dition, one of the probes detected a 2.6 kbp transcript in a number of tumors but not in normal mammary glands. None of the novel proviruses interrupted this coding sequence, which was named *int-1* (for MMTV *int*egration site). With one exception, all were found clustered on either side of the transcribed region. Most had integrated in the opposite transcriptional orientation, suggesting that they were activating *int-1* by enhancer insertion.

Since its discovery, the involvement of *int-1* in mammary tumor development has been confirmed by gene transfer experiments both in tissue culture and transgenic animals. What about the function of *int-1* in normal development? Studies on the pattern of transcription of *int-1* have revealed highly localized expression in the central nervous system (CNS) of embryos examined at 8 to 14 days post coitum (dpc), suggesting a role in development of the CNS. In the adults, expression has only been found in the testis, suggesting an additional role in spermatogenesis. Also suggesting a role in mouse development is the finding that *int-1* is about 54% identical at the amino acid sequence level to *Wingless* (*wg*), a gene that plays a functional role in the control of *Drosophila* development.

29.4.2 Identification of Other Proto-Oncogenes[8]

Several other unrelated genes which are activated in some mammary tumors (*int-2*, *int-3*, *int-4*, and *int-4l*) have been cloned using a similar approach. One of these genes, *int-2*, is structurally related to the fibroblast growth factor family of genes and is specifically expressed in the embryo. A number of other potential oncogenes have been cloned from other kinds of tumors, though none has yet been studied to the same extent as *int-1* and *int-2*.

The identification and characterization of novel transforming genes, may well have important implications for the study of mouse development. The homology between a mammalian proto-oncogene and a developmentally important gene in *Drosophila* makes this a compelling argument. Over the next few years we can expect to see a much more detailed examination of genes cloned from tumors. The full characterization of these genes should benefit the study of neoplastic growth and differentiation as well as normal developmental processes.

29.5 Transgenic Mice and Experimentally Induced Insertional Mutations[9]

While the above efforts have depended on events occurring spontaneously in the mouse, there are also exciting prospects for experimentally inducing new insertional mutations. These have been made possible by the development of several methods to introduce DNA stably into the mouse germ line.

The most widely used method to produce transgenic mice is to inject cloned DNA into a pronucleus of a fertilized egg. Although used primarily for studies of gene expression, an unexpected benefit has been the frequent integration of the injected DNA into genes, causing insertional mutations. Experimental retroviral infection of embryos has also been developed as a method for gene transfer. Just like spontaneous infections, this approach has resulted in insertional mutations as well. An additional approach to making transgenic mice involves genetic modification of tissue culture lines of embryonic stem cells (ES cells), followed by reincorporation into growing embryos. By applying somatic cell genetic techniques to these cells while in culture, both random and selected insertional mutations have been produced.

Approximately 10% of the random integration sites for experimentally introduced foreign DNA cause insertional mutations. The frequency resulting from experimental retroviral infection of embryos or ES cells is similar. At present, over 20 induced insertional mutations have been described and at least partially characterized (see Table 2). Some of these mutations lead to death in utero, while others exhibit mutant phenotypes after birth. Very few have shown a dominant phenotype. Most have proven to be recessive, presumably reflecting loss of function of the affected gene. In most cases, random insertional mutations in transgenic mice have identified previously unknown genes. Interestingly, a number of insertional mutations have actually turned out to be new alleles of genes previously identified by classical genetic analysis. Similar to *d* and *hr*, these insertional mutations provide direct access to genes already studied at the phenotypic level.

29.5.1 Identifying and Characterizing New Insertional Mutations [10]

In all mutation studies, when a new variant is discovered standard genetic tests are employed to demonstrate the heritability of the novel phenotype. For mutations arising in transgenic animals, an additional requirement is to prove that the mutation really is caused by integration of the transgene. It is always possible that the mutation is spontaneous and only coincidentally arose in the transgenic strain. Another possibility is that expression of the transgene, and not integration into an endogenous gene, causes the mutant phenotype. In the following sections, the genetic and molecular methods used to prove that mutations are due to transgene insertion are examined. The analysis of the S strain is used as an illustration. This is one of the first transgenic strains in which a mutation with a visible phenotype arose, a recessive mutation producing a developmental defect in the limbs.

Table 2. Random insertional mutations in transgenic mice

Strain	Prenatal lethal insertional mutations		
	Method	Characterization	Ref.
Mov13	Retroviral infection of postimplantation embryo	Gene affected is α1(I) collagen; see text for details on phenotype	12
Mov34	Retroviral infection of preimplantation embryo	Early postimplantation lethal; gene is ubiquitously expressed in embryos and adults	17
413d	Retroviral infection of ES cells	Early postimplantation lethal	14
HUGH3,4	DNA injection	DNA rearrangements in the flanking regions	18
line 4	DNA injection	Mapped to chromosome 3; gene cloned	19
T81-3	DNA injection	Underdevelopment of embryonic ectoderm	19
HB58	DNA injection	Mapped to chromosome 10; gene cloned	20
HE46	DNA injection	Neural tube defect	20
BS12	DNA injection	Mapped to chromosome 1	20

Strain	Visible insertional mutations		
	Method	Phenotype	Gene affected	Ref.
Mpv17	Retroviral infection of preimplantation embryo	Kidney failure; death before 9 months	Unknown	17
H4	Retroviral infection of ES cells	Dopamine depletion	*Hprt*	13
line 2	DNA injection	Fused toes	Unknown	11
S	DNA injection	Defects in limbs; absence of kidneys	Limb deformity (*ld*)	10
PHT1-1	DNA injection	Defects in brain and hindlimbs	Unknown; named legless	11
358-3	DNA injection	Anterior digit deformity	Dominant extra toes (*Xt*)	11
Fo. 14	DNA injection	Cranio-facial abnormalities	Unknown	11
OVE3A	DNA injection	Abnormal spermatid development; male infertility	Unknown; named symplastic spermatids (*sys*)	21
GF49	DNA injection	Infertility; low germ cell numbers	Unknown; named germ cell deficiency (*gcd*)	22
p432	DNA injection	Abnormal neurological development; male infertility	Purkinje cell degeneration (*pcd*)	23
p447	DNA injection	Motor disorder; male sterility	Hotfoot (*ho*)	23
Tg4	DNA injection	Limb incoordination by 12 days	Dystonia musculorum (*dt*)	24
A/B	DNA injection	Growth defect; 60% normal size	pygmy (*pg*)	22
OVE1B	DNA injection	Absence of certain hair types	downless (*dl*)	21

The most basic step in demonstrating the involvement of the transgene in a newly identified mutation is to show that the mutant phenotype and the transgene are genetically inseparable. Standard genetic crosses are complemented by DNA analysis to determine which animals are transgenic. In this way, the pattern of inheritance of the phenotype is determined at the same time that co-segregation of the transgene with the phenotype is investigated. Genetic analysis showed that the limb deformity phenotype in the S strain was inherited as an autosomal recessive trait. Proving that the mutation was caused by transgene insertion involved two steps. All animals that were known to be heterozygous for the mutation (carriers) were shown to carry the transgene by DNA analysis. These animals were then crossed and the progeny displaying the mutant phenotype were shown to be homozygous for the transgene.

In the discussion of *Srev-5*, we saw how the intensity of bands on Southern blots can be used to determine the gene dosage of an inserted retrovirus. The same approach can be applied to injected DNA. However, for the limb deformity mutation, a less ambiguous method was used to prove homozygosity of the transgene. The site of integration was cloned and a molecular probe was obtained from the flanking region that allowed the insert-containing allele and the wild-type allele to be distinguished (see Fig. 5). Animals homozygous for the transgene had only one band on Southern blots, the one diagnostic for the insert-containing allele. Nontransgenic animals had only the wild-type band while heterozygotes had both bands. Using this probe, complete concordance between the limb deformity phenotype and homozygosity for the transgene was demonstrated.

If preliminary analysis shows that all animals expressing the mutant phenotype are homozygous for the transgene, a strong case can be made for disruption of a recessive endogenous gene. However, this finding alone is not sufficient proof. There is still the possibility that the apparent recessive mutation is due to increased levels of expression of the homozygous transgene. One approach used to resolve this question is based on the fact that transgenes integrate at different sites in each strain. The same genetic and molecular tests can be applied to these

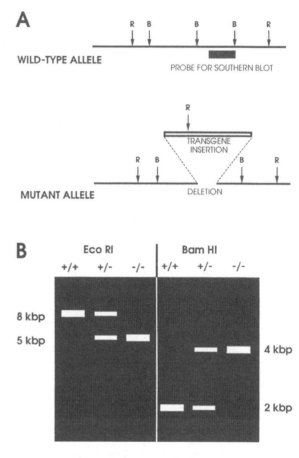

Fig. 5A, B. Structure of a transgene insertion. A schematic representation of the structures of the wild type and mutant alleles of an insertionally mutated gene is shown in A. Insertion of the transgene was accompanied by a small deletion of chromosomal sequences, as was seen for *ld*. The arrangement of sites for restriction enzymes Eco RI (R) and Bam HI (B), used to compare the structures of the wild-type and mutant allele, is shown. A schematic of a Southern blot hybridized to a probe corresponding to sequences *above the thick black bar* is shown in B. The probe recognizes distinct restriction enzyme fragments for the two alleles. DNA from wild-type animals (+/+) has a 8 kbp Eco RI fragment and 2 kbp Bam HI fragment. DNA from heterozygotes (+/−) have both the normal fragments as well as a novel 5 kbp Eco RI fragment and 4 kbp Bam HI fragment coming from the mutant allele. The removal of a Bam HI site due to deletion combined with addition of the transgenic sequences, creates a larger Bam HI fragment in the mutant allele for hybridization to the probe. The transgene introduces a novel Eco RI site, creating a smaller fragment for hybridization to the probe. Homozygous mutants (−/−) have only the novel fragments. The similar use of a probe on Southern blots of animals in the S strain allowed the absolute concordance of the *ld* phenotype and homozygosity of the transgene to be demonstrated

other transgenic strains carrying the same foreign DNA. If they do not show the mutant phenotype, yet express the transgene in an identical manner to the mutant strain, an indirect case can be made for the integration site in the original strain causing the mutation. The strain segregating the limb deformity mutation was one of a total of four transgenic strains made with the same DNA. Genetic analysis of the other three strains revealed no mutant phenotypes in animals homozygous for the transgene. In addition, one of the other strains had an identical pattern of expression of the transgene. The fact that homozygotes of this strain are normal is further proof that expression of the transgene did not cause the limb deformity phenotype.

An even more convincing way to prove insertional mutagenesis as the cause of a phenotype is to show that the mutation is a new allele of a known gene. If a phenotype resembles that of a previously described mutation, test crosses can be carried out to determine genetic complementation. The test animals are compound heterozygotes: heterozygous for the transgene and heterozygous for the original allele. Lack of genetic complementation would result in these animals exhibiting the mutant phenotype, and mean that the mutation is a new allele. In addition, finding the mutant phenotype in animals heterozygous for the transgene, would further confirm that transgene dosage or level of expression are unrelated to phenotype.

The initial indication that the mutation in the S strain might be allelic to a classical mutation was the striking similarity of phenotypes of homozygous S transgenics and mutants of the limb deformity (*ld*) gene. Besides deformed limbs and fused digits, the homozygotes often are missing one or both kidneys. Other evidence suggesting allelism was obtained by mapping the integration site to chromosome 2, where *ld* maps. However, firm evidence came from complementation tests in which transgenic mice and *ld* mutants were crossed. Animals that were compound heterozygotes for the transgene insertion and for *ld* were phenotypically identical to homozygous *ld* animals or homozygous S transgenics. This result provided very strong evidence for transgene disruption of the *ld* locus as the cause of the phenotype. This prompted an extensive molecular analysis of the region surrounding the insertion site so that the *ld* gene could be isolated and its role in limb development studied.

29.5.2 Isolating Genes Identified by Random Insertion of Injected DNA[11]

The structure of many of the transgenes produced by injection have complicated the task of obtaining molecular clones corresponding of the pre-integration site and, in turn, identifying a presumptive gene mutated by the integration. Injected DNA is often present in several copies at the integration site, usually in a head-to-tail concatenated arrangement. The long blocks of tandemly repeated copies often have a size that is larger than the capacity of standard cloning vectors meaning that most clones will contain only transgene DNA. Without the junction fragment, access to the region around the insertion is impossible. In the cases where it has been possible to clone the junction, analysis has shown that extensive rearrangements of the chromosomal sequences around the site of integration can occur. In addition to the transgene, the concatenated arrays often contain mouse DNA derived from chromosomal locations other than the integration site. In case where complex rearrangements or large deletions have occurred, several genes could in fact be affected by the integration event.

Fortunately, analysis of the *ld* gene was not hindered by rearrangements at the site of insertion. Only a small deletion of about 1.5 kbp of chromosomal sequences had accompanied insertion. Cloned sequences around the transgene provided flanking probes for an analysis of transcription in the region. The approach taken illustrates another general strategy used to search for genes within defined chromosomal regions. As described for CpG islands, the structure of the cloned flanking DNA can indicate potential transcribed regions. In this case, the flanking DNA was searched for sequences also found in a variety of other organisms. Such sequences have a potential to be expressed, the logic being that a sequence conserved over evolution probably codes for a functional domain of a protein.

A search for conserved sequences netted two regions with ORFs and correct splicing signals. These were used to probe cDNA libraries made from mRNA extracted from kidney, a tissue likely on the basis of the mutant phenotype to express the *ld* gene. The cDNA clones that were isolated have allowed an extensive molecular characterization of the *ld* gene. The gene is the template for a family of transcripts that arise through the use of alternative splicing and polyadenylation sites. These transcripts have a complex pattern of expression spatially, within the organism, and temporally, during development. Transcription occurs early in embryogenesis, consistent with the fact that onset of the *ld* mutant phenotype occurs by midgestation. In the adult, expression is found in several organs besides the kidney.

Of the many genes that have now been identified by insertional mutations (see Table 2), several cause limb dysmorphisms. Besides *ld*, these include a previously unknown gene named legless. In the mutant, the hindlimbs do not develop. In addition, these mice have craniofacial and brain abnormalities. Another mutant also has a phenotype of craniofacial malformation. Effects on the digits have been seen in several insertional mutants. Hopefully, the analysis of all these genes and their gene products should contribute to our overall understanding of the development of the body plan.

29.5.3 Isolating Genes Identified by Random Insertion of Exogenous Retroviruses[12]

A number of points make retroviruses attractive tools for experimental insertional mutagenesis. There is growing evidence that a substantial fraction of retroviral integrations occur in transcriptionally active regions of the genome. If true, this suggests that proviral integration might be targeted to genes active in the infected cell. When insertional mutations have occurred, the lack of chromosomal rearrangements accompanying retroviral integration has made the task of finding a gene in the flanking region easier. Retroviruses have been engineered to carry foreign genes, including ones for the express purpose of making the subsequent cloning of integration sites even easier.

The insertional mutation carried in the *Mov-13* transgenic line was one of the first described and remains one of the most thoroughly characterized to date. The *Mov-13* strain was generated by retroviral infection of a postimplantation embryo. It carries a single proviral insertion and is a recessive lethal, with death occurring in homozygous embryos by 14 dpc. A deductive approach was used to identify the gene at the *Mov-13*-retroviral integration site. The integration site was readily cloned and a region flanking the proviral insertion site was found to be transcribed. Based on the spatial and temporal patterns of normal expression of the flanking gene, probes for candidate genes whose products are found in the extracellular matrix were tested for identity to the *Mov-13* locus. An $\alpha 1(I)$ collagen gene probe and the *Mov-13* specific probe were shown to hybridize to the same bands on Southern and Northern blots. Analysis of the DNA sequence around the site of integration showed that the provirus had inserted in the first intron of the $\alpha 1(I)$ collagen gene. This provides another example of integrations of mobile elements into a noncoding regions causing mutations.

Analysis of the phenotype of the *Mov-13* strain has lead to a new appreciation of the role of $\alpha 1(I)$ collagen in normal development. *Mov-13* mutant embryos were found to develop normally until around 11 dpc, at which point sudden embryonic death occurs. Examination of the heart and large blood vessels revealed the cause of death to be extensive bleeding, apparently due to the rupturing of a major blood vessel. This finding revealed for the first time that collagen I is essential for the establishment and maintenance of the mechanical stability of the developing circulatory system. The role of collagen I in organ development was also investigated, even though most organs develop long after mutant embryos die. For these experiments, organ rudiments were taken from still viable homozygous embryos and placed in culture. The organ cultures reached developmental stages beyond that possible in intact embryos. These studies revealed that collagen I is not essential for the normal development of a variety of epithelial-mesenchymal organs, such as the lung, where a role had been previously postulated.

Further investigation of the effect of the mutation focused on tooth development. Normal odontoblasts excrete dentin, an extracellular matrix containing only type I collagen. Unexpectedly, development of homozygous mutant tooth rudiments in an in vivo organ culture system was completely normal and indistinguishable from the development of teeth from heterozygous or wild-type donors. Most surprisingly, a normal layer of type I collagen was found in the dentin of the developing teeth taken from homozygous embryos. Further analysis showed that in odontoblasts, unlike all other cell types studied, the presence of the provirus does not interfere with expression of the $\alpha1(I)$ collagen gene. This unexpected finding has shown that proviral insertions can be conditionally mutagenic in a cell-specific manner. One explanation is that different cis-regulatory sequences are used in different cell types and that the insertion affects these sequences differently. In any case, Mov-13 has proven to be an invaluable tool for defining the role of collagen in development and for probing the complexity of transcriptional regulation of the $\alpha1(I)$ collagen gene.

29.5.4 Inducing and Selecting Specific Insertional Mutations[13]

All of the mutations that have been described so far have come about from the random integration of DNA or a retrovirus into genes. Even in cases where integrations have disrupted known genes, the mutational events were the products of chance occurrences. Randomness is desirable when the goal is to identify unknown genes. What about the possibilities for introducing mutations into specific genes? ES cells have the unique ability to differentiate in vivo when reintroduced into embryos and contribute to all cell lineages, including the germline. Studies have shown that gene transfer into ES cells provides an indirect yet equally efficient route for transgenesis. In contrast to transgenesis via microinjection of DNA, somatic cell genetic methods can be applied to ES cells in tissue culture in order to induce and select specific mutations that can then be imported into the mouse itself. One of the first such mutations was produced by retroviral infection of ES cells. The successful incorporation of this mutation into the germ line showed that this indirect approach to specific mutagenesis was possible.

Mutations in the gene for the purine salvage pathway enzyme hypoxanthine-guanosine phosphoribosyltransferase (HPRT) are unique in that they can be easily selected in tissue culture. In addition, the Hprt gene is on the X-chromosome, so in ES cells derived from male embryos only one copy of the gene is present. Therefore in these cells the mutant phenotype [resistance to the purine analog 6-thioguanine (6-TG)] could be scored after only one mutagenic event. Although the gene is probably no more likely a target for insertional mutagenesis by retroviruses than any other gene, somatic cell genetic techniques make it possible to select for cells in which this particular insertion event has occurred. As it turned out, the random integration of one retrovirus into this specific gene was an extremely rare event, occurring only twice in over a million infected ES cells. ES cells from both of these mutant lines were incorporated into the germ lines of chimeric mice, enabling two different HPRT mutant strains to be derived (see Fig. 6). Mutant males from one of the strains have undetectable levels of HPRT enzyme activity. The association of a particular human genetic disease with HPRT inactivity prompted metabolic and behavioral studies of these mice.

Lesch-Nyhan (L-N) syndrome is a rare X-linked human genetic disease caused by mutations at the Hprt locus that result in loss of HPRT activity. The disease predominantly strikes boys, with symptoms appearing during early childhood. The most drastic effect is self-mutilation, mostly of the lips and fingers. It is still not understood how a primary deficiency in purine metabolism can have a pleiotropic effect on behavior. One relatively recent finding is that L-N children have severely depleted levels of the neurotransmitter dopamine (DA) in their basal ganglia, up to 90% less than normal. The relationship between HPRT activity and DA metabolism in the brain is not known. One way to approach these intriguing questions is through the use of animal models that reproduce the human disorder in some or all aspects. The HPRT-deficient mice were investigated in order to determine their usefulness as models of L-N syndrome.

Although the mutant mice lack detectable levels of HPRT activity, they display none of the peculiar behaviors that distinguish L-N children. Further

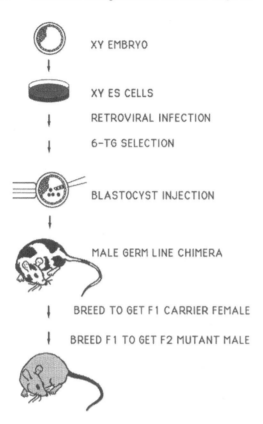

XY EMBRYO

XY ES CELLS

RETROVIRAL INFECTION

6-TG SELECTION

BLASTOCYST INJECTION

MALE GERM LINE CHIMERA

BREED TO GET F1 CARRIER FEMALE

BREED F1 TO GET F2 MUTANT MALE

Fig. 6. Producing HPRT mutant mice from ES cells. The strategy for generating a germ line mutation using ES cells is shown. An ES cell line derived from a male embryo was infected by culturing in the presence of another cell line that produced a retrovirus. The infected ES cell culture was put under selection for resistance to 6-TG. 6-TG resistant clones were shown to be HPRT mutant cells and were injected into embryos to form chimeric mice. Male germ line chimeras were bred to obtain an F_1 generation of carrier females and these were bred to obtain hemizygous mutant male F_2 animals

tests found no impairments in general motor skills or ability to learn. Lacking an overt phenotype, the mice were examined neurochemically and found to have a moderate depletion of DA in the basal ganglia, about 25% less than normal control mice. Further characterization revealed deficiencies in the serotonin neurotransmitter system as well. These results reveal a pattern of biochemical changes in *Hprt* mutant mice, with similarities and differences to the human syndrome. For instance, the change in DA is qualitatively similar to but quantitatively different from the situation in L-N patients. On the other hand, serotonin levels are normal in L-N brains, but reduced in HPRT mutant mice. The similarities suggest that these mice might model certain aspects of the L-N syndrome. However, the differences demonstrate the potential limitations of animal models, as well as raising new questions. Why does the same genetic lesion have such dramatically different developmental consequences in mice and humans? Resolving this question may eventually provide therapeutic strategies for circumventing the neurochemical changes that are secondary to HPRT deficiency.

29.6 Outlook [14]

Molecular characterization is still in the initial stages for most of the genes identified by insertional mutagenesis. Although *Mov-13* has already provided important insights into some of the mechanisms underlying mammalian development, the work on this mutant has been going on for most of the past decade. Similarly detailed analyses for other mutants will take an equally long time. In the years to come, insertional mutagenesis is likely to continue to identify novel genes and new alleles of known genes. One can expect that with the broadening availability of transgenic technology, more and more experimentally induced mutants will be discovered. The hybrid mouse strains that undergo high frequencies of spontaneous reintegrations of endogenous retroviruses will also provide a rich source of new mutants. Further investigation of the association of endogenous retroviruses and old mutations identified by classical genetic analysis, will require the identification of new specific probes. Coupled with improvements in the resolution of blotting techniques, it should be possible eventually to carry out molecular genetic analysis of even the 1000-member IAP class of sequences.

In addition to these efforts, new approaches are being developed which utilize unique aspects of ES cells and the transgenic mice derived from them. The power of embryonic stem cell technology should bring us closer to two important goals. Scientists working on mouse development are eager to duplicate the feat achieved by *Drosophila* geneticists who have used saturation mutagenesis to identify every gene involved in specific developmental processes. Saturation insertional mutagenesis is a long way off but the approaches utilizing ES cells described below provide the groundwork for such experiments. Equally desirable is the ability to generate mutations in genes that have already been cloned, but for which no mutants are currently available. The study of mutant phenotypes in all of the genes that have recently come to light through studies on development in lower organisms, mainly *Drosophila* (see Lobe and Gruss, Chap. 26, this Vol.), would be extremely informative. As we describe below, the application of novel somatic cell genetic methods to ES cells is allowing this once undreamed of goal to be attained.

29.6.1 New Strategies for Random Insertional Mutagenesis in Transgenic Mice[15]

Infection of ES cells in culture has proven to be a very efficient way to introduce retroviruses into the germ line, much more efficient than infecting intact embryos. In fact, retroviruses can be introduced at high copy number into individual cells by carrying out infection over an extended culture period. This provides a unique method for integrating large numbers of potentially mutagenic transgenes into the germ line of a single transgenic mouse. In the first application of this approach, 25 transgenic strains were derived. At least 200 new retroviral integration sites were established into the germ line in these mice. In effect, a new class of multi-copy endogenous retroviruses was created. Genetic analysis of these new proviruses has required similar approaches to those already described for the various classes of endogenous viruses present in multiple copies in inbred mice. Although this analysis has been long and tedious, it has been fruitful. Of the 200 new proviruses, several have integrated next to genes causing insertional mutations affecting development. The isolation and analysis of these genes is just beginning.

ES cells are also being used for a less random approach to identify genes involved in development. This approach, first developed for studies on *Drosophila* development, relies on the ability to de-

tect whether an introduced retroviral element or DNA construct has integrated next to or within a gene. The vectors carry the *Escherichia coli* gene for β-galactosidase (lacZ). Expression of the gene can be easily scored by staining with X-gal, a substrate for β-galactosidase that is converted into a visible deep blue product. In "enhancer traps", lacZ is linked to a weak promoter and expression is dependant on the vector integrating near an enhancer for a cellular gene. In "gene traps", the lacZ gene lacks regulatory sequences except for a splice acceptor site. The gene can only be expressed if the vector integrates into an intron of a cellular gene and splicing results in a chimeric mRNA that produces a functional fusion protein. In addition, the integration should produce a mutation. ES cells infected or transfected with vectors of these sorts have been introduced into chimeras and patterns of lacZ expression have been studied, and mutant mice have been derived. Large-scale application of this approach, already underway in several laboratories, should identify many new genes which are of developmental interest.

29.6.2 Nonrandom Insertional Mutagenesis: Targeting Mutations by Homologous Recombination[16]

Unlike the case for *Hprt*, direct selection of mutations in the majority of genes in ES cells is impossible. Fortunately, a more general approach has been developed to allow a wide range of mutations to be created. The only requirement is for the gene to be already cloned. The approach utilizes the ES cell's ability to carry out homologous recombination between introduced cloned DNA and the endogenous locus. Although DNA will for the most part integrate randomly at nonhomologous chromosomal positions, at a certain frequency integration will occur by homologous recombination with the chromosomal sequence. For this approach to result in mutations, the cloned copy of the gene must first be modified in vitro. In most applications, the coding region is disrupted by insertion of DNA that also functions as a selectable marker, usually the bacterial neomycin resistance gene (*neo*) that can be selected in mouse cells using the aminoglycoside antibiotic G418. Cloned DNA that has been insertionally mutated in vitro will transfer the mutation to the endogenous gene following homologous recombination in vivo. Both somatic cell genetics and molecular techniques can then be used to screen for the rare clones of cells in which these so-called targeted events have occurred (see Fig. 7).

Targeted mutagenesis is currently one of the most exciting fields in mammalian developmental biology. Already, mutations in a number of genes have been introduced into the mouse germ line. It is conceivable that in the next few years, an ever-increasing list of candidate developmental genes will be accompanied by targeted mutations generated in ES cells. One of the most striking examples of the power and scope of insertional mutagenesis is the identification and characterization of the *int-1* gene. As described in a previous section, the gene was first identified through insertional activation by retroviruses. The original insertional mutation allowed the gene to be cloned and studies on its normal role to be initiated. These studies showed that the gene is expressed in the developing brain. Now these studies are being extended using mutant mice derived from ES cells carrying targeted insertional mutations in the *int-1* gene. In fact, the null mutant produced by targeting displays dramatic developmental abnormalities of the brain. Hopefully in the near future, similar studies on other genes will contribute to our deepening understanding of fundamental mechanisms in development.

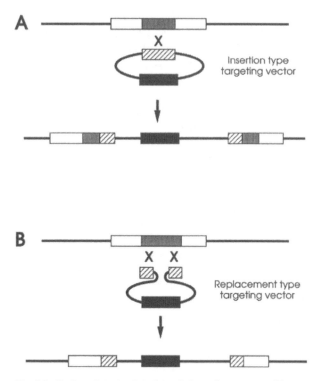

Fig. 7 A, B. Gene inactivation through homologous recombination. A general strategy for targeted disruption of a gene is shown. Two types of targeting vectors are shown, each with homology to the middle of the gene being targeted. The section of the gene shown in *vertical stripes* is homologous to the area in the targeting vectors shown in *diagonal stripes*. The *thick black bar* in the targeting vectors represents a selectable gene such as the bacterial neomycin resistance gene, *neo*. Homologous recombination in each case inserts the vector into the gene and disrupts it. Cells in which these events have taken place can be enriched for due to the presence of the selectable gene in the vector. In **A**, inactivation occurs by a single crossover event leading to duplication of the sequences shared by targeting vector and gene (*insertion*). In **B**, a double crossover event inactivates the gene without leading to duplication of any sequences (*replacement*)

29.7 Summary

Insertional mutagenesis is both an approach to molecularly clone and understand existing mutations as well as an experimental tool to create new mutations. Two mutations, well known from earlier genetic studies, have been shown to be due to insertions of endogenous retroviruses. Cloning these genes has allowed molecular studies into their function during development to be undertaken. In the past few years, over 30 genes have been identified and cloned following induced insertional mutation in transgenic mice (Table 2 lists only the best characterized ones). The characterization of these genes is also providing valuable insights into developmental processes. New strategies for insertional mutagenesis will not only allow many new genes to be identified and isolated, but also allow mutant alleles to be created in a wide variety of cloned genes.

References

General

Gridley T, Soriano P, Jaenisch R (1987) Insertional mutagenesis in mice. Trends Gent 3:162–166

Howe M, Berg D (eds) (1989) Mobile DNA. ASM, Washington

Lyon MF, Searle AG (eds) (1989) Genetic variants and strains of the laboratory mouse, 2nd edn. Oxford University Press, Oxford, Gustav Fischer Verlag, Stuttgart

Shapiro J (ed) (1983) Mobile genetic elements. Academic Press, New York

Weiss R, Teich N, Varmus H, Coffin J (eds) (1985) Molecular biology of tumor viruses: RNA tumor viruses, 2nd edn. Cold Spring Harbor Laboratory, Cold Spring Harbor, New York

[1]Bhattacharyya MK, Smith AM, Ellis THN, Hedley C, Martin C (1990) The wrinkled-seed character of pea described by Mendel is caused by a transposon-like insertion in a gene encoding starch-branching enzyme. Cell 60:115–122

Kazazian HH, Wong C, Youssoufian H, Scott AF, Philipps DG, Antonarakis SE (1988) Haemophilia A resulting from de novo insertion of L1 sequences represents a novel mechanism for mutation in man. Nature 332:164–166

McClintock B (1984) The significance of responses of the genome to challenge. Science 226:792–801

Moerman DG, Waterston RH (1984) Spontaneous unstable unc-22 IV mutations in C. elegans var. Bergerac. Genetics 108:859–877

Mount SM, Rubin GM (1985) Complete nucelotide sequence of the Drosophila transposable element Copia: homology between Copia and retroviral proteins. Mol Cell Biol 5:1630–1638

O'Hare K, Rubin GM (1983) Structures of P transposable elements and their sites of insertion and excision in the Drosophila melanogaster genome. Cell 34:25–35

Roeder GS, Farabaugh PJ, Chaleff DT, Fink GR (1980) The origins of gene instability in yeast. Science 209:1375–1380

Varmus H, Brown P (1989) Retroviruses. In: Howe M, Berg D (eds) Mobile DNA. ASM, Washington, pp 53–108

[2]Stoye JP, Coffin JM (1985) Endogenous retroviruses. In: Weiss R, Teich N, Varmus H, Coffin J (eds) Molecular biology of tumor viruses: RNA tumor viruses, vol 2. Cold Spring Harbor Laboratory, Cold Spring Harbor, NY, pp 357–404

Stoye JP, Coffin JM (1987) The four classes of endogenous murine leukemia virus: structural relationships and potential for recombination. J Virol 61:2659–2669

Stoye JP, Coffin JM (1988) Polymorphism of murine endogenous proviruses revealed by using virus class-specific oligonucleotide probes. J Virol 62:168–175

[3]Copeland NG, Hutchinson KW, Jenkins NA (1983) Excision of the DBA ecotropic provirus in dilute coat-color revertants of mice occurs by homologous recombination involving the viral LTRs. Cell 33:379–387

Eibel H, Philippsen P (1984) Preferential integration of yeast transposable element Ty into a promoter region. Nature 307:386–388

Jenkins NA, Copeland NG, Taylor BA, Lee BK (1981) Dilute (d) coat color mutation of DBA/2J mice is associated with the site of integration of an ecotropic MuLV genome. Nature 293:370–374

Jenkins NA, Copeland NG, Taylor BA, Lee BK (1982) Organization, distribution, and stability of endogenous ecotropic murine leukemia virus DNA in chromosomes of Mus musculus. J Virol 43:26–36

Rinchik EM, Russell LB, Copeland NG, Jenkins NA (1985) The dilute-short ear (d-se) complex of the mouse: lessons from a fancy mutation. Trends Genet 1:170–176

Snyder MP, Kimbrell D, Hunkapiller R, Hill R, Fristrom J, Davidson N (1982) A transposable element that splits the promoter inactivates a Drosophila cuticle protein gene. Proc Natl Acad Sci USA 79:7430–7434

Strobel MC, Seperack PK, Copeland NG, Jenkins NA (1990) Molecular analysis of two mouse dilute locus deletion mutations: spontaneous dilute lethal[20J] and radiation-induced dilute prenatal lethal Aa2 lethal alleles. Mol Cell Biol 10:501–509

[4]Bird AP (1986) CpG-rich islands and the function of DNA methylation. Nature 321:209–213

Stoye JP, Fenner S, Greenoak GE, Moran C, Coffin JM (1988) Role of endogenous retroviruses as mutagens: the hairless mutation of mice. Cell 54:383–391

[5]Jenkins NA, Copeland NG (1985) High frequency acquisition of ecotropic MuLV proviruses in SWR/J-RF/J mice. Cell 43:811–819

Kidwell M (1986) P-M mutagenesis. In: Roberts DB (ed) Drosophila: a practical approach. IRL Press, Oxford, pp 59–81

Lock LF, Keshet E, Gilbert DJ, Jenkins NA, Copeland NG (1988) Studies of the mechanism of spontaneous germline ecotropic provirus acquisition in mice. EMBO J7:4169–4177

Steffen DL, Taylor BA, Weinberg RA (1982) Continuing germ line integration of AKV proviruses during the breeding of AKR mice and derivative recombinant inbred strains. J Virol 42:165–175

[6]Hayward WS, Neel BG, Astrin SM (1981) Activation of a cellular oncogene by promoter insertion in ALV-induced lymphoid leukosis. Nature 290:475–480

Nusse R, Varmus HE (1982) Many tumors induced by the mouse mammary tumor virus contain a provirus integrated in the same region of the host genome. Cell 31:99–109

Nusse R, Berns A (1988) Cellular oncogene activation by insertion of retroviral genes: genes identified by provirus tagging. In: Klein G (ed) Cellular oncogene activation. Marcel Dekker, New York, pp 95–119

[7]Nusse R, van Ooyen A, Cox D, Fung YKT, Varmus H (1984) Mode of proviral activation of a putative mammary oncogene (int-1) on mouse chromosome 15. Nature 307:131–136

Rijsewijk F, Schuerman M, Wagenaar E, Parren P, Weigel D, Nusse R (1987) The Drosophila homolog of the mouse mammary oncogene int-1 is identical to the segment polarity gene wingless. Cell 50:649–957

[8]Moore R, Casey G, Brookes S, Dixon M, Peters G, Dickson C (1986) Sequence, topography and protein coding potential of mouse int-2: a putative oncogene activated by mouse mammary tumor virus. EMBO J 5:919–924

Nusse R (1988) The int genes in mammary tumorigenesis and in normal development. Trends Genet 4:291–295

[9]Gordon J (1989) Transgenic animals. Int Rev Cytol 115:171–229

Palmiter RD, Brinster RL (1986) Germ-line transformation of mice. Annu Rev Genet 20:465–499

[10]Woychik RP, Stewart TA, Davis LG, D'Eustachio P, Leder P (1985) An inherited limb deformity created by insertional mutagenesis in a transgenic mouse. Nature 318:36–40

[11]Covarrubias L, Nishida Y, Mintz B (1986) Early postimplantation embryo lethality due to DNA rearrangements in a transgenic mouse strain. Proc Natl Acad Sci USA 83:6020–6024

McNeish JD, Scott WJ, Potter SS (1988) Legless, a novel mutation found in PHT-1 transgenic mice. Science 241:837–839

Overbeek PA, Sing-Ping L, Van Quill KR, Westphal H (1986) Tissue-specific expression in transgenic mice of a fused gene containing RSV terminal sequences. Science 231:1574–1577

Pohl TM, Mattei M-G, Rüther U (1990) The recessive insertional mutation add is allelic to the dominant mouse mutation extra-toes (Xt). Development 110:1153–1158

Wakasugi S, Iwanaga T, Inomoto T, Tengan T, Maeda S, Uehira M, Araki K, Miyazaki J, Eto K, Shimada K, Yamamura K-I (1988) An autosomal dominant mutation of facial development in a transgenic mouse. Dev Genet 9:203–212

Wilkie TM, Palmiter RM (1987) Analysis of the integrant in MyK-103 transgenic mice in which males fail to transmit the integrant. Mol Cell Biol 7:1646–1655

Woychik RP, Maas RL, Zeller R, Vogt TF, Leder P (1990) Formins: proteins deduced from the alternative transcripts of the limb deformity gene. Nature 346:850–853

Zeller R, Jackson-Grusby L, Leder P (1989) The limb deformity gene is required for apical ectodermal ridge differentiation and anteroposterior limb pattern formation. Genes Dev 3:1481–1492

[12]Cepko CL, Roberts BE, Mulligan RC (1984) Construction and applications of a highly transmissible murine retroviral shuttle. Cell 37:1053–1062

Kratochwil K, Dziadek M, Löhler J, Harbers K, Jaenisch R (1986) Normal epithelial branching morphogenesis in the absence of collagen I. Dev Biol 117:596–606

Kratochwil K, von der Mark K, Kollar EJ, Jaenisch R, Mooslehner K, Schwarz M, Haase K, Gmachl I, Harbers K (1989) Retrovirus-induced insertional mutation in Mov13 mice affects collagen I expression in a tissue-specific manner. Cell 57:807–816

Löhler J, Timpl R, Jaenisch R (1984) Embryonic lethal mutation in mouse collagen I gene causes rupture of blood ves-

sels and is associated with erythropoietic and mesenchymal cell death. Cell 38:597–607

Mooslehner K, Karls K, Harbers K (1990) Retroviral integration sites in transgenic Mov mice frequently map in the vicinity of transcribed DNA regions. J Virol 64:3056–3058

Reik W, Weiher H, Jaenisch R (1985) Replication-competent Moloney murine leukemia virus carrying a bacterial suppressor tRNA gene: selective cloning of proviral and host sequences. Proc Natl Acad Sci USA 82:1141–1145

Scherdin U, Rhodes K, Breindl M (1990) Transcriptionally active genome regions are preferred targets for retrovirus integration. J Virol 64:907–912

Schnieke A, Harbers K, Jaenisch R (1983) Embryonic lethal mutation in mice induced by retrovirus insertion into the $\alpha 1$(I) collagen gene. Nature 304:315–320

Shih C-C, Stoye JP, Coffin JM (1988) Highly preferred targets for retrovirus integration. Cell 53:531–537

Soriano P, Gridley T, Jaenisch R (1989) Retroviral tagging in mammalian development and genetics. In: Howe M, Berg D (eds) Mobile DNA. ASM, Washington, pp 927–937

Vijaya S, Steffen DL, Robinson HL (1986) Acceptor sites for retroviral integrations map near DNase I-hypersensitive sites in chromatin. J Virol 60:683–692

[13]Dunnett SB, Sirinathsinghji DJS, Heavens R, Rogers DC, Kuehn MR (1989) Monoamine deficiency in a transgenic (Hprt$^-$) mouse model of Lesch-Nyhan syndrome. Brain Res 501:401–406

Finger S, Heavens RP, Sirinathsinghji DJS, Kuehn MR, Dunnett SB (1988) Behavioral and neurochemical evaluation of a transgenic mouse model of Lesch-Nyhan syndrome. J Neurol Sci 86:203–213

Kelley WN, Wyngaarden JB (1983) Clinical syndromes associated with hypoxanthine-guanine phosphoribosyltransferase deficiency. In: Stanberry YB, Wyngaarden JB, Fredrickson DS, Goldstein JL, Brown MS (eds) The metabolic basis of inherited disease. McGraw-Hill, New York, pp 1115–1143

Kuehn MR, Bradley A, Robertson EJ, Evans MJ (1987) A potential animal model for Lesch-Nyhan syndrome through introduction of HPRT mutations into mice. Nature 326:295–298

Robertson E, Bradley A, Kuehn M, Evans M (1986) Germline transmission of genes introduced into cultured pluripotential cells by retroviral vector. Nature 323:445–458

[14]Jackson IJ (1989) The mouse. In: Glover DM, Hames BD (eds) Genes and embryos. IRL Press at Oxford University Press, Oxford, pp 204–211

Rossant J, Joyner AJ (1989) Towards a molecular genetic analysis of mammalian development. Trends Genet 5:277–283

[15]Brenner DG, Lin-Chao S, Cohen SN (1989) Analysis of mammalian cel genetic regulation using retrovirus-derived "portable exons" carrying the *Escherichia coli lacZ* gene. Proc Natl Acad Sci USA 86:5517–5521

Conlon FL, Barth KS, Robertson EJ (1991) A novel retrovirally induced embryonic lethal mutation in the mouse: assessment of the developmental fate of embryonic stem cells homozygous for the 413d proviral integration. Development 111:969–981

Gossler A, Joyner AL, Rossant J, Skarnes WC (1989) Mouse embryonic stem cells and reporter constructs to detect developmentally regulated genes. Science 244:463–465

O'Kane CJ, Gehring WJ (1987) Detection in situ of genomic regulatory elements in *Drosophila*. Proc Natl Acad Sci USA 84:9123–9127

[16]Capecchi MR (1989) Altering the genome by homologous recombination. Science 244:1288–1292

Thomas KR, Capecchi MR (1990) Targeted disruption of the murine int-1 proto-oncogene resulting in severe abnormalities in midbrain and cerebellar development. Nature 346:847–850

McMahon AP, Bradley A (1990) The Wnt-1 (int-1) proto-oncogene is required for development of a large region of the mouse brain. Cell 62:1073–1085

[17]Soriano P, Gridley T, Jaenisch R (1987) Retroviruses and insertional mutagenesis in mice: proviral integration at the Mov 34 locus leads to early embryonic death. Genes Dev 1:366–375

Weiher H, Noda T, Gray DA, Sharpe AH, Jaenisch R (1990) Transgenic mouse model of kidney disease: insertional inactivation of ubiquitously expressed gene leads to nephrotic syndrome. Cell 62:425–434

[18]Wagner EF, Covarrubias L, Stewart TA, Mintz B (1983) Prenatal lethalities in mice homozygous for human growth hormone sequences integrated in the germ line. Cell 35:647–655

[19]Mark WH, Signorelli K, Lacy E (1985) An insertional mutation in a transgenic mouse line results in developmental arrest at day 5 of gestation. Cold Spring Harbor Symp Quant Biol 50:453–463

[20]Constantini F, Radice G, Lee JJ, Chada KK, Perry W, Son HJ (1989) Insertional mutations in transgenic mice. Prog Nucleic Acid Res Mol Biol 36:159–169

[21]MacGregor GR, Russell LD, Van Beek MEAB, Hanten GR, Kovac MJ, Kozak CA, Meistrich ML, Overbeek PA (1990) Symplastic spermatids (sys): a recessive insertional mutation in mice causing a defect in spermatogenesis. Proc Natl Acad Sci USA 87:5016–5020

Shawlot W, Siciliano MJ, Stallings RL, Overbeek PA (1989) Insertional inactivation of the downless gene in a family of transgenic mice. Mol Biol Med 6:299–307

[22]Pellas TC, Ramachandran B, Duncan M, Ran SS, Marone M, Chada K (1991) Germ cell deficient (gcd), a novel insertional mutation manifested as infertility in transgenic mice. Proc Natl Acad Sci USA (in press)

Xiang X, Benson KF, Chada K (1990) Mini-mouse: disruption of the pygmy locus in a transgenic insertional mutant. Science 244:967–969

[23]Gordon JW, Uehlinger J, Dayani N, Talansky BE, Gordon M, Rudomen GS, Neumann PE (1990) Analysis of the hotfoot (ho) locus by creation of an insertional mutation in a transgenic mouse. Dev Biol 137:349–358

Krulewski TF, Neumann PE, Gordon JW (1989) Insertional mutation in a transgenic mouse allelic with Purkinje cell degeneration. Proc Natl Acad Sci USA 86:3709–3712

[24]Kothary R, Clapoff S, Brown A, Campbell R, Peterson A, Rossant J (1988) A transgene containing *lacZ* inserted into the dystonia locus is expressed in neural tube. Nature 335:435–437

Chapter 30   The Introduction of Genes into Mouse Embryos and Stem Cells

ERWIN F. WAGNER and GORDON KELLER

30.1 Introduction [1]

The elucidation of the molecular mechanisms which regulate mammalian embryonic development and cellular differentiation is one of the most challenging tasks in modern biology. Since it is believed that regulation of gene expression is fundamental to the appropriate development of specific cell lineages, a considerable effort has gone into understanding how individual genes function and how they are regulated.

A very fruitful approach to studying gene controls and gene function in organisms such as *Drosophila* and *C. elegans* has been the use of naturally occurring or experimentally induced mutations. Because of the scarcity of developmental mutants in mammalian systems, other approaches have been followed to address problems related to differential gene expression during development. One of these approaches is the transfer of cloned genes into murine stem cells and embryos.

The advent of gene cloning has made possible the analysis of eukaryotic gene regulation with the aid of powerful techniques such as the introduction of viruses, viral vectors or cloned genes into mammalian tissue culture cells. Because of the limitations of such in vitro studies for analyzing tissue-specific regulation of gene expression and for investigating the function of complex systems such as the immune or nervous system, various attempts were made to introduce cloned genes into mouse embryos and into stem cells of embryonic (ES) and hematopoietic (HS) origin. The latter two cell populations (ES and HS cells) exhibit a distinct developmental program that can partly be followed in vitro and, more importantly, in vivo. Some of the

key issues to be addressed and answered with gene transfer using these experimental systems are the following: (1) analyzing elements which control the correct cell-specific and stage-specific expression of genes in vivo; (2) studying the consequences of deregulated gene expression leading to an altered pattern of differentiation and development; (3) expanding rare precursor populations of early embryonic and hematopoietic lineages through the use of v-*onc* genes and growth control genes; (4) testing the function of genes either by their overexpression in appropriate and inappropriate cell types or by gene inactivation in ES cells and chimeras and (5) mapping lineages in early development and differentiation through the use of the DNA integration site or by virtue of cell-specific expression of marker genes.

In this review we intend to summarize the studies that have addressed the issues raised above. We will confine ourselves to experiments dealing with the stable introduction of genes into fertilized eggs and the subsequent analysis of their expression in transgenic mice. In addition, we will also outline the recent advances that have been made through the approach of transferring genes into ES and HS cells. Before reviewing the results obtained, we will briefly describe the available techniques for introducing genes into mouse embryos.

30.2 Methods for Introducing Genes into Mouse Embryos

In this section, the methods available for introducing cloned genes into mouse embryos will be briefly outlined (Fig. 1).

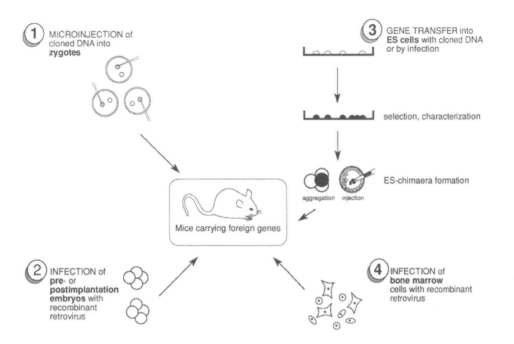

Fig. 1. Four ways of introducing genes into mice

30.2.1 DNA Injection into Fertilized Eggs[2]

The injection of recombinant DNA into the pronuclei of fertilized mouse eggs is at present the
most frequently used technique for generating
transgenic mice. Briefly, freshly isolated fertilized
eggs are cultured for $1-2$ h until the pronuclei are
visible. A DNA solution is then microinjected
directly into one of the pronuclei (Fig. 2A). Successful injection is evidenced by pronuclear swelling (Fig. 2B). The injected eggs are then transferred
to a pseudopregnant female where they develop to
term. Once the mice are born, they are analyzed as
outlined in Fig. 3. Identification of the transgenic
mice is necessary since only $10-40\%$ of the pups
born carry the injected DNA sequence. Further
analysis of such transgenic animals is carried out to

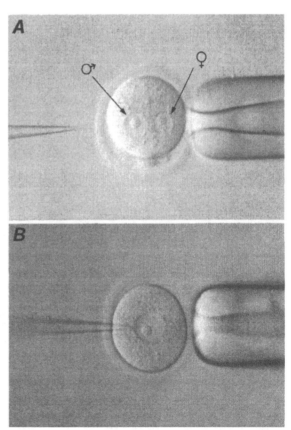

Fig. 2A, B. DNA injection into the male pronucleus of a fertilized mouse egg. **A** The egg displaying the female and male pronucleus is immobilized on a holding pipet. **B** Injection showing
swelling of male pronucleus; female pronucleus not visible
(Nomarski optics, $\times 1200$); egg diameter is approx. 70 μm; tip of
injection pipet approx. 1 μm

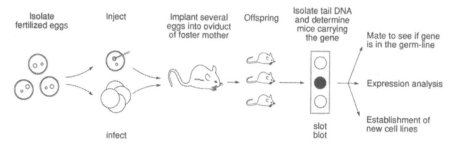

Fig. 3. Flow chart for generating transgenic mice

determine whether the genes are transmitted to off-spring and whether or not they are expressed (Fig. 3).

30.2.2 Infection of Embryos with Retroviral Vectors[3]

The second technique for introducing genes into mouse embryos uses viruses, most frequently recombinant retroviral vectors. One of the advantages of this system is its simplicity. Preimplantation embryos, denuded of their zona pellucida which blocks virus entry, can be easily infected by culturing them on cells that are producing the vector (Fig. 1). The infected embryos are then returned to a foster mother and, when born, are analyzed ex-

actly as outlined in Fig. 3. One of the disadvantages of using these vectors has been the inefficient expression of the proviral genome in embryonic cells; therefore, this technique cannot be considered as a routine method at present.

30.2.3 Gene Transfer Using Embryonic Stem (ES) Cells[4]

A third technique for introducing foreign genetic material into embryos has been to use the stem cells of teratocarcinomas and more recently of embryos (ES cells, Fig. 4). ES cells can be isolated from explanted mouse blastocysts and then established as permanent cell lines. When injected into blastocysts, these ES cells have the unique capacity to differentiate and contribute to all lineages of the developing mouse including the germ-line (Fig. 4). These characteristics of ES cells offer the possibility of manipulating the mouse genome by first isolating a clone of stem cells in vitro that had been selected to carry a particular alteration in its genome. This alteration can then be introduced into the mouse germ-line by reintroducing the ES cells into blastocysts by microinjection or into morulae by embryo-aggregation (Figs. 1, 4).

30.3 Expression of Introduced Genes in Transgenic Mice

Initial experiments demonstrated that the foreign DNA which is stably integrated in various copies in

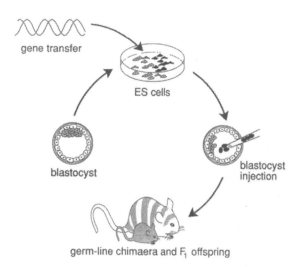

Fig. 4. The embryonic stem (ES) cell cycle

both somatic and germ-line cells in a random fashion was transmitted to offspring, thereby yielding homozygous strains of transgenic mice. A serious problem has been and still is the inability to predict tissue-specific expression of the injected genes. It is now realized that the presence of prokaryotic vector sequences as well as the use of cDNAs is usually inhibitory to efficient and reproducible expression. From the enormous number of genes injected into fertilized eggs for different purposes, we will discuss a few selected examples, which highlight the potential of the transgenic technology.

30.3.1 To Study Gene Control[5]

A great effort has been directed towards understanding the molecular basis for developmental and cell-type specific activation of genes. Transgenic mice are the experimental model of choice for localizing cis-acting sequence elements responsible for tissue-specific gene regulation. Such elements localized in the 5' and/or 3' flanking sequences as well as within genes were identified for a large number of genes which are expressed in many different cell types. The fusion of such elements to various

genes became an important tool in transgenic experiments (Fig. 5). Here we will limit ourselves to one relatively simple example of a cell-type specific gene, the rat elastase-I gene (rE), which encodes a serine protease expressed predominantly in the exocrine pancreas. When a DNA fragment containing the entire rE gene and 7.2 kb of the 5' flanking sequence was tested in transgenic mice, many efficiently expressed elastase mRNA (Fig. 5A). Deletions that removed all but 205 bp of the 5' flanking sequence were also expressed in a pancreas-specific manner (Fig. 5B). This 205 bp of the 5' flanking sequence was also able to direct the expression of human growth hormone (hGH) or SV40 T antigen exclusively to the acinar cells of the pancreas (Fig. 5C, D). Deletions that left only 72 bp of the 5' flanking sequence were inactive. This finding implies that the DNA sequence located between -205 and -72 is necessary for the tissue-specific expression of the rE gene. This sequence has all the properties of a tissue-specific enhancer element as it functions when moved 3 kb upstream of its normal location or when inserted into an intron, and it can activate heterologous promoters appropriately. These experiments strongly suggest that this 133-bp region contains all of the information necessary to direct normal levels of expression of an associated structural gene exclusively to the acinar cells of the pancreas.

30.3.2 To Change the Physiology of Mice[6]

The potential for using specific promoters or enhancers to direct expression of heterologous genes to a specific cell type stimulated numerous attempts to experimentally alter the physiology of mice. One example of these studies, designed to address aspects of growth control, again employed the human growth hormone (hGH) gene for overexpression in both normal and mutant dwarf mice. The efficient expression of a mouse metallothionine hGH fusion gene (mMT-hGH, Fig. 5E) in transgenic mice caused a dramatic growth stimulation resulting in the development of mice that were twice their normal size. Expression of this transgene in the dwarf mice corrected their genetic defect, stimulating their growth to that comparable

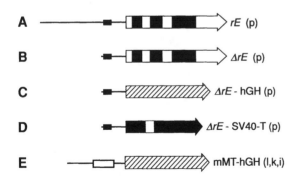

Fig. 5A–E. Fusion gene constructs used for gene expression studies in transgenic mice. **A** The rat elastase I gene (*rE*) found specifically expressed in the exocrine pancrease (*p*). **B** The truncated *ΔrE* construct exhibiting the same specificity of expression. **C, D** The tissue-specific element *ΔrE* (205 bp) directing the expression of reporter genes, human growth hormone (*hGH*) or SV40 T antigen (*SV40-T*) to the exocrine pancreas. **E** Fusion gene employing the mouse metallothionein promoter linked to hGH to alter the growth of mice. This fusion gene was found to be expressed in liver (*l*), kidney (*k*), and intestine (*i*)

to normal animals. These types of experiments will help to elucidate the molecular deficiency causing the respective hereditary disorder in humans.

30.3.3 To Study Oncogene Function[7]

The study of oncogene action in transgenic mice is directed towards indentifying the functions of their products and dissecting the mechanism underlying oncogene activation and tissue specificity of tumor formation. To investigate the consequences of oncogene expression in transgenic mice, three types of genes have been used: (1) oncogenes with their own regulatory elements which allow the examination of both the tissue specificity of the regulatory elements and the effect(s) of the oncogene in those tissues; (2) hybrid oncogenes designed for targeted expression of the oncoprotein to specific cell types in the transgenic mouse (Fig. 6); and (3) hybrid oncogenes with regulatory elements of broad specificity for addressing the issue of differential oncogene activity (Fig. 7). Here we would like to give two examples of the latter two approaches and will then attempt to summarize what can be learned from such studies.

30.3.3.1 T-Antigen Expression in the Pancreatic β Cells

Studies using the SV40 T antigen in combination with different tissue-specific regulatory elements demonstrated that almost all cell types are susceptible to transformation. In one of the first studies, transgenic mice were produced that carried the SV40 T antigen under the control of the 5′ flanking region of the rat insulin gene (RIP-SV40-T, Fig. 6). These animals developed normally but died prematurely at 9–12 weeks of age. T-antigen expression was confined to the insulin-producing β cells of the endocrine pancreas in the islets of Langerhans, indicating that the upstream sequences of the insulin gene are sufficient for cell-type-specific expression in vivo. In young mice, the islets of Langerhans were of normal size, with all β cells expressing T antigens; however, with increasing age the islets

became disorganized and hyperplastic, eventually developing into solid β-cell tumors (Fig. 6). Although tumor formation was very frequent, it was not absolute, as only a small fraction of the hyperplastic islets became tumorigenic.

30.3.3.2 Consequences of c-fos Expression

Several c-*fos* constructs (MT-c-*fos*LTR, H2-c-*fos*LTR) with broad range regulatory elements were used to generate transgenic mice. A consistent observation was that only one tumor type (bone tumors) developed despite the fact that the gene was expressed in several tissues (Fig. 7). The kinetics of tumor development suggests that *fos* expression alone is not sufficient for their formation

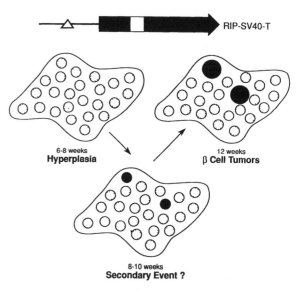

Fig. 6. Targeted expression of SV40 T antigen to the pancreatic β cells using the rat insulin promoter (RIP-SV40-T). Transgenic mice express SV40 T antigen in all of the insulin-producing β cells, which are localized into approximately 400 focal clusters of endocrine cells (the islets of Langerhans). Generalized hyperplasia reproducibly develops in approximately 50% of the islets between 6–8 weeks of age. By 10 weeks of age, >80% of the islets contain rapidly proliferating β cells. At 12 weeks, a few (2–6) highly vascularized tumors are evident. This pattern indicates that during the period of 5–10 weeks of age, an infrequent, localized secondary event converts a few of these focal clusters of proliferating cells into solid tumors

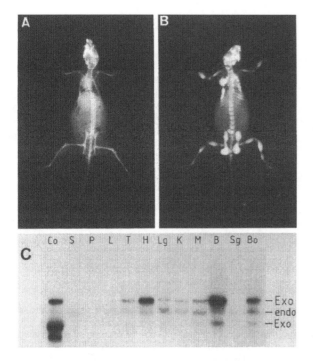

Fig. 7A–C. Phenotype in *fos* transgenic mice. X-ray and expression analysis of a H2-c-*fos* LTR transgenic mouse. **A** Control mouse. **B** Transgenic mouse with drastic bone alterations. **C** The same mouse was analyzed by Northern blotting. *Exo* exogenous; *endo* endogenous c-*fos* transcripts. *Letters on top of the figure* are: *Co* control cell line; *S* spleen; *P* pancreas; *L* liver; *T* thymus; *H* heart; *Lg* lung; *K* kidney; *M* muscle; *B* brain, *Sg* salivary gland; *Bo* bone

and that additional, yet undefined factors must be involved.

Together, these and other studies support the view that neoplastic transformation is a multi-step process. The presence of an oncogene in the germ-line is not sufficient to produce a tumorigenic phenotype in all cells expressing the gene; the different oncogenes exert a distinct specificity in different cell types.

30.3.4 To Investigate Complex Systems[8]

In the previous section we have outlined the study of oncogenesis using transgenic mice. Now, we want to briefly consider a series of experiments aiming at defining the role of genes within complex systems, such as the immune or nervous system. The goal of these studies is to determine how these systems respond to single genetic changes, thereby elucidating the properties of the system rather than of the gene. The introduction of rearranged immunoglobulin genes into the germ-line of mice demonstrated that these genes are correctly expressed and that their expression altered the endogenous immunoglobulin repertoire. Likewise, the transfer of a functionally rearranged T-cell receptor β-chain gene showed that the transgene expression inhibited the rearrangement of the endogenous β genes. Such approaches, designed to interfere with complex biological systems, were extended to the study of self-tolerance by B and T cells and to autoimmunity using a variety of foreign genes including those encoding MHC class I and class II antigens.

Less instructive until now have been the investigations into the function of genes in the central nervous system (CNS). Only one study will be mentioned, which deals with the expression of the myelin basic protein (MBP) gene, an essential gene for myelin formation in the CNS. The shiverer phenotype in a naturally occurring mouse mutant strain was characterized as a deficiency in MBP synthesis. Transgenic mice expressing an anti-sense MBP mini-gene were converted from a normal to mutant shiverer phenotype. This strongly argues for a tight regulation of MBP synthesis, in particular since the antisense mRNA reduced normal MBP synthesis by only 70%. More sophisticated experiments into the function of genes in the CNS will certainly be the goal of the future.

30.4 The ES Cell System

We have already briefly described the outstanding features of ES cells in Section 30.2.3 and will now concentrate on results obtained by overexpressing genes or by deleting genes from the ES cell genome with the aim to gain insights into mouse postnatal life.

30.4.1 The Gain of Function Approach[9]

One possible route for assaying the role of a gene product in early development is the overexpression

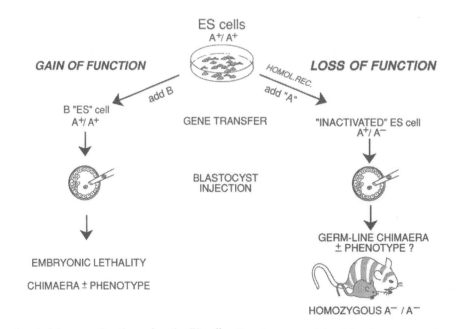

Fig. 8. Scheme for studying gene function using the ES cell system. The *right panel* describes the loss-of-function approach involving the inactivation of one allele (e.g., A^-) via homologous recombination and subsequent transmission through the germline of chimeric mice. The *left panel* describes the gain-of function approach involving the overexpression of a gene (e.g., *B*) in ES cells which might lead to an altered "*ES*" cell clone. Following blastocyst injection the consequences of B expression in chimeras could be embryonic lethality or the development of chimeras with or without a visible phenotype

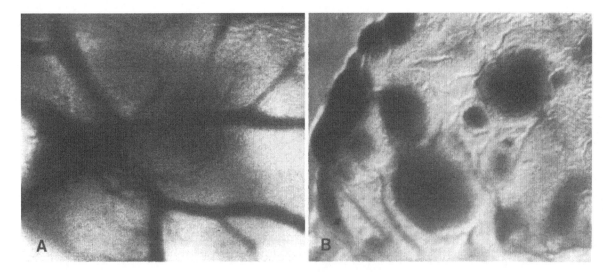

Fig. 9 A, B. Polyoma middle T expression disrupts blood vessel formation at mid gestation in chimeric embryos derived from selected ES cells (**B**). Blood vessel formation in the yolk sac of a control mouse (**A**)

in ES cells and subsequent analysis in chimaeras (Fig. 8). The consequences of exogenous gene expression at the onset of development can be embryonic lethality or an altered phenotype which becomes recognizable later in development.

To investigate the effects of the Polyoma middle T antigen (mT) on early mouse development, we have isolated mT expressing ES cell clones and used them to generate chimeras. All chimeric fetuses were found to be specifically arrested in development at midgestation displaying extensive disruption of the blood vessels due to the appearance of multiple cavernous hemangiomas (endothelial tumors, Fig. 9). Permanent endothelial cell lines, readily established from these tumors, provided a model system for further studies into the role of this oncogene. Such experiments clearly illustrate that ES cells are a valuable system for the study of growth control during embryogenesis.

30.4.2 The Loss of Function Approach [10]

An even better method for assaying the function of a gene would be to eliminate it from the genome of the whole organism. Very recent studies using strategies to inactivate genes by homologous recombination in ES cells with the subsequent generation of germ-line chimeras, make such a scenario possible (Fig. 8) and thereby establish a new era in mammalian developmental biology.

We have chosen the *int*-1 gene as an example for demonstrating the power of this technology. This gene is expressed during CNS development in a temporally and spatially restricted fashion. Following inactivation of the gene by homologous recombination in ES cells, germ-line chimeras were obtained and bred to derive homozygous *int*-1 deficient mice. These homozygous mice were not viable since specific regions of the brain were missing

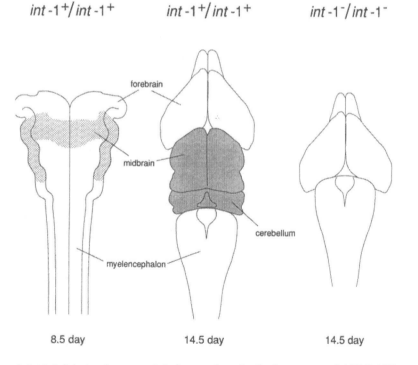

Fig. 10. Phenotype of *int*-1-deficient mice generated from selected *int*-1⁻/*int*-1⁺ ES cells. On the *left* the expression pattern of *int*-1 in the open cephalic neural plate of an 8.5 day mouse embryo and the corresponding regions depicted at 14.5

days. In the homozygous *int*-1⁻/*int*-1⁻ mouse the midbrain and cerebellum fails to develop, whereas the forebrain and myelencephalon seem to be unaffected

(Fig. 10). This result indicates that *int*-1 is an essential gene for early development and further suggests that its expression might determine the fate of a cell. These studies clearly open up new avenues for studying mammalian development at the molecular level and will certainly be instrumental in unraveling the molecular networks responsible for the functioning of a muticellular organism.

30.5 Gene Transfer into Hematopoietic Stem Cells

30.5.1 The Hematopoietic System [11]

The hematopoietic system develops early during embryogenesis and continues to function to generate mature blood cells throughout adult life. Continuous hematopoietic activity is necessitated by the fact that a large majority of mature blood cells have a short life span and need to be replaced as they die. The major sites of blood cell production are the embryonic yolk sac, the fetal liver, and the adult bone marrow. The structure of the hematopoietic system has been best defined experimentally in the mouse, and most evidence now suggests that the various developing cell populations within the system are organized in a hierarchy represented schematically in Fig. 11.

30.5.1.1 The Hierarchy of Hematopoietic Cells

The most immature cells which are found at the head of this hierarchy are the hematopoietic stem (HS) cells. Several characteristics of these cells distinguish them from virtually all other cells within the system. First, stem cells are able to generate more stem cells through a process which is often referred to as self-renewal. It is this self-renewal capacity which is pivotal to the stem cell's ability to maintain a functional hematopoietic system throughout life. Second, in addition to this unique ability to self-renew, stem cells also have the capacity to generate progeny of all of the diverse myeloid

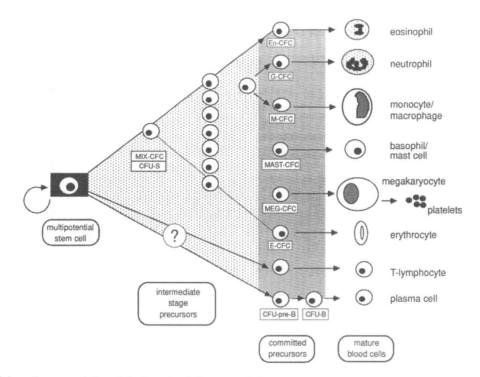

Fig. 11. Schematic representation of the hematopoietic system of the mouse

and lymphoid blood cell lineages and hence they are known as multipotential stem cells. As the stem cells proliferate, they generate differentiated progeny which are unable to self-renew and tend to be more restricted in their developmental potential. These progeny are collectively referred to as precursor cells and represent a continuum of cells at all stages of development between the primitive HS cell population and the mature blood cells. The most primitive of these precursor cells have extensive proliferative potential and are able to generate progeny of more than one lineage. In contrast, the most mature are restricted to a single lineage and show limited proliferative potential.

30.5.1.2 Assays for HS Cells in Vivo

Hematopoietic stem cells are identified and routinely assayed by their capacity to generate and maintain a functional hematopoietic system over extended periods of time in vivo. This is most easily accomplished experimentally by removing bone marrow from a "donor" mouse and transplanting it either into a recipient animal whose own hematopoietic system has been ablated by irradiation, or into a mutant animal from the W series whose own hematopoietic cells are at a growth disadvantage to those of the normal donor. The mutation within these W mice has been shown to reside within the c-kit proto-oncogene which encodes a growth factor receptor with tyrosine kinase activity. As the normal stem cells have a growth advantage over those of the W mouse, it is assumed, but not yet shown experimentally, that these primitive cells utilize this receptor. Repopulation of the hematopoietic system of these recipient animals over long periods of time is considered a true measurement of HS cell function. Hematopoietic regeneration per se does not demonstrate the multipotent nature of the stem cell population. The capacity of these cells to generate multiple hematopoietic lineages has been directly demonstrated in experiments involving the use of unique markers.

The first of these markers to be used was radiation-induced chromosomal translocations. When bone marrow cells carrying these unique chromosomal translocations were transplanted into reci-

pient mice, differentiated blood cells from multiple lineages were found to contain the same translocation, demonstrating that a single stem cell was able to generate these different lineages. Subsequent to these early studies, a number of groups have utilized the integration site of recombinant retroviruses as a unique marker for stem cells (see Sect. 30.5.4).

30.5.1.3 The Colony Assay in Vivo and in Vitro[12]

The various precursor populations have been identified and can be assayed by short-term colony assays. The first of these assays measures a population of precursors with the capacity to generate a clone of hematopoietic cells in the spleen of an irradiated mouse. Originally it was thought that these spleen colony forming cells or CFU-S represented the most primitive hematopoietic stem cells and to date the term stem cell is often used in reference to CFU-S. However, most evidence would now suggest that this assay measures a heterogeneous population of precursors that is more mature than and can be physically separated from the HS cells.

The second assay detects precursors that are able to form colonies of hematopoietic cells in semisolid cultures in vitro in response to specific growth factors. This assay detects a broad spectrum of precursors, ranging from primitive ones able to generate large colonies consisting of all myeloid lineages to relatively mature ones which form small colonies consisting of a single lineage. As with the CFU-S, these in vitro colony-forming cells or CFC represent a population that is more mature than the reconstituting stem cells.

Despite the fact that the in vitro assay does not detect the most primitive cells within the hematopoietic system, its use has been instrumental in identifying precursor populations and in defining various parent-progeny relationships within these populations. Consequentially, the term precursor is often used interchangeably with colony-forming cell or CFC and precursors of the various lineages are frequently identified by a prefix representing the lineages to which they are restricted. For example, a precursor restricted to erythroid development is referred to as an erythroid colony-forming cell or

E-CFC, one restricted to the megakaryocyte lineages as MEG-CFC and one restricted to eosinophil development as Eo-CFC. Those precursors which are able to generate multiple lineages are referred to as mixed-CFC. Unfortunately, the terminology is not all this straightforward and is at times confusing. For historical reasons the bipotential precursor restricted to neutrophil and macrophage development is known as a granulocyte-macrophage colony-forming cell or GM-CFC rather than a NM-CFC and is at times even referred to as a CFU-C for colony-forming cell in culture. Primitive erythroid precursors which generate relatively large colonies of erythroid cells were initially defined as BFU-E or burst-forming unit erythroid, whereas the more mature precursors which give rise to much smaller colonies are known as CFU-E or colony-forming unit-erythroid.

30.5.1.4 Hematopoietic Growth Factors[13]

In addition to providing a system for defining various precursor populations, the in vitro colony assay has been used extensively to identify a number of the hematopoietic growth factors which, in part, are responsible for regulating the system. Although a detailed analysis of these growth factors is clearly beyond the scope of this chapter, we will briefly describe a few of the common ones that are used routinely in experimental hematology. Those factors which were identified by their ability to stimulate the growth of colonies in vitro are referred to as colony-stimulating factors or CSFs. The specificity of the different CSFs varies widely. Granulocyte-CSF (G-CSF) and macrophage CSF (M-CSF or CSF-1) stimulate the growth of precursors from the neutrophil and macrophage lineages respectively, whereas granulocyte-macrophage CSF (GM-CSF) can act on both of these lineages. Interleukin-3 (IL-3), on the other hand, has a much broader range of activities and can stimulate the growth of precursors from all of the myeloid lineages. In addition to these colony-stimulating factors, a large number of interleukins (IL-1 through 10) have been identified, some of which have either direct or indirect effects on cells of the hematopoietic system.

Many of these well characterized factors act at the level of later stage precursors and not on the multipotential stem cell. Most evidence suggests that these early cells are regulated by molecules that function over short range or by direct contact with poorly defined regulatory populations, often referred to as stromal cells. Recently, a novel growth factor has been identified which is produced by bone marrow-derived fibroblasts, a component of the marrow stroma. This factor is encoded by the *steel* (*Sl*) gene and has been shown to be the ligand for the c-*kit* encoded receptor, mentioned previously. Mutations at this *Sl* gene affect the environment of the hematopoietic tissue such that it is unable to support the development of transplanted hematopoietic cells. Although the role of this factor within hematopoiesis is not yet clear, the fact that transplanted stem cells engraft poorly into these mice suggests that it will act on this population.

The in vitro colony assay together with the introduction of molecular biology to the field of hematopoiesis has greatly increased our understanding of the various precursor populations and the factors which control their growth and differentiation. In contrast, our understanding of the mechanisms which control the proliferation and differentiation of the primitive stem cell population has progressed at a much slower pace. This lack of progress in gaining a better understanding of the biology of the stem cell population is directly related to the fact that the assay for these cells is long and tedious, that they represent an extremely small minority of any hematopoietic population (approx. 1 in 30 000 normal marrow cells), and that it is not yet possible to grow them as a pure population in culture.

30.5.1.5 Approaches to Studying HS Cells[14]

A number of different approaches addressing questions of HS cell development have been taken by various investigators. One, which has been pursued by many over the years, has been the attempt to physically purify the stem cells from various hematopoietic tissues. Clearly, this is a formidable task given the frequency of stem cells in these tissues and, as a consequence, even the most enriched stem cell populations are not pure and consist of a mixture of precursor cells and stem cells. A second approach is to define conditions that will allow one to selectively grow these cells in culture. With the discovery of new growth factors such as the ligand for the c-*kit*-encoded receptor, this goal might soon be reached. A third approach that has been used to address issues of stem cell development in vivo and one that has great potential for future studies is the use of recombinant retroviruses to transduce genes into multipotential stem cells. The following sections outline a rationale for this approach, describe the method for infecting hematopoietic cells and highlight some of the new information that has been obtained using this approach.

30.5.2 Rationale for Retrovirus-Mediated Gene Transfer

The development of retroviral vectors that can efficiently transduce genes has made it possible to introduce genes into cells at most stages of development within the hematopoietic hierarchy, including the primitive multipotential HS cells. The questions that can be addressed using this approach have been outlined in the Introduction. The two lines of experimentation that we will focus on in the following sections are the use of retroviruses as unique markers to label individual stem cells and follow their fate in vivo and the use of retroviruses to transduce growth control genes into various precursor and stem cell populations with the intent of altering their growth and developmental potential.

The hematopoietic system is well suited to gene transfer experiments since various precursor and stem cell populations are readily accessible as single cell suspensions from either adult bone marrow or fetal liver. These cells can be manipulated to some extent in culture and the consequences of expression of the introduced gene can be analyzed both in culture and in transplanted animals.

30.5.3 Protocol for Infection

For infection of bone marrow, the best results are obtained with cells from 5-FU-treated animals

(Fig. 12). This treatment serves several purposes. First, 5-FU destroys a large number of intermediate stage cycling precursor cells but does not affect the stem cell population, and thus serves as a very simple method of enriching for these primitive cells. Second, it is assumed, but not yet proven, that depletion of these precursors will induce the more primitive ones, including the stem cells to divide. This is an important consideration, as the provirus will only integrate into dividing cells.

The marrow cells are infected by exposing them to high concentrations of virus either by coculture with the virus-producing cells or by culturing them in supernatant from the virus producing cells. Although the coculture protocol is easier and more popular, infection with a virus-containing supernatant offers several advantages and is the method of choice. First, a large stock of virus can be produced, titered, and then frozen ($-80\,°C$) for use in a number of different experiments. This eliminates the variability introduced by preparing fresh virus-producing cells for each experiment and it also allows one to use a constant virus to cell ratio. Second, recent evidence would suggest that, at least a subpopulation of stem cells are adherent, and thus would be lost during the coculture step.

Following infection of marrow with a virus which contains the selectable *neo* gene, the cells can be preselected for 48–72 h in the presence of G418 (Fig. 12). This preselection step significantly enriches the population for those cells which are infected and expressing the *neo* gene. For example, immediately following infection with a high titer virus such as N2, between 20% and 50% of all in vitro colony-forming cells are G418 resistant (G418^r). Following 72 h of preselection, the proportion of G418^r CFC increases to 60–95%.

The most serious difficulty encountered when attempting to infect stem cells with recombinant retroviruses is survival of these cells throughout the infection and preselection steps. As indicated earlier, conditions which promote the survival of hematopoietic stem cells in culture have not yet been defined and as a consequence, a large proportion of the initial stem cell population is likely to be lost throughout the various procedures. Although there is no evidence that IL-3 and IL-1 enhance the survival of HS cells, they are routinely included in

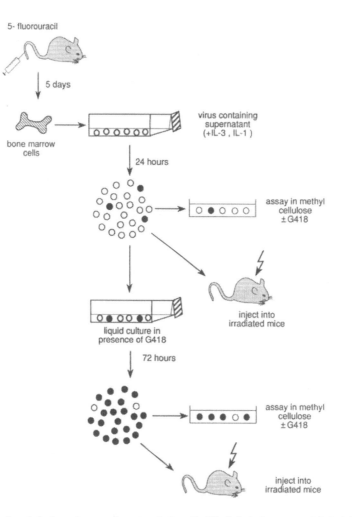

Fig. 12. Protocol for retrovirus infection of mouse hematopoietic cells. *Shaded circles* represent infected cells

the cultures as it has been shown that they do enhance the survival of precursor cells. Despite the fact that conditions for stem cell survival have not yet been optimized, a proportion of the population does survive the infection and preselection protocol and are able to function when transplanted into recipient animals.

30.5.4 Clonal Analysis of Retrovirus-Transduced Stem Cells[15]

Following infection of a cell by the recombinant retrovirus, a cDNA copy of the retroviral RNA is made through the process of reverse transcription. The viral DNA is then integrated into the genome of the cell in what can be considered a random site. This integration site is different in each infected cell and thus can be used as a convenient unique marker to follow their fate (Fig. 13).

To be able to identify the progeny of uniquely marked stem cells, DNA prepared from tissues and cell populations from mice reconstituted with virus-infected bone marrow cells is first digested with an enzyme that does not cleave within the recombinant virus and then analyzed by standard Southern blotting techniques. The size of each of the hybridizing bands is determined by the distance

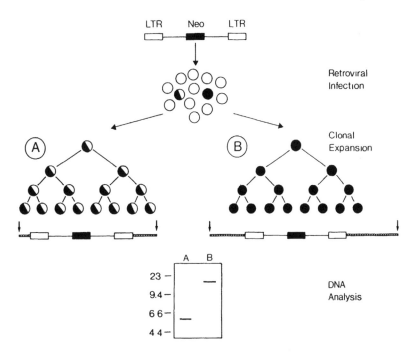

Fig. 13. The retrovirus integration site as a unique lineage marker. Two different infected stem cells within the hematopoietic population shown are depicted by a *half shaded* (A) and a *fully shaded* (B) *circle*. Following transplantation into recipient mice, these stem cells proliferate and differentiate giving rise to a large clone of hematopoietic cells (clonal expansion). At an appropriate time following reconstitution, DNA prepared from tissues and cell populations is digested with an enzyme that does not cleave within the integrated provirus. The distance between the restriction enzyme sites (indicated by the *arrows*) and the integrated provirus is unique for each of the clones. The different sized DNA fragments, representing the two clones, can easily be identified by Southern analysis

between the gene within the viruses to which the probe hybridizes and the nearest restriction enzyme sites. This distance will be different for each viral integration site and provided that only one virus has infected each cell, each band can be regarded as a single clone (Fig. 13). Using this approach of tracing stem cell-derived clones in reconstituted animals, the following information on the development potential of HS cells has been obtained.

The first observation from experiments involving the reconstitution of mice with retrovirus-marked stem cells was to essentially confirm the earlier studies using chromosomal translocations in demonstrating the existence of multipotential stem cells. In addition to demonstrating the existence of stem cells, these early retroviral marking studies also indicated that relatively few stem cells are sufficient for maintaining the hematopoietic system of an irradiated reconstituted mouse, thereby highlighting their extensive proliferative potential.

The next series of studies involved a temporal analysis of the clonal make-up of the developing hematopoietic system in irradiated reconstituted mice. The outcome of these experiments indicated that early following reconstitution, the hematopoietic system is in a state of clonal flux, with some new clones appearing and some of the existing ones disappearing. With time, however, stable clones emerge and persist for the lifetime of the animal. Those clones which disappear were most likely derived from infected precursors which exhibit no capacity to self-renew and thus are unable to sustain long-term hematopoiesis in these animals. The stable clones which eventually emerge and become predominant are probably derived from stem cells with long-term reconstituting capacity. The question which arises from these studies is at what time following reconstitution can one reliably detect the progeny of the multipotential stem cells? Although it is difficult to provide an estimate for all stem cells, most evidence would indicate that within 4 months of reconstitution, the majority of the hematopoietic cells in a reconstituted animal are derived from multipotential stem cells. The findings from these experiments would also indicate that hematopoiesis in a bone marrow-reconstituted mouse is maintained by a set of long-lived stem cells which can function for the lifetime of the animal.

Perhaps the most intriguing observation made from the most recent marking experiments is that the stem cell population can actually expand during the course of hematopoietic reconstitution. The fact that some expansion of this population does occur suggests that, at least in the environment of regenerating bone marrow, stem cells can proliferate without differentiating. If this potential is substantial it might be possible, with appropriate growth factors, to significantly expand this pool of cells in vivo and ultimately in culture. In conclusion, the findings from these marking experiments clearly demonstrate that multipotent stem cells can be infected with recombinant retroviruses and, following infection, can function upon transplantation into recipient animals. Clonal analysis of the hematopoietic system of these reconstituted animals has indicated that a few long-lived stem cells are sufficient to maintain the hematopoietic system of an irradiated reconstituted mouse.

The demonstration that retroviruses efficiently transduce genes into the stem cell population made it possible to design and carry out experiments aimed at transferring growth control genes into these cells with the aim of altering their developmental potential and ultimately expanding their number. To date, this goal has not been achieved. However, a number of studies have demonstrated that expression of growth control genes in hematopoietic cells can significantly alter their growth and differentiation potential. In the following section we will review a selected number of these studies focusing on those in which genes transduced by retroviruses have been shown to alter hematopoietic development in vivo.

30.5.5 Alteration of Hematopoiesis by Retrovirus-Mediated Gene Transfer[16]

In recent years much effort has focused on the effects of the v-*src* oncogene expression on hematopoietic development. Although its cellular counterpart, c-*src*, is expressed only in a limited number of lineages, findings from both in vitro and in vivo studies indicate that expression of v-*src* is able to alter the growth and differentiation of hematopoietic populations at various stages of development. To determine if expression of v-*src* would influence the development of a regenerating hematopoietic system, bone marrow cells infected with a v-*src*-containing virus were used to reconstitute irradiated mice. It is likely that the infected precursor cells do express the v-*src* oncogene, since colonies generated from them in culture were found to express the v-*src*-encoded kinase activity (Fig. 14A). Six to 10 weeks following reconstitution, most tissues within these animals also contained the pp60^{v-src} kinase, demonstrating that the introduced gene was efficiently expressed in vivo as well as in vitro (Fig. 14B). Many of the animals which expressed the v-*src* oncogene suffered from a severe and often lethal myeloproliferative disease, characterized by a shift of the developing precursor populations from the bone marrow to the spleen and peripheral blood. This dramatic increase in splenic hematopoiesis led to significant increases in spleen size. Despite this alteration, the precursors within these animals appeared normal with respect to their in vitro growth factor responsiveness and developmental potential. However, when spleen cells from these primary reconstituted recipients were passaged to secondary hosts, erythroid precursors with a transformed phenotype arose. These transformed cells, which appeared to be blocked at an intermediate stage of erythroid development, grew rapidly in culture, giving rise to continuously growing, factor-dependent cell lines. The outcome of these studies indicates that expression of the v-*src* oncogene can significantly disrupt the development of the hematopoietic system of a reconstituted mouse. The primary effect of v-*src* expression on the regenerating hematopoietic system is one of hyperplasia and not rapid transformation. Transformed populations arose only when the spleen cells were

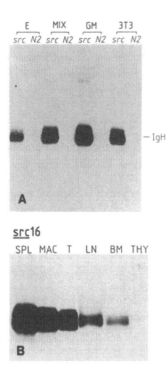

Fig. 14 A, B. In vitro IgH phosphorylation assay of pp60$^{v\text{-}src}$ tyrosine kinase activity in **A** cells from in vitro hematopoietic colonies and **B** tissues from reconstituted mice. **A** Following infection with either N-TK-*src*, the v-*src* expressing virus or N2, the control virus which expresses only *neo*, bone marrow cells were plated in methyl cellulose cultures in the presence of IL-3, IL-1, Epo and G418. Colonies containing mature hematopoietic cells were picked, pooled, and assayed for the presence of kinase activity. *E* erythroid colonies; *MIX* multilineage colonies consisting of cells from at least three different lineages; *GM* bilineage colonies containing both neutrophils and macrophages. Kinase activity is readily detected in the colonies derived from the *src*-infected precursors but not in those derived from the control N2-infected cells. **B** Mouse *src*-16 was reconstituted for 7 weeks with bone marrow cells infected with N-TK-*src*. At this time it was sacrificed and tissues as well as cell populations derived from them were assayed for kinase activity. *SPL* spleen; *MAC* bone marrow-derived macrophages; *T* splenic T cells; *LN* lymph node; *BM* bone marrow; *THY* thymus

passaged from one recipient to another, suggesting that secondary events were required for the development of this phenotype.

A second series of studies has analyzed the effects of *bcr-abl* expression on the hematopoietic system of bone-marrow reconstituted mice. *Bcr-abl*, a hybrid gene found in virtually all cases of human chronic myelogenous leukemia (CML), is generated at the level of the multipotential stem cell by a translocation that juxtaposes the coding sequences of the *bcr* gene (a gene with unknown function) with the coding sequences for the c-*abl* proto-oncogene. This translocation most likely confers some growth advantage to the stem cell since, even in the preleukemic state, the entire, hematopoietic system of the CML patients is dominated by the clone expressing the *bcr-abl* gene. The possible role of this gene in the initiation of CML at the level of the stem cell makes it an ideal candidate to introduce it into mouse HS cells. Mice reconstituted with bone marrow cells infected with a retrovirus containing the *bcr-abl* gene develop a variety of hematological disorders. These include a CML-like disease characterized by the predominance of granulocytes in the spleen and peripheral blood of the recipients, a pre-B cell leukemia, macrophage tumors, an erythroleukemia, and mast cell tumors. In some animals tumors from only a single lineage developed, whereas in others multiple tumors from different lineages arose within the same clone, indicating that a multipotential cell had been infected with the recombinant retrovirus. These findings demonstrate that *bcr-abl* expression can influence the development of multiple lineages in these reconstituted animals. Whether or not the stem cells themselves are also affected in these animals remains to be determined. Many of the tumors and leukemic populations were clonal and developed after a long latency period suggesting that expression of *bcr-abl* alone was not sufficient to lead to the transformation of these populations.

The third set of experiments along these lines involves the use of retroviruses to transduce a growth factor gene, the IL-3 gene, into hematopoietic cells. As indicated in Section 30.5.1.4, IL-3 is able to stimulate the growth of a wide range of myeloid lineages in culture. Mice reconstituted with bone marrow cells infected with a retrovirus which expresses the IL-3 gene were found to contain vastly elevated levels of cells from the granulocytic lineage. The eosinophilic, monocytic, and basophilic lineages were affected to a lesser extent. This overgrowth of granulocytic cells resulted in enlargement of both the spleen and liver. Precursor cells from these animals were able to generate continuously growing, factor-independent cell lines in culture, a number of which had the morphology of mast cells, a lineage which responds to IL-3. These findings indicate that overexpression of a normal hematopoietic growth factor gene can lead to alterations of the hematopoietic system and immortalization of hematopoietic precursor populations. An interesting extension of this approach would be to introduce genes for factors such as the newly identified *kit* ligand, which might act on more primitive populations.

In conclusion, these studies highlight the potential of this technology and clearly demonstrate that it is possible to significantly alter hematopoietic development by overexpression of different types of growth control genes. They also pave the way for future experiments in which combinations of genes or genes that act specifically on the early stages of hematopoiesis can be transduced.

30.6 Outlook

We would like to conclude by projecting our ideas on what we consider to be the important areas where rapid advances should occur in the near future:

1. It is necessary to better characterize stage- and tissue-specific elements as well as inducible elements regulating gene expression. The identification of such elements should guarantee efficient expression preferably independent of the chromosomal integration site. This will become very important for studies where efficient expression is required, e.g., in transgenic studies using commercial livestock.

2. In contrast to *Drosophila* and *C. elegans* there are only few mouse mutants known that affect early development and it is not clear if they play an important role in regulating developmental decisions. The prospect of generating mutations in preselected genes using ES cells will mark the beginning of a systematic genetic dissection of gene function in development; this will dramatically change the future of experimental mammalian genetics. Furthermore, this approach will allow the derivation of precise animal models for human hereditary diseases.

3. Transgenic mice will undoubtably provide important insights into the function of complex systems such as the immune and nervous systems. Gene ablation as well as overexpression experiments will continue to be extremely useful for describing the molecular structure and for gaining a better understanding of how these systems work.

4. With respect to oncogenesis a great deal of novel information regarding the function of the gene in question will be obtained. Genes acting downstream of the oncogene as well as co-operating genes responsible for secondary events will certainly be discovered and their function assessed. The chain of molecular events responsible for the transition from hyperplasia to neoplasia will be reconstructed, thereby allowing the development of novel strategies to prevent tumor formation.

5. The ability to reproducibly differentiate ES cells in vitro along specific pathways will provide a novel approach for defining the molecular events involved in the establishment of a particular lineage. The capacity to generate mature hematopoietic cells from ES cells in culture has already been demonstrated and is currently the subject of intense investigation. Defining the conditions which allow one to reproducibly generate the most primitive HS cells from ES cells in culture is the next major step. This ability will combine the advantages of the ease of manipulating ES cells in culture with that of studying the effects of this manipulation on a well-defined developmental system. Questions concerning the effects of overexpression or inactivation of a gene, as well as the causal relationship between expression of a specific growth factor or growth factor receptor gene and the direction of differentiation and development, can be addressed.

6. With respect to hematopoietic stem cells, protocols for their enrichment and ultimate purification need to be defined using long-term reconstitution as a measurement of stem cell function. Access to enriched populations of stem cells would enable one to transfer genes specifically to this population and would be instrumental in identifying the conditions which pro-

mote their survival and growth in culture. Furthermore, such approaches will provide ample scope for application in gene therapy experiments, which are just beginning and which could represent the start of a new era in medicine.

pulation of HS cells opens up new possibilities to unravel the molecular mechanisms of cellular differentiation and will certainly provide a wealth of new applications for use in therapeutic studies.

30.7 Summary

The introduction of genes into early embryos and ES cells leading to the generation of transgenic mice allows one to study the consequences of specific alterations in gene expression on the development of the whole organism. The introduced genes have the potential to be expressed in all somatic cells and can be transmitted through the germ-line to offspring, providing a rich and reproducible source of experimental material. On the other hand, the transfer of genes into HS cells enables one to investigate specifically the effects of individual genes on the well characterized hematopoietic system. The results from both approaches will undoubtably lead to a better understanding of the molecular mechanisms controlling mammalian differentiation and development.

Over the past years, we have observed an extraordinary increase in the use of transgenic mice and stem cell systems. Methods of manipulating embryos and transferring genes into stem cells have been developed and now constitute "standard" procedures used for a variety of approaches. Each of the four methods described have advantages for some and disadvantages for other applications. Pronuclear injection of recombinant DNA is the method of choice for dissecting the intricate regulatory elements governing gene regulation and for obtaining expression of a given gene in almost any specific tissue. Retroviruses and retroviral vectors to be used for embryo infection are problematic when expression is required, but they are superior agents when genetic tagging of chromosomal loci by insertional mutagenesis or when marking of cell lineages in stem cell differentiation and embryo development is desired. The most recently developed method of generating transgenic mice from ES cells allows in principle the derivation of mice with pre-designed genetic composition and phenotypic characteristics. Finally, the mani-

References

[1] Jaenisch R (1988) Transgenic animals. Science 240: 1468-1474
Palmiter RD, Brinster RL (1986) Germ-line transformation of mice. Annu Rev Genet 20:465-499
Wagner EF (1990) On transferring genes into stem cells and mice. EMBO J 9:3025-3032
[2] Brinster RL, Chen HY, Trumbauer ME, Yagle MK, Palmiter RD (1985) Factors affecting the efficiency of introducing foreign DNA into mice by microinjecting eggs. Proc Natl Acad Sci USA 52:4438-4442
Gordon JW, Ruddle FH (1983) Gene transfer into mouse embryos: production of transgenic mice by pronuclear injection. Methods Enzymol 101:411-433
[3] Stewart CL, Schuetze S, Vanek M, Wagner EF (1987) Expression of retroviral vectors in transgenic mice obtained by embryo infection. EMBO J 6:383-388
Wagner EF, Stewart CL (1986) Integration and expression of genes introduced into mouse embryos. In: Rossant J, Pedersen RA (eds) Experimental approaches to mammalian embryonic development. Cambridge Univ Press, Cambridge, pp 509-549
[4] Robertson EJ (1987) Teratocarcinomas and embryonic stem cells: a practical approach. IRL Press, Oxford
[5] Brinster RL, Palmiter RD (1986) Introduction of genes into the germ line of animals. Harvey Lect 80:1-38
[6] Palmiter RD, Brinster RL, Hammer RE, Trumbauer ME, Rosenfeld MG et al. (1982) Dramatic growth of mice that develop from eggs microinjected with metallothionein-growth hormone fusion genes. Nature 300:611-615
[7] Hanahan D (1985) Heritable formation of pancreatic β-cell tumors in transgenic mice expressing recombinant insulin/simian/virus 40 oncogenes. Nature 315:115-122
Hanahan D (1988) Dissecting multistep tumorigenesis in transgenic mice. Annu Rev Genet 22:479-519
Rüther U, Garber C, Komitowski D, Müller R, Wagner EF (1987) Deregulated c-fos expression interferes with normal bone development in transgenic mice. Nature 325:412-416
Wagner EF (1990) Oncogenes and transgenic mice. In: Habenicht A (ed) Growth factors, differentiation factors and cytokines. Springer, Berlin Heidelberg New York, pp 662-683
[8] Hanahan D (1989) Transgenic mice as probes into complex systems. Science 246:1265-1275
Katsuki M, Sato M, Kimura M, Yokoyama M, Kobayashi K, Nomura T (1988) Conversion of normal behavior to shiverer by myelin basic protein antisense cDNA in transgenic mice. Science 241:593-595

[9]Wagner EF, Wang Z-Q, Grigoriadis AE, Möhle-Steinlein U, Aguzzi A, Risau W (1990) Analysis of oncogene function in ES cells and chimaeric mice. In: Origins of human cancer. CSH Lab Press (in press)

Williams RL, Courtneidge SA, Wagner EF (1988) Embryonic lethalities and endothelial tumours in chimaeric mice expressing Polyoma virus middle T oncogene. Cell 52:121–131

[10]Capecchi MR (1989) Altering the genome by homologous recombination. Science 244:1288–1292

McMahon AP, Bradley A (1990) The *Wnt*-1 (*int*-1) proto-oncogene is required for development of a large region of the mouse brain. Cell 62:1073–1085

[11]Chabot B, Stephenson DA, Chapman VM, Besmer P, Bernstein A (1988) The proto-oncogene c-*kit* encoding a transmembrane tyrosine kinase receptor maps to the mouse *W* locus. Nature 335:88–89

Keller G, Paige C, Gilboa E, Wagner EF (1985) Expression of a foreign gene in myeloid and lymphoid cells derived from multipotent haematopoietic precursors. Nature 318:149–154

Metcalf D (1988) The molecular control of blood cells. Harvard Univ Press, Cambridge MA, London UK

Russel ES (1979) Hereditary anemias of the mouse: a review for the geneticist. Adv Genet 20:357–459

[12]Jones RJ, Wagner JE, Celano P, Zicha MS, Sharkis SJ (1990) Separation of pluripotent haematopoietic stem cells from spleen colony-forming cells. Nature 347:188–189

Suda T, Suda J, Ogawa M (1983) Single cell origin of mouse hemopoietic colonies expressing multiple lineages in variable combinations. Proc Natl Acad Sci USA 81:7151–7155

Till JE, McCulloch EA (1961) A direct measurement of the radiation sensitivity of normal mouse bone marrow cells. Radiat Res 14:213–222

[13]Nicola NA (1989) Hemopoietic cell growth factors and their receptors. Annu Rev Biochem 58:45–63

Witte ON (1990) *Steel* locus defines new multipotent growth factor. Cell 63:5–6

[14]Iscove N, Xiao-Qiang Y (1990) Precursors (pre-CFCmulti) of multilineage hemopoietic colony-forming cells quantitated in vitro. J Immunol 145:190–195

Spangrude GJ, Johnson GR (1990) Resting and activated subsets of mouse multipotent hematopoietic stem cells. Proc Natl Acad Sci USA 87:7433–7437

[15]Keller G, Snodgrass R (1990) Life span of multipotential hematopoietic stem cells in vivo. J Exp Med 171:1407–1418

[16]Daley GQ, Van Etten RA, Baltimore D (1990) Induction of chronic myelogenous leukemia in mice by the $P210^{bo-abl}$ gene of the philadelphia chromosome. Science 247:824–830

Elefanty AG, Hariharan IK, Cory S (1990) bcr-abl, the hallmark of chronic myeloid leukaemia in man, induces multiple haemopoietic neoplasms in mice. EMBO J 9:1069–1078

Keller G, Wagner EF (1989) Expression of v-*src* induces a myeloproliferative disease in bone marrow-reconstituted mice. Genes Dev 3:827–837

Wong PMC, Chung S, Dunbar CE, Bodine DM, Ruscetti S, Nienhuis AW (1989) Retrovirus-mediated transfer and expression of the interleukin-3 gene in mouse hematopoietic cells result in a myeloproliferative disorder. Mol Cell Biol 9:798–808

Chapter 31 Skeletal Muscle Differentiation

Deborah F. Pinney and Charles P. Emerson, Jr.

31.1 Introduction

Skeletal muscle is one of the best characterized cell types in multicellular eucaryotic animals and is a striking example of the relationship between structural organization and differentiated function. Skeletal muscle has been the subject of intense cell biological and biochemical investigations that have led to a detailed understanding of the molecular basis of contraction. In recent years, there has been progress in studies of the embryological origin of skeletal muscle cells and in related studies on the molecular genetic mechanisms that regulate skeletal muscle cell differentiation and the expression of contractile protein genes. This recent progress has made the skeletal muscle a paradigm for understanding the fundamental molecular and genetic mechanisms of cellular differentiation.

31.2 Differentiated Skeletal Muscle

31.2.1 Physiology and Histology[1]

Discussion of the process of muscle differentiation requires a brief review of skeletal muscle as a cell type apparatus. The sarcoplasmic reticulum is a storage site for calcium ions, and depolarization leads to the release of calcium into the cytoplasm near the contractile apparatus. Calcium ions activate contraction through binding to specific calcium binding proteins that facilitate enzymatic and physical interactions required for force generation.

The force of contraction is generated via shortening of myofiber length through the molecular interactions among polarized arrays of filamentous myofibrillar proteins of the contractile apparatus.

Myofibrils are a structurally organized set of interdigitating thick and thin filaments that physically interact to generate asymmetric force (Fig. 1). The thick filaments are composed primarily of about 400 myosin molecules organized in opposite polarity around a central zone. The myosin molecule is a dimer of a large 200 kDa protein, referred to as myosin heavy chain (mhc). The mhc dimer is stabilized by a coiled-coil interaction between alpha helical domains that extend along the carboxyl terminal half of the protein. The helical structure of mhc stabilizes the interactions among myosin dimers to form the higher order thick filament, which can self assemble in vitro. The thick filaments of some invertebrate muscles are assembled on a core protein, paramyosin. The amino terminal half of the mhc protein has a globular structure which binds two classes of smaller 20 kDa proteins, myosin light chains 1/3 (mlc1/3) and myosin light chain 2 (mlc2). Skeletal muscle is adapted for unidirectional shortening through the coordinated contraction of bundles of muscle cells (Fig. 1). The skeletal muscle cell is an elongate fiber and often a multinucleated syncytium (myofiber or myotube) formed during embryonic development by the fusion of mononucleated myogenic cells. Fibers are surrounded by a sheath of collagen-related proteins that attach fibers to fixed skeletal structures through specialized myotendinous junctions. Fibers are innervated by axons of motor neurons at single and multiple synapse sites. Contraction is initiated by depolarization of the muscle plasma membrane, usually induced by activity of the nerve axon. Activation of a motor nerve impulse causes the release of chemical signals, neurotransmitters, from vesicles at the nerve synapse to initiate depolarization of the muscle fiber membrane. These neurotransmitters are acetylcholine in vertebrates and glutamate in insects and arthro-

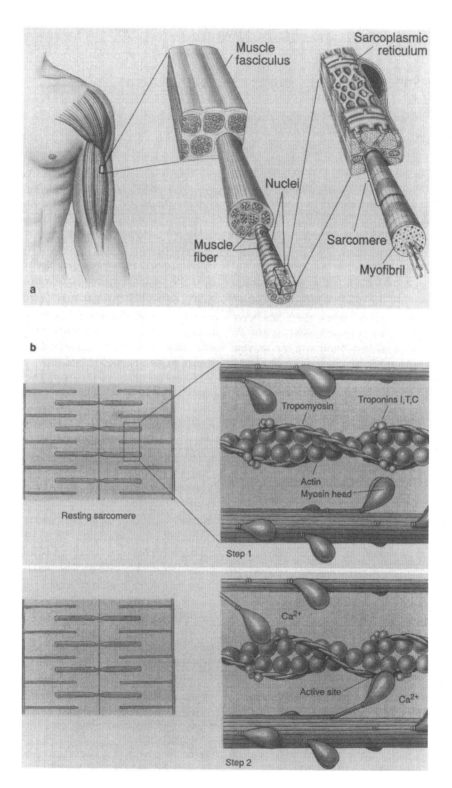

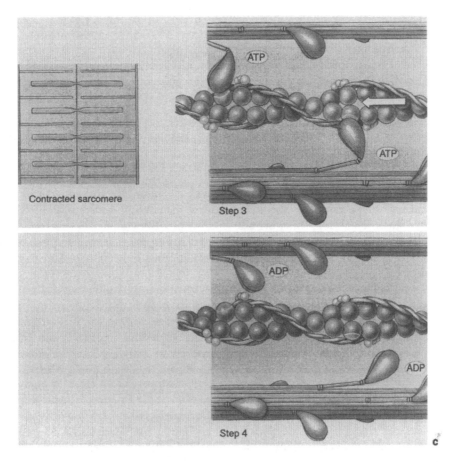

Fig. 1. a The organization of human skeletal muscle (illustration by W. Ober, in *Biology of Animals*, Mosby-Yearbook, 1989). The sliding filament mechanism for muscle contraction. **b** In the resting sarcomere (*Step 1*), troponin I inhibits the interaction between the myosin head and actin. Calcium is released from the sarcoplasmic reticulum and binds to troponin C resulting in a conformational change that allows the myosin head to interact with actin (*Step 2*). The thick and thin filaments move past each other because of a conformational change in the myosin head that requires the hydrolysis of ATP (*Steps 3* and *4*) resulting in muscle shortening. (Martini 1989)

pods. In some cases, such as insect indirect flight muscles, depolarization is initiated by mechanical stretch caused by the contraction of other muscle fiber groups. The neurotransmitters released from the axon synapse bind to specific receptor proteins clustered in the post-synaptic muscle fiber membrane. Binding of these transmitters to receptor proteins causes membrane depolarization, which is transmitted internally into the muscle fiber through specialized membranes, T tubules, that penetrate the muscle fiber and are coupled to the sarcoplasmic reticulum, which surrounds the myofibrillar contractile apparatus of heavy chains and light chains. The globular domains, referred to as the globular heads, protrude, in staggered arrays and with opposite polarities, from either side of the thick filament. The mhc globular head has a sequence domain that binds ATP which is the chemical energy for contraction. Other domains are involved in the cyclic interactions with thin filament actin when the contraction cycle is initiated by calcium flux from the sarcoplasmic reticulum. The thin filament is composed of actin protein subunits (G-actin) polymerized into two intertwined F-actin filaments. Sets of thin filaments are organized at regular intervals around each thick filament, these intervals differing in number for functionally different types of skeletal muscles. Proteins that regulate mhc interactions and muscle contraction are associated with the actin thin filament. These proteins are tropomyosin, Tm, and the troponin complex of proteins, TnC, TnI, and TnT. Specialized structural proteins, myomesin and alpha actinin, anchor and tie together the thick and thin filaments and anchor these filaments to the Z line to provide a scaffold for mechanical force generation. In addition to the myofibrillar proteins directly involved in force generation, other cytoskeletal and membrane structural proteins, including desmin and dystrophin, play an essential role in the structural and functional integrity of the muscle fiber.

The myofibrillar proteins interact to generate force when calcium is released into the myofibrillar space (Fig. 1). Calcium binds avidly to a calcium-binding protein, troponin C, which is part of the troponin complex spaced periodically along with the thin filament and is bound to tropomyosin by the troponin T subunit. Troponin I then undergoes a conformational change in response to the binding of calcium to troponin C. In the absence of calcium bound to troponin C, troponin I inhibits actin from interacting with the mhc globular heads. In the presence of calcium, tropomyosin apparently is displaced conformationally to allow mhc heads and actin to interact cyclically with adjacent actins and to move progressively along the thin filament accompanied by ATP hydrolysis. Relaxation of the contracted muscle occurs when calcium is released from troponin C and transported back into the sarcoplasmic reticulum via the activity of a membrane-associated protein pump, calcium ATPase, reestablishing the inhibition of actin myosin interaction.

While the myofibrils provide the mechanical structures needed for contractility and the sarcoplasmic reticulum coordinates muscle contraction and relaxation, contraction requires abundant ATP as the energy source. Differentiated muscle has specialized mechanisms for ATP production by anaerobic and aerobic metabolism. Muscle has numerous mitochondria called sarcosomes where oxidative phosphorylation occurs, the primary source of ATP. In addition, the sarcoplasm contains glycogen and creatine phosphate as well as glycolytic enzymes, creatine phosphokinase, and other specialized proteins as myoglobin needed to regenerate ATP from ADP following contraction.

31.2.2 Muscle Diversity and Contractile Protein Isoforms [2]

Skeletal muscle is the most prevalent tissue for force generation and movement in the animal kingdom. Skeletal muscles can have distinct physiological properties related to their specialized contractile functions. For instance, vertebrates have fast twitch and slow twitch skeletal muscle fibers involved in rapid motor activity and postural maintenance, respectively. Insects such as the dipteran, *Drosophila*, have a set of rapidly contracting indirect flight muscles (Ifm) specialized for flight, specialized muscles for walking and jumping, egg laying and proboscis protrusion, as well as body wall muscles for larval movement. Other less complex and smaller organisms, such as the well-studied worm, *Caenorhabditis elegans*, have two major muscle groups, pharynx muscles for feeding and body wall muscles for coordinated movement.

The molecular cloning of contractile protein genes and genetic studies of mutant alleles of these genes demonstrate that specific isoforms of contractile proteins are present in different muscles and that subtle differences in the structure and function of the isoforms are a basis for the different contractile properties of specific skeletal muscle types as well as other muscle types such as cardiac and smooth muscle in vertebrates (Fig. 2). Molecular cloning and DNA sequencing techniques have made it possible to characterize a large number of contractile protein genes and cDNAs. The cloning of these genes has provided a wealth of primary amino acid sequence data on contractile proteins, has led to the identification and characterization of additional muscle proteins, and has provided unique experimental tools to study the structure and function of these proteins and the developmental regulation of their genes.

FAST   SLOW

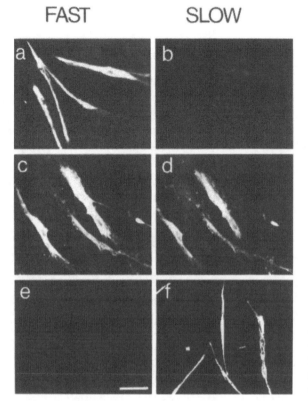

Fig. 2a–f. Avian muscle fibers express different isoforms of myosin heavy chain. Fluorescent double staining with antibodies that specifically recognize fast twitch mhc isoforms (**a, c, e**) or slow twitch mhc isoforms (**b, d, f**) shows that clonal populations of cultured primary chick embryo limb muscle express fast (**a, b**), slow (**e, f**), or both (**c, d**) isoforms. (Miller and Stockdale 1987)

The contractile protein genes and cDNAs that have been characterized by molecular cloning include subunits of the acetylcholine receptor, the sarcoplasmic reticulum Ca ATPase, myofibrillar proteins including myosin heavy chain, myosin light chains, troponin complex proteins troponin I, T, C; tropomyosin, actin, C protein, nebulin, α-actinin and desmin, myomesin, titan, and enzymes involved in energy metabolism including creatine phosphokinase, aldolase, and glycerol phosphokinase. One important conclusion that has been derived from these cloning studies is that skeletal muscle contractile proteins are members of large families

of structurally and functionally related proteins. Each has its own developmental pattern of expression and presumed specialized function in the contraction of specialized cells such as skeletal, cardiac, and smooth muscle and in cellular motility processes in nonmuscle cells and eukaryotic organisms such as *Acanthamoeba* and slime molds. The different members of these families encode isoforms that have closely related, but divergent sequences. For instance, eukaryotic cells have cytoplasmic myosin, tropomyosin, and actin protein isoforms whose primary amino acid sequences are related but not identical to the thick and thin filament protein isoforms in skeletal muscle. Furthermore, an even more striking finding has been that physiologically distinct muscles express different isoforms that are assumed to reflect different contractile properties of these muscles. For example, vertebrates have the slow twitch skeletal muscles important for posture and the fast twitch oxidative muscles necessary for mobility. Fast and slow muscles express different isoforms of many if not all muscle protein isoforms, including fast and slow isoforms of myosin heavy and light chains, troponin I, C, and T, and calcium ATPase. Interestingly, many of the isoforms that are expressed in slow vertebrate muscles are also expressed in cardiac muscle, even though these muscle cell types have distinct developmental origins. Cardiac and skeletal contractile protein isoforms, for example actin, are coexpressed in embryonic skeletal muscle, and specific isoforms are selectively repressed in their expression later in development. In contrast, embryonic isoforms of other contractile proteins such as myosin heavy chain are initially expressed during development and adult isoforms are specifically activated in different muscles later in development. The current evidence suggests that the muscle-specific expression of different families of isoform proteins in vertebrates is regulated by complex developmental and/or physiological control processes. In vertebrates, isoform expression is established during development and maintained in adults, but can change in response to processes of innervation and muscle activity. Such plasticity of isoform expression may be an adaptation to the long life span of these organisms. The expression of fast and slow isoforms can be switched in adult fibers by changes in whether the muscle is inner-vated by a fast or slow firing nerve or by the con-tractile activity imposed on the muscle fiber itself. In short-lived organisms such as *C. elegans* and *Drosophila*, expression of isoforms is established and may be fixed during development.

31.2.3 Multigene Families and Alternative RNA Splicing[3]

An important conclusion of the recent molecular cloning studies of contractile protein genes is that isoforms of contractile proteins are produced by two regulatory mechanisms: either by the differen-tial transcription of members of multigene families or by regulated alternative RNA splicing of trans-cripts produced by a single gene. In a variety of organisms, many muscle contractile protein iso-forms are produced by regulated alternative RNA splicing including tropomyosin, troponin T, and myosin light chains. Mammals have multiple tropo-myosin genes, but these genes can give rise to a complexity of isoforms, including both cytoplasmic and muscle-specific isoforms, by alternative splic-ing of the RNA transcripts derived from a single tropomyosin gene. In contrast, there are as many as ten muscle-specific myosin heavy chain genes that are differentially expressed in different muscle types including embryonic muscle, fast twitch and slow twitch skeletal muscles, cardiac muscle, and

Table 1. Myosin heavy chain isoform diversity in some animals

	Number of MHC genes	Number of MHC isoforms
Drosophila	1	10 – 480 isoforms generated by alternative splicing
C. elegans	4	4 isoforms – MHC A and B expressed in body wall – MH C and D expressed in pharyngeal muscles
Chicken and quail	As many as 31	≥ 5 embryonic, adult, and muscle-type specific isoforms
Man	Unknown	≥ 4 embryonic, adult, and muscle-type specific isoforms
Rat	≥ 7	7 genes cloned including embryonic neonatal, extra-ocular, fast II A, fast II B, cardiac ventricular/atrial, skeletal ventricular/slow

smooth muscles (Table 1). *C. elegans* has four myosin heavy chain genes, two expressed in body wall muscle, mhcA and mhcB, and two expressed in the pharynx, mhcC and mhcD. However, *Drosophi-la* has only one muscle-specific myosin heavy chain gene (and a distinct cytoplasmic myosin gene) that is expressed in all larval and adult muscle types. The *Drosophila* mhc gene produces a diversity of muscle-specific myosin isoforms in functionally different muscle types by regulated alternative RNA splicing. Specific functional domains of the myosin protein are encoded by the alternatively spliced exons (Fig. 3). The molecular mechanisms

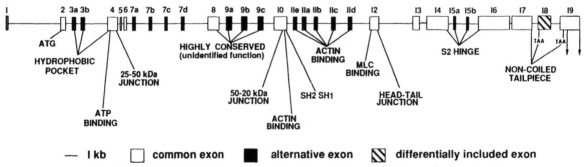

Fig. 3. The myosin heavy chain gene of *Drosophila*. Isoform diversity of *Drosophila* muscle mhc is generated by alternative splicing that is muscle-type specific. Functional domains of the protein can be assigned to specific exons so that alternative splicing creates differences in protein activity. One exon of each set of exons 3, 7, 9, and 11 are included in all mhc mRNA. Exon 18 is spliced into mRNA of some adult muscles

that regulate the muscle-specific alternative splicing of myosin and other muscle genes remain to be determined. However, these findings establish that different organisms have evolved different molecular strategies, differential expression of gene family members and alternative RNA splicing, to generate a complexity of isoforms for many if not all of the contractile proteins. The muscle specificity of the regulated expression of different myosin isoforms in different muscles and correlations between isoform protein sequence and functional domains of these proteins provide evidence for the importance of isoforms in the functional differentiation of muscle isoforms in different muscles.

31.2.4 Genetic Studies of Skeletal Muscle Differentiation[4]

Genetic studies on contractile protein genes in *C. elegans* and *Drosophila* provide evidence for the importance of isoforms in the assembly of muscle structures such as myofibrils and for the function of specialized muscles. Genetic studies of muscle are feasible because mutations in contractile protein genes often do not compromise organisms' viability, but selectively disrupt the specialized functions of specific muscles that are nonessential for embryogenesis or fertility. The phenotypes of contractile protein gene mutations are associated with specific motor dysfunctions that are easily identified visually, thus facilitating the use of large-scale genetic screens and selection schemes to recover mutant strains. Dominant and recessive mutations of muscle contractile protein genes have been identified that disrupt locomotion of *C. elegans, unc* mutations, and that disrupt flight in *Drosophila,* Ifm mutations. These genetic studies in combination with molecular probe mapping with cloned muscle genes has led to the chromosome mapping of a collection of muscle genes and contractile protein muscle genes in *Drosophila* and *C. elegans* and mammals, including humans. One important conclusion from these chromosome mapping studies is that muscle genes are not clustered in the genome, but are widely scattered throughout these genomes with a few exceptions including linkage of three mammalian mhc genes.

Mutations that disrupt or inactivate the expression and functions of myosin, actin, troponin T, troponin I and tropomyosin have been identified based on mutagenesis screens for flightless *Drosophila*. About 30 uncoordinated *C. elegans* mutants have been identified although the protein products of only a few loci have been identified including the *unc54* and *myo3* myosin heavy chain proteins, mhcB and mhcA. Mutations of the *unc54* mhcB of *C. elegans* have provided important information to identify functional domains of the myosin protein. Mutations of the *myo3* gene, which encodes mhcA in body wall muscle, are lethal and disrupt assembly of mhcB into thick filaments, as do mutations in the paramyosin *unc15* locus. Mutations of the *Drosophila* mhc, actin, tropomyosin, and troponin T genes expressed in Ifm reveal the importance of gene dosage and level of expression in the assembly and stability of myofibrils and demonstrate the different functional requirements of muscles for specific isoforms. Mutations of the *Drosophila* mhc gene, which encodes myosin for all skeletal muscles, have been identified based on their specific disruption of flight muscle function (Fig. 4). The flight muscle specificity of these mhc mutations is based on lesions that disrupt alternative RNA splicing of Ifm-specific exons or reduce the functional levels of mhc expressed in Ifm, which, in contrast to other muscle types, requires two myosin alleles to achieve, accumulate, and assemble functional levels of myosin protein into myofibrils.

Genetic and molecular studies of human genetic diseases, specifically the Duchenne muscular dystrophy locus, have led to the identification of a new muscle protein, dystrophin and to an investigation of the molecular and cellular basis of the progressive degeneration of skeletal muscle that accompanies this disease. The *dmd* Duchenne muscular dystrophy locus on the human X chromosome is extraordinarily large, 2000 kb, which is nearly the size of the entire *E. coli* genome. The dystrophin gene has multiple exons that encode a previously unknown muscle protein, dystrophin. Dystrophin is an integral membrane protein, and *dmd* mutations that disrupt the structure of this protein cause defects in muscle fiber membrane integrity that lead to muscle degeneration. Different mutations

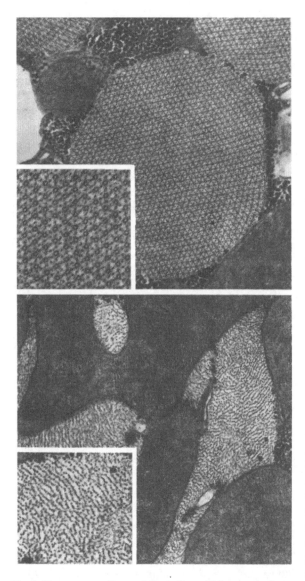

Fig. 4. Transverse sections of the indirect flight muscle from wild type (*upper panel*) and a flightless mhc mutant (Mhc 10) (*lower panel*). *Insets* show the myofibrils of each. Wild-type flies have an orderly array of thick and thin filaments while Mhc10 flies lack wild-type thick filaments and show no obvious arrangement of thin filaments. (After Collier et al. 1990)

of the *dmd* locus, usually deletions, have different times of onset and degrees of severity in an individual's life. Another human genetic disease, sudden death cardiac arrest syndrome, has been shown to be the result of a point mutation of a cardiac

myosin heavy chain isoform. Based on these early successes with the characterization of the *dmd* locus and the more extensive genetic studies of muscle mutations in *C. elegans* and *Drosophila*, we can anticipate that additional human diseases that cause muscle weakness or degeneration will be identified as mutations of muscle protein isoforms.

31.3 Differentiation of Embryonic Skeletal Muscle

31.3.1 Transcriptional Regulation During Myoblast Differentiation[5]

Vertebrate skeletal muscle fibers are formed by the cellular fusion of progenitor myoblasts, embryonic cells that proliferate and populate the muscle forming regions of embryos. The process of myoblast differentiation into muscle fibers has been investigated in cell culture of clonal embryonic myoblasts and established myoblast cell lines (Fig. 5). Myoblasts were first isolated from avian embryos. In dispersed cell cultures, myoblasts can proliferate clonally in the presence of medium rich in growth factors, but retain their potential to differentiate and fuse to form muscle fibers. These early studies established that myoblasts are a stably determined cell type, and progeny of extensive cell division faithfully inherit their myoblast identity and can express their potential to differentiate into muscle fibers. The growth and the differentiation of myoblasts is controlled by extracellular factors, specifically growth factors such as FGF and TGF-β. In the presence of growth factors, myoblasts proliferate, whereas in reduced concentrations of such factors, myoblasts can exit the cell cycle in G_1, fuse and differentiate into contractile fibers.

Myoblast cell culture has been a powerful experimental system to study the molecular processes that accompany and regulate myoblast differentiation. Myoblasts have a bipolar morphology that distinguishes them from fibroblast-like cells. As proliferative cells, myoblasts express distinctive gene products including an intermediate filament

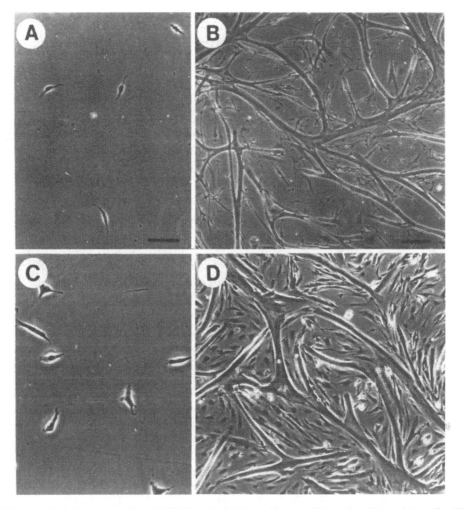

Fig. 5 A–D. Cultures of quail primary (**A**) and 10T1/2-derived (**C**) myoblasts proliferate in mitogen-rich media. When mitogens are depleted, quail (**B**) and 10T1/2-derived (**D**) myoblasts fuse to form multinucleated myofibers

protein, desmin, and cell surface marker antigens, such as H36 in human myoblasts. When myoblasts respond to growth factor signals to cease division and to differentiate, these cells coordinately activate the transcription of a large set of contractile protein genes that encode isoforms of the major contractile proteins of functional muscle. The proteins encoded by these mRNAs rapidly accumulate and assemble into myofibrillar structures. Thus, although myoblasts are stably determined to differentiate into muscle fibers and activate contractile protein genes, contractile protein genes are silent until differentiation is initiated. Myoblasts

have regulatory mechanisms that specify a defined set of contractile protein genes for activation in response to growth factor and cell division differentiation signals.

The cloning of contractile protein genes has led to characterization of *cis*-acting regulatory elements that control their transcriptional activation during myoblast differentiation (Fig. 6). Such regulatory elements have been identified by DNA transfection of myoblasts with cloned genes and analysis of the transcriptional activity of these genes in response to myoblast differentiation. These studies have led to the identification of mus-

-1200
TAAGGAGGCAAGGCCTGGGGACACCCGAGATGCCTGGTTATAATTAA-

MyoD1   **MyoD1**
CCCAGA**CATGTG**GCTGCTCCCCCCCCCAAC**ACCTG**CTGCTGCCTGAG-

AP2
CCTCAC**CCCCCACCCCGG**TGCCTGGGTCTTAGGCTCTGTACACCATGG-

CArG   -1030
AGGAGAAGCTCCTCTCT**CTAAAAATAA**CCCTGTCCCTGGTGGATCC-

Fig. 6. Muscle-specific enhancer of the mouse muscle creatine kinase gene. This sequence is located in the 5′ flanking sequence of the creatine kinase gene and shares extensive homology with the rat and human creatine kinase gene enhancers. Binding sites for MyoD1 class of helix/loop/helix myogenic regulatory proteins and the AP2 and CArG transcription factors are shown in *bold*

cle-specific control sequences that reside upstream of the transcription initiation sites of actin, myosin heavy chain, acetylcholine receptor, and creatine phosphokinase genes, within the intron of a troponin I gene, and downstream of the transcription termination site of a myosin light chain 1/3 gene. These transcriptional regulatory elements are relatively small, 60–120 bp, and share some similar, consensus sequence motifs related to AP-2, serum response element (CArG), and helix/loop/helix (HLH)-protein binding sites (Fig. 6). In vivo footprinting techniques reveal that these sequence motifs are binding sites for nuclear proteins. In vivo footprinting studies of the creatine kinase gene reveal that three presumptive *cis*-acting regulatory sites become occupied only when myoblast differentiation is initiated. These findings imply that the transcriptional activation of contractile protein genes is regulated directly by the association of regulatory proteins with these control sequences. The similarity of consensus sequence motifs suggests that similar control factors and processes coordinate the transcriptional activation of contractile protein genes when myoblasts differentiate.

31.3.2 Identification of Some Myogenic Regulatory Genes: *myd* and the MyoD Family of Genes[6]

Regulatory genes that control myoblast differentiation and the transcriptional activation of contractile protein genes have been identified by DNA transfection analysis of a multipotential mouse embryo cell line, C3H/10T1/2, clone 8 (10T1/2). The discovery of these regulatory genes has led to a new framework for understanding both the myoblast as a determined myogenic cell and the regulation of myoblast differentiation.

The 10T1/2 cell provides a cell culture model for molecular genetic studies of the embryological origin of myoblasts and myoblast differentiation. The cloning of several regulatory genes was possible because the cell line has special properties. The 10T1/2 cell is multipotential, as revealed by treatment of cultures of 10T1/2 cells with the DNA hypomethylating agent, 5-azacytidine (5-Aza-CR). 10T1/2 cells respond to 5-Aza-CR by undergoing clonal conversions to form proliferative lineages of myoblasts, adipoblasts, and chondroblasts. Cells of each of these lineages can be induced to differentiate into appropriate differentiated cell types, muscle, fat and cartilage, in response to culture conditions. The lineage specificity and high frequency of the conversion of 10T1/2 cells into myoblasts suggested a simple genetic model for myoblast conversion: that the 10T1/2 cell is genetically blocked from conversion to a myoblast because an essential myogenic regulatory gene is repressed by DNA methylation. This myogenic regulatory gene is derepressed by treatment of 10T1/2 cells with 5-Aza-CR, which demethylates a large fraction of the methylated DNA and demethylates this myogenic regulatory locus leading to activation of this myogenic gene. This model has been tested by DNA transfection of 10T1/2 cells with myoblast cDNAs and demethylated genomic DNA cloned as recombinant cosmids. These studies demonstrate that a single cosmid genomic DNA locus, *myd*, as well as four muscle-specific cDNAs, MyoD1, myogenin, myf5 and MRF4, can convert 10T1/2 cells at high frequency into muscle as transfected DNAs. The genes associated with the *myd*-genomic DNA segment have not yet been characterized, but the myogenic cDNAs encode related DNA-binding proteins of the helix/loop/helix family of transcription factors (Fig. 7).

MyoD1 homologs have been identified in other vertebrate and invertebrate species (Table 2). It is uncertain at this time whether all four of the mammalian MyoD1 homologs are expressed in all vertebrate species or whether additional MyoD1 homologs exist. Significantly, MyoD1 homologs have been discovered in *C. elegans* and *D. melanogaster*, arguing that these genes are highly conserved and important regulators of myogenesis.

The myogenic regulatory cDNAs encode members of the *myc* gene family of helix-loop-helix (HLH) genes, so called because of their prominent conserved protein domain features: a region of basic amino acids, involved in DNA-binding, an amphipathic α-helical domain (helix I), a loop region undistinguished by any obvious secondary structure, and a second amphipathic α-helical domain (helix II). The HLH regions of these proteins are preceded by basic regions that are demonstrably DNA binding domains (Fig. 7). Immunostaining of myoblasts and myofibers with antibodies raised to MyoD1 suggest that the myogenic HLH proteins localize to nuclei. While all the members of the *myc* family HLH genes share these common structural features, homology at the primary sequence level within these domains is striking only for the myogenic HLH proteins, suggesting that the conserved regions are important for muscle-specific activities. Deletion analysis and domain swapping experiments have shown that the basic amino acid domain is necessary for myogenic conversion and confers muscle-specific transcriptional activating function. In addition, invitro-binding studies together with mutational analysis have shown that helix II is involved in hetero-oligomerization of the MyoD1 homologs with other HLH proteins. Specifically, MyoD1 homologs oligomerize with the ubiquitously expressed E12/E47 class of proteins transcribed from the human E2 gene and a potential negative regulator of myogenic HLH proteins, Id. Id lacks a DNA binding basic amino acid domain, but is capable of heterodimerizing with MyoD1 and of attenuating MyoD1 DNA binding in vitro. Id is expressed in a variety of cell types, but its expression is decreased coincident with differentiation suggesting that Id negatively regulates other HLH proteins in undifferentiated cells. Id may be a member of a class of HLH proteins that includes the extramachrochaetae (emc) gene of *Drosophila* that oligomerizes with basic HLH proteins and inhibits their DNA binding and transcription activation function.

Table 2. Identification of MyoD1 and some MyoD1 homologs

MyoD1 homolog	Animal	Homology
MyoD1	Mouse	MyoD1
Myf-3	Human	MyoD1
XMyoD	Frog (*X. laevis*)	MyoD1
CMD1	Chicken	MyoD1
qmf1	Quail	MyoD1
CeMyoD	Worm (*C. elegans*)	MyoD1
nau	Fly (*D. melanogaster*)	MyoD1
Myogenin	Rat/mouse	myogenin
Myf-4	Human	myogenin
qmf2	Quail	myogenin
myf-5	Human	myf-5
qmf3	Quail	myf-5
MRF-4	Rat	MRF-4
Herculin	Human	MRF-4

31.3.3 Some Properties of MyoD1, myf5, Myogenin, and MRF4 Genes[7]

The four mammalian myogenic HLH cDNAs, MyoD1, myf5, myogenin, and MRF4, are domi-

```
            80        90        100       110       120
MyoD1     EHVRAPSGHHQAGRCLLWACKACKRKTTNADRRKAATMRERRRL
qmf1      EHVRAPSGHHQAGRCLLWACKACKRKTTNADRRKAATMRERRRL
XMyoD     EHVRAPSGHHQAGRCLLWACKACKRKTTNADRRKAATMRERRRL
nau       LAPLVCSSAQSSRPCLTWACKACKKKSVTVDRRKAATMRERRRL
CeMyoD    VDSQHEDTTTSTAGGAGVGGPRRTKLVLSVDRRKAATMRERRRL
Myf-5     EHVRAPTGHHQAGHCLMWACKACKRKSTTMDRRKAATMRERRRL
myogenin  KGLGTP--EHCPGQCLPWACKVCKRKSVSVDRRRAATLREKRRL
MRF4      PPGLQP--PHCPGQCLIWACKTCKRKSAPTDRRKAATLRERRRL

E12       KELKAPRARTSPDEDEDDLLPPEQKAEREKERRVANNARERLRV
E47       KELKAPRARTS---STDEVLSLEEKDLRDRERRMANNARERVRV
da        GKGTKRPRRYCSSADEDDDAEPAVKAIREKERRQANNARERIRI
twist     GSDAGGKAFRKPRRLKRKPSKTEETDEFSNQRVMANVRERQRT
AS-C T4   GMALTRCSESVSSLSPGSSPAPYNVDQSQSVQRR--NARERNRV
AS-C T5    MAL----GSENHSVFNDDEESSSAFNGPSVIRR--NARERNRV

N-myc     VTLPIHQQ-HNYAAPKAKSLSPRNSDSEDSERRRNHNILERQRR
L-myc     VSISIHQQQHNYAAPKP--VS---SDTEDVTKRKNHNFLERKRR
c-myc     RVKLDSVRVLRQISNNRKCTSPRSSDTEENVKRRTHNVLERQRR
```

```
         HELIX 1              LOOP           HELIX 2
         ------------                        ------------

            130       140                 150       160
MyoD1     SKVNEAFETLKRCTSSNP--------------NQRLPKVEILRNAIRYIEGLQALLR
qmf1      SKVNEAFETLKRCTSTNP--------------NQRLPKVEILRNAIRYIESLQALLR
XMyoD     SKVNEAFETLKRYTSTNP--------------NQRLPKVEILRNAIRYIESLQALLR
nau       RKVNEAFEILKRRTSSNP--------------NQRLPKVEILRNAIEYIEGLEDLLQ
CeMyoD    RKVNEAFEVVKQRTCPNP--------------NQRLPKVEILRSAIDYINNLERMLQ
Myf-5     KKVNQAFETLKRCTTTNP--------------NQRLPKVEILRNAIRYIESLQELLR
myogenin  KKVNEAFEALKRSTLLNP--------------NQRLPKVEILRSAIQYIERLQALLS
MRF4      KKINEAFEALKRRTVANP--------------NQRLPKVEILRSAINYIERLQDLLH

E12       RDINEAFKELGRMCQLHLNSEKP-------------QTKLLILHQAVSVILNLEQQVR
E47       RDINEAFRELGRMCQMHLKSDKA-------------QTKLLILQQAVQVILGLEQQVR
da        RDINEALKELGRMCMTHLKSDKP-----------QTKLGILNMAVEVIMTLEQQVR
twist     QSLNDAFKSLQQIIPTL--------------PSDKLSKIQTLKLATRYIDFLCRMLS
AS-C T4   KQVNNSFARLRQHIPQSIITDLTKG--G-GRGPHKKISKVDTLRIAVEYIRSLQDLVD
AS-C T5   KQVNNGFSQLRQHIPAAVIADLSNGRRGIGPGANKKLSKVSTLKMAVEYIRRLQKVLH

N-myc     NDLRSSFLTLRDHVP-----ELVK--------NEKAAKVVILKKATEYVHSLQAEEH
L-myc     NDLRSRFLALRDQVP-----TLAS--------CSKAPKVVILSKALEYLQALVGAEK
c-myc     NELKRSFFALRDQIP-----ELEN--------NEKAPKVVILKKATAYILSVQAEEQ
```

Fig. 7. Amino acid sequences of the basic and helix/loop/helix regions of some *myc* gene family members. Primary sequence comparison shows that these proteins share motifs in the basic (a.a. 102–124) and helix/loop/helix regions. Residues *in bold type* represent shared amino acid residues showing greater homology among the muscle-specific HLH proteins. Daughterless (*da*), twist, and the achaete-scute complex (*AS-C T4* and *T5*) are other HLH genes that are important in *Drosophila* development

nantly acting regulatory genes that have the remarkable power to convert 10T1/2 cells into differentiated muscle. In vitro DNA-binding assays show that these myogenic HLH proteins, as well as other HLH proteins, bind to CANNTG sequence motifs. These DNA-binding studies also suggest that oligomerization of these myogenic proteins with the general HLH proteins, E12/E47, is necessary for high affinity binding to DNA. This sequence motif is present in the regulatory elements of all muscle contractile protein genes characterized to date (see above and Fig. 6). Mutagenesis studies show that this sequence motif is necessary for transcriptional activity of the muscle genes. Furthermore, MyoD1, myogenin, and myf5 can transactivate reporter constructs containing regulatory elements of creatine kinase, myosin light chain, acetylcholine receptor, actin, and troponin I genes. The functions of these myogenic regulatory proteins, however, may not be precisely equivalent because the kinetics of transactivation of different regulatory elements differ among the various regulatory proteins, and MRF4 fails to transactivate regulatory elements of the troponin I and creatine kinase genes. Furthermore, the temporal and developmental expression of the MyoD1 homologs differ; MyoD1 and myf5 mRNA are expressed in myoblasts and in differentiated muscle, myogenin expression commences as differentiation begins, and MRF4 is expressed later in development during muscle maturation. The unique and/or redundant functions of the individual myogenic regulatory gene homologs await gene knockout experiments to disrupt gene function to create null mutations by homologous recombination or antisense RNA inhibition. The current evidence, however, clearly establishes the important regulatory functions of myogenic HLH genes in skeletal muscle differentiation, and specifically in the coordinated transcriptional activation and expression of contractile protein genes.

The HLH myogenic genes have additional regulatory functions in skeletal muscle differentiation. DNA transfection and gene expression studies show that MyoD1 plays a role as a regulator of myoblast cell division. MyoD1 controls cessation of myoblast cell division and the initiation of differentiation in response to the depletion of growth factors through a probable signal transduction pathway. The myogenic HLH genes also can autoactivate and cross-activate to maintain their transcriptional activities. Transfection of 10T1/2 cells with each of the MyoD1 homologs and with *myd* shows that endogenous MyoD1, myf5, and myogenin are expressed in response to *myd* transfection while the MyoD1 homologues show cross- and auto-activation of the endogenous genes. This highly interactive regulatory loop of regulatory gene control can explain the stability of the myoblast phenotype and its differentiation to form fibers. These myogenic regulatory genes provide molecular genetic mechanisms to ensure maintenance and inheritance of the differentiation capacity of myoblasts and the coordinated transcriptional activation of the set of contractile protein genes.

31.4 Embryological Origins of Skeletal Muscle

31.4.1 Somite Origins of Vertebrate Muscle [8]

Vertebrate skeletal muscle is derived from the mesodermal layer of the gastrulating embryo. Prospective mesodermal cells invaginate as a sheet of cells through the blastopore (amphibians) or migrate as single cells through the primitive streak (birds and mammals) into spaces between the ectoderm and the endoderm, eventually forming continuous bilateral layers of mesenchymal cells on either side of the notochord. The dorsal portion of this mesodermal layer, that part of the mesoderm adjacent to the notochord, becomes segmented into a series of paired structures termed somites (Fig. 8). Newly formed somites are masses of mesodermal cells surrounding a small cavity. Segmentation of the dorsal mesoderm into somites occurs in a cranial to caudal sequence so that fully formed somites exist cranially at the same time as unsegmented dorsal mesoderm caudally. As development proceeds, the somites compartmentalize so that the ventral cells disperse as sclerotomal cells that will eventually form cartilage (Fig. 9). The remaining cells of the somite form a bilayered structure consisting of dorsal dermatomal cells which contribute to the dermis and underlying myotomal cells which give rise to the skeletal muscles.

Convincing evidence that skeletal muscle is derived exclusively from somitic mesoderm has been obtained using chick and quail as model systems. Avian embryos provide easily manipulable and affordable systems in which to observe early development. Early experiments investigating the origin of muscle cells depended upon marking cells with fine particles of carbon. Carbon particles adhere to the cell surface and identify cells at their eventual location. These experiments were infor-

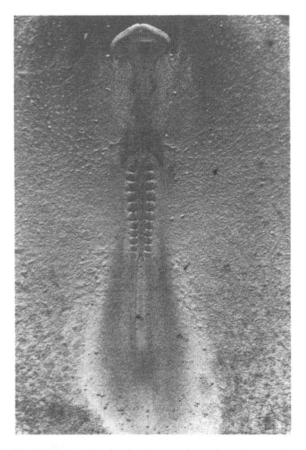

Fig. 8. Micrograph of a developing avian embryo showing the paired somites on either side of the neural tube. (After Drake and Little 1991)

mative, but suffer from technical limitations of the carbon marking technique. Definitive marking experiments have taken advantage of the histological differences between quail nuclei, which contain densely clumped chromatin, and chick nuclei, which contain dispersed chromatin. Quail and chick cells can be easily distinguished following heterospecific grafts of quail somites into a chick host or vice versa. These experiments showed that donor somitic cells migrate to muscle forming regions and participate in myofiber formation at the appropriate time (Fig. 10). Consistent with results of these chick/quail grafting experiments, surgical removal or X-irradiation of wing or leg level somites results in muscle-less limbs. Inter-

estingly, normal muscle results from transplantation of any somite along the cranial-caudal axis, suggesting that the information signalling muscle type specialization resides in the muscle forming regions themselves and is not a predetermined characteristic of the somite myoblasts. Migration of cells from the dermomyotome into the muscle-forming regions has been demonstrated using the quail-chick marker system and ^{3}H-thymidine-labeled tracer cells. Thus, during normal development, muscle cells migrate from the somites to muscle forming regions where local environmental signals determine muscle type and the timing of differentiation.

31.4.2 Determination of Myoblasts: Role of Cell-Cell Interaction[9]

While grafting and marking experiments establish the somite origin of the muscle-forming cells within the embryo, they do not address the questions of when and where the cells become determined myoblasts that autonomously maintain the capacity to differentiate. It has been shown that somites in organ culture require interaction with adjacent embryonic tissue (neural tube) to form muscle, but that somites removed from embryos at slightly later stages alone will form muscle. This result suggests that presumptive myoblasts exist within the somites at very early stages of development, but that these presumptive somite myoblasts require other tissue types, specifically the neural tube, to promote their myogenic determination. Because the MyoD1 family of myogenic regulatory genes is expressed exclusively by skeletal myogenic lineages, they have been used as molecular markers to identify determined myogenic cells in embryos. In situ hybridization of mouse and quail embryos using labeled probes of MyoD1 or MyoD1 homologs show that these regulatory genes are expressed very early in development prior to contractile protein gene expression in the somites. Activation of these genes follows the cranial-caudal gradient of somite formation. Furthermore, different MyoD1 family members are expressed with different temporal patterns, suggesting either that the somite gives rise to

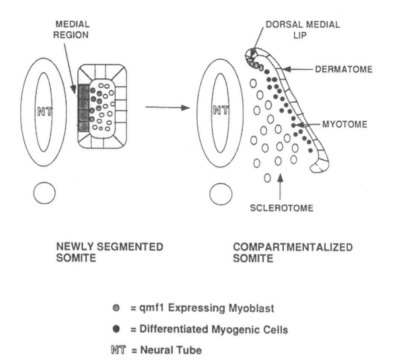

= qmf1 Expressing Myoblast

= Differentiated Myogenic Cells

NT = Neural Tube

Fig. 9. Schematic representation of newly segmented and compartmentalized somites. *Stippled regions* indicate qmf1 expression detected by in situ hybridization

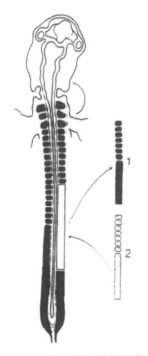

Fig. 10. Diagram representing the chick-quail somite grafting experiments

different lineages of myogenic cells or that myoblasts express these genes sequentially. In the quail, the MyoD1 homolog, qmf1, is expressed exclusively in the dorsal medial lip of the developing somite, where no contractile protein gene expression is detected, suggesting that there is strict spatial control of myogenic determination and differentiation (Fig. 11).

The experiments outlined using avian embryos suggest the importance of cell-cell or inductive interactions in myogenic determination. Tissue induction has become a central theme in developmental biology and has been most fully investigated in amphibian eggs because of their large size and accessibility to experimental manipulation. Interactions between vegetal and animal cells in the early frog embryo induce mesoderm formation, while signals emitted by a portion of the dorsal mesoderm induce proximal mesodermal cells to form somites (Fig. 12). A search for mesoderm-inducing factors reveals an agent, XTC-MIF, that induces animal cells to produce a full range of meso-

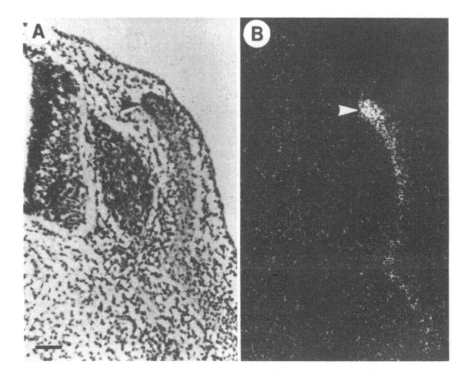

Fig. 11 A, B. Brightfield (**A**) and darkfield (**B**) micrographs, of a quail embryo transverse section showing hybridization of a qmf1 probe to the dorsal medial lip and the myotomal region of a compartmentalized somite (*arrows*)

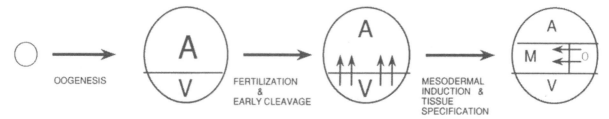

Fig. 12. Schematic representation of early frog development. *A*, animal pole; *V* vegetal pole; *M* mesoderm; *o* inducing or organizer region of the dorsal mesoderm. *Arrows* suggest the direction of inducing signal

dermal tissues, except blood. XTC-MIF-mesodermal inducing activity is partially neutralized by antibodies to transforming growth factor, TGF-β_2. Both TGF-β_2 and fibroblast growth factor (FGF) act synergistically to induce mesoderm. FGF mRNA and a TGF-β_2-like protein, Vg1, are synthesized in frog embryos. It has been hypothesized that these inducing molecules are secreted and present in polarized gradients within the embryo so that cells respond dependent on their position within the embryo.

The *Xenopus* homolog of MyoD1, XmyoD, is expressed in the mesoderm of early gastrulae and becomes restricted to the developing somites of neurula stage embryos. XmyoD expression can be detected prior to contractile protein gene expression so that it is the earliest known mesodermal marker expressed. When animal pole explants are induced to convert to mesoderm by vegetative pole explants, XMyoD expression is also induced. It is unclear whether activation of XMyoD expression and conversion of cells to muscle is direct or whether there are intervening steps in the induction process.

31.4.3 Expression of MyoD in Invertebrates[10]

Embryological as well as cell and molecular biological studies using vertebrate model systems have established a framework for understanding myogenic determination and differentiation. Still many questions remain concerning myogenic determination and the regulatory mechanisms operating to allow myoblasts to respond to environmental cues for growth, differentiation, and muscle type-specific gene expression. Two invertebrate model systems, *Drosophila melanogaster* and *Caenorhabditis elegans*, are proving valuable organisms for studies that address these questions because they are amenable to genetic as well as cell biological and biochemical studies. Genetic studies of *Drosophila* have revealed a hierarchy of developmental regulatory genes that control first embryonic pattern formation, then position and region-specific identity, and finally tissue specificity. Consistent with the theory that factors involved in developmental processes are often conserved over diverse species, the expression of *Drosophila* MyoD, *nau*, coincides with the early commitment of a subset of cells to somatic muscles. *Nau* gene expression peaks first in early embryos preceding expression of the very earliest muscle-specific genes and again in pupae at the time when adult muscle is being formed. *Nau*-positive cells detected by in situ hybridization may correspond to the early muscle precursor cells that appear to preform muscle location and specification in the *Drosophila* embryo. In addition to this HLH gene, a homeobox gene, S59, is expressed in mesodermal cells that are precursors to specific somatic muscles in *Drosophila*. The pattern of expression of the *nau* and S59 genes are similar and future genetic experiments will demonstrate the relationship between these genes.

C. elegans provides another unique system for studies of muscle cell determination and differentiation as cell lineages are invariant and are known for the entire organism. The rigidly determined cleavage pattern of the *C. elegans* zygote most likely results from the segregation of egg cytoplasm that may contain determinants that instruct cells in their differentiation pathway. The *C. elegans* MyoD homolog, CeMyoD, is expressed very early in lineages that give rise to body wall muscles but not to pharyngeal muscle (Fig. 13). As body wall muscle and pharyngeal muscle have different developmental programs, it may be that cytoplasmic determinants act to regulate CeMyoD expression. Like the *Drosophila nau* gene, CeMyoD is not expressed in all muscle types, even though all the muscles express some of the same differentiation-specific genes, suggesting that there are as yet undiscovered myogenic regulatory genes. Identification of mutations that inactivate the function and alter the expression of these myogenic genes will provide understanding of the specific regulatory functions of genes during myogenesis and a basis for identifying new genes that control the expression and functions of the CeMyoD and *nau* genes.

31.5 Outlook

Continued molecular and genetic investigations of skeletal muscle differentiation using tissue culture cells and avian and mammalian embryos will most likely provide more refined understanding and insight into specific functions of the different HLH myogenic regulatory genes and the complexity of genetic interactions involved in cell type-specific gene regulation. In addition, these genes also are most likely involved in some aspects of the mechanisms involved in the determination of embryonic cells to the lineage of skeletal myogenic progenitor cells and to initiation of the coordinated activation of the set of muscle-specific genes. Genetic studies, now possible in *Drosophila* and *C. elegans*, and transgenic and genetic engineering studies using homologous recombination in mammalian embryos offer new and complementary approaches to investigate and define the functions of these myogenic genes and to identify additional regulatory genes that activate or repress the myogenic regulatory genes in the skeletal muscle and non-muscle cell lineages in the embryo. In this regard, it is interesting to consider that the currently known repertoire of known HLH myogenic regulatory genes encodes nuclear proteins. Almost nothing is yet known about other genes and proteins that con-

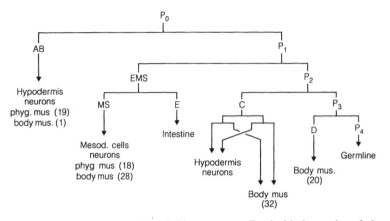

Fig. 13. *C. elegans* cell lineages. The fates of each of the six blastomeres are listed with the number of pharyngeal (*phyg.*) or body wall muscle cells noted in *parentheses*. (Krause et al. 1990)

trol the cellular signaling and cellular interactions that control the activation and nuclear functions of these myogenic regulatory genes, that specify the myogenic cell lineages in the mesoderm of early embryos, and that initiate the differentiation response of myogenic progenitor cells in the muscle-forming regions. Embryological and cellular studies in amphibian embryos provide compelling evidence that such cellular interactions and signaling processes direct and modulate the expression of the myogenic genes and their encoded proteins. Future embryological and molecular experiments in amphibian and avian embryos and genetic studies using *C. elegans* and *D. melanogaster* will most likely provide new insights into these mechanisms and identify genes and proteins involved in these signaling and signal transduction processes. Even less is known about the physiological differentiation processes that influence and modulate the expression of muscle genes and the synthesis and turnover of contractile proteins in muscle in response to activity and exercise, although the recent progress in understanding the molecular basis of skeletal muscle differentiation during embryogenesis makes even such complex physiological problems approachable. Skeletal muscle continues to be an excellent cell type for studies of cell differentiation using molecular, genetic, cellular and embryological approaches in a diversity of organisms, and continued exciting progress towards un-

derstanding fundamental mechanisms of cellular differentiation is certain.

31.6 Summary

Rapid progress is being made in understanding the fundamental mechanisms of skeletal muscle differentiation. The differentiated functions of skeletal muscle in contractile activity and motility have been the subject of investigation by physiologists and biochemists and have provided a basis for the current embryological, genetic, cellular and molecular studies of the developmental origins of differentiated muscles in both vertebrate and invertebrate embryos. In common with many other differentiated cell types, the process of skeletal muscle differentiation involves the coordinated transcriptional activation of a large set of cell type-specific genes by transcriptional regulatory mechanisms involving interactions of enhancer control sequences localized within or in the vicinity of these cell type-specific genes with specific protein transcription factors. In the specific case of skeletal muscle differentiation, the helix/loop/helix (HLH) myogenic regulatory proteins function as muscle specific transcription factors involved in the coordinated activation of the cell type-specific contractile protein genes.

References

[1]Emerson CP Jr, Bernstein SI (1987) Molecular genetics of myosin. Annu Rev Biochem 56:695–726

Engel AG, Banker BQ (eds) (1986) Myology, vol 1. McGraw-Hill, New York

Epstein HF, Fischman DA (1991) Molecular analysis of protein assembly in muscle development. Science 251:1039–1044

Harold FM (1986) The vital force: a study of bioenergetics. WH Freeman, New York

MacIntyre RS, Davis MB (1987) A genetic and molecular analysis of the α-glycerophosphate cycle in *Drosophila melanogaster*. In: Ratazzi MC, Scandalios JG, Whitt GS (eds) Isozymes: Current topics in biological and medical research, vol 14. Molecular and cellular biology. Alan R Liss, New York, p 195

Martini R (1989) Fundamentals of anatomy and physiology. Prentice-Hall, Englewood Cliffs, NJ

[2]Buckingham ME (1985) Actin and myosin multigene families: their expression during the formation of skeletal muscle. Essays Biochem 20:77–109

Dhoot GK, Perry SV (1979) Distribution of polymorphic forms of troponin components and tropomyosin in skeletal muscle. Nature 278:714–718

Miller JB, Stockdale FE (1987) What muscle cells know that nerves don't tell them. Trends Neurosci 10:325–329

Peckham M, Molloy JE, Sparrow JC, White DCS (1990) Physiological properties of the dorsal longitudinal flight muscle and the tergal depressor of the trochanter muscle of *Drosophila melanogaster*. J Muscle Res Cell Motil 11:203–215

Pette D, Vrbova G (1985) Neural control of phenotypic expression in mammalian muscle fibers. Muscle Nerve 8:676–689

Waterston RH (1989) Molecular genetic approaches to the study of motility in *Caenorhabditis elegans*. Cell Motil Cytoskel 14:136–145

Whalen RG (1985) Myosin isoenzymes as molecular markers for muscle physiology. J Exp Biol 115:43–53

[3]Bucher EA, Charles de la Brousse F, Emerson CP Jr (1989) Developmental and muscle specific regulation of avian fast skeletal troponin T isoform expression by mRNA splicing. J Biol Chem 264:12482–12491

Dibb NJ (1989) Sequence analysis of the complete *Caenorhabditis elegans* myosin heavy chain gene family. J Mol Biol 205:603–613

George EL, Ober MB, Emerson CP Jr (1989) Functional domains of the *Drosophila melanogaster* muscle myosin heavy chain gene are encoded by alternatively spliced exons. Mol Cell Biol 9:2957–2974

Nadal-Ginard B (1990) Muscle cell differentiation and alternative splicing. Curr Opinion Cell Biol 2:1058–1064

[4]Bejsovec A, Anderson P (1990) Functions of the myosin ATP and actin binding sites are required for *C. elegans* thick filament assembly. Cell 60:133–140

Collier VL, Kronert WA, O'Donnell PT, Edwards KA, Bern-

stein SI (1990) Alternative myosin hinge regions are utilized in a tissue-specific fashion that correlates with muscle contraction speed. Genes Dev 4:885–895

Dibb NJ, Brown DM, Karn J, Moerman DG, Bolton SL, Waterston RJ (1985) Sequence analysis of mutations that affect the synthesis, assembly, and enzymatic activity of the *unc-54* myosin heavy chain of *Caenorhabditis elegans*. J Mol Biol 183:543–551

Engel AG, Banker BQ (eds) (1986) Myology, vol 2. McGraw-Hill, New York

Fyrberg E, Fyrberg CC, Beall C, Saville DL (1990) *Drosophila melanogaster* troponin-T mutations engender three distinct syndromes of myofibrillar abnormalities. J Mol Biol 216:657–675

Homyk T Jr, Emerson CP Jr (1988) Functional interactions between unlinked muscle genes within haploinsufficient regions of the *Drosophila* genome. Genetics 119:105–121

Karlik CC, Fyrberg EA (1985) An insertion within a variably spliced tropomyosin gene blocks accumulation of only one encoded isoform. Cell 41:57–66

Karlik CC, Coutu MD, Fyrberg EA (1984) A nonsense mutation within the *Act*88F actin gene disrupts myofibril formation in *Drosophila* indirect flight muscles. Cell 38:711–719

Koenig M, Hoffman EP, Bertelson CJ, Monaco AP, Feener C, Kunkel L (1987) Complete cloning of the Duchenne muscle dystrophy (DMD) cDNA and preliminary genomic organization of the DMD gene in normal and affected individuals. Cell 50:509–517

Tanigawa G, Jarcho JA, Kass S, Solomon SD, Vosberg H-P, Seidman J, Seidman CE (1990) A molecular basis for familial hypertrophic cardiomyopathy: an α/β cardiac myosin heavy chain hybrid gene. Cell 62:991–998

Waterston RH (1989) The minor myosin heavy chain, mchA, of *Caenorhabditis elegans* is necessary for the initiation of thick filament assembly. EMBO J 8:3429–3436

[5]Clegg CH, Linkart TA, Olwin BB, Hauschka SD (1987) Growth factor control of skeletal muscle differentiation: commitment to terminal differentiation occurs in G₁ phase and is repressed by fibroblast growth factor. J Cell Biol 105:949–956

Horlick RA, Benfield PA (1989) The upstream muscle-specific enhancer of the rat muscle creatine kinase gene is composed of multiple elements. Mol Cell Biol 9:2396–2413

Konigsberg IR (1963) Clonal analysis of myogenesis. Science 140:1273–1284

Rosenthal N (1989) Muscle cell differentiation. Curr Opinions Cell Biol 1:1094–1101

[6]Benezra R, Davis RL, Lockshon D, Turner DL, Weintraub H (1990) The protein Id: a negative regulator of helix-loop-helix DNA binding proteins. Cell 61:49–59

Olson EN (1990) MyoD family: a paradigm for development? Genes Dev 4:1454–1461

Pinney DF, Pearson-White SH, Konieczny SF, Latham KE, Emerson CP Jr (1988) Myogenic lineage determination and differentiation: evidence for a regulatory gene pathway. Cell 53:781–793

Pinney DF, Charles de la Brousse F, Emerson CP Jr (1990)

Molecular genetic basis of skeletal muscle determination and differentiation. In: Mahowald A (ed) Genetics of pattern formation and growth control. Wiley-Liss, New York, pp 65–89

Weintraub H, Davis R, Tapscott S, Thayer M, Krause M, Benezra R, Blackwell TK, Turner D, Rupp R, Hollenberg S, Zhuang Y, Lassar A (1991) The myoD gene family: nodal point during specification of the muscle cell lineage. Science 251:761–766

[7]Brennan TJ, Olson EN (1990) Myogenin resides in the nucleus and acquires high affinity for a conserved enhancer element upon heterodimerization. Genes Dev 4:582–595

Brennan TJ, Edmondson DG, Olson EN (1990) Aberrant regulation of MyoD1 contributes to the partially defective myogenic phenotype of BC_3H1 cells. J Cell Biol 110:929–937

Buskin JN, Hauschka SD (1989) Identification of a myocyte-specific nuclear factor which binds to the muscle-specific enhancer of the mouse muscle creatine kinase gene. Mol Cell Biol 9:2627–2640

Konieczny SF, Drobes BL, Menke SL, Taparowsky EJ (1989) Inhibition of myogenic differentiation by the H-ras oncogene is associated with the down regulation of the MyoD1 gene. Oncogene 4:473–481

Mueller PR, Wold B (1989) In vivo footprinting of a muscle specific enhancer by ligation mediated PCR. Science 246:780–786

Murre C, McCaw PS, Vaessin H, Caudy M, Jan LY, Jan YN, Cabrera CV, Buskin JN, Hauschka SD, Lassar AB, Weintraub H, Baltimore D (1989) Interactions between heterologous helix-loop-helix proteins generate complexes that bind specifically to a common DNA sequence. Cell 58:537–544

Sorrentino V, Pepperkok R, Davis RL, Ansorge W, Philipson L (1990) Cell proliferation inhibited by MyoD1 independently of myogenic differentiation. Nature 345:813–815

Vaidya TB, Rhodes SJ, Taparowsky EJ, Konieczny SF (1989) Fibroblast growth factor and transforming growth factor β repress transcription of the myogenic regulatory gene MyoD1. Mol Cell Biol 9:3576–3579

Yutzey KE, Rhodes SJ, Konieczny SF (1990) Differential trans-activation associated with the muscle regulatory factors MyoD1, myogenin and MRF4. Mol Cell Biol 10:3934–3944

[8]Balinski BI, Fabian BC (1981) An introduction to embryology. Saunders College Publ, Philadelphia

Drake CJ, Little CD (1991) Integrins play an essential role in somite adhesion to the embryonic axis. Dev Biol 143:418–421

Kenny-Mobbs T, Thorogood P (1987) Autonomy of differentiation in avian brachial somites and the influence of adjacent tissues. Development 100:449–462

Konigsberg IR (1986) The embryonic origin of muscle. In: Engel A, Banker BQ (eds) Myology, vol 1. McGraw-Hill, New York, pp 39–71

[9]Charles de la Brousse F, Emerson CP Jr (1990) Localized expression of a myogenic regulatory gene, qmf1, in the somite dermatome of avian embryos. Genes Dev 4:567–581

Hopwood ND, Pluck A, Gurdon JB (1989) MyoD expression in the forming somite is an early response to mesoderm induction in Xenopus embryos. EMBO J 8:3409–3417

Sassoon D, Wright WE, Lin V, Lassar A, Weintraub H, Buckingham ME (1989) Expression of two myogenic regulatory factors myogenin and MyoD1 during mouse embryogenesis. Nature 341:303–307

Scalles JB, Olson EN, Perry M (1990) Two distinct Xenopus genes with homology to MyoD1 are expressed before somite formation in early embryogenesis. Mol Cell Biol 10:1516–1524

Smith JC (1989) Induction and early amphibian development. Curr Top Cell Biol 1:1061–1070

Wright WE, Sassoon DA, Lin VK (1989) Myogenin, a factor regulating myogenesis, has a domain homologous to MyoD. Cell 56:607–617

[10]Bate M (1990) The embryonic development of larval muscles in Drosophila. Development 110:791–804

Cowan AE, McIntosh JR (1985) Mapping the distribution of differentiation potential for intestine, muscle, and hypodermis during early development in Caenorhabditis elegans. Cell 41:923–932

Dohrmann C, Azpiazu N, Frasch M (1990) A new Drosophila homeobox gene is expressed in mesodermal precursor cells of distinct muscle during embryogenesis. Genes Dev 4:2098–2111

Krause M, Fire A, Harrison SW, Priess J, Weintraub H (1990) CeMyoD accumulation defines the body wall muscle cell fate during C. elegans embryogenesis. Cell 63:907–919

Michelson AM, Abmayr SM, Bate M, Arias AM, Maniatis T (1990) Expression of a MyoD family member prefigures muscle pattern in Drosophila embryos. Genes Dev 4:2086–2097

Sulston JE, Schierenberg E, White JG, Thomson JN (1983) The embryonic cell lineage of the nematode Caenorhabditis elegans. Dev Biol 100:64–119

Chapter 32   Hepatocyte Differentiation

CALVIN J. KUO and GERALD R. GRABTREE

32.1 Introduction

Despite a somewhat debatable taste and a rather mundane appearance, the liver is unquestionably essential to life. Indeed, among other tasks, the liver secretes plasma proteins, clotting factors and bile, detoxifies drugs and harmful by-products of cellular metabolism, and through gluconeogenesis supplies the body with glucose in times of severe stress. Although performing these diverse functions, the adult liver is predominantly composed of a single cell type – the hepatocyte – allowing the abstract and rather intimidating developmental question of "how does the liver form" to be roughly simplified to "how do hepatocytes form". In this chapter, we will address the question of hepatocyte differentiation with specific reference to how hepatocytes acquire the specialized proteins, often unique to the liver, that enable these individual cells to perform the many tasks attributed to the liver as an organ (Table 1).

32.2 Embryologic Origin of the Liver[1,2]

After fertilization, the zygote undergoes a series of cell divisions forming in turn a morula, and the hollow blastocyst. Subsequently, gastrulation occurs, with concomitant formation of the three germ layers: endoderm, mesoderm, and ectoderm. The unspecialized endoderm thus formed differentiates into the liver, lungs, as well as the entire gastrointestinal tract, including esophagus, stomach, large and small intestines, and exocrine pancreas.

Clearly, different signals must act upon the endoderm, indicating whether it should form, for instance, stomach rather than intestine. In the partic-

ular case of the liver, classical embryological studies have provided some insight into the nature of the "liver-inducing" signal. Apparently, at least two independent inductive events are required for proper liver organogenesis. The post-gastrulation endoderm soon assumes the form of a long tube, spanning the length of the embryo, with its termini roughly defining the locations of the future mouth and anus. Because of a rapid increase in the number of mesodermal somites, the embryo undergoes bending, placing the nascent heart – initially at the most anterior end of the embryo – in close proximity to the endodermal tube. It is this close association of heart mesoderm with the previously

Table 1. Representative hepatic genes

Plasma proteins
  Albumin
  α-Fetoprotein
  α-2-Macroglobulin
  Transthyretin
  Apolipoprotein CIII
Clotting factors
  Fibrinogen
  Prothrombin
  Factor VIII
Protease inhibitors
  α-1-Antitrypsin
  Antithrombin III
Gluconeogenic enzymes
  Phosphoenolpyruvate carboxykinase
  Pyruvate carboxylase
Acute phase reactants
Hemopexin
C-reactive protein
Kininogen
Cholesterol synthetic enzymes
  HMG-CoA reductase
  Lecithin-cholesterol acetyltransferase
Detoxification enzymes
  Cytochrome P450
  Glutathione S-acetyltransferase

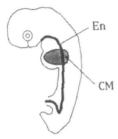

Fig. 1. Schematic representation of the overlap between cardiac mesenchyme (*CM*) and undifferentiated endoderm (*En*) in the early embryo. This mesoderm:endoderm interaction is the first inductive event of liver organogenesis

uncommitted endoderm which represents the first inductive event in liver organogenesis. At the 5-somite stage in the chick, the heart mesoderm exerts an inductive influence upon the endodermal tube, causing cells at the point of "contact" to become fated to form liver, although these cells (hepatic endoderm) do not yet synthesize liver-specific markers (Fig. 1).

The differentiation of these hepatic endoderm cells into mature hepatocytes producing liver-specific proteins is dependent upon a second inductive event. As with the first induction, this also consists of an interaction between mesoderm and endoderm. By the 20–22-somite stage in the chick, the hepatic endoderm and the cardiac mesenchyme have re-segregated. However, the hepatic endoderm has now become invested with hepatic mesodermal cells which will later form the extensive endotheli-

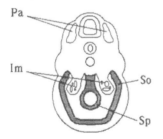

Fig. 2. Mesodermal derivatives competent to perform the second inductive event of liver organogenesis. The lateral plate mesoderm, represented by the somatopleure (*So*) and the splanchnopleure (*Sp*), can induce terminal hepatocyte differentiation, while the para-axial (*Pa*) and intermediate (*Im*) mesoderm can not. This is a schematic transverse section of an early embryo at the level of the small intestine. (After Le Dourain 1975)

um of the well-vascularized adult liver. This distinct association between hepatic endoderm and hepatic mesoderm constitutes the second inductive event of liver organogenesis, during which the endoderm is triggered to differentiate along the hepatocyte lineage, with subsequent synthesis of liver markers. The requirement for a second induction has been demonstrated by Le Dourain and Houssaint, who demonstrated that only when hepatic endoderm is co-cultured with mesoderm do hepatocytes develop. Interestingly, this effect may be obtained with any lateral plate mesoderm, even from heterologous species (Fig. 2).

32.3 Cell Lineages in the Embryonic Liver[3]

Having received two distinct inductive stimuli, the hepatic endoderm proliferates to form a stratified structure bulging from the gastrointestinal endodermal tube, known as the hepatic diverticulum (HD), or liver bud. This process occurs during the 10th day of gestation in the mouse, at the 20-somite stage. During formation of the liver bud, terminal differentiation into hepatocytes occurs, as measured by synthesis of α-fetoprotein, a marker for embryonal liver cells. The liver bud sends forth cords of hepatocytes rostrally, invading the septum transversum, or presumptive diaphragm, thereby fixing the position of the liver anlage (Fig. 3).

The emerging hepatic diverticulum may be subdivided into two regions, cranial and caudal, a distinction rooted in the unequal capacity of these regions to differentiate into hepatocytes. The caudal HD cannot form hepatocytes; rather, these cells are

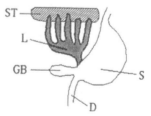

Fig. 3. Invasion of the septum transversum (*ST*) by proliferating liver cords (*L*). *GB* Gall bladder; *D* duodenum; *S* stomach

destined to form the extrahepatic bile duct. Only cells of the cranial HD eventually form hepatocytes, since immunoreactive α-fetoprotein is only detected in this region (Fig. 4a). Based upon the capacity of regions of the hepatic diverticulum to form hepatocytes and other cell types when transplanted to ectopic sites, Shiojiri has proposed a cell lineage schematic for the embryonic liver (Fig. 4b).

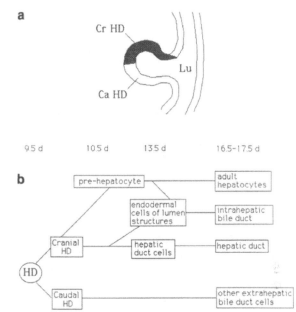

Fig. 4. a Schematic parasaggital section of the hepatic diverticulum, 9.5 d p. c. The hepatic diverticulum bulges from the lumen (*Lu*) of the endodermal tube. The cranial hepatic diverticulum (*Cr HD*) demonstrates intense alpha-fetoprotein synthesis (*shaded*), while the caudal hepatic diverticulum (*Ca HD*) does not. (After Shiojiri 1981). **b** Cell lineage schematic of the embryonic liver. (After Shiojiri 1981)

32.4 Environment Versus Autonomy in Hepatocyte Differentiation

In theory, the achievement of a given differentiated state may proceed by a combination of two antithetical mechanisms. On one extreme, a cell may be totally dependent upon its environment to provide stimuli which drive it from one intermediate developmental step to the next. These environmental stimuli may take the form of soluble growth/differentiation factors, cell contacts with extracellular matrix (cell-matrix), or direct cell contact with other cells (cell-cell). On the other extreme, a cell may differentiate in a totally autonomous manner. It may possess an internal program of development enabling it to progress from one differentiation stage to the next, in a manner totally ignorant of environmental cues. A more appealing middle ground is that the cell's internal program must match that of other cells with which it interacts to achieve multicellular organization (see Garcia-Bellido and Mari-Beffa, Chap. 22, this Vol.).

The role of environmental stimuli in hepatocyte differentiation has already been suggested by the description of specific inductive events occurring during early liver development. However, evidence which we will review argues that both cell-autonomous and environmental processes prominently influence hepatocyte differentiation, cooperating, and indeed synergizing, to drive liver organogenesis. The specific example of the hepatocyte embodies a dilemma central to developmental biology: in the initiation and maintenance of a given differentiated state, what are the relative contributions of external environmental stimuli and internal cell-autonomous programs of development?

32.5 Cell-Autonomous Genetic Cascades in Hepatocyte Differentiation[4]

After transduction of an environmental differentiation signal, perhaps mediated by molecules such as receptor tyrosine kinases and G-proteins, what internal machinery might a hepatocyte possess that would allow it to progress throught a succession of developmental stages without regard to extracellular influences? Such a cell-autonomous developmental juggernaut must include molecules which confer a new pattern of gene expression upon the cell, such as transcription factors (see Errington, Chap. 2, this Vol.). Simplistically, one might envision a master hepatocyte transcription factor which activated the promoters for several other genes which themselves encoded transcription factors. In turn, these newly synthesized transcription factors would then activate transcription from yet another generation of promoters for genes lying lower in the genetic cascade. Eventually, transcription factors would reach the final genes in the hierarchy, the genes encoding the proteins and enzymes of the adult hepatocyte – the proteins and enzymes which actually execute the liver's diverse physiological functions (Fig. 5)! Such a genetic cascade could potenially drive the development of an uncommitted endodermal precursor into an adult hepatocyte, synthesizing such characteristic proteins as albumin, α-1-antitrypsin, and fibrinogen.

Several precedents exist describing developmental pathways regulated by transcription factor cascades. The ability of single genes encoding putative transcription factors. such as MyoD1 or myogenin, to convert fibroblasts into muscle cells suggests that a single master gene, by transcriptionally activating genes lower in the hierarchy, can indeed drive an entire differentiation program to completion (see Pinney and Emerson, Chap. 31, this Vol.). In addition, the mutations responsible for gross developmental abnormalities in Drosophila and other organisms occur with foremost frequency in genes encoding transcription factors, indicating the widespread and extensive use of genetic cascades in the construction of wild-type phenotypes.

The achievement of any differentiated state may be divided into two distinct phases. First, events must occur which lead to the initial acquistion of phenotype, events occurring during *initiation*. After the expression of differentiated traits, however, events must occur which insure the continuous perpetuation of phenotype, events occurring during *maintenance*. How might transcription factor-mediated genetic cascades contribute to these two phases of differentiation in the hepatocyte? The action of transcription factors is most easily appreciated during the initiation phase, as genetic cascades could drive the presumptive liver cell through a series of stages culminating with the first expression of adult hepatocyte markers (Table 1).

However, the same transcription factors responsible for initiation may also confer maintenance of the hepatocyte phenotype. An emerging theme among developmentally important transcription factors, such as *even-skipped*, *ultrabithorax*, and *myoD1*, is positive autoregulation, the ability of a factor to stimulate its own transcription by binding

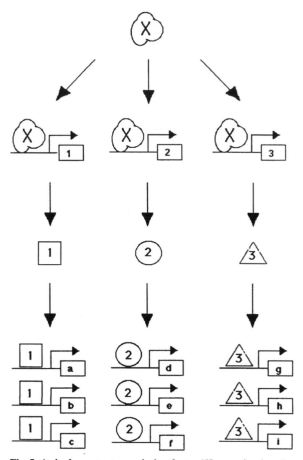

Fig. 5. A single master transcription factor (X) can stimulate the expression of numerous end-target genes (a−i) through the use of genetic cascades employing intermediate, subservient transcription factors (1−3)

its own regulatory regions. If hepatocyte transcription factors underwent positive autoregulation as well, they would promote their own expression in addition to differentiated cell markers such as albumin and α-1-antitrypsin. In this manner, positive autoregulation could irreversibly commit a cell to the hepatic phenotype. Additionally, the hepatocyte transcription factors might positively regulate each other, further reinforcing the differentiated state.

Recently, advances in the analysis of DNA-binding proteins have greatly facilitated the identification of transcription factors participating in the genetic cascades of hepatocyte differentiation. The general approach has been to begin with the final genes in the regulatory pathway — extensively studied liver markers such as albumin and fibrinogen — and then to characterize the unknown transcription factors controlling their expression, thereby advancing "backwards" towards genes acting at earlier, more fundamental stages in the hierarchy. Fittingly, this approach has yielded several novel transcription factors which are expressed at high levels in hepatocytes but not other cell types, and which activate transcription of genes characteristic of the adult liver.

32.5.1 Hepatocyte Nuclear Factor-1α[5-7]

Hepatocyte Nuclear Factor-1α (HNF-1α, also LF-B1,APF) was originally described as a liver protein capable of binding to the promoters of α-1-antitrypsin and α- and β-fibrinogen. Since these genes are predominantly expressed in liver, and since HNF-1α itself is highly enriched in hepatocytes versus other cell types, it was predicted that HNF-1α might be required for the liver-specific transcription of its target genes. Subsequently, HNF-1α was found to bind sequences in the promoters of over 15 additional genes transcribed in the adult liver, including those of albumin and α-fetoprotein, while not binding to promoters of genes not transcribed in liver. A comparison of various HNF-1α promoter-binding sites revealed a consensus 13 base pair inverted repeat motif which was conserved among an extremely large number of liver genes, as well as across species (Table 2).

HNF-1α was further demonstrated to be crucial to the liver-specific expression of its target genes. The addition of a single HNF-1α binding site could confer upon non-liver gene promoters the ability to be selectively activated in liver cells but not in other cell types. Moreover, if HNF-1α sites in various liver gene promoters were deleted, or mutated to a sequence incapable of binding HNF-1α, these genes lost their capacity for expression in liver cells. In addition, if a spleen nuclear extract, which did not contain HNF-1α and did not transcribe liver genes, was then supplemented which purified HNF-1α protein, liver promoters could be demonstrated to undergo dramatic activation in vitro. Finally, transient insertion (transfection) of the HNF-1α gene

Table 2. Regulatory sequences interacting with HNF-1 α or β

Promoter	Species	Sequence	Position
Fibrinogen α	Rat	GGTGATGATTAAC	-47
Fibrinogen β	Rat	GTCAAATATTAAC	-84
Fibrinogen β	Human	ATTAAATATTAAC	-77
Albumin	Mouse	GTTAATGATCTAC	-52
Albumin	Rat	GTTAATGATCTAC	-53
Albumin	Human	GTTAATAATCTAC	-51
Albumin	Xenopus	GTTAATAATTTTC	-53
α-Fetoprotein	Mouse	GTTACTAGTTAAC	-50
α-Fetoprotein	Mouse	GTTAATTATTGGC	-116
α-Fetoprotein	Rat	GTTACTAGTTAAC	-49
α-Fetoprotein	Rat	GTTAATTATTGGC	-115
α-Fetoprotein	Human	GTTACTAGTTAAC	-47
α-Fetoprotein	Human	GATTAATAATTAC	-3400
α-Fetoprotein	Human	GTTAATTATTGGC	-118
Transthyretin	Mouse	GTTACTTATTCTC	-118
Transthyretin	Rat	GTTACTTATTCTC	-116
Transthyretin	Human	GTACTTATTCTC	-116
α1-Antitrypsin	Mouse	GTTAAT-ATTCAT	-63
α1-Antitrypsin	Human	GTTAAT-ATTTCAC	-63
Aldolase B	Human	GTGTTGAATAAAC	-74
Aldolase B	Chicken	AGGGAGAATAAAC	-71
L-pyruvate kinase	Rat	GTTATACTTTAAC	-79
HBV pre S	Human	GTTAATCATTACT	-75
PEPCK	Rat	AACATTCATTAAC	-182
Vitellogenin A2	Xenopus	GGTAATTGTTTAC	-98
α2$-$6 Transferase	Rat	GTTAATGTTTAAC	-66
CYP2E1	Rat	GCTAATAATAAAC	-95

HNF-1 Consensus $G_{24}T_{21}T_{23}A_{23}A_{19}T_{21}N\ A_{22}T_{26}T_{21}A_{11}A_{18}C_{25}$ (27 sequences).

into nonliver cells allowed co-transfected HNF-1α-dependent promoters to be transcribed. These studies strongly argued that the binding of HNF-1α to its cognate site in liver gene promoters had functional consequences — the conferral of a liver-specific expression pattern onto its target genes.

Unexpectedly, HNF-1α transcripts exist at high levels in stomach, intestine, and kidney, in addition to liver. This apparently diverse group of tissues is unified by the peculiar property that they all express genes normally considered "liver-specific". For instance, the kidney contains low levels of mRNA encoding α-1-antitrypsin, albumin, and the α- and β-chains of fibrinogen. The detection of HNF-1α mRNA in these tissues is therefore consistent with their pattern of gene expression, and suggests a causal relationship between the presence of HNF-1α and hepatic gene transcription. An understanding of the basis for the transcription of the HNF-1α gene in these tissues will allow greater insight into their individual organogeneses, as well as reveal whether these organs receive a common developmental stimulus.

The cloning and sequencing of the HNF-1α gene has allowed its functional dissection, and has revealed that the HNF-1α protein contains three major regions: a dimerization domain, a transcription activation domain, and a DNA-binding domain. The HNF-1α DNA-binding domain possesses homology with the homeodomains found in developmentally significant transcription factors of *Drosophila* and other organisms, consistent with HNF-1α's active participation in the process of hepatocyte differentiation. Among homeodomain transcription factors, however, HNF-1α is notable for its ability to associate with itself in solution, or to homodimerize, and to therefore interact with DNA as a bimolecular complex. The significance of this property resides in the potential ability of HNF-1α to interact with *non*-HNF-1α molecules, or to heterodimerize, and in so doing to acquire an altered DNA-binding specificity or trans-activation strength. As a result, heterodimeric HNF-1α could theoretically determine cell fate in organs besides liver by associating with a variety of tissue-specific partners, and subsequently binding to promoters of nonhepatic genes.

What molecules regulate the expression of HNF-1α in liver and other organs? The answer to this question would surely grant insight into antecedent events in hepatocyte differentiation, enabling identification of intermediate molecules linking the inductive events of liver organogenesis to HNF-1α expression. One manner in which this has been approached has been to analyze the promoter for HNF-1α itself, searching for transcription factors acting at a higher level in the HNF-1α genetic cascade. This analysis has revealed the binding of two non-HNF-1α transcription factors in adult liver to the HNF-1α promoter − HNF-4, a liver-enriched member of the steroid hormone receptor superfamily (see Sect. 32.5.6), and HNF-3 molecules, vertebrate homologues of the *Drosophila fork head* protein (see Sect. 32.5.5). Intriguingly, two factors associate with the HNF-1α promoter in hepatocyte cell lines but not primary liver tissue:

AP-1 and an as yet unidentified molecule; this peculiar expression pattern likely reflects either the transformed state of these hepatomas or their entrapment in a primitive developmental stage [9]. Because HNF-1α expression is presumably initated during organogenesis by an inductive event, it is potentially significant that HNF-4, as a putative steroid hormone receptor, and AP-1, as a protein kinase C-responsive transcription factor, both possess the capacity to transduce extracellular signals.

32.5.2 Hepatocyte Nuclear Factor-1β (HNF-1β, also vHNF, vAPF) [8]

The rat Fao hepatoma is extensively studied, well-differentiated cell line which has maintained both hepatocyte gene expression and morphology. M. Weiss isolated a rare, spontaneously arising clone of the Fao line, designated C2, which exhibited an altered morphology, and which had ceased to express several liver-specific genes. In a inspired strategy, Weiss and Deschatrette cultured the C2 mutant line in glucose-free medium, killing the vast majority of the cells, and in rare survivors forcing obligate re-expression of hepatic genes encoding gluconeogenic enzymes such as PEP carboxykinase and pyruvate carboxylase. Through this means, a revertant cell line was obtained, designated Rev7, which had regained liver gene expression in parallel with re-expression of gluconeogenic genes.

The presence of HNF-1α was investigated in Fao, C2, and Rev7 cells to determine whether HNF-1α expression correlated with liver gene expression in these cells. Not surprisingly, cross-linking studies indicated that the differentiated Fao and Rev7 cells, as well as primary hepatocytes possessed wild-type 88 kDa HNF-1α. However, the dedifferentiated C2 cells did not contain HNF-1α, instead expressing a variant 75 kDa HNF-1β (vHNF-1). HNF-1β, and not HNF-1α was also discovered in somatic hybrids between hepatocytes and heterologous cells which extinguished liver gene expression, furthering the correlation between HNF-1β expression and hepatic dedifferentiation. The HNF-1β gene was isolated on the basis of homology to HNF-1α, and HNF-1β antisera reacted with C2 cell HNF-1β but not

HNF-1α in Fao or Rev7 cells, confirming that the 75 and 88 kDa forms of HNF-1 were indeed encoded by distinct genes. Western blotting and cross-linking analyses have demonstrated that HNF-1β protein exists at low levels in normal liver, at amounts much lower than HNF-1α.

Based upon the negative correlation between HNF-1β expression (in Fao/C2/Rev7/somatic hybrids) and the hepatic phenotype, it was expected that HNF-1β would possess weak, if any, trans-activation activity. Surprisingly, however, transfected HNF-1β potently activated transcription from reporter genes bearing HNF-1 binding sites in a variety of cell types. Why, then, the lack of hepatic gene expression in cell lines expressing HNF-1β and not HNF-1α? One possible explanation is that the amount of HNF-1β in dedifferentiated hepatomas – much less than the HNF-1α in normal hepatocytes, differentiated liver cell lines, or transfected cells – is too small to effectively activate hepatic transcription. Alternatively, since promoters generally depend on the simultaneous presence of numerous factors for maximal activity (see Sect. 32.5.8), the ability of HNF-1β to trans-activate its target genes may be compromised by the lack of other hepatic DNA-binding proteins (c.f. HNF-4) in dedifferentiated hepatomas.

The HNF-1α and HNF-1β molecules share great homology in the DNA-binding homeodomain, consistent with both proteins being able to recognize the same target sequences, and also share the myosin-like HNF-1α dimerization domain. However, while HNF-1α and HNF-1β readily mix and heterodimerize after individual translation in vitro or after co-transfection into lymphocytes, native HNF-1α from liver cells does not heterodimerize when presented with HNF-1β. As a result of these studies, it is evident that co-expression of both α and β HNF-1 does not necessarily result in heterodimer formation, and suggests that heterodimerization can be regulated in a tissue-specific fashion.

HNF-1β mRNA exhibits a broader tissue distribution than HNF-1α: both transcripts are present in liver, intestine, stomach, and kidney, but HNF-1β mRNA is additionally expressed in lung and ovary. De Simone and Cortese have noted that all HNF-1β-expressing tissue contain an epithelial layer, and have confirmed this as the site of HNF-1β expres-

sion by in situ hybridization studies. Consequently, it has been proposed that HNF-1β participates in an "epithelial" differentiation program, as does the *Drosophila* gene *crumbs*.

In mice the appearance of HNF-1β transcripts (6.5 days p.c.) precedes that of HNF-1α transcripts (9.5 days p.c.), the latter occurring concomitantly with the onset of liver organogenesis. This same temporal pattern of appearance of DNA-binding activity, β preceding α, has been observed in F9 teratocarcinoma cells induced by retinoic acid to differentiate along an α-fetoprotein-expressing lineage. Taken together, these similar observations indicate that HNF-1β participates in developmental decision-making at an earlier stage than HNF-1α, and that these two molecules possess distinct, nonredundant embryologic functions, despite their many structural and biochemical similarities.

The expression of HNF-1β protein can be strongly repressed, apparently at a post-transcriptional level. Despite abundant amounts of HNF-1β mRNA in adult liver, vanishingly little HNF-1β DNA-binding activity is detectable, suggesting an extremely effective post-transcriptional inhibition. This negative regulation is most likely mediated by extracellular signals, since when hepatocytes are removed from the organ environment and placed in tissue culture, they reactivate high-level expression of HNF-1β DNA-binding activity. Consequently, adult hepatocytes, which normally express predominantly HNF-1α, appear to retain a latent capacity to express HNF-1β that is actively repressed by extracellular signals.

32.5.3 CAAT/Enhancer Binding Protein[9]

The CAAT/Enhancer Binding Protein (C/EBP) was originally identified as a liver protein capable of binding to the CCAAT homologies of several viral promoters, as well as to the enhancer cores of the SV40, MSV, and polyoma tumor viruses. Why might viral enhancers possess binding sites for a cellular DNA-binding protein? Frequently, viral genomes have become evolutionarily streamlined through the use of overlapping and cistronic genes and alternative splicing patterns in an effort to encode the greatest number of proteins with a mini-

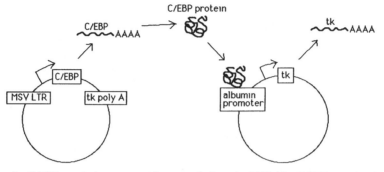

Fig. 6. Co-transfection of a C/EBP expression vector and a reporter gene construct illustrates a general strategy for assay of transcription factor activity. The C/EBP gene is driven by a strong, constitutive promoter (MSV long terminal repeat), allowing high level expression of C/EBP mRNA. After the translation of mRNA into C/EBP protein, the C/EBP transcription factor can trans-activate the thymidine kinase (*tk*) reporter gene by binding to a site in the upstream albumin promoter. The presence of tk mRNA thus represents the ability of C/EBP to trans-activate the albumin promoter. (After Friedman et al. 1989)

mally sized genome. Conceivably, during evolution the SV40, MSV and polyoma viruses might have successfully eliminated the need to encode a transcription factor capable of transcribing viral genes, by incorporating bindes sites for a *cellular* transcription factor into their enhancers. In this parsimonious manner, such viruses could save precious genome size by co-opting for their own purposes a transcription factor already present in host cells and already performing a physiological function.

Apparently subverted for the transcription of viral genes, what might be the normal, physiological function of the C/EBP protein? Its capacity to bind enhancers, and its relatively restricted expression in liver and fat suggested that it could represent a transcription factor involved in hepatocyte differentiation. Consistent with such a role, the albumin promoter has been described to possess a high-affinity C/EBP binding site whose deletion or mutation greatly reduces albumin transcription in liver cells. In addition, the introduction of a C/EBP expression vector into hepatoma cell lines markedly stimulates the albumin promoter contained on a co-transfected reporter plasmid, directly demonstrating its transcription factor activity (Fig. 6). Other C/EBP binding sites have been described in the albumin enhancer, as well as in control regions of the transthyretin and α-1-antotrypsin genes, suggesting a prominent role for C/EBP during hepatocyte differentiation.

The molecular cloning of the C/EBP gene, achieved by McKnight and colleagues, has revealed several noteworthy features. The C/EBP gene is located on mouse chromosome 7, confirming that C/EBP is of cellular, not viral origin. Furthermore, like HNF-1α and -1β, C/EBP does not exhibit an absolutely liver-specific expression pattern. Instead, C/EBP mRNA is detectable in a heterogeneous group of tissues including liver, adipose tissue, lung, intestine, adrenal glands, and placenta, apparently unified by a high level of lipid synthesis or metabolism. Darnell and associates have rigorously shown that the major regulation of C/EBP mRNA levels occurs at the level of transcription initiation, rather than mRNA stability.

In addition, an analysis of the C/EBP nucleotide sequence unveiled a fascinating homology with the cellular and viral *myc*, *fos*, and *jun* proteins. This homology predicted an amphipathic α-helix in which five invariant leucine residues were each separated by six nonconserved amino acids. The regular periodicity of the leucine residues, occurring at every seventh position, indicated that the leucines all appeared on a single face of the putative α-helix, creating a highly hydrophobic surface ideal for protein-protein interactions. McKnight subsequently proposed that the leucines in two adjacent C/EBP proteins could interact, zipper-like, thereby allowing the molecules to interact, and to dimerize (Fig. 7). The importance of the "leucine zipper" as a dimerization motif, first recognized in C/EBP, has been extended for a variety of other molecules in which it occurs, including members of the *fos* and *jun* families.

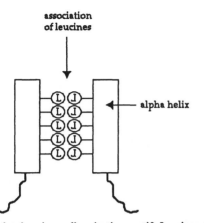

Fig. 7. The leucine zipper dimerization motif. Leucines on two adjacent alpha-helices undergo hydrophobic interaction, allowing dimerization of C/EBP molecules

At the same time that C/EBP promotes development, it apparently simultaneously inhibits growth – an additional characteristic of terminal differentiation. The inverse correlation between C/EBP levels and cell growth was initially suggested by the prominent C/EBP expression in growth-arrested, fully differentiated hepatocytes from mouse or rat liver, in distinction to the very weak C/EBP expression in constantly proliferating hepatoma cell lines. Moreover, attempts to stably express C/EBP in transfected cell lines were uniformly unsuccessful, indicating a potential lethality. This was circumvented by the construction of a chimeric estrogen receptor-C/EBP molecule which associated with C/EBP binding sites but which activated transcription in an estrogen-dependent manner. The stable expression of this fusion protein produced hormone-dependent growth arrest and acceleration of differentiation, revealing a dual function.

32.5.4 DBP and LAP, Related Members of the C/EBP Family[10]

Unexpectedly, the C/EBP binding sites in the albumin promoter are recognized not only by C/EBP, but also by two related transcription factors present in liver cells. DBP (D-Binding Protein) and LAP (Liver-enriched Activator Protein). Sequence analysis of the DBP and LAP genes, isolated in the lab-

oratory of U. Schibler, reveals a strong homology with C/EBP in the basic DNA recognition domain, consistent with their common binding specificity. LAP, moreover, shares the C/EBP leucine zipper dimerization motif, and consequently forms heterodimers with C/EBP in vitro. The existence of three distinct factors recognizing the same DNA sequence may, instead of representing a functional redundancy, perhaps be indicative of complex regulation of albumin transcription. In fact, Schibler and colleagues have demonstrated that DBP and LAP transcripts and protein follow circadian expression patterns, and that this is likely responsible for the circadian variation in albumin gene transcription. The contribution of these factors to the initial events of liver organogenesis is unclear, however, as their first expression occurs post-natally.

Whereas DBP and LAP proteins are liver-specific, their transcripts are detectable in brain, lung, and testis among other tissues. The ability of DBP and LAP mRNA to exist in the absence of protein product indicates that these genes are regulated post-transcriptionally, perhaps in response to physiological, metabolic, or developmental stimuli. The 5′ untranslated region of the DBP gene contains a short open reading frame with homology to a 5′ leader which confers translational regulation on the yeast GCN2 mRNA, thus suggesting a potential mechanism for DBP post-transcriptional regulation.

32.5.5 Hepatocyte Nuclear Factor-3αβγ (HNF-3αβγ)[11]

The Hepatocyte Nuclear Factor-3α (HNF-3α) transcription factor was originally identified by its association with the promoters of the hepatic genes transthyretin and α-1-antitrypsin. An analysis of the sequence of the HNF-3α gene revealed a remarkable 125 amino homology between the HNF-3α DNA-binding domain and functionally indispensible regions of the *Drosophila* developmental protein *fork head*. Moreover, HNF-3α contains two additional areas of *fork head* homology, indicating that HNF-3α and *fork head* are true evolutionary homologues (Fig. 8). Compelling evidence for this theory is provided by the phenotypes of *fork head*

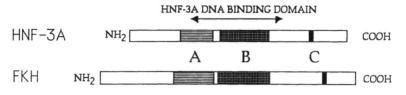

Fig. 8. Homology between HNF-3A and *fork head*. (After Weigel and Jackle 1990)

mutant flies, which exhibit lesions in endodermal development (midgut, hindgut, yolk sac, and salivary gland), as well as in morphogenesis of the Malphighian tubules – the *Drosophila* "kidney". It should prove interesting to determine whether molecules which regulate *fork head* in *Drosophila*, such as the gap gene *huckebein*, are conserved in form and function in vertebrates.

HNF-3α binding sites are occupied by not one, but three liver-enriched nuclear proteins, as demonstrated by gel mobility shift assays. In agreement with these results, two additional, highly related genes sharing the *fork head* motif were identified through low-stringency screening – HNF-3β and HNF-3γ. The transcripts of the HNF-3 family members are all expressed in liver, with differential expression in small intestine, lung, and testis. In contrast to HNF-1α and -1β, which do not exhibit differences in binding site preference, HNF-3α, -3β, and -3γ differ greatly in affinity for various sequences, strongly implying that their functions are nonredundant.

32.5.6 Hepatocyte Nuclear Factor-4 (HNF-4)[12]

Hepatocyte Nuclear Factor-4 (HNF-4) is a liver-enriched transcription factor which stimulates the hepatic expression of the transthyretin, apolipoprotein CIII, and α-1-antitrypsin genes. As typical of hepatic trans-activators. HNF-4 mRNA is detectable in liver, kidney, and intestine, but not in tissues which do not express liver genes. Sequence analysis of the HNF-4 gene indicated that HNF-4 is a member of the steroid hormone receptor superfamily, unified by nuclear translocation upon ligand binding, by ligand-dependent activation of transcription, and by a "zinc-finger" DNA-binding domain in which a zinc atom is tetrahedrally coordinated by four histidine and/or cysteine amino acids. The

possibility therefore exists that HNF-4 is activated by an unidentified extracellular ligand participating in developmental and/or metabolic processes. However, transient transfection of the HNF-4 cDNA into heterologous cells such as HeLa or Jurkat T-lymphocytes indicates that HNF-4 does not exhibit cell-type-specific requirements for function; perhaps HNF-4 is constitutively active or its ligand is ubiquitous or acts to repress trans-activation.

A striking similarity exists between the expression patterns of HNF-4 and HNF-1α. Indeed, both mRNAs are found in the same adult tissues, and both DNA-binding activities are absent in dedifferentiated hepatomas such as C2, H5, and R8, as well as in somatic hybrids between hepatocytes and heterologous cells which have extinguished liver gene expression. A high-affinity HNF-4 binding site is present in the HNF-1α promoter, and HNF-4 can potently activate a basal promoter via this site in transient transfection assays. Consequently, HNF-4 is a strong candidate to directly regulate HNF-1α expression in a genetic hierarchy. However, it is equally plausible that both molecules may be initially activated during development by a common, yet to be identified, higher-order stimulus and that HNF-4 acts merely to maintain, rather than initiate, HNF-1α expression.

32.5.7 Combinatorial Models of Liver Determination

The existence of multiple transcription factors relevant to hepatocyte gene expression points to the existence of multiple transcription factor-mediated genetic pathways. Conceivably these genetic cascades might all be subservient to a master transcription factor whose appearance would comprise the *single* initiating event of liver organogenesis. A single stimulus "X" would induce the master hepatocyte transcription factor only in the region of the embryo destined to form liver, and would be itself sufficient to confer the hepatocyte phenotype via several ensuing genetic cascades utilizing different liver transcription factors (e.g., HNF-1α, C/EBP, HNF-4). The spatial restriction of stimulus X would define the embryonic area capable of forming liver, by limiting the appearance of hepatic transcription factors, and the expression of genes regulated by these factors. (Fig. 9).

However, virtually none of the liver transcription factors nor their target genes are uniquely expressed in liver, as might be predicted by the above model. Rather, the distinct, yet overlapping expression patterns of these proteins suggest a combinatorial model, in which *several* stimuli "X", "Y", and "Z" might all be required as *multiple* initiating events of liver organogenesis. These stimuli could each occur in distinct, yet overlapping regions of the embryo, with the regions of their occurrence intersecting only in the embryo's presumptive liver region. Contrasting with the previous model, in which a single stimulus X could trigger expression of all hepatic transcription factors and their respective target genes, now each stimulus X, Y, and Z could only induce a *partial* subset of hepatic trans-activators. Accordingly, the embryonic area fated to form liver would be demarcated by the overlapping occurrence of liver-promoting stimuli, and therefore also by the subsequent overlapping expression of tran-

scription factors and their respective target genes. If the essence of being a hepatocyte, the essence of "liver-ness", is defined as the possession of the entire compliment of liver genes, then multiple stimuli could prevent the development of "liver-ness" in undesirable regions by acting in a combinatorial, cooperative fashion, dictating that the full roster of hepatic transcription factors and hepatic genes only occur in a defined region where all liver-inducing signals overlap (Fig. 9).

32.5.8 Combinatorial Models of Transcription Factor Action[13]

Moreover, transcription factors themselves may also act combinatorially to impose an additional level of spatial restriction on liver organogenesis. Only rarely is a single transcription factor sufficient to induce maximal transcription of a given gene. Rather, eukaryotic promoters generally contain DNA-binding sites for multiple transcription factors, with the simultaneous, interactive presence of all factors being necessary for maximal transcription. The albumin promoter, for example, contains six binding sites, denoted A-F, each interacting with a cognate transcription factor. Deletion or mutation of *any* of sites A-F inhibits binding of the respective factor, and markedly impairs albumin transcription (Fig. 10). By inference, the complete integrity of sites A-F is essential to the binding of

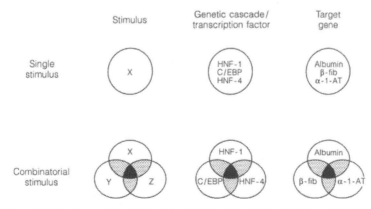

Fig. 9. Single versus combinatorial stimulus models of liver determination. *Top* single stimulus model. *Bottom* combinatorial stimulus model. Note that the *blsck shaded area* (presumptive liver region) is the only area to express all the transcription factors and end-target genes. See text for further details

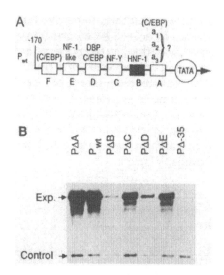

Fig. 10 A,B. The albumin promoter illustrates the principle of combinatorial transcription factor action. **A** Binding sites for DNA-binding proteins in the albumin promoter. (Maire et al. 1989). **B** In vitro transcription analysis demonstrating that the mutation of albumin promoter sites A−E greatly impairs transcription (*Exp.*) relative to the wild-type promoter P_{wt}. The constitutive expression of the adenovirus major-late promoter has been included as an internal control. (After Maire et al. 1989)

their respective transcription factors, to the subsequent interaction of these factors in forming an active transcription complex, and to the ensuing high level expression of albumin. The combinatorial, synergistic action of several transcription factors in stimulating albumin expression is therefore manifested in the assembly of an active multiprotein complex on the albumin promoter. Obviously, such combinatorial action creates an additional level of spatial restriction on liver organogenesis, as hepatocyte genes such as albumin will only be maximally expressed in regions where the correct combination of transcriptional activators exist.

32.5.9 Multi-Tiered Regulation of Liver Transcription Factors[6−12,14]

In order for a given transcription factor to trans-activate its target genes, and to therefore contribute to specifying the identity of a certain tissue, its gene must be transcribed, its mRNA must be translated, and its protein product must undergo appropriate covalent modification and noncovalent association with accessory proteins. All of these essential steps, of course, constitute potential levels of regulation which may be exploited to tightly regulate the expression of liver transcription factors and to thus prevent inappropriate organogenesis. The existence of transcriptional control of hepatic trans-activators is strongly suggested by the restricted expression patterns of their mRNAs and/or by nuclear run-on experiments. Furthermore, strong evidence exists for post-transcriptional regulation of DBP, LAP, and HNF-1β, and HNF-1α is additionally subject to post-translational glycosylation and phosphorylation. Certainly, the developmental regulation of the liver transcription factors may prove to be quite complex.

32.6 Environmental Influences on Hepatocyte Differentiation

Transcription factors such as HNF-1α and C/EBP mediate cell-autonomous processes potentially involving both the initiation and maintenance of the hepatocyte phenotype. However, a large body of evidence strongly suggests that environmental influences play an equally significant role in hepatocyte development. These environmental factors may be broadly grouped into three main categories: circulating differentiation factors, cell-cell interactions, and cell-matrix interactions. In general, the precise molecular events elicited by environmental stimuli have not been as thoroughly characterized as for the transcription factors, rendering this an area of great current interest.

When adult, fully differentiated hepatocytes are removed from their natural milieu within the liver organ and grown in primary culture, they rapidly *de*differentiate, and cease to express liver-specific genes. This simple experiment demonstrates that the hepatocyte phenotype is not cell-autonomous in nature, and that constant environmental stimulation of liver cells, as occurs within the intact liver organ, is required to keep hepatocytes in a fully mature form. Consequently, effects of specific environmental factors on hepatocyte development may be tested by adding various substances (soluble factors, cells, matrix) to primary hepatocyte cultures, and examining their ability to preserve heaptocyte morphology, physiology, and gene expression. Assayed in this manner, a positive effect would indicate that a given variable participates in *maintenance* of the hepatocyte phenotype (Fig. 11). Moreover, if a factor can be added to a dedifferentiated primary hepatocyte culture with the subsequent restoration of liver functions, then a role in the *initiation* of the hepatocyte phenotype would be inferred.

32.6.1 Circulating Differentiation Factors[15]

By virtue of the liver's dual efferent circulation, receiving input from both the portal vein and the hepatic artery, the hepatocyte is richly perfused by blood abundant in soluble factors such as ions, steroid hormones, and polypeptide factors. These molecules may act to maintain the hepatocyte phenotype in vivo, as suggested by the ability of tissue culture medium containing rigorously defined amounts of various hormones and factors to help preserve the differentiation of primary hepatocyte cultures.

Unfortunately, no mammalian factor has yet been described capable of driving hepatocyte differentiation by inducing hepatic transcription factors and their respective target genes. Instead, the two circulating factors which do alter liver gene expression, glucocorticoids and interleukin-6(IL-6), act upon *already mature* liver cells, transiently increasing the expression of certain subsets of hepatocyte genes in response to metabolic conditions such as stress, starvation, and inflammation. The glucocorticoids and IL-6 exert their effects by interacting with high affinity receptors displayed by hepatocytes, generating a signal which is transduced into the necleus, and which stimulates transcription of selected liver-specific genes.

32.6.1.1 Glucocorticoids[16]

The earliest characterized example of circulating factor action on hepatocytes, and one of the first demonstrations of hormonal enzyme regulation in

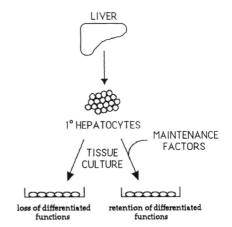

Fig. 11. An assay for environmental influences based upon the ability of a given factor to sustain the differentiated functions of primary hepatocyte cultures

a mammalian system, is represented by the ability of glucocorticoids to induce expression of gluconeogenic enzymes. Glucocorticoids markedly induce the transcription of gluconeogenic genes such as tyrosine aminotransferase (TAT), pyruvate carboxylase, and phosphoenolpyruvate carboxykinase (PEPCK), thereby maintaing glucose homeostasis during periods of exertion or starvation. Initially released by the adrenal cortex, glucocorticoids associate with high affinity receptors present in the cytoplasm of hepatocytes, as well as many other cell types. Within the hepatocyte, in a manner typical of steroid hormone action, the cytoplasmic hormone:receptor complex then translocates to the nucleus, binding promoter sequences containing consensus glucocorticoid response elements (GRE). When associated with its cognate GRE, the glucocorticoid receptor in the agonist-occupied form induces transcription of genes encoding various gluconeogenic enzymes, leading to an increased level of these proteins in the hepatocyte, and to the enhanced synthesis of glucose from metabolic intermediates.

32.6.1.2 Interleukin-6[17]

Hepatocyte gene expression may also be modulated by the polypeptide lymphokine interleukin-6 (IL-6). Elaborated by cells of the monocyte/macrophage lineage in response to physiological insults such as infection, trauma, or neoplasia, IL-6 induces a dramatic increase in transcription of a subset of hepatocyte genes encoding what are known as the acute phase proteins. These proteins, including α-2-macroglobulin, α-1-acid glycoprotein, fibrinogen, and α-1-antitrypsin, are secreted by the liver into the bloodstream, and participate in the systemic response directed against the tissue injury prominent in trauma, infection, and many disease processes.

The molecular action of IL-6 presumably commences with the interaction of IL-6 with a high-affinity plasma membrane receptor. Subsequently, through a signal transduction cascade possibly involving protein kinase C, the transcriptional pattern of the hepatocyte is drastically reorganized, in favor of the acute phase genes, and at the expense of genes such as albumin and transthyretin. The acute phase gene promoter sequences conferring positive transcriptional regulation by IL-6 have been mapped in a variety of genes, and a hexanucleotide motif highly conserved among several IL-6 response elements has been described (Table 3). This CTGGGA motif, interestingly, is identical to that recognized by the protein kinase C-responsive transcription factor NF-κB. The involvement of NF-κB in IL-6 signal transduction is uncertain, however, as several groups have recently characterized and isolated cDNA clones encoding an IL-6-inducible, C/EBP-like molecule which activates transcription via IL-6 response elements in the haptoglobin, hemopexin, and C-reactive protein promoters.

32.6.2 Cell-Cell Interactions [2, 18]

The realization that cell-cell interactions play crucial roles in embryogenesis is as old as the discipline of embryology itself. Classically, Spemann first demonstrated in the 1930s the capacity of a transplanted blastopore lip to induce a second developmental axis, spawning the modern science of embryology. With the advent of molecular biology, the detailed elaboration of cell-cell interactions has become the province of the molecular embryologist, with a modern paradigm represented by the elegant descriptions by Rubin, Benzer, and others of cell-cell interactions required for morphogenesis of the *Drosophila* retina. Analysis of mutations in *Drosophila* retinal development has revealed lesions in loci encoding the *sevenless* and *ellipse* transmembrane tyrosine kinases, as well as the putative tyrosine kinase ligand *bride of sevenless*.

In general, cell-cell interactions may assume various forms. For instance, cell-cell contact may involve highly specific interaction between proteins on adjacent cell membranes, as exemplified by the T-cell receptor: major histocompatibility complex interaction, or the postulated association between the *sevenless* and *bride of sevenless* proteins. In these particular cases, protein:protein (ligand:receptor) contact underlies the apparent cell-cell contact, initiating a signal which is transduced through the plasma membrane and into the nucleus, with accompanying changes in gene transcription and phenotype of the receptor-bearing cell. In a slightly different example, one cell may secrete a soluble factor which only binds receptors on immediately

Table 3. Interleukin-6 response elements

Gene	Position[a]	Sequence
Rat α2-macroglobulin	−169	TCCTTCTGGGAATTCTG
Rat α1-acid glycoprotein	−5221	GGCTTCTGGGAAAAACT
Rat α-fibrinogen	−134	AATTTCTGGGATGCCGT
Rat β-fibrinogen	−154	GCTTGCTGGGAAGATGT
Rat γ-fibrinogen	−156	AAAATCTGGGAATCCCT
Rat kininogen	−85	CCCAGCTGGGTACCTGC
Human haptoglobin	−130	TGTTACTGGAAAAGATA
Human C-reactive protein	−81	CAATGTTGGAAAATTAT
Hexanucleotide consensus:		CTGGGA

[a] Position is the number of nucleotide 3′ upstream of +1 (transcription start).

neighboring cells, thereby exhibiting a paracrine mode of action. A less specific type of cell-cell interaction involves cell adhesion, the binding of cells to each other to form a physical unit, tissue type, or organ. This type of cellular association can utilize cytoskeletal elements in adjacent cells, such as desmin, actin, and intermediate filaments, which form the so-called tight junctions and gap junctions, or generalized adhesion molecules such as L-CAM (Liver cell adhesion molecule) which form homtypic dimers to bridge cells. Finally, what is apparently cell-cell interaction may actually be mediated by proteins situated between cells, comprising an intervening basement membrane, or extracellular matrix.

The cell-cell interactions important in initiating hepatocyte differentiation have been only vaguely characterized with respect to precise molecular mechanisms. The two inductive events by which mesodermal cells determine which endodermal cells are fated to form liver suggest a paramount role for cell-cell contacts in *initiating* the hepatocyte phenotype. However, it is unknown whether ligand: receptor, paracrine factor: receptor, or cell: matrix contacts underlie these inductive events which are so crucial to the initial phases of liver organogenesis. Interestingly, Le Dourain has demonstrated that the interposition of a 1 μm pore Millipore filter between endodermal and mesodermal layers permits hepatic induction, while a 0.6 μm filter does not; yet, this does not distinguish between the effects of minute cell processes, secreted factors, or extracellular matrix.

Various cell-cell interactions of equally vague character have been implicated as participating in *maintenance* of the hepatocyte phenotype. The culture of primary hepatocytes as a monolayer rapidly leads to an 80–99% decrease in the levels of steady state mRNA encoding liver genes. However, Darnell and colleagues have shown that liver-specific transcription rates and steady state mRNA levels may be maintained if primary hepatocytes are cultured as a slice of whole liver thin enough to allow nutrient diffusion, thus arguing for a cell-cell contact of undefined nature acting to maintain the hepatocyte phenotype. In addition, treatment of cultured slices with circulating factors or serum does not further boost liver-specific transcription,

emphasizing the primacy of cell-cell interaction in eliciting this effect.

The nature of this cell-cell interaction might be homotypic, only involving hepatocytes, or heterotypic, involving hepatocytes and another cell type, possibilities which are not distinguishable by culture of whole liver slices in situ. Suggesting that heterotypic interactions are at least capable of maintaining liver functions, the co-cultivation of cells of the biliary apparatus with primary hepatocyte monolayers can partially preserve albumin secretion and steady state mRNA levels. On the other hand, the ability of cultured primary hepatocytes to retain liver markers appears to increase with cell density, indicating a role for homotypic interactions.

32.6.3 Cell-Matrix Interactions [19]

Most epithelial cells possess a fundamental asymmetry by displaying both an apical surface, through which substances are secreted and absorbed, and a basolateral surface, through which the cell is anchored to an underlying basement membrane/extracellular matrix. However, while the liver performs diverse secretory functions, its classification as a true epithelium has been problematic since it demonstrates no morphological polarity, displaying instead two identical surfaces which face blood sinusoids on either side. Thus, traveling from the hepatocyte toward the sinusoids in either direction, one first traverses the space of Disse, and then a discontinuous endothelial layer, finally encountering blood (Fig. 12). Apparently empty by ultrastructural criteria, it is the space of Disse which nevertheless constitutes the hepatocyte equivalent of basement membrane. Indeed, the Disse space has been recently demonstrated to contain archetypal basement membrane components such as laminin, type-IV collagen, and heparin sulfate proteoglycan. Moreover, instead of merely representing a passive structural support, the extracellular matrix proteins of the Disse space may actively promote hepatocyte differentiation. Many of the effects of cell-cell interactions in maintaining liver phenotype may be easily elicited by extracellular

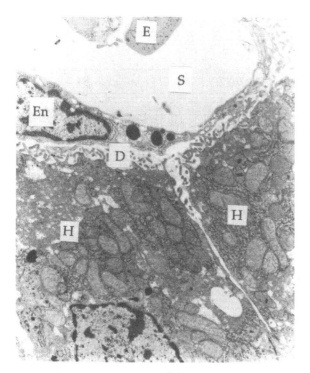

Fig. 12. Transmission electron micrograph demonstrating relationship of hepatocytes (*H*) to Disse space (*D*), endothelial cells (*En*) and blood sinusoids (*S*). An immature erythrocyte (*E*) is present in the sinusoid. The Disse space contains extracellular matrix components believed to promote hepatocyte differentiation. (Courtesy of Lynne Mercer)

matrix components, when these are employed as a substratum for primary hepatocyte culture.

An impressive preservation of the liver phenotype, extending over 3 weeks, may be obtained by culturing primary hepatocytes on a combination of matrix proteins known as EHS gel. Extracted from the Engelbreth-Holm-Swarm mouse tumor, the EHS gel contains laminin, type-IV collagen, heparin sulfate proteoglycan, as well as other minor proteins such as growth factors and cytokines. The ability of matrix to maintain liver functions occurs at the transcriptional level, as EHS gel potently supports albumin transcriptional rates, steady-state mRNA levels and secretion in primary cultures. In addition, EHS gel effects may be obtained with an EHS fraction highly enriched in laminin, suggesting that the active component may be either laminin or a laminin-associated protein.

Several more indirect lines of evidence support a role for extracellular matrix in hepatocyte differentiation. The deleterious effects of serum on the differentiated phenotype of primary hepatocytes can be negated by culture on matrix substratum. In addition, when liver chunks are disaggregated with EDTA, the individual hepatocytes detach from a still intact matrix. However, with the addition of serum in place of EDTA, the detached hepatocytes quickly reassociate with the matrix, reforming an organ structure and resuming hepatocyte gene expression.

The effects of matrix may, of course, extend as well to the initiation of the hepatocyte phenotype. Certainly, the inductive events of liver organogenesis, ascribed very broadly to cell-cell contacts, may be in reality mediated heterotypically by the mesodermal secretion of matrix which could in turn stimulate an endodermal pre-hepatocyte. The extracellular matrix retains growth factors such as FGF und TGF-β which may convey this effect. Moreover, if a primary hepatocyte culture, having undergone dedifferentiation by culture without a substratum, is then cultured with EHS gel, a spectacular recovery of albumin transcriptional rate, steady-state mRNA levels, and secretion may be observed. The ability of EHS gel to restore albumin expression to such dedifferentiated cells strongly suggests that matrix components may actively induce liver-specific functions during organogenesis.

32.7 Outlook

The hepatic transcription factors influence liver organogenesis on several levels. Without these potent trans-activators, liver-specific gene transcription would be greatly impaired, resulting in the absence of the proteins which execute the liver's daily physiological functions, as well as a decidedly nonhepatic phenotype. However, while the tissue-restricted expression of these factors is certainly not unexpected, it raises a fascinating question: what confers tissue specificity upon the transcription factors? Since several, if not all, of the hepatocyte transcription factors are regulated transcriptionally (Table 4), a logical approach will be to analyze the promoters of the transcription factors themselves,

Table 4. Hepatocyte transcription factors

Transcription factor	DNA-binding motif	Transcriptional regulation	Post-transcriptional regulation	Dimerization
HNF-1	Homeodomain	+	+	+
C/EBP	Basic	+	ND	+
DBP	Basic	−	+	−
LAP	Basic	±	+	+
HNF-3	Forkhead	+	ND	ND
HNF-4	Zinc finger	+	ND	+

ND, Not determined.

to determine the "transcription factors which regulate the transcription factors", enabling their liver-specific expression. In addition, as the tissue specificity of certain hepatic transcription factors is determined post-transcriptionally (Table 4), it should be feasible to define for each factor whether "post-transcriptional regulation" involves more precisely mRNA stability, translational control, or post-translational modification/nuclear translocation, and subsequently to identify the proteins mediating this regulation.

Given the inductive events which initiate liver organogenesis, as well as the many extracellular cues implicated in maintaining the hepatic phenotype, it is obvious that a complete description of hepatic development will not be obtained within the hepatocyte, and that the molecular basis for environmentally significant influences must be determined. What extracellular proteins convey circulating, cell-cell and cell-matrix differentiation signals to the hepatocyte? What receptors do hepatocytes display to receive these signals? Which non-hepatocyte cells elaborate these environmental cues during liver induction and during maintenance of phenotype? As these environmental signals potently stimulate hepatic gene expression, it seems extremely likely that they promote the liver-specific expression of hepatocyte transcription factors at both transcriptional and post-transcriptional levels. The greatest challenge to those studying hepatocyte development may well prove to be the unraveling of the signal transduction cascades linking environmental inductive signals to liver-specific gene transcription.

32.8 Summary

Great recent progress has been made in identifying transcription factors permitting the liver-specific expression of adult liver markers such as albumin, α-1-antitrypsin, and fibrinogen. These hepatic trans-activators (HNF-1$\alpha\beta$, C/EBP, DBP, LAP, HNF-3$\alpha\beta\gamma$, and HNF-4) function combinatorially and synergistically to stimulate gene transcription, and are themselves regulated at transcriptional and post-transcriptional levels. In addition, far from possessing an exclusively cell-autonomous developmental program, driven uncompromisingly by transcription factor cascades, the hepatocyte receives extracellular signals which initiate and maintain the liver phenotype. Such environmental signals, in the form of circulating, cell-cell and cell-matrix influences, are believed to initiate the nuclear events of liver organogenesis through the promotion of hepatocyte-specific transcription.

References

[1] Hogan B, Constantini F, Lacy E (1986) Manipulating the mouse embryo. Cold Spring Harbor Laboratory Press, New York, pp 19–70

[2] Greengard O, Federman M, Knox WE (1972) Cytophotometry of developing rat liver and its application to enzymic differentiation. J Cell Biol 52:261–262

Houssaint E (1980) Differentiation of the mouse hepatic primordium. I. An analysis of tissue interactions in hepatocyte development. Cell Differ 9:269–279

Le Dourain NM (1975) An experimental analysis of liver development. Med Biol 53:427–455

[3] Shiojiri N (1981) Enzymo- and immunocytochemical analyses of the differentiation of liver cells in the prenatal mouse. J Embryol Exp Morphol 62:139–152

[4] Bienz M, Tremml G (1988) Domain of ultrabithorax expression in *Drosophila* visceral mesoderm from autoregulation and exclusion. Nature 333:576–578

Davis RL, Weintraub H, Lassar AB (1987) Expression of a single transfected cDNA converts fibroblasts to myoblasts. Cell 51:987–1000

Gehring WJ (1987) Homeoboxes in the study of development. Science 236:1245–1252

Hiromi Y, Gehring WJ (1987) Regulation and function of the *Drosophila* segmentation gene fushi tarazu. Cell 50:963–974

Scott MP, Tamkun JW, Hartzell GW III (1989) The structure and function of the homeodomain. BBA Rev Cancer 989:25–49

Weintraub H, Davis R, Tapscott S, Thayer M, Krause M, Benezra R, Blackwell TK, Turner D, Rupp R, Hollenberg S, Zhuang Y, Lassar A (1991) The myoD gene family: nodal point during specification of the muscle cell lineage. Science 251:761–766

Wright WE, Sassoon DA, Lin VK (1989) Myogenin, a factor regulating myogenesis, has a domain homologous to MyoD. Cell 56:607–617

[5] Courtois G, Morgan JG, Campbell LA, Fourel G, Crabtree GR (1987) Interaction of a liver-specific factor with the fibrinogen and α-1-antitrypsin promoters. Science 238:688–692

Courtois G, Baumheuter S, Crabtree GR (1988) Purified hepatocyte nuclear factor 1 interacts with a family of liver-specific promoters. Proc Natl Acad Sci USA 85:7937–7941

Baumheuter S, Courtois G, Crabtree GR (1989) The role of HNF-1 in liver-specific gene expression. Ann N Y Acad Sci 69:272–279

[6] Baumheuter S, Mendel DB, Conley PB, Kuo CJ, Turk C, Graves MK, Edwards CA, Courtois G, Crabtree GR (1990) HNF-1 shares three sequence motifs with the POU domain proteins and is identical to LF-B1 and APF. Genes Dev 4:372–379

Frain M, Swart G, Monaci P, Nicosia A, Stampfli S, Frank R, Cortese R (1989) The liver-specific transcription factor LF-B1 contains a highly diverged homeobox DNA-binding domain. Cell 59:145–157

Kuo CJ, Conley PB, Hsieh C-L, Francke U, Crabtree GR (1990) Molecular cloning, functional expression and chromosomal localization of mouse hepatocyte nuclear factor 1. Proc Natl Acad Sci USA 87:9838–9842

Lichtsteiner S, Schibler U (1989) A glycosylated liver-specific transcription factor stimulates transcription of the albumin gene. Cell 57:1179–1187

[7] Kuo CJ, Conley PB, Chen L, Sladek F, Darnell J, Crabtree GR (1992) A transcriptional hierarchy involved in mammalian cell-type specification. Nature 355:457–461

Nicosia A, Monaci P, Tomei L, De Francesco R, Nuzzo M,

Stunnenberg H, Cortese R (1990) A myosin-like dimerization helix and an extra-large homeodomain are essential elements of the tripartite DNA binding structure of LF-B1. Cell 61:1225–1236

[8] Baumheuter S, Courtois G, Crabtree GR (1988) A variant nuclear protein in dedifferentiated hepatoma cells binds to the same functional sequences in the β-fibrinogen gene promoter as HNF-1. EMBO J 7:2485–2493

Cereghini S, Blumenfeld M, Yaniv M (1988) A liver-specific factor essential for albumin transcription differs between differentiated and dedifferentiated rat hepatoma cells. Genes Dev 2:957–974

Deschatrette J, Weiss M (1974) Characterization of differentiated and dedifferentiated clones of a rat hepatoma. Biochimie 56:1603–1611

Deschatrette J, Moore EE, Dubois M, Weiss M (1980) Dedifferentiated variants of a rat hepatoma: reversion analysis. Cell 19:1043–1051

DeSimone V, De Magistris LD, Lazzaro D, Gertsner J, Monaci P, Nicosia A, Cortese R (1991) LFB3, a heterodimer-forming homeoprotein of the LFB1 family, is expressed in specialized epithelia. EMBO J 10:1435–1443

Kuo CJ, Mendel DB, Hansen LP, Crabtree GR (1991) Independent regulation of HNF-1α and HNF-1β by retinoic acid in F9 teratocarcinoma cells. EMBO J 10:2231–2236

Kuo CJ, Caron J, Crabtree GR unpublished observations

Mendel D, Hansen L, Graves M, Conley P, Crabtree GR (1991) HNF-1α and HNF-1β share dimerization and homeo domains, but not activation domains, and form heterodimers in vitro. Genes Dev 5:1042–1056

Rey-Campos J, Chouard T, Yaniv M, Cereghini S (1991) vHNF1 is a homeoprotein that activates transcription and forms heterodimers with HNF1. EMBO J 10:1445–1457

[9] Birkenmeier EH, Gwynne B, Howard S, Jerry J, Gordon JI, Landschultz W, McKnight SL (1989) Tissue-specific expression, developmental regulation, and genetic mapping of the gene encoding CCAAT/enhancer binding protein. Genes Dev 3:1146–1156

Costa RH, Grayson DR, Xanthopoulous KG, Darnell JE (1988) A liver-specific DNA-binding protein recognizes multiple nucleotide sites in the regulatory regions of transthyretin, α-1-antitrypsin, albumin and SV40 genes. Proc Natl Acad Sci USA 85:3840–3844

Friedman AD, Landschultz WH, McKnight SL (1989) CCAAT/Enhancer binding protein activates the promoter of the serum albumin gene in cultured hepatoma cells. Genes Dev 3:1314–1322

Johnson PF, Landschultz WH, Graves BJ, McKnight SL (1987) Idenification of a rat liver nuclear protein that binds to the enhancer core of three animal viruses. Genes Dev 1:133–146

Landschultz WH, Johnson PF, Adashi EY, Graves BJ, McKnight SL (1988) Isolation of a recombinant copy of the gene encoding C/EBP. Genes Dev 2:786–800

Landschultz WH, Johnson PF, McKnight SL (1988) The leucine zipper: a hypothetical structure common to a new class of DNA-binding proteins. Science 240:1759–1764

Maire P, Wuarin J, Schibler U (1989) In vitro dissection of

the albumin gene promoter reveals positive cis-acting elements for tissue-specific and ubiquitous transcription factors. Science 244:343–346

Umek RM, Friedman AD, McKnight SL (1991) CCAAT-Enhancer binding protein: a component of a differentiation switch. Science 251:288–292

Xanthopoulos KG, Mirkovitch J, Decker T, Kuo CF, Darnell JE (1989) Cell-specific transcriptional control of the mouse DNA-binding protein mC/EBP. Proc Natl Acad Sci USA 86:4117–4121

[10] Descombes P, Chojkier M, Lichtsteiner S, Falvey E, Schibler U (1990) LAP, a novel member of the C/EBP gene family, encodes a liver-enriched transcriptional activator protein. Genes Dev 4:1541–1551

Mueller CR, Maire P, Schibler U (1990) DBP, a liver-enriched transcriptional activator is expressed late in ontogeny and its tissue specificity is determined post-transcriptionally. Cell 61:279–291

Wuarin J, Schibler U (1990) Expression of the liver-enriched transcriptional activator protein DBP follows a stringent circadian rhythm. Cell 63:1257–1266

[11] Costa RH, Grayson DR, Darnell JE (1989) Multiple hepatocyte-enriched nuclear factors function in the regulation of transthyretin and α-1-antitrypsin genes. Mol Cell Biol 9:1415–1425

Grayson DR, Costa RH, Darnell JE (1989) Regulation of hepatocyte-specific gene expression. Ann N Y Acad Sci 557:243–256

Lai E, Prezioso VR, Smith E, Litvin O, Costa R H, Darnell JE (1990) HNF-3A, a hepatocyte-enriched transcription factor of novel structure is regulated transcriptionally. Genes Dev 4:1427–1436

Lai E, Prezioso VR, Tao W, Chen WS, Darnell JE (1991) Hepatocyte nuclear factor 3α belongs to a gene family in mammals that is homologous to the *Drosophila* homeotic gene *fork head*. Genes Dev 5:416–427

Weigel D, Jackle H (1990) The *fork head* domain: a novel DNA binding motif of eukaryotic transcription factors? Cell 63:455–456

Weigel D, Seifert E, Reuter D, Jackle H (1990) Regulatory elements controlling expression of the *Drosophila* homeotic gene *fork head*. EMBO J 9:1199–1207

[12] Kuo CJ, Conley PB, Chen L, Sladek F, Darnell J, Crabtree GR (1992) A transcriptional hierarchy involved in mammalian cell-type specification. Nature 355:457–461

Sladek FM, Zhong W, Lai E, Darnell JE (1990) Liver-enriched transcription factor HNF-4 is a novel member of the steroid hormone receptor superfamily. Genes Dev 4:2353–2365

[13] Lichtsteiner S, Wuarin J, Schibler U (1987) The interplay of DNA-binding proteins on the promoter of the mouse albumin gene. Cell 51:963–973

Maire P, Wuarin J, Schibler U (1989) In vitro dissection of the albumin gene promoter reveals positive cis-acting elements for tissue-specific and ubiquitous transcription factors. Science 244:343–346

[14] Lichtsteiner S, Schibler U (1989) A glycosylated liver-specific

transcription factor stimulates transcription of the albumin gene. Cell 57:1179–1187

Mendel DB, Crabtree GR unpublished observations

[15] Enat R, Reid LM (1984) Hepatocyte proliferation in vitro: its dependence on the use of serum-free hormonally defined medium and substrata of extracellular matrix. Proc Natl Acad Sci USA 81:1411–1415

Jefferson DM, Clayton DF, Darnell JE Jr, Reid LM (1984) Post-transcriptional modulation of gene expression in cultured rat hepatocytes. Mol Cell Biol 4:1929–1934

[16] Beato M (1989) Gene regulation by steroid hormones. Cell 56:335–344

Granner DK, Hargrove JL (1983) Regulation of the synthesis of tyrosine aminotransferase: the relationship to $mRNA^{TAT}$. Mol Cell Biochem 53:113–128

Yamamoto KR (1985) Steroid receptor regulated transcription of specific genes and gene networks. Annu Rev Genet 12:209–252

[17] Castell JV, Andus T, Kunz D, Heinrich PC (1989) Interleukin 6: the major regulator of acute-phase protein synthesis in man and rat. Ann NY Acad Sci 557:87–101

Evans E, Courtois G, Killiasn PL, Fuller GM, Crabtree GR (1987) Induction of fibrinogen and a subset of acute phase response genes involves a novel monokine which is mimicked by phorbol esters. J Biol Chem 262:10850–10854

Fey G (1989) Regulation of rat liver acute phase genes by interleukin-6 and production of hepatocyte stimulating factors by rat hepatoma cells. Ann NY Acad Sci 557:317–331

Fowlkes DM, Mullis NT, Comeau CM, Crabtree GR (1984) Potential basis for regulation of the coordinately expressed fibrinogen genes: homology in the 5' flanking regions. Proc Natl Acad Sci USA 81:2313–2316

Poli V, Mancini FP, Cortese R (1990) IL-6DBP, a nuclear protein involved in interleukin-6 signal transduction, defines a new family of leucine zipper proteins related to C/EBP. Cell 63:643–653

[18] Ben-Ze'ev A, Robinson GS, Bucher NR, Farmer SR (1988) Cell-cell and cell-matrix interactions differentially regulate the expression of hepatic and cytoskeletal genes in primary cultures of rat hepatocytes. Proc Natl Acad Sci USA 85:2161–2165

Clayton DF, Harrelson AL, Darnell JE Jr (1985) Dependence of liver-specific transcription on tissue organization. Mol Cell Biol 5:2623–2632

Guguen-Guillouzo C, Clement B, Baffet G, Beaumont C, Morel-Chany E, Glaise D, Guillouzo A (1983) Maintenance and reversibility of active albumin secretion by adult rat hepatocytes co-cultured with another liver epithelial cell type. Exp Cell Res 143:47–54

Reinke R, Zipursky SL (1988) Cell-cell interaction in the *Drosophila* retina: the bride of *sevenless* gene is required in photoreceptor cell R8 for R7 cell development. Cell 55:321–330

Rubin GM (1989) Development of the *Drosophila* retina: inductive events studied at single cell resolution. Cell 57:519–520

Sorkin BC, Hemperly JJ, Edelman GM, Cunningham BA

(1988) Structure of the gene for the liver cell adhesion molecule, L-CAM. Proc Natl Acad Sci USA 85: 7617–7621

[19] Bissel DM, Arenson DM, Maher JJ, Roll FJ (1987) Support of cultured hepatocytes by a laminin-rich gel. J Clin Invest 79:801–812

Caron JM (1990) Induction of albumin gene transcription in hepatocytes by extracellular matrix proteins. Mol Cell Biol 10:1239–1243

Michealopoulos G, Pitot HC (1975) Primary culture of parenchymal liver cells on collagen membranes. Exp Cell Res 94:70–78

Chapter 33 Cellular Differentiation in the Hematopoietic System: an Introduction

M. INTRONA, J. GOLAY, and S. OTTOLENGHI

33.1 A Brief Description of the Cells and Their Functions[1]

The blood contains many types of cells with very different functions, with the common characteristics of having a limited life span ranging from a few hours (for example granulocytes) to several months (for example lymphocytes). Due to their limited life span, all these cells need to be continuously produced to keep their number constant in the blood stream and in the tissues. This function is accomplished in the bone marrow in the adult. In the fetus, the major hematopoietic organs are the liver and spleen and in the embryo, the yolk sac. Blood cells can be distinguished into red cells (erythrocytes) and white cells (leukocytes). Leukocytes are further subdivided in lymphoid (T and B lymphocytes) and myeloid cells (granulocytes, monocytes and megakaryocytes). The granulocytes comprise neutrophils, eosinophils, and basophils (Fig. 1). All these cell types derive from the same cell, called stem cell, through a series of intermediate stages that are described below.

33.1.1 Hematopoietic Stem Cells[2]

During early embryonic development, mesodermal cells give rise to pluripotent stem cells (PSC), which are the most immature hematopoietic cells and can give rise to mature cells of all hematopoietic lineages (erythroid, myeloid, and lymphoid). Indeed, PSC can reconstitute the whole hematopoietic system of irradiated animals.

Stem cells represent approximately 0.01% of the bone marrow cells, are long-lived and may last the entire life span of an individual. By definition, they have the capacity for self-renewal: upon each divi-

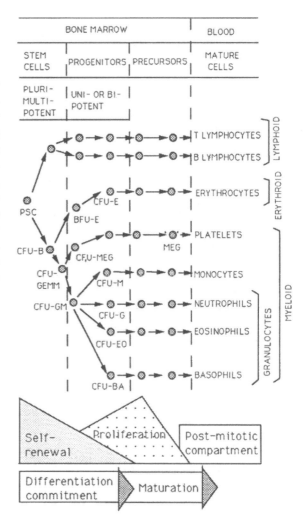

Fig. 1. Schematic representation of hematopoietic differentiation pathways. *PSC* pluripotent stem cells; *CFU* colony-forming unit; *B* blast; *GEMM* granulocyte-erythroid-monocyte-megakaryocyte; *BFU* burst-forming unit; *E* erythroid; *MEG* megakaryocyte; *GM* granulocyte-monocyte; *EO* eosinophil; *BA* basophil. Multi-potent stem cells and progenitors progressively lose the ability to divide into cells maintaining multipotentiality (self-renewal). Variation in the frequency of cell division (proliferation) during maturation is also shown (*bottom*)

sion, each daughter cell can either remain a stem cell and retain full potential for differentiation, or it can embark on a course leading irreversibly to terminal differentiation towards one particular lineage, i.e., become committed. The ability to self-renew without undergoing commitment is typical, and exclusive of the stem cells. However, under normal conditions stem cells rarely divide. Differentiation to mature cell types is in fact mainly sustained by the committed progenitors (see below). Under pathological conditions such as blood loss or infections, the normal balance between commitment and self-renewal may be modified, to favor differentiation to erythrocytes or granulocytes respectively. Conversely, stem cell self-renewal may be favored following bone marrow ablation due to X-irradiation or chemotherapy.

The pluripotentiality of stem cells has been demonstrated in several ways. In man, chronic myelogenous leukaemia (CML) results from spontaneous transformation of stem cells (see Introna and Golay, Chap. 34, this Vol.). These transformed cells and their differentiated progeny can easily be identified since they carry a abnormal chromosome (the Philadelphia chromosome). It could thus be demonstrated that in CML the transformed stem cell gives rise to clonal lines (as defined by isoenzyme analysis) of myeloid as well as lymphoid cells. In mice, bone marrow transplantation experiments have shown that a single cell, identifiable by radiation-induced chromosomal abnormalities, can give long-term reconstitution of the whole hematopoietic system of irradiated animals. This finding has been recently confirmed using stem cells marked by retroviral introduction of the neo-resistance gene (for details, see Wagner and Keller, Chap. 30, this Vol.). In these experiments, each stem cell and its clonal progeny can be recognized from the typical pattern of bands of integrated retroviral DNA in Southern blots. In this way, individual stem cells were shown to reconstitute all different lineages. In addition, it was demonstrated that a single stem cell, though totipotent, could repopulate a given lineage (for example T lymphocytes) in one hematopoietic organ (for example in the thymus) but not in another (spleen). Similarly, a given stem cell may repopulate an organ with various hematopoietic cell types, but not with T cells.

These phenomena may reflect an early stochastic "decision" of a stem cell to differentiate into a given lineage (see below).

33.1.2 Progenitor Cells

Progenitor cells are the descendents of stem cells that have become committed to one or more hematopoietic lineages (Fig. 1). Unlike stem cells, progenitors divide frequently and have a limited life span. They cannot give long-term reconstitution of the hematopoietic system in irradiated animals.

Intermediate steps between stem cells and progenitors are shown in Fig. 1. It is important to note that, in the early stages, cell proliferation is not antagonistic to progressive cellular differentiation. Only during the late maturation stages does differentiation occur in the absence of proliferation (Fig. 1).

All these cells, with the exception of PSC can be assayed in vitro by their ability to give rise to monoclonal colonies in semi-solid media (colony assay) and are hence called colony-forming units (CFUs). The CFU-blast (CFU-B) generate colonies of multipotent undifferentiated cells. They have therefore a high self-renewal capacity like true stem cells but have undergone a first commitment step (Fig. 1). All other progenitors generate colonies containing differentiated cells.

33.1.3 Morphologically Identifiable Immature Cells or Precursors

Precursors divide much less frequently than progenitors; they are already committed to a single lineage of differentiation and give rise to the fully mature cells through few rounds of cell division. Several differentiation steps have been identified in the erythroid and myeloid pathways by morphological and biochemical analysis of the differentiating cells and will be described in more detail in the subsequent chapters of this Volume (Ottolenghi, Chap. 35, for erythroid differentiation and Introna and Golay, Chap. 34, for myeloid differentiation).

33.2 Stochastic Versus Deterministic Mechanisms in the Commitment to Individual Hematopoietic Lineages[3]

The various stages of differentiation depicted in Fig. 1 could be the result of either a programmed (deterministic) series of ordered events or random (stochastic) changes occurring in the stem cell and its progeny and leading to differentiation. Experiments to distinguish between these two possible mechanisms have been performed at the single cell level. Single cells were picked from CFU-B colonies, allowed to divide into two cells which were then individually replated. This process was repeated once more and the four daughter cells obtained were allowed to differentiate in a CFU-assay. As shown in Fig. 2, multipotent cells become progressively restricted in their ability to choose specific lineages, but no preferential order can be observed; thus commitment towards a given lineage is stochastic and not preprogrammed in the cell. As mentioned above, some stem cells, although multipotent in the whole animal, give rise to only a limited range of cell types within a given organ. Although in this case the stem cell's "environment" may have directed its early commitment decision, it is possible that this phenomenon reflects the stochastic events observed in vitro.

33.3 Growth Factors in Hematopoiesis[4]

In vitro differentiation of hematopoietic cells can be obtained in simple semi-solid media (colony assay) in the presence of growth factors (GFs). Several GFs have been identified (and their genes cloned) which act on progenitor cells at various stages of development (Fig. 3). Some GFs induce proliferation and differentiation of very early pluripotent cells leading to the development of colonies containing mature cells of all lineages (e.g., IL-3). Other GFs (e.g., G-CSF) act both on early progenitors and more mature cells with limited differentiation potential. Finally, some GFs (e.g., M-CSF or IL-5) act exclusively on mono- or bi-potent

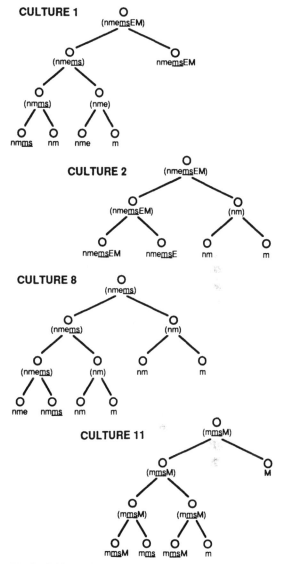

Fig. 2. Evidence for stochastic mechanisms in the process of commitment decision in in vitro culture of blast cells. The presumptive genealogical trees and developmental potential of progenitors (in parentheses) have been reconstructed on the basis of the observed cell types generated by descendants of individual progenitors. (After Suda et al. 1984). *n* neutrophils; *m* monocytes; *ms* basophils-mast cells; *e* eosinophils; *E* erythrocytes; *M* megakaryocytes. Four different experiments are shown

precursor cells. These differences lead to a hierarchy in the effects of GFs on specific cells. For example, GM-CSF, IL-6, and G-CSF all stimulate CFU-GEMM cells; however, GM-CSF induces the formation of colonies containing neutrophils, macro-phages, eosinophils, and megakaryocytes; IL-6 generates neutrophilic and megakaryocytic colonies, while G-CSF leads to the appearance of neutrophilic colonies only (Fig. 3). It is also clear from Fig. 3 that more than one GF is necessary for the differentiation of cells within each specific lineage. In general, a factor with a broad range of targets like IL-3 will synergize with a factor that has a restricted targets to induce a specific cell type (e.g., IL-3 and erythropoietin are required for erythroid differentiation, IL-3 and G-CSF for neutrophil differentiation). Finally, in addition to their proliferative and differentiative action, GFs may stimulate cell survival and the functional activation of fully mature cells.

33.4 Cell-Cell Contacts in Hematopoiesis

In vivo hematopoiesis takes place within a microenvironment represented by a heterogeneous population of "stromal" cells composed of macrophages, endothelial cells, fat cells, and fibroblasts associated with extracellular matrix. Stromal cells locally produce GFs which may then become absorbed to the extracellular matrix, where they are protected from proteolysis and can be presented to their target cells immobilized on the stroma. The major glycosaminoglycan of the stroma, heparan sulfate, is able to absorb both GM-CSF and IL-3 and to present it to stromal cells. On the other hand, cell adhesion molecules (CAM, fibronectin, trombospondin, and other molecules) may be instrumental to mediating the attachment of hematopoietic cells to the stroma, thus enabling localized regulation by absorbed GFs. Cells belonging to different lineages and at different maturation stages may exhibit variable requirements for adhesion. These observations suggest possible compartmentalization of hematopoietic cells within the bone marrow stroma.

It is important to note that, in addition to GFs acting on progenitor and precursor cells, stromal cells produce a GF specific for the pluripotent stem cells. This factor called stem cell growth factor (SGF) (Fig. 3) is a ligand for a membrane receptor

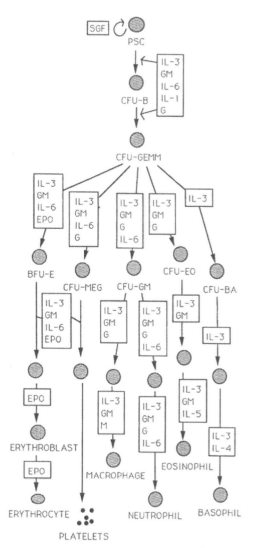

Fig. 3. Sites of action of GFs in hematopoietic differentiation. The abbreviations for stem and progenitor cells are identical to those indicated in Fig. 1. Several growth factors (indicated *within boxes*) synergize at various stages to induce optimal proliferation and differentiation of target cells. *IL* interleukin; *SGF* stem cell growth factor; *GM* GM-CSF *G* G-CSF; *M* M-CSF; *EPO* erythropoietin

encoded by the protooncogene c-kit. Its importance is demonstrated by observations on mutant mouse strains which are deficient either in SGF production (*Steel* mutants) or in c-*kit* synthesis (W⁻). Both mutant mice suffer to a variable degree from anaemia, mast cell deficiency and other defects. *W⁻* mice can be cured by transplantation of bone marrow stem cells from normal or even *Steel* mice. On the contrary, the defects of *Steel* mice are not reversed by transplantation of stem cells, suggesting that in these mice, the stem cell defect is secondary to a defect involving the environment (i.e., the stromal cells producing SGF).

References

[1]Metcalf D (1989) The molecular control of cell division, differentiation commitment and maturation in haemopoietic cells. Nature 339:27–30

[2]Iscove N (1990) Searching for stem cells. Nature 347:126–127

Keller G, Palge C, Gilboa E, Wagner EF (1985) Expression of a foreign gene in myeloid and lymphoid cells derived from multipotent haematopoietic precursors. Nature 318:149–154

Lenischka IR, Raulet DH, Mulligan RC (1986) Developmental potential and dynamic behaviour of haematopoietic stem cells. Cell 45:917–927

Martin PJ, Najfeld V, Hansen JA, Penfold GK, Jacobson RJ, Fialkow PJ (1980) Involvment of the lymphoid system in chronic myelogenous leukaemia. Nature 287:49–50

Mintz B, Palge C, Litwin S (1984) Monoclonal derivation of mouse myeloid and lymphoid lineages from totipotent hematopoietic stem cells experimentally engrafted in fetal hosts. Proc Natl Acad Sci USA 81:7835–7839

Torok-Storb B (1988) Cellular interactions. Blood 72:373–385

[3]Ogawa M, Porter PN, Nakahata T (1983) Renewal and commitment to differentiation of hematopoietic stem cells (an interpretative view). Blood 61:823–829

Suda T, Suda J, Ogawa M (1984a) Analysis of differentiation of mouse hematopoietic stem cells in culture by sequential replating of paired progenitors. Blood 64:393–399

Suda T, Suda J, Ogawa M (1984b) Disparate differentiation in mouse hemopoietic colonies derived from paired progenitors. Proc Natl Acad Sci USA 81:2520–2524

[4]Clark SC, Kamen R (1987) The human hematopoietic colony-stimulating factors. Science 236:1229–1237

Coulombel L, Vullet MH, Leroy C, Tchernia G (1988) Lineage- and stage-specific adhesion of human hematopoietic progenitor cells to extracellular matrices from marrow fibroblasts. Blood 71:329–334

Gimble JM (1990) The function of adipocytes in the bone marrow stroma. New Biol 2:304–312

Gordon MY, Ford AM, Greaves MF (1990) Cell interaction and gene expression in early haematopoiesis. Int J Cell Clon 8:11–25

Gordon MY, Riley GP, Watt SM, Greaves MF (1987) Compartmentalisation of a hematopoietic growth factor (GM-CSF) by glycosaminoglycans in the bone marrow microenvironment. Nature 326:403–405

Long MW, Dixit VM (1990) Thrombospondin functions as a cytoadhesion molecule for human hematopoietic progenitor cells. Blood 75:329–334

Spooncer E, Heyworth CM, Dunn A, Dexter TM (1986) Self-renewal and differentiation of interleukin-3-dependent multipotent stem cells are modulated by stromal cells and serum factors. Differentiation 31:111–118

Walker F, Nicola NA, Metcalf D, Burgess AW (1985) Hierarchical down-modulation of hemopoietic growth factor receptors. Cell 43:269–276

Williams GT, Smith CA, Spooncer E, Dexter TM, Tayler DR (1989) Haemopoietic colony stimulating factors promote cell survival by suppressing apoptosis. Nature 343:76–78

Witte ON (1990) Steel locus defines new multipotent growth factors. Cell 63:5–6

Chapter 34   The Role of Oncogenes in Myeloid Differentiation

M. INTRONA and J. GOLAY

34.1 Introduction

All the different cell types that constitute the immune hematopoietic system derive from a monomorphic population of totipotent stem cells which reside in the bone marrow through a complex process that is extensively reviewed in a separate chapter in this Volume (Introna et al., Chap. 33). Many biological aspects are still unknown, including the complete characterization of all the genes which contribute to the process. In this chapter we will focus only on a family of genes (proto-oncogenes and oncogenes) which may play a role in hematopoietic differentiation. Furthermore, we will limit the discussion to the branch of differentiation that leads to monocytes and granulocytes. Before summarizing the information on the potential role of oncogenes in myeloid differentiation, we will describe the different model systems available to study this process and their drawbacks.

34.1.1 The Normal Myeloid Pathway of Differentiation [1]

The pathway of differentiation of normal bone marrow cells to granulocytes and monocytes has been studied through analysis of the changes in morphological, biochemical, and immunological markers of the differentiating cells and the different steps recognized are schematized in Fig. 1. The major difficulty encountered in studying the mechanisms and the genes that regulate the normal pathway of differentiation is that this complex process takes place in the bone marrow, which contains many different cell types at various stages of differentiation so that any one population is scarcely represented.

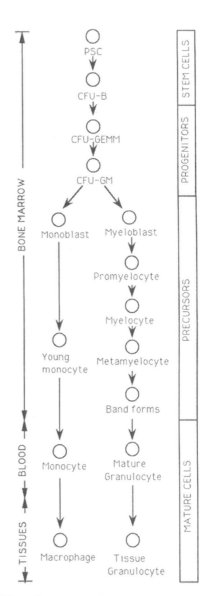

Fig. 1. Schematic representation of myeloid differentiation steps. *PSC* pluripotent stem cell; *CFU* colony-forming-unit; *B* blast; *GEMM* granulocyte-erythroid-megakaryocyte-monocyte. *GM* granulocyte-monocyte

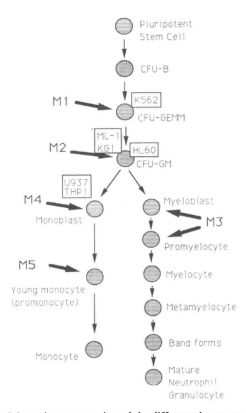

Fig. 2. Schematic representation of the different phenotypes of AML with respect to normal differentiation, according to the FAB classification. *M1* myeloblastic undifferentiated leukemia; *M2* myeloblastic differentiated; *M3* promyelocytic; *M4* myelo-monoblastic; *M5* monoblastic. The *thick arrows* indicate the cell type most commonly found in the blood and bone marrow of patients with these subtypes of leukemia. The most commonly used cell lines are also indicated in *boxes* which are placed in correspondence to the stage of differentiation of their normal cellular counterpart

34.1.2 The Transformed Myeloid Cells to Study Differentiation [2]

Human myelogenous leukemias are neoplastic diseases of myeloid cells which can be subdivided in acute and chronic forms. Both acute and chronic myelogenous leukemias (AML and CML) belong to a vast family of clonal diseases known as Hematopoietic Stem Cell Disorders, in that the basic genetic lesions leading to malignancy are believed to take place in a single hematopoietic stem cell. AML is morphologically very heterogeneous and at least

five different myeloid subtypes have been characterized according to the most widely used French-American-British (FAB) classification (Fig. 2). The morphology of the neoplastic cells present in the blood represent more or less closely one of the differentiation steps normally observed in the bone marrow (Fig. 2). In other words, most neoplastic cells appear to be "blocked" or "arrested" at a single step, which varies from patient to patient, of the normal differentiation cascade so that they accumulate in enormous amounts in the blood and other organs. The arrested cell has, in addition, increased self-renewal capacity so that cell division gives rise to daughter cells all identical to each other.

The chronic myelogenous leukemia (CML) is characterized by a dramatic increase in the blood of all essentially mature cell types, although one type (usually neutrophils) predominates. Unlike in AML, therefore, in the chronic phase of CML, no differentiation arrest is evident. Rather, it appears as a proliferative disease of the stem cell. The monoclonality of the disease has been used to demonstrate the differentiating capacity of the stem cell as explained in Chapter 33 (Introna et al., this Volume). CML inevitably undergoes transformation to an acute disease, called blast crisis, in which a clone of immature, differentiation-arrested, blast cells of myeloid or lymphoid origin invades the organism.

Since the leukemic cells in both AML and CML appear to represent the transformed counterpart of normal stages of myeloid differentiation and are available often in large numbers, they have been used as models to study differentiation. In addition, the genetic abnormalities detected in these leukemias can give a clue about the role of particular genes for the proliferation and/or differentiation of myeloid cells.

34.1.3 The Cell Lines as Models to Study Differentiation [3]

A model system that has been extensively used to study hematopoietic differentiation is that of the in vitro established, immortalized myeloid cell lines. These are derived from in vivo leukemias, and

Table 1. Characteristics and differentiation potential of the most commonly used human myeloid leukemia lines

Cell line	Stage of differentiation	Chemical inducers	Differentiated phenotype
HL60	Bipotent myeloid precursor	Retinoic acid Dimethyl sulfoxide	Granulocytes
		Vitamin D_3	Monocytes
		Phorbol esters	Macrophages
KG1	Myeloid precursor	Phorbol esters	Macrophages
		Teleocidins	Macrophages
U937	Monoblast	Phorbol esters	Macrophages
		Teleocidins	
THP1	Monoblast	Phorbol esters	Macrophages
ML1	Myeloid precursor	Phorbol esters	Macrophages
		Teleocidins	

therefore may only partially represent what takes place during normal differentiation. Nevertheless, they are a useful tool. Many of these cell lines do no require any specific growth factor (GF) for their survival and growth. It is interesting to note that fresh leukemic cells are not immortalized. It is not clear whether their occasional immortalization in vitro (establishment as lines) results from a further genetic alteration not present in the original leukemic cells or to the occasional presence and selection in vitro of a rare immortalized and GF-independent leukemic cell. The characteristics of the most frequently used myeloid cell lines are briefly described in Table 1. Since they are mostly derived from AML patients, they essentially represent progenitor or precursor cells blocked at different stages of differentiation (Fig. 2). For many of these lines (HL60, U937, KG1) the differentiation block can be at least partially released by exposure to several agents (phorbol esters, retinoic acid, vitamin D, etc., Table 1) which generally cause arrest in the proliferation of the cells and concomitantly differentiation to more mature cells. Differentiation can be detected by morphological changes and/or the appearance of some of the markers and functions normally present in the corresponding mature circulating blood cells. U937 and KG1 can differentiate towards monocytes while HL60 is unique in that it can differentiate towards both monocytes and granulocytes depending on the inducing agent used (Table 1).

The cell lines differ from their normal immature counterparts in at least one important aspect: cell

lines differentiate following or in conjunction with a decrease in proliferation: this is the opposite of what has been observed in normal bone marrow cells where proliferation is accompanied by differentiation (see: Introna et al., Chap. 33, this Vol.). The considerations have to be taken into account when interpreting results obtained with cell lines.

34.1.4 Retroviruses and Oncogenes[4]

The first oncogenes were discovered in a subfamily of retroviruses (RNA viruses) called acutely transforming defective retroviruses. The characteristic of retroviruses is that after infection on the host cell, the viral RNA can be "transcribed" into DNA which then becomes stably integrated in the host genome and replicates with it. A subgroup of these viruses induces tumors of different histiotypes in various animal species with a very short latency. These acutely transforming retroviruses have acquired from their host cells, and integrated into their genome, cellular genes which have suffered structural alterations (deletions, point mutations) during this "transduction" process. These mutated genetic elements of cellular origin are responsible for the high tumorigenicity of these viruses and have been called viral oncogenes when present in the viral transforming conformation (v-onc) and cellular oncogenes or proto-oncogenes when present in their normal cellular configuration (c-onc).

The genome of most of the spontaneously occurring, acutely transforming retroviruses have been sequenced and their oncogenes are listed in Table 2. They can be classified according to the main biochemical characteristics of their protein products (Table 2), and the significance of this is discussed in more detail in another chapter of this Volume (Philipson and Sorrentino, Chap. 36).

Only some of these retroviruses rapidly transform hematopoietic targets in vivo and in vitro.

Table 2. List of the functions and cellular localisation of the oncogenes mentioned

Oncogenes	Function/homology	Cellular localization
myc myb fos jun ets-1	DNA-binding transcription factors	Nucleus
evi-1	Zinc-finger protein	?
raf	Ser/Threo kinase	Cytoplasmic
abl	Tyrosine kinase	
src hck yes fgr	Tyrosine kinases src-like	Cytoplasmic, at the internal side of the plasma membrane
fes/fps	Tyrosine kinase	
fms	M-CSF receptor	Plasma membrane
erbB	EGF-receptor, Tyr kinase	Plasma membrane
ros	Tyr kinase, insulin receptor-like	Plasma membrane
H-ras K-ras N-ras	G-protein-like	Cytoplasmic
sis	PDGF β chain	Secreted

Therefore in such models the oncogenes directly alter the proliferative and differentiating capacity of the hematopoietic target cell which gives excellent models to understand the genetic mechanisms underlying the hematopoietic differentiation.

34.2 The Role of Oncogenes in Myeloid Differentiation

34.2.1 The Avian Defective Retroviruses Which Cause Acute Myeloid Leukemias

In chickens, several naturally occurring viruses giving myeloid leukemias have been isolated. Some of these have been studied in much detail and have given useful information about the role of their respective oncogenes in myeloid transformation and differentiation. The best studied of these avian retroviruses are listed in Table 3. They can be divided into two groups that carry either the v-myb or v-myc gene. In addition, two of the viruses (E26 and MH2) carry two different oncogenes (the v-myb/v-ets, and the v-myc/v-mil oncogenes, respectively).

34.2.1.1 The myb-Containing Retroviruses[5]

Within a few days or weeks after injection of the acute myeloblastosis virus (AMV) or the E26 virus into very young animals or in fertilized eggs, the

Table 3. Avian acutely transforming retroviruses and phenotypes of the chicken cells transformed in vitro and in vivo

Virus	Viral oncogene	Phenotype in vitro	Effect in vivo	GF dependence of transformed cells
AMV	myb	Monoblasts	Monoblastic leukemia	cMGF[a]
E26	myb + ets	Erythroblasts + myeloblasts	Erythroblastic + myeloblastic leukemia	EPO[b] or cMGF
OK10	myc	Macrophages	Myelocytomatosis	cMGF
MC29	myc	Macrophages	Myelocytomatosis	cMGF
CMII	myc	Macrophages	Myelocytomatosis	cMGF
MH2	myc + mil	Macrophages	Monocytic leukaemia	GF-independent

[a] cMGF: chicken myeloid growth factor.
[b] EPO: Erythropoietin.

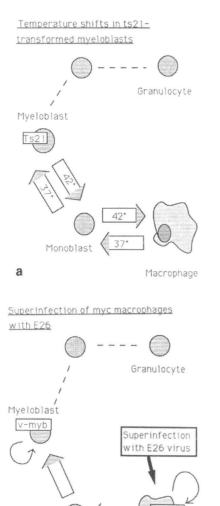

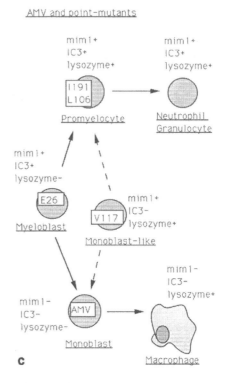

Fig. 3a–c. Effect of *v-myb* on myeloid differentiation in various experimental conditions. **a** Schematic representation of the effect of temperature shifts on ts21-transformed myeloblasts. **b** Effect of superinfection of *v-myc*-transformed macrophages with E26 virus. **c** Phenotypes of myeloid cells transformed by E26, AMV, and the AMV point mutants I191, L106, and V117. Mim1 is a gene and IC3 an antigen, both present during some stages of granulocyte differentiation. The *continuous arrows* represent the presumed normal differentiation pathways of the avian myeloblast towards granulocytes or macrophages, by analogy to the known differentiation pathways in man. The *dashed arrows* indicate that the phenotype of the V117 transformed cell is similar but not identical to that of both the promyelocyte and the monoblast

blood, spleen and marrow become invaded with blasts and the animals die rapidly. As shown in Table 3, AMV causes a strong acute myeloid leukemia, while E26 causes mostly an erythroid leukemia with a concomitant and weaker myeloid leukaemia ("strong" and "weak" refer to the numbers of transformed cells detectable in the periphery). The phenotype of the transformed myeloid cells is that of immature precursors of macrophages in the case of AMV (monoblasts) and that of even more immature and less differentiated myeloid precursors in the case of E26 (myeloblasts) (Fig. 3 C).

The leukemias that arise in these animals are oligoclonals, implying that the viruses are directly oncogenic and that transformation does not require any additional secondary event, such as a somatic mutation. The short latencies support this hypothesis. Furthermore, the sites of viral integration in the cellular genome are randomly distributed, thus excluding a pathogenic role for the cellular genes located in the vicinity of the integrated virus (unlike what is found in the case of the chronic retroviruses, Sect. 34.2.3).

It is also possible to infect normal chicken bone marrow cells in vitro and obtain the specific outgrowth of transformed myeloid or erythroid cells or both, depending on the culture conditions used. It is important to note that although the transformed cells have an increased self-renewal capacity with respect to their normal counterparts they are still dependent on myeloid or erythroid GFs for growth and are not immortalized. Finally and most importantly, an absolute correspondence was found between the phenotypes and GF-dependence of the cells transformed by AMV and E26 in vitro and in vivo.

34.2.1.2 v-myb Transforms Myeloblasts and v-ets Transforms Erythroblasts[6]

Since E26 can transform cells of both myeloid and erythroid phenotypes and carries two different genes of cellular origin (v-myb and v-ets), it was asked whether they were each responsible for a different transformed phenotype. A relation between v-myb and myeloid transformation was suspected from the fact that AMV, which carries v-myb alone, transforms only myeloid cells. More direct evidence in support of this hypothesis came from studies of mutants of E26. This virus was mutagenized in vitro, and extensive screening allowed to isolate temperature-sensitive (ts) mutants which had retained the ability to transform myeloblasts at 37 °C (permissive temperature) but not at 42 °C (nonpermissive temperature). One of these mutants, called ts21, was further studied, and it could be shown to be ts only for myeloid and not erythroid transformation. Furthermore, when myeloblast colonies grown at 37 °C were shifted to 42 °C, the cells stopped dividing and differentiated to mature macrophages within several days. Conversely, when these macrophages were then backshifted to 37 °C, proliferation resumed, and the immature myeloblastic phenotype was simultaneously reacquired (Fig. 3A). Complete sequencing of the ts21 virus revealed that it carries a single point mu-

tation in the v-myb oncogene while v-ets has remained unaltered. This point mutation in v-myb was unambiguously demonstrated to be responsible for the ts phenotype. This mutation was located at the 5' end of the v-myb protein, in the region involved in DNA binding (see below). These results imply that an active v-myb oncogene is required not only to induce the self-renewal of the cells but also to maintain their undifferentiated phenotype.

Recently a similar mutant of E26 virus was isolated, called ts1.1, which is ts for erythroid but not myeloid transformation: erythroblastic colonies transformed by ts1.1 at 37 °C differentiate to mature erythrocytes when shifted to 42 °C. The finding of a single point mutation in the v-ets gene of ts1.1 relative to wild-type E26 demonstrated a role of this oncogene in the transformation of chicken erythroid cells. It is important to remember that although a virus carrying only v-myb exists (AMV), no virus containing v-ets alone has ever been found or constructed that could still transform erythroid cells by itself. This may be either because v-myb is required in addition to v-ets for erythroid transformation, or because the function of the v-ets protein is altered when dissociated from v-myb to which it is normally fused in E26.

34.2.1.3 v-myb Affects Differentiation[7]

Clones of v-myc-transformed macrophages (see below) can be "superinfected" with E26. These superinfected clones were found to have "reverted" phenotype and to resemble the immature myeloblasts normally transformed by E26 (Fig. 3B). In other words, cells that are already highly proliferating through the action of the v-myc oncogene and have a differentiated (macrophage) phenotype can nevertheless de-differentiate upon addition of the v-myb oncogene. Additional evidence for an effect of v-myb on differentiation independently from proliferation comes from studies of AMV and of three different in vitro constructed point mutants of this virus (Fig. 4). Normally, AMV-transformed cells resemble monoblasts (Fig. 3C). Two of the AMV point-mutants, however (L106 and I191), transform cells with a promyelocytic phenotype (precursor of granulocytes), whereas the third mutant (V117)

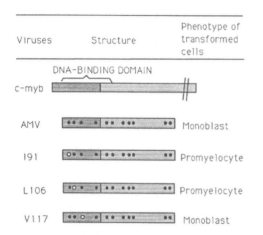

Fig. 4. Structure of AMV and its point mutants and phenotype of the cells transformed by these viruses. The *black dots* represent the mutations present in AMV relative to the chicken *c-myb* proto-oncogene. The *white dots* show the location of the amino acids that were mutated in the AMV mutants back to the *c-myb* configuration

gives monoblasts apparently identical to those obtained with the parental AMV (Fig. 3C). The different phenotypes could be deduced from the morphology of the cells (e.g., the presence of granules typical of chicken promyelocytes) and from the pattern of expression of a cell surface marker (IC3) and of lysozyme (Fig. 3C). The fact that only slight changes in the *v-myb* protein can apparently alter the differentiation programme of myeloid cells strongly suggests that the *v-myb* gene product can directly regulate myeloid differentiation. It has been suggested that *v-myb* may affect myeloid differentiation by mimicking at the transcriptional level the activity of *c-myb*, its normal counterpart. One hypothesis is that *c-myb* may work as a master gene for a genetic program of differentiation. In support of such a hypothesis is the fact that the three point mutations of AMV described above are located in the DNA-binding region of the *v-myb* protein and therefore the specificity of the recognition of DNA sequences by these transcriptional factors may be slightly different (Fig. 4). Furthermore, studies of the only gene known to be directly regulated by *myb* are even more interesting in this respect. This gene, called *mim-1*, is induced by the *v-myb* of E26 and by the three AMV point mutants, but not by the parental AMV *v-myb*

(Fig. 3C). This indicates that various forms (mutants) of *v-myb*, differing only by single amino acid, can differentially regulate the expression of a gene. In addition, these data suggest that transformed cells can have a great degree of heterogeneity since the V117 point mutant which gave an apparent monoblastic phenotype like AMV was able to induce the transcription of *mim-1* while AMV does not (Fig. 5C). Whether this heterogeneity reflects normal differentiation or is specific to these virally transformed cells remains to be determined. The *mim-1* gene is specifically expressed in normal chicken granulocytes and their precursors which further stresses the hypothesis that the *myb* gene may be a master gene for hematopoietic differentiation.

34.2.1.4 The v-myc-Containing Avian Retroviruses[8]

Four major isolates have been studied which carry the *v-myc* oncogene and are able to induce acute leukemias in the chicken (Table 3). Only one of them, the MH2 virus, carries a second oncogene in addition to *v-myc*, called *v-mil*, which has serine/threonine kinase activity. The MC29, CMII, and OK10 viruses cause myelocytomatosis in vivo, that is a poorly characterized accumulation in the blood of granulocyte precursors, and they do this only in some chicken flocks. The *myc/mil* MH2 virus, on the other hand, reproducibly gives an acute monocytic leukemia.

All *myc*-containing viruses, upon in vitro infection of bone marrow cells, transform these cells into macrophages but do not immortalize them. As in the case of the *v-myb*-transformed myeloid precursor cells, the macrophages produced by in vitro transformation with *v-myc* are dependent on GF for growth, with the exception of the MH2-transformed cells which grow in the absence of exogenous GFs.

To summarize, studies of the avian retroviral models of myeloid leukemia suggest that: (1) *v-myb* induces immature myeloblasts; (2) *v-myc* induces mature macrophages; (3) in both cases a second oncogene with kinase activity can render the transformed cells GF-independent; (4) finally, *v-myb*

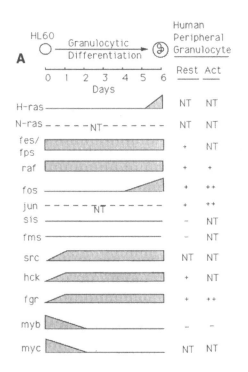

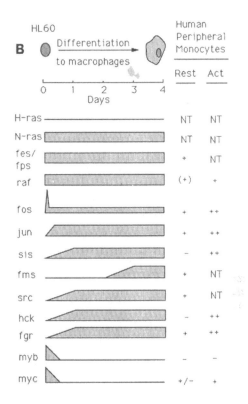

Fig. 5 A, B. Pattern of proto-oncogene expression during differentiation of HL60 cells. **A** Granulocytic differentiation of HL60 cells, induced by DMSO or retinoic acid. For comparison, the pattern of expression of the same oncogenes in normal resting or phorbol ester activated mature human granulocytes is indicated, when known. **B** Differentiation of HL60 cells to macrophages induced by phorbol esters. The expression pattern of proto-oncogenes in resting and lipopolysaccharide-activated human peripheral monocytes is also indicated. NT: not tested

may regulate the differentiation of myeloid cells independently of its effect on cell proliferation.

34.2.2 The Murine Acute Retroviruses Which Transform Myeloid Cells[9]

The two best studied natural isolates of acute murine retroviruses are the murine histiocytic sarcoma virus (MHSV) which carries the *v-H-ras* as oncogene and the myeloproliferative sarcoma virus (MPSV) that carries the *v-mos* oncogene. In the first case, the virus induces the transformation of mature tissue macrophages (histiocytes) which form localized solid tumors (sarcomas) with no sign of blood or marrow invasion. When these tumours are explanted in vitro they give rise to im-

mortalized macrophage cell lines which do not require any GFs to grow. The second virus, on the contrary, only slightly affects the proliferation of hematopoietic cells leading to a benign myeloproliferative disease. Finally, a third virus has been created in the laboratory carrying the *c-myc* oncogene which also gives localized intraperitoneal macrophage tumors. Upon in vitro explant, these tumors also give rise to GF-independent immortalized macrophage cell lines.

The murine data suggest that the *myc* oncogene alone can induce macrophages that are GF-dependent. A second genetic event is required in these cells to render them GF-independent. Furthermore the *v-raf* gene may have the capacity to induce a drastic lineage shift (pre-B lymphocytes to macrophages).

34.2.3 The Murine Chronic Retroviruses that Cause Myeloid Leukemias [10]

Chronic retroviruses induce tumors with long latencies. The reason is that they lack the exogenously acquired oncogenes. Unlike the acute viruses, they cause tumors by integrating in the vicinity of a cellular proto-oncogene altering its structural integrity and/or its transcriptional regulation through the action of the viral sequences called long-terminal repeats (LTRs), which have strong promoter/enhancer activities. Since the integration sites of all retroviruses is random, the probability of a chronic virus integrating near a proto-oncogene is relatively low, which explains that an active infection has to last several weeks or months until an appropriate integration takes place leading to transformation (see Kuehn and Stoye, Chap. 29, this Vol.). The resulting tumors are consequently monoclonal, unlike those induced by the acute viruses, even though the actual transforming event, that is the alteration of a cellular proto-oncogene, is very similar to that induced by the acute transforming retrovirus, as will be seen below.

The great majority of chronic retroviruses that cause tumors have been isolated from mice. Three major retroviruses have been implicated in murine myeloid leukemogenesis: the Moloney murine leukemia virus (MoMLV), the Friend murine leukemia virus (Friend MLV) and the wild ecotropic Cas-Br-M virus. For all three, appropriate infection regimens in selected strains of mice lead to a relatively low incidence of myeloid leukemias. In a relatively high proportion of these myeloid tumors, viral integration takes place within the *c-myb* gene and disrupts the coding sequence at the 5' or 3' end of the gene. In fewer cases, integration takes place near the *evi-1* gene (a transcription factor) or near the *c-fms* proto-oncogene (M-CSF receptor). The analysis of these integration sites has therefore confirmed a likely role for *myb* in the transformation of immature myeloid cells and has led to the finding of new potential oncogenes.

34.2.4 Potential Oncogenes in Human Myeloid Leukemias

It has already been said that the chronic and acute forms of human myeloid leukemias arise as a consequence of a disregulation of the differentiation/proliferation program of stem and progenitor cells leading to a number of different leukemic phenotypes. The search for the genetic changes that lead to tumorigenesis has been the focus of most studies on this disease. Since human leukemias frequently carry nonrandom chromosomal rearrangements or translocations, one of the main approaches has been to investigate whether known proto-oncogenes were present at the sites of these translocations and became activated during this process. The second approach has been to search for small alterations, such as point mutations, already known to be able to activate specific oncogenes in the animal models of leukemia. Finally, a third line of investigation has been to study the expression of known oncogenes in human leukemias to determine whether their transcripts were abnormal and/or overexpressed.

34.2.4.1 Translocations [11]

Several proto-oncogenes have been found at sites of chromosomal translocations in human leukemias (*myc, abl,* see below), while in other cases previously unknown sequences have been characterized. A role of the translocated gene in the leukemia becomes likely when it can be demonstrated that the breakpoint falls within the proto-oncogene in a way to alter its structure or the regulation of its transcription, and that such alterations are always or very frequently observed in a particular type of leukemia. In myeloid leukemias, the t9/22 (*bcr/abl*) is the best known translocation that satisfies these criteria and will be described in detail. The t9/22 translocation results in an abnormal chromosome called the Philadelphia chromosome that can be detected cytogenetically in almost all CML patients as well as in a subgroup of Acute Lymphoblastic Leukemias (ALL). Characterization of the Philadelphia chromosome revealed that the *c-abl* oncogene is translocated from chromosome 9 to chromosome 22. The region on chromosome 22 to which c-abl has translocated has been called bcr (break-point cluster region). Knowledge of the genetic elements involved in the translocation event has led to the demonstration that in some cases of CML negative for the Philadelphia Chromosome, the c-abl gene had nonetheless translocated to the bcr site. The virtual complete overlapping between bcr/abl rearrangement and CML diagnosis makes it a good candidate as a causative event in the leukemogenesis leading to CML.

The break point in c-abl is variable but usually takes place within the first intron such that the first exon is absent from the bcr-abl hybrid. The break points on chromosome 22 (bcr) cluster in three major regions of the bcr gene. The resulting bcr-abl hybrids contain different numbers of bcr exons so that in most CML, the hybrid codes for a 210 kDa protein, while in almost all Philadelphia positive ALL, the hybrid protein has a molecular weight of 190 kDa. The details of such genetic rearrangements fall beyond the limit of this chapter. Many aspects of the activity of the bcr-abl gene remain obscure. The most relevant question to this chapter is how such small differences in the bcr-abl hybrids apparently lead to different leukemias. Neither the cell in which the translocation has taken place nor the specific molecular abnormality completely determine leukemia phenotypes, since no absolute correlation has been found between a particular type of translocation and the phenotype of the cells. Thus, if this lesion may be a primary transforming event in the stem cell, other genetic alterations may play a role in determining the phenotype and other aspects of the biology of the transformed cells and therefore may be able to interfere with some genetic aspect to the differentiation program.

34.2.4.2 Point Mutations in Oncogenes[12]

Detailed studies of the viral oncogenes have demonstrated that only few specific point mutations within the coding region can lead to oncogenic activation of proto-oncogenes, such as the three members of the ras family (H-, K-, and N-ras) and c-fms (the M-CSF receptor). A systematic search for the presence of such point mutations in fresh human

leukemia cells has been carried out. Point mutations in all three ras genes are frequent in AML and in the related Myelodisplastic Syndromes (MDS, often referred to as pre-leukemic conditions, since they evolve in acute myeloid leukemias). The same point mutations are infrequent in CML, either in the stable phase of the disease or during the blast crisis. Several reports suggest that these mutations may not be part of a primary transforming event, but rather secondary genetic lesions that contribute to some of the biological properties of the leukemic cells.

Recently, point mutations in the c-fms gene have been detected in about 15% of AML and MDS patients. These mutations are located in domains of the c-fms protein that regulate its kinase activity so that the mutated protein is constitutively activated. As for the ras mutations, they may well represent secondary genetic abnormalities which confer some yet undetermined biological property to the leukemic cells.

Finally, in a subset of CML patients undergoing blast crisis, genomic abnormalities in the p53 gene have been described. p53 belongs to a group of genes called suppressor genes or anti-oncogenes. Oncogenicity is thought to result from the inactivation (via mutations) or deletion of both alleles. It remains therefore to be demonstrated whether in these patients both p53 alleles are altered and contribute to the disease progression.

34.2.4.3 Alteration in the Expression of Proto-Oncogenes[13]

Genetic amplification, that is the abnormal presence of multiple copies of a gene, can lead to overexpression of the amplified gene. Several investigators have detected amplification of some oncogenes (myb and more frequently myc) in cell lines, predominantly of myeloid origin. These data, however, have rarely been reproduced with fresh samples, suggesting that amplification arose as a consequence of in vitro selective pressures. The pattern of expression of a number of proto-oncogenes has been examined in human myeloid leukemias. Except in the case of the bcr-abl, abnormal sizes of mRNAs have never been consitently detected. The only data available are on the quantitative differences in the different diseases. The interpretation of these results is very difficult due to the heterogeneity of the leukemic samples (different phenotypes, proliferative capacity and contamination with normal cells) and the relative paucity of information about the transcriptional behavior of most proto-oncogenes in normal myeloid precursor cells. Whether some proto-oncogenes are involved in the transformation or in determining the phenotypes of human leukemias through deregulation of their transcription or translation remains therefore an open question.

34.3 The Role of Proto-oncogenes in Myeloid Differentiation

The observations described above that several oncogenes can increase self-renewal, block differentiation, or alter the program of differentiation of myeloid cells suggested that this may reflect the normal function of their cellular proto-oncogene counterparts, which may be important genes that control hematopoietic growth and differentiation. Many studies have consequently aimed at determining what may be the role of proto-oncogenes during myeloid differentiation. The approaches used in vitro were two: (1) To follow temporally the changes in proto-oncogene expression, either at the RNA and less frequently at the protein levels, during myeloid differentiation either in normal bone marrow cells or using the established cell lines as "normal" models. (2) A more direct approach was to use anti-sense oligonucleotides specific for particular proto-oncogenes in order to block translation of their transcripts into protein and therefore their function.

34.3.1 Studies in Normal Cells[14]

High levels of c-myb transcript are detectable in true human progenitor cells (CD34$^+$), and other marrow subpopulations show lower but detectable levels of c-myb expression. Since the CD34$^+$ cells are relatively quiescent compared to later progeni-

tors, the data has been taken to indicate that high levels of c-myb relate more to the immaturity of myeloid precursor cells rather than to their rate of proliferation. Incubation of bone marrow cells with anti-c-myb oligonucleotides drastically reduce the number of colonies that subsequently form in soft agar. The interpretation is that c-myb is required for the proliferation of hematopoietic progenitor cells. Such an effect of anti-c-myb was, however, not observed when the experiment was repeated on CD34$^+$ cells which contrasts with the high expression of c-myb in these cells. All the data could be reconciled by the hypothesis that c-myb has at least two functions, one related to immaturity in CD34$^+$ cells, and one related to proliferation in many hematopoietic cells.

Little information is available on other oncogenes. c-myc is relatively highly expressed in myeloid progenitors, while c-fes/fps is transcribed in every cell type of myeloid differentiation with a slight decrease towards the more mature forms. Finally, in some experiments on murine bone marrow, the c-src and the c-fms genes were found to be equally transcribed in proliferating monocytic precursors and nonproliferating macrophages, while the fgr kinase (src-related) was specific for the mature forms, whereas c-fos gene was expressed only in the proliferating precursors.

Finally, antisense oligonucleotides specific for c-abl efficiently block the formation of colonies in soft agar, whether total bone marrow or CD34$^+$ cells are used, suggesting an important function of c-abl in sustaining the proliferation of myeloid progenitors.

34.3.2 Studies of Differentiating Cell Lines[15]

The kinetics of expression of proto-oncogenes during differentiation of many cell lines induced by various agents have been extensively studied. The data on HL60 are summarized in Fig. 5A and B and are representative of what has been observed in other lines as well (U937, KG1, THP1, M1). It is clear from Fig. 5 that some proto-oncogenes are constitutively expressed (fes/fps, N-ras, raf/mil) and never change during differentiation. Other genes, such as the src-related tyrosine kinases (src, *fgr,* and *hck*) are not transcribed in immature cells but are induced following differentiation towards either granulocytes or macrophages. They may therefore be involved in the differentiation process but presumably not in lineage determination in this model.

Both the *c-fms* gene (M-CSF receptor) and the *c-sis* gene, that codes for the PDGF β chain, can be regarded as a molecular marker for normal macrophages since it is known that macrophages express M-CSF receptors and secrete PDGF, particularly upon activation. In agreement with this, they are both induced late during macrophage differentiation of HL60 cells and not during granulocytic differentiation. Furthermore, differentiation of HL60 cells to macrophages is abolished by anti-sense oligonucleotides specific for *c-fms*, confirming the absolute requirement for M-CSF and its receptor for monocytic differentiation.

Finally, the data on the behavior of the nuclear proto-oncogenes are worth discussing. It is now generally accepted that the *c-myc* gene in many systems is constitutively transcribed in all phases of the cell cycle and that a drop in *c-myc* transcription is closely associated with growth arrest in hematopoietic and other cell types. *c-myc* has hence been suggested to serve an important function in inducing the G_0 to G_1 transition. Indeed, when the hematopoietic cell lines are induced to differentiate, the accompanying block in proliferation is associated with a decline of the *c-myc* gene expression, although with variable kinetics. A decline in *c-myc* may therefore be a prerequisite for growth arrest to take place. An additional role of *c-myc* in driving differentiation is suggested by studies with anti-*c-myc* oligonucleotides that can inhibit the proliferation and induce partial differentiation of HL60 cells. The effect on differentiation is, however, limited and has not been fully characterized, suggesting that other signals are required in addition to drop in *c-myc.*

The *fos* gene has been extensively investigated in this context because it was originally thought to be involved in the control of myeloid differentiation. More recently, however, *c-fos* has been shown to be rapidly and transiently induced in many cell types and by many different signals, suggesting that it has a role other than in differentiation. It is still

possible that the presence of *c-fos* may be required in relatively mature myeloid cells, since, following its transient induction, detectable levels of *c-fos* are still observed during both the monocytic and granulocytic differentiation of HL60 and detected in purified populations of mature macrophages and granulocytes isolated from human blood. A similar behavior is ascribed to the *c-jun* gene which is also detectable in either mature monocytes and granulocytes. This is consistent with the hypothesis that *c-fos* and *c-jun* are always concomitantly expressed since their protein products form a heterodimer that functions in regulating the transcription of other genes.

Finally, *c-myb* is actively transcribed in immature myeloid cell lines, and its levels decline sharply during differentiation, but it is not clear in these models whether the decrease in *c-myb* relates to the growth arrest or to differentiation. In favor of the former hypothesis is the finding that anti-*c-myb* oligonucleotides inhibit the proliferation of HL60 but do not induce their differentiation, suggesting that other events are required for the latter. Furthermore, purified populations of monocytes and granulocytes isolated from human blood do not show any transcript for the *c-myb* gene.

As a conclusion it can be said that although many reports have appeared on the potential role of proto-oncogenes during the differentiation of cell lines, the information collected is essentially indirect in suggesting a not better specified role for some proto-oncogene in driving the process of differentiation. Only the data with the use of antisense oligonucleotides may indicate a necessary role for the *c-myb, c-fms,* and *c-myc* genes in some step of the differentiation pathway, but these studies clearly require more work in order to directly identify the genetic target for their action.

34.4 Transgenic Mice [16]

As far as the study of the role of proto-oncogenes in myeloid differentiation is concerned, few transgenic experiments of interest have been performed. The potential role of the *bcr-abl* fusion products were studied in transgenic mice. Introduction of the p210 *bcr/abl* in these mice resulted in T or B lymphomas. Recently, the p190 *bcr/abl* was tested and preliminary data suggest that the transgenic mice developed acute B lymphoblastic leukemias. The problem with these studies is that the results depend on the promoters used (in this case the immunoglobulin enhancer or retroviral promoters) and thus on the cell type in which the gene is expressed, and may not reflect the function of the rearranged *abl* gene in CML. In fact, these data with transgenic mice imply a transforming activity by the *bcr/abl* chimera rather different from the one observed with the same construct upon bone marrow infection as reported previously.

34.5 Outlook and Summary

Normal hematopoietic differentiation takes place by a complex interplay between self-renewal of stem cells and progenitors and differentiation of these cells and their progeny, a process that is accompanied by commitment to specific lineages. This complex series of events is at least in part regulated by GFs and by the ordered appearance of the receptors for these factors on hematopoietic cells during differentiation. Except in the case where an oncogene has been demonstrated to be the receptor for a hematopoietic GF (*v-fms*, the M-CSF receptor), evidence for the role of oncogenes during myeloid differentiation is still circumstantial. Nevertheless, a large body of evidence in several systems indicates that some oncogenes can affect one or more of the elements that regulate myeloid differentiation, that is self-renewal, differentiation, commitment and GF dependence. In particular, the strongest evidence for a function during self-renewal of myeloid cells has been obtained for the *v-myc*, *c-abl*, and *v-myb* oncogenes. The *v-myb* oncogene is particularly relevant, since its oncogenicity appears to be very specific for myeloid cells in several model systems and it is the gene for which there is the best evidence for a function during myeloid differentiation and perhaps also for commitment. Another oncogene implicated in commitment is *v-raf/v-mil*, although its major function seems to be that of inducing GF independence, a role shared by the tyro-

sine kinases. The greatest challenge that remains is to determine what precise functions have their respective proto-oncogenes during normal myeloid differentiation.

References

[1] Arai KI, Lee F, Miyajima A, Miyatake S, Arai N, Yokota T (1990) Cytokines: coordinators of immune and inflammatory responses. Annu Rev Biochem 59:783–836
Cannistra SA, Griffin JD (1988) Regulation of the production and function of granulocytes and monocytes. Semin Hematol 25:173–188
Clark SC, Kamen R (1987) The human hematopoietic colony-stimulating factors. Science 236:1229–1237
Gonda TJ, Dunn AR (1986) A molecular basis for growth regulation in normal and neoplastic hemopoiesis. Cancer Rev 3:58–90
Metcalf D (1989) The molecular control of cell division, differentiation commitment and maturation in haemopoietic cells. Nature 339:27–30
Nicola NA (1989) Hemopoietic cell growth factors and their receptors. Annu Rev Biochem 58:45–77
Sachs L (1987) The molecular control of blood cell development. Science 238:1374–1378

[2] Drexler HG (1987) Classification of acute myeloid leukemias: a comparison of FAB and immunophenotyping. Leukemia 1:697–705
Griffin JD, Lowenberg B (1986) Clonogenic cells in acute myeloblastic leukemia. Blood 68:1185–1195
Henderson ES, Lister TA (eds) (1990) Leukemia, 5th edn. WB Saunders, London
Stamatpyannopoulos G, Nienhuis AW, Leder P, Majerus PW (eds) (1987) The molecular basis for blood diseases. WB Saunders, London

[3] Butturini A, Gale RP (1990) Oncogenes and leukaemia. Leukemia 4:138–160
Collins SJ (1987) The HL-60 promyelocytic leukemia cell line: proliferation, differentiation, and cellular oncogene expression. Blood 70:1233–1244

[4] Graf T, Beug H (1978) Avian leukemia viruses, interaction with their target cells in vivo and in vitro. Biochim Biophys Acta 516:269–299
Weiss R, Teich N, Varmus H, Coffin J (1985) RNA tumor viruses, molecular biology of tumor viruses, 2nd edn. Cold Spring Harbor Laboratory, New York

[5] Beug H, Leutz A, Kahn P, Graf T (1984) Ts mutant of E26 leukemia virus allow transformed myeloblasts, but not erythroblasts or fibroblasts to differentiate at the nonpermissive temperature. Cell 39:579–588
Radke K, Beug H, Kornfeld S, Graf T (1982) Transformation of both erythroid and myeloid cells by E26 an avian leukemia virus that contains the *myb* gene. Cell 31:643–653

[6] Frykberg L, Metz T, Brady G, Introna M, Beug H, Vennstrom B, Graf T (1988) A point mutation in the DNA binding domain of the v-myb oncogene of E26 virus confers temperature sensitivity for transformation of myelomonocytic cells. Oncogene Res 3:313–322

Golay J, Introna M, Graf T (1988) A single point mutation in the v-ets oncogene affects both erythroid and myelomonocytic cell differentiation. Cell 55:1147–1158

[7] Introna M, Golay J, Frampton J, Nakano T, Ness SA, Graf T (1991) Mutations in v-myb after the differentiation of myelomonocytic cells transformed by the oncogene. Cell 63:1287–1297

Ness SA, Marknell A, Graf T (1989) The v-myb oncogene product binds to and activates the promyelocyte-specific mim-1 gene. Cell 59:1115–1125

[8] Adkins B, Leutz A, Graf T (1984) Autocrine growth induced by src-related oncogenes in transformed chicken myeloid cells. Cell 39:439–445

Graf T, Weizsaecker FW, Grieser S, Coll J, Stehelin D, Patschinsky T, Bister K, Bechade C, Calothy G, Leutz A (1986) V-mil induces autocrine growth and enhances tumorigenicity in v-myc-transformed avian macrophages. Cell 45:357–364

[9] Baumbach WR, Stanley ER, Cole MD (1987) Induction of clonal monocyte-macrophage tumors in vivo by a mouse c-myc retrovirus: rearrangement of the CSF-1 gene as a secondary transforming event. Mol Cell Biol 7:664–671

Klingler K, Johnson GR, Nicola NA, Arman G, Kluge N, Ostertag W (1988) Transformation of single myeloid precursor cells by the malignant histocytosis sarcoma virus (MHSV): generation of growth factor-independent myeloid colonies and permanent cell lines. J Cell Physiol 135:32–38

Klinken SP, Alexander AW, Adams JM (1988) Hemopoietic lineage switch: v-raf oncogene converts Eu-myc transgenic B cells into macrophages. Cell 53:857–867

[10] Gonda TJ, Ramsay RG, Johnson GR (1989) Murine myeloid cell lines derived by in vitro infection with recombinant c-myb retroviruses express myb from rearranged vector proviruses. EMBO J 8:1767–1775

Shen-Ong GL, Wolff L (1987) Moloney murine leukemia virus-induced myeloid tumors in adult Balb/c mice: requirement of c-myb activation but lack of v-abl involvement. J Virol 61:3721–3725

[11] Croce CM (1987) Role of chromosome translocations in human neoplasia. Cell 49:155–156

Daley GQ, Van Etten RA, Baltimore D (1990) Induction of chronic myelogenous leukemia in mice by the p210 bcr/abl gene of the Philadelphia chromosome. Science 247: 824–830

Elefanty AG, Hariharan IK, Cory S (1990) Bcr-abl, the hallmark of chronic myeloid leukemia in man, induces multiple hemapoietic neoplasms in mice. EMBO J 9:1069–1078

Goldman JM, Grosveld G, Baltimore D, Gale RP (1990) Chronic myelogenous leukemia, the unfolding saga. Leukemia 4:163–167

Longo L, Pandolfi PP, Biondi A, Rambaldi A, Mencarelli A, Lo Coco F, Diverio D, Pegoraro L, Avanzi G, Tablio A, Zangrilli D, Alcalay M, Donti E, Grignani F, Pelicci PG (1990) Rearrangement and aberrant expression of the retinoic acid receptor α gene in acute promyelocytic leukemias. J Exp Med 172:1571–1576

[12] Ahuja HG, Foti A, Bar-Eli M, Cline MJ (1990) The pattern of mutational involvement of RAS genes in human hematologic malignancies determined by DNA amplification and direct sequencing. Blood 75:1684–1690

Ridge SA, Worwood M, Oscier D, Jacobs A, Padua RA (1990) Fms mutations in myelodysplastic, leukemic, and normal subjects. Proc Natl Acad Sci USA 87:1377–1380

[13] Mashal R, Shtalrid M, Talpaz M, Kantarjian H, Smith L, Beran M, Cork A, Trujillo J, Gutterman J, Deisseroth A (1990) Rearrangement and expression of p53 in the chronic phase and blast crisis of chronic myelogenous leukemia. Blood 75:180–189

Rambaldi A, Wakamiya N, Vallenga E, Horiguchi J, Warren MK, Kufe D, Griffin JD (1988) Expression of the macrophage colony-stimulating factor and c-fms genes in human acute myeloblastic leukemia cells. J Clin Invest 81:1030–1035

[14] Caracciolo D, Valtieri M, Venturelli D, Peschle C, Gewirtz AM, Calabretta B (1989) Lineage specific requirement of c-abl function in normal hematopoiesis. Science 245:1107–1110

Caracciolo D, Venturelli D, Valtieri M, Peschle C, Gewirtz AM, Calabretta B (1990) Stage related proliferative activity determines c-myb functional requirements during normal human hematopoiesis. J Clin Invest 85:55–61

Emilia G, Donelli A, Ferrari S, Torelli U, Selleri L, Zucchini P, Moretti L, Venturelli D, Ceccherelli G, Torell G (1986) Cellular levels of mRNA from c-myb, c-myc, and c-fes onco-genes in normal myeloid and erythroid precursors of human bone marrow: an in situ hybridization study. Br J Haematol 62:287–292

Gewirtz AM, Calabretta B (1988) A c-myb antisense oligodeoxynucleotide inhibits normal human hematopoiesis in vitro. Science 242:1303–1306

Golay J, Capucci A, Arsura M, Castellano M, Rizzo V, Introna M (1991) The expression of c-myb and B-myb, but not A-myb correlates with proliferation in human hematopoietic cells. Blood 77:149–158

Kastan MB, Slamon DJ, Civin Cl (1989) Expression of proto-oncogene c-myb in normal human hematopoietic cells. Blood 73:1444–1451

[15] Bertani A, Polentarutti N, Sica A, Rambaldi A, Mantovani A, Colotta F (1989) Expression of c-jun protooncogene in human myelomonocytic cells. Blood 74:1811–1816

Calabretta B (1987) Dissociation of c-fos induction from macrophage differentiation in human myeloid leukemic cell lines. Mol Cell Biol 7:769–774

Colotta F, Wang JM, Polentarutti N, Mantovani A (1987) Expression of c-fos protooncogene in normal human peripheral blood granulocytes. J Exp Med 165:1224–1229

Introna M, Hamilton TA, Kaufman RE, Adams DO, Bast RC

(1986) Treatment of murine peritoneal macrophages with bacterial lipopolysaccharide alters expression of *c-fos* and *c-myc* oncogenes. J Immunol 137:2711–2715

Introna M, Bast RC, Tannenbaum CS, Hamilton TA, Adams DO (1987) The effect of LPS on expression of the early "competence" genes JE and JC in murine peritoneal macrophages. J Immunol 138:3891–3896

Introna M, Bast RC, Johnston PA, Adams DO, Hamilton TA (1987) Homologous and heterologous desensitization of proto-oncogene *c-fos* expression in murine peritoneal macrophages. J Cell Physiol 131:36–42

Resnitzky D, Yarden A, Zipori P, Kimchi A (1986) Autocrine β-related interferon controls *c-myc* suppression and growth arrest during hematopoietic cell differentiation. Cell 46:31–40

Sariban E, Mitchell T, Kufe D (1985) Expression of the *c-fms* proto-oncogene during human monocyte differentiation. Nature 316:64–66

Chapter 35 Developmental Regulation of Human Globin Genes: a Model for Cell Differentiation in the Hematopoietic System

SERGIO OTTOLENGHI

35.1 Introduction

The differentiation of the erythroid cells begins in an early progenitor, representing a minimal proportion (<0.1%) of the cell population in hematopoietic organs, like bone marrow or fetal liver. To understand the mechanism of erythroid differentiation, it is therefore necessary to start from the analysis of the regulation of genes involved in the terminal phases of differentiation, like globin genes, and then to move back to investigate how genes regulating globin expression are themselves regulated. This chapter will mainly describe information obtained in the human system. This system has the obvious disadvantage that human beings cannot be experimentally manipulated like other animal species; however, many inherited mutations affecting globin gene regulation exist in the 5 billion individuals representing the human population, and these "mutants" often spontaneously refer themselves to doctors and then to investigators. Inferences derived from the study of these "mutants" can then be tested by introducing mutant globin genes into cells in culture or by generating mice transgenic for manipulated human genes.

35.2 The Human Globin Genes: a Paradigm for Cell Type-Specific Gene Expression in the Erythroid Lineage

35.2.1 Distant DNA Sequences 5' to the Globin Gene Clusters Regulate the Expression of All Globin Genes

The human globin genes are clustered in two different regions, on chromosome 11 (ε, $^G\gamma$, $^A\gamma$, δ, and β globin genes: the β-like globin cluster) and 16 (ζ, α_2, and α_1; the α-like globin cluster). These genes are subject to two different types of regulation: they must all be expressed specifically in erythroid cells, and they must individually be expressed at specific times during development (i.e. ζ and ε globin genes during the embryonic period; $^G\gamma$, $^A\gamma$ globin genes during the fetal period; α globin genes from the fetal period onwards, and β globin genes predominantly during the adult period). Thus HbF (fetal) is represented by the tetramers $\alpha_2{}^G\gamma_2$ and $\alpha_2{}^A\gamma_2$, respectively. HbA is the major adult Hb (97%), HbA$_2$ the minor species (2.5% of total Hb). Furthermore, the balance between synthesis of α- and β-like globins must always be close to 1; in humans, genetic defects in the expression of either α-globin or β-globin cause intracellular precipitation of the excess globin within erythroid cells, leading to premature cell death and disease, usually of severe gravity in homozygosis. These diseases are called thalassemias; the very severe homozygous type of β-thalassemia is also known as Cooley's anemia. Hundreds of thousands of patients are affected overall in the world by various types of thalassemic diseases.

35.2.2 Deletions Causing γ-δ-β-Thalassemia May Remove a Region Controlling Globin Gene Expression[1]

Large deletions within the β-globin cluster may simultaneously remove several β-like globin genes, leading to multiple defects; in γ-δ-β-thalassemia both the two fetal and the two adult globin genes are inactive (Fig. 1). Surprisingly, in a few cases the β-globin gene appeared to be intact, and it was possible to show that the cloned β-globin gene, when transfected into cells grown in culture, behaves as a

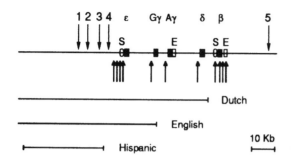

Fig. 1. Deletions within the β-like globin regions in γδβ-thalassemia remove several DNAse-hypersensitive sites. *Arrows pointing downwards* show the erythroid specific, nondevelopmentally regulated HS sites; *small arrows pointing upwards* indicate erythroid specific-developmentally regulated sites. γ-globin sites are ~120 nt upstream to the CAP site; and β-globin promoter sites are 75–175 nt upstream to the CAP site. Three additional sites in the β-globin gene map in the second intron, in the third exon (~200 nt 5' to the poly A addition site) and ~800 nt 3' to the poly A addition site. *E* enhancer; *S* silencer. The two β-globin silencer regions are between −610 to −490 and −338 to −233 relative to the transcription initiation (CAP) site; the ε silencer region is between −177 and −392 relative to the CAP site. The extent of the deletions is shown by *horizontal bars. E* enhancer; *S* silencer. Note that HS *sites 1–4* are numbered by some authors in the reverse order

normal β-globin gene. The common defect in all of these types of γ-δ-β-thalassemia is the loss of large DNA regions upstream to the ε-globin gene. This observation suggest that the deletions remove a DNA region (upstream to the ε-globin gene) that somehow controls globin gene expression in *cis*.

35.2.3 The Locus Control Region (LCR) Allows High-Level, Erythroid-Specific, Position-Independent Gene Expression[2,3]

Globin genes transfected into erythroid cells and integrated at random positions in chromatin are expressed at relatively low levels, while exogenous globin genes integrated (by homologous recombination) within the globin cluster are expressed at essentially normal levels. Thus, regions flanking (at a distance) the β-globin gene might affect its behavior. To evaluate the possible regulatory role of distant DNA regions in *cis* to the β-like globin cluster, DNA fragments including the region upstream to

the ε-globin gene were linked to a human β-globin gene, and used to generate transgenic mice (for methods see Wagner and Keller, Chap. 30, this Vol) carrying a low copy number of the transgene. All of the mice in which the transgene remained intact, expressed β-globin in the appropriate cells at levels comparable (on a per-copy basis) to those expected for the endogenous mouse β-globin gene; in contrast, β-globin genes not linked to the distant DNA region were silent or expressed at very low levels (~1%). Many similar experiments have now been carried out, and support several general notions: the region upstream to the ε-globin gene (Dominant Control Region, DCR or Locus Control Region, LCR) contains elements capable of activating erythroid specific (globin) or non-erythroid-specific genes in an erythroid specific fashion in transgenic mice and in cell culture. The expression level is relatively independent of the location (upstream or downstream) and orientation of the LCR relative to the tested gene. These properties are those of a typical enhancer; however, genes linked to the LCR (but not genes unlinked to it) are expressed at high levels independently of the site of integration in the mouse DNA. This property is not usually observed with conventional enhancers and might reflect unique features of the LCR.

35.2.4 Crucial Elements of the LCR Are Marked by DNase Hypersensitive Sites (HS Sites)[4]

High sensitivity of specific DNA sequences within chromatin or nuclei to DNase digestion has long been known to correlate with expression of adjacent genes. When erythroid cell nuclei are analyzed using very low DNase levels, short regions of selectively very high sensitivity (HS sites) are identified few kb pairs upstream to the ε-globin gene and (one site) ~18 kb downstream to the β-globin gene (Fig. 1). The upstream HS sites map within the LCR. The HS sites are present exclusively in erythroid cells and, significantly, at all stages of development; in contrast, minor erythroid specific DNase-sensitive sites present in the immediate vicinity of γ- and β-globin genes are developmentally regulated together with the respective genes.

35.2.5 LCR Activity May Be Mediated by Long-Range Effects on Chromatin Structure [2,5]

The observations reported above, taken together with the functional effects of the LCR in transgenic mice experiments, have suggested that the LCR may "open" (in erythroid cells only) the chromatin in its vicinity, allowing the globin genes to be regulated by their individual transacting factors in a stage-specific manner. The significance of the correlation between erythroid-specific HS sites, their location in the LCR, and function is further stressed by other observations. In mice made transgenic for LCR-linked globin genes, the chromatin in the region of the HS sites is organized, in erythroid cells, so as to reconstitute a DNase I sensitive structure. Similarly, if normal nonerythroid human cells (not showing HS sites in their chromatin) are fused, in vitro, to mouse erythroid cells, the DNase I sensitive sites are formed concurrently with activation of the human globin genes (induced by mouse erythroid cell transacting factors). In these cells, DNase I sensitivity extends as far as ~ 100 kb 3' to the β-globin gene. However, nonerythroid cells from the patient with the Hispanic type of γ-δ-β-thalassemia (who lacks the region including HS 1 to HS 3) fail to both activate the globin genes and to develop DNase sensitivity on the intact HS 4 and HS 5 when fused to mouse erythroid cells. In these cells the whole region extending from ~ 100 kb 5' to the ε-globin gene to ~ 100 kb 3' to the β-globin gene remains in a DNase I insensitive configuration.

35.2.6 Individual HS Sites Are Sufficient for Position-Independent High-Level-Expression [6-8]

Recent work has shown that the functional effect of the whole LCR can be reproduced by short (a few hundred nucleotides) regions surrounding the individual HS sites; apparently, the full effect (or almost so) only requires sequences within HS 2 or HS 3.

The analysis of the sequences within HS sites has disclosed binding motifs for two erythroid-specific factors (GATA-1 and NFE 2, see below) and a po-

tent conventional enhancer segment in HS 3, also containing NFE 2 binding motifs. None of these elements alone is sufficient for allowing position-independent expression of linked genes. Mutations within the NFE 2 binding sites of the HS 3 enhancer still allow position-independent expression, albeit at a greatly decreased level. Thus, the LCR may be constituted by the combination of a strong enhancer with other elements mediating position-independent expression. While the enhancer can stimulate gene expression in transient assays, other individual subfragments mediating position-independent expression do so only in stable assays, indicating a need for integration into chromatin.

Finally, the concept of LCR may be of general importance. An LCR region has been described upstream to the α-globin cluster; examples of similar mechanisms have also been provided for genes expressed in other cell types, like lymphocytes and hepatocytes.

35.3 Erythroid and Stage-Specific Elements Mediating Expression of Individual Globin Genes

35.3.1 The Hemoglobin Pattern Is Modulated Throughout the Whole Development Within a Single Cell Population [9]

In humans, there are two main globin switches; the embryonic → fetal and the fetal → adult switches. In the former switch, occurring between the 5th and the 8th week of gestation, embryonic ε and ζ globin gene expression is suppressed, while α-, γ-, and to a minor extent β-globin synthesis are activated; in the latter, occurring at a fixed period coinciding with the expected (not the actual) time of birth, γ-globin synthesis is almost completely repressed and full δ- and β-globin synthesis is activated.

Progenitor cells obtained at the time of the switches can be allowed to differentiate in vitro to erythroblasts and individual clonal erythroid colonies can be analyzed for their hemoglobin synthetic patterns. Individual embryonic colonies can synthesize variable levels of ε-, γ-, and very low levels of β-globin chains, showing that a single type of embryonic progenitor generates mature cells capable of both embryonic and fetal-adult globin synthesis. Similarly, individual colonies obtained at birth from cord blood progenitors synthesize both γ- and β-globin. If two different populations of progenitors existed (one with fetal and the other with adult properties) individual colonies should show only one of two types (fetal or adult) of γ/β-globin synthetic ratios. In contrast, the data show a continuum of ratios, ranging between typical fetal (i.e., ~80% γ, ~20% β) and adult levels (Fig. 2). These data show that the switches are not mediated by changes in the relative proportions of different cell populations (embryonic, fetal, and adult), but rather by gene regulation within a single cell population, from the embryonic throughout the adult period. For these reasons, the search for erythroid specific elements has always been closely linked to that of stage specific elements.

35.3.2 Inherited Defects of the Regulation of Fetal Hemoglobin Synthesis in Humans (Hereditary Persistence of Fetal Hemoglobin, HPFH) Provide Clues to the Identification of Regulatory Elements[10,11]

Individuals heterozygous for one of the several types of HPFH express, in the adult period, one or

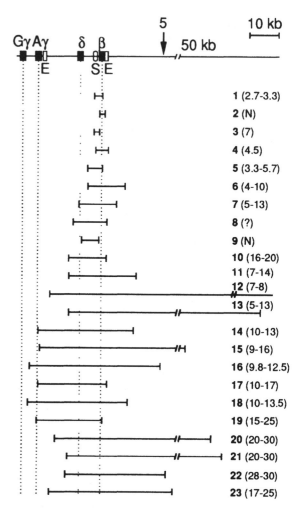

Fig. 3. Deletions within the β-like globin cluster in β-thalassemia, δβ-thalassemia and Hereditary Persistence of HbF. HbF levels, in heterozygous carriers, are indicated within parentheses when available; *N* indicates nonincreased levels, ? indicates increased, but not well-defined levels. *Horizontal bars* indicate the deleted DNA regions. The *numbers (1 to 23)* refer to individual types of mutation. *1* Turkish; *2* Indian; *3* Black; *4* Thai; *5* Czech; *6* Dutch; *7* Sicilian; *8* Black; *9* Lepore, *10* Italian 2; *11* Macedonian; *12* Japanese; *13* Spanish; *14* Black; *15* Chinese; *16* German; *17* Indian 2; *18* Turkish; *19* Kenya; *20* HPFH 2; *21* HPFH 1; *22* Italian 1; *23* Indian 3. Note that the simple loss of the β-globin promoter and enhancer regions is sufficient for generating increased HbF levels (deletions *1, 3–5*), although larger deletions (*7*) have greater effects. In addition, compare the δ-β fusion Lepore gene (deletion *9*), containing a functional β-like globin promoter and the β-globin enhancer, with deletion *7*, that has lost the β-globin gene; only the latter case shows high HbF. Note that HbF levels in Kenya patients include the contribution of the hybrid α_2 $^A\gamma$-β_2 molecule (Hb Kenya) (deletion *19*)

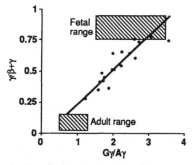

Fig. 2. Erythroid colonies developed in vitro from cord blood progenitors collected at birth show a continuum of γ/γ + β and Gγ/Aγ globin synthetic ratios, ranging between typical fetal and typical adult levels. Thus the switch occurs by gene regulation in a single cell line, not by substitution of an adult cell clone for a fetal one. Note that the γ/γ + β ratio usually obtained in in vitro culture of adult cells is higher than that shown in vivo (<0.01)

both of the nonallelic $^G\gamma$- and $^A\gamma$-globin genes at level exceeding by 15–200-fold the expected one. In these individuals fetal hemoglobin (HbF, $\alpha_2\gamma_2$) may represent more than 30% of total hemoglobin, in contrast to the normal adult level of less than 1% HbF. Two types of defect have been demonstrated; in the first group, deletions within the 3' end of the β-like globin cluster are almost invariably associated with increased expression of both γ-globin genes (or of the $^G\gamma$-globin gene, if the $^A\gamma$-globin gene is eliminated by the deletion) in cis to the deletion (Fig. 3). The analysis of γ-globin expression in individuals heterozygous for polymorphic γ-globin genes shows that this effect occurs strictly on the γ-globin genes lying on the deleted chromosome, not on those on the normal chromosome. In some cases the deletions bring into close proximity of the γ-globin genes an enhancer element lying 3' to the β-globin gene (see, for example, HPFH in Hb Kenya carriers and Italian-2 HPFH); however, the breakpoints of the deletions are widespread, and deletions increasing HbF production exist, in which all the putative enhancer elements are deleted (see Fig. 3 for explanation). An additional possibility is suggested by the observation that complete deletion

of the β-globin gene is invariably associated with elevated HbF; it is proposed that regulatory elements within the adult β-globin promoter and enhancer (see below) may normally compete with γ-globin elements in cis for interaction with a common enhancer. This model would explain why a type of δ-β-thalassemia (Sicilian) having its 5' breakpoint in the δ-globin gene and the 3' breakpoint just 3' to the β-globin gene shows high HbF, while the deletion generating the hybrid β-like 5'δ-3'β globin Lepore has normal HbF levels (Fig. 3).

A different group of HPFHs is characterized by point mutations in the promoter of the overexpressed $^G\gamma$ or $^A\gamma$-globin gene (Table 1); the same mutations have been observed repeatedly in HPFH cases from disparate populations, and within different polymorphic gene frameworks, but never in normal individuals; these observations strongly support a causal role of these mutations in the overexpression of the mutated gene. One of these mutations ($-175\,T \rightarrow C$) greatly decreases the in vitro binding to the promoter of the ubiquitous transcription factor OTF1, and creates a better binding site for an erythroid specific factor, GATA-1; the latter change appears to be responsi-

Table 1. Point mutations in the γ-globin promoter in HPFH. Their effects on HbF synthesis and on the activity of the γ-globin gene

Mutation	Geographic origin	Chain type	HbF%	cis/trans
$-202\,C \rightarrow G$	Africa	G	15–25	$-0.89\,(\beta^A/\beta^S)$
$-202\,C \rightarrow T$	Africa	A	3	
$-198\,T \rightarrow C$	England	A	8	
$-196\,C \rightarrow T$	Italy	A	15	$-0.7\,(\beta^A/\beta^A)$
$-196\,C \rightarrow T$	Italy (Sardinia)	A	15–20	
$-196\,C \rightarrow T$	China	A	15	
$-196\,C \rightarrow T$	China	A	15	
$-175\,T \rightarrow C$	Italy	G	20	
$-175\,T \rightarrow C$	Africa	G	30	$-0.67\,(\beta^A/\beta^S)$
$-175\,T \rightarrow C$	Africa	A	30	$-0.30\,(\beta^A/\beta^C)$
$-161\,G \rightarrow A$	Africa	G	1–2	
$-158\,C \rightarrow T$	Several regions	G	1–3	
$-117\,G \rightarrow A$	Italy	A	12	$0.8\,(\beta^A/\beta^A)$
$-117\,G \rightarrow A$	Greece	A	15	$0.6-0.7\,(\beta^A/\beta^A)$
$-117\,G \rightarrow A$	Africa	A	15	
$-114\,C \rightarrow T$	Japan	G	10–14	
-115 to -103 Deletion	Africa	A	30	$0.45\,(\beta^A/\beta^S)$

Identical mutations in different populations probably have independent origins, as they are linked to different restriction enzyme polymorphic sites. The activity of the β-globin gene adjacent (in cis) to the HPFH mutation relative to that of the β-globin gene on the non-HPFH chromosome (in trans) is calculated from data on families carrying structurally mutant or $\beta°$-thalassemnic globin genes and HPFH. For more details, see refs. 11 and Ottolenghi et al. (1989).

ble (on the basis of transfection experiments) for at least part of the overactivity of the mutated γ-globin gene. A different mutation, $-198\,T\to C$, generates a relatively strong binding site for the ubiquitous factor Sp1; once again, this mutation accounts for increased erythroid specific activity of the mutated promoter, possibly by cooperation with downstream GATA-1-binding sites (Fig. 4). In conclusion, these inherited defects point to different long- and short-range mechanisms mediating alterations of the relative proportions of fetal and adult globins. In no case, however, was any evidence

found for stage-specific factors capable of explaining the high level of fetal globin expression in adults. It would rather appear, instead, that the relative proportion of γ- and β-globin synthesis depends, to some extent at least, on competition between the two sets of genes. Indeed, genetic analysis (based on the study of the expression of "marker" mutant β-globin genes in *trans* to HPFH mutations) shows (Table 1) that several HPFH point mutations are correlated with significantly decreased expression of the β-globin gene in *cis*, an effect mirroring the increased γ-globin expression observed in *cis* to deletions in the β-globin gene region. Experiments in transgenic mice (see below) support these ideas.

35.3.3 Transgenic Mice. Globin Gene Regulatory Elements Are Located Both 5′, Within, and 3′ to the Genes[12]

To identify regulatory elements, chimeric genes have been constructed by fusing the promoter of either β-, γ- and α-globin genes or of heterologous genes, to the body or 3′ parts of different globin genes; these chimeric constructs have been used for generating transgenic mice and in transfection experiments.

Studies in transgenic mice in particular indicate that human γ- and β-globin genes can be correctly expressed in erythroid cells at the appropriate stages. A variety of studies show that both the γ- and β-globin promoters and enhancers elements are responsible for erythroid specific expression of the globin genes. In particular, an enhancer (possibly containing erythroid specific elements) is located $0.5-1$ kb downstream to the $^A\gamma$-globin gene; two erythroid specific enhancers are located within the second β-globin gene intron, the third exon, and 3′ to the β-globin gene (Fig. 5). However, the sequences within these broad domains which are responsible for stage-specific expression of the globin genes are not yet clearly identified. Sequences in the γ-globin promoter comprised between -201 and -136 allow fetal-like expression of the β-globin gene when placed upstream to the β-globin promoter; nothing is known about sequences regulating β-globin expression in the adult period.

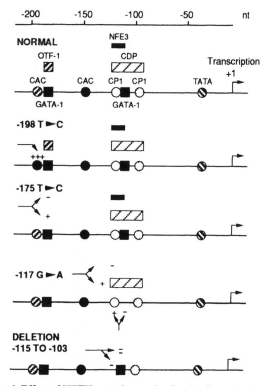

Fig. 4. Effect of HPFH mutations on in vitro binding of nuclear proteins to the γ-globin promoter. The positions of binding sites are shown; factors binding at overlapping sites are indicated by *symbols on the top of the drawing*. The effect of the mutation is indicated by *arrows*; + means better binding, − decreased binding. Note that the only common effect of the -117 and -115 to -103 deletion is the decreased binding of the protein NFE3; it has been inferred that this erythroid-specific protein may be an inhibitor of γ-globin expression. *CDP* is the CCAAT displacement protein, and competes for binding with CP1 (CCAAT box-binding protein). *OTF-1* is a ubiquitous protein binding the ATGCAAAT octamer. For *CACCC* and *GATA-1* see text, Section 35.3.5)

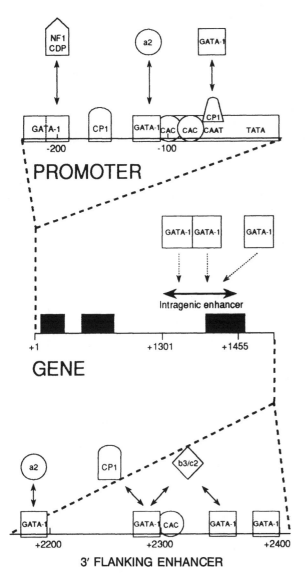

Fig. 5. Schematic representation of the human β-globin gene and its binding sites for transcriptional factors (modified from deBoer et al. 1988). *Double arrows* indicate overlapping binding sites for two or more factors. For *GATA-1, CACC, CDP,* and *CP1* see Fig. 4 and text; *NF1* nuclear factor 1; *a2* and *b3/c2* are not yet characterized proteins

The 3' β-globin enhancer, when fused to the γ-globin gene, induces its expression in the adult period (mimicking the HPFH effect of the natural Hb Kenya mutation, see Fig. 3). However, this 3' enhancer cannot be responsible, per se, for the adult only expression of the β-globin gene, as the enhancer is also active during the embryonic period (in transgenic mice).

35.3.4 In Transgenic Mice the LCR Linked to Individual Globin Genes Overrides Their Stage-Specific Regulation [13,14]

Constructs containing the γ- and β-globin genes, with their promoter and enhancer elements, but withouf LCR, are expressed at the correct developmental stages in transgenic mice, though at low levels. However, it LCR sequences are linked to individual γ- or β-globin genes, each is expressed at both the expected and inappropriate stages. However, if γ- and β-globin genes are both linked, within the same construct, to LCR sequences, the pattern of expression observed in transgenic mice is again reminiscent of the expected one, γ being expressed mainly in the embryonic-fetal period and β in the adult period. Further experiments have shown that individual globin genes may be differentially susceptible to the induction of inappropriate developmental activation by LCR. The developmental regulation of the β-globin gene is easily overridden by a linked LCR, while the expression of the ε-globin gene remains strictly embryonic, possibly due to its strong silencer (Fig. 1). The γ-globin gene shows intermediate sensitivity; in fact, certain, but not all, LCR-γ-globin constructs can be almost completely silenced in the adult period, even in the absence of a linked β-globin gene. Additional experiments, in which the order of the γ- and β-globin genes relative to the LCR has been reversed, show that the gene closer to the LCR is the one more easily activated (polar effect). These results directly confirm the existence of a competition (already inferred on genetic grounds, see Sect. 35.3.2) between γ- and β-globin genes for cooperation with sequences within the LCR.

35.3.5 Binding Sites for the Same Erythroid-Specific and Ubiquitous Factors Are Present in the Globin Promoter, Enhancer, and LCR Regions [7,11,13,15,16]

The β-globin promoter contains binding sites for the erythroid specific factor GATA-1 and for ubiq-

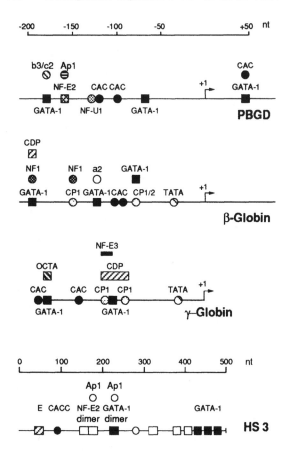

Fig. 6. Binding sites for the same erythroid and ubiquitous nuclear factors are present in erythroid-specific promoters and in HS sites. The prevalent binding factors (from in vitro experiments) are shown as bound to the DNA; factors binding at overlapping sites are indicated on *top*. (For *abbreviations* see Figs. 4 and 5.) *AP-1* indicates a site capable of binding Jun-Fos homo- or heterodimers; *E* a site for an erythroid protein (not characterized). *PBGD* porpholibinogen deaminase gene

uitous factors (the CCAAT-binding protein CP 1 and the CACCC-box binding proteins, among which is Sp 1) (Fig. 5). Remarkably, the β-globin enhancers contain several GATA-1-binding sites, as well as CCAAT and CACCC boxes; additionally, the γ-globin promoter contains duplicated GATA-1-binding sites, duplicated CCAAT boxes, and a CACCC box; note that the two HPFH mutations discussed above strengthen the distal GATA-1-binding site and create a new CACCC box, respectively. Finally, HS 2 and HS 3 in the LCR contain several binding sites for GATA-1 and an addi-

tional erythroid specific factor NFE 2 at a short distance and alternating with CACCC-box like sequences (Figs. 4–6). The similarity of binding sites in promoter, enhancer, and LCR regions of the β-globin cluster has stimulated the development of models based on competition for the regulation of the β-globin cluster (Fig. 7) during normal development (Fig. 7A) and in HPFH (Fig. 7B).

35.3.6 GATA-1 and NFE-2 Binding Sites Are Present in Several Erythroid-Specific Genes[17-19]

The identification of the sequences capable of binding GATA-1 and NFE 2, rapidly led to the discovery of similar motifs in several nonglobin genes expressed in an erythroid specific manner (Table 2). A particularly interesting case is represented by the erythroid-specific promoter of porphobilinogen deaminase (PBGD); both a GATA-1 and a NFE-2 site are present in this promoter and are essential for its activity.

35.3.7 Elements Regulating Globin Gene Expression Are Conserved in Evolution[20]

The general scheme outlined above for the human globin genes has interesting similarities in several animal species. For example, a 3' β-globin enhancer has been found in species as evolutionarily distant from man as chicken and duck. Sequences resembling those in the HS sites of the human LCR are found in mouse and goat β-globin cluster regions, as well as upstream to the human α-globin cluster; as the β-globin gene diverged from the α-globin gene ~ 500 million years ago, this observation possibly suggests a very ancient origin of the HS regions. Targets for transcriptional factors are themselves highly conserved; the AGATAG motif recognized by GATA-1 is present and functional in chicken globin promoters and enhancers, as well as (together with the NFE 2 binding site) in the β-globin LCR. It is not surprising then that some general mechanisms for regulation are shared between distant species; competition between the adjacent chicken ε- and β-globin genes for interaction with a common enhancer lying between the two genes

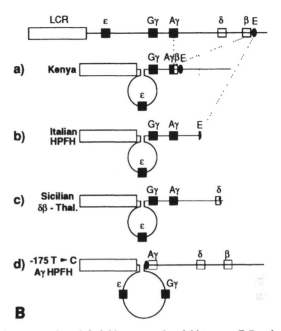

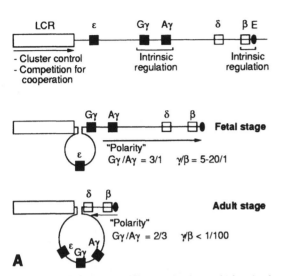

Fig. 7. A Regulation of the β-like globin cluster during development. The γ- and β-globin genes are inherently regulated by sequences within the genes themselves or in their immediate vicinity (see experiments in transgenic mice). The LCR, linked to either the γ- or the β-globin gene, may override this control; however, when both γ- and β-globin genes are linked to the LCR, developmental control is restored. It is speculated that during development the embryonic, fetal, and adult globin genes may each preferentially "interact" with the LCR, at their individually appropriate times; this results in preferentially enhanced activity of the interacting gene, and relative inhibition of the others by competition. "Polarity" in the effects of LCR is suggested by the observation that precise γ/β and $^G\gamma/^A\gamma$ ratios (correlating with their positions) exist during development; moreover, distance from LCR and relative positions of linked genes affect their expression in transgenic mice (the gene closer to LCR is more easily activated). Since the natural gene order is ε-γ-β-globin, the LCR-mediated activation of the β-globin gene may require preliminary inhibition of the γ-globin genes. Complete suppression of γ-globin gene activity may then be mediated by competition between activated β-globin gene and γ-globin genes. **B** Regulation of the β-like globin cluster in HPFH. Correct developmental control may be subverted by deletions, which remove the competing β-globin region (*a*, *b*, and *c*) and in some cases (*a* and *b*) bring new enhancer elements into the γ-globin region; these result in HPFH. Conversely, point mutations in the γ-globin promoter generating new binding sites for transcription factors allow adult expression of the mutated γ-globin gene and inhibit δ-β globin expression in *cis* (*a*). The fact that normal γ-globin genes may be expressed at almost full levels in deletion HPFH implies that stage-specific factors may not be required for adult expression of a γ-globin gene; the same conclusion is suggested by HPFH due to point mutations creating binding sites for nongene-specific factors (GATA-1 and Sp1). In mice, and possibly in humans, GATA-1 levels are lower in fetuses than in adults; if the γ-globin genes have higher affinity for GATA-1 than the β-globin gene, the latter would only cooperate with the LCR in adult age, but not in fetal age. Thus, changing levels of a nongene-specific factor might affect developmental regulation of globin genes

has been reported. In chicken, globin promoter function may be regulated primarily by Eryf1 (a factor homologous to GATA-1, see below), and by CACCC binding and other proteins, which show changes in abundance during development; notably, however, an apparently adult stage-specific protein (NFE4) has been detected, that is thought to bind to the β-globin promoter, activating it, and allowing it to preferentially interact with the enhancer. As one might expect, AGATAG-recognizing factors are also conserved. The mouse and human GATA-1 factors are extensively homologous through their whole sequence; the chicken sequence, however, greatly differs at the 5' and 3' regions, but is highly homologous to the mammalian cDNAs in its central part, the DNA-binding domain represented by a region encoding zinc fingers. Interestingly, two additional cDNAs (GATA-2 and GATA-3) have recently been cloned from chicken and mammalian cells. They show over 90% protein

Table 2. Presence of the GATA-1 binding AGATAG (or related) motifs in erythroid- and/or megakaryocytic-specific genes or in genes up-regulated during erythroid development[a]

– Globin genes	α- and β-like promoter and/or	Erythroid
	enhancer regions (man, mouse,	Erythroid
	chicken, duck)	Erythroid
– Erythroid porphobilinogen deaminase promoter[a] (man)		Erythroid
– Glycophorin A promoter (man)		Erythroid
– R-type pyruvate kinase promoter (rat)		Erythroid
– Glutathione peroxidase promoter (man, mouse)		Erythroid
– Band 4.2 (man)		Erythroid
– 5′-Aminolevulinate synthase, ALAS promoter[a] (man)		Erythroid
– Erythropoietin receptor promoter (mouse)		Erythroid and megakaryocytic
– GATA-1 promoter and first intron (man, mouse)		Erythroid and megakaryocytic
– GP II b promoter (man)		Megakaryocytic
– PF 4 promoter (rat)		Megakaryocytic

[a] Asterisks indicate genes showing also NFE-2 binding sites.

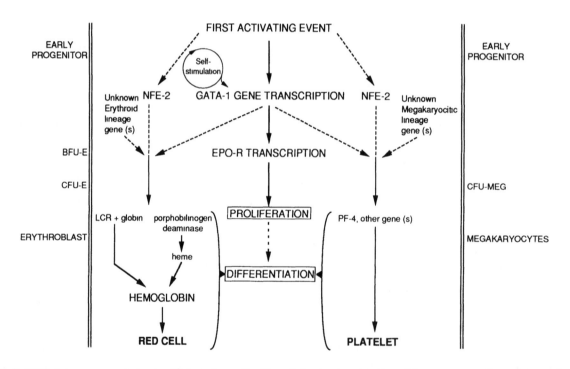

Fig. 8. GATA-1 is necessary, but not sufficient, for erythroid-specific gene expression. The model assumes that GATA-1 is able to self-stimulate its synthesis leading to progressive increase of GATA-1 levels; this stimulates transcription of genes involved in control of proliferation (Epo-Receptor gene) and of differen-tiation and maturation. Other genes are also necessary for erythroid- or megakaryocytic-specific expression. Finally, heme synthesis is dependent on porphobilinogen deaminase activity, and is required for both translational control and for hemoglobin assembly

sequence homology to the finger region of GATA-1, and recognize the same binding motif. GATA-2 and GATA-3, although expressed in erythroid cells at varying levels during development, are also expressed in nonerythroid cells.

35.4 Several Hematopoietic Cell Types Express GATA-1

35.4.1 Genes Expressed in Megakaryocytes Bind to and Are Regulated by GATA-1 [21]

At least three genes expressed in megakaryocytes show GATA-1-binding motifs in their promoters; of these, the PF4 gene appears to require GATA-1 for expression (Table 2). Interestingly, the promoter of the erythropoietin receptor gene also shows a potential GATA-1-binding site; as the erythropoietin receptor gene is expressed in both the erythroid and megakaryocytic lineages, GATA-1 might regulate the same gene in these two cell types. In addition, NFE2 (as well as GATA-1) is present in megakaryocytes.

35.4.2 The Gene Encoding GATA-1 May Be Activated in a Common Progenitor of Several Lineages [21,22]

In addition to erythroid and megakaryocytic cells, basophilic cell lines and mast cells (the tissue counterparts of basophilic granulocytes) have been found to express the GATA-1 gene. These observations suggest that the gene may be first activated in a common progenitor to all of these cell types. In addition, GATA-1 expression has been detected in mouse myeloid cell lines which are dependent for proliferation (and possibly for differentiation) on either GM-CSF or G-CSF; the dependence on these growth factors is typical of bipotent granulo-monocytic and committed granulocytic progenitors, respectively. Interestingly, switching the GM-CSF-dependent cells to IL-3 increases GATA-1 expression; moreover, the more differentiated G-CSF-dependent cell lines have much lower levels of GATA-1 than the GM-CSF-dependent lines. These data lead

to the suggestion that GATA-1 expression may be first activated in a common progenitor of the erythroid-megakaryocytic-basophilic and myeloid lineages: however, differentiation along the myeloid lineage (and possibly response to G-CSF and GM-CSF) lead to the extinction of GATA-1 expression. Thus GATA-1 shows restricted expression in mature blood cells. It is also of interest that the GATA-1 gene shows several GATA-1-binding sites, which may contribute to its activity; thus, following its initial activation in the unknown progenitor, the GATA-1 gene may be further upregulated by its own product (Fig. 8).

35.4.3 GATA-1 Is Necessary, but not Sufficient, for Erythroid Differentiation [23]

The gene encoding GATA-1 is located on the X-chromosome. It has been possible to destroy the gene by targeted mutation (by homologous recombination) in male embryonic stem cells (ES cells) (see Chap. 29, this Vol., Küehn and Stoye for details of the methods), which have therefore become unable to synthesize any GATA-1. When participating in the development of chimeric mice, these cells give rise to apparently normal myeloid cells, but not to erythroid cells. This result demonstrates that GATA-1 is necessary for the normal differentiation (and/or proliferation) of erythroid cells. However (Fig. 8), neither GATA-1 nor NFE-2 is sufficient for erythroid or megakaryocytic differentiation, as both factors are present in these cell types, yet these cells show clearly distinct differentiation markers. Other unknown genes must be necessary; these observations are in favor of combinatorial models for differentiation (discussed in Kuo and Crabtree, Chap. 32, this Vol.).

35.4.4 The Same Transcription Factors Coordinately Regulate Genes Which Are Involved in Several Aspects of Cell Differentiation as Well as Proliferation [17-19, 23,24]

Figure 8 and Table 2 show that GATA-1 and NFE-2 regulate the genes encoding the major protein products of the cell (globin) together with key enzymes (PBGD, ALAS) for the synthesis of the prosthetic group (heme) necessary for hemoglobin assembly. In addition, the expression of other enzymes and major membrane proteins (glycophorins, band 4.2) is also under the control of GATA-1 and/or NFE-2. Finally, the erythropoietin receptor, mediating erythroid cell proliferation, is also regulated by GATA-1. The inability of ES cells, in which the GATA-1 gene has been destroyed, to participate in the erythroid development is likely to reflect, at least in part, a defect in erythropoietin receptor synthesis. This scheme may be of general significance; mutant mice unable to synthesize Pit-1, a pituitary transcription factor regulating the expression of the genes encoding prolactin, the thyroid stimulating, and the growth hormones, neither produce the relevant hormones nor show the respective pituitary cell types responsible for their synthesis. Thus, Pit-1 may regulate a growth factor receptor in a similar way as GATA-1 regulates the erythropoietin receptor.

35.4.5 Which Mechanisms Are Responsible for Establishing the Cell-Specific Pattern of Expression of Transcription Factors? [18, 21, 22]

Although there is some evidence that GATA-1 may be regulated by its own product (Fig. 8), but not by NFE-2, the mechanism initially establishing its transcription in a progenitor cell is still unknown. Conceivably, a master gene, like MyoD in muscle (see Pinney and Emerson, Chap. 31, this Vol.), might start a cascade of activation of genes encoding transcription factors, like GATA-1 and NFE-2. However, as GATA-1 and NFE-2 are present in several hematopoietic cell types, and are probably activated in a common progenitor (see Sect. 35.4.2), the main problem is now to understand how lineage specificity is achieved within the hematopoietic system.

35.5 Growth Factors Are Involved in the Differentiation of Hematopoietic Cells [22,25]

35.5.1 Transfection into Hematopoietic Cells of Genes Encoding Growth Factor Receptors May Stimulate Proliferation and Allow Survival

Recent experiments in which genes encoding growth factor receptors have been introduced and expressed into hematopoietic cells suggest that the differentiation of hematopoietic cells can be affected by growth factors in two possible ways. According to the first possibility, some growth factors act on progenitors that are already committed to one or few differentiation fates. These cells might exhibit only one or two types of receptors (for example, Epo-R in BFU-Es and CFU-Es, GM-CSF and/or M-CSF receptor in mono or bipotential myeloid-progenitors), whose expression might itself be part of the cell type-specific gene expression pattern. In this instance, growth factor stimulation could be necessary only to allow selective proliferation and survival (Fig. 9A) of the cells exhibiting the appropriate receptor but not necessarily for inducing lineage-specific transcription factors mediating expression of cell type-specific proteins; such a model has been suggested for BFU-Es and CFU-Es. In addition to promoting cell proliferation and survival, growth factors might also induce a subset of genes essential for completing the preestablished differentiation program of committed cell; these genes might already be poised for expression. For example, the EGF-receptor gene was introduced and constitutively expressed in a premast cell line that is unable to differentiate spontaneously and that does not normally express this receptor. EGF treatment allowed the transfected cells to differentiate to mast cells only (Fig. 9B). Presumably, EGF delivered a signal converging onto the same pathway as a physiological signal, bypassing the block to terminal differentiation.

35.5.2 Growth Factors
May Cause Differentiation

An alternative possibility is that progenitors, still uncommitted to specific lineages, have receptors

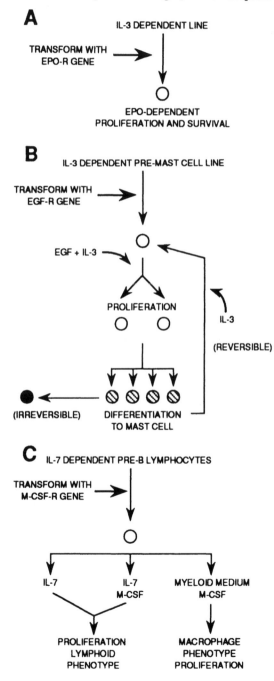

for several growth factors, and that stimulation with a given factor may selectively activate a new differentiation program. For example, IL-3 promotes the proliferation of multipotent progenitors and their differentiation into several types of mature cells, including myeloid, erythroid, megakaryocytic, and other cells. However, combined treatments with IL-1/G-CSF and GM-CSF or IL-1/G-CSF and M-CSF induce the same multipotent progenitors to differentiate only into neutrophils or macrophages, respectively, diverting some progenitors from erythroid or other possible fates. In a recent experiment (Fig. 9C), the M-CSF receptor was expressed (via gene transfection) in B-lymphocytes; M-CSF treatment then caused transdifferentiation of lymphocytes to a monocyte-like phenotype, showing that a new program of gene expression can be activated by certain growth factors.

The two models are not mutually exclusive. For example, it is known that stimulation of one type of receptor may lead to "down-regulation" of other receptors, by mechanisms affecting either the stability of the growth factor receptor mRNA or post-translational events. Additionally, early myeloid cell lines, expressing low levels of GATA-1 when grown in IL-3, can further decrease this level of expression when treated with GM-CSF, suggesting that expression of transcription factors inappropriate for a given lineage (myeloid) may be turned off by growth factors involved in the differentiation of cells belonging to that lineage.

As discussed in the introduction, progenitors have an intrinsic ability to become committed to various developmental fates, in the apparent ab-

Fig. 9A–C. Hematopoietic cells engineered via gene transfection to express the receptor for an inappropriate growth factor, may respond to that growth factor by proliferation and/or differentiation. **A** Lymphocytes (or other cell types) respond to inappropriate erythropoietin stimulation by proliferation, but do not induce any erythroid markers; the cells retain their original phenotype. **B** Undifferentiated mast cells proliferate and complete their original differentiation program in response to inappropriate EGF stimulation; in the majority of the cells, the differentiation is reversible upon EGF withdrawal. **C** Pre-B lymphocytes respond to inappropriate M-CSF stimulation by transdifferentiation to macrophages, which become dependent on M-CSF for further proliferation; transdifferentiation requires optimal environmental conditions ("myeloid medium") and is antagonized by IL-7, a lymphoid growth factor

sence of any extrinsic clues. However, in vivo, progenitors are somewhat compartimentalized within bone marrow stromal cells producing various growth factors. It is thus possible that, in vivo, a stem cell and its progeny may be induced and/or further selected for a specific differentiation pathway and for proliferation according to the nature of their neighboring cells.

35.6 A Complex Network of Interactions at Several Distinct Levels Underlies Terminal Differentiation [19,26]

The differentiation of hematopoietic cells cannot simply be due to cascades of transcription factors. For example, erythroid expression of porphobilinogen deaminase requires selective use of an erythroid-specific promoter (rather than of an alternative ubiquitous promoter) and alternative splicing. The stability of globin mRNA is greatly increased (half-life over 50 h) in erythroleukemic cells induced to differentiate along the erythroid pathway; however, induction of overt megakaryocytic differentiation in a bipotent erythroid-megakaryocytic cell line leads to destabilization and disappearance of γ-globin mRNA. Effects at other less defined levels are exemplified by the need for heme to induce differentiation in erythroid cell lines. Additionally, a specific mRNA binding protein regulates the iron-dependent stability of transferrin receptor mRNA as well as the translation of the mRNAs encoding ferritin and 5'-aminolevulinate synthase (ALAS), the enzyme catalyzing the first step of heme synthesis. This mechanism allows for a tight control of iron intake and utilization for hemoglobin synthesis. Finally, it is of interest that, in chicken erythroblasts, a drug-induced block of the function of ALA-synthetase or of band 3 (an anion transporter) mimics, in part, the transforming effect of the v-erbA oncogene, preventing complete differentiation; v-erbA is a repressor of the transcription of some erythroid genes, like carbonic anhydrase, ALA-S and band 3. Thus, heme and changes in intracellular ionic concentration might affect terminal differentiation.

35.7 Summary and Outlook [27]

The analysis of the structure of normal and mutant globin genes and of other erythroid-specific genes, has allowed the identification of some erythroid transcription factors that are involved in the regulation of multiple erythroid-specific genes. Cloning of the regulatory regions of the genes encoding these transcription factors might make it possible to understand how these genes are primarily regulated in early erythroid progenitors. Techniques enabling the creation, by homologous recombination, of mice deficient in the expression of these genes will make it possible to investigate how important is their activity in the development of the erythroid system. However, additional unidentified genes already appear to be necessary to allow the differentiation of erythroid cells from other cell types, like megakaryocytes, deriving from a common progenitor. The recent cloning of the receptors of most known growth factors should also improve our understanding of how growth factors affect the expression of transcription factors and other cellular proteins.

From a practical standpoint, the information collected in the last decade may be important for the development of new therapeutical approaches to widespread and serious diseases like thalassemia. Short DNA sequences from the LCR will be inserted close to the β-globin gene in vectors designed for efficient introduction of foreign DNA into stem cells; it is hoped that LCR sequences will greatly stimulate the expression of the foreign β-globin gene in the progeny of the stem cells, leading to effective "gene therapy". On the other hand, better understanding of the mechanisms regulating γ-globin gene expression might allow investigation of treatments capable of increasing HbF synthesis in thalassemic patients, to replace HbA.

References

[1] Driscoll MC, Dobkin CS, Alter BP (1989) $\gamma\delta\beta$-thalassemia due to a de novo mutation deleting the 5'ε-globin gene activation-region hypersensitive sites. Proc Natl Acad Sci USA 86:7470–7474

[2] Grosveld F, van Assendelft GB, Greaves DR, Kollias G (1987) Position-independent, high-level expression of the human β-globin gene in transgenic mice. Cell 51:975–985

Nandi AK, Roginski RS, Gregg RG, Smithies O, Skoultchi AI (1988) Regulated expression of genes inserted at the human chromosomal β-globin locus by homologous recombination. Proc Natl Acad Sci USA 85:3845–3849

[3] Talbot D, Collis P, Antoniou M, Vidal M, Grosveld F, Greaves DR (1989) A dominant control region from the human β-globin locus conferring integration site-independent gene expression. Nature 338:352–354

van Assendelft GB, Hanscombe O, Grosveld F, Greaves DR (1989) The β-globin dominant control region activates homologous and heterologous promoters in a tissue-specific manner. Cell 56:969–977

[4] Forrester WC, Thompson C, Edler JT, Groudine M (1986) A developmentally stable chromatin structure in the human β-globin gene cluster. Proc Natl Acad Sci USA 83:1359–1363

Groudine M, Kohwi-Shigematsu T, Gelinas R, Stamatoyannopoulos G, Papayannopoulou T (1983) Evidence for a locus activating region; the formation of developmentally stable hypersensitive sites in globin expressing hybrids. Proc Natl Acad Sci USA 80:7551–7555

Tuan D, London IM (1984) Mapping of DNase I-hypersensitive sites in the upstream DNA of human embryonic ε-globin gene in K562 leukemia cells. Proc Natl Acad Sci USA 81:2718–2722

Tuan D, Solomon W, Li Q, London IM (1985) The "β-like-globin" gene domain in human erythroid cells. Proc Natl Acad Sci USA 82:6384–6388

[5] Forrester WC, Epner E, Driscoll MC, Enver T, Brice M, Papayannopoulou T, Groudine M (1990) A deletion of the human β-globin locus activation region causes a major alteration in chromatin structure and replication across the entire β globin locus. Genes Dev 4:1637–1649

Forrester WC, Takegawa S, Papayannopoulou T, Stamatoyannopoulos G, Groudine M (1987) Evidence for a locus activating region: the formation of developmentally stable hypersensitive sites in globin expressing hybrids. Nucleic Acids Res 10:10158–10177

[6] Ney PA, Sorrentino BP, Lowrey CH, Nienhuis AW (1990) Inducibility of the HS II enhancer depends on binding of an erythroid specific nuclear protein. NAR 18:6011–6017

Talbot D, Grosveld F (1991) The 5′HS2 of the globin locus control region enhances transcription through the interaction of a multimeric complex binding at two functionally distinct NF-E2 binding sites. EMBO J 10:1391–1398

Tuan DYH, Solomon WB, London IM, Pee DP (1989) An erythroid-specific, developmental-stage-independent enhancer far upstream of the human "βlike globin" genes. Proc Natl Acad Sci USA 86:2554–2558

[7] Caterina JJ, Ryan TM, Pawlik KM, Palmiter RD, Brinster RL, Behringer RR, Townes TM (1991) Human β-globin locus control region: analysis of the 5′ DNase I hypersensitive site HS2 in transgenic mice. Proc Natl Acad Sci USA 88:1626–1630

Philipsen S, Talbot D, Fraser P, Grosveld F (1990) The β-globin dominant control region: hypersensitive site 2. EMBO J 7:2159–2167

Pruzina S, Hanscombe O, Whyatt D, Grosveld F, Philipsen S (1991) Hypersensitive site 4 of the human β globin locus control region. NAR 19:1413–1419

Talbot D, Philipsen S, Fraser P, Grosveld F (1990) Detailed analysis of the site 3 region of the human β-globin dominant control region. EMBO J 9:2169–2178

[8] Eccles S, Sarner N, Vidal M, Cox A, Grosveld F (1990) Enhancer sequences located 3′ of the mouse immunoglobulin λ locus specify high-level expression of an immunoglobulin λ gene in B cells of transgenic mice. New Biol 2:801–811

Greaves DR, Wilson FD, Lang G, Kioussis D (1989) Human CD2 3′-flanking sequences confer high-level, T cell-specific, position-independent gene expression in transgenic mice. Cell 56:979–986

Higgs DR, Wood WG, Jarman AP, Sharpe J, Lida J, Pretorius I-M, Ayyub H (1990) A major positive regulatory region located far upstream of the human α-globin gene locus. Genes Dev 4:1588–1601

[9] Comi P, Giglioni B, Ottolenghi S, Gianni AM, Polli E, Barba P, Covelli A, Migliaccio G, Condorelli M, Peschle C (1980) Globin chain synthesis in single erythroid bursts from cord blood: studies on the γ→β and $^{G}γ→^{A}γ$ switches. Proc Natl Acad Sci USA 77:362–365

Peschle C, Migliaccio AR, Migliaccio G, Petrini M, Calandrini M, Russo G, Mastroberardino G, Presta M, Gianni AM, Comi P, Giglioni B, Ottolenghi S (1984) The embryonic fetal Hb switch in humans: studies in erythroid colonies derived from fetal liver and yolk sac BFU-E. Proc Natl Acad Sci USA 81:2416–2420

[10] Camaschella S, Serra A, Saglio G, Baiget M, Malgaretti N, Mantovani R, Ottolenghi S (1987) The 3′ ends of the deletions of Spanish δβ°thalassemia and Black HPFH1 and 2 lie within 17 kilobases. Blood 70:593–596

Camaschella C, Serra A, Gottardi E, Alfarano A, Revello D, Mazza U, Saglio G (1990) A new hereditary persistence of fetal hemoglobin deletion has the breakpoint within the 3′ β-globin gene enhancer. Blood 75:1000–1005

Elder JT, Forrester WC, Thompson C, Mager D, Henthorn P, Peretz M, Papayannopoulou T, Groudine M (1990) Translocation of an erythroid-specific hypersensitive site in deletion-type hereditary persistence of fetal hemoglobin. Mol Cell Biol 10:1382–1389

Fritsch EF, Lawn RM, Maniatis T (1979) Characterization of deletions which affect the expression of fetal globin genes in man. Nature 279:598–603

Ottolenghi S, Giglioni B, Comi P, Gianni AM, Polli E, Ayquaye CTA, Oldham JH, Masera G (1979) Globin gene deletions in HPFH, δ°β°thalassaemia and Hb Lepore disease. Nature 278:654–656

Saglio G, Camaschella C, Serra A, Bertero T, Rege Cambrin C, Guerrasio A, Mazza U, Izzo P, Terragni F, Giglioni P, Comi P, Ottolenghi (1986) Italian type of deletional hereditary persistence of fetal hemoglobin. Blood 68:646–651

Sanguansermsri T, Pape M, Laig M, Hundrieser J, Flatz G

(1990) β°Thalassemia in a Thai family is caused by a 3.4 kb deletion including the entire β-globin gene. Hemoglobin 14:157–168

11 Giglioni B, Casini C, Mantovani R, Merli S, Comi P, Ottolenghi S, Saglio G, Camaschella C, Mazza U (1984) A molecular study of a family with Greek hereditary persistence of fetal hemoglobin and β thalassemia. EMBO J 3:2641–2645

Mantovani R, Malgaretti N, Giglioni B, Comi P, Cappellini N, Nicolis S, Ottolenghi S (1987) A protein factor binding to an octamer motif in the γ-globin promoter disappears upon induction of differentiation and hemoglobin synthesis in K562 cells. NAR 15:9349–9364

Mantovani R, Malgaretti N, Nicolis S, Ronchi A, Giglioni A, Giglioni B, Ottolenghi S (1988) The effects of HPFH mutations in the human γ-globin promoter on binding of ubiquitous and erythroid specific nuclear factors. NAR 16:7783–7806

Mantovani R, Superti-Furga G, Gilman J, Ottolenghi S (1989) The deletion of the distal CCAAT box region of the ^{A}γ-globin gene in Black HPFH abolishes the binding of the erythroid specific protein NFE3 and of the CCAAT displacement protein. NAR 17:6681–6691

Martin DIK, Tsai SF, Orkin SH (1989) Increased γ-globin expression in a nondeletion HPFH mediated by an erythroid-specific DNA-binding factor. Nature 338:435–438

Nicolis S, Ronchi A, Malgaretti N, Mantovani R, Giglioni B, Ottolenghi S (1989) Increased erythroid-specific expression of a mutated HPFH γ-globin promoter requires the erythroid factor GATA-1. NAR 17:5509–5516

Ottolenghi S, Nicolis S, Taramelli R, Malgaretti N, Mantovani R, Comi P, Giglioni B, Longinotti M, Dore F, Oggiano L, Pistidda P, Serra A, Camaschella C, Saglio G (1988) Sardinian ^{G}γ-HPFH: a T→C substitution in a conserved "octamer" sequence in the ^{G}γ-globin promoter. Blood 71:815–817

Ronchi A, Nicolis S, Santoro C, Ottolenghi S (1989) Increased Sp1 binding mediates erythroid-specific overexpression of a mutated (HPFH) γ-globin promoter. NAR 17:10231–10241

12 Bodine DM, Ley TJ (1987) An enhancer element lies 3' to the human ^{A}γ globin gene. EMBO J 2:2997–3004

Kollias G, Wrighton N, Hurst J, Grosveld F (1986) Regulated expression of human ^{A}γ-, β-, and hybrid γ/β globin genes in transgenic mice: manipulation of the developmental expression patterns. Cell 46:89–94

Magram J, Chada K, Costantini F (1985) Developmental regulation of a cloned adult β-globin gene in transgenic mice. Nature 315:338–340

Magram J, Niederreither K, Costantini F (1989) β-globin enhancers target expression of a heterologous gene to erythroid tissues of transgenic mice. Mol Cell Biol 9:4581–4584

Perez-Stable C, Costantini F (1990) Roles of fetal ^{G}γ-globin promoter elements and the adult β-globin 3' enhancer in the stage-specific expression of globin genes. Mol Cell Biol 10:1116–1125

Trudel M, Magram J, Bruckner L, Costantini F (1987) Up-

stream ^{G}γ-globin and downstream β-globin sequences required for stage-specific expression in transgenic mice. Mol Cell Biol 7:4024–4029

13 Behringer RR, Ryan TM, Palmiter RD, Brinster RL, Townes TM (1990) Human γ- to β-globin gene switching in transgenic mice. Genes Dev 4:380–389

Dillon N, Grosveld F (1991) Human γ-globin gene silenced independently of other genes in the β-globin locus. Nature 350:252–254

Enver T, Raich N, Ebens AJ, Papayannopoulou T, Costantini F, Stamatoyannopoulos G (1990) Developmental regulation of human fetal-to-adult globin gene switching in transgenic mice. Nature 344:309–314

Hanscombe O, Whyatt D, Fraser P, Yannoutsos N, Greaves D, Dillon N, Grosveld F (1991) Importance of globin gene order for correct developmental expression. Genes Dev 5:1387–1394

Orkin SH (1990) Globin gene regulation and switching: circa 1990. Cell 63:665–672

Ottolenghi S, Mantovani R, Nicolis S, Ronchi A, Giglioni B (1989) DNA sequences regulating human globin gene transcription in non-deletional hereditary persistence of HbF. Hemoglobin 13:523–541

Raich N, Enver T, Nakamoto B, Josephson B, Papayannopoulou T, Stamatoyannopoulos G (1990) Autonomous developmental control of human embryonic globin gene switching in transgenic mice. Science 250:1147–1149

Townes TM, Behringer RR (1990) Human globin locus activation region (LAR): role in temporal control. Tig 6:219–223

14 Berg PE, Williams DM, Quian RL, Cohen RB, Cao SX, Mittelman M, Schechter AN (1989) A common protein binds to two silencers 5' to the human β-globin gene. Nucleic acid Res 17:8833–8852

Cao SX, Gutman PD, Dave HPG, Schechter AN (1989) Identification of a transcriptional silencer in the 5'-flanking region of the human ε-globin gene. Proc Natl Acad Sci USA 85:5306–5309

15 Antoniou M, deBoer E, Habets G, Grosveld F (1988) The human β-globin gene contains multiple regulatory regions: identification of one promoter and two downstream enhancers. EMBO J 7:377–384

deBoer E, Antoniou M, Mignotte V, Wall L, Grosveld F (1988) The human β-globin promoter; nuclear protein factors and erythroid specific induction of transcription. EMBO J 7:4203–4212

Wall L, deBoer E, Grosveld F (1988) The human β-globin gene 3' enhancer contains multiple binding sites for an erythroid-specific protein. Genes Dev 2:1089–1100

16 Whitelaw E, Tsai SF, Hogben P, Orkin SH (1990) Regulated expression of globin chains and the erythroid transcription factor GATA-1 during erythropoiesis in the developing mouse. Mol Cell Biol 10:6596–6606

17 Mignotte V, Wall L, deBoer E, Grosveld F, Romeo PH (1989) Two tissue-specific factors bind the erythroid promoter of the human porphobilinogen deaminase gene. Nucleic Acid Res 17:37–54

Plumb M, Frampton J, Wainwright H, Walker M, Macleod

K, Goodwin G, Harrison P (1989) GATAAG: a cis-control region binding an erythroid-specific nuclear factor with a role in globin and non-globin gene expression. Nucleic Acids Res 17:73–92

Youssoufian H, Zon LI, Orking SH, D'Andrea AD, Lodish HF (1990) Structure and transcription of the mouse erythropoietin receptor gene. Mol Cell Biol 10:3675–3682

[18] Nicolis S, Bertini C, Ronchi A, Crotta S, Lanfranco L, Moroni E, Giglioni B, Ottolenghi S (1991) An erythroid specific enhancer upstream to the gene encoding the cell-type specific transcription factor GATA-1. NAR 19:5285–5291

Tsai S-F, Strauss E, Orkin SH (1991) Functional analysis and in vivo footprinting implicate the erythroid transcription factor GATA-1 as a positive regulator of its own promoter. Genes Dev 5:919–931

[19] Mignotte V, Eleouet JF, Raich N, Romeo PH (1989) Cis- and trans-acting elements involved in the regulation of the erythroid promoter of the human porphobilinogen deaminase gene. Proc Natl Acad Sci USA 86:6548–6552

[20] Behringer RR, Hammer RE, Brinster R, Palmiter RD, Townes TM (1987) Two 3' sequences direct adult erythroid-specific expression of human β-globin genes in transgenic mice. Proc Natl Acad Sci USA 84:7056–7060

Chol OB, Engel JD (1988) Developmental regulation of β-globin gene switching. Cell 55:17–26

Efstratiadis A, Posakony JW, Maniatis T, Lawn RM, O'Connell C, Spritz RA, deRiel JK, Forget BG, Weissman SM, Slightom JL, Blechl AE, Smithies O, Baralle FE, Shoulders CC, Proudfoot NJ (1980) The structure and evolution of the human β-globin gene family. Cell 21:653

Emerson BM, Nickol JM, Fong TC (1989) Erythroid-specific activation and derepression of the chick β-globin promoter in vitro. Cell 57:1189–1200

Evans T, Reitman M, Felsenfeld G (1988) An erythrocyte-specific DNA-binding factor recognizes a regulatory sequence common to all chicken globin genes. Proc Natl Acad Sci USA 85:5976–5980

Gallarda JL, Foley KP, Yang Z, Engel JD (1989) The β-globin stage selector element factor is erythroid-specific promoter enhancer binding protein NF-E4. Genes Dev 3:1845–1859

Ho I-C, Vorhees P, Marin N, Oakley BK, Tsai S-F, Orkin SH, Leiden JM (1991) Human GATA-3: a lineage-restricted transcription factor that regulates the expression of the T cell receptor α gene. EMBO J 10:1187–1192

Jackson PD, Evans T, Nickol JM, Felsenfeld G (1989) Developmental modulation of protein binding to β-globin gene regulatory sites within chicken erythrocyte nuclei. Genes Dev 3:1860–1873

Li Q, Zhou B, Powers P, Enver T, Stamatoyannopoulos G (1990) β-Globin locus activation regions: conservation of organization, structure, and function. Proc Natl Acad Sci USA 87:8207–8211

Martin DIK, Orkin SH (1990) Transcriptional activation and DNA binding by the erythroid factor GF-1/NF-E1/Eryf1. Genes Dev 4:1886–1898

Moon AM, Ley TJ (1990) Conservation of the primary structure, organization, and function of the human and mouse β-gene locus-activating regions. Proc Natl Acad Sci USA 87:7693–7697

Reitman M, Felsenfeld G (1988) Mutational analysis of the chicken β-globin enhancer two positive-acting domains. Proc Natl Acad Sci USA 85:6267–6271

Trainor CD, Evans T, Felsenfeld G, Boguski MS (1990) Structure and evolution of a human erythroid transcription factor. Nature 343:92–96

Tsai SF, Martin DIK, Zon LI, D'Andrea AD, Wong GG, Orkin SH (1989) Cloning of cDNA for the major DNA-binding protein of the erythroid lineages through expression in mammalian cells. Nature 339:446–451

Yamamoto M, Ko LJ, Leonard MW, Beug H, Orkin S, Engel DJ (1990) Activity and tissue-specific expression of the transcription factor NF-E1 multigene family. Genes Dev 4:1650–1662

Zon LI, Tsai S, Burgess S, Matsudaira P, Bruns GAP, Orkin SH (1990) The major human erythroid DNA-binding protein (GF-1): primary sequence and localization of the gene to the X chromosome. Proc Natl Acad Sci USA 87:668–672

[21] Martin DIK, Zon LI, Mutter G, Orkin SH (1990) Expression of an erythroid transcription factor in megakaryocytic and mast cell lineages. Nature 344:444–447

Romeo PH, Prandini ME, Joulin V, Mignotte V, Prenant M, Vainchenker W, Marguerie G, Uzan G (1990) Megakaryocytic and erythrocytic lineages share specific transcription factors. Nature 344:447–449

[22] Crotta S, Nicolis S, Ronchi A, Ottolenghi S, Ruzzi L, Shimada Y, Migliaccio AR, Migliaccio G (1990) Progressive inactivation of the expression of an erythroid transcriptional factor in GM- and G-CSF dependent myeloid cell lines. NAR 18:6863–6869

[23] Pevny L, Simon MC, Robertson E, Klein WH, Tsai S-F, D'Agati V, Orkin SH, Costantini F (1991) Erythroid differentiation in chimaeric mice blocked by a targeted mutation in the gene for transcription factor GATA-1. Nature 349:257–260

[24] Li S, Crenshaw III EB, Rawson EJ, Simmons DM, Swanson LW, Rosenfeld MG (1990) Dwarf locus mutants lacking three pituitary cell types result from mutations in the POU-domain gene pit-1. Nature 347:528–533

[25] Borzillo GV, Ashmun A, Sherr CJ (1990) Macrophage lineage switching of murine early pre-B lymphoid cells expressing transduced fms genes. Mol Cell Biol 10:2703–2714

Cross M, Dexter TM (1991) Growth factors in development, transformation, and tumorigenesis. Cell 64:271–280

D'Andrea AD, Zon LI (1990) Erythropoietin receptor. Subunit structure and activation. J Clin Invest 86:681–687

Dexter TM, Heyworth CM, Spooncer E, Ponting I (1990) The role of growth factors in self renewal and differentiation of haemopoietic stem cells. Philos Trans R Soc Lond Ser B 327:85–98

Gliniak BC, Rohrschneider LR (1990) Expression of the M-CSF receptor is controlled posttranscriptionally by the dominant actions of GM-CSF or multi-CSF. Cell 63:1073–1083

Gordon MY, Ford AM, Greaves MF (1990) Cell interactions and gene expression in early hematopoiesis. Int J Cell Cloning 8:11–25

Metcalf D (1989) The molecular control of cell division, differentiation commitment and maturation in haemopoietic cells. Nature 339:27–30

Walker F, Nicola NA, Metcalf D, Burgess AW (1985) Hierarchical down-modulation of hemopoietic growth factor receptors. Cell 43:269–276

Wang HM, Collins M, Arai K, Miyajima A (1989) EGF induces differentiation on an IL-3 dependent cell line expressing the EGF receptor. EMBO J 8:3677–3684

26 Chretien S, Dubart A, Beaupain D, Raich N, Grandchamp B, Rosa J, Goossens M, Romeo PH (1988) Alternative transcription and splicing of the human porphobilinogen deaminase gene result either in tissue-specific or in housekeeping expression. Proc Natl Acad Sci USA 85:6–10

Cox TC, Bawden MJ, Martin A, May BK (1991) Human erythroid 5-aminolevulinate synthase: promoter analysis and identification of an iron-responsive element in the mRNA. EMBO J 10:1891–1902

Li JP, D'Andrea AD, Lodish HF, Baltimore D (1990) Activation of cell growth by binding of Friend spleen focus-forming virus gp55 glycoprotein to the erythropoietin receptor. Nature 343:762–764

Lumelsky NL, Forget BG (1991) Negative regulation of globin gene expression during megakaryocytic differentiation of a human erythroleukemic cell line. Mol Cell Biol 20:3528–3536

Müllner EW, Neupert B, Kühn LC (1989) A specific mRNA binding factor regulates the iron-dependent stability of cytoplasmic transferrin receptor mRNA. Cell 58:373–382

Zenke M, Kahn P, Disela C, Vennström B, Leutz A, Keegan K, Hayman MJ, Choi HR, New N, Engel JD, Beug H (1988) v-erbA specifically suppresses transcription of the avian erythrocyte anion transporter (band 3) gene. Cell 52:107–119

Zenke M, Muñoz A, Sap J, Vennström B, Beug H (1990) v-erbA oncogene activation entails the loss of hormone-dependent regulator activity of c-erbA. Cell 61:1035–1049

27 Novak U, Harris EAS, Forrester W, Groudine M, Gelinas R (1990) High-level β-globin expression after retroviral transfer of locus activation region-containing human β-globin gene derivatives into murine erythroleukemia cells. Proc Natl Acad Sci USA 87:3386–3390

Chapter 36 Growth Control in Animal Cells

LENNART PHILIPSON and VINCENZO SORRENTINO

36.1 Introduction

Multicellular organisms contain a vast number of cells with different functions. It has been estimated that each human contains 10^{13} cells all originating from the zygote produced by fertilization of an egg by a sperm. At each division the cells duplicate their DNA content, but some cytoplasmic constituents may be differentially distributed between the two daughter cells. Cells in different tissues, in different species, and in different stages of embryonic development have division or cell cycles that vary enormously in duration from less than 1 h in the early frog embryo to more than 1 year in the adult human liver. An understanding of the controls operating during growth of animal cells may therefore provide insight into the basic mechanisms of embryonic and organ development and tissue-specific gene expression, as well as tissue regeneration in adult life. It is clear that an adult animal must manufacture a million new cells each second simply to maintain the status quo and if the cell cycle is halted by chemicals or ionizing radiation it will lead to major pathological manifestations and ultimately death. All the diseases involving either increased cell divisions like the benign and malignant tumors or atherosclerotic plaques, or those involving defective cell divisions like anemia, sclerotic ulcers, and tissue necrosis might be prevented or treated if we had a detailed knowledge of the growth controls operating in animal cells.

Several recent contributions in this research field − the discovery of *onc* genes and the identification of growth factors and growth factor receptors − have significantly contributed to our understanding of the molecular machinery of growth control. The studies of retroviral *onc* genes have established that they derive from cellular genes and are all controlling induction of growth of mammalian and chicken cells. The biochemical properties and the subcellular localization of these proteins as well as the interaction between them and their signal transducers has led to a network model where numerous components are involved in transmitting a signal from the plasma membrane of the animal cell across the cytoplasm to the nucleus, to affect the transcription machinery. The growth-stimulating factors bind to specific receptors, generating a mitogenic response which activates other intracellular components in this signaling cascade, ultimately leading to transcriptional activation and DNA synthesis. Several different intracellular pathways can be activated by the same stimulatory growth factor, and different growth factors can mediate a signal through the same pathway. Furthermore, different cell types and tissues exhibit different response mechanisms. More recently, negative control of cell growth has been investigated and growth inhibitory polypeptides with antiproliferative properties have been identified. It has also recently been established that a few human tumors arise because of deletion within specific genes whose protein products are required for suppression of the tumorigenic phenotype.

This chapter will review current thinking about growth control mechanisms in mammalian cells, emphasizing both the factors stimulating growth as well as those suppressing growth and their polymorphic intracellular effects. In our opinion the interplay between positive and negative regulation of cell growth, including feedback controls as well as antagonistic and synergistic effects in the complex molecular machinery governing cell growth, forms the basis for understanding many other more sophisticated aspects of biology.

36.2 The Cell Cycle[1]

During mitotic division of animal cells, which lasts 1–2 h, the content of the nucleus condenses to form visible chromosomes which, through an elaborate mechanism, are pulled apart into two equal sets. This is followed by a separation of all other components in the cytoplasm, called cytokinesis, which leads to two daughter cells each receiving one of the two sets of the chromosomes. The mitosis phase or M-phase in the cell cycle was early recognized because it could be observed in the light microscope (Fig. 1). The period between each M-phase, referred to as the interphase, has been more difficult to characterize. The interphase is actually the period in which preparation for division occurs and involves a carefully ordered sequence of events. With the help of radioactive precursor labeling, the phase in the cell cycle devoted to DNA synthesis called the S-phase was soon identified. Between the M- and the S-phase there is an interval or a gap, known as the G1 phase, in which the cell prepares itself for DNA synthesis. Another gap phase, G2, separates the end of the S-phase from the beginning of the next M-phase. The eukaryotic cell cycle is therefore divided into the four phases G1, S, G2, and M. For most animal cells in culture the time to complete the entire cell cycle is between 10 and 30 h, and the variation in the cell cycle length is most often due to variation of the G1 phase, while the

length of the other phases appears fairly constant. A restriction point in the middle of G1 seems to control the decision whether a cell will go further into DNA synthesis and division or alternatively enter quiescence. Cells that do not divide, i.e., quiescent cells are considered to be in a special niche called the G0 state. Cells in the G0 state usually have a G1 content of DNA, but whether this phase is qualitatively or quantitatively different from the G1 phase leading to division is not yet clear. The decision, however, to enter proliferation or quiescence appears to reside in the G1 phase. Several parameters like availability of nutrients, cell size, cell density, cell adhesion, or the presence of growth factors can influence this decision. Many transformed cells seem to have lost the ability to regulate progression at this point and cannot, therefore, revert to G0. Thus quiescence is considered to be a distinct event within the G1 phase.

The quiescent cell requires a longer time to enter the S-phase when stimulated to grow than its proliferative counterpart. In mouse NIH 3T3 cells that have been used as a model, G0-S transition takes 12 h, while in continuously growing cells the time between M and S is only 6 h. Several models may explain the prolonged G1 phase of quiescent cells. It may be due to a prolonged lag phase required to build up the signal molecules required to enter DNA synthesis. Alternatively, quiescent cells may have lower numbers of receptors for growth factors or adhesion molecules; this could account for the delay in reaching the S-phase.

36.3 Controls in the Cell Cycle[2]

It is difficult to analyze the steps of the cell cycle in tissues of an intact animal, and most of these events have, therefore, been delineated in cell cultures. Through time-lapse photography the duration of the M-phase has been accurately determined and cells undergoing DNA synthesis can be detected by ^{3}H-thymidine autoradiography. Cells progressing through the cycle can also be followed by measuring their DNA content directly by fluorescent activated cell sorting. An important aspect for studying the cell cycle has been the ability to

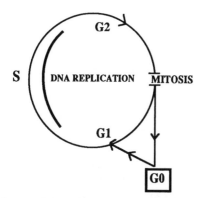

Fig. 1. Diagram of the cell cycle. At completion of mitosis, cells have two options; (1) to enter a non-proliferating G0 state, from which they can reenter the proliferative pool if an appropriate stimulus is provided; (2) to cycle again, through G1, S-phase (DNA replication), G2 and mitosis

synchronize the events. This can be done by shaking off mitotic cells from a solid support and then following the events through the G1, S, and the G2 phase.

Fusion of synchronized cells has been useful to identify the dominant principle in the different phases of the cell cycle. Fusion of G1- and S-phase cells leads to DNA synthesis in the G1 nucleus but the G2 nucleus which has 4n DNA content is unresponsive to this S-phase activator. Likewise, mitotic cells fused to G2 cells can induce premature breakdown of the nuclear envelope and condensation of the chromosomes suggesting that a mitosis activator exists. This putative activator has no effect on either G1 or S cells, suggesting that another cytoplasmic signal delays preparation for mitosis until DNA replication is completed.

Some of the genes controlling cell division have been characterized in yeast which is a useful model system since haploid cells are available and homologous recombination is easily performed. By using temperature-sensitive mutants affecting the cell cycle both in budding and in fission yeast, some of the genes controlling the start of the cell cycle, as well as the different steps in mitosis leading to partition of chromosomes, cytoplasm, and centrosomes have been identified. A 34 kDa phosphoprotein, the product of the cdc 2 gene (cell division cycle) of fission yeast, corresponding to the CDC28 gene in budding yeast, has been shown to be the catalytic subunit in the complex of proteins known as the maturation-promoting factor (MPF). The cdc2 phosphoprotein which provides the start signal in yeast also provides the start signal for the events in fertilized *Xenopus* eggs leading to rapid synchronous cell division generating 4000 cells from the original one. The evolutionary preservation of this gene has allowed cloning of the human gene by direct functional complementation of a yeast mutant with a human cDNA. The similarity between the yeast and the human genes is around 60% at the amino acid level, suggesting a considerable evolutionary preservation of the set of genes involved in the control of the cell cycle.

Normal human cells in vitro appear to have a finite life time, only allowing up to 50 population doublings, thereafter the division cycle becomes more asynchronous and the cells ultimately cease dividing and exhibit a stage similar to senescence of the whole organism. The total number of division cycles prior to senescence varies depending on the composition of the medium, and increases three- or fourfold if the concentration of some growth factors is increased. Cell senescence in culture has a pathological counterpart. In a human genetic disease, Werner's syndrome, the cells become prematurely senescent in culture. The fibroblasts from these patients are obviously unresponsive to some growth factors, including platelet-derived growth factor, PDGF, and fibroblast growth factor, FGF, although they are able to proliferate vigorously in response to other growth factors. The studies on cell senescence demonstrate that cells that otherwise seem identical are heterogenous in their ability to divide.

36.4 Cell-to-Cell Signaling [3]

The molecular nature of the controls involved in the cell cycle is gradually emerging from studies of cancer cells which are immortal and not subject to senescence. The cell cycle in multicellular organisms depends on complex controls from the outside, often referred to as social controls, and proliferation is governed by different combinations of growth stimulatory and inhibitory factors, as well as cell-cell contact within the multicellular organ. Most normal cells are unable to divide unless they are anchored to an extracellular matrix, or in a multicellular compartment with other cells from which they receive signals.

Signaling mechanisms between animal cells vary mostly with regard to the distance over which they operate (Fig. 2). At long distances hormones secreted from special endocrine cells travel through the blood stream to influence target cells which are distributed widely throughout the body. At much shorter distances neighboring cells may secrete local chemical mediators which are rapidly taken up, destroyed, or immobilized, so the signals can only work on cells in the immediate environment, referred to as paracrine signals. The extreme case of paracrine signaling is when a cell secretes signaling mediators that directly influence the receptors on the same cell. This is usually referred to as auto-crine signaling.

In most target cells, extracellular signals are recognized by a specific protein, a receptor, that with great specificity binds the signaling molecule and then initiates the intracellular response. Depending on the structure of the ligand, the receptor molecule is either located intracellularly or anchored in the plasma membrane. Most protein hormones and some chemical ligands are water-soluble and can thus not pass directly through the lipid bilayer of the target cell membrane. They therefore must bind to a specific receptor on the cell surface. The water-insoluble steroid and thyroid hormones, on the other hand, are lipid-soluble and must be transported in the blood system bound to carrier molecules; once released from these proteins they can pass through the lipid bilayer of the plasma membrane and bind to specific receptors inside the cell.

A mechanistically different type of cell-cell signaling also exists through direct channels between neighboring cells leading to an exchange of cytoplasm and specific mediators within a cell compartment. Cell-to-cell contact, mediated through such gap junctions, appears to control abnormal cell growth. Transformed cells that do not form these junctions can escape growth suppression transmitted by normal surrounding cells. The molecular details of such channel signaling have not yet been clarified but this also applies to many of the other pathways.

36.5 Growth Stimulatory Factors[4]

Most of the growth stimulatory factors appear to operate through paracrine or autocrine mechanisms. However, since some of them are found in the blood, they may also act in an endocrine fashion. More than 30 different growth factors have been identified.

Nerve growth factor (NGF) and epithelial growth factor (EGF) were the first to be discovered some 30 years ago. The cDNA and genes have now been cloned and the sequences have been determined for a multitude of growth stimulating factors. EGF, a 6 kDa polypeptide, is a strong mitogen for both ep-

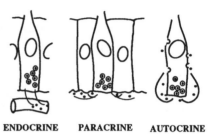

ENDOCRINE   PARACRINE   AUTOCRINE

Fig. 2. Different modes of mediating signals from one cell to another by the endocrine, paracrine, and autocrine routes. *Black points* refer to growth factors and the *semi-circular regions* on the cell membrane represent their receptors

ithelial and mesenchymal cells. It is released from the cells by cleavage of a large precursor protein which has characteristics of a membrane protein. A structurally related protein, tumor growth factor a (TGFα) of the same size is secreted from some tumor cells and binds to the EGF receptor with the same affinity as EGF itself. Overexpression of TGFα results in cell transformation and it is abundantly produced by some transformed cells but it has also been found in normal cells. TGFα, like EGF, is synthesized as a membrane-bound precursor protein which is cleaved to yield the soluble factor.

Platelet-derived growth factor (PDGF) was identified as the main mitogen present in human serum and it is predominantly synthesized in megakaryocytes. It is mainly a mitogen for mesenchymal cells. Two genes coding for the A and B chains respectively of PDGF have been identified. The active PDGF molecule, with a molecular weight of around 30 kDa, is a dimer of these two chains either in the form of homo- or heterodimers linked by disulfide bonds. The specific biological response to the AA, AB, and BB forms of PDGF seems to differ, but the exact function of the isomeric forms has not been clarified. The v-*sis* oncogene, isolated from Simian sarcoma virus, encodes the PDGF B chain, suggesting that transformation by v-*sis* is mediated by an autocrine mechanism. The v-*sis* protein is in most cases secreted and binds the receptor on the outside of the plasma membrane. A small fraction of the protein may, however, interact with the PDGF receptor inside the cells. By linking the v-*sis* protein with a retention sequence, an intracellular autocrine loop was recently demonstrated. Although expression of this v-*sis* PDGF B chain has been observed in human tumors its involvement in tumorigenesis is still uncertain.

Growth factors of another class, that support proliferation of endothelial cells, were initially purified on the basis of their affinity for heparin. The two prototypes of this group are endothelial cell growth factor (ECGF) and fibroblast growth factor (FGF). A gene coding for the human ECGF has been characterized and its sequence shows similarity to the FGF sequence. Two different forms of FGF, one acidic and the other basic, were initially identified, and several genes related to the basic form have been discovered. Some of them bear sequence similarity with the int-2 *onc* gene. Recently a platelet-derived endothelial cell growth factor (PDECGF) that selectively stimulates endothelial cell growth was purified and cloned. Because of its selected target cell specificity, this growth factor may have a role in maintaining the integrity of blood vessels.

36.6 Growth Factors and Cell Multiplication[5]

Although several growth factors have been identified and characterized it still remains to be established whether they are all needed for cell proliferation and if they are interchangeable. In the best characterized system for cell proliferation, the Balb 3T3 cells from mice, PDGF is only required transiently to render quiescent cells competent to progress towards the S-phase. The treated cells will not proceed to S-phase unless other growth factors like EGF or the insulin-like growth factor (IGF-1) are supplied. It appears, therefore, that two distinct steps operate. PDGF makes the cell competent to reenter the cell cycle from the G0-state and EGF and IGF-1 function as progression factors allowing the cells to proceed towards the S-phase. In late G1, IGF-1 appears to be the important growth factor since cells arrested 6 h before the S-phase can progress further only when IGF-1 is provided. Although Balb 3T3 cells have helped to subdivide the G1-phase, other cells respond differently to several

of the growth factors. PDGF is not active in epithelial cells, while EGF is a mitogen for both fibroblastic and epithelial cells. For hemopoietic cells specific growth factors, many of them discovered as cell-specific growth factors, so-called cytokines, are required, and most of them appear to be cell lineage-specific.

The multitude of growth factors and cytokines and their cell specificity suggest that several processes in stem cell renewal and differentiation, as well as cell division during development or adult life, require a combined action of several of them. They provide together a signal network that indicates to the cell when and how to resume proliferation.

36.7 Growth Factors and Transformation[6]

A connection between stimulatory growth factors and transforming genes, which led to the concept of autocrine growth regulation, was established by two different approaches. First some mouse and human tumor cells in cultures produced TGFα which could be purified. As long as the growth factor was present in the medium of normal cells, the purified TGFα induced a transformed phenotype. When it was removed, the cells reverted to their normal untransformed state. The cells in culture thereby sustained a higher growth rate through an autocrine control mechanism. The second approach was to clone EGF and FGF under a strong promoter and introduce them by transfection into normal cells. Selecting the cells with high growth potential revealed that such cells had amplified the signal and strongly expressed the stimulating growth factor at a high level. These results again suggest that the receptors were available on the cells and increased growth factor production was enough to establish a high growth potential. Such cells, in addition, could also induce tumors in nude mice, suggesting that some of the cells had a capacity to progress to malignancy. To understand cancer, it is essential to establish the molecular mechanism involved when a polypeptide growth stimulating factor applied from the outside of the cell delivers an internal signal leading to an increased growth potential and even allows some of the cells to become tumorigenic.

36.8 Growth Factor Receptors[7]

Many of the water-soluble growth stimulators of cells must be recognized by receptors anchored in the plasma membrane. There are essentially three types of such receptors. The first class is channel-linked receptors mainly involved in rapid synaptic signaling between electrically excitable cells. They respond to neurotransmitters and transiently open or close an ion channel, leading to a change in the ion permeability of the membrane.

The second class is G protein-associated receptors. (G proteins refer to GTP binding proteins with GTPase activity). An extracellular ligand alters the conformation of the cytoplasmic domain of the receptor so that it can bind cytoplasmic G proteins, which in turn activates a plasma membrane enzyme. Through introduction of a G protein with a bound GTP in the cascade, the adenylate cyclase may become activated, leading to an increase in cyclic AMP. Activation persists until the ligand leaves the receptor.

The third class of receptors involved in growth control are usually classified as catalytic receptor proteins. Several of these receptors contain a tyrosine-specific protein kinase on the cytoplasmic side of the plasma membrane (Fig. 3). Most of the growth stimulatory factors recognize catalytic receptors which are proteins with the following characteristics:

1. A single polypeptide divided into an extracellular, a transmembrane, and an intracellular domain. The insulin receptor, however, contains two polypeptide chains generated by cleavage of a single precursor receptor protein.
2. The extracellular domain is often glycosylated and rich in disulfide bonds and contains the binding site for the growth factor or hormones.
3. The transmembrane domain is composed of a simple hydrophobic polypeptide segment followed by polar basic sequence which probably blocks further extrusion of the chain through

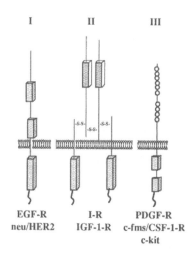

Fig. 3. Structural organization of receptors with tyrosine kinase activity. The *upper boxes* represent the cystein-rich repeat regions in the extracellular domain; the *lower boxes* represent the tyrosine kinase region of the cytoplasmic domain. *Open circles* in the type III subclass of the tyrosine kinase receptors represent conserved cysteines among these receptors

the membrane. For both the EGF and the PDGF receptor it appears that a ligand-induced dimerization of the receptor is an important step in activation of the receptor kinase activity.
4. The cytoplasmic domain of the receptor provides the signal to the cytoplasmic side and may express kinase activity.

Two classes of receptor for growth stimulating factors with or without an intracellular tyrosine kinase domain have been characterized. Those with kinase activity include different *onc* genes. The *erb*-B gene present in an avian leukemia virus is derived from the EGF receptor. It is truncated in the extracellular domain, suggesting that it may express tyrosine kinase activity on the cytoplasmic side even without a growth factor ligand: The c-*fms* gene, originally discovered in a variant sarcoma virus, encodes the receptor for the macrophage-colony stimulating factor, MCSF or CSF I. The activation of the transforming potential of this receptor results probably from point mutation in the C-terminal part of the molecule, which leads to a changed regulation of the kinase activity. Other *onc* genes may, based on structural comparison to known receptors, also fall in this category although the ligand is still unknown. When the catalytic receptors bind a specific ligand, the tyrosine kinase is activated leading to autophosphorylation of the receptor, which in turn may give a cytoplasmic signal that can ultimately change the expression pattern of several cellular genes. Contact between the growth stimulating factor and its receptor often leads to downregulation of remaining receptors on the cell surface suggesting that the cell can protect itself from overstimulation by external growth factors.

Receptors with tyrosine kinase activity represent, however, only one class of growth factor receptors. In the recently cloned receptors for some cytokines like GMCSF, IL6, and IL2, the cytoplasmic fragment contains no tyrosine or serine/threonine kinase domain, indicating that the signals relayed by these receptors are transmitted to other molecules by other mechanisms.

36.9 The Intracellular Signal Pathway[8]

The most intriguing part of the signal pathway is the intracellular events. The signal from the receptor can change the cellular gene expression pattern, leading to an increased growth potential and ultimately, if overstimulated, progress through several

steps to tumorgenesis. Several observations have focussed on the drastic changes in the intracellular environment which occur shortly after growth factor-receptor contact (Fig. 4). Within a minute an increase in cytoplasmic pH is registered, probably mediated through the ATP-Na proton pump in the plasma membrane. This is followed within minutes by an increased turnover of a special type of lipid molecules, phosphatidyl inositol phosphates (PIP), closely associated with the internal side of the plasma membrane. Catabolism is probably triggered by a specific phospholipase C (PLC) leading to a release of inositol phosphates (IP) and an accumulation of diacylglycerol (DAG). DAG can, in turn, activate another kinase, the phosphoprotein kinase C (PKC) which appears to require this lipid as a cofactor. It is of special interest that some lipid-soluble phorbol esters called tumor promoters can replace DAG and activate PKC. This kinase phosphorylates several different target proteins at serine or threonine residues. It is therefore distinct from the tyrosine kinase residing at the intracellular region of the receptors. The accumulation of IP or signals from other receptors appears in addition to release calcium ions in the cytoplasm of stimulated cells either from calcium-binding proteins like calmodulin or from mitochondria. The hydrolysis of PIP to IP and DAG, which are instrumental in the release of Ca^{2+} from intracellular stores and activation of protein kinase C respectively, is catalyzed by one member of a large family of enzymes called phospholipase C (PLC). Three forms of PLC have been characterized and cloned. PLCγ has re-

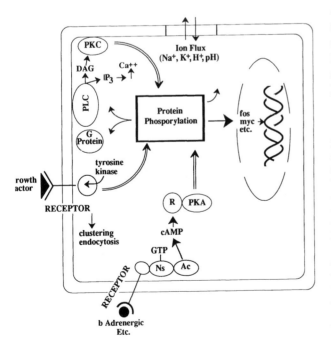

Fig. 4. A schematic diagram revealing the two main pathways of intracellular signal transduction. The growth factor pathway is recognized by growth stimulatory factors but also by insulin. The β-adrenergic pathway recognized by adrenaline couples via a G-regulatory protein (*Ns*) to adenyl cyclase (*Ac*) and cyclic cAMP binding to the regulatory subunit of protein kinase A (*PKA*). Various agonists can activate phospholipase C (*PLC*), leading to the release of diacylglycerol (*DAG*) and inositol 1,4,5-triphosphate (*IP$_3$*). DAG activates protein kinase (*PKC*) and IP$_3$ causes the release of calcium from the endoplasmic reticulum. These events probably lead to a cascade of protein phosphorylation that alters the function of membrane associated receptors, ion channels, and cytoplasmic proteins. Undefined signals from this cascade probably engage the nucleus to induce the expression of various genes including c-*fos* and c-*myc*

cently been shown to be phosphorylated on tyrosines and serines within a minute after PDGF and EGF treatment of cells. Purified PDGF receptor can also phosphorylate PLCγ in vitro.

An analogous but different situation occurs when cells are stimulated at β-adrenergic receptors leading to activation of cyclic AMP-dependent protein kinase inside the cell. In this pathway one of three different G proteins serve as an intermediary in the activation or inhibition of the adenyl cyclase required to generate cyclic AMP, which operates as a mediator for transferring phosphates to other proteins through protein kinase A (PKA).

36.10 Oncogenes in the Signal Pathway[9]

There are among the *onc* genes with tyrosine kinase activity some which lack an extracellular receptor moiety. The prototypes of these membrane-associated tyrosine kinases are the *src* and the *abl* genes (see Table 1). The mechanism that activates c-*src* is obviously complex, involving alteration in its tyrosine kinase activity. Although little is known about the function and physiology of the v-*abl*, the c-*abl* gene is rearranged in about 95% of chronic myeloid leukemia cases which carry the Philadelphia chromosome. As a consequence of this rearrangement, the *Abl*protein expressed in leukemic cells has an uncontrolled protein kinase activity. The v-*mos* and the v-*raf* oncogenes both encode cytoplasmic proteins with serine or threonine protein kinase activity. Studies, aimed towards dissecting the interaction between oncogenes in the intracellular signal pathway that ultimately leads to DNA synthesis, indicate that the Mos and Raf proteins are located in a different pathway from the *fms, src, fos* and *sis* oncogenes. The latter genes are probably dependent on the presence of a functional Ras protein for induction of cell proliferation. After PDGF treatment of cells the 74-kDa Raf protein becomes associated with the PDGF receptor. The PDGF receptor phosphorylates the Raf protein on tyrosines leading to activation of a serine/threonine kinase in the Raf. Other proteins of the *src* family also bind the PDGF receptor and are phosphorylated after addition of PDGF.

The *ras* oncogene family belongs to the second major class of *onc* genes associated with the cytoplasmic side of the plasma membrane. They are G proteins with GTP binding and GTPase activity and are also the family of *onc* genes most frequently activated in human tumors. This activation results from point mutations which decrease the ability of Ras proteins to hydrolyze GTP. Overexpression of some of the Ras proteins can also transform cells. The structural and biochemical similarity between the Ras proteins and the signal-transducing G proteins is of special interest. Recently a protein called GAP that can stimulate the GTPase activity of Ras has been isolated. It fails, however, to stimulate GTP hydrolysis by mutant oncogenic *ras*. Although the *ras* genes in yeast seem to participate in the regulation of cyclic nucleotide metabolism, the role of *ras* in mammalian cells is still unresolved.

Table 1 identifies the relationship between several *onc* genes and their corresponding cellular gene, and Fig. 5 demonstrates schematically the position of the *onc* genes in the signal pathway.

The relationship between all these intracellular events and the specific interaction between a growth factor and its receptor has not yet been established. In fact, some growth-stimulating factors do not cause calcium release, or will not activate the PKC, while others will. Alternative pathways might therefore exist for different growth-stimulating factors and the signal pathway will probably be much more complex than currently assumed. It is reassuring, however, that most of the genes involved in stimulating human cells in cultures are also present in several other species. The *ras* oncogene has been found in all species from yeast to man and there is considerable sequence homology between species. It is of high priority to dissect all the alternative branches of the signal pathway from the receptor through the cytoplasm into the nucleus of the mammalian cell.

36.11 Nuclear Response Elements[10]

Ultimately the signal for growth must reach the nucleus of the animal cell to cause a change in gene expression, leading to a high growth potential

Table 1. The relationship between *onc* genes and their corresponding cellular genes

Acronym	Original source		Subcellular location of protein	Cellular gene
	Retroviruses	Nonviral tumor		
Growth factors				
sis	Simian sarcoma		Secreted	PDGF
Protein kinases				
abl	Abelson murine leukemia	Chronic myelogenous leukemia Acute lymphocytic leukemia	Plasma membrane	Tyrosine kinase
erbB	Avian erythroblastosis		Plasma membrane	EGF receptor
fms	Feline sarcoma		Plasma membrane	CSF-1 receptor
kit	Feline sarcoma		Plasma membrane	Receptor-like tyrosine-specific protein kinase
mos	Moloney murine sarcoma		Cytoplasm	Serine/threonine kinase
neu		Neuroblastoma	Plasma membrane	Receptor-like tyrosine-specific protein kinase
raf (mil)	Murine sarcoma		Cytoplasm	Serine/threonine kinase
src	Rous avian sarcoma		Plasma membrane	Tyrosine kinase
G-Like proteins				
Ha-*ras*	Murine sarcoma	Bladder, mammary skin carcinomas	Plasma membrane	GTP-GDP binding protein with GTPase activity
Ki-*ras*	Kirsten murine	Lung, colon and pancreas carcinomas, sarcoma	Plasma membrane	GTP-GDP binding protein with GTPase activity
N-*ras*		Neuroblastoma, leukemias, melanoma	Plasma membrane	GTP-GDP binding protein with GTPase activity
Nuclear proteins				
erbA	Avian erythroblastosis		Nucleus	Thyroid hormone receptor
fos	Mouse osteosarcoma virus		Nuclear matrix	DNA binding protein
myb	Avian myeloblastosis	Leukemia	Nuclear matrix	DNA-binding protein
myc	Avian myelocytomatosis	B cell tumors	Nuclear matrix	DNA-binding protein
N-*myc*		Neuroblastoma	Nuclear matrix	DNA-binding protein

which may progress into malignant transformation. Around 100 cellular genes, and among them the cellular homologous of several viral *onc* genes, especially the c-*fos*, c-*myc*, and c-*myb*, are transcribed more actively upon growth induction than during the normal cell cycle. In fact, stimulation of c-*fos* transcription starts within 10 min after growth factors are supplied to the outside of the cell and the stimulation is transient and disappears within 1–2 h. The stimulation of c-*myc* transcription is delayed, but also abates after several hours. All of these *onc* genes have been cloned and sequenced and the normal cellular genes have also been identified. Several other *onc* genes that encode proteins with nuclear localization, like *myb*, *ski*, and *jun*, have also been identified as part of the nuclear re-

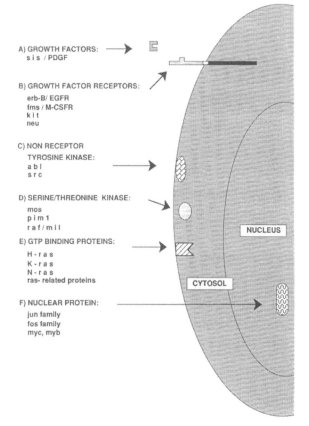

A) GROWTH FACTORS:
s i s / PDGF

B) GROWTH FACTOR RECEPTORS:
erb-B/ EGFR
fms / M-CSFR
k i t
neu

C) NON RECEPTOR
TYROSINE KINASE:
a b l
s r c

D) SERINE/THREONINE KINASE:
mos
p i m 1
r a f / m i l

E) GTP BINDING PROTEINS:
H - r a s
K - r a s
N - r a s
ras- related proteins

F) NUCLEAR PROTEIN:
jun family
fos family
myc, myb

NUCLEUS

CYTOSOL

Fig. 5. The position of several oncogenes in the signal transduction pathways. The cellular location of the oncogenes corresponding to growth factors, growth factor receptors, various kinases, G-proteins, and nuclear proteins have been schematically identified

sponse elements. Recently v-*jun* has been shown to encode an altered form of a transcription factor (AP1), containing a complex of c-*fos* and c-*jun*. Interactions among nuclear oncogenes like *fos* and *jun* have also been demonstrated, and this may modulate the ability of these proteins to interact with specific DNA regulatory sequences. Studies with antisense RNA have established that the expression of *fos, myc,* and *myb* is required for cell proliferation and the injection of the *myc* protein into cells can induce DNA synthesis. The difference in structure between the transforming moiety contained in retroviruses and the cellular gene is not as extensive for nuclear as for the cytoplasmic oncogenes. Transfection of the nuclear response ele-

ments in a vector which allows amplification of their expression leads to a transformed phenotype and in some cases similar events may occur through translocation of DNA from a weak to a strong promoter or to a site that can readily be amplified. Several of the DNA tumor viruses also contain oncogenes which can activate transcription but their cellular counterparts have not yet been identified.

Transgenic mice may ultimately help to clarify some of the normal functions of the transforming genes and especially their tissue-specific expression, as well as their expression during development. By introducing the transforming genes into the germ line of the mouse, one can easily follow their expression throughout development and in different tissues of the adult mice by including a specific reporter sequence to identify the introduced gene. Such experiments have revealed that the promoter sequences of the gene may determine the tissue specificity of the tumor. A construct containing the promoter of the insulin gene fused to an oncogene from a DNA virus caused only β-cell insulomas and the immunoglobulin promoter fused to c-*myc*-generated B-cell lymphomas.

It is clear that the signal pathway from the growth stimulatory factor at the outside of the cell through the receptor traversing the membrane and the final switch in the nucleus is far from resolved but many of the cellular *onc* genes reside in this pathway and correspond to stimulatory growth factors, their receptors, mediators of phosphate transfer, or activators of nuclear transcription. The activity of all these components must, of course, be tuned precisely to achieve growth control and both positive and negative regulatory elements must coexist. Only an interplay between all regulatory elements can provide the normal phenotype.

36.12 Cell Cycle-Regulated Genes[11]

In dividing cells, division follows an increase of cellular components, replication of DNA, and redistribution of the organelles. To accomplish this series of events, a temporally well regulated organization of gene expression is required. Identification of cell cycle-regulated genes is, therefore, paramount to understand cell growth regulation and neoplastic transformation.

The expression of several genes involved in DNA synthesis in animal cells is known to be cell cycle-regulated. Histones, for example, are predominantly expressed in the S-phase. Genes encoding enzymes linked to the metabolism of nucleotides and polyamines, like thymidine kinase (TK), ribonucleotide reductase, DNA polymerase, and ornithine decarboxylase, are also differentially expressed during the cell cycle.

Regulation of expression of cell cycle genes operates at several levels. The gene coding for TK is transcriptionally regulated in S-phase, via a specific enhancer region binding specific transcription factors. Controls also exist at the translational level since the rate of synthesis of the TK proteins increases more than can be accounted for by the increase in mRNA concentration during the S-phase. The stability of the protein is also dramatically decreased after cell division. These controls appear also to operate for other genes like topoisomerase II.

While expression of certain genes is limiting for cell growth, for example the tRNA synthetase genes, expression of others is important for cell progression through specific phases of the cell cycle, thus when *ts* mutants are grown at the nonpermissive temperature, cells will only arrest at a specific phase of the cell cycle. Two genes necessary for the G1 progression in mammalian cells have been identified by complementation of such mutants. One complements the *ts*11 mutant in BHK cells which is blocked at the nonpermissive temperature in the G1 phase. Expression of this gene starts 5 h after serum stimulation of quiescent cells. This gene codes for asparagine synthetase, but its role in cell cycle control remains to be established. After the initial discovery that several genes, including *fos* and *myc*, are transcriptionally activated following serum stimulation of quiescent cells, much attention was focused on the isolation of other genes which are induced under similar conditions. A systematic differential screening of a c-DNA library from serum-stimulated cells with cDNA probes from quiescent and activated cells identified about 100 genes which are activated between 15 min and 4 h after serum addition to quiescent cells. Several of the genes isolated code for nuclear proteins such

as *fos, myc, jun,* and DNA-binding proteins containing zinc fingers. Two of these genes, *fra*-1 and *fos* B, are related to c-*fos,* while two others, *jun*-B and *jun*-D, are related to the c-*jun* gene. Other genes were encountered which resemble growth factors or encode cytoskeletal or extracellular matrix proteins. Many of the 100 genes isolated after serum stimulation of quiescent fibroblasts do not correspond to known proteins and their function remains to be elucidated.

36.13 Gene Expression at Quiescence [12]

Much less is known about genes preferentially expressed in quiescent cells. The first protein specific for quiescent cells that has been identified is statin, a protein of 57 kDa, which is located in the nucleus of nonproliferating human fibroblasts, but not in growing or transformed cells. Expression of statin is also lost when arrested cells are induced to reenter the cell cycle. Another protein, with an apparent MW of about 20 kDa, specifically expressed in nondividing cells, has been identified in chicken fibroblasts. The gene coding for this protein has recently been cloned. Expression of this protein is repressed when cells are induced by stimulatory growth factors or by constitutive high expression of the *src* oncogene. In parallel to the activation of transcription of new genes, serum stimulation also results in the decrease of the expression of some genes and proteins. Experiments aimed at identifying genes expressed in quiescent cells, which are downregulated when cells are induced to reenter the cycle by serum stimulation, revealed a group of such genes in mouse NIH 3T3 cells. The expression of these growth arrest specific (*gas*) genes is depressed by serum stimulation. Some of these genes are also expressed at low levels in exponentially growing cells but RNA is more abundant when cells reach confluency, or become quiescent for other reasons. These genes are, however, expressed in vivo during embryo development and are distributed in a tissue-specific manner in both newborn and adult mice organs. They may therefore play a role in cell or organ physiology and not directly control growth arrest. More recently a new set of growth arrest-specific genes has been identified through cloning the genes induced upon UV-irradiation or after chemical mutagenesis. Some of these appear also to be downregulated when cells traverse the G0/G1 boundary. It will be important to identify the function of the growth arrest genes and verify whether they have a role in controlling cell proliferation.

36.14 Growth Inhibitory Factors[13]

Numerous other negatively controlling elements have also recently been identified both at the level of growth inhibitory factors as well as growth inhibitory receptors, and we might soon unravel a signal pathway that will negatively control cell growth as a counterpart to the positive elements which induce growth.

In contrast to stimulatory growth factors that act from the outside of cells and trigger cell division, other polypeptides have an inhibitory effect on cell proliferation. In many respects interferon (IFN) and transforming growth factor beta (TGFβ) behave as the antagonists of mitogenetic growth factors. Other factors have dual effects. Amphiregulin has both stimulatory and inhibitory activity on cell proliferation. Likewise peptide factors that have growth inhibitory activity, like tumor necrosis factor (TNF), TGFβ, interleukin 6 (IL6), act both through stimulation and inhibition of cell proliferation, as well as on many other cellular functions not involved in cell growth. Some cytokines, like IL4, may exert an antiproliferative effect in vivo by stimulating release of other mediators from surrounding cells in a paracrine fashion. A better characterization of the biochemical properties of these peptides will probably lead to an understanding of the mechanism of negative growth regulation. The antigrowth properties of IFN and TGFβ, two of the best studied proteins with inhibitory properties, may offer suitable models for this category of controlling elements.

36.14.1 Interferons (IFNs)

Many studies have analyzed the antiproliferative capacity of IFNs to gain insight into the mechanism-regulating normal cell proliferation. Addition of IFN to quiescent cells inhibits serum-induced transition from G0 to the S phase of the cell cycle and if added to exponentially growing cells, all the phases of the cell cycle are prolonged. This does not result from an arrest of cellular biosynthesis as judged by total RNA or protein synthesis. Furthermore, IFN-growth arrested cells undergo an increase in size. IFNs appear to inhibit the expression of growth-induced genes, suggesting that they may regulate proliferation by interfering with expression of *onc* genes. Production of IFNs has been observed after mitogenic stimulation of cells. This led to the suggestion that IFNs might be part of an autocrine feedback mechanism that may help to stop proliferation after a few initial rounds of divisions triggered by mitogenic stimuli. Some of the IFN-induced proteins have been implicated in induction of growth arrest. Evidence indicates that the 2−5 adenylate oligonucleotide (2−5A) synthetase, one of the enzymes specifically induced by IFN, may be involved in growth control. In fact micromolar amounts of 2−5A can decrease the number of cells that enter the S-phase after serum stimulation of quiescent fibroblasts. However, the levels of 2−5A synthetase do not correlate with growth inhibition in all cell systems. The double-stranded RNA-dependent protein kinase, another enzyme induced by IFN, is also associated with growth arrest in some cell lines. In addition an IFN-induced protein of 15 kDa may have a role in inhibition of cell growth.

36.14.2 Transforming Growth Factor Beta (TGFβ)

TGFβ belongs to a new family of genes, which can inhibit growth in epithelial cells and primary fibroblasts. TGFβ also modulates the differentiated phenotype, and leads to reorganization of the cytoskeletion and synthesis of extracellular matrix proteins. It has been purified to homogeneity, and is a dimer protein of 25 kDa resulting from the association of two polypeptide chains. Both homo- and heterodimers can form by association of the products of the TGFβ_1 and TGFβ_2 genes. Recently, two more genese called TGFβ_3 and TGFβ_4 have been

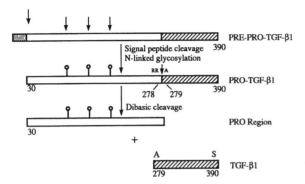

Fig. 6. Schematic representation of processing events of pre-pro-TGF-β_1, glycosylation sites $\downarrow \rightarrow$; signal peptide *black box*; pro-region *empty box*; mature TGF-β_1 *striped box*

cloned. Sequence analysis reveals that TGFβ is cleaved from a large precursor (Fig. 6). Genes for TGFβ are almost identical between man, mouse, pig, and cow, and mRNA for TGFβ is expressed in many cell types and tissues. The production of TGFβ appears partially to be controlled at the translational level. Furthermore the protein is released from cells as an inactive precursor which is unable to bind to the receptor and is, therefore, biologically inactive (Fig. 6). Activation of the TGFβ protein can be achieved by acidification, alkalinization, and denaturing agents such as urea. The inactive form of TGFβ can also form complexes with a carrier protein. A glycosylated domain of the TGFβ precursor appears to be involved in the release of active TGFβ molecules. Several cell lines produce high levels of TGFβ in its latent inactive form. The activation of TGFβ may therefore negatively control cell proliferation, but there are obviously several different mechanisms for making active TGFβ available to the target cell.

The activity of TGFβ is mediated by specific receptors that appear to be refractory to downregulation by high concentration of the ligand. The binding constant is around 20 pM. TGFβ binds strongly to three surface proteins of 50, 70, and 300 kDa. The 50 kDa protein appears to be the functional receptor, since it is expressed in almost all cells that respond to TGFβ, and is selectively lost in mutant cells resistant to growth inhibition by TGFβ. At variance with most growth factor receptors, no tyrosine kinase is associated with the TGFβ receptors.

TGFβ stimulates or inhibits cell proliferation, depending on the cell target, culture conditions, or the presence of other growth factors. Anchorage-independent growth induced by TGFβ may depend on the ability of TGFβ to induce production and secretion of extracellular matrix components. In cells growing in monolayer, TGFβ appears to be mitogenic mainly in fibroblasts and mesenchymally derived cells. In primary rodent fibroblasts TGFβ is predominantly growth inhibitory, although this effect can be overcome by high cell number or concomitant presence of stimulatory growth factors. In contrast to primary cells, established fibroblastic cell lines are stimulated to grow by TGFβ. This effect is in some cases mediated by production of PDGF. Loss of the growth inhibitory response to TGFβ may accompany the acquisition of a transformed phenotype. In other cell types including B and T lymphocytes, hepatocytes, endothelial and epithelial cells, TGFβ is a strong inhibitor of proliferation. As little as 0.1 ng, corresponding to 4 pM TGFβ, can inhibit cell growth up to 95% after 20 h of treatment. TGFβ appears to suppress expression of genes like *myc* and other *onc* genes in endothelial cells and keratinocytes. However, in primary rodent fibroblasts inhibition of growth is independent of the induction of these genes and most of the signals known to be induced by mitogen stimulation are unaffected. TGFβ, in fact, does not appear to interfere with induction of tyrosine phosphorylation, protein kinase C activity, Na/H antiproton pump activity nor transcription of several nuclear genes in stimulated hamster cells. These data indicate that the signals transduced by the TGFβ receptors are distinct from those induced by stimulatory growth factors.

36.15 Control of Growth Arrest[14]

Evidence for negative control of cell growth stems mainly from genetic analysis. Somatic cell genetics taking advantage of hybrid cells has been useful to establish the presence of negative elements. Many tumor cells may become less tumorigenic after fusion with normal fibroblasts, lymphocytes, or macrophages. Studies with stable hybrids have demonstrated that reappearance of tumorigenicity only occurs after chromosome losses. In the best studied cases with human intraspecies hybrids, reexpression of tumorigenecity between HeLa and normal human fibroblasts was associated with the loss of one copy of chromosome 11. Taking all data together, the cell hybrid studies suggest that tumor or transformed cells can be normalized after fusion with fibroblasts. It appears therefore that negatively controlling elements may be lost during transformation and tumor formation. Since hybrid cells in most cases take on the phenotype of the normal parental cell regardless of the origin of the malignant cell, the normalizing factor is probably dominant over the tumorigenic element.

Recessive tumor suppressor genes exist also in other species. Careful genetic studies in *D. melanogaster* established the existence of several tumor suppressor genes. At least 24 such genes have been identified. All of these genes seem to play a role in normal development since when the mutant allele is homozygotic it is often lethal, probably due to arrested development at a specific stage. This finding in *Drosophila* might explain why humans with a defect in tumor suppressor genes are heterozygotic at birth and the defect in the second allele must arise from a subsequent somatic event. The homozygotic state may be lethal during development.

36.15.1 Tumor Suppressor Genes

Genes that may act as suppressors of normal cell proliferation have recently been identified. Both cytogenetic studies and *r*estriction *f*ragment *l*ength *p*olymorphism (RFLP) analyses have identified several human cancer loci. Genes in this loci appear to contribute to neoplastic transformation of cells if both alleles have been eliminated. Table 2 identifies some of these loci on specific human chromosomes.

The negative controlling element involved in the development of retinoblastoma (RB) has been cloned. A cDNA derived from the retinoblastoma locus encodes a protein of 928 amino acids, called p105-RB. The RB protein can associate with the DNA virus *onc* gene products, i.e., the adenovirus E1A protein, the SV40 large T (SV40 LT) and the papillomavirus E7 proteins. These findings point to a connection at the protein level between the genes that act positively and negatively in cell proliferation. The 105-RB protein is present in all phases of the cell cycle and may be modified by phosphorylation. It becomes phosphorylated when cells approach the S- and G2-phases of the cell cycle, and is unphosphorylated in quiescent and differentiated cells. The SV40 LT antigen which only binds the unphosphorylated form may consequently induce growth by sequestering the RB protein thereby preventing its suppressor function. Introduction of a normal RB protein in a retinoblastoma cell line obviously decreases the transformed phenotype of the recipient cells.

Altered alleles of the RB gene have been detected in many tumors other than retinoblastoma. In all such neoplastic cells the RB protein is unexpressed or altered. It appears therefore that, although originally discovered because of its association with a specific type of neoplasia, the RB gene is expressed in almost all cells and it probably plays a role in growth control.

Elimination of a tumor suppressor gene appears also to be involved in Wilms' tumor. Wilms' tumor is a pediatric nephroblastoma often associated with other malformations such as aniridia, genito-urinary abnormalities, and mental retardation. The Wilms' tumor is linked to a gene complex on chromosome 11 p13 region, referred to as the WAGR locus. Reintroduction of a normal chromosome 11 in Wilms' tumor cells via a microcell transfer technique leads to suppression of the ability to form tumors in the hybrid cells. This suggests that Wilm's tumor cells are defective for a gene on chromosome 11 p13 that can control the expression of the malignant phenotype in kidney cells. A gene has been cloned that maps on this region of chromosome 11. It seems to be altered in many patients with Wilms' tumors. The encoded protein shows properties of a DNA-binding protein containing a zinc-finger motif. Its role in tumorigenesis has not been unequivocally established.

The nuclear phosphoprotein p53 is also involved in cell growth control and transformation. It was discovered for its ability to complex with SV40 LT. Mutants of the p53 gene, cloned from transformed cells, can, when overexpressed, immortalize primary cells and cooperate with activated Ras protein to transform primary rat embryo fibroblasts. However, wild type p53 is incapable of transforming cells and, in fact, inhibits transformation by mutant p53 and other *onc* genes. Wild-type p53 can furthermore suppress growth of some human carcinoma cells. P53 is thus probably a tumor suppressor gene, and when mutated, it leads to transformation, acting as a recessive *onc* gene. Alterations involving allelic deletions and missense mutations have been observed in many human tumors.

36.16 Outlook

Identification and cloning of growth factors, growth factor receptors, and the discovery of onco-genes over the last years have provided some direct evidence of the molecular and biochemical basis for control of normal and neoplastic cell growth. Further biochemical dissection of some of these proteins has helped to delineate a network of proteins that interact in stimulating DNA synthesis. Studies of anti-*onc* genes or suppressor genes are now providing the first evidence for biochemical interactions between proteins that stimulate growth and proteins that counteract uncontrolled cell proliferation. We can next expect a better resolution of the mitogenic signaling pathways. More precise localization of the positive effectors of the mitogenic signal network may help to introduce a block to prevent illegitimate activation.

Alternative ways of preventing stimulation of cell division may also be identified when the signal pathways for growth inhibitory factors have been

Table 2. Chromosomal location of genes in humans associated with various tumors

Cancer type	Chromosome location									
	1	2	3	5	10	11	13	17	18	22
Neuroblastoma, melanoma	p									
Breast carcinoma	q									
Uveal melanoma		+								
Small-cell lung carcinoma			p				q			
Adenocarcinoma of the lung			p							
Renal cell carcinoma			p							
Von Hippel-Lindau disease (familial RCC)			p							
Colorectal carcinoma				q				p	q	
Familial polyposis coli				q						
Multiple endocrine neoplasia type 2A					+					
Glioblastoma					+					
Multiple endocrine neoplasia type 1						q				
Wilms' tumor	p									
Hepatoblastoma						p				
Rhabdomyosarcoma						p				
Bladder carcinoma						p				
Breast carcinoma						p	q			
Ductal breast carcinoma							q			
Retinoblastoma							q			
Osteosarcoma							q			
Soft tissue sarcomas							q			
Colon carcinoma								p		
Colon carcinoma									q	
Acoustic neuroma										+
Meningioma										+
Bilateral acoustic neurofibromatosis										+

p and q refer to chromosome arms.
+ means that the gene has not yet been located to a chromosome arm.

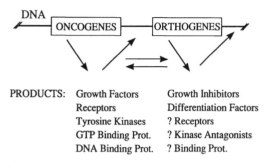

DNA

ONCOGENES → ORTHOGENES

PRODUCTS:
Growth Factors	Growth Inhibitors
Receptors	Differentiation Factors
Tyrosine Kinases	? Receptors
GTP Binding Prot.	? Kinase Antagonists
DNA Binding Prot.	? Binding Prot.

Fig. 7. A model for possible interaction between oncogene and orthogene products. Several of the oncogenes have been identified but the function of their products is still uncertain. Some growth inhibitors and differentiation factors have been identified among the orthogene products. Others may be putative growth inhibitory receptors, kinase antagonists, or inhibitory DNA-binding proteins

characterized. Advances in understanding the role of the genes involved in growth control are essential to initiate rational attempts to identify growth factor antagonists which in turn may open new ways for treatment of abnormal cell proliferation. All the genes involved in the inhibitory pathway have been referred to as orthogenes, derived from the Greek word for straight, and they may be as diversified as the oncogenes, derived from the Greek word for mass or tumor (Fig. 7).

The numerous steps involved in changing normal cells to fully malignant metastasizing tumor cells, through the tumor initiation and progression pathways, may be more understandable if we remember that these changes result from an interplay between the products of at least a hundred genes involved in positively regulating cell growth and probably an equal number of negatively regulatory elements.

36.17 Summary

Since the introduction of the cell cycle concept, two approaches to study growth regulation of cells have been proposed: one claims that cells are naturally quiescent, requiring a stimulatory encounter with growth factors for induction of cell division. The other considers cellular multiplication as the natural steady state; cessation of multiplication is thus

a restriction imposed on the system. In the latter case, emphasis is mainly on the signals involved in arrest of multiplication.

The present chapter gives an introductory overview of the events involved both in growth induction and growth arrest. A signal network obviously exists in mammalian cells governed by an interplay between multiple genes involved in positive regulation of cell growth and an equal number of negatively regulatory elements.

References

[1] Baserga R (1985) The biology of ell reproduction. Harvard University Press, Cambridge, Massachusetts

[2] Goldstein S (1990) Replicative senescence: the human fibroblast comes of age. Science 249:1129−1133

Lee MG, Nurse P (1987) Complementation used to clone a human homologue of the fission yeast cell cycle control gene cdc2. Nature 327:31−35

Moria AO, Draetta G, Beach D, Wang JY (1989) Reversible tyrosine phosphorylation of cdc2: dephosphorylation accompanies activation during entry into mitosis. Cell 58:193−203

Pardee AB (1989) Events and regulation of cell proliferation. Science 246:603−608

[3] Sporn MB, Roberts AB (1988) Peptide growth factors are multifunctional. Nature 332:217−219

[4] Delarco JE, Todaro GJ (1978) Growth factors from murine sarcoma virus-transformed cells. Proc Natl Acad Sci USA 75:4001−4005

Deuel TF (1987) Polypeptide growth factors: Roles in normal and abnormal cell growth. Annu Rev Cell Biol 3:443−492

[5] Pledger WJ, Stiles CD, Antoniades HN, Scher CD (1987) An ordered sequence of events is required before *BALB/c*-3T3 cells become committed to DNA synthesis. Proc Natl Acad Sci USA 75:2839−2843

[6] Armelin HA, Armelin MCS, Kelly K, Stewart T, Leder P, Cochran BH, Stiles CD (1984) Functional role for c-*myc* in mitogenic response to platelet-derived growth factor. Nature 310:655−660

Waterfield MD, Scrace GT, Whittle N, Stroobant P, Johnsson A, Wasteson A, Westermark B, Heldin C-H, Hunag HS, Deuel TF (1983) Platelet-derived growth factor is structurally related to the putative transforming protein p28*sis* of simian sarcoma virus. Nature 304:35−39

[7] Ullrich A, Coussens L, Hayflick JS, Dull TJ, Gray A, Tam AW, Lee J, Yarden Y, Libermann TA, Schlessinger J, Downward J, Mayes ELV, Whittle N, Waterfield MD, Seeburg PH (1984) Human epidermal growth factor receptor cDNA sequence and aberrant expression of the amplified gene in A431 epidermoid carcinoma cells. Nature 309:418−425

Yarden Y, Ullrich A (1988) Growth factor receptor tyrosine kinases. Annu Rev Biochem 57:443–478

[8] Rozengurt E (1986) Early signals in the mitogenic response. Science 234:161–166

[9] Dalla-Favera R, Bregni M, Erickson J, Patterson D, Gallo RC, Croce CM (1982) Human c-*myc onc* gene is located on the region of chromosome 8 that is translocated in Burkitt lymphoma cells. Proc Natl Acad Sci 79: 7824–7827

Levinson AD, Oppermann H, Levintow L, Varmus HE, Bishop JM (1978) Evidence that the transforming gene of avian sarcoma virus encodes a protein kinase associated with a phosphoprotein. Cell 15:561–572

Trahey M, McCormick F (1987) A cytoplasmic protein regulates GTPase of normal N-*ras* p21, but does not affect oncogenic mutants. Science 238:542–545

Weinberg RA (1989) Oncogenes and the molecular origins of cancer. Cold Spring Harbor Laboratory Press, New York

[10] Bohmann D, Bos TJ, Admon A, Nishimura T, Vogt PK, Tjian R (1987) Human proto-oncogene c-*jun* encodes a DNA binding protein with structural and functional properties of transcription factor AP-1. Science 238: 1386–1392

Cochran BH, Reffel AC, Stiles CD (1983) Molecular cloning of gene sequences regulated by platelet-derived growth factor. Cell 33:939–947

Greenberg ME, Ziff EB (1984) Stimulation of 3T3 cells induces transcription of the c-*fos* proto-oncogene. Nature 311:433–438

[11] Almendral JM, Sommer D, MacDonald-Bravo H, Burckhardt J, Perera J, Bravo R (1988) Complexity of the early genetic response to growth factors in mouse fibroblasts. Mol Cell Biol 8:2140–2148

Heintz N (1989) Temporal regulation of gene expression during the mammalian cell cycle. Curr Opinion Cell Biol 1:275–278

[12] Philipson L, Sorrentino V (1991) From growth arrest to growth suppression. J Cell Biochem 46:95–101

Wang E (1987) Contact-inhibition-induced quiescent state is marked by intense nuclear expression of statin. J Cell Physiol 133:151–157

[13] De Maeyer E, De Maeyer-Guignard J (eds) (1988) Interferons and other regulatory cytokines. John Wiley & Sons, New York

Massagué J (1990) The transforming growth factor-β family. Annu Rev Cell Biol 6:597–641

Resnitzky D, Yarden A, Zipori D, Kimchi A (1986) Autocrine β-related interferon controls c-*myc* suppression and growth arrest during hematopoietic cell differentiation. Cell 46:31–40

Roberts AB, Sporn MB (1988) Transforming growth factor b. Adv Cancer Res 51:107–145

[14] Cavenee WK, Dryja TP, Phillips RA, Benedict WF, Godbout R, Gallie BL, Murphee AL, Strong LC, White R (1983) Expression of recessive alleles by chromosomal mechanisms in retinoblastoma. Nature 305:779–784

Finlay CA, Hinds PW, Levine AJ (1989) The p53 proto-oncogene can act as a suppressor of transformation. Cell 57:1083–1093

Friend SH, Bernards R, Rogelj S, Weinberg RA, Rapaport JM, Albert DM, Dryja TP (1986) A human DNA segment with properties of the gene that predisposes to retinoblastoma and osteosarcoma. Nature 323:643–646

Sager R (1989) Tumor suppressor genes: the puzzle and the promise. Science 246:1406–1411

Stanbridge EJ (1987) Genetic regulation of tumorigenic expression in somatic cell hybrids. Adv Viral Oncol 6:83–101

Weissman BE, Saxon PJ, Pasquale SR, Jones GR, Geiser AG, Stanbridge EJ (1987) Introduction of a normal human chromosome 11 into a Wilms' tumor cell line controls its tumorigenic expression. Science 236:175–180

Whyte P, Bucjkovich KJ, Horowitz JM, Friend SH, Raybuck M, Weinberg RA, Harlow E (1988) Association between an oncogene and an anti-oncogene: the adenovirus E1A proteins bind to the retinoblastoma gene product. Nature 334:124–129

Glossary

β-Adrenergic receptor:
Receptor on the cell surface for adrenaline-like ligands.

Allantois:
One of the membranes associated with the vertebrate embryo (see amnion, chorion).

Alu repeat:
A DNA sequence about 300 base pairs long found in multiple copies in the genomes of many species. There are more than 100 copies, randomly distributed, in the human genome. A member of the SINE family (q.v.).

Amnion:
One of the membranes associated with the vertebrate embryo (see allantois, chorion).

Amphiregulin:
A regulator with both positive and negative effects.

Anchor cell:
A somatic cell (q.v.) from the gonad of *Caenorhabditis elegans* that induces adjacent cells to undergo development to generate the functional vulval opening.

Anlage:
A group of cells in an embryo from which a particular organ will develop.

Anterior group genes:
Maternal genes (q.v.) involved in determining the anterior segmentation pattern of the *Drosophila* embryo.

Antisense RNA:
RNA complementary to the normally expressed RNA transcript. Antisense RNA can, in some cases, be used to block the expression of a gene.

Apical dominance:
Suppression by plant apical meristems (q.v.) of the division of distal meristems.

Ascogonium:
The female sexual organ in certain ascomycete fungi.

Assembly initiation complex:
The minimum defined complex, in virus assembly, which is capable of directing the formation of the final product by addition of further subunits. Once such a complex has formed the remaining viral components associate with it until assembly is complete.

Autocrine:
Regulation of cellular processes by substances produced within the same cell.

Bacteroid:
Differentiated form of rhizobial bacteria that exists as an intracellular symbionts within the legume nodule.

Band cloning:
Cloning of a DNA fragment from size-fractionated DNA by conventional or pulse-field gel electrophoretic (q.v.) methods.

Biological clock:
An endogenous mechanism in cells or organisms that generates a pattern of activity varying regularly with time.

Blastocyst:
An early stage in the development of an animal embryo, having outer cells (the trophoblast, q.v.), an inner cell mass (which will make up the embyro proper) and a hollow cavity (the blastocoel).

Blastoderm:
An embryo comprised of a single germ layer.

Blastoma:
An early cell stage in terminal differentiation (q.v.).

Blastomere:
Any cell produced during cleavage.

Bristle:
A cuticular structure in insects, generally innervated by one or more sensory neurons, therefore a sensory organ.

Capsids:

(= viral shells). A term used to describe the macromolecular container of viral nucleic acid. For simple viruses the capsid is often composed only of proteins, but in more complex systems the complex of protein and nucleic acid (nucleocapsid) can be surrounded by a membrane containing virally encoded glycoproteins.

Carcinoma:

A tumor in an epithelial cell (q.v.) lineage.

Catabolite repression:

Decrease in gene expression due to the presence, in the growth medium, of glucose or another easily useable source of carbon.

Caulonema:

A type of cell found in the filamentous stage of the gametophyte (q.v.) in many species of moss.

Cell commitment:

See commitment.

Cell lineage:

The progeny of a defined cell.

Centromere:

The region of eukaryote chromosome involved in its attachment to the spindle at mitosis and meiosis.

Cephalic furrow:

A groove in the transverse axis of the *Drosophila* embryo towards its anterior end.

Chemotaxis:

Reaction of motile cells or microorganisms to chemical stimuli by moving towards or away from source of chemical.

Chloronema:

A type of cell found in the filamentous stage of the gametophyte (q.v.) in many species of moss.

Chondrocyte:

A type of animal cell responsible for secreting the special collagens and glucosaminoglycans (such as chondroitin sulfate) characteristic of cartilage.

Chorion:

An extra-embryonic membrane which encloses both the amnion (q.v.) and the allantois (q.v.), consisting of inner and outer mesoderm.

Chromosome jumping:

Technique that permits the cloning of both ends but not the middle of a long piece of genomic DNA.

Chromosomal walk:

Technique used to isolate a gene beginning with a linked region which has already been cloned, through the use of a set of overlapping cloned DNA fragments.

Cleistothecium:

A type of fruiting body (q.v.), having no opening, produced by some ascomycete fungi (see ostiole, perithecium).

Clock-affecting gene:

A gene encoding a product that affects, either directly or indirectly, the biological clock (q.v.).

Commitment:

Determination, not normally reversible, of a cell or tissue to a path of development.

Compaction:

A change in cell cohesiveness, resulting in cells adhering tightly to each other.

Conidium:

Fungal asexual spore (see macroconidium, microconidium).

Conidiophore:

A specialized hypha (q.v.) which bears conidia.

Contig:

A member of a set of recombinant clones containing overlapping segments of DNA which define a large segment of chromosomal DNA.

Copia element:

A member of the retrotransposon (q.v.) family of mobile elements (q.v.). Up to 100 virtually identical copies are present in the *Drosophila* genome.

Cosmid:

A vector containing the DNA packaging site (cos) of bacteriophage lambda, which can be used to clone up to 40 kbp of DNA.

CpG island:

An unmethylated genomic region, 0.5 – 1 kbp long, frequently found at the 5′ end of genes, in which the dinucleotide CpG is present at the same frequency as the dinucleotide GpC.

Crown gall:

Plant tumor caused by the bacterium, *Agrobacterium tumefaciens.*

Cryptogene:

A gene encoding an RNA that is edited post-transcriptionally.

Cys-Cys finger motif:
Type of zinc finger domain (q.v.).

Cys-Hys motif:
Type of zinc finger domain (q.v.).

Cytokine:
A factor that changes the developmental phenotype of an animal cell.

Dermatome:
In vertebrate development, the dorsal layer of the dermamyotome (q.v.), which generates the mesenchymal (q.v.) connective tissue of the skin.

Dermomyotome:
In vertebrate development, the epithelial cells remaining in a somite (q.v.), after the cells of the scleretome (q.v.) have migrated away.

Developmental gene:
(= morphogene) A gene, the mutation of which may alter development.

Direct repeat:
A repeat of a nucleic acid sequence in the same orientation (see inverted repeat).

DNA imprinting:
Epigenetic change (i.e. a change not involving the DNA sequence) to genomic DNA (e.g. by methylation or bound proteins) which is dependent on cell history or parentage, which is transmitted to progeny cells and which may affect later cell function or differentiation (see genomic imprinting).

Domain:
A sub-structure within a protein usually involved in a particular function such as DNA binding.

Dormancy:
Physiological state in which metabolism is reduced to a very low level.

Dorso-ventral group genes:
Maternal genes (q.v.) involved in determining the pattern along the dorso-ventral axis of the developing *Drosophila* embryo.

Ectopic expression:
Expression of a gene in a tissue which is different from the tissue in which the gene is normally expressed.

Ecotropic virus:
Mouse retrovirus which encodes an envelope glycoprotein, interacting with a cellular receptor found predominantly on mouse cells (see polytropic virus, xenotropic virus).

Electroporation:
A technique for transformation (q.v.) using a pulsed electrical charge to facilitate DNA uptake.

Endocrine:
The regulation of cellular processes by a substance produced by cells at a remote site (= hormone, q.v.).

Endocrine pancreas:
The part of the pancreas which secretes hormones (e.g. insulin, glucagon) (see exocrine pancreas).

Endospore:
Spore which develops within the cytoplasm of another cell, called the mother cell (q.v.), or sporangium (q.v.).

Endothelial cell:
A type of animal cell, derived from the endoderm, which lines an internal surface, e.g. blood vessels.

Enhancer:
A DNA sequence which interacts with transcription factors to increase the rate of transcription of adjacent genes (see silencer).

Enhancer-trapping:
A technique employing a vector containing a transposable element, which can be used to find tissue-specific enhancer elements.

Entelechy problem:
The problem of how an organism is programmed to develop to a size characteristic of its species.

Epiblast:
In vertebrate development, the part of the inner cell mass which will form the embryo (also called primitive ectoderm).

Epidermoblast:
A type of animal cell which gives rise to the epidermis, including the sensory organs.

Epistasis:
The masking of the phenotype associated with one gene as a result of the phenotype caused by a different gene (e.g. headlessness is epistatic to eye colour!).

Epithelial cell:
A type of animal cell, derived from the ectoderm, lining the external or interior surfaces of the body.

Etiolated plant:
A plant grown in the absence of light, characterized by a highly elongated shoot, unexpanded leaves, and immature chloroplasts.

Exocrine pancreas:
The part of the pancreas, drained by ducts, which secretes digestive enzymes (see endocrine pancreas).

Fate map:
Cells that have become committed (q.v.) are said to have acquired a particular fate. Fate maps relate fate to position. They may be constructed, for some organisms, at an early stage of development using methods which mark or destroy particular groups of cells, and observing the effects at later stages of development.

Fibroblast:
A type of animal cell derived from the mesoderm which gives rise to interstitial tissue.

Flagellum:
An extracellular organelle used for cell locomotion.

Floral evocation:
Commitment (q.v.) of the vegetative meristem to flower formation.

Floral induction:
The process in which endogenous or exogenous stimuli trigger floral evocation (q.v.).

Fruiting body:
A multicellular spore-containing structure on colonies of certain bacteria and fungi.

Gametophyte:
The phase of the life cycle of a plant which gives rise to the gametes. The cells of gametophytes are normally haploid (see sporophyte).

Gap genes:
Zygotic gene (q.v.) required for the formation of a group of adjacent segments in the developing *Drosophila* embryo.

Gel retardation assay:
A technique to identify cellular proteins binding to a specific DNA sequence.

Gene conversion:
The alteration of one of two near-homologous DNA sequences to the sequence of the other.

Gene disruption:
Targeted inactivation of a gene by homologous recombination.

Gene family:
A number of similar or related genes found in a single genome.

Genetic cascade:
Successive waves of gene expression; the expression of one or more genes in each wave activates the next, culminating with the final cellular phenotype.

Genetic economy:
The evolutionary pressure to minimize the size of a virus genome, and so increase its replication rate. As a result, viral shells are made up of multiple copies of just one, or a few proteins.

Genetic mosaic:
An organism containing cells of different genotypes.

Genetic mosaic analysis:
A technique to determine in which cell(s) a gene functions, which employs genetic mosaics (q.v.).

Genomic imprinting:
A process by which genes are differentially expressed from the maternal or paternal genome (see DNA imprinting).

Germ line:
Those cells having the potential to contribute genetic material to subsequent generations (see soma).

Growth factor:
A chemical, in animals usually a protein or peptide, which is involved in growth regulation.

Guide RNA:
An RNA molecule that provides sequence information during the process of RNA editing.

Gynandromorph:
A genetic mosaic (q.v.) containing cells of both sexes.

Haemangioma:
Endothelial (q.v.) tumor.

Haematapoiesis:
Developmental process which gives rise to blood cells.

Helix-loop-helix motif:
(= helix-turn-helix motif) A protein domain, probably involved in dimerization, characteristic of certain transcription factors.

Hepatocyte:
A type of animal cell which gives rise to liver cells.

Hepatoma:
Hepatocyte (q.v.) tumor; used also to indicate a transformed (q.v.) hepatocyte cell line.

Heterochronic gene:
A gene that controls the developmental state or time at which a particular developmental event occurs.

Heterochronic mutant:
A mutant in which a specific event occurs at an inappropriate stage of development.

Heterokaryon:
A cell containing nuclei of more than one genotype (see homokaryon).

Heterokaryon incompatibility:
Property of two strains of the same species which are unable to form a stable heterokaryon.

Heterothallism:
Property of a species which has different mating types but no distinct sexes.

Histotype:
The type of cell or tissue into which cells differentiate.

Homeobox:
An amino acid sequence which forms a homeodomain (q.v.).

Homeodomain:
A protein domain (q.v.) involved in DNA binding and recognition, characteristic of certain transcription factors. First detected in homeotic genes of *Drosophila* but now detected in genes from other organisms.

Homeotic gene:
A gene which can mutate to cause a structure to be produced in an inappropriate position.

Homokaryon:
A cell containing nuclei of only one genotype (see heterokaryon).

Homothallism:
Property of a species in which any individual can mate with any other, i.e. having no mating types or sexes (see heterothallism).

Hormone:
A substance regulating cellular activity in animals which has its effect distant from its site of synthesis.

Host range:
The range of organisms with which a symbiont (q.v.) or parasite (q.v.) can form a relationship.

Housekeeping gene:
A gene, the product of which is required in many tissues and at many developmental times, e.g. a gene coding for an enzyme of intermediary metabolism.

Hybrid dysgenesis:
Condition leading to reduced fertility. Caused, in *Drosophila*, by the activation of mobile elements (q.v.) and characterized by male recombination, enhanced mutation rates, chromosomal recombination and segregation distortion.

Hypermorphic mutant:
A mutant displaying overproduction of a normal feature.

Hyperplasia:
Increased replication of morphologically normal cells. May result in the production of an oversized organ.

Hypha:
A type of filamentous cell found in some fungi and actinomycetes.

Hypoplasia:
Reduced replication of morphologically normal cells. May result in the production of an undersized organ.

Imaginal disk:
A compact group of cells in the *Drosophila* larva commited (q.v.) to produce specific adult structures (see transdetermination).

Immortalization:
A change in a cell allowing it to generate progeny that will divide indefinitely in culture, escaping from the general rule that a finite number of cell divisions is allowed. In vivo, some cancer cells may exhibit this property (see transformation).

Infection thread:
Tunnel-like structure within a plant that conveys bacteria towards the nodule meristem.

Insuloma:
Tumor in the endocrine pancreas (q.v.).

Interferon:
A protein which inhibits virus infection and multiplication.

Interleukin:
A subgroup of cytokines (q.v.) exhibiting multiple effects in lymphoid cells and macrophages.

Inverted repeat:

Repeat of a nucleic acid sequence in the opposite orientation (see direct repeat).

Karyogamy:

Fusion of two genetically different nuclei during a sexual cross.

Laser ablation:

A technique to kill cells by using a microscopically focussed lased beam. Can be used to construct fate maps (q.v.).

Leucine zipper:

A leucine-rich amphipathic alpha helix structure within a protein, which functions as a dimerization interface. First recognized in the C/EMP molecule.

LINE family:

The Long Interspersed Nuclear Element family of DNA sequences, 6–7 kbp retroposon (q.v.) sequences present about 50000 times in mammalian genomes.

Lymphocyte:

A type of blood cell engaged in antibody response and other defence mechanisms.

Lymphoma:

A tumor in lymphocytic cells.

Macroconidium:

A large conidium (q.v.) containing more than one (typically two or three) nuclei (see microconidium).

Macrophages:

A type of mammalian cell involved in the immune response and cellular defence mechanisms.

MADS box:

A region which binds DNA, found in a number of proteins including yeast MCM1, Arabidopsis AG, Antirrhinum DEF A and mammalian SRF.

Mannopine:

N2-(1′-deoxy-D-mannitol-1′-yl)-L-glutamine.

Mat cassette:

A yeast mating type locus sequence which can be switched for an alternative sequence by gene conversation (q.v.).

Maternal gene:

A gene, the product of which is required for embryo development, but which is transcribed from the maternal genome (see anterior group genes, dorso-ventral group genes, posterior group genes, terminal group genes, zygotic gene).

Meristem:

A plant tissue, the primary function of which is the formation of new cells by division.

Mesenchyme:

A type of animal cell involved in lining the abdomen and intestines.

Metacyclic trypanosome:

A stage of the trypanosome (q.v.) life cycle which lives in salivary glands of insects and which is infectious to its vertebrate host.

Metula:

A cell type interposed between the conidiophore vesicle and the phialide (q.v.) in *Aspergillus*.

Microconidium:

A small conidium (q.v.) containing only one nucleus (see macroconidium).

Mobile element:

(= transposon) A DNA sequence capable of moving (or transposing) from one chromosomal position to another. Can be divided into two classes; those in which transposition occurs via an RNA intermediate and those in which transposition is DNA mediated (see retroposon, retrotransposon).

Morphogen:

A general term referring to a substance which plays a role in the regulation of development.

Morphogene:

See developmental gene.

Morphogenetic gradient:

An asymmetrically distributed morphogen (q.v.) which acts as a differential regulator according to its local concentration.

Mother cell:

A cell involved in endospore formation. The mother cell engulfs the prespore soon after a division that separates it from its sister prespore cell (see endospore, spore, sporangium).

Mycelium:

A network of branching hyphae (q.v.).

Myelencephalon:

The part of the developing vertebrate hindbrain, which will form the medulla oblongata.

Myotome:

The ventral part of the dermamyotome (q.v.), which will form the body wall and limb muscles.

Neoplasia:

Cellular events changing a normal cell into a tumor cell.

Nephrotome:

The intermediate mesoderm which will ultimately form the kidney.

Neuroblast:

A type of insect cell which gives rise to the central nervous system.

Neuroectoderm:

The region of the ectoderm from which the central nervous system develops.

Neurogenic gene:

A gene, the loss of function of which leads to the developmental misrouting of cells which otherwise would develop as epidermoblasts, to give rise to nerve cells.

Nodule:

A specialized root organ within which bacteria grow and fix nitrogen that is made available to the plant.

Nopaline:

N2-(1,3-dicarboxypropyl)-L-arginine, an opine (q.v.), composed of arginine and a ketoglutaric acid.

Nucleoid:

Bacterial chromosome with associated cell envelope, protein, and RNA; isolated by lysis of the cell with lysozyme and detergent.

Octopine:

N2-(D-1-carboxyethyl)-L-arginine, an opine (q.v.), composed of arginine and pyruvic acid.

Odontoblast:

Connective tissue cell which forms the outer surface of the dental pulp and secretes dentin.

Oncogene:

Cellular gene which may cause tumors if overexpressed.

Opine:

Amino acid and/or sugar derivate, the synthesis of which in planta is encoded by genes in the Agrobacterium T DNA.

Oscillator:

A mechanism that leads to a recurring pattern of changing cellular activity between fixed limits.

Ostiole:

Small opening in the fruiting body of a fungus.

Oxidative stress:

Exposure of an organism to levels of oxygen that are harmful.

Pair rule genes:

Zygotic genes (q.v.) required for the formation of alternating segments in the developing *Drosophila* embryo.

Pan editing:

Extensive post-transcriptional changes in the sequence of an RNA molecule.

Paracrine:

Regulation of cell activity by substances produced in neighbouring cells.

Parasite:

An organism that lives within or upon another organism (the host), in a relationship that is harmful to the host.

Perithecium:

A type of fruiting body having an opening or ostiole (q.v.) produced by some ascomycete fungi (see cleistothecia).

PFG mapping:

The construction of long-range restriction maps in the Mbp size range from hybridization analysis of DNA separated by pulse-field gel electrophoresis (q.v.).

Pheromones:

Signalling molecules released into the environment to allow communication with organisms, usually of the same species.

Phialide:

A flask-shaped cell in *Aspergillus nidulans*, which acts as a stem cell from which conidia (q.v.) are repeatedly budded.

Photomorphogenesis:

Those processes in development that are regulated by the presence, spectral composition, and duration of incident light.

Phytochrome:

A plant photoreceptor pigment that absorbs red and far-red light and controls the timing of a large number of plant developmental processes.

Phytohormone:

A substance which influences plant growth and development at very low concentrations. The term is used more loosely in plant biology than hormone (q.v.) is used in animal biology and is not necessarily confined to substances which have their effect distant from their site of synthesis.

Pinna:
The external ear.

Plasmodium:
A single large cell containing many nuclei which are derived from nuclear divisions without concomitant cell division. The early *Drosophila* embryo is an example of a plasmodium (see syncytium).

Pleiotropy:
The process whereby a gene can affect a number of aspects of the phenotype not always obviously related.

Pluripotency:
Capacity of a cell to give rise to a number of cell types (see totipotency).

Polyketide antibiotic:
An antibiotic synthesized by the sequential condensation of carboxylic acid residues, such that the addition of each residue adds a keto group to the growing chain (some of the keto groups may subsequently be reduced or eliminated).

Polytropic virus:
Retrovirus which recognizes receptors found on cells of both mouse and non-mouse species (see ecotropic virus, xenotropic virus).

Positional information:
Development information which allows position to be sensed, possibly by the presence of morphogenetic gradients (q.v.).

Positional value:
The positional information (q.v.) determining a particular developmental position.

Posterior group genes:
Maternal genes (q.v.) involved in determining the posterior segmentation pattern of the developing *Drosophila* embryo. Many posterior group genes are also involved in forming the polar plasm which is required for germ line (q.v.) differentiation.

POU domain:
A protein domain (q.v.) which mediates DNA binding and which is characteristic of certain transcription factors. About 1/3 of the domain is composed of one class of homeodomain.

Primary hepatocyte:
A non-cancerous hepatocyte (q.v.) isolated from the liver and grown in culture; in contrast to a transformed hepatoma.

Primitive streak:
A region along the middle of the ectoderm of a vertebrate embryo where cells migrate through to form the endoderm and mesoderm.

Procyclic trypanosome:
A stage in the trypanosome (q.v.) life cycle that is not infectious to its vertebrate host. In *Trypanosoma brucei* this is found in the insect midgut.

Proneural gene:
A gene, the product of which promotes neural development.

Protoperithecium:
A spherical structure composed of hyphae (q.v.) produced by the female parent of *Neurospora*. After fertilization it develops into a perithecium (q.v.).

Pseudogene:
A non-functional gene present in the normal genome.

Pulse-field gel electrophoresis:
A technique that allows the separation of DNA fragments of up to 6 Mbp, by subjecting DNA to two field pulses applied sequentially at an angle of greater than 90° to each other.

Quiescence:
A resting state of mammalian cells, after arrest induced by depletion of a component of the medium (often serum). Quiescent cells have unreplicated DNA and are therefore in G1, although they are physiologically quite different from cycling cells in G1. The state is sometimes called G0.

Redundant gene:
A gene, of which there is more than one copy in the haploid genome. Thus, all copies are not likely to be essential and mutation of one copy may not give a mutant phenotype.

Restriction fragment length polymorphism:
(= RFLP) Inherited differences between individuals of a species, detected by variation in the lengths of DNA fragments generated by digestion with restriction enzymes.

Retroposon:
Mobile elements (q.v.) believed to move via an RNA intermediate but which have a structural

organization unlike retroviruses. Many of these elements must transpose passively since they do not encode an enzyme capable of reverse transcription (see retrotransposon).

Retrotransposon:
Mobile element (q.v.) with the same structural organization as an integrated retrovirus including the presence of long terminal repeats. They transpose by a modified retroviral life cycle lacking the extracellular phase characteristic of viruses.

Rhombencephalic isthmus:
Indentation at the caudal limit of the rhombencephalon in the developing vertebrate hindbrain.

Ribozyme:
RNA molecule with specific RNase activity.

RNA editing:
Post-transcriptional processing of an RNA sequence.

S-phase:
The period during the cell cycle when DNA replication takes place.

Sarcoma:
Malignant tumor composed of cells resembling embryonic connective tissue.

Sclerotome:
Mesenchyme (q.v.) cells, derived from somites, which migrate towards the notochord and form the vertebral column.

Segment polarity genes:
Zygotic genes (q.v.) required for the formation of the single segment units in the developing *Drosophila* embryo.

Segmentation genes:
Zygotic genes (q.v.) required for the formation of the anterior-posterior segment pattern, in the developing *Drosophila* embryo (see also gap genes, homeotic genes, maternal genes, pair rule genes, segment polarity genes).

Self-assembly:
The spontaneous association of macromolecules to form a defined complex (e.g. a viral shell). In the strict sense it implies that all the components of the assembly reaction become part of the final complex.

Self-renewal:
The ability of a progenitor cell to divide into two daughter cells, at least one of which will retain the properties of its progenitor.

Serum response element:
DNA sequence required for a transcriptional response to serum growth factors.

Silencer:
DNA sequence which interacts with transcription factors to decrease the rate of transcription of adjacent genes (see enhancer).

SINE family:
The Short Interspersed Nuclear Element family of DNA sequences, short retroposons (q.v.) present in up to a million copies in mammalian genomes (see LINE family).

Soma:
Those cells which do not normally have the potential to contribute genetic information to the next generation (see germ line).

Somatic cell:
A cell which is part of the soma (q.v.).

Somite:
One member of a series of paired segments into which the thickened dorsal zone of mesoderm is divided.

Sporangium:
Structure containing one or more spores (q.v.).

Spore:
A differentiated dormant cell type specialized to survive harsh conditions and/or dispersal.

Spore maturation:
The later stages of spore development during which the dormancy and resistance properties appear.

Sporophore:
Structures that bear spores.

Sporophyte:
The phase of the plant life cycle which gives rise to spores by meiosis. The cells of sporophytes are normally diploid (see gametophyte).

Stalk cell:
A *Caulobacter* cell that is non-motile and contains an appendage that is continuous with the cell envelope (see swarmer cell).

Stem cell:
A cell which produces one daughter cell like itself while the other daughter cell undergoes differential development.

Stem cell line:
Totipotent (q.v.) cells derived from the inner cell mass of a mammalian embryo.

Sterigma:

Alternative name for both metula (q.v.) and phialide (q.v.).

Stromal cells:

Heterogeneous complex of cells, mostly not hematopoietic (q.v.), which gives rise to the bone marrow micro-environment by synthesizing the extracellular matrix and releasing growth factors.

Swarmer cell:

A *Caulobacter* cell that is motile and has a single polar flagellum (q.v.) (see stalk cell).

Symbiont:

An organism that lives within or in close association with another organism of a different species, in a relationship that is mutually beneficial.

Syncytium:

Single cell containing many nuclei which is derived from a fusion of previously independent cells (see plasmodium).

Telomere:

Specialized structure which terminates an eukaryotic chromosome.

Teratocarcinoma:

A malignant cell line derived from a mammalian germ cell or an early blastomere.

Teratogen:

Agent which causes malformations by interfering with the process of embryo development.

Terminal differentiation:

A differentiated state to which a cell becomes irreversibly committed. Dedifferentiation or resumption of proliferative growth is no longer possible.

Terminal group genes:

Maternal genes (q.v.) involved in determining the terminal regions of the developing *Drosophila* embryo

Totipotency:

The capacity of a cell to give rise to a whole organism (see pluripotency).

Transdetermination:

Modification of the commitment (q.v.) of *Drosophila* imaginal disks (q.v.) which may be achieved by the culture of disks in the abdominal cavity of an adult fly.

Transfection:

See transformation.

Transformation:

Used in two senses. In animal cell culture, transformation describes a change which allows a cell line to grow indefinitely (= immortilization, q.v.). More generally, transformation describes an experimental procedure involving the treatment of a cell or organism with DNA, so as to modify its genotype. To avoid confusion, animal cell biologists often use the term transfection to describe this procedure.

Transposable element:

See mobile element.

Transposase:

An enzyme which catalyzes the transposition of mobile elements (q.v.) to new sites.

Transposon:

See mobile element.

Transposon tagging:

A technique for the isolation of a gene in which a mobile element (q.v.) is inserted into a gene thereby mutating it and providing a molecular marker for its isolation.

Transposition:

See mobile element.

Trans-splicing:

The splicing together of independently transcribed RNA molecules.

Trichogyne:

The receptive hyphal structure of the protoperithecium (q.v.), produced by some ascomycete fungi, to which conidia (q.v.) from the male parents are attached prior to fertilization.

Trophoblast:

Outer embryonic cells of the blastocyst stage of a mammalian embryo.

Tropism:

Growth of plant cells or tissues in response to an external stimulus, e.g. gravity (gravitropism) or light (phototropism).

Trypanosome:

Member of the family Trypanosomatidae, which includes the genera *Trypanosoma, Leishmania, Crithidia, Blastocrithidia, Leptomonas,* and *Herpetomonas.*

Variant surface glycoprotein:

(= VSG) A glycoprotein which forms a protective surface coat on a mammalian blood stage trypanosome. Variations in VSG proteins allow some evasioin from the host immune response.

Vegetative growth:
Growth not involving sexual reproduction.

Xenotropic virus:
Retrovirus which recognizes receptors found on many non-mouse species but which are apparently absent in most mice (see polytropic virus, ecotropic virus).

Yeast artificial chromosome:
Cloning construct, incorporating a yeast centromere (q.v.) and telomeres (q.v.) that allows very long pieces of DNA to be cloned in yeast.

Zinc finger:
A protein domain (q.v.), characteristic of certain transcription factors, that allows interaction with DNA, and that requires a single atom of zinc to maintain its structural integrity.

Zone of polarizing activity:
A region of the posterior part of a vertebrate limb bud that gives the limb its positional information (q.v.) along the anterior-posterior and ventro-dorsal axes.

Zoo blot:
Southern blot employing a collection of genomic DNAs from representative animal species and used to assess the degree of evolutionary conservation of a DNA probe.

Zygotic gene:
A gene which is transcribed from the genome of the zygote or early embryo, the product of which is required in the developing embyro (see gap genes, maternal gene, pair rule genes, segment polarity genes, segmentation genes).

Subject Index

Abbreviations given in parentheses refer to the particular organism in which the entries are explained.

Printed by Publishers' Graphics LLC USA
MO20120905-328